# Communications in Computer and Information Science 2918

**Rationale**

The CCIS series is devoted to the publication of proceedings of computer science conferences. Its aim is to efficiently disseminate original research results in informatics in printed and electronic form. While the focus is on publication of peer-reviewed full papers presenting mature work, inclusion of reviewed short papers reporting on work in progress is welcome, too. Besides globally relevant meetings with internationally representative program committees guaranteeing a strict peer-reviewing and paper selection process, conferences run by societies or of high regional or national relevance are also considered for publication.

**Topics**

The topical scope of CCIS spans the entire spectrum of informatics ranging from foundational topics in the theory of computing to information and communications science and technology and a broad variety of interdisciplinary application fields.

**Information for Volume Editors and Authors**

Publication in CCIS is free of charge. No royalties are paid, however, we offer registered conference participants temporary free access to the online version of the conference proceedings on SpringerLink (http://link.springer.com) by means of an http referrer from the conference website and/or a number of complimentary printed copies, as specified in the official acceptance email of the event.

CCIS proceedings can be published in time for distribution at conferences or as post-proceedings, and delivered in the form of printed books and/or electronically as USBs and/or e-content licenses for accessing proceedings at SpringerLink. Furthermore, CCIS proceedings are included in the CCIS electronic book series hosted in the SpringerLink digital library at http://link.springer.com/bookseries/7899. Conferences publishing in CCIS are allowed to use our online conference service (Meteor) for managing the whole proceedings lifecycle (from submission and reviewing to preparing for publication) free of charge.

**Publication process**

The language of publication is exclusively English. Authors publishing in CCIS have to sign the Springer CCIS copyright transfer form, however, they are free to use their material published in CCIS for substantially changed, more elaborate subsequent publications elsewhere. For the preparation of the camera-ready papers/files, authors have to strictly adhere to the Springer CCIS Authors' Instructions and are strongly encouraged to use the CCIS LaTeX style files or templates.

**Abstracting/Indexing**

CCIS is abstracted/indexed in DBLP, Google Scholar, EI-Compendex, Mathematical Reviews, SCImago, Scopus. CCIS volumes are also submitted for the inclusion in ISI Proceedings.

**How to start**

To start the evaluation of your proposal for inclusion in the CCIS series, please send an e-mail to ccis@springer.com

Hongyang Yan · Yu Wang · Goi Bok Min
Editors

# Artificial Intelligence Security and Privacy

## Third International Conference, AIS&P 2025 Guangzhou, China, December 5–7, 2025 Proceedings

*Editors*
Hongyang Yan
Guangzhou University
Guangzhou, China

Yu Wang
Guangzhou University
Guangzhou, China

Goi Bok Min
Universiti Tunku Abdul Rahman (UTAR)
Kuala Lumpur, Malaysia

ISSN 1865-0929 ISSN 1865-0937 (electronic)
Communications in Computer and Information Science
ISBN 978-981-95-9883-0 ISBN 978-981-95-9884-7 (eBook)
https://doi.org/10.1007/978-981-95-9884-7



This Springer imprint is published by the registered company Springer Nature Singapore Pte Ltd.
The registered company address is: 152 Beach Road, #21-01/04 Gateway East, Singapore 189721, Singapore

# Preface

The 3rd International Conference on Artificial Intelligence Security and Privacy (AIS&P 2025) was successfully held in Guangzhou, China, from December 5 to 7, 2025. The conference brought together scholars, researchers, and industry practitioners from around the world, reflecting the growing global attention paid to artificial intelligence security, privacy protection, and trustworthy intelligent systems.

AIS&P 2025 aimed to promote international cooperation and academic exchange in the rapidly evolving fields of AI security and privacy. The conference covered a wide range of topics, including privacy-preserving machine learning, federated learning, adversarial robustness, trustworthy AI, secure data sharing, privacy-enhancing technologies, and interdisciplinary applications of secure and reliable intelligent systems. The event provided an important platform for presenting the latest research achievements, discussing emerging challenges, and exploring future directions in AI security and privacy.

A highlight of AIS&P 2025 was its distinguished series of keynote speeches delivered by leading experts in the field. Qiao Yan (Shenzhen University, China) opened the keynote session with an in-depth exploration of "Vanishing Privacy: Fast Gradient Leakage Threat to Federated Learning," shedding light on the vulnerabilities of gradient-sharing mechanisms and the evolving landscape of privacy attacks. Shaowei Wang (Guangzhou University, China) presented "Privacy Preserving Combinatorial Optimization and Data Contribution Estimation," introducing a novel paradigm that expands differential privacy in the shuffle model to support personalized computation. Thippa Reddy Gadekallu (Zhongda Group, China) delivered a forward-looking talk on "Federated Learning for Big Data—Opportunities, Applications, and Future Directions," offering valuable insights into decentralized learning frameworks and their role in secure data ecosystems. Together, these keynote presentations provided a comprehensive overview of the theoretical foundations, practical challenges, and future directions of AI security and privacy research.

AIS&P 2025 also included a series of oral presentations covering diverse research topics, ranging from intrusion detection systems and trustworthy AI to optimization algorithms, intelligent threat response, and secure reinforcement learning. These contributions highlight the interdisciplinary nature of AI security and the increasing demand for robust, transparent, and privacy-preserving intelligent systems.

A total of 106 manuscripts were submitted to AIS&P 2025, of which 41 were accepted after rigorous peer review; submissions received two double-blind reviews each, on average. The papers included in this proceedings volume represent high-quality research and innovative methodologies contributed by participants from academia and industry. The collected works reflect current trends and advancements in AI security and privacy, offering valuable references for researchers, engineers, and professionals working in related fields.

We sincerely thank all authors for their contributions, the reviewers for their careful evaluations, and the keynote speakers for sharing their expertise. We also extend our gratitude to the organizing committee, technical program committee, and all supporting teams for their dedication and support, which made AIS&P 2025 a successful and inspiring event.

AIS&P 2025 Conference Committee

# Organization

## Conference Chairs

| | |
|---|---|
| Chunsheng Yang | National Research Council, Canada |
| Goi Bok Min | Universiti Tunku Abdul Rahman, Malaysia |
| H. Herbert Song | University of Maryland, Baltimore County, USA |
| Weizhi Meng | Lancaster University, UK |
| Jin Li | Guangzhou University, China |

## Technical Program Committee Chairs

| | |
|---|---|
| Changyu Dong | Guangzhou University, China |
| Charles Morisset | Durham University, UK |

## Publication Chairs

| | |
|---|---|
| Hongyang Yan | Guangzhou University, China |
| Md Liakat Ali | Rider University, USA |

## Organizing Chairs

| | |
|---|---|
| Shaowei Wang | Guangzhou University, China |
| Yu Wang | Guangzhou University, China |

## Publicity Chairs

| | |
|---|---|
| Yun Peng | Guangzhou University, China |
| Zahid Akhtar | State University of New York Polytechnic Institute, USA |

## Technical Program Committee

| | |
|---|---|
| Kai Chen | Hong Kong University of Science & Technology, China |
| Xiaofeng Chen | Xidian University, China |
| Feng Hao | University of Warwick, UK |
| Tong Li | Nankai University, China |
| Weisheng Li | Chongqing University of Posts and Telecommunications, China |
| Joseph Liu | Monash University, Australia |
| Chunhua Su | University of Aizu, Japan |
| Cong Wang | City University of Hong Kong, China |
| Qian Wang | Wuhan University, China |
| Zhendong Wu | Hangzhou Dianzi University, China |
| Tao Xiang | Chongqing University, China |
| Yang Xiang | Swinburne University of Technology, Australia |
| Qiben Yan | Michigan State University, USA |
| Min Yang | Fudan University, China |
| Xu Yuan | University of Delaware, USA |
| Hua Zhang | Beijing University of Posts and Telecommunications, China |
| Albert Zomaya | University of Sydney, Australia |

## Organizing Committee

| | |
|---|---|
| Zhili Zhou | Guangzhou University, China |
| Fei Peng | Guangzhou University, China |
| Yan Pang | Guangzhou University, China |
| Xiaojun Ren | Guangzhou University, China |
| Yuling Chen | Guizhou University, China |
| Debiao He | Wuhan University, China |
| Yuanyuan He | Huazhong University of Science and Technology, China |
| Deqing Zou | Huazhong University of Science and Technology, China |
| Haibo Hu | Hong Kong Polytechnic University, China |
| Nan Jiang | East China Jiaotong University, China |
| Kaishun Wu | Hong Kong University of Science and Technology, China |

# Contents

# Integrating Large Language Models with IDS for Analyst-Guided Automated Threat Response

Elijah M. Maseno[1](✉), Yanxia Sun[1], and Zenghui Wang[2]

[1] Department of Electrical and Electronic Engineering Science, University of Johannesburg, Johannesburg, South Africa
masenoelijah@gmail.com, ysun@uj.ac.za

[2] Department of Electrical and Smart Systems Engineering, University of South Africa, Johannesburg, South Africa

**Abstract.** The increasing sophistication and volume of cyberattacks demand intrusion detection systems (IDS) that combine speed, accuracy, and operational trust. To address this, we present a hybrid detection and response system combining IDS with GPT to improve threat detection, investigation, and response, ultimately lowering Mean Time to Respond (MTTR). Suricata IDS is integrated with GPT in a virtualised environment, continuously monitoring network traffic and generating alerts for suspicious activities. GPT classifies alerts as Low, Medium, High, or Critical based on intelligence. The system automatically queues a recommended response action, such as IP blocking, port filtering, or service shutdown, into a human-in-the-loop approval panel for medium and high-severity alerts. Approved actions are enforced using a host-based firewall (ufw), with automatic rollback after a configurable duration. The architecture incorporates Slack notifications for rapid analyst engagement and ensures auditability through a persistent state store. Experimental testing demonstrates reduced response latency and improved triage accuracy. This method combines AI-driven analysis with SOC procedures to provide a trustworthy automated cyber defence solution.

**Keywords:** Intrusion Detection System (IDS) · GPT-4 · Cyber Threat Intelligence · Security Operations Centre (SOC) · Automated Threat Response

## 1 Introduction

With the ever-increasing number of cybersecurity incidents, analysts are overwhelmed by the volume of alerts requiring rapid response. These incidents could lead to catastrophic consequences for the organisation if left unaddressed. Security Operations Centres (SOCs) have played a critical role in addressing these issues over the decade. However, due to the rapid changes in the cyber world, they have struggled to respond promptly (Chamkar et al., 2024). To enhance the operations of SOCs, there is a growing demand for faster and more efficient tools to support detection, investigation, and response (Freitas et al., 2025). To address these demands, researchers have proposed the adoption of machine learning algorithms as an aid to the SOCs' operations (Sedjelmaci, 2022) (Khayat et al., 2025).


H. Yan et al. (Eds.): AIS&P 2025, CCIS 2918, pp. 1–14, 2026.
https://doi.org/10.1007/978-981-95-9884-7_1

(Khayat et al., 2025), It has been observed that modern SOCs' operations have evolved from traditional roles, such as identification, to more complex ones, including threat intelligence and response. These systems can comb through vast volumes of data and unearth possible patterns of attacks (Gupta et al., 2019). Research by (Freitas et al., 2025) observed that Extended Detection and Response (XDR) products provide precise, context-rich guided responses, a capability that is crucial in tackling the modern complex threats. Towards this noble task of developing superior SOC tools, researchers have been investigating the possibility of integrating large language models (LLMs) in the operations of SOC.

This research is motivated by the urgent need to strengthen defences against persistent threats like SSH brute force attacks, which remain a common entry point for unauthorised access and system compromise. Traditional IDS tools such as Suricata effectively detect these attempts but often overwhelm analysts with raw alerts lacking contextual insight. By integrating Large Language Models (LLMs) like GPT-4, this study aims to enhance alert interpretation, automate threat classification, and recommend tailored remediation actions in real time. The goal is to bridge the gap between detection and intelligent response, ultimately reducing analyst fatigue and response time while advancing the development of AI-powered Security Operations Centres (SOCs). The contributions for this work include:

1) We propose a novel integration of Suricata IDS with GPT-4 to generate contextual threat intelligence, improving the interpretability of alerts for SOC analysts.
2) The framework integrates a policy-driven response mechanism that includes analyst approval checkpoints, where GPT-based risk classification guides decisions such as IP blocking, enabling secure automation while maintaining human oversight.
3) An interactive Security Operations Centre (SOC) dashboard is developed using Streamlit, offering real-time alert visualisation with features such as dynamic filtering, GPT-powered threat explanations, CVE enrichment, and Slack-based notifications, all designed to enhance analyst situational awareness and operational efficiency.
4) We demonstrate a complete and practical AI-enhanced SOC workflow, from detection and GPT enrichment to approval-based response, validated in a real-time testbed to reduce Mean Time to Respond (MTTR).

This paper is organised as follows: the next section discusses the related works. Section 3 covers the methodology. In Sect. 4, we present the results and discussion, and finally, we conclude the paper in Sect. 5.

## 2 Related Works

Recent advancements in cybersecurity research have explored the integration of large language models (LLMs) into traditional security frameworks, with an emphasis on enhancing interpretability and response capabilities. A report released by Microsoft demonstrated the potential of generative AI tools, specifically Microsoft Security Copilot, to impact the efficiency of security operations centres (SOCs) (Bono et al., 2024). Generative AI tool Microsoft Security Copilot reduced incident resolution time through several key features that assist security analysts during investigations and responses, including incident summation, guidance on response actions, interpreting malicious scripts, automating queries, and retrieving threat intelligence. These features enhanced the operational efficiency of security operations centres by reducing manual effort and speeding up the mean time to resolution (MTTR) for security incidents. The disadvantages of this system are that first, it is a proprietary black-box system with limited ability to audit, customise, or explain its internal decision-making. Second, it requires licensing, which may be costly to some organisations. Lastly, integration depends on Microsoft's roadmap and access permissions. (Asfour and Murillo, 2023) Researchers have explored how GPT-4 can simulate realistic human responses to social engineering attacks like phishing, by modelling different personality traits (the Big Five). The researchers looked at how these simulated reactions could help security experts understand which traits are most vulnerable and suggested ways to improve cybersecurity defences. In the future, researchers propose exploring the applicability of findings to other forms of social engineering attacks beyond phishing, as the current study focused on phishing scenarios only.

(Al-Hawawreh et al., 2023) This study explored ChatGPT's implications in cybersecurity, presenting practical applications and a case study on false data injection attacks. Similarly, researchers in (Beckerich et al., 2023) demonstrated how large language models (LLMs) like ChatGPT can be turned into hidden intermediaries for malware attacks. The authors created a proof-of-concept where ChatGPT, combined with plugins, acts as a proxy to spread malicious software and control a victim's system without raising alarms. In their work (McKee and Noever, 2023), researchers applied ChatGPT for the creation of malware and cyberattacks. Using ChatGPT, the researchers generated complex code for keyloggers, worms, and ransomware. These works primarily focused on the application of Large Language Models (LLMs) in offensive cybersecurity contexts.

(Gupta et al., 2019) The authors proposed a novel approach that utilises graph analysis to derive new features from event logs, followed by the application of a deep neural network for automatic classification and prioritisation of these security events. (Mugeshwaran et al., 2025) applied deep learning for recognising complex attack patterns, improving detection accuracy while lowering false positives. In addition, the architecture used natural language processing for contextual alert generation. Both works failed to address the issue of response automation.

CYGENT, a GPT-powered agent, demonstrated how LLMs can summarise raw security logs into analyst-friendly narratives, significantly reducing cognitive load and investigation time (Balasubramanian et al., 2024). In parallel, (Kheddar, 2025) provided a comprehensive review of transformer-based models, including BERT and GPT in intrusion detection systems (IDS), revealing their capacity to model complex traffic patterns and improve threat classification. Furthermore, (Ferrag et al., 2025) emphasised both the opportunities and potential risks associated with generative AI in cybersecurity, advocating for frameworks that maintain human oversight and secure deployment.

These reviewed studies highlight a growing trend toward intelligent and context-aware detection systems. However, most focus on log summarisation or predictive classification, whereas our approach advances the field by coupling GPT-driven reasoning with real-time Suricata alert interpretation, prioritisation, and actionable response orchestration, a novel fusion that strengthens situational awareness and shortens the response loop in Security Operations Centres (SOCs).

# 3 Methodology

This study proposes a hybrid intrusion detection and response system that integrates Suricata, a signature-based IDS, with a GPT-4 assistant to enhance threat interpretation, risk prioritization, and analyst support. The architecture delivers real-time detection, guided triage, and semi-automated mitigation, reinforced by state tracking and human oversight. The system is implemented in Python with Streamlit and Slack APIs and deployed on Ubuntu Linux. The methodology consists of five components: Suricata sensor deployment, auto-responder engine, analyst dashboard, approval enforcement, and state management as shown in Fig. 1 below:

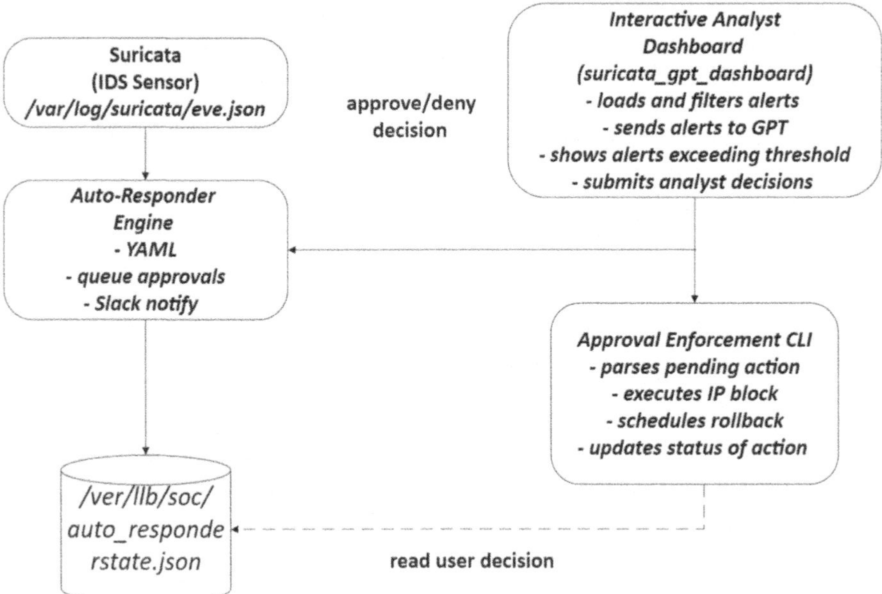


**Fig. 1.** Hybrid Intrusion Detection System and Response System Integrating Suricata IDS with GPT-4 Powered Assistance

## 3.1 Suricata IDS Sensor Deployment

Suricata acts as the primary intrusion detection engine. It inspects both live and simulated network traffic and logs alerts in JSON format to the file /var/log/suricata/eve.json. Each alert record includes key fields such as timestamp, source/destination IPs, protocol, signature, category, and severity. These serve as the input to downstream modules. To demonstrate the practical effectiveness of the proposed hybrid framework, we conducted a controlled proof-of-concept experiment simulating an SSH brute force attack. In this scenario, a Kali Linux machine was configured to initiate repeated unauthorised login attempts against an Ubuntu machine over port 22, as shown in Fig. 2 below.

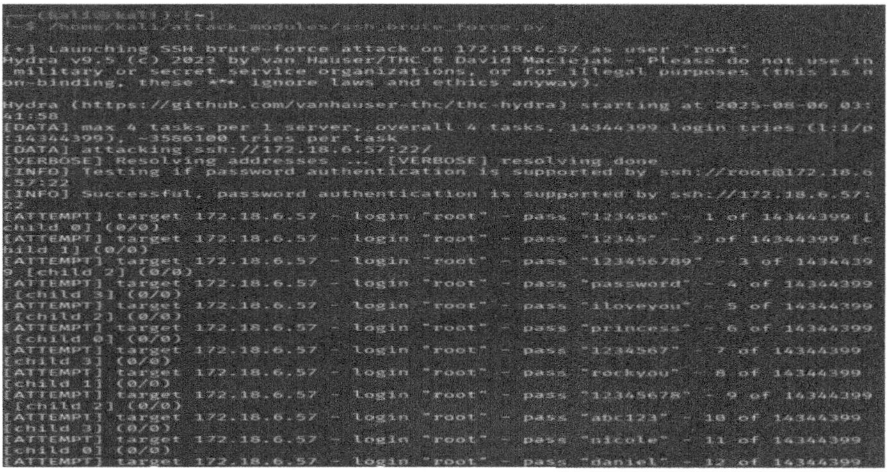

**Fig. 2.** Initiated SSH brute force attack against the Ubuntu machine

### 3.2 Auto-Responder Engine

This Python script continuously tails the eve.json file to extract and filter security alerts based on predefined YAML-based matching rules, such as IPs, categories, or severities. The flagged alerts are queued, with actionable responses pending in a persistent state store located at /var/lib/soc/auto_responder_state.json. The queued action is uniquely identified using a hash of key fields and saved with a status of pending, along with metadata like creation time.

### 3.3 Interactive Analyst Dashboard

The web-based Streamlit dashboard serves as the analyst interface, offering real-time visibility and enhanced interpretability of alerts. Key functionalities include:

a. Loading and filtering alerts from eve.json based on severity, IP, or signature, as shown in Fig. 3 below. The Streamlit-based dashboard visualised the generated SSH brute force attacks in real-time.
b. Using a dropdown menu, an alert of interest is selected, sending context-rich prompts to OpenAI's GPT-4 via API to classify severity, explain potential threats, and recommend mitigations, as shown in Figs. 4 and 5.
c. Highlighting cases where GPT severity exceeds a predefined threshold, suggesting analyst attention. GPT classified the threat as "medium" severity based on attack behaviour and known threat intelligence, as shown in Fig. 6 below.
d. Enabling analysts to submit approve/deny decisions for queued actions directly from the interface.

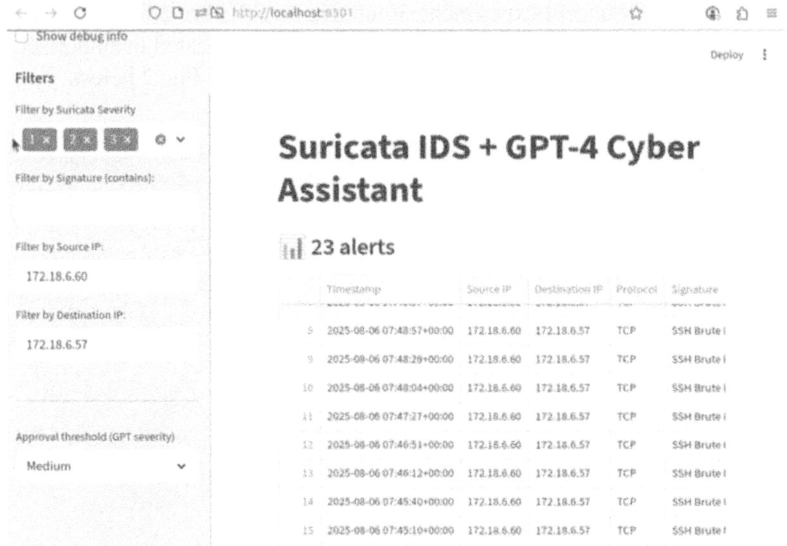


**Fig. 3.** The generated SSH brute force attacks in real-time

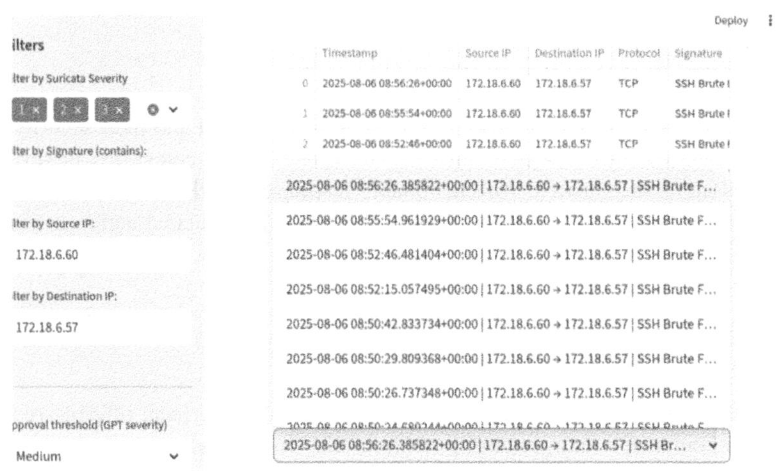


**Fig. 4.** A dropdown menu for alert selection

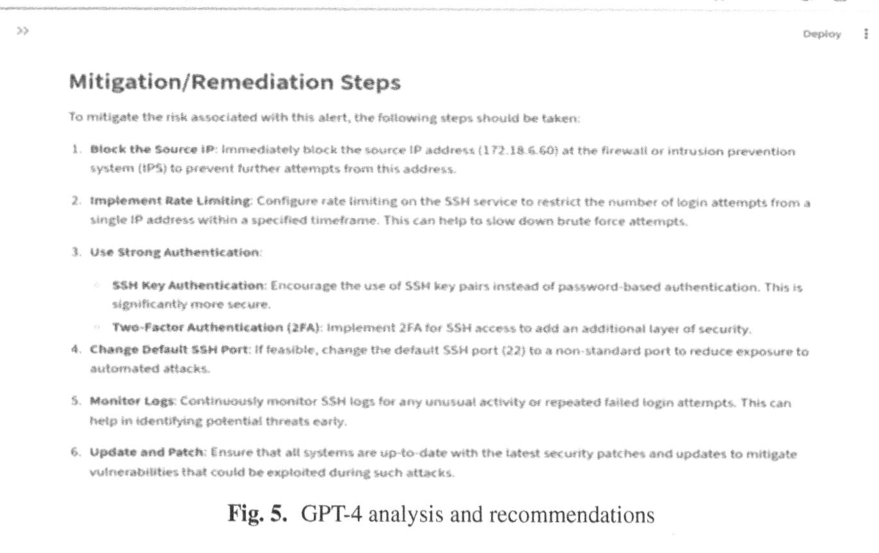


**Fig. 5.** GPT-4 analysis and recommendations

Deploy

**GPT Severity:** Medium

Queued pending approval 8ee800e35313dbcf7da68ba11e440945 (target src_ip=172.18.6.60).

Visual Summary Charts

**Pending Response Approvals**

Actions are proposed by the auto-responder (or GPT suggestions) and require analyst approval.

**Pending items: 1**

| | Pending ID | Signature | SID | Src → Dst | Proposed | Duration |
|---|---|---|---|---|---|---|
| 0 | 8ee800e35313dbcf7da68ba11e440945 | SSH Brute Force Attempt | None | 172.18.6.60 → 172.18.6.57 | block_ip | 10m |

SSH Brute Force Attempt (sid:None) — 8ee800e35313dbcf7da68ba11e440945

**Fig. 6.** GPT classification of the attack

### 3.4 Approval Enforcement CLI

To enforce blocking decisions, a command-line tool (approve_action.py) is invoked with sudo privileges. It parses the pending action from the shared state file, executes the corresponding mitigation, such as IP block via ufw, schedules an automatic rollback after the action duration lapses, and updates the action status to approved or denied in the state store. This CLI is triggered manually by the analyst, via the dashboard interface, as shown in Fig. 7.

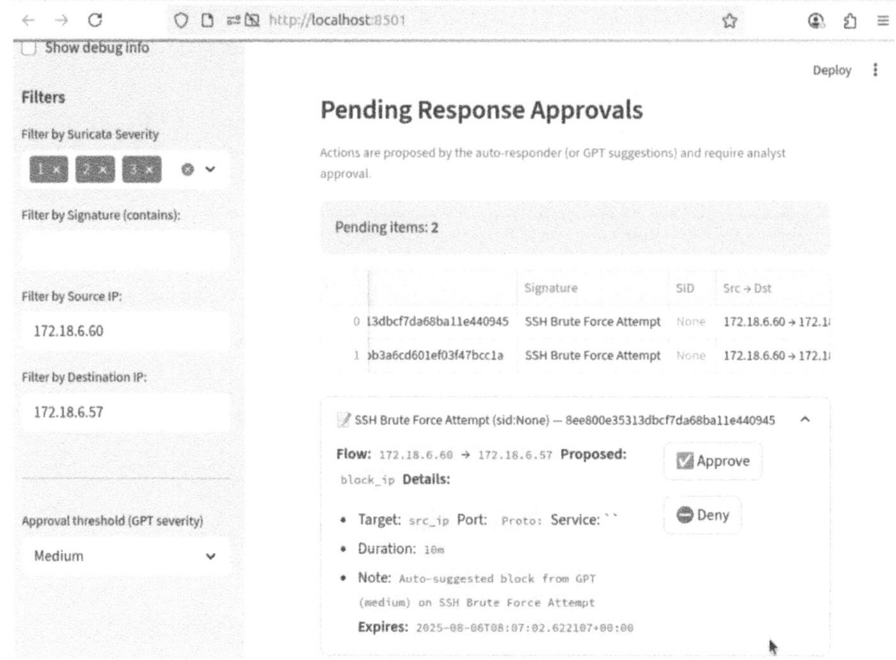


**Fig. 7.** Approval enforcement CLI

### 3.5 Persistent State Management

All queued actions and their metadata are stored in a timestamp-safe JSON file (auto_responder_state.json). The state store ensures that no duplicate actions are executed, analyst decisions are tracked, and timed-out actions are rolled back cleanly.

The proposed system, an analyst-in-the-loop pipeline, facilitates timely detection and contextual response to network threats, bridging the gap between traditional IDS tools and modern AI-powered threat intelligence.

## 4 Results and Discussion

To evaluate the effectiveness of the proposed hybrid intrusion detection and response framework, a proof-of-concept test was conducted using an SSH brute force attack originating from a Kali Linux machine targeting an Ubuntu system. Suricata successfully detected the malicious traffic based on a custom rule signature, generating real-time alerts in the eve.json output. These alerts were processed by the auto_responder.py module, which applied YAML-defined matchers to classify the incident and queue it for review. GPT-4 was then tasked with providing human-readable threat intelligence, including contextual interpretation and remediation steps.

Upon detecting the SSH brute force attempt, the GPT module classified the alert as "Medium" severity, citing the prevalence and automation of such attacks as justification.

Table 1 presents a summary of the threat intelligence generated by GPT-4 for the SSH brute force attempt.

**Table 1.** SSH Brute Force Threat Analysis Summary

| Section | Content |
|---|---|
| Threat Classification | Classification: Medium |
| Justification for Classification | The alert indicates a detected SSH brute force attempt, which is a common attack vector used by malicious actors to gain unauthorised access to systems. The classification as Medium is based on several factors:<br>• **Commonality:** SSH brute force attacks are prevalent and often automated, making them a routine threat<br>• **Potential Impact:** If successful, such an attack could lead to unauthorised access, data breaches, or system compromise. However, the actual risk depends on the strength of the SSH credentials and the security measures in place<br>• **Detection:** The detection of the attempt indicates that the system is actively monitoring for such threats, which mitigates the risk somewhat |
| Mitigation/Remediation Steps | To reduce the risk of successful SSH brute force attacks, consider implementing the following mitigation strategies:<br>• Change Default SSH Port: Alter the default SSH port (22) to a non-standard port to reduce the number of automated attacks<br>• Implement Strong Password Policies: Ensure that all SSH accounts use strong, complex passwords. Consider enforcing password expiration and complexity requirements<br>• Use Public Key Authentication: Disable password authentication and enforce the use of SSH keys for authentication. This significantly increases security as it requires possession of a private key<br>• Limit SSH Access: Restrict SSH access to known IP addresses or ranges using firewall rules or security groups. This can significantly reduce exposure to brute force attacks<br>• Enable Fail2Ban or Similar Tools: Utilise tools like Fail2Ban to monitor SSH logs and automatically block IP addresses exhibiting suspicious behaviour, such as multiple failed login attempts |

The system then auto-generated a pending response action recommending a block on the attacker's IP address (src_ip) via ufw. A notification was sent to Slack, as shown in Fig. 8. The analyst approved this action through the integrated Streamlit dashboard, triggering the approve_action.py script, which inserted the ufw deny from rule and scheduled its automatic removal after the expiry time. As shown in Fig. 9, the SSH brute force attempt dropped immediately. Further testing confirmed that approved blocks took immediate effect, with successful denial of traffic from the malicious IP and observable changes in ufw status as shown in Fig. 10.

**Fig. 8.** Slack notifications

```
[ATTEMPT] target 172.18.6.57 - login "root" - pass "agustin" - 3309 of 14344400 [child 0] (0/1)
[ATTEMPT] target 172.18.6.57 - login "root" - pass "webster" - 3310 of 14344400 [child 2] (0/1)
[ATTEMPT] target 172.18.6.57 - login "root" - pass "pussy1" - 3311 of 14344400 [child 2] (0/1)
[ATTEMPT] target 172.18.6.57 - login "root" - pass "pucca" - 3312 of 14344400 [child 1] (0/1)
[ATTEMPT] target 172.18.6.57 - login "root" - pass "penis" - 3313 of 14344400 [child 2] (0/1)
[ATTEMPT] target 172.18.6.57 - login "root" - pass "button" - 3314 of 14344400 [child 0] (0/1)
[ATTEMPT] target 172.18.6.57 - login "root" - pass "billiejoe" - 3315 of 14344400 [child 3] (0/1)
[ATTEMPT] target 172.18.6.57 - login "root" - pass "bandit1" - 3316 of 14344400 [child 2] (0/1)
[ATTEMPT] target 172.18.6.57 - login "root" - pass "56789" - 3317 of 14344400 [child 0] (0/1)
[ATTEMPT] target 172.18.6.57 - login "root" - pass "123456789123456" - 3318 of 14344400 [child 3] (0/1)
[ATTEMPT] target 172.18.6.57 - login "root" - pass "turner" - 3319 of 14344400 [child 2] (0/1)
[ATTEMPT] target 172.18.6.57 - login "root" - pass "sexychick" - 3320 of 14344400 [child 0] (0/1)
[ATTEMPT] target 172.18.6.57 - login "root" - pass "queens" - 3321 of 14344400 [child 3] (0/1)
[ATTEMPT] target 172.18.6.57 - login "root" - pass "iminlove" - 3322 of 14344400 [child 0] (0/1)
[ATTEMPT] target 172.18.6.57 - login "root" - pass "greeneyes" - 3323 of 14344400 [child 3] (0/1)
[ATTEMPT] target 172.18.6.57 - login "root" - pass "cinthia" - 3324 of 14344400 [child 0] (0/1)
[ERROR] could not connect to target port 22: Timeout connecting to 172.18.6.57
[ERROR] ssh protocol error
[VERBOSE] Retrying connection for child 2
[RE-ATTEMPT] target 172.18.6.57 - login "root" - pass "turner" - 3324 of 14344400 [child 2] (0/1)
[ERROR] could not connect to target port 22: Timeout connecting to 172.18.6.57
[ERROR] ssh protocol error
[VERBOSE] Retrying connection for child 3
[RE-ATTEMPT] target 172.18.6.57 - login "root" - pass "greeneyes" - 3324 of 14344400 [child 3] (0/1)
[ERROR] could not connect to target port 22: Timeout connecting to 172.18.6.57
[ERROR] ssh protocol error
[VERBOSE] Retrying connection for child 0
[RE-ATTEMPT] target 172.18.6.57 - login "root" - pass "cinthia" - 3324 of 14344400 [child 0] (0/1)
[ERROR] could not connect to target port 22: Timeout connecting to 172.18.6.57
[ERROR] ssh protocol error
[VERBOSE] Retrying connection for child 2
[RE-ATTEMPT] target 172.18.6.57 - login "root" - pass "turner" - 3324 of 14344400 [child 2] (0/1)
[ERROR] could not connect to target port 22: Timeout connecting to 172.18.6.57
[ERROR] ssh protocol error
[VERBOSE] Retrying connection for child 3
[RE-ATTEMPT] target 172.18.6.57 - login "root" - pass "greeneyes" - 3324 of 14344400 [child 3] (0/1)
[ERROR] could not connect to target port 22: Timeout connecting to 172.18.6.57
[ERROR] ssh protocol error
[VERBOSE] Retrying connection for child 0
[RE-ATTEMPT] target 172.18.6.57 - login "root" - pass "cinthia" - 3324 of 14344400 [child 0] (0/1)
```

**Fig. 9.** Blocked SSH brute force attack

```
┌──(kali㉿kali)-[~]
└─$ nc -zv 172.18.6.57 22
172.18.6.57: inverse host lookup failed: Host name lookup failure
(UNKNOWN) [172.18.6.57] 22 (ssh) : Connection timed out
```

**Fig. 10.** Proof of the blocked attack source IP

To measure the responsiveness of the proposed system, we applied Latency Metrics. We first measured the Detection to Approval Latency, which is the time taken between when the alert is first detected and when a human approves the mitigation action. Secondly, we measured the Detection to Mitigation Latency, which is the time from detection to actual enforcement of the response. This includes approval time and any system overhead. Good systems have only a few seconds' gap between approval and mitigation. Figure 11 presents a line graph illustrating the latency metrics captured at three critical stages of the automated intrusion response workflow: detection, approval, and mitigation. The x-axis represents the sequence of unique alert instances, while the y-axis quantifies the elapsed time in seconds. Two distinct latency lines are plotted: Detection to Approval and Detection to Mitigation, enabling comparative analysis of human responsiveness versus enforcement delay. The graph highlights that approval latency closely aligns with mitigation latency in most cases, confirming immediate action post-approval. Across ten alert instances (A001–A010), the system consistently demonstrates efficient performance, with approval latency typically ranging between 80 and 140 s, and mitigation latency closely tracking it with a minimal overhead. Notably, alert A005 achieved the fastest response time (~35s to approval, ~ 45s to mitigation), indicating optimal pipeline execution. While alerts A004 and A006 exhibited higher latencies (~170–180s), likely due to increased GPT reasoning or delayed analyst feedback, the system overall maintains low variance and tightly coupled automation. The small delta between approval and mitigation confirms the effectiveness of the integrated response mechanism. These results affirm the system's suitability for real-time threat triage in live SOC environments.

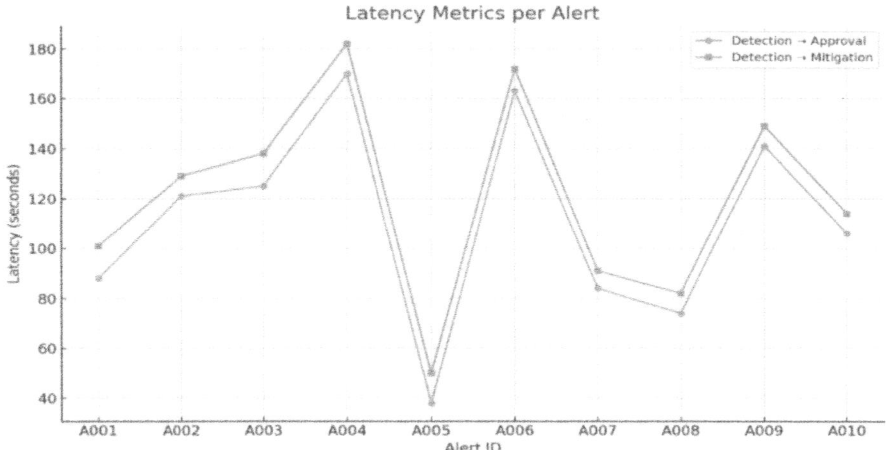


**Fig. 11.** Latency Metrics Visualisation

To enable analysts to evaluate and improve the quality of GPT-generated threat explanations continuously, we implemented a feedback mechanism, shown in Fig. 12 below. This feedback mechanism allows users to provide simple positive or negative feedback on each recommendation. The system captures valuable insights about the accuracy and usefulness of the AI's responses in real-world security contexts. This feedback not only helps to identify patterns of strengths and weaknesses in GPT's threat assessments but also establishes a foundation for iterative refinement of prompts, severity classification, and automated response rules. Ultimately, incorporating analyst feedback ensures that the tool remains aligned with operational needs and enhances the overall effectiveness of the intrusion detection and response process.

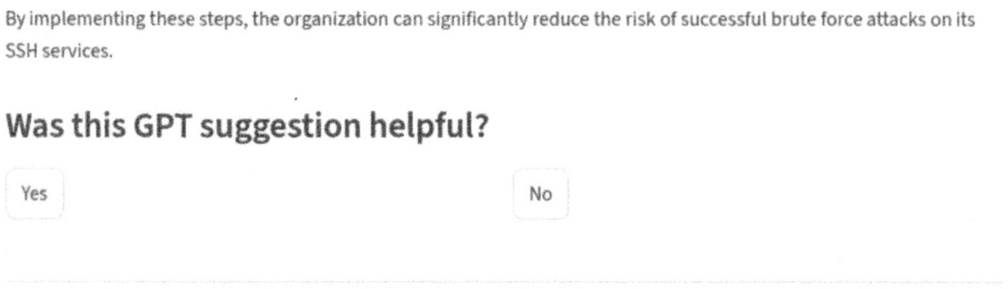

By implementing these steps, the organization can significantly reduce the risk of successful brute force attacks on its SSH services.

Was this GPT suggestion helpful?

Yes No

**Fig. 12.** The feedback buttons

This integration significantly improves operational efficiency in SOC environments by reducing Mean Time to Respond (MTTR). Unlike traditional IDS setups, where human triage and response introduce delays, the proposed system automates detection, contextual analysis, and enforcement, while still preserving human oversight at the decision-making step. Compared to existing commercial tools such as Microsoft Security Copilot, which requires cloud-based licensing and tight ecosystem integration, our solution remains lightweight, self-hosted, and modular, offering transparency, extensibility, and fine-grained control over automated actions. The proposed system was deployed in a controlled lab environment simulating real-world SOC operations. Through the SSH brute force attack scenario, we observed that GPT-4 consistently generated context-rich explanations that aligned with publicly known CVEs and threat intelligence frameworks such as MITRE ATT&CK. For example, Suricata alerts labelled generically as "Potential Protocol Command Decode" were correctly interpreted by GPT as brute force login attempts, complete with associated risk classification, and tailored mitigation guidance.

## 5 Conclusion

This study presents a practical and extensible intrusion detection and response framework that combines Suricata IDS with GPT-4-powered threat analysis and a human-in-the-loop approval mechanism. The system bridges the gap between alert generation and actionable remediation by automating interpretation and proposing response actions, while preserving analyst control. Through the integration of a Streamlit dashboard, Slack notifications, and ufw-based enforcement, the solution significantly reduces Mean Time to Respond (MTTR) to network threats without sacrificing transparency or flexibility.

The system was successfully validated using a simulated SSH brute-force attack originating from a Kali Linux machine, with alerts detected by Suricata, contextualised by GPT-4, reviewed via the dashboard, and enforced through dynamic firewall rules.

The results demonstrate the system's capacity to provide real-time, explainable, and auditable cybersecurity defence within constrained environments. Compared to proprietary systems such as Microsoft Security Copilot, this approach offers a lightweight, modular, and open-source alternative, better suited for institutions seeking affordable, interpretable solutions. While SSH brute-force was used as a proof of concept, future tests will extend to other attack vectors. In addition, we plan to evaluate the system's scalability by deploying it within a distributed SOC environment, leveraging asynchronous event ingestion and parallel GPT processing, while simulating high-traffic conditions using large-scale PCAP replay to assess performance degradation and alert handling efficiency.

**Credit Authorship Contribution Statement**

EMM: Writing Original draft, Data Curation, Formal Analysis, Investigation, Methodology, Simulation, Validation YS: Formal analysis, Investigation, Resources, Supervision, Fund Acquisition, Review and Editing, ZW: Formal, analysis, Methodology, Supervision, Validation, Review and Editing.

**Acknowledgments.** This work was supported in part by the South African National Research Foundation under Grants numbers CHN231229201835, CPRR240327211161, RA22112976288, KIC250321305600 and AJCR230704126719.

**Disclosure of Interests.** The authors declare that they have no known competing financial interests or personal relationships that could have appeared to influence the work reported in this paper.

## References

Al-Hawawreh, M., Aljuhani, A., Jararweh, Y.: ChatGPT for cybersecurity: practical applications, challenges, and future directions. Clust. Comput. **26**(6), 3421–3436 (2023). https://doi.org/10.1007/s10586-023-04124-5

Asfour, M., Murillo, J.C.: Harnessing large language models to simulate realistic human responses to social engineering attacks: a case study. Int. J. Cybersec. Intel. Cybercr. **6**(2), 21–49 (2023). https://doi.org/10.52306/2578-3289.1172

Balasubramanian, P., Seby, J., Kostakos, P.: CYGENT: a cybersecurity conversational agent with log summarisation powered by GPT-3. In: Proceedings of the 3rd International Conference on Artificial Intelligence and Internet of Things (AIIoT 2024), p. 101021911. IEEE (2024). https://doi.org/10.1109/AIIoT58432.2024.10574658

Beckerich, M., Plein, L., Coronado, S.: RatGPT: turning online LLMs into proxies for malware attacks (2023). arXiv Preprint arXiv:2308.09183. http://arxiv.org/abs/2308.09183

Bono, J., Grana, J., Xu, A.: Generative AI and Security Operations Center productivity: evidence from live operations (2024). arXiv Preprint arXiv:2411.03116. http://arxiv.org/abs/2411.03116

Chamkar, S.A., Maleh, Y., Gherabi, N.: Security Operations Centers: use case best practices, coverage, and gap analysis based on MITRE adversarial tactics, techniques, and common knowledge. J. Cybersec. Privacy **4**(4), 777–793 (2024). https://doi.org/10.3390/jcp4040036

Ferrag, M.A., et al.: Generative AI in cybersecurity: a comprehensive review of LLM applications and vulnerabilities. Inter. Things and Cyber-Phys. Sys. **5**, 1–46 (2025). https://doi.org/10.1016/j.iotcps.2025.01.001

Freitas, S., Kalajdjieski, J., Gharib, A., McCann, R.: AI-driven guided response for Security Operation Centers with Microsoft Copilot for Security. In: Proceedings of the 2025 ACM Conference, pp. 191–200. ACM (2025). https://doi.org/10.1145/3701716.3715209

Gupta, N., Traore, I., De Quinan, P.M.F.: Automated event prioritisation for Security Operation Center using deep learning. In: Proceedings of the 2019 IEEE International Conference on Big Data, pp. 5864–5872. IEEE (2019). https://doi.org/10.1109/BigData47090.2019.9006073

Khayat, M., Barka, E., Serhani, M.A., Sallabi, F., Shuaib, K., Khater, H.M.: Empowering Security Operation Center with artificial intelligence and machine learning: a systematic literature review. IEEE Access **13**, 19162–19197 (2025). https://doi.org/10.1109/ACCESS.2025.3532951

Kheddar, H.: Transformers and large language models for efficient intrusion detection systems: a comprehensive survey. Information Fusion **124**, 103347 (2025). https://doi.org/10.1016/j.inffus.2025.103347

McKee, F., Noever, D.: Chatbots in a botnet world. Int. J. Cybernet. Info. **12**(2), 77–95 (2023). https://doi.org/10.5121/ijci.2023.120207

Mugeshwaran, K., Maharajan, K., Nithish, D., Dinesh, S., Uday, N.: An integrated approach to AI-enhanced Security Information and Event Management. In: Proceedings of the International Conference on Computer, Robot, Testing, Engineering and Evaluation (ICCRTEE 2025), pp. 1–6. IEEE (2025). https://doi.org/10.1109/ICCRTEE64519.2025.11053120

Sedjelmaci, H.: Attacks detection and decision framework based on generative adversarial network approach: case of vehicular edge computing network. Trans. Emerg. Telecomm. Technol. **33**(10), e4073 (2022). https://doi.org/10.1002/ett.4073

# GBAA-GA: General Black-Box Adversarial Attack Method for Audio Recognition Based on Genetic Algorithm

Huan Tan, Sai Wang, Zhanghua Zhu(✉), and Chengze Li

Communication NCO Academy, ChongQing 400000, China
1060264392@qq.com

**Abstract.** Automatic Speech Recognition (ASR) has brought great convenience, whereas security issues are not supposed to be neglected. Universal adversarial attack techniques, which are known for their generalizability, have received wide-spread attention in academia. Current research primarily focuses on white-box attacks, where the attacker has full knowledge of the target model. However, obtaining a wide range of information is difficult at present. Therefore, a general black-box adversarial example attack algorithm, GBAA-GA, is designed based on a genetic algorithm. Specifically, this study first trained a speech command classification model based on the Speech Commands dataset as the attack target. Under the assumption that only the model's input and output information is accessible, the study incorporated the concept of genetic algorithm and introduced an adaptive mutation probability strategy to address the issue of local optima in the algorithm. This approach successfully generated universal audio adversarial examples in a black-box setting. Finally, comparative experiments with the white-box universal perturbation attack algorithm PGD-UAP demonstrate that GBAA-GA achieves performance comparable to PGD-UAP in key metrics such as attack success rate and stealthiness, thereby fully validating the effectiveness and practicality of the proposed algorithm.

**Keywords:** speech recognition · adversarial attack · universal adversarial examples · genetic algorithm · black-box attack

## 1 Introduction

With the rapid development of artificial intelligence technology, automatic speech recognition (ASR), as one of the key technologies in human computer interaction, has been widely applied in numerous fields such as virtual assistants, smart homes, in-vehicle systems, and financial services. However, as the application scenarios of speech recognition systems continue to expand and deepen, the security threats they face are becoming increasingly prominent. Adversary sample at-tacks represent one of the most challenging issues in this regard. Adversary samples(Kurakin et al., 2016) are created by adding carefully crafted, imperceptibly slight perturbation noises to clean data, resulting in new data samples that may cause the original recognition model to produce incorrect identification


H. Yan et al. (Eds.): AIS&P 2025, CCIS 2918, pp. 15–28, 2026.
https://doi.org/10.1007/978-981-95-9884-7_2

results. Similarly, in ASR, attackers can construct adversarial speech samples to trick ASR into misinterpreting malicious commands as legitimate ones or misrecognizing normal speech, thereby leading to a series of security issues.

Current research on adversarial attacks against ASR systems has been categorized by academia from various dimensions. Based on the attacker's level of knowledge about the target model, attacks can be classified into black-box attacks and white-box attacks (Wei et al., 2022; Li et al., 2022; Song, 2023). Among these, white-box attacks assume that the attacker has complete knowledge of the target model's network architecture, parameters, and other information. Leveraging this information, the attacker deliberately adds adversarial perturbations to the original samples to generate adversarial examples. The Fast Gradient Sign Method (FGSM) (Goodfellow et al., 2015) is one of the most commonly used and classic techniques for white-box adversarial attacks. It was initially applied in the field of image recognition adversarial attacks, and subsequent re-search has shown its effectiveness in the domain of speech adversarial attacks as well. This algorithm works by obtaining the gradient information of the original model and iteratively moving a certain step size along the gradient direction, adding this to the original sample until the target model produces an incorrect recognition result, thereby generating an adversarial example. Gong et al. (2018) first designed a method based on the concept of FGSM to generate adversarial examples for speech. This approach directly adds adversarial noise to the raw audio waveform, thus avoiding the losses associated with converting between acoustic features and waveform features. In the same year, Iter et al. (2017) combined FGSM with the deceptive gradient method, modifying and adding adversarial perturbations at the MFCC feature level using a pretrained WaveNet model, and then inversely transforming the MFCC features back into the waveform audio signal to achieve the adversarial attack. Experiments demonstrated that although the inverse trans-formation of MFCC features incurs some loss, the original speech and the adversarial samples still maintain consistent characteristics.

Black-box attacks assume that the adversary possesses little to no knowledge of the target model's internal information when conducting adversarial attacks. However, in real world scenarios, obtaining detailed internal information about the target model is often extremely difficult. Therefore, black-box attacks are more challenging compared to white-box attacks, yet they possess inherent advantages. Previous research outcomes encompass: Cisse et al. (2017) designed a method named Houdini. They first performed untargeted white-box attacks on the DeepSpeech2 (Amodei et al., 2016) model, then applied the generated adversarial samples to the Google Voice system. Experimental results demonstrated that the adversarial samples also exhibited significant attack effectiveness against Google Voice, indicating that white-box adversarial samples possess a certain degree of transferability. Additionally, Alzantot et al. (2022) proposed a method based on a genetic algorithm approach. Relying solely on the probability vector output by the target model, they iteratively searched for powerful adversarial samples under black-box conditions to deceive the speech classification model. Khare et al. (2019) proposed a multi-objective evolutionary approach to conduct adversarial attacks against ASR systems under black-box conditions. Their method aimed to cause transcription errors in the ASR system while ensuring high similarity between the adversarial audio and the original speech. The approach simultaneously optimized multiple objectives: the edit distance between the targset and original text, and the Euclidean distance between the MFCC features of the original and adversarial audio. By incorporating evolutionary principles and simulating biological evolution theory, the method used an environmental fitness function to screen for the optimal populations and individuals.

Furthermore, from the perspective of the attack scope of adversarial examples, they can be categorized into individual adversarial attacks and universal adversarial attacks. As the names imply, individual adversarial attacks target only a single audio instance, meaning the adversary must find and add distinct perturbations for each audio sample. In contrast, universal adversarial attacks are effective against all or a certain category of audio samples, eliminating the need to generate unique pertur-bations for each instance. It is evident that universal adversarial attacks are more suitable for realworld scenarios, offering greater generality and flexibility when targeting models. Early research on universal adversarial attacks primarily focused on the image domain. In 2017, Moosavi Dezfooli et al. (2017) first introduced the concept of universal perturbations. They proposed a method named the vanilla universal attack, which iteratively creates image specific perturbations for data points and accumulates them into a universal adversarial perturbation. Subsequently, Neekhara et al. (2019) discovered that universal perturbations are not limited to the image domain but are also applicable in the audio domain. Later, building on the work of Neekhara, Li et al. (2020) proposed a universal perturbation generation method targeted at speaker verification ASR systems. This approach utilizes a generative model to map from a low-dimensional normal distribution to a universal perturbation subspace, producing universal perturbations capable of targeted attacks and generating multiple similar yet distinct universal perturbations.

In real world scenarios, particularly in military-related domains, obtaining extensive information about the target model for white-box attacks is highly challenging. Moreover, an attack method that generates adversarial samples for only individual audio instances is difficult to implement on edge devices and is also unacceptable in terms of time cost. To address these issues, this paper proposes a General Black-Box Adversarial Attack Method for Audio Recognition Based on Genetic Algorithm (GBAA-GA). This method can generate universal perturbations that are effective only against a specific category of voice commands without affecting the recognition of other command categories, relying solely on the input-output information of the target model. Thereby, it achieves both black-box attack and universal attack effects.

The main contributions of this paper are as follows:

a) A voice command classification and recognition model was designed and implemented, providing a foundational target model for subsequent universal adversarial example attack algorithms.
b) A genetic algorithm based black-box universal adversarial example generation algorithm, GBAA-GA, was proposed, along with an improved adaptive mutation probability strategy.
c) Comparative experiments on the Speech Command dataset evaluating attack effectiveness and stealth demonstrated that, compared to the white-box PGD-UAP algorithm, the GBAA-GA algorithm achieves performance levels close to those of the white-box approach in both metrics, even with minimal information about the target model.

## 2 Research Methods

First, this paper constructing and training a speech recognition model using the Speech Commands dataset. Subsequently, under the assumption that only the input and output information of the model can be accessed, a universal adversarial attack is performed on multiple speech data samples of a specific command category based on a genetic algorithm. Through iterative processes of crossover, mutation, and selection, the study ultimately achieves the effect of a universal adversarial attack under black-box conditions. The primary models and algorithms involved in the adversarial attack are as follows:

### 2.1 Speech Command Recognition Model

This paper proposes a nine class speech command recognition model based on a Bidirectional Long-Short-Term Memory (BiLSTM) network. By integrating BiLSTM (Minami, 2018) with fully connected neural network layers, the model achieves effi-cient speech command classification. The specific implementation workflow is as follows: first, the training set audio data is preprocessed by extracting Mel-Frequency Cepstral Coefficients (MFCCs), which are then normalized; subse-quently, the processed features are fed into a hidden layer composed of three BiLSTM layers, followed by further processing through two fully connected layers, ultimately outputting the classification result and corresponding confidence score for each speech command. As shown in Fig. 1, it illustrates the detailed network architecture of the model (Tan, 2021).

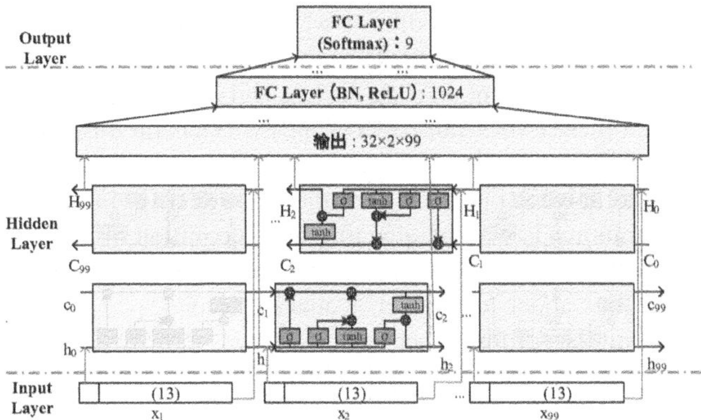


**Fig. 1.** Network Architecture for Audio Command Classification and Recognition Model

### 2.2 General Black-Box Adversarial Attack Method Based on Genetic Algorithm (GBAA-GA)

Assumptions: The algorithm operates under the assumption that only the input and output information of the speech command recognition model is accessible.

**Objective:** The goal is to generate a universal perturbation that causes the recognition model to misclassify only the speech samples from the targeted attack category, while leaving its performance on samples from other categories unaffected.

**Procedure:** As outlined in the pseudocode of Algorithm 1, the process begins by initializing a population of noise vectors (noise_pop) and other relevant parameters. The algorithm then enters an iterative genetic cycle. A subset of elite individuals (elite_size) is chosen directly based on these fitness scores. Subsequently, new offspring are generated through pairwise crossover and gene mutation, forming the new noise population (noise_pop) for the next generation. This cycle of selection, crossover, and mutation repeats until convergence, ultimately yielding the optimal universal noise (best_pop_noise).

**Algorithm 1. Black-Box Universal Adversarial Audio Example Generation Algorithm**

```
Input: All categories of audio data (all_audios); Target category audio data (tar
        get_audios); Other categories of audio data (other_audios).
Output: The current best universal perturbation (best_pop_noise).
  1) Initialize noise_pop; pop_size; elite_size; mutation_p
  2) target_audios,other_audios = audio_load()
  3) while iter >= 300 or attack_rate >= 0.7
  4)    for noise in noise_pop:
  5)      out_put=speechModel(all_audios)
  6)      scores=get_fitness_score(out_put) // Evaluate fitness scores
  7)    end
  8)    noise_pop = get_new_pop(noise_pop , scores, elite_size, pop_size)
  9)    Statistical Attack Effectiveness; iters++
 10) end
```

The core of Algorithm 1 lies in its fitness function, crossover, and mutation strategies, which are detailed below.

**Fitness Function.** The genetic algorithm requires a fitness function to simulate the natural selection process. According to the algorithm's objective, the fitness function should be designed such that the target audio samples (target_audios) are misclassified as much as possible, while the other audio samples (other_audios) are correctly classified to the greatest extent. Since the fitness function must balance these two competing objectives, it is formulated as shown in Algorithm 2.

**Algorithm 2. Fitness Function**

```
Input: Adversarial speech recognition outputs (output_batch), ground-truth audio
       labels (batch_y), loss function weight (loss_w).
Output: Fitness score
  1) def get_fitness_score(output_batch,batch_y):
  2)  losses = CrossEntropyLoss(output_batch, batch_y)
  3)  for i in range(batch_size):
  4)   if Target Category:
  5)     loss1 += - losses[i]
  6)   else
  7)     loss2 += losses[i]
  8)  score = loss1 * loss_w[0] + loss2 * loss_w[1]
  9)  return - score
```

Algorithm 2 employs the cross-entropy loss to quantify the discrepancy between the recognition model's output vector and the ground truth labels. The final fitness score comprises two components: for the target category samples, the objective is to maximize the deviation from the original label (i.e., a larger cross-entropy loss is desirable), hence this term is assigned a negative value; for the other category samples, the goal is to maintain recognition accuracy, so this term is assigned a positive value.

**Adaptive Mutation Probability** Genetic Algorithm is a process of heuristic search for the global optimum. Within this process, the fitness function acts as the environmental selection mechanism, while crossover and mutation serve as the sources of variation. Typically, crossover involves only the exchange of genetic material between two individuals. Without mutation, repeated crossover operations would lead to a fixed gene pool, ceasing to produce new variations. Therefore, mutation is crucial for the search capability of GAs, as it determines the influx of new variables into the population. Consequently, setting an appropriate mutation probability is critical for the algorithm's rapid convergence. This paper proposes an adaptive mutation probability strategy designed to achieve the following objectives:

1) The mutation probability is in [0, Pmax], where Pmax $< 1$.
2) The mutation probability is dynamically adjusted in each iteration, being inversely proportional to the magnitude of change in the population's fitness. A smaller change triggers a higher mutation rate, whereas a larger change results in a lower rate.
3) The strategy must provide fine grained discrimination during periods of evolutionary stagnation to facilitate escape from local optima.

Based on the above objectives, the strategy is designed as follows:

$$mutation\_p = p_{base} \times (\frac{1}{1 + e^{kL_{score}}}) \tag{1}$$

In Eq. (1), $\boldsymbol{P_{base}}$ is the base upper limit of the mutation rate, $\boldsymbol{L_{score}}$ is the difference in fitness score of the best individual between the current iteration and the previous one, and $\boldsymbol{k}$ is a scaling constant. The adaptive mutation probability is automatically adjusted based on the difference between the optimal fitness values of the last two iterations. By leveraging a sigmoid style function to map this difference to an appropriate mutation probability, the strategy elegantly satisfies the objectives outlined above.

**Crossover.** After the population is screened by the fitness function from the previous generation, the elite individuals undergo a pairwise crossover operation, with the probability of crossover weighted by their fitness scores. In this experiment, we employ gene point crossover rather than gene segment crossover, as detailed in Algorithm 3.

**Algorithm 3: Genetic Crossover Algorithm**

Input: elite population (elite_pop), fitness scores of the elite population (elite_pop_scores), total population size (pop_size).
Output: next generation population (next_pop).

```
1) def get_new_pop(elite_pop,elite_scores,pop_size):
2)  elite_probs = elite_scores / elite_scores.sum()
3)  cand1,cand2=choice(elite_probs,elite_pop, pop_size)
4)  mask = rand(pop_size,elite_pop.size)<0.5
5)  next_pop = mask*cand1+(1-mask)*cand2
6)  return next_pop
```

# 3 Experiment

## 3.1 Dataset and Evaluation Metrics

**Dataset Construction.** The attack model in this study primarily targets the classification and recognition of voice control command files. The target model utilizes the Speech Commands dataset, jointly developed by Google TensorFlow and the AIY team, as training data. This dataset comprises approximately 65,000 one-second-long WAV format english short audio clips with a sampling rate of 16 kHz, covering 30 common categories including words, numbers, and directional commands. To better simulate real world application scenarios, this study selects eight directional commands ("down", "go", "left", "off", "on", "right", "stop", "up") and randomly samples uniformly from the remaining categories to form an "other" class. The detailed composition of the final experimental dataset is provided in Table 1.

**Table 1.** Overview of the Speech Command Dataset Construction

| Split subset | down | go | left | off | on | right | stop | up | Other |
|---|---|---|---|---|---|---|---|---|---|
| Data set | 2359 | 2372 | 2353 | 2357 | 2367 | 2367 | 2380 | 2375 | 2500 |
| Target model's training set | | | | | 19287 | | | | |
| Target model's test set | | | | | 2143 | | | | |

**Evaluation Metrics.** To quantitatively evaluate the effectiveness of the algorithm, this study employs several metrics to measure the performance of universal adversarial examples, including the Targeted Attack Success Rate (T-ASR) and Untargeted Accidental

Attack Rate (U-AAR). T-ASR is defined as the ratio of successfully attacked adversarial examples to the total number of target-class samples. U-AAR refers to the proportion of non-target audio samples that are incorrectly classified due to the adversarial perturbation, relative to the total number of non-target samples.

Simultaneously, the stealthiness of the universal adversarial examples is assessed using the Average Similarity metric. The Average Similarity is represented by the mean of the correlation coefficients calculated between the original audio samples and their adversarial counterparts. The formula for the correlation coefficient is as follows:

$$r(X, Y) = \frac{Cov(X, Y)}{\sqrt{var|X| * var|Y|}} \tag{2}$$

In Formula (2), $Cov(X, Y)$ represents the covariance of X and Y, $var|X|$ denotes the variance of X, and $var|Y|$ denotes the variance of Y. By comparing the correlation coefficient between the original audio and the adversarial audio sample, the overall similarity between the two can be quantitatively assessed.

### 3.2 Speech Command Classification Model Training

During the training phase of the speech command classification and recognition model, the relevant parameter configurations are outlined in Table 2. The training set was constructed using the method described in Table 1. The number of training epochs was set to 10, and the batch size (BATCH_SIZE) was configured as 64. Furthermore, the cross-entropy loss function was employed as the loss metric, and the Adam optimization algorithm was utilized for gradient based parameter updates to achieve efficient convergence.

**Table 2.** Training Parameters for the Speech Command Classification Model

| Parameter name | Parameter value |
|---|---|
| BATCH_SIZE | 64 |
| epochs | 10 |
| Learning Rate | 1 × 10–3 |
| The optimization algorithm Adam | 64 |

After model training was completed, its classification performance was evaluated using the test set. An early stopping strategy was adopted during training, retaining the model with the best performance from the 10 epochs as the final classifier. Testing confirmed that the model achieved a classification accuracy of 96.17%.

### 3.3 Speech Command Classification Model Training

Regarding the selection of the mutation rate for the genetic algorithm, a comparison was conducted between a fixed mutation rate and varying mapping constants with different k values. The k values selected for comparison were 5, 10, and 20. As shown in Fig. 2, the magnitude of k determines the steepness of the mapping curve. Furthermore, by analyzing the variation pattern of scores per generation under a fixed mutation probability, it was observed that during the early iterations, the mutation probability generally remained below 0.5, while it dropped below 0.1 during the later stages of deep optimization. Selecting a mapping constant of $k = 10$ effectively distinguishes the original probability distribution and aligns well with the practical experimental requirements.

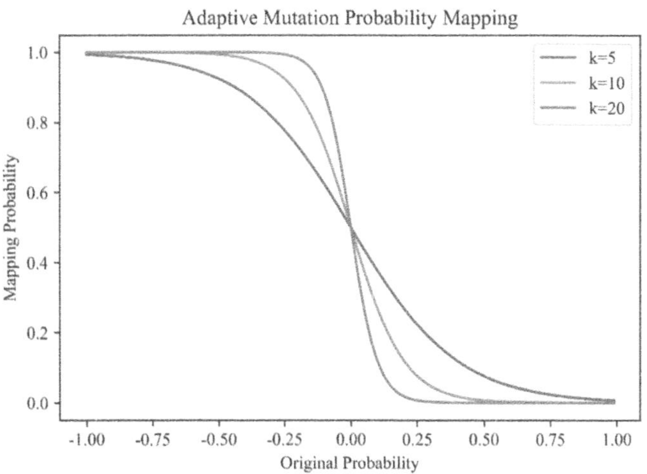


**Fig. 2.** Adaptive Mutation Probability Mapping Diagram

Simultaneously, based on a mapping constant of $k = 10$, we compared and analyzed the change trends of the best individual fitness during the iterative process when using fixed mutation probability versus adaptive mutation probability. As shown in Figs. 3 and 4, when a fixed mutation probability is employed, the genetic algorithm transitions from a relatively rapid ascent phase into a slow-rising bottleneck phase after approximately 50 generations, making the optimization process prone to becoming trapped in local optima. In contrast, the adaptive mutation probability strategy successfully escapes these local optima. Furthermore, when both algorithms were iterated for 300 generations, as illustrated in Fig. 4, the adaptive method mapped the mutation rate to the interval of approximately (0, 0.3). This resulted in an improvement of about 47.1% in the highest fitness score of the loss function (measured by the best score in the final iteration). In terms of attack success rate, the adaptive mutation probability achieved a 16.9% improvement over the fixed mutation probability. The performance of the two mutation strategy algorithms on the test set is summarized in Table 3.

**Table 3.** Comparison of Mutation Probability Strategy Effectiveness

| Metric | Fixed Mutation Rate | Adaptive Mutation Rate |
|---|---|---|
| Highest Fitness Score | 20.809 | 30.609 |
| Attack Success Rate | 28.58% | 45.48% |

Therefore, it can be concluded that the adaptive mutation probability method offers significant advantages and improvements over the fixed mutation rate approach.

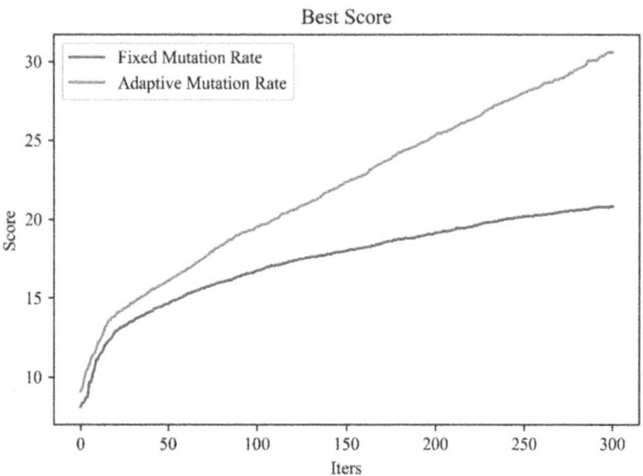


**Fig. 3.** Optimal Score Trend under Fixed Mutation Rate Iteration

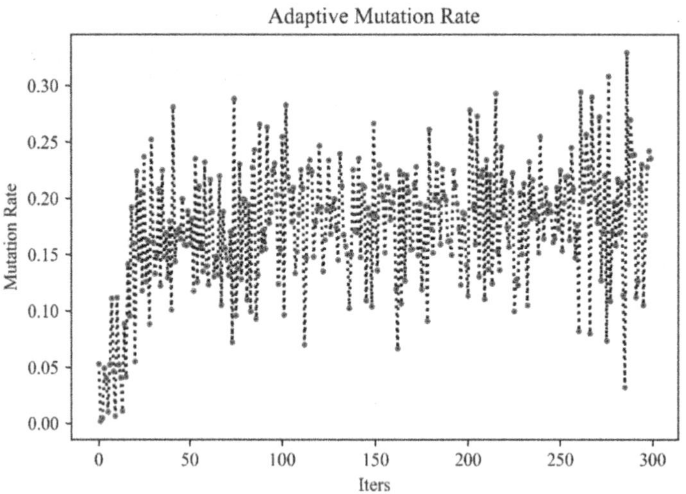


**Fig. 4.** Optimal Score Trend with Adaptive Mutation Rate Iteration

### 3.4 Universal Adversarial Examples Algorithm: Results and Analysis

Given that utilizing all audio samples for the training set would exceed the capacity of conventional computational resources, 100 audio samples per category were selected to form the training set for the black-box universal adversarial attack training. Furthermore, the quality of the algorithm was evaluated based on three metrics: (Attack Effectiveness) T-ASR and U-AAR; and (Attack Stealthiness) Average Similarity. Since there are currently no directly comparable schemes for generating black-box universal adversarial examples in the audio domain. Therefore, a comparative analysis was conducted with the white-box universal perturbation attack algorithm PGD-UAP (Zhang, 2022). PGD-UAP is a universal perturbation attack based on the PGD algorithm. Its primary strategy involves applying the concept of gradient integration across each batch of the entire known training set. Specifically, it minimizes the objective loss function by utilizing the average of the gradients obtained after performing gradient descent on each batch of data.

The parameter configurations for the black-box universal adversarial example algorithms are detailed in Table 4.

**Table 4.** Selection of GBAA-GA Algorithm Parameters

| Parameter name | Parameter value |
|---|---|
| pop_size | 100 |
| elite_size | 10 |
| epsMax | 5*10–4 |
| Maximum Absolute Amplitude of the Noise | 0.01 |
| Fitness Weight | [10, 1] |
| iters | 300 |
| k | 10 |

**Attack Effectiveness Evaluation.** This study employed the PGD-UAP algorithm and the GBAA-GA algorithm for training on the training set, respectively. The main parameters for PGD-UAP were selected as follows: a single perturbation step size Eps = 0.0004, number of iterations epoch = 50, with other main parameters consistent with those of GBAA-GA. After testing on the same test set, the effectiveness results are shown in Table 5:

**Table 5.** Performance Metrics Evaluation Results

| Effectiveness Metric | PGD-UAP | GBAA-GA |
|---|---|---|
| T-ASR | 49.89% | 45.48% |
| U-AAR | 17.07% | 23.36% |

It can be observed that T-ASR of GBAA-GA can reach approximately 91% of that achieved by PGD-UAP, while the U-AAR only increases by 6.29% compared to PGD-UAP. However, considering that white-box attacks require complete knowledge of the target model, whereas black-box attacks only need access to the model's inputs and outputs, the optimization search difficulty of the algorithm is notably higher in the black-box setting. Therefore, from the perspective of attack effectiveness metrics, the GBAA-GA algorithm demonstrates validity in attacking the target model, and its attack performance is largely comparable to that of white-box attacks.

**Assessment of Attack Stealthiness.** As can be seen from Table 6, the average similarity of both the PGD-UAP and GBAA-GA algorithms exceeds 90%. Furthermore, As shown in Fig. 5, the similarity of adversarial examples generated by GBAA-GA is predominantly distributed above 90%, with a distribution pattern that shows little difference from that of white-box attacks. This indicates that the black-box adversarial examples are highly similar to the original audio.

**Table 6.** Stealthiness Metrics Evaluation Results

| Stealthiness Metric | PGD-UAP | GBAA-GA |
|---|---|---|
| Average Similarity | 95.59% | 92.45% |

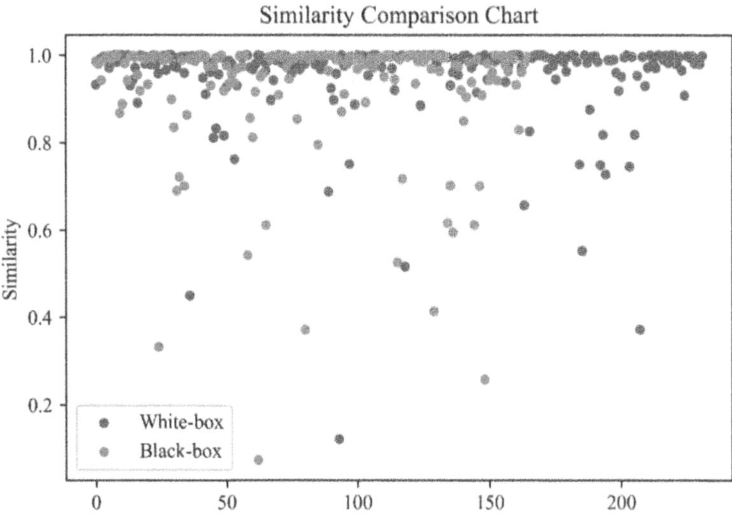


**Fig. 5.** Comparative Analysis of White-box vs. Black-box Adversarial Sample Similarity

Meanwhile, as shown in Fig. 6, it reveals that the black-box universal adversarial noise primarily consists of low-frequency background noise, which has minimal impact on key audio information and human auditory sensitivity. In summary, the adversarial examples generated by GBAA-GA exhibit high similarity to the original audio, demonstrating strong attack stealthiness.

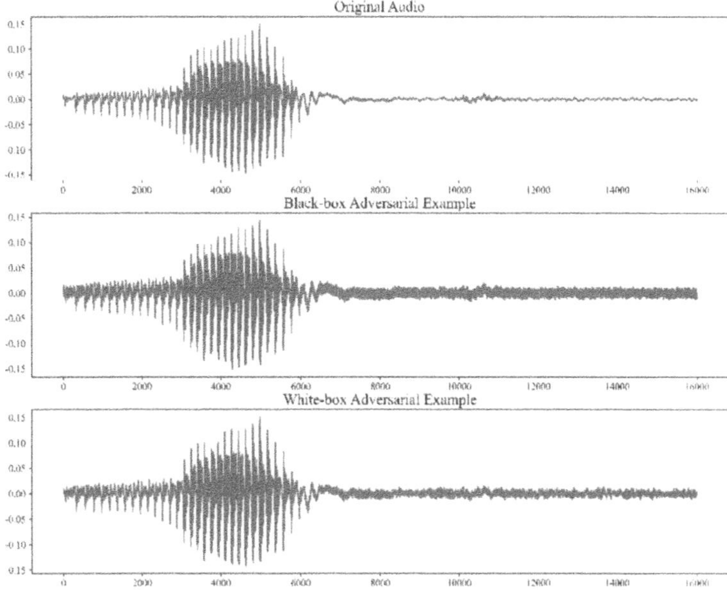


**Fig. 6.** Waveform Comparison of Adversarial Examples

## 4 Conclusion

This paper proposes a General Black-box Adversarial Example Generation algorithm based on Genetic Algorithms (GBAA-GA). Leveraging the concept of genetic algorithms integrated with an adaptive mutation probability strategy, GBAA-GA successfully generates universal adversarial perturbations using only the input-output information of the target model. Comparative experiments with the white-box PGD-UAP algorithm demonstrate that GBAA-GA achieves significant effectiveness in both attack success and perturbation stealthiness.

For future work, research can be extended to generating universal black-box adversarial examples in the physical world. It would be valuable to account for various physical world interference factors, such as environmental noise and device specific noise (Xie et al., 2020; Yakura and Sakuma, 2018; Chen et al., 2020). This direction would contribute to developing a more practical and applicable technical framework for adversarial attacks against universal speech intelligence systems.

## References

Alzantot, M., Balaji, B., Srivastava, M.: Did you hear that? Adversarial examples against automatic speech recognition (n.d.). Retrieved 11 August 2022, from https://arxiv.org/pdf/1801.00554.pdf

Amodei, D., Ananthanarayanan, S., Anubhai, R., et al.: Deep speech2 end-to-end speech recognition in English and mandarin. In: Proceedings of the 33rd International Conference on Machine Learning, pp. 173–182 (2016)

Chen, T., Shangguan, L., Li, Z., et al.: Metamorph: Injecting inaudible commands into over-the-air voice controlled systems. In: Network and Distributed Systems Security (NDSS) Symposium (2020)

Cisse, M., Adi, Y., Neverova, N., et al.: Houdini: Fooling deep structured visual and speech recognition models with adversarial examples. In: 31st International Conference on Neural Information Processing Systems, pp. 6980–6990 (2017)

Gong, Y., Poellabauer, C.: Crafting adversarial examples for speech paralinguistics applications. In: Proceedings of the Dynamic and Novel Advances in Machine Learning and Intelligent Cyber Security Workshop, pp. 1–8 (2018)

Goodfellow, I.J., Shlens, J., Szegedy, C.: Explaining and harnessing adversarial examples. In: 3rd International Conference on Learning Representations, pp. 1–11 (2015)

Iter, D., Huang, J., Jermann, M.: Generating adversarial examples for speech recognition (Stanford Technical Report) (2017)

Khare, S., Aralikatte, R., Mani, S.: Adversarial black-box attacks on automatic speech recognition systems using multi-objective evolutionary optimization. In: Interspeech 2019, pp. 3208–3212 (2019)

Kurakin, A., Goodfellow, I., Bengio, S.: Adversarial examples in the physical world (2016). Retrieved from https://arxiv.org/abs/1607.02533

Li, K.Z., Xu, Y., Zhang, S.C., Yan, J.L.: Survey on adversarial example attack and defense technology for automatic speech recognition. Comput. Eng. Appl. **58**(14), 1–15 (2022)

Li, J.G., Zhang, X.F., Jia, C.M., et al.: Universal adversarial perturbations generative network for speaker recognition. In: 2020 IEEE International Conference on Multimedia and Expo, pp. 1–6. IEEE (2020)

Minami, S.: Predicting equity price with corporate action events using LSTM-RNN. J. Math. Fin. **8**(1), 58–63 (2018)

Moosavi-Dezfooli, S.M., Fawzi, A., Fawzi, O., et al.: Universal adversarial perturbations. In: 2017 IEEE Conference on Computer Vision and Pattern Recognition (CVPR), pp. 86–94. IEEE Computer Society (2017)

Neekhara, P., Hussain, S., Pandey, P., et al.: Universal adversarial perturbations for speech recognition systems. In: Interspeech 2019, pp. 481–485 (2019)

Zhang, Q.B.: Research and implementation of universal perturbation technique for intelligent voice control systems (Doctoral dissertation). Xidian University (2022). https://doi.org/10.27389/d.cnki.gxadu.2022.002430

Han, S.S.: Research on speech adversarial example generation techniques for universal and stealth (Doctoral dissertation). PLA Strategic Support Force Information Engineering University (2023). https://doi.org/10.27188/d.cnki.gzjxu.2023.000065

Tan, H.: Research and implementation of intelligent voice control system adversarial sample attack and defense technology (Doctoral dissertation). Xidian University (2021). https://doi.org/10.27389/d.cnki.gxadu.2021.003477

Wei, C.Y., Meng, S., Zhou, X., et al.: Reviews on the attack and defense methods of voice adversarial examples. Journal of Cyber Security **7**(01), 100–113 (2022). https://doi.org/10.19363/J.cnki.cn10-1380/tn.2022.01.07

Xie, Y., Shi, C., Li, Z.H., et al.: Real-time, universal, and robust adversarial attacks against speaker recognition systems. In: ICASSP 2020 - 2020 IEEE International Conference on Acoustics, Speech and Signal Processing, pp. 1738–1742. IEEE (2020)

Yakura, H., Sakuma, J.: Robust audio adversarial example for a physical attack (2018). arXiv Preprint arXiv:1810.11793

# ConsistRAG: Defending Retrieval-Augmented Generation of Large Language Models Against Knowledge Corruption Attacks

Shumin Liu and Xiaohui Yang(✉)

School of Software Engineering, Jiangxi University of Science and Technology, Nanchang 330013, China
1633608107@qq.com

**Abstract.** Retrieval-Augmented Generation (RAG) demonstrates outstanding performance in open-domain question-answering tasks. However, its heavy reliance on external knowledge bases makes it vulnerable to Knowledge Corruption Attacks. Recent studies like PoisonedRAG demonstrate that attackers can effectively mislead retrieval and generation processes by injecting malicious documents into knowledge bases, highlighting the practicality and severity of such attacks. To address this threat, this paper proposes a two-stage defense mechanism named ConsistRAG. This mechanism first employs knowledge expansion strategies to introduce supplementary documents, enhancing retrieval diversity and robustness while addressing the extreme case of empty results after filtering. It then uses K-Means clustering to identify and filter out semantically homogeneous document clusters. Finally, it performs inter-document consistency checks to detect semantic conflicts among candidate documents, further eliminating potential logical poisoning content. Experimental results demonstrate that ConsistRAG effectively reduces the attack success rate (ASR) while maintaining high accuracy (ACC) and validates the effectiveness and generalization capability of the ConsistRAG defense mechanism against knowledge corruption attacks.

**Keywords:** LLM · Retrieval-Augmented Generation · Knowledge Corruption Attacks · ConsistRAG

## 1 Instruction

Large language models (LLMs), such as GPT-3.5 (Brown et al., 2020), GPT-4 (Achiam et al., 2023), and PaLM 2 (Anil et al., 2023), have found extensive real-world applications due to their exceptional generative capabilities. However, these models still face inherent limitations: First, due to the temporal truncation of pre-training data, they lack up-to-date knowledge. Second, they are prone to "hallucination" (Ji et al., 2023), generating inaccurate responses. Third, in specialized domains like medicine, models may exhibit knowledge gaps. These shortcomings pose significant challenges when deploying such models in critical sectors such as healthcare (Brown et al., 2020; Al Ghadban et al., 2023; Wang et al., 2024), finance and law (Kuppa et al., 2023; Mahari, 2021), and scientific research (Kumar et al., 2023; Prince et al., 2024; Boyko et al., 2023).


H. Yan et al. (Eds.): AIS&P 2025, CCIS 2918, pp. 29–40, 2026.
https://doi.org/10.1007/978-981-95-9884-7_3

To mitigate challenges like knowledge update difficulties and unstable responses in large models, Retrieval-Augmented Generation (RAG) (Karpukhin et al., 2020; Lewis et al., 2020; Borgeaud et al., 2022; Thoppilan et al., 2022) has emerged as a mainstream solution. RAG enhances answer reliability and timeliness by retrieving relevant documents from external knowledge bases to provide additional context for LLMs.

However, this architecture introduces new security vulnerabilities. Since RAG relies on external knowledge bases, attackers can inject malicious documents to execute knowledge corruption attacks or corpus poisoning attacks without altering model parameters (Zhong et al., 2023). A prime example is PoisonedRAG (Zou et al., 2024). Such attacks can prioritize carefully crafted "poisoned documents" during the retrieval phase, feeding them as context into the generation stage to induce the model to output predetermined incorrect answers. For instance, attackers can maliciously edit Wikipedia entries (Carlini et al., 2024), publish fake news, host malicious websites, or even insert poisoned documents into private corporate knowledge bases.

PoisonedRAG pioneered systematic research on knowledge corruption attacks against RAG systems, demonstrating that a small number of fabricated documents can achieve extremely high attack success rates (ASR$\approx$97%) without significantly impacting normal performance, while remaining difficult to detect. In a PoisonedRAG attack, an attacker first selects one or more target questions (Target Questions) and assigns arbitrary target answers (Target Answers) to them. By injecting highly relevant poisoned documents into the knowledge base, attackers can induce RAG to output predetermined answers when queried on these topics. This method not only propagates misinformation (e.g., erroneous leader names) and steers commercial preferences but also generates financial disinformation, posing severe threats to high-risk scenarios such as cybersecurity, finance, and healthcare.

To counter knowledge corruption attacks (e.g., PoisonedRAG) in Retrieval-Augmented Generation (RAG) systems, this paper proposes a two-stage defense framework——that combines clustering filtering with cross-document consistency detection. Unlike methods relying solely on similarity or perplexity metrics, ConsistRAG fully leverages two distinct characteristics of poisoned documents: homogenization and logical inconsistency. In the first stage, we employ K-Means clustering to identify and filter out clusters of documents with high semantic homogeneity, thereby suppressing attacker poisoning achieved through repeated injection. In the second stage, we introduce inter-document consistency detection to identify semantic conflicts among candidate documents, further eliminating potential logically poisoned content. Simultaneously, during the retrieval phase, ConsistRAG employs knowledge expansion strategies to introduce supplementary documents. This enhances the diversity and robustness of retrieval results while addressing the extreme case of empty results after filtering. Through this two-stage collaboration, ConsistRAG effectively improves the robustness of RAG systems against various attack scenarios.

The main contributions of this paper are as follows:

1. We designed a two-stage defense method combining K-Means clustering with cross-document consistency detection, systematically mitigating the threat of knowledge base injection backdoor attacks against RAG systems.

2. ConsistRAG significantly reduces attack success rates (by up to 80%) while maintaining high and stable accuracy (improved by up to 30%) across different large language models, outperforming all existing defenses.

# 2 Design of Consistrag

## 2.1 Overview of ConsistRAG

ConsistRAG is a defense framework designed to counter knowledge corruption attacks targeting RAG systems. It integrates K-Means clustering filtering, knowledge expansion mechanisms, and inter-document consistency detection to generate more robust and trustworthy responses. As illustrated in Fig. 1, attackers optimize and inject malicious documents tailored to specific queries and target answers. The retriever first obtains candidate documents from the knowledge base and performs preliminary filtering using K-Means clustering to exploit the homogeneity of toxic documents. If all Top-k documents fall within toxic clusters or few documents remain after this step, the system introduces additional candidates via knowledge expansion to increase corpus diversity and weaken the clustering effect of toxic documents. Subsequently, consistency detection is employed to identify logical conflicts among candidate documents, further eliminating potential malicious content. Finally, the system generates responses based on the most reliable document set, significantly enhancing robustness against adversarial attacks.

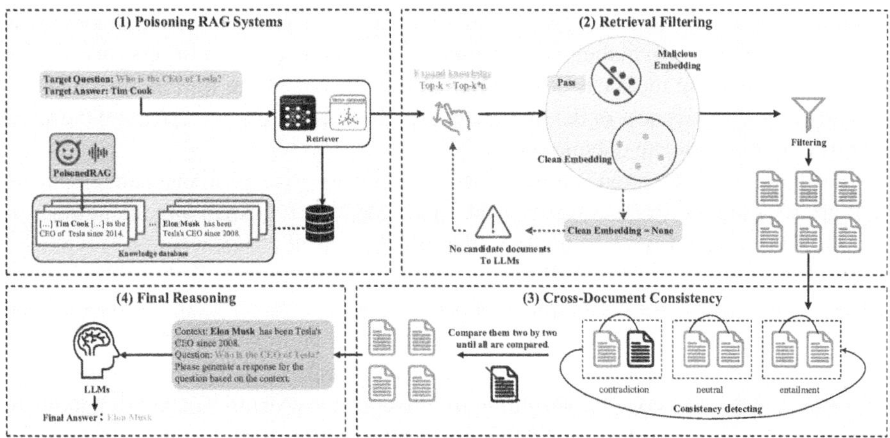


**Fig. 1.** ConsistRAG Defense Structure.

## 2.2 Retrieval Filtering - Stage 1

In the attacker strategies, for each query, attackers may generate either a single malicious document or multiple malicious documents to disrupt the entire RAG process. These are termed Single Injection Attacks and Multiple Injection Attacks, respectively.

The Retrieval Filtering stage addresses two distinct attack scenarios: single-injection attacks and multiple-injection attacks. Under multiple-injection attacks, it aims to filter out malicious documents, ensuring only clean documents are passed to the next stage as retrieval content. In single-injection scenarios, although malicious documents lurk as isolated entries within the document candidate set, ConsistRAG's second stage provides secondary defense against both malicious and clean documents delivered from the Retrieval Filtering stage. Malicious documents are constructed as $S \oplus I$, where attackers optimize component S to maximize query-malicious document similarity, thereby elevating retrieval priority. As shown in Eqs. 4 and 5, I denotes attacker-predefined initial text designed to misguide LLM responses. Under these constraints, generated malicious documents typically exhibit higher similarity in the embedding space than clean documents.

Simultaneously, methods like PoisonedRAG (Zou et al., 2024) leverage LLMs to generate initial text I and produce multiple malicious documents for the same query by adjusting temperature parameters, enabling repeated injection attacks. Since these documents share the same generation process, they inherently exhibit high similarity. Further optimized under the constraints of Eq. 3, these malicious documents form tighter clusters in the embedding space than clean documents. Leveraging this characteristic, ConsistRAG employs K-means clustering ($k = 2$) during the Retrieval Filtering stage to distinguish clean documents from potentially malicious ones based on their embedding distributions. In scenarios with multiple injections, this defense strategy effectively filters out the most malicious document set due to their extreme similarity.

In practical scenarios, the number and diversity of retrieved candidate documents significantly impact defense effectiveness. Our experiments demonstrate that when the candidate set is small (e.g., Top-k = 5), individual malicious documents can easily blend into normal clusters and evade detection. Conversely, when the candidate set is large (e.g., Top-k = 20), malicious documents form distinct small clusters due to high homogeneity, making them easier to eliminate via the clustering filtering mechanism. Therefore, the effectiveness of defense mechanisms depends to some extent on the scale and diversity of candidate documents.

Based on this observation, we introduce a knowledge expansion mechanism during the Retrieval Filtering stage to increase the candidate set size (expanding the number of candidate documents from k to n during the initial retrieval phase, where n is typically an integer multiple of k, such as 4k). This enhances the robustness and diversity of the clustering filtering stage. Furthermore, in extreme scenarios where the top-k retrieval results are completely overridden by toxic content and entirely filtered out during the Retrieval Filtering stage, leaving no candidate documents to be passed to subsequent RAG stages, the knowledge expansion mechanism re-retrieves from a larger candidate set to ensure the RAG system continues functioning.

When addressing single-injection attacks, the N-gram mechanism employs the ROUGE-L score (Renze, 2024) to measure intra-cluster similarity. This approach preserves as many clean documents as possible while establishing a defensive foundation for the second stage of ConsistRAG.

### 2.3 Cross-Document Consistency - Stage 2

After the initial clustering and filtering in Stage 1, the system has preliminarily eliminated large-scale homogeneous poisoned candidate documents. However, relying solely on semantic clustering remains insufficient to fully eliminate potential malicious interference, as attackers may exploit diverse phrasing to inject semantically contradictory poisoned documents. To address this, ConsistRAG introduces a Cross-Document Consistency (CDC) mechanism in Stage 2 to further screen, identify, and eliminate candidate documents exhibiting contradictions or poisoning characteristics.

Within a normal knowledge base, documents retrieved for the same query typically exhibit high consistency—meaning core facts should mutually corroborate across documents. Poisoned documents, however, primarily feature logical or factual contradictions with most normal documents, attempting to mislead the model into generating incorrect answers.

The document consistency detection mechanism employs a Natural Language Inference (NLI) model (Liu et al., 2021; Chen et al., 2021; Laban et al., 2022) to measure semantic relationships between documents. Given two document fragments $d_i$ and $d_j$, the NLI model outputs probability distributions for three relationship types: entailment, neutral, and contradiction. If a document $d_i$ exhibits a high contradiction probability ($\tau$) relative to most other documents, it is flagged as a potential poison document and discarded. Only the candidate set with high inter-document consistency is retained as input for the generation model.

This consistency detection mechanism effectively mitigates the impact of knowledge corruption attacks. In such attacks, malicious documents often exhibit significant conflicts with original, legitimate knowledge in the corpus. By statistically aggregating reasoning relationships among documents, ConsistRAG extracts more consistent knowledge support from multiple sources, thereby enhancing the reliability and robustness of generated answers. Furthermore, this mechanism does not rely on an internal model knowledge base. Instead, it defends by analyzing the semantic structure and consistency relationships among candidate documents, making it more suitable for countering external knowledge base poisoning scenarios.

# 3 Results and Analysis

## 3.1 Evaluation Metrics

To comprehensively evaluate the effectiveness of the ConsistRAG defense method against PoisonedRAG attacks in RAG scenarios, the following metrics are adopted:

Attack Success Rate (ASR). ASR measures whether the generated answer for a target question matches the attacker's predefined target answer. Following (Rizqullah et al., 2023; Huang et al., 2023), we employ a substring matching criterion: an attack is considered successful when the target answer is a substring of the model's generated answer. This method is suitable for closed-ended questions (e.g., "Sam Altman" and "The CEO of OpenAI is Sam Altman" are considered consistent in substring matching), avoiding false positives from exact-match requirements. Zou et al. validated this approach through human evaluation, demonstrating high consistency between ASR results from substring matching and human assessments.

Accuracy (ACC). Measures the proportion of correct responses generated by the RAG system under normal conditions, reflecting the reliability of its retrieval and generation.

## 3.2 Comparison of Experimental Results

We conducted comprehensive experiments under two scenarios: Single Injection Attack and Multiple Injection Attack.

**Single Injection Attack.**
In this scenario, all attack methods can inject only one malicious document into the retrieval database to manipulate the RAG system's responses to specific queries. As shown in Table 1, most existing defense methods perform poorly when only a single malicious document is injected. For instance, in attacks on the MS-MARCO dataset, ASR for other defense frameworks varied significantly from 12.0% to 56.0%, while ConsistRAG maintained a stable ASR between 11.0% and 13.0% while achieving the highest ACC.

**Multiple Injection Attack.**
As shown in Table 2, we evaluated performance at various poisoning rates (20% to 100%) on the open-source model $\text{Llama}_{3.1\text{-}8B}$. Experimental results demonstrate that ConsistRAG maintains stable performance in both ASR and ACC across all datasets, achieving average accuracy exceeding 85% on the NQ and MS-MARCO datasets. Notably, the RobustRAG defense framework—which employs an aggregation and voting strategy fails when malicious documents outnumber clean ones. In contrast, ConsistRAG's knowledge expansion mechanism and K-means clustering filter significantly reduce both the quantity and probability of retrieving malicious documents during the retrieval phase. Only a small fraction of these documents are subsequently used in the Cross-Document Consistency verification stage. These results clearly demonstrate that ConsistRAG significantly enhances the robustness and stability of RAG systems against repeated injection attacks.

**Table 1.** The result table presents the performance of various defense frameworks applied to different LLMs integrated with RAG systems, primarily against single injection attacks. The best performance is in bold.

| Models | Defense | NQ | | HotpotQA | | MS-MARCO | |
|---|---|---|---|---|---|---|---|
| | | LM_targeted | hotflip | LM_targeted | hotflip | LM_targeted | hotflip |
| | | ACC↑/ ASR↓ | ACC↑/ ASR↓ | ACC↑/ ASR↓ | ACC↑/ ASR↓ | ACC↑/ ASR↓ | ACC↑/ ASR↓ |
| $\text{Llama}_{3.1\text{-}8B}$ | Vanilla RAG | 26.0/73.0 | 25.0/71.0 | 27.0/81.0 | 26.0/80.0 | 28.0/70.0 | 26.0/71.0 |
| | RobustRAG | 51.0/27.0 | 50.0/26.0 | 40.0/50.0 | 42.0/49.0 | 67.0/19.0 | 68.0/18.0 |
| | InstructRAG | 58.0/37.0 | 59.0/35.0 | 54.0/41.0 | 55.0/39.0 | 63.0/33.0 | 64.0/32.0 |
| | ASTUTE RAG | 77.0/11.0 | 75.0/10.0 | 65.0/**16.0** | 64.0/**15.0** | 54.0/41.0 | 53.0/40.0 |
| | TrustRAG | 83.0/**9.0** | 81.0/12.0 | 68.0/19.0 | 67.0/21.0 | 81.0/12.0 | 78.0/15.0 |
| | ConsistRAG | **89.0**/10.0 | **89.0/10.0** | **75.0/16.0** | **73.0**/21.0 | **86.0/11.0** | **88.0/11.0** |
| $\text{Mistral}_{7B}$ | Vanilla RAG | 21.0/82.0 | 20.0/80.0 | 20.0/89.0 | 22.0/90.0 | 23.0/81.0 | 22.0/80.0 |
| | RobustRAG | 50.0/16.0 | 51.0/17.0 | 43.0/47.0 | 44.0/45.0 | 68.0/15.0 | 67.0/15.0 |

(*continued*)

**Table 1.** *(continued)*

| Models | Defense | NQ | | HotpotQA | | MS-MARCO | |
|---|---|---|---|---|---|---|---|
| | | LM_targeted | hotflip | LM_targeted | hotflip | LM_targeted | hotflip |
| | | ACC↑/ ASR↓ | ACC↑/ ASR↓ | ACC↑/ ASR↓ | ACC↑/ ASR↓ | ACC↑/ ASR↓ | ACC↑/ ASR↓ |
| | InstructRAG | 44.0/45.0 | 43.0/44.0 | 33.0/62.0 | 34.0/64.0 | 64.0/55.0 | 63.0/54.0 |
| | ASTUTE RAG | 69.0/**10.0** | 68.0/**11.0** | 67.0/26.0 | 65.0/28.0 | 66.0/28.0 | 67.0/27.0 |
| | TrustRAG | 83.0/15.0 | 81.0/18.0 | **84.0/16.0** | **83.0/18.0** | 81.0/18.0 | 78.0/21.0 |
| | ConsistRAG | **85.0**/14.0 | **83.0**/16.0 | 67.0/20.0 | 68.0/25.0 | **86.0/11.0** | **87.0/11.0** |
| $Qwen_{2.5\text{-}7B}$ | Vanilla RAG | 24.0/80.0 | 23.0/78.0 | 25.0/87.0 | 26.0/90.0 | 22.0/83.0 | 20.0/82.0 |
| | RobustRAG | 47.0/14.0 | 43.0/16.0 | 48.0/44.0 | 50.0/47.0 | 69.0/17.0 | 67.0/18.0 |
| | InstructRAG | 47.0/42.0 | 42.0/46.0 | 36.0/60.0 | 37.0/58.0 | 62.0/56.0 | 62.0/58.0 |
| | ASTUTE RAG | 71.0/**10.0** | 68.0/**14.0** | 70.0/22.0 | 71.0/26.0 | 68.0/29.0 | 67.0/30.0 |
| | TrustRAG | 80.0/18.0 | 78.0/21.0 | **75.0**/25.0 | **72.0**/27.0 | 79.0/21.0 | 76.0/23.0 |
| | ConsistRAG | **88.0**/11.0 | **84.0/14.0** | 71.0/**17.0** | **72.0/21.0** | **85.0/13.0** | **84.0/15.0** |

**Table 2.** The result table presents the performance of various defense frameworks applied to $Llama_{3.1\text{-}8B}$ integrated with RAG systems, primarily against LM_targeted attack method of PoisonedRAG with different poison rates. The best performance is in bold.

| Datasets | Defense | Poison-(100%) | Poison-(80%) | Poison-(60%) | Poison-(40%) | Poison-(20%) | Poison-(0%) |
|---|---|---|---|---|---|---|---|
| | | ACC↑/ ASR↓ | ACC↑/ ASR↓ | ACC↑/ ASR↓ | ACC↑/ ASR↓ | ACC↑/ ASR↓ | ACC↑ |
| NQ | Vanilla RAG | 2.0/98.0 | 2.0/98.0 | 3.0/97.0 | 4.0/93.0 | 26.0/73.0 | 71.0 |
| | RobustRAG | 11.0/83.0 | 15.0/75.0 | 23.0/63.0 | 37.0/46.0 | 51.0/27.0 | 61.0 |
| | InstructRAG | 27.0/69.0 | 38.0/56.0 | 40.0/56.0 | 51.0/45.0 | 58.0/37.0 | 68.0 |
| | ASTUTE RAG | 61.0/29.0 | 64.0/24.0 | 68.0/19.0 | 69.0/18.0 | 77.0/11.0 | 75.0 |
| | TrustRAG | 82.0/**2.0** | 84.0/**2.0** | 83.0/**2.0** | 82.0/**1.0** | 83.0/**9.0** | 82.0 |
| | ConsistRAG | **89.0/2.0** | **88.0/2.0** | **87.0/2.0** | **90.0**/2.0 | **89.0**/10.0 | **89.0** |
| HotpotQA | Vanilla RAG | 1.0/99.0 | 2.0/97.0 | 6.0/94.0 | 5.0/94.0 | 27.0/81.0 | 71.0 |
| | RobustRAG | 8.0/89.0 | 10.0/87.0 | 19.0/76.0 | 33.0/57.0 | 40.0/50.0 | 54.0 |
| | InstructRAG | 26.0/73.0 | 40.0/57.0 | 50.0/47.0 | 52.0/44.0 | 54.0/41.0 | **83.0** |
| | ASTUTE RAG | 48.0/41.0 | 53.0/38.0 | 59.0/30.0 | 59.0/31.0 | 65.0/**16.0** | 65.0 |
| | TrustRAG | 66.0/**4.0** | 70.0/**4.0** | 70.0/8.0 | 68.0/6.0 | 68.0/19.0 | 74.0 |
| | ConsistRAG | **72.0**/6.0 | **75.0/4.0** | **75.0/2.0** | **77.0/2.0** | **75.0/16.0** | 75.0 |
| MS-MARCO | Vanilla RAG | 3.0/97.0 | 3.0/96.0 | 5.0/94.0 | 7.0/93.0 | 28.0/70.0 | 79.0 |
| | RobustRAG | 25.0/68.0 | 28.0/66.0 | 37.0/54.0 | 57.0/34.0 | 67.0/19.0 | 73.0 |
| | InstructRAG | 44.0/54.0 | 47.0/51.0 | 49.0/45.0 | 60.0/36.0 | 63.0/33.0 | 89.0 |
| | ASTUTE RAG | 26.0/73.0 | 40.0/57.0 | 50.0/47.0 | 52.0/44.0 | 54.0/41.0 | 83.0 |
| | TrustRAG | 85.0/**5.0** | 83.0/7.0 | 84.0/8.0 | 85.0/7.0 | 81.0/12.0 | 85.0 |
| | ConsistRAG | **87.0**/6.0 | **89.0/4.0** | **90.0/4.0** | **91.0/4.0** | **86.0/11.0** | **90.0** |

### 3.3 Ablation Study

We conducted an ablation study on Llama$_{3.1\text{-}8B}$ to evaluate the impact of three key components in the ConsistRAG defense mechanism: (1) Using knowledge expansion strategies during the retrieval phase to augment the candidate set size, thereby enhancing the robustness and diversity of the clustering filtering phase and addressing the extreme case of empty results after retrieval filtering (denoted as "w/o Expand knowledge"); (2) the use of the K-Means filtering strategy in Phase 1 for clean retrieval (denoted as "w/o K-Means"); (3) the application of cross-document consistency detection in Phase 2 to further screen, identify, and eliminate contradictory or poisoned candidate documents (denoted as "w/o Cross-Document Consistency").

As shown in Table 3 and Fig. 2, each component of the ConsistRAG framework significantly contributes to its robustness against poisoning attacks. Notably, the defenses in the K-Means and Cross-Document Consistency stages do not appear to significantly impact the accuracy of RAG system responses. This is because the Expand Knowledge mechanism substantially reduces the probability of poisoned documents being retrieved by the RAG system before these two stages. Consequently, these subsequent defenses focus on more granular exclusion and filtering. The effectiveness of the Expand knowledge mechanism is further validated in subsequent experiments examining the expansion factor—specifically, the scale of the Top-K documents—within Knowledge Expansion.

The combination of K-Means filtering and Cross-Document Consistency detection has proven to be a powerful approach for reducing ASR.

**Table 3.** Ablation study results for the Llama$_{3.1\text{-}8B}$ model across three datasets (NQ, HotpotQA, and MS-MARCO), w/o E-k means w/o Expand knowledge; w/o KM means w/o K-Means; w/o CDC means w/o Cross-Document Consistency

| Datasets | Defense | Poison-100% | Poison-80% | Poison-60% | Poison-40% | Poison-20% | Poison-0% |
|---|---|---|---|---|---|---|---|
| | | **ACC↑/ ASR↓** | **ACC↑/ ASR↓** | **ACC↑/ ASR↓** | **ACC↑/ ASR↓** | **ACC↑/ ASR↓** | **ACC↑** |
| NQ | Vanilla RAG | 2.0/98.0 | 2.0/98.0 | 3.0/97.0 | 4.0/93.0 | 26.0/73.0 | 71.0 |
| | ConsistRAG$_{w/o\ E\text{-}k}$ | 66.0/5.0 | 59.0/15.0 | 67.0/5.0 | 69.0/5.0 | 69.0/28.0 | 79.0 |
| | ConsistRAG$_{w/o\ KM}$ | 88.0/12.0 | 87.0/13.0 | 87.0/9.0 | 90.0/8.0 | 89.0/10.0 | 89.0 |
| | ConsistRAG$_{w/o\ CDC}$ | 87.0/3.0 | 86.0/4.0 | 86.0/**2.0** | 88.0/5.0 | 90.0/9.0 | 90.0 |
| | ConsistRAG | **89.0/2.0** | **88.0/2.0** | **91.0/2.0** | **92.0/2.0** | **92.0/8.0** | **92.0** |
| HotpotQA | Vanilla RAG | 1.0/99.0 | 2.0/97.0 | 6.0/94.0 | 5.0/94.0 | 27.0/81.0 | 71.0 |
| | ConsistRAG$_{w/o\ E\text{-}k}$ | 50.0/9.0 | 46.0/18.0 | 51.0/5.0 | 53.0/5.0 | 54.0/35.0 | 65.0 |
| | ConsistRAG$_{w/o\ KM}$ | 67.0/16.0 | 74.0/15.0 | 73.0/9.0 | 74.0/8.0 | 72.0/18.0 | 72.0 |
| | ConsistRAG$_{w/o\ CDC}$ | 70.0/7.0 | 73.0/5.0 | 74.0/3.0 | 76.0/6.0 | 74.0/**15.0** | 73.0 |
| | ConsistRAG | **72.0/6.0** | **75.0/4.0** | **75.0/2.0** | **77.0/2.0** | **75.0**/16.0 | **75.0** |
| MS-MARCO | Vanilla RAG | 3.0/97.0 | 3.0/96.0 | 5.0/94.0 | 7.0/93.0 | 28.0/70.0 | 79.0 |
| | ConsistRAG$_{w/o\ E\text{-}k}$ | 63.0/9.0 | 59.0/16.0 | 67.0/7.0 | 69.0/7.0 | 63.0/30.0 | 77.0 |
| | ConsistRAG$_{w/o\ KM}$ | 86.0/16.0 | 87.0/15.0 | 87.0/11.0 | 90.0/10.0 | 83.0/13.0 | 87.0 |
| | ConsistRAG$_{w/o\ CDC}$ | 85.0/7.0 | 88.0/5.0 | 88.0/**4.0** | 89.0/7.0 | 84.0/12.0 | 89.0 |
| | ConsistRAG | **87.0/6.0** | **89.0/4.0** | **90.0/4.0** | **91.0/4.0** | **86.0/11.0** | **90.0** |

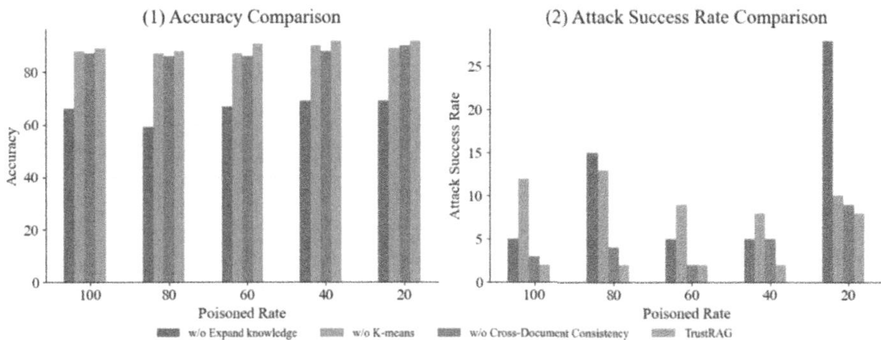


**Fig. 2.** Figure showing the impact of different modules on attack success rate and accuracy in ablation study. (1) The bar plot of ablation study on ACC in NQ based on the $\text{Llama}_{3.1\text{-}8B}$. (2) The bar plot of ablation study on ASR in NQ based on the $\text{Llama}_{3.1\text{-}8B}$

## Impact of Different Retrievers

Table 4 demonstrates ConsistRAG's performance across different retrievers (Contriever, Contriever-ms, ANCE) under default experimental settings. Results indicate that the ConsistRAG defense mechanism reliably identifies and filters malicious text, ensuring answer accuracy and credibility, while exhibiting stable and broad applicability across various retrievers.

**Table 4.** Impact of different retrievers.

| Models | Defense | NQ | | HotpotQA | | MS-MARCO | |
|---|---|---|---|---|---|---|---|
| | | LM_targeted | hotflip | LM_targeted | hotflip | LM_targeted | hotflip |
| | | **ACC↑/ ASR↓** | **ACC↑/ ASR↓** | **ACC↑/ ASR↓** | **ACC↑/ ASR↓** | **ACC↑/ ASR↓** | **ACC↑/ ASR↓** |
| **Contriever** | | | | | | | |
| $\text{Llama}_{3.1\text{-}8B}$ | Vanilla RAG | 2.0/98.0 | 2.0/97.0 | 1.0/99.0 | 2.0/98.0 | 3.0/97.0 | 1.0/98.0 |
| | ConsistRAG | 89.0/2.0 | 89.0/2.0 | 72.0/6.0 | 73.0/6.0 | 87.0/6.0 | 87.0/7.0 |
| $\text{Mistral}_{7B}$ | Vanilla RAG | 1.0/99.0 | 1.0/97.0 | 1.0/99.0 | 1.0/98.0 | 2.0/98.0 | 1.0/97.0 |
| | ConsistRAG | 79.0/5.0 | 77.0/5.0 | 70.0/3.0 | 71.0/4.0 | 83.0/7.0 | 83.0/6.0 |
| $\text{Qwen}_{2.5\text{-}7B}$ | Vanilla RAG | 1.0/99.0 | 1.0/97.0 | 1.0/99.0 | 1.0/99.0 | 2.0/98.0 | 2.0/97.0 |
| | ConsistRAG | 87.0/2.0 | 85.0/2.0 | 69.0/4.0 | 70.0/4.0 | 84.0/8.0 | 85.0/8.0 |
| **Contriever-ms** | | | | | | | |
| $\text{Llama}_{3.1\text{-}8B}$ | Vanilla RAG | 6.0/93.0 | 5.0/91.0 | 4.0/96.0 | 3.0/95.0 | 23.0/76.0 | 21.0/74.0 |
| | ConsistRAG | 88.0/1.0 | 87.0/1.0 | 70.0/5.0 | 72.0/5.0 | 90.0/4.0 | 90.0/6.0 |
| $\text{Mistral}_{7B}$ | Vanilla RAG | 5.0/93.0 | 4.0/92.0 | 1.0/99.0 | 1.0/99.0 | 25.0/77.0 | 24.0/78.0 |
| | ConsistRAG | 79.0/4.0 | 78.0/4.0 | 68.0/2.0 | 70.0/4.0 | 85.0/5.0 | 84.0/6.0 |
| $\text{Qwen}_{2.5\text{-}7B}$ | Vanilla RAG | 5.0/95.0 | 4.0/94.0 | 2.0/99.0 | 1.0/99.0 | 27.0/79.0 | 27.0/78.0 |
| | ConsistRAG | 86.0/1.0 | 85.0/2.0 | 70.0/5.0 | 70.0/4.0 | 87.0/5.0 | 85.0/6.0 |
| **ANCE** | | | | | | | |
| $\text{Llama}_{3.1\text{-}8B}$ | Vanilla RAG | 9.0/91.0 | 9.0/92.0 | 3.0/97.0 | 3.0/98.0 | 17.0/82.0 | 16.0/83.0 |

*(continued)*

**Table 4.** *(continued)*

| Models | Defense | NQ | | HotpotQA | | MS-MARCO | |
|---|---|---|---|---|---|---|---|
| | | LM_targeted | hotflip | LM_targeted | hotflip | LM_targeted | hotflip |
| | | **ACC↑/ ASR↓** | **ACC↑/ ASR↓** | **ACC↑/ ASR↓** | **ACC↑/ ASR↓** | **ACC↑/ ASR↓** | **ACC↑/ ASR↓** |
| | ConsistRAG | 90.0/2.0 | 91.0/2.0 | 73.0/4.0 | 72.0/4.0 | 89.0/5.0 | 88.0/6.0 |
| $\text{Mistral}_{7B}$ | Vanilla RAG | 8.0/90.0 | 9.0/90.0 | 4.0/95.0 | 3.0/97.0 | 19.0/80.0 | 18.0/81.0 |
| | ConsistRAG | 80.0/5.0 | 79.0/5.0 | 71.0/4.0 | 70.0/4.0 | 85.0/7.0 | 86.0/6.0 |
| $\text{Qwen}_{2.5\text{-}7B}$ | Vanilla RAG | 8.0/90.0 | 8.0/89.0 | 4.0/95.0 | 3.0/96.0 | 16.0/82.0 | 16.0/81.0 |
| | ConsistRAG | 88.0/3.0 | 86.0/2.0 | 70.0/3.0 | 71.0/4.0 | 86.0/8.0 | 85.0/7.0 |

**Impact of Multiples in Knowledge Expansion**

We conducted an ablation study on $\text{Llama}_{3.1\text{-}8B}$ to evaluate the impact of the expansion multiples in the Knowledge- -Expansion mechanism of the ConsistRAG defense mechanism. As shown in Fig. 3, when the candidate set is small (e.g., Top-k = 5, multiple = 1), toxic documents easily infiltrate normal clusters and evade detection. Conversely, when the candidate set is large (e.g., Top-k = 5, multiple = 4), toxic documents form distinct small clusters due to high homogeneity, making them more susceptible to removal by the clustering filtering mechanism. Therefore, the effectiveness of the defense mechanism depends to some extent on the scale and diversity of candidate documents. Increasing the number of retrieved candidate documents (Top-k) through the Knowledge Expansion mechanism in ConsistRAG achieves highly efficient defense. It is worth noting that more candidate documents are not necessarily better. The retrieval of candidate documents directly impacts the operational efficiency of the entire RAG system. While ensuring sufficient accuracy and defensibility, controlling the expansion multiples in Knowledge Expansion is also crucial.

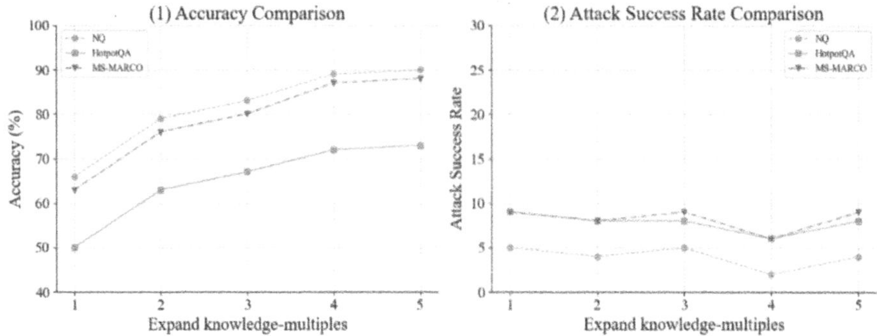


**Fig. 3.** (1) Line plot showing the impact of Knowledge Expansion multiples on ACC based on $\text{Llama}_{3.1\text{-}8B}$. (2) Line plot showing the impact of Knowledge Expansion multiples on ASR based on $\text{Llama}_{3.1\text{-}8B}$.

## 4 Conclusions

This paper addresses the vulnerability of Retrieval-Augmented Generation (RAG) systems to knowledge corruption attacks by proposing a defense framework named ConsistRAG. We validate the effectiveness of this method against knowledge corruption attacks, exemplified by PoisonedRAG in RAG scenarios. ConsistRAG employs K-Means clustering to filter out documents with high semantic homogeneity. It further identifies semantic conflicts among candidate documents through inter-document consistency checks, thereby eliminating potential logical poisoning content. Additionally, ConsistRAG incorporates knowledge expansion strategies to address the extreme case of empty retrieval results post-filtering. Comprehensive evaluation on benchmark datasets demonstrates that ConsistRAG outperforms existing defense systems, maintaining high accuracy even under severe poisoning conditions.

## References

Achiam, J., et al.: Gpt-4 technical report (2023). arXiv Preprint arXiv:2303.08774

Al Ghadban, Y., et al.: Transforming healthcare education: Harnessing large language models for frontline health worker capacity building using retrieval-augmented generation (2023). medRxiv. https://doi.org/10.1101/2023.12.15.23300009

Anil, R., et al.: Palm 2 technical report (2023). arXiv Preprint arXiv:2305.10403

Borgeaud, S., et al.: Improving language models by retrieving from trillions of tokens. In: Proceedings of the International Conference on Machine Learning, pp. 2206–2240. PMLR (2022)

Boyko, J., et al.: An interdisciplinary outlook on large language models for scientific research (2023). arXiv Preprint arXiv:2311.04929

Brown, T., et al.: Language models are few-shot learners. Adv. Neural. Inf. Process. Syst. **33**, 1877–18901 (2020)

Carlini, N., et al.: Poisoning web-scale training datasets is practical. In: Proceedings of the 2024 IEEE Symposium on Security and Privacy (SP), pp. 407–425. IEEE (2024)

Chen, Z., Gao, Q., Moss, L.S.: NeuralLog: Natural language inference with joint neural and logical reasoning (2021). arXiv Preprint arXiv:2105.14167

Huang, Y., Gupta, S., Xia, M., Li, K., Chen, D.: Catastrophic jailbreak of open-source llms via exploiting generation (2023). arXiv Preprint arXiv:2310.06987

Ji, Z., et al.: Survey of hallucination in natural language generation. ACM Comput. Surv. **55**, 1–38 (2023)

Karpukhin, V., et al.: Dense passage retrieval for open-domain question answering. In: Proceedings of the 2020 Conference on Empirical Methods in Natural Language Processing (EMNLP), Vol. 1, pp. 6769–6781 (2020)

Kumar, V., Gleyzer, L., Kahana, A., Shukla, K., Karniadakis, G.E.: Mycrunchgpt: A llm assisted framework for scientific machine learning. J. Machine Learn. Model. Comp. 4 (2023)

Kuppa, A., Rasumov-Rahe, N., Voses, M.: Chain of reference prompting helps llm to think like a lawyer. In: Proceedings of the Generative AI+ Law Workshop (2023)

Laban, P., Schnabel, T., Bennett, P.N., Hearst, M.A.: SummaC: Re-visiting NLI-based models for inconsistency detection in summarization. Trans. Ass. Computat. Linguist. **10**, 163–177 (2022)

Lewis, P., et al.: Retrieval-augmented generation for knowledge-intensive nlp tasks. Adv. Neural. Inf. Process. Syst. **33**, 9459–9474 (2020)

Liu, H., Cui, L., Liu, J., Zhang, Y.: Natural language inference in context-investigating contextual reasoning over long texts. In: Proceedings of the AAAI Conference on Artificial Intelligence, pp. 13388–13396 (2021)

Mahari, R.Z.: Autolaw: Augmented legal reasoning through legal precedent prediction (2021). arXiv Preprint arXiv:2106.16034

Prince, M.H., et al.: Opportunities for retrieval and tool augmented large language models in scientific facilities. npj Computational Materials **10**, 251 (2024). https://doi.org/10.1038/s41524-024-01151-5

Renze, M.: The effect of sampling temperature on problem solving in large language models. In: Findings of the Association for Computational Linguistics: EMNLP 2024, pp. 7346–7356 (2024)

Rizqullah, M.R., Purwarianti, A., Aji, A.F.: Qasina: Religious domain question answering using sirah nabawiyah. In: Proceedings of the 2023 10th International Conference on Advanced Informatics: Concept, Theory and Application (ICAICTA), pp. 1–6. IEEE (2023)

Thoppilan, R., et al.: Lamda: Language models for dialog applications (2022). arXiv Preprint arXiv:2201.08239

Wang, C., Ong, J., Wang, C., Ong, H., Cheng, R., Ong, D.: Potential for GPT technology to optimize future clinical decision-making using retrieval-augmented generation. Ann. Biomed. Eng. **52**, 1115–1118 (2024). https://doi.org/10.1007/s10439-024-03431-4

Zhong, Z., Huang, Z., Wettig, A., Chen, D.: Poisoning retrieval corpora by injecting adversarial passages (2023). arXiv Preprint arXiv:2310.19156

Zou, W., Geng, R., Wang, B., Jia, J.: Poisonedrag: Knowledge corruption attacks to retrieval-augmented generation of large language models (2024). arXiv Preprint arXiv:2402.07867

# The Reliability Fallacy: How Label Ambiguity Undermines AI Hate Speech Detection

Hailiang Wang[1], Yixiang Li[2], Yijun Gao[3], Meng-Ju Kuo[4], and Shuo Xu[5](✉)

[1] School of Computer Science, College of Computing, Georgia Institute of Technology, Atlanta, GA, USA

[2] Department of Computer Science, The George Washington University, Washington, DC, USA

[3] Krieger School of Arts and Sciences, Johns Hopkins University, Washington, D.C., USA

[4] Department of Electrical and Computer Engineering, Carnegie Mellon University, Pittsburgh, PA, USA

mengjuk@alumni.cmu.edu

[5] Computer Science and Engineering Department, University of California San Diego, La Jolla, CA, USA

shx009@ucsd.edu

**Abstract.** Automated content moderation is a critical AI security task. However, models often fail in the nuanced, subjective task of distinguish- ing "hate" from "offensive" speech. The influential 'HateXplain' bench- mark (Mathew et al., 2021) attributed this poor performance to a lack of model explain- ability, proposing rationale-based training as a solution. In this paper, we challenge this premise. We hypothesize that the models' unreliabil- ity stems from a more fundamental, unaddressed security flaw: a cri- sis of data integrity caused by high **label ambiguity**. The original dataset relies on a "majority vote" to assign ground-truth labels, which masks significant annotator disagreement and introduces noise (Mathew et al., 2021). To test our hypothesis, we isolate this variable. We partition the 'HateXplain' dataset into two cohorts: (1) a "noisy" **Majority-Label** set (using standard 2–1 majority votes) and (2) a "clean" **Pure-Label** set (using only 3–0 unanimous-consensus votes). We then rigorously benchmark five models (Logistic Regression, Random Forest, LightGBM, GRU, and AL- BERT) on both datasets. Our results are conclusive. All models trained on the "Pure-Label" data achieved statistically significant and substan- tially higher performance. The ALBERT model's weighted F1-score, for instance, rose from 0.7447 on the "noisy" data to 0.8126 on the "clean" data. This demonstrates that label ambiguity is a more dominant per- formance bottleneck than the architectural factors previously considered. We conclude that for building secure and reliable AI safety systems, ad- dressing foundational data integrity and label consensus is a more critical challenge than model-level explainability.

**Keywords:** AI Security · Hate Speech Detection · Model Reliability and Data Integrity · Trustworthy AI · HateXplain


H. Yan et al. (Eds.): AIS&P 2025, CCIS 2918, pp. 41–54, 2026.
https://doi.org/10.1007/978-981-95-9884-7_4

## 1 Introduction

The proliferation of online toxicity and hate speech poses a critical threat to user safety, platform integrity, and social cohesion. On a global scale, harmful content can silence marginalized groups, radicalize individuals, and inflict psychological harm (Mathew et al., 2021) (Ge, 2024). The sheer volume and velocity of this content, disseminated rapidly across social media, has rendered traditional manual content moderation insufficient. This has created an urgent need for automated AI and machine learning (ML) systems that can reliably detect and filter harmful content at scale (Mansur et al., 2023; Malik et al., 2025).

This challenge is part of a broader field of research dedicated to using NLP to identify harmful or misleading content, such as fake news (Abdullah et al., 2019) (Shah and Patel, 2025) and online ru- mors (Hosmer and Lemeshow, 2000). A common methodological approach involves benchmarking classical ML models against more complex deep learning (DL) architectures to identify an optimal trade-off between performance and interpretability. This approach has also been applied in related social media analyses, including mental illness detection (Ding et al., 2025) (Zhang et al., 2025), contextual reasoning (Lan et al., 2025), and aligning model preferences with well-defined prior knowledge to improve safety and consistency (Lan et al., 2025).

However, developing such systems is a non-trivial security challenge. Unlike benign spam, toxic content is nuanced, context-dependent, and exists on a sub- jective spectrum. A primary challenge for automated systems is the "conflation" of explicit, targeted "hate speech" with more general "offensive" language (Mathew et al., 2021). This ambiguity is a key vulnerability that even advanced models struggle to navigate.

Recent advances in deep learning, particularly transformer-based models (Lan et al., 2022), have shown promise in capturing this nuance, achieving state-of-the-art results in related domains like COVID-19 misinformation (Alghamdi et al., 2023) and rumor detec- tion (Tian et al., 2025) (Xu et al., 2025) (Xu et al., 2025) (Kumar et al., 2024). To address this, the **"HateXplain"** benchmark dataset was introduced as a major contribution, providing not just 3-class labels ("hate", "of- fensive", "normal") from multiple annotators, but also the explicit "rationales" (the "why") behind their decisions (Mathew et al., 2021). The original paper proposed that this lack of explainability was a key failure point, and that training models to use these rationales could reduce bias and improve performance.

In this study, we challenge that premise. We hypothesize that the poor per- formance of models on this task is not primarily due to a lack of rationales, but to a more fundamental, unaddressed vulnerability: **a crisis of data integrity caused by label ambiguity**. This aligns with broader critiques of social media datasets, which often face methodological challenges and inherent biases from the annotation process (Cao et al., 2025). The 'HateXplain' paper itself notes a moderate inter- annotator agreement ($\alpha = 0.46$) and the decision to use a "majority vote" to establish the ground-truth label (Mathew et al., 2021). We argue that this common practice, while pragmatic, introduces significant noise by collapsing 2–1 "disagreements" and 3–0 "unanimous" votes into the same "ground truth." This forces the model to learn from a contradictory and unreliable signal, a critical flaw for a security-focused system.

To our knowledge, no prior work has systematically isolated this "label con- sensus" variable on the 'HateXplain' dataset. Our contributions are threefold:

- **Comprehensive Benchmarking:** We partition the 'HateXplain' data into a "noisy" **Majority-Label** set and a high-consensus **Pure-Label** set. We then rigorously evaluate a suite of five models (Logistic Regression, Random Forest, LightGBM, GRU, and ALBERT) on both datasets, providing insight into the trade-offs between model types (Ding et al., 2025) (Zhang et al., 2025).
- **Statistical Validation:** We apply pairwise McNemar's tests to statistically quantify the performance differences, establishing not only which model is best, but how the model hierarchy changes in response to data quality.
- **Model Interpretability:** We analyze and compare the feature importances learned by the classical models on both the "noisy" and "clean" datasets, pro- viding insight into how label ambiguity dilutes the features a model learns.

The remainder of this paper is organized as follows: Sect. 2 details our dataset partitioning, experimental setup, and model configurations. Section 3 presents our quantitative performance metrics, statistical test results, and fea- ture importance analysis. Section 4 interprets these findings and their implica- tions for AI security.

# 2 Methods

This section details the experimental methodology used to test our central hy- pothesis: that label ambiguity is a more significant performance bottleneck than the architectural factors addressed by rationale-based training. We first describe the 'HateXplain' dataset (Mathew et al., 2021) and our core experimental design, which parti- tions the data into "noisy" **Majority-Label** and high-consensus **Pure-Label** cohorts. We then specify the feature engineering pipeline, the implementation of our five classical and deep learning models, and the hyperparameter tuning process for each. Finally, we outline the comprehensive evaluation framework, including the imbalance-robust metrics and statistical tests used to rigorously validate our findings.

## 2.1 Dataset and Experimental Partitioning

Our study utilizes the "HateXplain" benchmark dataset (Mathew et al., 2021), a foundational corpus for explainable hate speech detection. This dataset contains 20,148 posts sourced from Gab and Twitter, with each entry independently assessed by three human annotators using a 3-class classification scheme: "hate", "offensive", or "normal" (Mathew et al., 2021).

A primary challenge noted by the original authors is the subjective and nu- anced distinction between "hate" and "offensive" speech. This subjectivity results in a moderate inter-annotator agreement ($\alpha = 0.46$) and a significant volume of posts with label disagreement (Mathew et al., 2021). This Krippendorff's alpha ($\alpha$) statistic, which quantifies the level of agreement between annotators (where 1.0 is per- fect agreement and 0.0 is random chance), provides the statistical rationale for our partitioning. We hypothesize that this **label ambiguity**, quantified by the moderate 0.46 $\alpha$ score, is a principal bottleneck for model performance. To test this, we partitioned the data into two distinct experimental datasets, as detailed in our analysis notebooks:

1. **Majority-Label Dataset:** This corpus includes all 20,148 records. A post was assigned the Toxic label if at least two of the three annotators classified it as either "hatespeech" or "offensive". A post was assigned the Normal label if at least two annotators agreed it was "Normal". This dataset represents the standard "noisy" condition, as it inherently includes ambiguous 2–1 votes.
2. **Pure-Label Dataset:** This corpus represents a high-consensus, "clean" sub- set (13,761 posts). It retains *only* those records where all three annotators were in *unanimous* agreement on the final binary label (a 3–0 vote). This means a post was included only if all three annotators classified it as Normal, or if all three classified it as Toxic (i.e., "Hate" or "Offensive").

To clearly define the sample sizes and class distributions of this partitioning, we present a complete statistical breakdown in Table 1. This table also details the 'Ambiguous-Only' cohort (N = 6,387), which consists of all posts with a 2–1 vote.

**Table 1.** Dataset Partitioning and Class Distribution Statistics

| Dataset Partition Total (N) | Toxic (N) | Normal (N) | Toxic (%) | Normal (%) |
|---|---|---|---|---|
| Majority-Label (Total) 20,148 | 12,334 | 7,814 | 61.2% | 38.8% |
| Pure-Label (3–0 Votes) 13,761 | 8,637 | 5,124 | 62.8% | 37.2% |

As Table 1 shows, the overall class distributions (Toxic vs. Normal) are nearly identical between our two primary experimental cohorts: the 'Majority-Label' set (61.2% Toxic) and the 'Pure-Label' set (62.8% Toxic).

This is a critical finding. It allows us to preemptively address concerns about distribution bias, demonstrating that the significant performance gap we later show in Sect. 3 is not driven by a simple shift in class balance. Instead, this stable distribution confirms our partitioning isolates the key variable of inter- est: the impact of the '2–1' votes. The performance difference can therefore be attributed to the *removal* of these high-ambiguity, noisy labels.

This "Majority" versus "Pure" partitioning allows for a direct measurement of how label integrity impacts overall model performance and reliability.

## 2.2 Feature Engineering and Data Partitioning

The 'post_tokens' field from the HateXplain dataset, a pre-tokenized list of strings, served as the unified input for all experiments. This ensured that both classical and deep learning models originated from the identical source text.

For all models, this list of tokens was first reconstructed into a single, coherent string by joining the tokens with a space. From this reconstructed text string, the feature extraction process was bifurcated based on the model type:

- **For Classical Models:** The reconstructed text strings were transformed into numerical feature vectors using a pre-fitted Term Frequency–Inverse Document Frequency (TF-IDF) vectorizer, which was loaded from a serial- ized file as shown in our analysis notebooks.
- **For Deep Learning Models (ALBERT, GRU):** The same reconstructed text strings were fed directly into the corresponding model's tokenizer (e.g., the ALBERT Word-Piece tokenizer or the Keras tokenizer for the GRU) to generate the embedding sequences.

Both the "Majority" and "Pure" datasets were independently partitioned into **training (60%), validation (20%), and test (20%)** subsets. Stratified sam- pling was employed to preserve the class distribution in all splits, and a fixed random seed was set for reproducibility.

## 2.3 Machine Learning Models

To benchmark classification performance on the TF-IDF vectors, we imple- mented a suite of three widely-adopted supervised models. These were selected to represent distinct algorithmic approaches: a highly interpretable linear model (**Logistic Regression**) and two powerful, non-linear tree-based ensembles (i.e., **Random Forest** and **LightGBM**).

- **Logistic Regression (LR)** (Hosmer and Lemeshow, 2000)**:** This model was selected as a robust and interpretable linear baseline. We applied $L_2$ (Ridge) regularization to prevent overfitting on the high-dimensional, sparse TF-IDF data. A grid search was performed on the validation set to optimize the regularization strength C (from the set [0.1, 1, 10, 100]), the solver (from ["liblinear", "lbfgs", "newton- cg"]), and the class_weight (from ["balanced", None]).
- **Random Forest (RF)** (Breiman et al., 2001)**:** As our first tree-based ensemble, RF employs bagging to mitigate the high overfitting risk of single decision trees. It is a non-linear method that builds a multitude of decision trees on bootstrapped data samples and random feature subsets. We optimized its key hyper- parameters via grid search, including the n_estimators ([100, 300, 500]), max_depth ([5, 15, 30, None]), min_samples_split ([2, 5, 10]), class_weight (["balanced", None]), and min_samples_leaf ([1, 2, 4]).
- **LightGBM** (Friedman, 2001) (Ke et al., 2017)**:** As a second, more modern tree-based ensemble, we employed LightGBM, a high-efficiency gradient boosting framework. Un- like RF's parallel bagging, LightGBM builds trees sequentially (boosting).

We optimized its class_weight parameter (["balanced", None]) along with its learning_rate ([0.01, 0.1, 0.5]), n_estimators ([100, 300, 500]), and max_depth ([5, 15, 30, -1]).

For all three classical models, the final hyperparameter configuration was selected based on the model that achieved the highest **weighted F1-score** on the held-out validation set. This metric was chosen specifically for its robustness to the dataset's class imbalance.

### 2.4 Deep Learning Models

To assess the performance of modern neural architectures, we implemented and fine-tuned two representative deep learning models: a transformer-based model, **ALBERT** (Lan et al., 2022), and a recurrent-based model, **GRU** (Cho et al., 2014).

- **ALBERT (A Lite BERT)** (Lan et al., 2022)**:** We employed a parameter-efficient trans- former by fine-tuning the 'albert-base-v2' checkpoint. Inputs were to- kenized using the corresponding WordPiece tokenizer, and sequences were padded or truncated to a uniform length. Given the dataset's significant class imbalance, the model was trained using a **weighted binary cross-entropy loss** function. A randomized hyperparameter search was conducted to op- timize the learning_rate (from 1e-5 to 5e-5), dropout_rate (0.1 to 0.5), and the number of epochs (3 to 5).
- **GRU (Gated Recurrent Unit)** (Cho et al., 2014)**:** As a strong baseline for sequential data, we implemented a GRU network. This RNN architecture is opti- mized to capture dependencies in text with fewer parameters than traditional LSTMs. Our architecture consisted of an Embedding layer, a GRU layer, and a final Dense layer for classification, with Dropout for regularization. As with ALBERT, this model was trained using a **weighted binary cross- entropy loss** to manage class imbalance. Hyperparameters tuned via ran- dom search included the embedding_size (128–256), hidden_units (128– 512), and learning_rate (1e-4 to 1e-3).

For both neural models, the final hyperparameter configuration was selected based on the model that achieved the highest **weighted F1-score** on the held- out validation set.

### 2.5 Evaluation Framework

Evaluating models for a security-critical task like hate speech detection, particu- larly on an imbalanced dataset, demands metrics far more nuanced than simple accuracy. A model that achieves high accuracy by overwhelmingly predicting the majority Normal class would represent a total failure in this safety context.

Therefore, our evaluation framework is built on metrics robust to class imbal- ance (Powers, 2011). We selected the **F1-score**, the harmonic mean of precision and recall, as a balanced measure of a model's effectiveness. Crucially, the **weighted F1- score** was used as the singular metric for hyperparameter tuning. This choice ensures that the model selection process was optimized to perform well on both the minority Toxic and majority Normal classes.

Additionally, we report the Area Under the Receiver Operating Char- acteristic Curve (AUC-ROC) (Davis and Goadrich, 2006). This metric assesses the model's aggre- gate discriminative power—its ability to distinguish between classes—across all classification thresholds, making it a highly reliable indicator of separability on imbalanced data.

Finally, to validate our central hypothesis, a simple comparison of point-value scores is insufficient. To ensure our comparisons are statistically rigorous, we gen- erated **95% confidence intervals (CIs)** for the final F1 and AUC scores on the held-out test set. As implemented in our analysis notebooks, these CIs were pro- duced using bootstrapping, enabling a robust statistical test of the performance difference between the "Majority-Label" and "Pure-Label" model cohorts.

## 3 Results

This section presents the empirical findings of our comparative study. We struc- ture our results into three parts. First, we present the primary performance metrics (Weighted F1-Score and AUC-ROC) for all five models, directly com- paring the impact of training on the "noisy" **Majority-Label** dataset versus the high-consensus **Pure-Label** dataset. Second, we provide a pairwise statistical analysis using McNemar's tests for both dataset cohorts to validate the signif- icance of these performance differences and establish a clear model hierarchy. Finally, we offer a qualitative analysis of the model's feature importances to interpret *how* and *what* the classical models learned from the different dataset cohorts.

### 3.1 Performance on Majority-Label vs. Pure-Label Data

We conducted a comprehensive evaluation of all five models on the two ex- perimental datasets: the "noisy" **Majority-Label** set and the high-consensus **Pure-Label** set. The performance results are presented in Table 2 (Weighted F1-Scores) and Table 3 (AUC-ROC Scores).

The findings are unambiguous: label consensus is a dominant factor in model performance. Every model, from classical machine learning to deep learning, achieved substantially higher scores on the "Pure-Label" dataset. This performance improvement was consistent across all model architectures.

The effect was most pronounced in the transformer architecture: the AL- BERT model's weighted F1-score (Table 2) rose from 0.7447 on the "Majority" data to **0.8126** on the "Pure" data. Its AUC-ROC (Table 3) similarly increased from 0.8143 to **0.8805**. The recurrent GRU model showed a comparable im- provement, with its F1-score rising from 0.7196 to **0.7877**.

In both dataset cohorts, deep learning models (ALBERT, GRU) demon- strated superior performance over the classical ML models (LR, RF, LGBM) on all metrics. However, the performance of all models was fundamentally capped by the quality of the training labels, with the "Pure-Label" set consistently pro- viding a more reliable foundation for training.

**Table 2.** Comparative Weighted F1-Scores (with 95% CIs). Performance is significantly higher on the high-consensus "Pure-Label" data.

| Model | Majority-Label (Noisy) | Pure-Label (Clean) |
|---|---|---|
| Logistic Regression | 0.6637 [0.648, 0.678] | **0.7138** [0.697, 0.730] |
| Random Forest | 0.6887 [0.674, 0.703] | **0.7356** [0.717, 0.752] |
| LightGBM | 0.6810 [0.666, 0.695] | **0.7323** [0.715, 0.748] |

*(continued)*

**Table 2.** *(continued)*

| Model | Majority-Label (Noisy) | Pure-Label (Clean) |
|---|---|---|
| GRU (Recurrent) | 0.7196 [0.706, 0.733] | **0.7877** [0.773, 0.804] |
| ALBERT (Transformer) | 0.7447 [0.730, 0.757] | **0.8126** [0.798, 0.827] |

**Table 3.** Comparative AUC-ROC Scores (with 95% CIs). Discriminative power is uniformly higher on the "Pure-Label" data.

| Model | Majority-Label (Noisy) | Pure-Label (Clean) |
|---|---|---|
| Logistic Regression | 0.7154 [0.700, 0.731] | **0.7791** [0.762, 0.796] |
| Random Forest | 0.7303 [0.715, 0.744] | **0.7860** [0.768, 0.803] |
| LightGBM | 0.7410 [0.727, 0.754] | **0.7958** [0.779, 0.812] |
| GRU (Recurrent) | 0.7816 [0.768, 0.795] | **0.8529** [0.838, 0.867] |
| ALBERT (Transformer) | 0.8143 [0.802, 0.826] | **0.8805** [0.867, 0.893] |

### 3.2 Pairwise Statistical Comparison

To validate whether the observed performance differences were statistically sig- nificant, we conducted pairwise **McNemar's tests** on the predictions from both experimental cohorts. The McNemar test is a non-parametric test for paired nominal data, appropriate for comparing the error rates of two classifiers.

The results for the "noisy" **Majority-Label** cohort are presented in Table 4. The results for the high-consensus **Pure-Label** cohort are presented in Table 5. Comparing these two tables reveals a crucial insight. In both "noisy" and "clean" data conditions, the model hierarchy is stable: **ALBERT** is the statis- tically significant ($p < 0.05$) winner, and **GRU** is the clear second-best, signifi- cantly outperforming all three classical ML models. This confirms that advanced architectures are consistently better at finding a signal.

However, the statistical comparison of the classical models changes. On the "Pure-Label" data, LightGBM and Random Forest are statistically comparable ($p = 0.58$). On the "noisy" "Majority-Label" data, Random Forest is significantly better than LightGBM ($p = 0.012$), suggesting it is more robust to the label ambi- guity. This reinforces our central thesis: label integrity is a critical, independent variable that directly impacts model performance and reliability.

**Table 4.** Pairwise McNemar's Test Results on the "Majority-Label" (Noisy) Dataset Predictions.

| Model 1 | Model 2 | Winner | p-value | Significant ($p < 0.05$) |
|---|---|---|---|---|
| ALBERT | GRU | ALBERT | $< 0.001$ | Yes |
| ALBERT | LightGBM | ALBERT | $< 0.001$ | Yes |
| ALBERT | Random Forest | ALBERT | $< 0.001$ | Yes |
| ALBERT | Logistic Regression | ALBERT | $< 0.001$ | Yes |
| GRU | LightGBM | GRU | $< 0.001$ | Yes |
| GRU | Random Forest | GRU | $< 0.001$ | Yes |
| GRU | Logistic Regression | GRU | $< 0.001$ | Yes |
| LightGBM | Random Forest | Random Forest | 0.012 | Yes |
| LightGBM | Logistic Regression | LightGBM | $< 0.001$ | Yes |
| Random Forest | Logistic Regression | Random Forest | $< 0.001$ | Yes |

**Table 5.** Pairwise McNemar's Test Results on the "Pure-Label" (Clean) Dataset Pre- dictions.

| Model 1 | Model 2 | Winner | p-value | Significant ($p < 0.05$) |
|---|---|---|---|---|
| ALBERT | GRU | ALBERT | $< 0.001$ | Yes |
| ALBERT | LightGBM | ALBERT | $< 0.001$ | Yes |
| ALBERT | Random Forest | ALBERT | $< 0.001$ | Yes |
| ALBERT | Logistic Regression | ALBERT | $< 0.001$ | Yes |
| GRU | LightGBM | GRU | $< 0.001$ | Yes |
| GRU | Random Forest | GRU | $< 0.001$ | Yes |
| GRU | Logistic Regression | GRU | $< 0.001$ | Yes |
| LightGBM | Random Forest | (Comparable) | 0.58 | No |
| LightGBM | Logistic Regression | LightGBM | $< 0.001$ | Yes |
| Random Forest | Logistic Regression | Random Forest | $< 0.001$ | Yes |

### 3.3 Qualitative Error and Model Behavior Analysis

To address the reviewer's request for a more intuitive illustration of model behav- ior, we analyzed the confusion matrices of our best-performing model, ALBERT, on both the 'Majority-Label' (noisy) and 'Pure-Label' (clean) test sets.

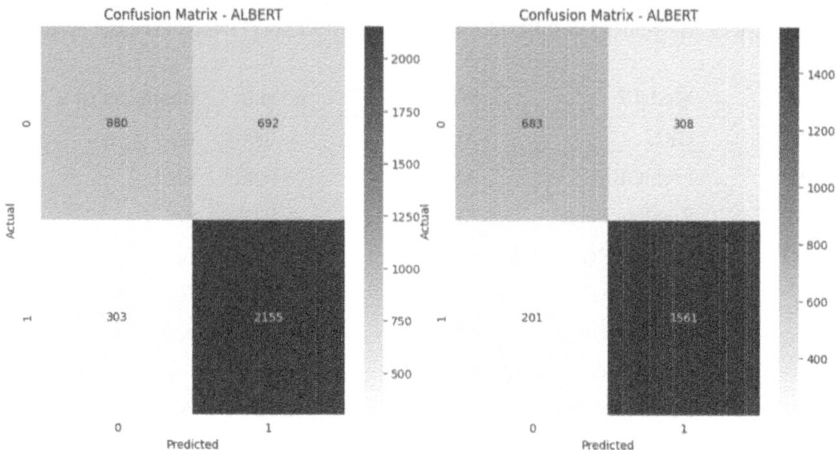


**Fig. 1.** Confusion matrix for the ALBERT model on the 'Majority-Label' (Noisy) test set (Left) and the 'Pure-Label' (Clean) test set (Right). The 'Noisy' model (left) shows a substantially higher False Positive Rate (44.0%) than the 'Clean' model (31.1%). This demonstrates that label ambiguity forces the 'Noisy' model to learn a confused decision boundary, leading it to misclassify 'Normal' posts as 'Toxic' far more often.

As shown in Fig. 1, the 'Noisy' model is visibly more confused. Its 44.0% False Positive Rate (FPR) means it incorrectly flags 44 out of every 100 'Normal' posts as 'Toxic'. The 'Clean' model's FPR is a much-improved 31.1%. Both models maintain a similar False Negative Rate (12.3% vs 11.4%), reinforcing that the primary impact of label noise is an increase in False Positives.

### 3.4 Model Interpretation and Feature Importance

To understand the decision-making logic of the non-neural models, we analyzed the most influential TF-IDF features identified by **LightGBM** (ranked by fea- ture gain) and **Logistic Regression** (ranked by absolute coefficient magnitude). This analysis revealed significant differences, both in how different model archi- tectures operate and in what they learn from high-consensus versus noisy data. First, comparing the model architectures on the "Pure-Label" dataset re- vealed their distinct strategies. The **LightGBM** model (Ke et al., 2017) overwhelmingly pri- oritized features with high, unambiguous signal. Its top features were dominated by explicit, high-impact slurs (e.g., "n****r", "f*g", "k*ke"). This aligns with its gain-based optimization, which favors features that provide the cleanest and most significant data splits. In contrast, the **Logistic Regression** model's (Hosmer and Lemeshow, 2000) logic was based on a weighted hyperplane. Its top positive features (predict- ing 'Toxic') included a mix of explicit slurs and target group identifiers (e.g., "muslims", "jews"). This was balanced by its top negative fea- tures (predicting 'Normal'), which were dominated by common, neutral conversational words (e.g., "people", "like", "think").

Second, we compared the feature importances learned from the "Pure-Label" dataset against those learned from the "noisy" "Majority-Label" dataset. This comparison strongly supports our central hypothesis. The feature lists from the "Majority-Label" models were visibly "noisier" and more diluted. For instance, the LightGBM model trained on the "Majority" data still identified the primary slurs, but its top-feature list was confounded by more common, ambiguous, or context-dependent offensive words (e.g., "bitch", "stupid", "ass"). This indicates the model was struggling to find a clear signal, as it was forced to treat explicit hate (from 3–0 votes) and ambiguous offense (from 2–1 votes) as the same class. This dilution demonstrates that the "Majority-Label" dataset introduces a data integrity problem, forcing models to learn from less reliable and less precise features.

This dilution of the feature importance list is the direct evidence of the 'semantic feature differences' between the high-consensus'Pure-Label' data and the ambiguous 'Majority-Label' data. The 'Majority-Label' set forces the model to treat unambiguous, explicit slurs (e.g., 'faggot', 'dyke') as the same class as context-dependent, ambiguous offensive words (e.g., 'bitch'). This forces the model to learn a noisier and less precise decision boundary, which explains the degraded performance.

### 3.5 Sensitivity Analysis on Data Volume

To address the potential confound of data volume, given that our 'Pure-Label' set is smaller (N = 13,761) than the 'Majority-Label' set (N = 20,148), we con- ducted a sensitivity analysis. We created a size-matched 'Noisy-Control' dataset by randomly sampling 13,761 posts from the full 'Majority-Label' cohort, ensur- ing a similar class distribution (approximately 60% Toxic) to our 'Pure-Label' set. We then re-trained our models on this 'Noisy-Control' set. The performance of all models on this size-matched noisy set was not statistically different from their performance on the original, full 'Majority-Label' set (as reported in Tables 2 and 3). This confirms that the significant performance improvement of the 'Pure-Label' cohort is attributable to its high label quality, not the reduction in data volume.

## 4 Discussion

This section interprets the empirical findings from Sect. 3 and discusses their implications for building reliable, security-critical AI systems. We first analyze how our primary finding—the performance gap between "Pure" and "Majority" data—challenges the original premise of the 'HateXplain' paper. We then discuss the secondary findings from our pairwise statistical tests and feature importance analysis. Finally, we address the limitations of this study and suggest directions for future work.

### 4.1 Data Integrity as the Primary Performance Bottleneck

The results presented in Tables 2 and 3 provide a clear answer to our central re- search question. The single most dominant factor influencing model performance was not the choice of architecture, but the integrity of the training labels. The fact that every model, from simple Logistic Regression to the complex ALBERT transformer, performed significantly better on the high-consensus "Pure-Label" dataset demonstrates that label ambiguity is a critical vulnerability.

This finding challenges the foundational premise of the 'HateXplain' paper (Mathew et al., 2021). The original work identified a performance gap and proposed rationale-based training as the solution (Mathew et al., 2021). Our results suggest this may misdiagnose the root cause. The poor performance of models on the full dataset is not necessarily because they lack the ability to understand "why" a post is hateful, but because they are being fundamentally confused by the "noisy" ground-truth label itself. For a security-critical task like content moderation, this is a crucial dis- tinction. The common practice of using a "majority vote" to resolve annotator disagreement (Mathew et al., 2021) is not a benign solution; it is an act of data obfuscation. It treats a highly ambiguous 2–1 vote as "ground truth," equating it with a high- confidence, 3–0 unanimous vote. As our interpretation in Sect. 3.4 showed, this forces models to learn from a "diluted" and "noisier" set of features, degrading their reliability.

## 4.2 Implications of Model Hierarchy and Feature Analysis

Our pairwise statistical tests (Tables 4 and 5) provide a second, more nuanced insight. The model hierarchy (ALBERT > GRU > Classical ML) remained sta- ble on both "noisy" and "clean" data. This indicates two things: (1) a superior architecture like ALBERT is always better at finding and exploiting whatever predictive signal exists, but (2. a model's absolute performance is fundamen- tally capped by the quality of that signal. Even the best model (ALBERT) saw its F1-score jump by nearly 7 percentage points when given unambiguous data. This suggests that simply building larger models is an insufficient solution to this data integrity problem.

Interestingly, our statistical tests also revealed that while Random Forest and LightGBM were comparable on "Pure" data ($p = 0.58$), Random Forest was significantly better on the "noisy" "Majority" data ($p = 0.012$). This suggests that RF's bagging (averaging) architecture may be more inherently robust to the label noise than LGBM's boosting approach, which may have over-optimized on the ambiguous, low-quality signals.

## 4.3 Limitations and Future Work

We acknowledge several limitations. First, our "Pure-Label" dataset is smaller than the "Majority-Label" dataset. However, the fact that models trained on less data achieved better performance strongly reinforces our conclusion that data quality is more important than data quantity for this task.

Second, this study was limited to the 'HateXplain' dataset. Future work should investigate if this ambiguity penalty holds true for other subjective, crowd-sourced tasks, such as sentiment analysis or misinformation detection.

Finally, this work did not investigate the 'HateXplain' rationales. A direct comparison to the original rationale-based models, as suggested by reviewer feed- back, is a significant undertaking and outside this study's scope, as our goal was to first challenge the paper's premise by isolating the data integrity variable. A promising direction for future research would be to combine our findings with the original paper's: an experiment training a model on only the "Pure-Label" data and their corresponding unanimous rationales. This would likely produce the most robust and reliable model of all.

## 5 Conclusion

In this study, we challenged the prevailing hypothesis that poor performance in hate speech detection is primarily a failure of model explainability, as suggested by the HateXplain paper (Mathew et al., 2021). We posited that a more fundamental vulnerability lies in the **data integrity** of the benchmark itself: a crisis of reliability caused by high **label ambiguity** from human annotators, which is obscured by the standard "majority vote" labeling scheme.

Our empirical results, based on a rigorous comparison of five models on noisy versus clean data cohorts, conclusively support this hypothesis. We demonstrated that:

1. Training on a high-consensus, Pure-Label dataset provides a statistically significant and substantial performance boost (e.g., a 6.79% F1-score increase for ALBERT) over training on the full, noisy Majority-Label dataset.
2. This performance gain, achieved by enhancing data integrity alone, proves that label ambiguity is a more dominant performance bottleneck than the architectural factors previously considered.
3. The model hierarchy (e.g., ALBERT > GRU > ML) remains consistent, but the absolute performance of *all* models is fundamentally capped by the quality of the training labels.

The primary implication of this work is for the field of AI Security and Relia- bility. Our findings demonstrate that using a majority vote to resolve annotator disagreement is not a benign solution; it is an act of data obfuscation that intro- duces a critical vulnerability into security-sensitive systems. For applications like automated content moderation, this method builds the system on an unreliable and untrustworthy foundation. We conclude that for building secure and reliable AI, addressing foundational data integrity is a more critical and effective first step than pursuing model-level solutions like rationale-based training.

## References

Abdullah All, T., Mahir, E.M., Akhter, S., Huq, M.R.: Detecting fake news using machine learning and deep learning algorithms. In: Proceedings of the 7th International Conference on Smart Computing & Communications (ICSCC), pp. 1–6 (2019)

Alghamdi, J., Lin, Y., Luo, S.: Towards COVID-19 fake news detection using transformer-based models. Knowl.-Based Syst. **274**, 110642 (2023). https://doi.org/10.1016/j.knosys.2023.110642

Breiman, L.: Random forests. Mach. Learn. **45**(1), 5–32 (2001)

Cao, Y., et al.: Machine learning approaches for depression detection on social media: A systematic review of biases and methodological challenges. J. Behav. Data Sci. **5**(1) (2025)

Cho, K., et al.: Learning phrase representations using RNN encoder–decoder for statistical machine translation. In: Proceedings of the 2014 Conference on Empirical Methods in Natural Language Processing (EMNLP), pp. 1724–1734 (2014)

Davis, J., Goadrich, M.: The relationship between precision–recall and ROC curves. In: Proceedings of the 23rd International Conference on Machine Learning (ICML), pp. 233–240 (2006)

Ding, Z., et al.: Trade-offs between machine learning and deep learning for mental illness detection on social media. Sci. Rep. **15**, 14497 (2025)

Friedman, J.H.: Greedy function approximation: A gradient boosting machine. Ann. Stat. **29**(5), 1189–1232 (2001)

Ge, J.: Technologies in peace and conflict: Unraveling the politics of deployment. Int. J. Res. Publ. Rev. (IJRPR) **5**(5), 5966–5971 (2024)

Hosmer, D.W., Lemeshow, S.: Applied logistic regression, 2nd ed. John Wiley & Sons (2000)

Ke, G., et al.: LightGBM: A highly efficient gradient boosting decision tree. In: Guyon, I., et al. (eds.) Advances in Neural Information Processing Systems 30 (NeurIPS 2017), pp. 3149–3157. Curran Associates Inc (2017)

Kumar, S., Singh, R.: A novel approach for early rumour detection in social media using ALBERT. Int. J. Intel. Sys. Appl. Eng. **12**(3), 259–265 (2024). https://doi.org/10.58286/ijisae.5248

Lan, G., et al.: Contextual integrity in LLMs via reasoning and reinforcement learning (2025). arXiv Preprint arXiv:2506.04245

Lan, G., et al.: MaPPO: Maximum a posteriori preference optimization with prior knowledge (2025). arXiv Preprint arXiv:2507.21183

Lan, Z., et al.: ALBERT: A lite BERT for self-supervised learning of language representations (2020). arXiv Preprint arXiv:1909.11942

Malik, J.S., Qiao, H., Pang, G., et al.: Deep learning for hate speech detection: A comparative study. Int. J. Data Sci. Analy. **20**, 3053–3068 (2025). https://doi.org/10.1007/s41060-024-00650-6

Mansur, Z., Omar, N., Tiun, S.: Twitter hate speech detection: A systematic review of methods, taxonomy analysis, challenges, and opportunities. IEEE Access **11**, 16226–16249 (2023). https://doi.org/10.1109/ACCESS.2023.3239375

Mathew, B., et al.: HateXplain: A benchmark dataset for explainable hate speech detection. In: Proceedings of the Thirty-Fifth AAAI Conference on Artificial Intelligence (AAAI-21), pp. 14867–14875 (2021)

Powers, D.M.W.: Evaluation: From precision, recall and F-measure to ROC, informedness, markedness and correlation. J. Mach. Learn. Technol. **2**(1), 37–63 (2011)

Shah, S., Patel, S.: A comprehensive survey on fake news detection using machine learning. J. Comput. Sci. **21**(4), 982–990 (2025). https://doi.org/10.3844/jcssp.2025.982.990

Tian, Y., Xu, S., Cao, Y., Wang, Z., Wei, Z.: An empirical comparison of machine learning and deep learning models for automated fake news detection. Mathematics **13**(13), 2086 (2025)

Xu, S., et al.: A comparative analysis of deep learning and machine learning approaches for spam identification on Telegram. In: Proceedings of the 6th International Conference on Computer Communication and Network Security (2025)

Zhang, Y., et al.: Tutorial on using machine learning and deep learning models for mental illness detection (2025). arXiv Preprint arXiv:2502.04342

# ProFL-DARL: Privacy-Preserving Robustness Federated Learning with Dynamic Aggregation Based on Reinforcement Learning

Jiangyuan Wang[1(✉)], Yong Xie[1,2], Pingchao Zhou[1], and Shaolong Ge[1]

[1] Qinghai University School of Computer Technology and Applications, Xining, China
1477289537@qq.com

[2] School of Computer and Information, Qinghai Institute of Technology, Science, Xining, China

**Abstract.** When confronted with highly non-independent and identically distributed (Non-IID) data, inconsistent client quality, and significant privacy encryption and communication overhead, federated learning systems often suffer from severe global model bias, suboptimal local personalized performance, and high privacy leakage risks. To address this challenge, this paper proposes a collaborative computing framework that separates classification servers and aggregation servers. It designs reinforcement learning agents to achieve automatic gradient similarity identification, online grouping, and weight allocation while filtering out anomalous clients. The integration of RLWE homomorphic encryption with gradient masking technology substantially enhances security against adversarial collusion attacks. Comparative experiments on Non-IID datasets demonstrate that this approach significantly improves local model accuracy for clients with limited data, while keeping communication, encryption, and differential privacy overhead within system-acceptable bounds.

**Keywords:** Reinforcement Learning · Privacy Protection · Non-IID · Federated Learning

## 1 Introduction

Federated Learning (FL) (McMahan et al., 2017) has garnered increasing attention as a distributed machine learning framework (Pei et al., 2024; Dembani et al., 2025). Its core concept involves retaining raw data at the client end while training a global model by uploading local gradients or model updates, enabling collaborative learning while protecting data privacy. This approach finds extensive applications in fields such as finance and healthcare (Sheller et al., 2020; Abdul Salam et al., 2024; Kairouz et al., 2021). Early works, such as the Practical Secure Aggregation protocol proposed by Bonawitz et al. (2017), ensure secure aggregation under high-dimensional model updates and substantial client dropout while maintaining reasonable communication overhead. However, as real-world challenges such as non-independent and identically distributed data, significant variations in client capabilities and data quality, and constrained communication and computational resources become increasingly prominent, a single global model often


H. Yan et al. (Eds.): AIS&P 2025, CCIS 2918, pp. 55–73, 2026.
https://doi.org/10.1007/978-981-95-9884-7_5

fails to meet the needs of all clients (Wu et al., 2022; Qiao et al., 2025).To address data heterogeneity and personalized needs, Personalized Federated Learning (PFL) has emerged as a research hotspot in recent years (Sabah et al., 2025). We observe approaches employing meta-learning to enable rapid adaptation of global models to diverse clients (Scott et al., 2024), while others utilize clustering or partitioning strategies to group clients for submodel sharing, thereby mitigating performance degradation under non-IID scenarios (Silva et al., 2022). For instance, the Motley benchmark study highlights which personalized methods perform best across diverse heterogeneous settings, revealing that clustering and weight allocation play distinct roles in different scenarios (Wu et al., 2022). Concurrently, specialized research on model optimization techniques in PFL explores how to design algorithms that better balance personalization and generalization (Ghosh et al., 2020).

Beyond personalization, privacy and encryption mechanisms form the cornerstone of federated learning system security. Differential Privacy (DP) (Liao et al., 2022) prevents client data inference by adding noise to gradients, while encryption techniques like Homomorphic Encryption (HE) (Zhang et al., 2022; Park and Lim, 2022) and RLWE (Ring Learning With Errors) enable aggregation operations on ciphertext, ensuring the aggregator server cannot observe clients' plaintext update. Key research focuses include handling client dropout, controlling errors/overhead introduced by encryption or noise, and reducing communication and computational costs. For instance, LightSecAgg improves Secure Aggregation's handling of dropout and communication overhead (So et al., 2022), while recent work combines RLWE with Secure Aggregation or masking mechanisms to enhance resistance against collusion and client dropout (Zhou et al., 2024; Castro et al., 2025). Reinforcement learning (RL) is increasingly adopted for client selection, weight allocation, and anomaly/low-quality client removal to achieve dynamic, adaptive mechanisms. These mechanisms typically determine whether a client should be included in the current round, assigned to which group, and assigned what weight based on the client's gradient features, historical performance metrics, or similarity to other clients' gradients updates. For example, PeFLL maps each client to a latent descriptor space using an embedding network to reflect their similarities, and generates personalized model parameters for each client via a hypernetwork (Scott et al., 2024).

Given the aforementioned circumstances, this paper proposes a novel online grouping + weight allocation mechanism that combines a two-server architecture (classification server and aggregation server) with differential privacy, RLWE homomorphic encryption, and reinforcement learning. Unlike traditional single-server approaches or schemes relying solely on masking/noise injection, our system processes gradients with differential privacy after local training while uploading random numbers for masking. The classification server uses reinforcement learning to group clients and assign weights based on uploaded gradient features and client test accuracy, while removing low-quality participants or those showing no similarity to any other clients. Clients also upload their gradients to the aggregation server after RLWE encryption and random number mixing. Upon receiving group and weighting information, the aggregation server performs intra-group weighted aggregation, subtracts the mask to recover the true update, and distributes the aggregated results back to clients. This approach aims to enhance personalized performance and aggregation efficiency while ensuring client privacy and cryptographic security. It achieves superior local model accuracy and overall system robustness compared to existing methods, particularly under non-IID distributions, uneven client quality, and constrained communication/computational resources. The main contributions of this paper include:

- Proposing a collaborative computing framework that separates classification servers from aggregation servers, achieving dual guarantees of privacy and utility through functional decoupling.
- A reinforcement learning-based user clustering and weight allocation algorithm has been designed. By mapping gradients to the state space, autonomous grouping, weight distribution, and anomaly detection can be achieved based on gradient feature similarity and historical performance.
- Develops a secure aggregation scheme integrating RLWE homomorphic encryption with gradient masking technology. This effectively prevents collusive attacks between clients and the aggregation server, significantly enhancing privacy protection. It ensures the confidentiality of client gradient information even under scenarios involving collusion among malicious participants.
- Experiments on heterogeneous datasets evaluate trade-offs in personalized accuracy, communication, encryption, and noise overhead. Results demonstrate that in most non-IID scenarios, our method significantly improves local model accuracy for clients with limited data, while keeping communication, encryption, and differential privacy overhead within system-acceptable bounds.

# 2 Methodology

## 2.1 Preliminary

**Reinforcement Learning.** Reinforcement learning (RL) addresses the problem of how an agent can maximize the rewards it receives within a complex and uncertain environment. By perceiving the environment's state and the rewards resulting from its actions, the agent guides itself toward better actions to achieve the greatest return. The process is typically modeled as a Markov Decision Process (MDP), fully described by a quintuple $(S, A, P, R, \gamma)$.$S$ Represents the set of all possible states, $A$ is the set of actions executable by the agent, The state transition probability $P(s'|s, a)$ defines the model for transitioning from state $s$ to state $s'$ after executing action $a$, the reward function $R(s, a)$ provides immediate feedback signals on the quality of actions, the discount factor $\gamma \in [0, 1]$ Balances the importance of current rewards versus future rewards. The objective is to maximize the expected cumulative discount reward:

$$G_t = \sum_{k=0}^{\infty} \gamma^k R_{t+k+1} \tag{1}$$

Traditional reinforcement learning suffers from a severe curse of dimensionality when dealing with high-dimensional or continuous state spaces, rendering it completely ineffective. Therefore, deep reinforcement learning introduces a parameterized deep neural network as a function approximator.Among numerous deep reinforcement learning algorithms, the Soft Actor-Critic (SAC) algorithm has garnered significant attention due to its outstanding performance in sample efficiency, exploration effectiveness, and training stability.SAC extends the traditional RL objective of expected cumulative reward by incorporating policy entropy $\mathcal{H}$. Its objective becomes maximizing the weighted sum of expected cumulative reward and policy entropy:

$$J_{\text{SAC}}(\pi) = \mathbb{E}_{\tau \sim \pi}[\sum_{t=0}^{\infty} \gamma^t (R(s_t, a_t, s_{t+1}) + \alpha \mathcal{H}(\pi(\cdot|s_t)))] \tag{2}$$

Here, $\tau = (s_0, a_0, s_1, a_1, \ldots)$ denotes a trajectory, and $\alpha > 0$ is the temperature parameter, which balances the importance of reward maximization and entropy maximization. The entropy term $\mathcal{H}(\pi(\cdot|s)) = -\mathbb{E}_{a \sim \pi(\cdot|s)}[\log\pi(a|s)]$ encourages policy randomness, thereby promoting exploration and preventing premature convergence to local optima. SAC simultaneously learns one policy (actor) $\pi_\theta$ and two Q-functions (critic)$Q_{\phi_1}, Q_{\phi_2}$.Update the Q-network parameters $\phi$ by minimizing the Soft Bellman error.

$$J_Q(\phi_i) = \mathbb{E}_{(s,a,r,s',d) \sim \mathcal{D}}\left[\left(Q_{\phi_i}(s, a) - \hat{Q}_{\text{target}}(s, a)\right)^2\right] \tag{3}$$

where $d$ is the Done Flag, and $\mathcal{D}$ is the Replay Buffer.Update the policy network parameters $\theta$ by minimizing the KL divergence, thereby updating the policy toward maximizing the expected Q-values and minimizing entropy.

$$J_\pi(\theta) = \mathbb{E}_{s \sim \mathcal{D}, a \sim \pi_\theta(\cdot|s)}\left[\alpha \log\pi_\theta(a|s) - \min Q_{\phi_j}(s, a)\right] \tag{4}$$

Similarly, action $a$ is sampled using a reparameterization trick: $a = f_\theta(\epsilon; s)$, where $\epsilon$ is the input noise.

**RLWE.** The Ring Learning with Errors (RLWE) problem stands as one of the core computational challenges in post-quantum cryptography (PQC). Building upon the Linear Equation with Errors (LWE) problem, it significantly enhances efficiency by leveraging the algebraic structure of polynomial rings in algebraic number theory. This makes it an ideal candidate for constructing practical post-quantum cryptographic schemes, such as key exchange, public-key encryption, and digital signatures. The RLWE problem is defined over polynomial rings. Let $R = \mathbb{Z}[x]/(f(x))$ denote a polynomial quotient ring, where $f(x)$ is an $n$-degree prime polynomial (typically chosen as the cyclotomic polynomial $f(x) = x^n + 1$, where $n$ is a power of 2). The elements of this ring are all polynomials of degree less than $n$:

$$a(x) = a_0 + a_1 x + a_2 x^2 + \cdots + a_{n-1} x^{n-1} \tag{5}$$

In cryptographic applications, we typically further consider the quotient ring of the coefficient field $q$:

$$R_q = R/(q) \cong \mathbb{Z}_q[x]/(f(x)) \tag{6}$$

Here, $\mathbb{Z}_q$ denotes the ring of integers modulo $q$. Elements in $R_q$ are polynomials with coefficients in $\mathbb{Z}_q$, and operations are polynomial operations modulo $f(x)$ and modulo $q$.RLWE encryption typically also requires sampling an error term from a discrete probability distribution $\chi$ concentrated around small integers near zero. Common distributions include:

- **Discrete Gaussian distribution**: $D_{\mathbb{Z},\sigma}$, with mean 0 and standard deviation $\sigma$
- **Uniform distribution**: Uniform sampling over the finite interval $[-B, B]$, where $B \ll q$

Given a polynomial ring $R_q$ and an error distribution $\chi$, the searchable RLWE problem is defined as follows, Sampling a random secret $\mathbf{s} \leftarrow R_q$ from the distribution and accessing an oracle $\mathcal{O}_{\mathbf{s},\chi}$ that outputs a sample $(\mathbf{a}, \mathbf{b})$ at each query, where:

$$\mathbf{a} \leftarrow R_q, \mathbf{b} = \mathbf{a} \cdot \mathbf{s} + \mathbf{e} \text{ with } \mathbf{e} \leftarrow \chi^n \tag{7}$$

The search-based RLWE problem requires recovering the secret $\mathbf{s}$ from multiple samples $\{(\mathbf{a}_i, \mathbf{b}_i)\}_{i=1}^{m}$. Its security rests on two core assumptions: (1) The computational complexity assumption holds for appropriately chosen parameters $(n, q, \chi)$, meaning no polynomial-time quantum or classical algorithm exists that solves the RLWE problem with a non-negligible advantage. (2) The average-case hardness of solving a random RLWE instance can be reduced to the worst-case hardness of solving the Approximate Shortest Vector Problem (approx-SVP) over ideal lattices.

RLWE can be very naturally applied to construct cryptographic protocols. A typical construction approach for a public-key encryption scheme is as follows:

- **Key Generation**: Secret key $sk = s \leftarrow \chi_s$ (sampled from a secret distribution). Public key $pk = (a, b = a \cdot s + e)$, where $a \leftarrow R_q$ and $e \leftarrow \chi_e$.
- **Encryption**: Encrypt message $m$ (encoded over the ring) using public key $pk$. The ciphertext takes the form $(c_1, c_2) = \left(a' \cdot s + e_1, b' \cdot s + e_2 + m\right)$, where $a'$, $b'$, and the error term are generated on-the-fly.
- **Decryption**: Using the secret key $sk = s$, compute $c_2 - c_1 \cdot s \approx m$, where the approximation is determined by the error terms and requires processing via error-correcting codes or similar methods.

The advantage of RLWE lies in its ability to conceal information equivalent to $n$ (the ring dimension) scalar LWE samples within a single sample $(a, b)$. Furthermore, since operations are based on polynomial multiplication and addition, they can be efficiently computed using number-theoretic transformations (NTT), achieving high computational efficiency.Compared to the original LWE problem, RLWE offers significant advantages, as shown in Table 1.

**Table 1.** Parameter Efficiency Comparison Between RLWE and LWE

| Property | LWE | RLWE |
|---|---|---|
| Public Key Size | $O(n^2 \log q)$ | $O(n \log q)$ |
| Computational Complexity | $O(n^2)$ | $O(n \log n)$ |
| Sample Size | $o\left(n^2\right)$ | $O(n)$ |

By leveraging the algebraic properties of ring structures, RLWE reduces both key size and computational complexity from quadratic to near-linear, making it more feasible for practical applications.

## 2.2 System Architecture

The system comprises four core entities that combine cryptography with deep reinforcement learning to achieve secure and efficient federated learning model training.

- **Clients**: As participants in federated learning, clients possess local private data. Their primary responsibilities include training on their local datasets using the current global model and computing local model gradients. To ensure privacy security, clients perform two parallel encryption processes on gradients and upload them separately to the Classification Weight Server (CS) and Aggregation Server (AS).
- **Classification Weight Server (CS)**: This server serves as the system's intelligent coordination center. Its core responsibilities include receiving differentially privacy-protected gradients uploaded by clients and executing client clustering and weight allocation based on this information. Additionally, CS interacts with the Aggregation Server (AS) to facilitate aggregation operations.
- **Aggregation Server (AS)**: This server serves as the system's secure computation node, specifically responsible for performing computational operations on ciphertext domains. Its core function is to execute secure weighted aggregation on encrypted gradients directly received from clients, leveraging the homomorphic addition and multiplication properties of the RLWE cryptosystem.
- **Trusted Authority (TA)**: As the system's initialization and trust anchor, the TA is responsible for establishing global system parameters and managing keys. Its primary duties include generating and distributing system parameters such as keys.

## 2.3 Threat Model

In this architecture, the Trusted Authority (TA) is assumed to be a fully trusted entity that does not collude with any other party and correctly executes initialization operations. Clients are categorized as "honest but curious" or partially malicious clients; honest clients comply with the protocol but may attempt to infer other clients' information from received messages, while malicious clients may submit erroneous gradients or attempt to disrupt the system. Servers (CS and AS) are treated as "honest but curious" non-colluding entities; they correctly execute specified computation and communication protocols but may attempt to infer clients' private data from received information.

## 2.4 ProFL-DARL

**Initialization.** The TA is responsible for performing a one-time global initialization, establishing the foundation for security and collaboration across the entire framework. Core parameters for RLWE are selected, including determining the polynomial ring dimension $n$, modulus $q$, and error distribution $\chi$. Subsequently, the TA runs the RLWE key generation algorithm $KeyGen(params) \rightarrow (pk, sk)$, producing an asymmetric key pair: the public key $pk$ for encryption and the private key $sk$ for decryption. The TA securely distributes the private key $sk$ to all registered clients via a secure channel, while publicly publishing the public key $pk$ to the CS and AS and all clients. Simultaneously, a unique symmetric key $K_{i,\text{CS}}$ is generated and distributed for each client $i$ and the CS. This key establishment can be achieved in two ways: first, the TA generates the key as a trusted third party and securely distributes it to both the corresponding client and CS. Second, the diffie-hellman key exchange protocol is used to directly negotiate a shared key between the client and CS, with the TA solely responsible for identity authentication during the exchange process.

After completing the above key distribution, the TA generates or designates an initial global model parameter $\boldsymbol{w}_0$. This model can serve as either a random initialization or a pre-trained base model. Finally, the TA distributes $\boldsymbol{w}_0$ to all clients via the broadcast channel.

**Local Training and Privacy Processing.** During the $t$ th training round, each client $i$ downloads the latest global model parameters $\boldsymbol{w}_t$ from the Aggregation Server (AS), then performs local training on its private dataset $\mathcal{D}_i$:

$$\boldsymbol{g}_i^t = \nabla_{\boldsymbol{w}} \mathcal{L}_i(\boldsymbol{w}_t) = \frac{1}{|\mathcal{B}_i|} \sum_{(\boldsymbol{x},y)\in\mathcal{B}_i} \nabla_{\boldsymbol{w}} \ell(\boldsymbol{w}_t; \boldsymbol{x}, y) \tag{8}$$

where $\mathcal{B}_i \subset \mathcal{D}_i$ denotes a randomly sampled batch from the local dataset, and $\ell(\cdot)$ represents the model's loss function on a single sample.

The client applies two parallel privacy-enhancing processes to the original gradient $\boldsymbol{g}_i^t$. To enable CS to perform effective client clustering and weight allocation while strictly preventing the inference of original data from gradients, client $i$ injects Gaussian noise into the original gradient, generating a perturbed gradient $\tilde{\boldsymbol{g}}_i^t = \boldsymbol{g}_i^t + \boldsymbol{\eta}_i^t$ that satisfies $(\varepsilon, \delta)$-differential privacy, where $\boldsymbol{\eta}_i^t \sim \mathcal{N}\left(0, \sigma_{\text{dp}}^2 \boldsymbol{I}\right)$, and the noise standard deviation $\sigma_{\text{dp}}$ is computed based on the predefined privacy budget $(\varepsilon, \delta)$ and gradient sensitivity $\Delta$. To achieve secure homomorphic aggregation and defend against AS eavesdropping, client $i$ homomorphically encrypts the original gradient using the RLWE public key $pk$ to obtain $\text{Enc}_{pk}\left(\boldsymbol{g}_i^t\right)$ using the RLWE public key $pk$ distributed by TA. A one-time high-strength random mask $\boldsymbol{r}_i^t$ is generated and homomorphically added to the encrypted gradient, yielding the final mixed ciphertext:

$$\boldsymbol{c}_i^t = \text{Enc}_{pk}\left(\boldsymbol{g}_i^t\right) \oplus \text{Enc}_{pk}\left(\boldsymbol{r}_i^t\right) = \text{Enc}_{pk}\left(\boldsymbol{g}_i^t + \boldsymbol{r}_i^t\right) \tag{9}$$

To ensure the random mask $\boldsymbol{r}_i^t$ can be securely transmitted to the CS for subsequent use, client $i$ encrypts the mask using the pre-established long-term shared symmetric key $K_{i,\text{CS}}$ between itself and the CS, generating the ciphertext $\text{Enc}_{K_{i,\text{CS}}}\left(\boldsymbol{r}_i^t\right)$. After completing all processing, client $i$ transmits the following to the CS: the current accuracy rate $\text{Acc}_i^t$ computed on the locally retained test set, and the encrypted random mask $\text{Enc}_{K_{i,\text{CS}}}\left(\boldsymbol{r}_i^t\right)$. Simultaneously, the encrypted gradient $\boldsymbol{c}_i^t$ mixed with the random mask is uploaded to the AS.

**Clustering and Weight Assignment.** AS sends the received list of client IDs to CS. CS then takes the intersection with its own received client list to obtain the tuples uploaded by online clients: $\left(\tilde{\boldsymbol{g}}_i^t, \text{Acc}_i^t, \text{Enc}_{K_{i,\text{CS}}}\left(\boldsymbol{r}_i^t\right)\right)$. CS then decrypts the received ciphertext $\text{Enc}_{K_{i,\text{CS}}}\left(\boldsymbol{r}_i^t\right)$ using the symmetric key $K_{i,\text{CS}}$ shared long-term with each client $i$, recovering the plaintext $\boldsymbol{r}_i^t$ of the random mask. Subsequently, the CS executes an online clustering and weight assignment algorithm based on reinforcement learning (detailed design and derivation are provided in Section E of this paper). By evaluating multi-dimensional features such as gradient similarity among clients, current and historical accuracy rates, and device status, the algorithm dynamically partitions all clients into $K$ clusters $\{C_1, C_2, ..., C_K\}$. Each cluster represents a group of clients with similar data distributions or learning characteristics. During this process, the algorithm simultaneously detects and excludes clients exhibiting abnormal behavior (such as excessive deviation of gradients from the cluster center, abnormal accuracy, or suspected poisoning attacks) from the current round of aggregation. For normal clients that pass screening, the algorithm assigns appropriate aggregation weights $\alpha_i^t$ satisfying the normalization constraint $\sum\limits_{i \in S^t} \alpha_i^t = 1$.

**Secure Aggregation and Mask Removal.** After completing client clustering and weight assignment, the CS transmits the following information to the AS for each cluster $C_k$: the membership details, the corresponding clustering weight $\alpha_k^t$, and the weighted sum of all client masks within that cluster, denoted as $\boldsymbol{R}_k^t = \sum\limits_{i \in C_k} \alpha_i^t \boldsymbol{r}_i^t$. The AS then performs secure aggregation operations independently for each cluster $C_k$. Leveraging the homomorphic properties of the RLWE cryptosystem, the received encrypted gradients $\boldsymbol{c}_i^t$ can be directly processed to compute the encrypted aggregation result

$$\boldsymbol{C}_k^t = \sum_{i \in C_k} \alpha_i^t \cdot \boldsymbol{c}_i^t = \mathrm{Enc}_{pk}\left(\sum_{i \in C_k} \alpha_i^t (\boldsymbol{g}_i^t + \boldsymbol{r}_i^t)\right) \tag{10}$$

Next, the AS must remove the previously added random mask from the aggregated result to recover the true gradient aggregation value. The AS encrypts the mask-weighted sum $\boldsymbol{R}_k^t$ received from the CS using the public key $pk$, yielding $\mathrm{Enc}_{pk}(\boldsymbol{R}_k^t)$ and subtracts it from $\boldsymbol{C}_k^t$ via homomorphic subtraction: $\hat{\boldsymbol{C}}_k^t = \boldsymbol{C}_k^t \ominus \mathrm{Enc}_{pk}(\boldsymbol{R}_k^t)$. The resulting ciphertext $\hat{\boldsymbol{C}}_k^t$ is the encrypted form of the true weighted gradient $\sum\limits_{i \in C_k} \alpha_i^t \boldsymbol{g}_i^t$ for cluster $C_k$, $\hat{\boldsymbol{C}}_k^t = \mathrm{Enc}_{pk}\left(\sum\limits_{i \in C_k} \alpha_i^t \boldsymbol{g}_i^t\right)$. Finally, AS distributes the encrypted aggregation result $\hat{\boldsymbol{C}}_k^t$ for each cluster to all clients within the corresponding cluster $C_k$.

**Local Updates.** For each client $i$ belonging to cluster $C_k$, upon receiving the encrypted aggregated result $\hat{\boldsymbol{C}}_k^t$ from the AS, it begins executing the local update process. The client decrypts the ciphertext using its private key $sk$ to recover the plaintext aggregated gradient $\boldsymbol{g}_{\mathrm{agg},k}^t = \mathrm{Dec}_{sk}\left(\hat{\boldsymbol{C}}_k^t\right)$ for its cluster $C_k$. Subsequently, using the stochastic gradient descent (SGD) update rule with learning rate $\eta$, it updates its locally stored previous-round global model $\boldsymbol{w}_t$ to $\boldsymbol{w}_i^{t+1} = \boldsymbol{w}_t - \eta \boldsymbol{g}_{\mathrm{agg},k}^t$. The updated model $\boldsymbol{w}_i^{t+1}$ serves as the initial model for the next round $(t+1)$ of local training.

**Algorithm 1** Reinforcement Learning-Based Client Clustering and Weight Allocation

**Require:** Client data $\{(i, \tilde{g}_i^t, Acc_i^t)\}_{i=1}^N$

**Ensure:** Clustering allocation results $\{C_1, C_2, ..., C_K\}$, weight vector $\alpha^t = \{\alpha_1^t, \alpha_2^t, ..., \alpha_N^t\}$

1. Initialize policy network and value network of reinforcement learning agent
2. Construct state representation: $\mathbf{s}_t = \text{Concat}(\mathbf{F}^t, \boldsymbol{\alpha}^{t-1}, \mathbf{C}^{t-1}, \mathbf{H}^t)$
3. Agent generates action $\mathbf{a}_t = (\boldsymbol{\pi}_a, \boldsymbol{\pi}_c, \boldsymbol{\pi}_w)$ based on current state $\mathbf{s}_t$
4. **for** each client $i = 1$ to $N$ **do**
5. Compute pruning flag:
6. **if** $\text{is\_pruned}_i = \mathbb{I}(\pi_c(i) > \tau) = \text{True}$ **then**
7. Mark client $i$ as an outlier node, excluded from this round's aggregation
8. Set $\alpha_i^t = 0$
9. **else**
10. Determine cluster assignment: $c_i = \text{argmax}_k \boldsymbol{\pi}_a(i, k)$
11. Compute aggregation weight: $\alpha_i^t = \text{Softmax}(\lambda \cdot Acc_i^t + (1 - \lambda) \cdot \boldsymbol{\pi}_w(i))$
12. **end if**
13. **end for**
14. Construct clusters based on assignments: $C_k = \{i | c_i = k \text{ and } \alpha_i^t > 0\}$
15. Normalize weights within each cluster: $\alpha_i^t = \alpha_i^t / \sum_{j \in C_k} \alpha_j^t$
16. Compute reward signal $r_t$ and update reinforcement learning agent
17. **return** Cluster assignments $\{C_1, C_2, ..., C_K\}$ and weight vector $\boldsymbol{\alpha}^t$

### 2.5 Reinforcement Learning Clustering and Weight Allocation Algorithm

In this section, we will provide a comprehensive overview of the training process based on deep reinforcement learning, with the workflow detailed in Algorithm 1. The core input data required for executing the algorithm primarily consists of the following two components:

- $\tilde{g}_i^t = g_i^t + \varepsilon_i^t$, $\varepsilon_i^t \sim \mathcal{N}(0, \sigma^2 I)$ denotes the noisy gradient satisfying $(\varepsilon, \delta)$-differential privacy.
- $Acc_i^t \in [0, 1]$ is the accuracy on the test set.

Gradients are preprocessed to reduce noise effects and extract meaningful features:

$$f_i^t = \text{Preprocess}(\tilde{g}_i^t) = \text{Normalize}(\text{PCA}(\tilde{g}_i^t, d')) \tag{11}$$

$\text{PCA}(\cdot, d')$ projects high-dimensional gradients onto the $d'$-dimensional principal component space ($d' \ll d$), while $\text{Normalize}(\cdot)$ standardizes them to unit norm. The feature matrix $F^t \in \mathbb{R}^{N \times d'}$ formed by all feature vectors models the task as a sequential decision problem, where reinforcement learning optimizes the clustering strategy.

**State.** The state $s_t$ comprehensively characterizes the current system status and is defined as:

$$s_t = \left(F^t, \alpha^{t-1}, C^{t-1}, H^t\right) \tag{12}$$

$\boldsymbol{F}^t$ denotes the gradient feature matrix for all clients in the current round. $\boldsymbol{\alpha}^{t-1} = \left[\boldsymbol{\alpha}_1^{t-1}, ..., \boldsymbol{\alpha}_N^{t-1}\right]$ represents the weight vector assigned to clients in the previous round. $\boldsymbol{C}^{t-1} = \left\{\boldsymbol{C}_1^{t-1}, ..., \boldsymbol{C}_K^{t-1}\right\}$ denotes the clustering results from the previous round. $\boldsymbol{H}^t = \{\mathrm{Acc}_i^1, ..., \mathrm{Acc}_i^t\}_{i=1}^N$ denotes the historical accuracy sequence for all clients. The state $\boldsymbol{s}_t$ is encoded as a fixed-length vector for compressed representation.

**Action.** Action $a_t$ defines the decision made in this round. Its output is a composite action:

$$a_t = (\boldsymbol{\pi}_a, \boldsymbol{\pi}_c, \boldsymbol{\pi}_w) \tag{13}$$

Allocation Action $\boldsymbol{\pi}_a$: An $N \times K$ soft assignment matrix where element $\pi_a(i, k)$ represents the probability of assigning client $i$ to cluster $k$. The final hard assignment is obtained by taking the maximum value in each row: $c_i^t = \mathrm{argmax}_k \pi_a(i, k)$, forming the new clusters $\left\{C_1^t, ..., C_K^t\right\}$. Pruning Action $\boldsymbol{\pi}_c$: An $N$-dimensional vector, where $\pi_c(i) \in [0, 1]$ represents the probability of classifying client $i$ as low-quality and excluding it from the current round of aggregation. Set a threshold $\tau$. If $\pi_c(i) > \tau$, the client is flagged as anomalous and excluded from all cluster aggregations. Weighting Action $\boldsymbol{\pi}_w$: An $N$-dimensional vector, $\pi_w(i) \in [0, 1]$, used to compute the aggregation weight $\alpha_i^t$ for client $i$. The final weight is jointly determined by accuracy and strategy: $\alpha_i^t = \mathrm{Softmax}\left(\lambda \cdot \mathrm{Acc}_i^t + (1 - \lambda) \cdot \pi_w(i)\right)$, where $\lambda$ is a hyperparameter balancing historical performance and current strategy.

**Reward Function.** The design of the reward function $r(s_t, a_t)$ is crucial, as it must guide the agent to learn effective clustering strategies. The reward is composed of the following components:

$$r(s_t, a_t) = r_{\text{intra}} + r_{\text{inter}} + r_{\text{acc}} + r_{\text{penalty}} \tag{14}$$

Intra-cluster Similarity Reward $r_{\text{intra}}$: Encourages high gradient similarity among clients within the same cluster. Calculated as the average cosine similarity of all client features within the cluster:

$$r_{\text{intra}} = \frac{1}{K}\sum_{k=1}^{K}\frac{1}{|C_k|(|C_k| - 1)}\sum_{i,j\in C_k, i\neq j}\mathrm{cosine}\left(\boldsymbol{f}_i^t, \boldsymbol{f}_j^t\right) \tag{15}$$

Inter-cluster Diversity Reward $r_{\text{inter}}$: Encourages greater diversity between different clusters. Calculate the average cosine distance between cluster centroids:

$$r_{\text{inter}} = \frac{1}{K(K-1)}\sum_{k=1}^{K}\sum_{l\neq k}(1 - \mathrm{cosine}(\boldsymbol{\mu}_k, \boldsymbol{\mu}_l)) \tag{16}$$

where $\boldsymbol{\mu}_k = \frac{1}{|C_k|}\sum_{i\in C_k}\boldsymbol{f}_i^t$ is the centroid of cluster $C_k$.

Accuracy Improvement Reward $r_{\text{acc}}$: Encourages clustering and weight allocation that enhance model performance. Compares the change in average test accuracy between the current and previous rounds:

$$r_{\text{acc}} = \beta \cdot \left( \frac{1}{N} \sum_{i=1}^{N} \text{Acc}_i^t - \frac{1}{N} \sum_{i=1}^{N} \text{Acc}_i^{t-1} \right) \tag{17}$$

where $\beta$ is a scaling factor.

Penalty term $r_{\text{penalty}}$: Prevents abnormal behavior, such as dropping too many clients or forming clusters of extreme size.

$$\begin{aligned} r_{\text{penalty}} &= -\gamma_1 \cdot \mathbb{I}(N_{\text{dropped}} > \phi) \\ &-\gamma_2 \cdot \text{StdDev}(|C_1|, ..., |C_K|) \end{aligned} \tag{18}$$

where $N_{\text{dropped}}$ is the number of dropped clients, $\phi$ is the threshold, StdDev is the standard deviation of cluster sizes, and $\gamma_1, \gamma_2$ are penalty coefficients.

**Policy Learning and Optimization.** This system employs the Soft actor-critic (SAC) algorithm to achieve intelligent policy learning and optimization. Based on the maximum entropy reinforcement learning framework, this algorithm balances exploration and exploitation. By leveraging state information such as noisy gradients and accuracy rates, it autonomously learns optimal clustering and weight allocation actions to maximize long-term cumulative rewards.

## 3 Convergence Analysis

ProFL-DARL incorporates three mutually coupled dynamic processes within its optimization flow: iterative updates of global model parameters, clustering and weighted decision-making for clients by the SAC-based reinforcement learning agent, and random perturbations introduced by differential privacy noise and homomorphic encryption mechanisms. Consequently, its overall convergence analysis requires simultaneous consideration of both optimization-level and policy-level perspectives. This section analyzes convergence from three dimensions: the convergence of the global model, the stability of the dynamic clustering mechanism, and the policy convergence of reinforcement learning agents within the federated learning loop. Particular emphasis is placed on how the RL module's convergence behavior in non-stationary FL environments enhances the effectiveness of the dynamic aggregation strategy.

First, we analyze the optimization convergence of the global model. Let the global objective function be defined as $L(w) = \sum_{i=1}^{N} p_i L_i(w)$, where $L_i$ represents the local loss of each client. In ProFL-DARL, the effective update at each training round can be expressed as:

$$w_{t+1} = w_t - \eta \sum_{k=1}^{K} \sum_{i \in C_k^t} \alpha_i^t g_i^t. \tag{19}$$

Since the reinforcement learning-derived clusters $C_k^t$ and weights $\alpha_i^t$ dynamically change during training, the update direction essentially represents directional descent with random fluctuations. However, under the following standard assumptions, the global model can still guarantee convergence in the desired sense: if each local loss $L_i$ is $\beta$-Lipschitz smooth, and the DP noise satisfies the variance bound $\mathbb{E}|\eta_i^t\|^2 \leq \sigma^2$, and the dynamic clustering mechanism maintains boundedness of intra-group gradient consistency error, i.e., there exists a constant $\delta$ such that

$$\mathbb{E}\|g_i^t - g_{\text{agg},k}^t\|^2 \leq \delta^2 \quad (20)$$

Then, based on stochastic nonconvex optimization theory, we obtain

$$\mathbb{E}\|\nabla L(w_t)\|^2 = O\left(\frac{1}{T}\right) + O\left(\sigma^2\right) + O\left(\delta^2\right) \quad (21)$$

This demonstrates that ProFL-DARL maintains an $O(1/T)$ descent rate under controllable noise and clustering bias, and converges to a bounded neighborhood near the optimal solution.

Next, we analyze the impact of the dynamic clustering mechanism on convergence. ProFL-DARL employs reinforcement learning agents to perform online clustering and weight computation for clients, effectively reducing inconsistencies in intra-group gradients. This alignment brings update directions closer to the true gradient direction of the global loss. Experiments reveal that intra-group gradient cosine similarity gradually increases during training before stabilizing in later stages, while the distance between inter-group clustering centers remains at a high level. This indicates that reinforcement learning has learned a relatively stable grouping structure that effectively reduces gradient conflicts. The stability of this structure diminishes the triangular non-equivalence bias in gradient directions across rounds, significantly suppressing the gradient variance term $\delta^2$ during global updates. This serves as a crucial factor accelerating the overall convergence of federated optimization.

Finally, we focus on discussing the policy convergence of RL agents in the FL cycle. The environment faced by reinforcement learning agents is highly non-stationary: on one hand, the global model $w_t$ is constantly changing, causing the client gradient distribution to drift across rounds; on the other hand, differential privacy noise and local data heterogeneity introduce random perturbations to input states. Consequently, traditional RL convergence theories cannot be directly applied. To ensure policy stability, this method introduces three steady-state mechanisms: (1) An experience replay buffer breaks the temporal correlation between gradients and states, enabling agents to learn from a "smoothed experience distribution" rather than the instantaneous distribution; (2) A target network reduces high-frequency jitter in Q-value estimation; (3) The maximum entropy objective $\alpha H(\pi(\cdot|s))$ ensures the policy retains sufficient exploration capability, thereby avoiding getting stuck in local optimal.

The convergence of reinforcement learning policies is critical to the effectiveness of the dynamic aggregation mechanism. Once the RL policy stabilizes, the clustering structure and weight distribution cease to fluctuate dramatically, significantly reducing aggregation volatility. This smoothens the global gradient update direction, ultimately accelerating the overall federated convergence rate. Simultaneously, the stability of the clustering structure maintains feature consistency within groups, substantially reducing gradient conflicts caused by heterogeneous data. It also continuously filters out low-quality or anomalous clients, making the effective gradients in each round more reliable. Thus, RL policy convergence and federated optimization convergence form a positive feedback loop: as the global model converges, state distributions become more regular, facilitating RL policy convergence. Conversely, policy convergence ensures stable clustering structures, further reducing gradient variance and accelerating global model convergence.

In summary, the ProFL-DARL optimization process exhibits an asymptotic progression: "dynamic policy stability → stable clustering structure → stable gradient direction → accelerated global model convergence." Even under privacy-preserving perturbations and encryption mechanisms, this framework maintains robust convergence performance while significantly enhancing the effectiveness and robustness of the dynamic aggregation mechanism through reinforcement learning agents.

## 4 Experiments

In this section, we present representative experimental results conducted on a computer equipped with a 2.100 GHz CPU and a GeForce A100 GPU, running on the Linux operating system. The experiments were implemented using the PyTorch package in the Python programming language. The proposed method was comprehensively evaluated on the standard benchmark datasets MNIST and CIFAR-10. The experimental setup employed varying scales of client populations. To simulate real-world data distribution characteristics, this paper adopted a non-iid data partitioning scheme based on the Dirichlet distribution. This approach ensures significant differences in category distributions across clients, with sample counts for each category following a long-tail distribution across different clients.Regarding the reinforcement learning agent, it is configured to execute FL communication once per round. The experience replay buffer capacity is set to 10,000 experience transfer tuples, a scale sufficient to cover diverse state-action pairs during long-term training. The learning rate is set to 0.001, and the episode length is set to 5.

We compare several approaches. KT-pFL (Zhang et al., 2021) proposes a novel training framework where each client maintains personalized soft predictions on the server to guide local training for other clients. It updates each client's personalized soft predictions using a knowledge coefficient matrix through linear combinations of all local soft predictions. PHP-FL (Pan et al., 2024) employs a knowledge matrix and client device performance for clustering, executing a cross-cluster knowledge transfer mechanism to enhance the personalized performance of weak clients. A ring aggregation algorithm is designed as a secure aggregation scheme for each group. FedMD (Li and Wang, 2019) utilizes transfer learning and knowledge distillation to design a universal framework enabling federated learning when each client possesses not only their private data but also uniquely designed models.The results are shown in Table 2.

**Table 2.** Performance Comparison of Different Federated Learning Methods

| Method | MNIST Accuracy (%) | CIFAR-10 Accuracy (%) |
|---|---|---|
| KT-pFL | 98.89 | 61.34 |
| PHP-FL | 98.58 | 62.65 |
| FedMD | 98.33 | 60.29 |
| FedAvg | 98.26 | 57.20 |
| ProFL-DARL | 98.69 | 73.00 |

We simulated a typical federated learning scenario with 50 clients, where each client's data was generated via a Dirichlet distribution, exhibiting high heterogeneity. As shown in Fig. 1, the data distribution across different classes (Class 0–9) is extremely imbalanced for each client (client ID 0–49). Some clients' samples are almost entirely concentrated in one or two classes, while other classes have few or no samples. This complex Non-IID distribution is the primary cause of model performance degradation and convergence difficulties in federated learning.

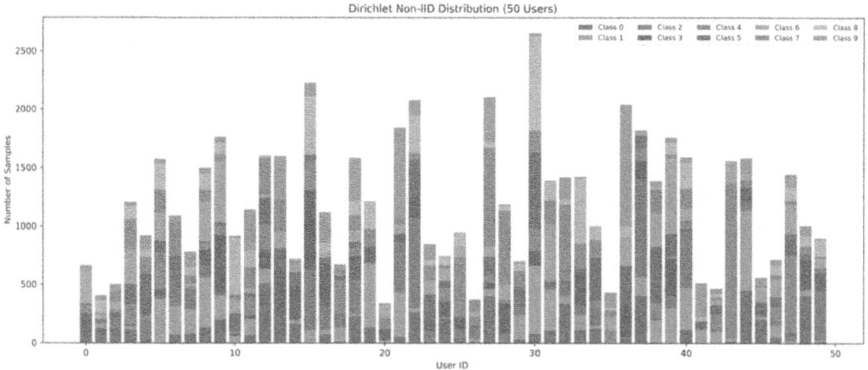


**Fig. 1.** Distribution of Non-IID Data Across Different Clients

We employ reinforcement learning agents to execute dynamic grouping tasks. The reward function is carefully designed as a composite metric balancing intra-group consistency, inter-group diversity, and final model performance. Through interaction with the environment, the agents learn and ultimately output a grouping strategy that assigns 50 chaotic clients into 5 distinct groups.

The experimental results are highly significant. As shown in Fig. 2a, after RL-optimized grouping, the originally chaotic client data is reorganized into five groups with distinct internal characteristics. Each group exhibits a dominant data distribution pattern. For example, Group 0 consists of a large number of clients with Class 0 samples, totaling 1,500 Class 0 samples—far exceeding other groups. Simultaneously, the pie chart in Fig. 2b further reveals the detailed composition within a single group. One category (likely Class 7, inferred by color) holds absolute dominance (56.9%), while other categories also occupy certain proportions. This indicates that the RL grouping strategy not only successfully clusters clients with similar data distributions, achieving high intra-group consistency, but also preserves necessary diversity to avoid overfitting. The reinforcement learning agent learned an effective grouping strategy that reorganizes highly heterogeneous raw data into a structure characterized by intra-group consistency and inter-group complementarity. Training federated learning based on this grouping structure can significantly improve the training efficiency of the global model and accelerate convergence.

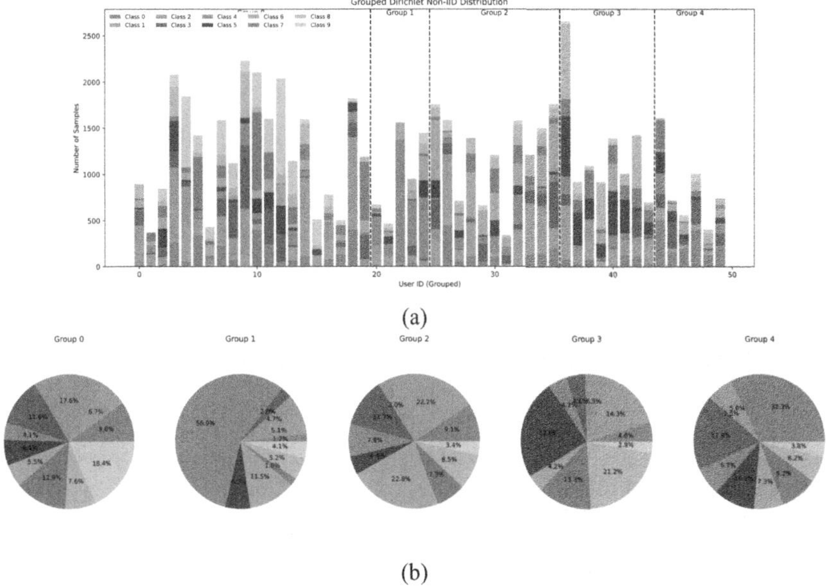


(a)

(b)

**Fig. 2.** Grouping Results

We investigated the impact of adding noise when clients transmit gradients to classification servers. Two widely used data sets, MNIST and CIFAR-10, were selected and neural network models were trained on each data set. Gradients computed by clients were sent to the server under two distinct settings:

- **No noise:** Gradients were transmitted to the server without any modification.
- **With noise:** Gaussian noise perturbations were applied to the gradients before transmission to improve the privacy of the client's data.

Experimental results shown in Fig. 3 indicate that adding noise has a negligible impact on test accuracy for both the MNIST and CIFAR-10 datasets. This demonstrates that adding reasonable noise to gradients can protect client privacy without significantly compromising model performance.

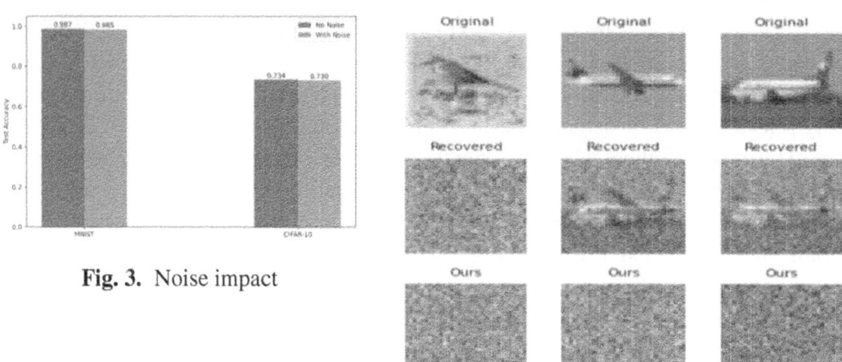


**Fig. 3.** Noise impact

**Fig. 4.** Results of model inversion attacks

In the empirical evaluation of privacy protection schemes, this paper conducted systematic experiments on the CIFAR10 dataset using a gradient perturbation privacy protection mechanism. Experimental results demonstrate that this approach significantly enhances privacy protection effectiveness: in model inversion attack assessments, the peak signal-to-noise ratio of reconstructed images decreased substantially, indicating that attackers struggle to recover original training data from model outputs. As shown in Fig. 4. The first row is the original image, the second row is the result after 10,00 iterations of attack without encryption, and the third row uses our secure aggregation scheme.

To more clearly demonstrate communication efficiency, we trained a three-layer fully connected network on the MNIST dataset and conducted a detailed analysis of the total runtime for each round of federated learning communication. This time encompasses client-side encryption, the RL decision process on the CS, local model training time, and homomorphic aggregation operations on the AS. The Table 3 presents the time distribution statistics for each component.

**Table 3.** Communication Phase Time Analysis

| Phase | Operation | Time(ms) |
|---|---|---|
| Client Encryption | Data Encryption | 108.91 |
| CS RL Decision | Client Batch and Weight Allocation | 4770.14 |
| Local Model Training | Training Process per Gradient Upload | 30320.06 |
| AS Aggregation | Aggregation Gradient | 6.05 |

We further conducted a comparative analysis between our proposed scheme and the classical federated learning algorithm FedAvg, training the MNIST dataset using a 3-layer fully connected network across 50 clients. The experiments primarily compared the total computational overhead on the client side per communication round and the overhead during the server-side aggregation phase, with specific data presented in Table 4.

**Table 4.** Computational Overhead Comparison

| Algorithm | Client Cost (ms) | Client Cost (ms) | Remarks |
|---|---|---|---|
| FedAvg | 30553.25 | 0.86 | Baseline Algorithm |
| ProFL-DARL | 35199.11 | 6.05 | Supports Secure Aggregation |

Regarding communication overhead, we conducted a detailed comparative analysis of different schemes. Experimental results show that the introduction of the privacy masking mechanism has a negligible impact on communication overhead. However, due to security requirements, RLWE encryption technology expands ciphertext size by approximately 4.1 times compared to plaintext. Additionally, to achieve secure aggregation, clients must transmit gradient data with added homomorphic noise to the AS, which increases the number of communication rounds per iteration compared to the FedAvg scheme. Specific communication overhead comparison data is shown in Table 5.

**Table 5.** Communication Overhead Comparison Analysis

| Scheme Type | Communication Cost (KB) | Expansion | Communication Rounds |
|---|---|---|---|
| Masked Scheme | 1990.4 | 5.1 × | 2 |
| Unmasked Scheme | 1990.4 | 5.1 × | 2 |
| FedAvg | 390.4 | 1.0 × | 1 |

Although our proposed scheme incurs increased computational and communication overhead compared to Fedavg, it achieves significant overall gains in performance and privacy protection, making it particularly suitable for real-world applications with stringent security requirements.

## 5 Summary

This paper analyzes existing approaches and proposes a novel collaborative architecture to address the bottlenecks in federated learning arising from highly non-iid data, client quality disparities, and privacy versus cryptographic overhead. First, we introduce functional decoupling between classification servers and aggregation servers to enhance dual safeguards for privacy and utility. Subsequently, we employ reinforcement learning strategies to automatically identify client gradient similarity within the end-to-end process, enabling dynamic grouping, weight allocation, and anomaly detection. Concurrently, we integrate RLWE homomorphic encryption with gradient masking techniques to strengthen secure aggregation capabilities and prevent collusion attacks. Experimental results demonstrate that our method significantly improves model performance in non-IID environments, while the overhead from encryption and differential privacy remains within system-acceptable limits. Future work should focus on testing and deploying this framework in larger-scale and more severely heterogeneous environments to further validate its practicality and robustness.

**Acknowledgements.** This study was funded by the Response of cvapotranspiration to carbon sink changes in typical alpine steppe and mcadow on the Tibetan Plateau(2025-GPKY 21) and National Natural Science Foundation of China (62462053).

## References

Abdul Salam, M., Fouad, K.M., Elbably, D.L., Elsayed, S.M.: Federated learning model for credit card fraud detection with data balancing techniques. Neural Comput. Appl. **36**(11), 6231–6256 (2024)

Bonawitz, K., et al.: Practical secure aggregation for privacy-preserving machine learning. In: Proceedings of the 2017 ACM SIGSAC Conference on Computer and Communications Security, pp. 1175–1191 (2017)

Castro, F., Impedovo, D., Pirlo, G.: An efficient and privacy-preserving federated learning approach based on homomorphic encryption. IEEE Open Journal of the Computer Society (2025)

Dembani, R., Karvelas, I., Akbar, N.A., Rizou, S., Tegolo, D., Fountas, S.: Agricultural data privacy and federated learning: A review of challenges and opportunities. Comput. Electron. Agric. **232**, 110048 (2025)

Ghosh, A., Chung, J., Yin, D., Ramchandran, K.: An efficient framework for clustered federated learning. Adv. Neural. Inf. Process. Syst. **33**, 19586–19597 (2020)

Kairouz, P., et al.: Advances and open problems in federated learning. Foundations and Trends® in Machine Learning **14**(1–2), 1–210 (2021)

Liao, J., Chen, Z., Larsson, E.G.: Over-the-air federated learning with privacy protection via correlated additive perturbations. In: 2022 58th Annual Allerton Conference on Communication, Control, and Computing (Allerton), pp. 1–8. IEEE (2022)

Li, D., Wang, J.: Fedmd: Heterogenous federated learning via model distillation (2019). arXiv Preprint arXiv:1910.03581

McMahan, B., Moore, E., Ramage, D., Hampson, S.: Communication-efficient learning of deep networks from decentralized data. In: Artificial Intelligence and Statistics, pp. 1273–1282. PMLR (2017)

Pan, Y., Su, Z., Ni, J., Wang, Y., Zhou, J.: Privacy-preserving heterogeneous personalized federated learning with knowledge. IEEE Trans. Netw. Sci. Eng. **11**(6), 5969–5982 (2024)

Park, J., Lim, H.: Privacy-preserving federated learning using homomorphic encryption. Appl. Sci. **12**(2), 734 (2022)

Pei, J., Liu, W., Li, J., Wang, L., Liu, C.: A review of federated learning methods in heterogeneous scenarios. IEEE Trans. Consum. Electron. **70**(3), 5983–5999 (2024)

Qiao, Y., et al.: Towards artificial general or personalized intelligence? A survey on foundation models for personalized federated intelligence (2025). arXiv Preprint arXiv:2505.06907

Sabah, F., et al.: FairD-PFL-SCS: Fair Dynamic Personalized Federated Learning with strategic client selection for improved accuracy and fairness. Information Fusion **115**, 102756 (2025)

Scott, J., Zakerinia, H., Lampert, C.H.: PeFLL: Personalized Federated Learning by Learning to Learn. In: The Twelfth International Conference on Learning Representations (ICLR) (2024)

Sheller, M.J., Edwards, B., Reina, G.A., Martin, J., Pati, S., Kotrotsou, A.: Federated learning in medicine: facilitating multi-institutional collaborations without sharing patient data. Sci. Rep. **10**(1), 12598 (2020)

Silva, A., Metcalf, K., Apostoloff, N., Theobald, B.J.: Fedembed: Personalized private federated learning (2022). arXiv Preprint arXiv:2202.09472

So, J., et al.: Lightsecagg: a lightweight and versatile design for secure aggregation in federated learning. Proceedings of Machine Learning and Systems **4**, 694–720 (2022)

Wu, S., et al.: Motley: Benchmarking heterogeneity and personalization in federated learning (2022). arXiv Preprint arXiv:2206.09262

Zhang, J., Guo, S., Ma, X., Wang, H., Xu, W., Wu, F.: Parameterized knowledge transfer for personalized federated learning. Adv. Neural. Inf. Process. Syst. **34**, 10092–10104 (2021)

Zhang, L., Xu, J., Vijayakumar, P., Sharma, P.K., Ghosh, U.: Homomorphic encryption-based privacy-preserving federated learning in IoT-enabled healthcare system. IEEE Trans. Netw. Sci. Eng. **10**(5), 2864–2880 (2022)

Zhou, Y., Zheng, P., Cao, X., Huang, J.: Two-Tier Data Packing in RLWE-based Homomorphic Encryption for Secure Federated Learning. In: Proceedings of the 2024 ACM SIGSAC Conference on Computer and Communications Security, pp. 2844–2858 (2024)

# A Deep Learning-Based Dual-Branch Feature Fusion Model for Network Intrusion Detection

Jianxing Yang(✉), Zixin Fang, Junna Gao, Taizhong Zhang, and Shiju Sun

School of Artificial Intelligence, Beijing Polytechnic University, Beijing, China
yangjianxing@bpu.edu.cn

**Abstract.** In order to meet the challenge of intrusion detection caused by the complexity and diversity of current network attacks, A deep learning model based on the fusion of two branch features is proposed. The shared convolution layer to extract basic spatial features is employed, then LSTM and GRU are introduced as parallel branches to model long-term dependence and short-term dynamics respectively. On this basis, the cross attention mechanism is used to realize the adaptive fusion of two branch features and enhance the discrimination ability of features. The six-category classification experiment are carried out on the strictly preprocessed CSE-CIC-IDS2018 data set. The verification results show that the accuracy rate of the proposed model is up to 97.96%, and the false positive rate is only 0.09%, which is significantly better than the traditional machine learning method and the single branch deep learning model. Ablation experiments further verified the effectiveness of the double branch structure and fusion mechanism. This work presents a novel technical solution for achieving high-precision, low false-positive intrusion detection in complex network environments.

**Keywords:** Intrusion Detection · Deep Learning · Dual-Branch Network · Feature Fusion

## 1 Introduction

### 1.1 Challenges in Current Intrusion Detection Systems

The rapid development of the new generation of information technology represented by artificial intelligence (AI) not only promotes the transformation of social innovation, but also makes the situation of network security increasingly severe. The China Internet Network Information Center (CNNIC) released the 55th Statistical Report on China's Internet Development on January 17, 2025. The report shows that the number of Internet users in China has reached 1.123 billion, and the penetration rate has increased to 79.7%. Greater risks have been posed to the system due to the continuous expansion of the network scale and the rapid growth in the number of users. The system vulnerabilities can be used by attackers to carry out intrusion, which poses a serious threat to citizens' privacy and network security system. In particular, the current network attack shows a large-scale and diversified development trend, and the traditional security measures, which rely on identity verification and perimeter protection, exhibit structural deficiencies, such as slow response and blind coverage in the dynamic and complex network environment.


H. Yan et al. (Eds.): AIS&P 2025, CCIS 2918, pp. 74–86, 2026.
https://doi.org/10.1007/978-981-95-9884-7_6

The intrusion detection system (IDS) is designed to detect abnormal behavior through real-time traffic monitoring. As a key link in the network security defense system, the importance of IDS is increasingly prominent, but there still remained several challenges. On the one hand, the limiting factors including the uneven distribution of network traffic, redundancy of feature space dimensions and insufficient feature extraction ability of the algorithm seriously restrict the improvement of classification accuracy of the detection model. On the other hand, traditional rule and feature-based detection methods are plagued by low detection rates, high false positive rates, and insufficient adaptability when confronted with new attacks. Especially, when the dimensions of network traffic characteristics exceed 500, the detection accuracy of the traditional machine learning model will decrease by 23%–35%, and the model update cycle is difficult to meet the needs of real-time attack and defense confrontation. More critically, the average exploit cycle for zero-day vulnerabilities had shrunk to 4.2 days by 2024, while the response cycle of traditional systems based on feature library update is usually more than 7 days, resulting in obvious protection lag. In order to solve these problems, intrusion detection technology is transforming to end-to-end detection architecture based on deep learning.

### 1.2 Research Status at Home and Abroad

In recent years, the intrusion detection technology based on Artificial Intelligence has made significant progress. Ayantayo et al. (2023) combined convolutional neural network (CNN) with fully connected network by introducing multi-layer feature fusion mechanism, which improved the detection accuracy of CIC-IDS2018 data set for complex attack types. Thaljaoui et al. (2025) proposed an intelligent intrusion detection system based on optimized CNN-LSTM structure, and verified the feasibility of convolution and recursion joint modeling in network traffic analysis. Ali et al. (2024) systematically analyzed the impact of multi-source features, sampling strategy and classifier selection on the performance of intrusion detection system from the perspective of machine learning algorithm design and feature extraction strategy. The CNN-GRU-FF two-layer feature fusion model proposed by imrana et al. (2024) shows excellent performance on NSL-KDD and unsw-nb15 data sets. Li et al. (2024) further summarized the research progress of intrusion detection system based on deep learning, and pointed out the potential and limitations of CNN, RNN and Transformer in the task of multi-dimensional traffic feature extraction and classification. In addition, the x-fuserlstm cross domain interpretable intrusion detection framework proposed by Alabbadi and bajaber (2025) achieves high-precision and low false positive detection performance in the IOT scenario. Overall, the research trends in recent years show that the deep fusion architecture based on CNN, RNN and attention mechanism has become the main direction to improve the performance of intrusion detection model. However, its generalization ability in the network environment and adaptability to unknown attack types still need to be further strengthened.

### 1.3 Main Contributions of This Article

A two-branch feature fusion model is proposed for network intrusion detection, which achieves 97.96% accuracy and 0.09% very low false alarm rate in complex multi classification tasks, and its performance is significantly better than the baseline method. The key of the model is to use parallel LSTM and GRU branches to learn the long-term and short-term spatio-temporal dependence characteristics in the network traffic sequence, and use the cross attention mechanism to realize the dynamic weighted fusion of features, so as to obtain a more robust feature representation. The research relies on the CSE-CIC-IDS2018 data set which has been strictly processed to ensure the reliability and reproducibility of the experiment. At the end, the effectiveness of the double branch structure and fusion mechanism has been further verified in the ablation experiments. The research not only verifies the effectiveness of the hybrid deep learning architecture in the field of intrusion detection, but also provides a new technical idea for building a new network security defense scheme.

## 2 Experimental Data and Evaluation Indicators

### 2.1 Experimental Data

**Data Set.** The CSE-CIC-IDS2018data set (Sharafaldin et al. 2018) released by the Canadian Institute of Cybersecurity Research was selected for this experiment, which can effectively eliminate the limitations of the early benchmark data set such as KDD99 and NSL-KDD by simulating the real network environment. The data set contains CSV files organized by date, and it has 80 network traffic characteristics in each record. The data set is composed of discrete network flow records rather than continuous time slice sequences.

In order to construct a modeling subset with consistent time sequence and balanced category, data preprocessing is carried out according to the following process as shown in Fig. 1. Firstly, read the original large-scale data by streaming block mode, and sort it stably according to the time stamp in each block, so as to strictly maintain the true order of network traffic. Then, the traffic tag is processed in a unified and standardized way, and the time stamp is parsed into a standard structured format, and a few abnormal records that cannot be parsed are eliminated.

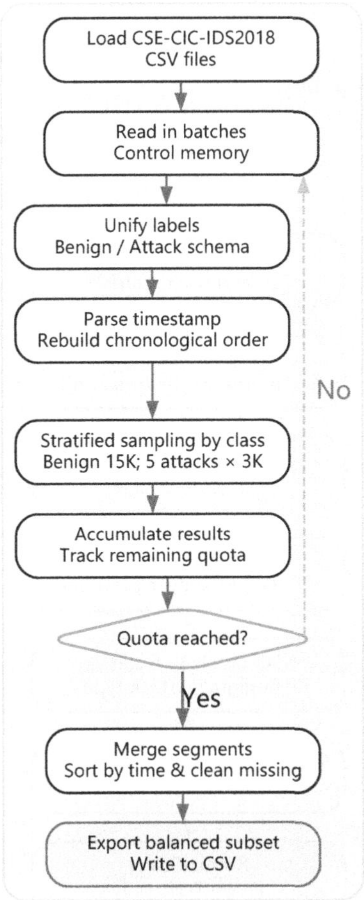


**Fig. 1.** Flow chart of experimental data set construction

**Sampling Stage.** In the process of sample collection, the sampling strategy based on time sequence priority was adopted to extract 15000 samples from "benign" traffic, and 3000 samples were collected from five types of target attacks including Slowhttptest DoS Attack, FTP Brute-force Cracking, SSH Brute-force Cracking, GoldenEye DoS and Slowloris DoS, respectively, forming a balanced data set with a total scale of 30000 records. To maintain the temporal consistency of the data set, all blocks are re merged and globally stable sorted after internal sorting and sampling.

**Data Preprocessing Stage.** In this stage, (see in Fig. 2) the identification fields unrelated to classification such as IP address and port number are firstly removed and only numerical features are retained. Subsequently, non finite values are cleaned up and 1%-99% quantile truncation is applied to each feature to suppress extreme outliers. On this basis, StandardScaler is used to standardize all features, transforming them into a distribution with a mean of 0 and variance of 1, thereby unifying feature dimensions and improving model training stability.

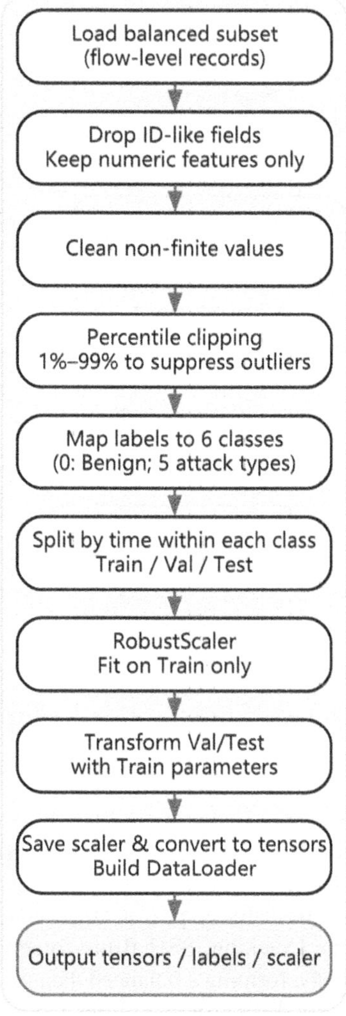


**Fig. 2.** Flow chart of data preprocessing.

The labels are uniformly mapped into a six category classification system, where 0 represents the "benign" category, and the remaining five categories correspond to the aforementioned attack types. To adapt to different model structures, the sample sequence can be encapsulated into a three-dimensional tensor structure including the number of samples, time steps, and feature numbers through a sliding window according to task requirements.

In terms of data partitioning strategy, each category is first divided into training set, verification set and test set in chronological order, and then each set is merged into a global data set. This method can ensure that each data subset has sufficient coverage of all categories while maintaining data temporal continuity.

## 2.2 Evaluation Metric

**Classification Accuracy.** As a core basic indicator, it can intuitively present the overall predictive accuracy of the model.

$$Accuracy = \frac{TP + TN}{TP + TN + FP + FN} = \frac{Number of\, correctly predicted\, samples}{Total sample size} \tag{1}$$

Among them, TP, TN, FP, and FN denote true positive,true negative,false positive,and false negative,respectively.

**Weighted Precision.** It measures the proportion of correctly predicted positives among all samples predicted as positive for a given class, reflecting the reliability of the predictions.

$$Precision_{weighted} = \frac{1}{N}\sum_{i=1}^{C} n_i \times \frac{TP_i}{TP_i + FP_i} \tag{2}$$

where $n_i$, $TP_i$, and $FP_i$ denote the number of test samples,the true positive count,and the false positive count for the i-th class,respectively,while C and N represent the total number of classes and the overall sample size.

**Weighted Recall Rate.** The weighted recall rate measures the proportion of samples actually belonging to a certain category that are correctly identified by the model,intuitively reflecting the model's completeness in identifying various categories.

$$Recall_{weighted} = \frac{1}{N}\sum_{i=1}^{C} n_i \times \frac{TP_i}{TP_i + FN_i} \tag{3}$$

where $TP_i$ and FNi denote the true positive and false negative instances for the i-th class, respectively.

**Weighted F1 Score.** The weighted F1 score, as the harmonic mean of weighted precision and weighted recall,can comprehensively evaluate the classification performance of a model.

$$F1_{weighted} = \frac{1}{N}\sum_{i=1}^{C} n_i \times \frac{2 \times TP_i}{2 \times TP_i + FP_i + FN_i} \tag{4}$$

where TPi, FPi and FNi represent the true positive,false positive,and false negative for the i-th class, respectively.

**False Alarm Rate.** False alarm rate (far) is used to evaluate the degree of misjudgment of the model on normal traffic, that is, the probability that the system mistakenly identifies normal behavior as abnormal behavior.

$$FAR = \frac{FP}{FP + TN} \tag{5}$$

where FP and TN denote the number of false positives and true negatives, respectively.

# 3 Experimental Results and Analysis

## 3.1 Intrusion Detection Experiment Based on Traditional Methods

**Performance Benchmark.** In order to build a reliable performance benchmark, the classification performance of several classical machine learning algorithms is systematically evaluated based on the preprocessed data set. Nine mainstream models including logistic regression, support vector machine and random forest were selected as the basic evaluation objects. In order to meet the input structure requirements of the traditional machine learning model, the original two-dimensional time series data are flattened into one-dimensional eigenvectors.

The experimental results show that KNeighbors, ExtraTrees, GradientBoosting and other models perform well, and the overall performance comparison is shown in Fig. 3. Among them, the accuracy rate of KNeighbors model on the test set reached 94.34%, and the accuracy rate, recall rate and F1 score were 95.39%, 94.30% and 94.41%, respectively, which was the best traditional model in this benchmark experiment. Although other models also show strong classification ability, their comprehensive performance does not exceed KNeighbors. However, models such as SVC-RBF and AdaBoost perform poorly in processing such high-dimensional time series signals.

**Limitations of Machine Learning Methods.** Generally speaking, it still has some limitations of traditional machine learning methods in dealing with high-dimensional time series classification tasks. Although the data flattening operation is adapted to the model input format, it inevitably destroys the temporal dependence and local structure information, which limits the ability of the model to extract deep dynamic features. The performance benchmark established not only reveals the shortcomings of traditional methods in processing complex time-series signals, but also provides a comparative basis for the performance advantage of deep learning model in such tasks.

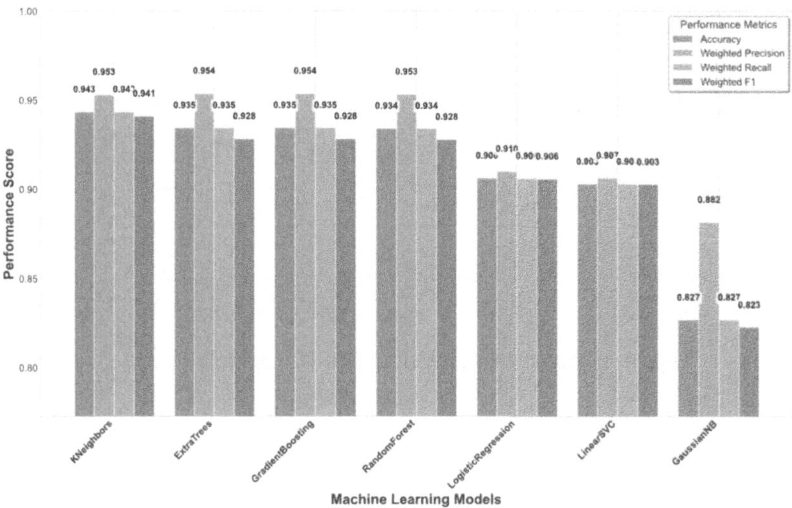


**Fig. 3.** Accuracy comparison in intrusion detection classification task

## 3.2 Intrusion Detection Experiment Based on Deep Learning Model

It is difficult for traditional machine learning methods to capture cross feature interactions in the process of flattening. In order to overcome this limitation, the application of deep learning in network intrusion detection task is then explored based on CSE-CIC-IDS2018 data set. Different from the method of explicitly reconstructing the complete time series, a serialized feature representation strategy is adopted, that is, the numerical features of each network flow record is organized into a two-dimensional structure in a fixed order as the model input. Based on this representation, the CNN-LSTM hybrid model and GRU network are constructed respectively, so that it can automatically learn cross feature dependencies and discrimination patterns at the feature sequence level, so as to realize the deep representation and recognition of intrusion behavior information.

**CNN-LSTM.** As illustrated in Fig. 4, the CNN-LSTM hybrid model processes the input by first employing convolutional layers to extract local discriminant patterns on the feature axis, which is then followed by a recurrent layer that captures global dependencies across these features. Specifically, the input data is first extracted with local features through two one-dimensional convolution layers (the number of filters is 64 and 128 respectively), and then down sampled through the maximum pool layer to compress the redundant information and enhance the robustness of the model. The processed feature sequence is further input to the LSTM layer containing 100 units to learn the long-range dependency across features. Finally, the network is regularized through the dropout layer to mitigate the risk of over fitting, and the final multi classification results are output through the softmax full connection layer.

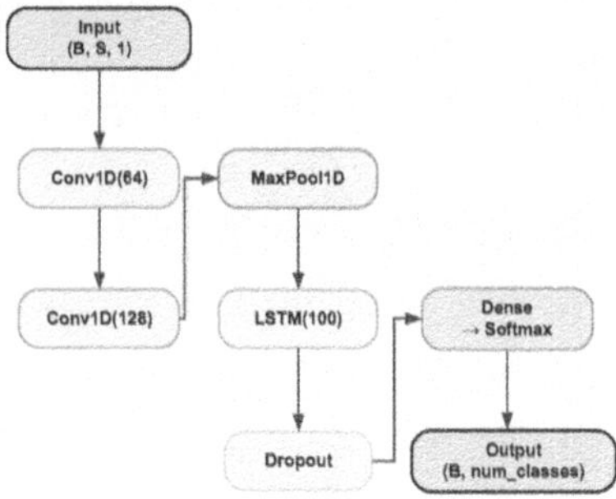


**Fig. 4.** Schematic diagram of CNN-LSTM model structure

**GRU.** The GRU network architecture is shown in Fig. 5. The single-layer GRU with 100 hidden units is considered as the core of the model. And the end-to-end learning is carried out on the input feature sequence to effectively capture the combination relationship and time-series dependency patterns between cross features. The sequence is processed by dropout layer and softmax classification layer in turn, and finally the multi classification results are output. Compared with LSTM, GRU achieves similar sequence modeling ability through a more simplified gating structure (update gate and reset gate), which significantly reduces model complexity and computational overhead while maintaining high classification performance.

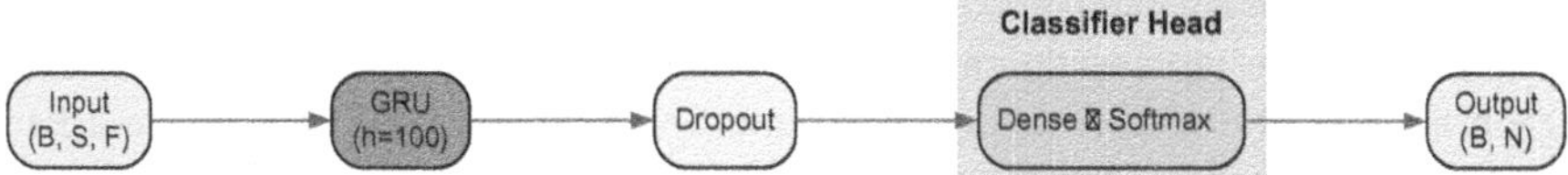


**Fig. 5.** Schematic diagram of GRU model structure

It shows in the experiment that the deep learning model is obviously superior to the traditional machine learning methods in intrusion detection tasks. Performance comparison shows that the CNN-LSTM hybrid model (accuracy 97.20%, F1 score 97.23%, false positive rate 0.56%) is better than GRU model (accuracy 95.96%, F1 score 96.05%, false positive rate 0.76%) (see in Fig. 6). The performance difference shows that under the modeling framework based on feature sequence, the convolution loop hybrid structure can more effectively extract the local discriminant patterns and the time-series dependency of modeling, so as to obtain higher detection accuracy and lower false alarm rate. Although GRU is slightly inferior in accuracy, it still shows good generalization ability and lower computational complexity, which is suitable for deployment scenarios with high real-time requirements.

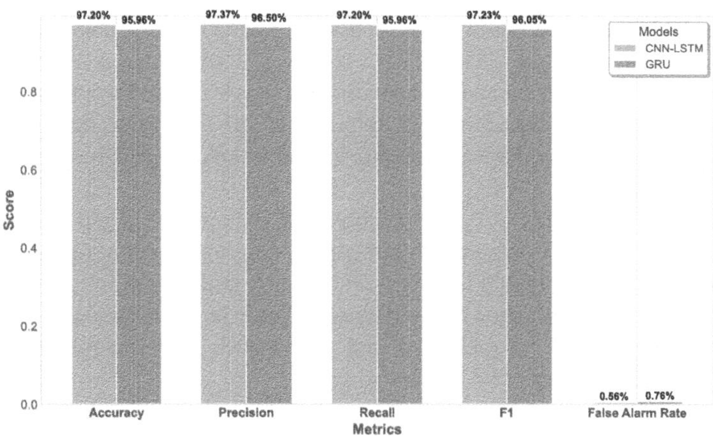


**Fig. 6.** Comparison of classification performance on test set

### 3.3 Intrusion Detection Experiment Based on Dual Branch Network Model

**Model Selection.** In order to further improve the accuracy of network intrusion detection, a dual branch feature fusion network is proposed (see **Error! Reference source not found.**). Firstly, the shared one-dimensional CNN is used to automatically extract discriminative local spatial patterns from the original input. Subsequently, the generated feature maps are sent into two recursive branches with complementary structures in parallel, LSTM and GRU, to model the long-range and short-range time dependencies in the traffic data respectively. Finally, through a cross attention fusion mechanism, the two branch outputs are integrated into a unified and more discriminative robust feature representation for accurate classification of intrusion behaviour (Fig. 7).

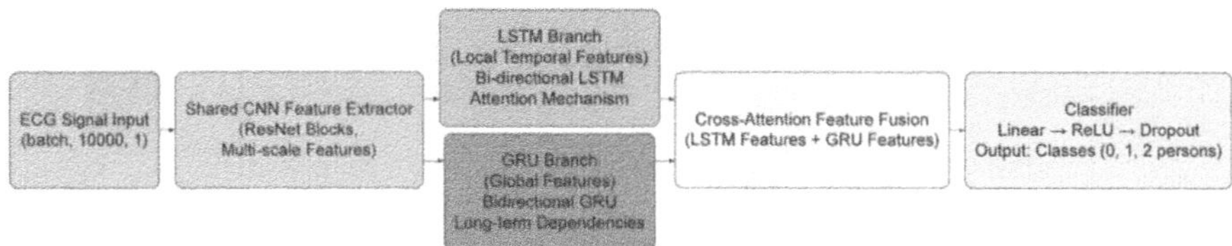


**Fig. 7.** Structure framework diagram of dual branch network model

Specifically, the local discriminant patterns are firstly extracted from the original input sequence through one-dimensional CNN, and a high-dimensional feature map is generated. The feature map is then input into the two branches of LSTM and GRU in parallel, and the dynamic dependencies under different time scales are modeled by using their respective gating mechanisms. In the feature fusion stage, this paper systematically compares various fusion strategies, and finally determines the cross attention mechanism as the optimal scheme. In the mechanism, the output features of the two branches are firstly aligned to the same semantic space through linear projection, and then the 8-head attention module is used for two-way interaction, including LSTM feature as query, GRU feature as key and value, and GRU feature as query, LSTM feature as key and value, so as to realize two-way context enhancement. Finally, the two attention outputs are spliced into 256 dimensional fusion vectors and compressed to 64 dimensions through two-layer fully connected networks to form a more discriminative enhanced feature representation for final classification.

The interaction weights of each branch can be dynamically learned according to the input characteristics. Compared with the traditional fixed weight fusion strategy, it shows stronger adaptability and flexibility in complex multimodal intrusion scenarios. However, due to the introduction of multi head attention mechanism, the computational complexity of the model has increased, and its detection performance and reasoning efficiency need to be weighed when deploying.

**Performance Comparison Experiment.** The effectiveness of the dual branch feature fusion network is fully verified in the performance comparison experiment (see Fig. 8). By introducing parallel GRU branches, The fusion of multi-view temporal features is achieved, thus effectively overcoming the limitation of single-network representation. A final accuracy of 97.96% and a false positive rate of 0.09% were attained, with performance significantly exceeding all single-branch benchmarks, thus demonstrating superior robustness and generalization.

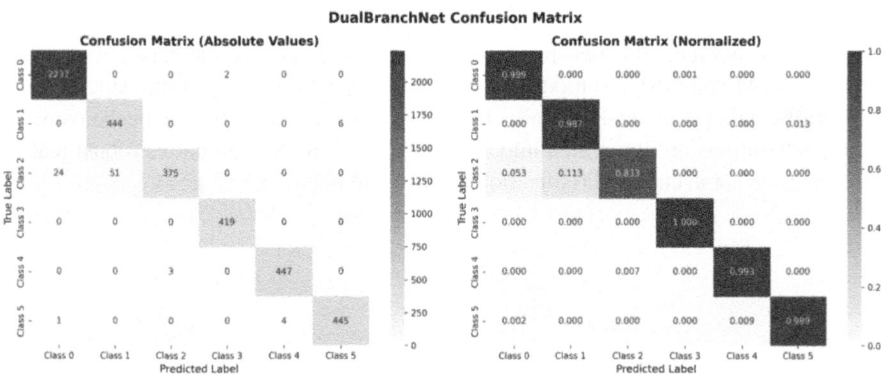


**Fig. 8.** Results of the Dual Branch Network Model

Ablation Experiment. According to the ablation experimental results shown in Table 1, the improvement of the performance of the dual branch structure model is attributed to the effective fusion of the LSTM and GRU branches. Specifically, LSTM is outstanding in capturing long-term dependence and maintaining memory stability, while GRU structure is lighter and more responsive to short-term dynamic changes. The integration of the above two structures significantly improves the modeling ability of the model for multi-scale temporal features, thus enhancing the overall robustness.

In contrast, the performance of the single branch model is significantly degraded. Although LSTM has advantages in long-term dependence modeling, it has weak ability to perceive the local features of rapid fluctuations, resulting in its insufficient sensitivity

**Table 1.** Comparison of ablation results

| Model | Accuracy | Compared to best |
|---|---|---|
| LSTM-only | 0.9720 | −0.0076 |
| GRU-only | 0.9596 | −0.0300 |
| Dual-Branch Network | 0.9796 | 0.0000 |

to some complex attack modes. Although the GRU model has a simpler structure and can effectively capture short-term dynamic patterns, its long-term memory capacity and parameter expression ability are relatively limited, so its performance in complex intrusion scenarios is inferior to that of the fusion model.

## 4 Conclusion

A deep learning-based dual branch network model with high-performance is proposed and systematically evaluated by comparing various traditional machine learning methods and deep learning architectures for network intrusion detection classification tasks. The experimental results show that the model is ranked highest in classification performance on the test set (see Table 2), with an accuracy rate of 97.96% and a false positive rate of only 0.09%, significantly surpassing CNN, CNN-LSTM, GRU and other single models. Its performance advantage is mainly due to the hybrid dual branch architecture, which can effectively integrate the complementary characteristics of different recursive units in time series modeling.

**Table 2.** Comprehensive performance comparison of various models

| Rank | Model | Accuracy | Precision | Recall | F1 Score | FAR | Type |
|---|---|---|---|---|---|---|---|
| 1 | DualBranchNet | 0.9796 | 0.9806 | 0.9796 | 0.9791 | 0.0009 | DL |
| 2 | CNN-LSTM | 0.9720 | 0.9737 | 0.9720 | 0.9723 | 0.0056 | DL |
| 3 | GRU | 0.9596 | 0.9650 | 0.9596 | 0.9605 | 0.0076 | DL |
| 4 | XGBoost | 0.9345 | 0.9537 | 0.9345 | 0.9283 | 0.0121 | ML |
| 5 | RandomForest | 0.9341 | 0.9532 | 0.9341 | 0.9279 | 0.0123 | ML |
| 6 | SimpleCNN | 0.9269 | 0.9294 | 0.9269 | 0.9258 | 0.0139 | DL |
| 7 | LogisticRegression | 0.9062 | 0.9101 | 0.9062 | 0.9060 | 0.0174 | ML |
| 8 | SVM_Linear | 0.9031 | 0.9065 | 0.9031 | 0.9029 | 0.0185 | ML |
| 9 | RandomForest | 0.8795 | 0.9027 | 0.8795 | 0.8656 | 0.0312 | DL |

In the future, the research will focus on lightweight design of model, exploring more advanced feature fusion strategies, doing experiments on more intrusion detection data sets and deploying to edge devices with limited resources to further improve the generalization ability in complex scenes.

## References

Alabbadi, Y., Bajaber, F.: X-FuseRLSTM: a cross-domain explainable intrusion detection framework in IoT using the attention-guided dual-path feature fusion and residual LSTM. Sensors **25**(12), 3693 (2025). https://doi.org/10.3390/s25123693

Ali, H., et al.: Unveiling machine learning strategies and considerations in intrusion detection systems: a comprehensive survey. Front. Comput. Sci. (2024). https://doi.org/10.3389/fcomp.2024.1387354

Ayantayo, A., et al.: Network intrusion detection using feature fusion with deep learning. J. Big Data **10**, 167 (2023). https://doi.org/10.1186/s40537-023-00834-0

China Internet Network Information Center (CNNIC). Statistical report on China's Internet development (55th ed.) (2025). http://www.cnnic.cn/

Li, Z., Fang, W., Zhu, C., Song, G., Zhang, W.: Toward deep learning-based intrusion detection system: a survey. In: Proceedings of the 6th International Conference on Big Data Engineering (BDE) (2024). https://doi.org/10.1145/3688574.3688578

Imrana, Y., et al.: CNN-GRU-FF: a double-layer feature fusion-based network intrusion detection system using convolutional neural network and gated recurrent units. Complex Intell. Syst. **10**, 3353–3370 (2024). https://doi.org/10.1007/s40747-023-01313-y

Sharafaldin, I., Lashkari, A.H., Ghorbani, A.A.: Toward generating a new intrusion detection dataset and intrusion traffic characterization. In: Proceedings of the 4th International Conference on Information Systems Security and Privacy (ICISSP). (2018). https://doi.org/10.5220/0006639801080116

Thaljaoui, M., Zouch, M., Bouallegue, B.: Intelligent network intrusion detection system using optimized deep CNN-LSTM with UNSW-NB15. Int. J. Intell. Netw. **6**, 153–165 (2025). https://doi.org/10.1007/s41870-025-02416-0

# Research on the Security Evaluation System for Sensitive Personal Information of ICV Terminals in Real Vehicle Usage Scenarios

Zhonghao Mi(✉), Lu Zhang, and Weinan Ju

China Automotive Technology and Research Center Co., Ltd., Tianjin 300300, China
mizhonghao@catarc.ac.cn

**Abstract.** This paper focuses on the field of evaluation of sensitive personal information of ICV terminals, establishing an evaluation system based on objective evaluation methods based on the real needs of user use. Through the use of the Delphi method and the Analytic Hierarchy Process, the selection of evaluation objects for sensitive personal information of ICV terminals, the establishment of a hierarchical evaluation index system, and the weight allocation are completed. This paper selects 18 mainstream models currently on sale, carries out experimental verification based on the established intelligent connected vehicle sensitive personal information security evaluation system, and analyzes the score loss from the overall dimension and the hierarchical index dimension according to the evaluation results.

**Keywords:** Privacy protection · Sensitive personal information · Evaluation system · Intelligent connected vehicle · Vehicle usage scenario

## 1 Introduction

With the deep integration of artificial intelligence, the Internet of Things, and big data technologies, Intelligent Connected Vehicles (ICV) have become a core driving force behind the transformation of the automotive industry (Qiu and Li 2022). Through advanced in-vehicle sensors, vehicle-to-everything(V2X) communication, and cloud platform technologies, ICV achieve comprehensive connectivity among vehicles, roads, and people, significantly enhancing driving safety, traffic efficiency, and user experience (Zhang et al. 2021). However, this highly interconnected nature also brings new challenges: ICV generate massive amounts of data during operation, including user behavior data, geographic location information, biometric data, and other sensitive personal information (Wang et al. 2022). If this data is leaked or misused, it not only threatens personal privacy and property safety but may even pose a risk to national security (Zhong et al. 2021).

In recent years, frequent privacy security incidents worldwide, such as Tesla's "privacy breach" and various hacking cases, have fully exposed the vulnerabilities in ICV personal information security management (Zhang et al. 2021). These incidents have


H. Yan et al. (Eds.): AIS&P 2025, CCIS 2918, pp. 87–101, 2026.
https://doi.org/10.1007/978-981-95-9884-7_7

not only drawn widespread attention from all sectors of society but have also prompted governments and regulatory agencies in various countries to strengthen relevant legislative efforts. Regulations such as the EU's General Data Protection Regulation (GDPR) (Zhang et al. 2021), China's Personal Information Protection Law (Zhu 2025), and the newly released national standard GB/T 44464-2024 "General Requirements for Vehicle Data" all explicitly emphasize the importance of protecting sensitive personal information. However, the complex real-world driving scenarios present dynamic and multi-dimensional challenges for data security protection (Wang et al. 2022).

Current ICV data security research exhibits a significant trend of "overemphasizing technology and underemphasizing scenarios, and overemphasizing compliance and underemphasizing user experience" (Qiu and Li 2022). From a technical perspective, although basic protective measures such as encrypted transmission (Zhang et al. 2023) and anonymization processing (GB/T 44464-2024) have become relatively mature, and relevant technical standards like GB/T 45574-2025 "Security Requirements for Processing Sensitive Personal Information" (Hu et al. 2025) provide detailed specifications for data processing, most existing evaluation methods remain confined to laboratory ideal environments and struggle to replicate the complexity of real-world driving scenarios (Wang et al. 2022). Through an in-depth analysis of existing research findings and practical cases, it can be observed that current research primarily focuses on hard metrics such as the strength of encryption algorithms, yet fails to effectively evaluate the impact of security measures on user experience in real-world driving scenarios (Wang et al. 2022).

The essence of these issues lies in the lack of evaluation dimensions centered around real user scenarios. This absence causes manufacturers' optimization directions to deviate from users' actual expectations, failing to truly meet market demands. Therefore, it is urgent to establish an evaluation system that covers real user scenarios, which should adhere to the core principles of scenario reproducibility, technological quantifiability, and experience mappability, advancing the security of sensitive personal information in intelligent connected vehicles from functional availability to experience optimization.

## 2 Construction of a Vehicle-Side Sensitive Personal Information Security Evaluation System

### 2.1 Determination of Evaluation Content for Sensitive Personal Information Security at Vehicle Terminals

With the gradual enrichment of in-vehicle system functions, the functions involving the collection and use of sensitive information are gradually increasing, as shown in Table 1. This paper collected opinions from experts of 22 mainstream automotive manufacturers in the industry, based on user attention, combined with the current existing classification and grading methods for sensitive personal information, and sorted out the sensitive personal information that users may be involved in real driving scenarios (as shown in Table 2). According to the mechanisms that can protect user privacy in three periods before (user notification, collection authorization), during (collection prompt), and after (access authorization, data deletion) data processing, the specific content of the evaluation system for sensitive personal information security at vehicle terminals was determined.

**Table 1.** Scope of Enterprise Expert Opinion Collection.

| Type of Vehicle Manufacturer | Recruited Enterprises |
|---|---|
| Independent Vehicle Manufacturers | FAW, SAIC Motor, Great Wall Motor, Chery Automobile, BYD, Geely, Beijing Auto, Changan Auto, Langtuo Auto |
| New Forces in Automotive Manufacturing | NIO, Xiaomi Auto, Lixiang Auto, Seres, XPeng Auto, Li Auto |
| Foreign-funded Automakers | smart, Volkswagen, Volvo, Tesla, FAW Auto |
| Tie 1 supplier | Huawei, Horizon |

**Table 2.** Vehicle-end sensitive personal information involved in real driving scenarios.

| Category | Example |
|---|---|
| In-vehicle video image information | Facial features, earlobes, irises, etc |
| Audio information | Voiceprints, recording content, etc |
| Vehicle location information | Location trajectories, precise positioning information, latitude and longitude, etc |
| Web browsing history | Browser website browsing history, etc |

## 2.2 Construction of the Vehicle-Side Sensitive Personal Information Security Evaluation System

This paper focuses on constructing an evaluation system for four types of vehicle-side sensitive personal information: in-vehicle video image information, audio information, vehicle location information, and web browsing history. Within this evaluation system, corresponding secondary indicators are established to conduct comprehensive evaluations of the vehicle-side sensitive personal information security evaluation system. (As shown in Fig. 1).

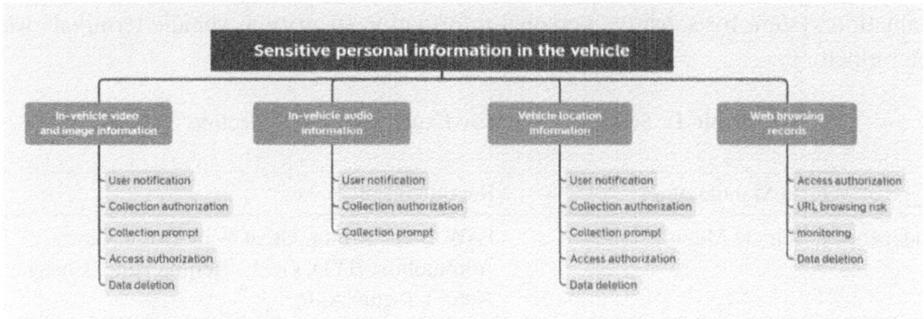


**Fig. 1.** Vehicle-side Sensitive Personal Information Security Evaluation System.

**In-vehicle video image information.** This paper is based on the key nodes of the vehicle processing in-car video image information in actual user driving scenarios, and under in-car video image information, it sets five sub-indicators: user notification, collection authorization, collection prompt, access authorization, and data deletion (as shown in Fig. 1).

Among them, regarding user notification, testing and evaluation are conducted on the methods for camera usage functionality and application privacy policy statements (including separate privacy policies or unified privacy policies); at the same time, testing and evaluation are also carried out on whether enhanced service processing notifications adopt separate consent or refusal methods.

Regarding authorization collection, testing and evaluation are conducted on whether the methods for turning the camera on and off adopt physical shielding or software control.

Regarding collection prompt, testing and evaluation are conducted on whether, when collecting in-vehicle video image information, there are icons or flashing information prompts on the vehicle interface or around the camera indicating the camera's usage status; additionally, when the function switches to background use, testing and evaluation are also conducted on whether the vehicle interface will have icons or flashing information prompts indicating the camera's usage status.

Regarding access authorization, testing and evaluation are conducted on whether data access authorization has implemented access account isolation, as well as whether data can be accessed directly from the vehicle end through an interface.

Regarding data deletion, the test evaluation checks whether it has the function to delete data, whether account isolation was performed during deletion, and whether files can still be accessed through other means after deletion.

**In-vehicle audio information.** This paper is based on the key nodes of the vehicle processing in-car audio information in actual user driving scenarios, and under in-car audio-visual information, it sets three sub-indicators: user notification, collection authorization, and collection prompt (as shown in Fig. 1).

Among them, regarding user notification, testing and evaluation are conducted on whether microphone usage features or applications use independent or unified privacy policy statements; at the same time, for enhanced service processing notifications (such as targeted push notifications, user experience programs, etc.), testing and evaluation are conducted on whether users can individually choose to agree or refuse.

Regarding authorization collection, testing and evaluation are conducted on whether the activation and deactivation of the microphone can be authorized through software (or functional settings), and the software method is assessed to determine if it supports individual or unified activation and deactivation functions.

Regarding collection prompt, testing and evaluation are conducted on whether the in-vehicle audio information collection system has icons or flashing information prompts indicating the microphone's usage status on the vehicle's infotainment system when the function is in use; additionally, when the function switches to background use, it is also necessary to test whether the infotainment system still has icons or flashing information prompts indicating the microphone's usage status.

**Vehicle location information.** This paper is based on the key nodes of vehicle processing vehicle location information in actual user scenarios. Under vehicle location information, there are 5 sub-indicators: user notification, collection authorization, collection prompt, access authorization, and data deletion (as shown in Fig. 1).

Among them, regarding user notification, testing and evaluation are conducted on whether the use of vehicle location information features or applications adopts independent or unified privacy policy statements; at the same time, for enhanced service processing notifications (such as precise push notifications, user experience programs, etc.), testing and evaluation are conducted on whether users can individually choose to agree or refuse.

Regarding authorization collection, testing and evaluation are conducted on whether the enablement and disablement of vehicle geographical location can be authorized through software (or functional settings), and whether this software method supports the enablement and disablement of individual features.

Regarding collection prompt, testing and evaluation are conducted on whether the vehicle interface has icons or flashing information prompts indicating the status of vehicle location information usage during the collection process; additionally, when the function switches to background operation, it is also necessary to test whether the vehicle interface still provides icons or flashing information prompts indicating the status of vehicle location information usage.

Regarding access authorization, testing and evaluation are conducted on whether data access authorization implements access account isolation, and whether data can be accessed directly from the vehicle terminal through an interface.

Regarding data deletion, test whether the evaluation system has the function to delete historical trajectory information, whether account isolation deletion has been implemented, and confirm whether files deleted afterward cannot be accessed again through other means.

**Web browsing history.** This article focuses on the key nodes in the vehicle processing vehicle location information in actual user scenarios, and under vehicle location information, sets up three sub-indicators: access authorization, browsing URL risk monitoring, and data deletion (as shown in Fig. 1).

Among them, regarding access authorization, the test evaluation should determine whether, when the browser has login functionality, users who are not logged in can only access content from the current session, and whether different users can only access their respective historical records after logging in; if there is no login functionality, it is necessary to test and evaluate whether the handling of historical records has other authorization methods.

Regarding the risk monitoring of browsing URLs, the test evaluation should determine whether the browser can identify malicious URLs and whether it prompts or prohibits users from accessing them after identification.

Regarding data deletion, the test evaluation should determine whether the web browsing history has a deletion function, and whether the deletion function implements account isolation.

## 2.3 Vehicle-End Sensitive Personal Information Security Assessment System Weight Distribution

Due to the three-tiered indicators covered by the vehicle-end sensitive personal information security evaluation system, in order to scientifically complete the weight allocation of each tiered indicator, this paper adopts the Analytic Hierarchy Process (AHP) (Dong and Chen 2004), constructs a comparison matrix for X indicators that belong to the same upper-level indicator, and determines the indicator weight allocation at each level through the Feld method (see Table 3).

**Table 3.** Example of Determining Indicator Weights Using the Analytic Hierarchy Process

| Vertical Horizontal | In-vehicle video image information | In-vehicle audio information | Vehicle location information | Web Browsing history |
|---|---|---|---|---|
| In-vehicle video image information | | | | |
| In-vehicle audio information | | | | |
| Vehicle location information | | | | |
| Web browsing history | | | | |

Note: 1-Equally important; 3-Slightly important; 5-Clearly important; 7-Strongly important; 9 Extremely important; 2, 4, 6, 8 represent intermediate values of varying degrees

After collecting opinions and results from experts at 22 enterprises, it was found that there is only one indicator in the first-level indicator, hence its weight is 100%. The weight allocation of second-level indicators and evaluation content are shown in Table 4. For second-level indicators containing multiple sub-indicators, the Analytic Hierarchy Process (AHP) is used to finally complete the weight assignment of these sub-indicators (as shown in Table 4).

**Table 4.** Vehicle-terminal sensitive personal information security evaluation indicator system and weight allocation

| First-level indicator | Second-level indicator | Third-level indicator |
|---|---|---|
| Vehicle-side sensitive personal information (100%) | In-vehicle video image information (30%) | User notification (25%) |
| | | Authorization collection (25%) |
| | | Prompt collection (15%) |
| | | Access authorization (20%) |
| | | Data deletion (15%) |
| | In-Car Audio Information (20%) | User Notification (40%) |
| | | Authorization collection (40%) |
| | | Collection prompts (20%) |
| | Vehicle location information (30%) | User notification (25%) |
| | | Authorization collection (25%) |
| | | Prompt collection (15%) |
| | | Access authorization (20%) |
| | | Data deletion (15%) |
| | Web browsing history (20%) | Access authorization (30%) |
| | | URL browsing risk monitoring (40%) |
| | | Data deletion (30%) |

### 2.4 Data Processing and Calculation for Vehicle-Side Sensitive Personal Information Security Assessment

Based on the vehicle-side sensitive personal information security assessment system constructed, this paper refers to industry expert opinions and accumulated experimental data of each individual indicator to formulate objective test data scoring rules for each four-level indicator, as shown in Fig. 2. After completing the test scoring, a hierarchical weighted method is used to calculate the total score for vehicle-side sensitive personal information security of ICVs.

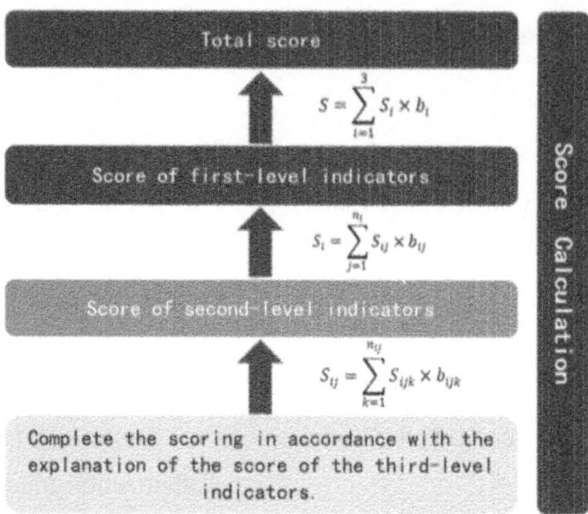


**Fig. 2.** Calculation of Vehicle-Side Sensitive Personal Information Security Assessment Score.

# 3 Application Cases of Vehicle-Side Sensitive Personal Information Security Assessment System

To verify the rationality of the constructed on-board sensitive personal information security assessment system, this paper selects 18 popular market vehicle models for testing and verification, as shown in Table 5. The test vehicles cover the mainstream sales range of (10–45) million yuan, involving three types of automotive brands: domestic independent brands, new forces in automotive manufacturing, and joint ventures. After completing the assessment of 18 models based on this paper's on-board sensitive personal information security assessment system, an analysis of their assessment scores and test results is conducted, specifically as follows.

**Table 5.** Vehicle-end sensitive personal information security assessment 18 trial verification vehicle information

| Selling Price<br>Brand type | Domestic Independent | New Forces in Automotive Manufacturing | Joint venture |
|---|---|---|---|
| (100,000–200,000) yuan | 2 | 2 | 2 |
| (200,000–350,000) yuan | 3 | 6 | 0 |
| Over 350,000 yuan | 2 | 1 | 0 |

## 3.1 Overall Analysis of the Security Assessment Results for Sensitive Personal Information on Vehicle Terminals

As shown in Fig. 3, the overall performance of the tested vehicles is excellent, but there are significant differences. Most vehicles (16 models) have a score rate concentrated above

80%, indicating that the industry's overall level of protection for personal information is acceptable; however, the score gap between different models is large, with the highest score rate (97.8%) and the lowest score rate (64.8%) differing by more than 30%, showing significant differences among manufacturers in protecting sensitive personal information at the vehicle end.

Figure 4 reveals a significant trend: the selling price of vehicle models is positively correlated with their average scores in sensitive personal information security evaluations. That is, the higher the price, the higher the overall level of information security protection for the vehicle. High-end models (above 350,000 yuan) perform the best, with an average score rate of 95.7%. Models in this price range achieve the highest score rates, indicating that high-end brands and models generally place greater emphasis on the security of users' sensitive personal information, possibly investing more in security hardware, software development, and privacy protection design. Mid-range models (200,000–350,000 yuan) and high-end models (above 350,000 yuan) show a very small gap, with mid-range models having an average score rate of 95.2% compared to high-end models at 95.7%, a mere 0.5% difference, almost on par. This suggests that in the car market above 200,000 yuan, the protection of sensitive personal information may have become a "standard feature" or core competitive advantage, with mainstream manufacturers achieving a high level. Mainstream-priced models (100,000–200,000 yuan) exhibit a significant gap, with an average score rate of 88.4%. The scores of models in this price range are notably lower than the previous two segments, with a gap exceeding 7%. This reflects that in the mainstream market with stricter cost control, personal information security may be one of the aspects compromised. Although an 88.4% score rate is not low, there is a clear hierarchical difference in protection levels compared to higher-priced models.

Figure 5 shows the average score rates of three different types of car brands. Overall, the average performance of various brands in protecting sensitive personal information on vehicle terminals is relatively close, all at a high level (with a score rate above 90%). Among them, the average score rate of emerging automotive forces is slightly higher, while traditional independent car manufacturers follow closely behind, with foreign-funded car manufacturers having the lowest average score rate, but the gap is not significant. In the current Chinese market, car manufacturers with different backgrounds have all reached a high level in protecting sensitive personal information on vehicle terminals. Emerging automotive forces have a slight advantage, but the overall difference is not substantial.

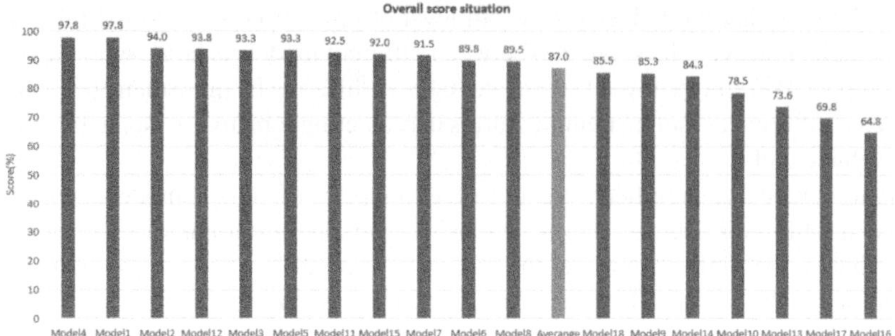


**Fig. 3.** Distribution of vehicle-end sensitive personal information security assessment scores for different models.

**Fig. 4.** Average score rates of vehicle-end sensitive personal information security assessments for different price segments.

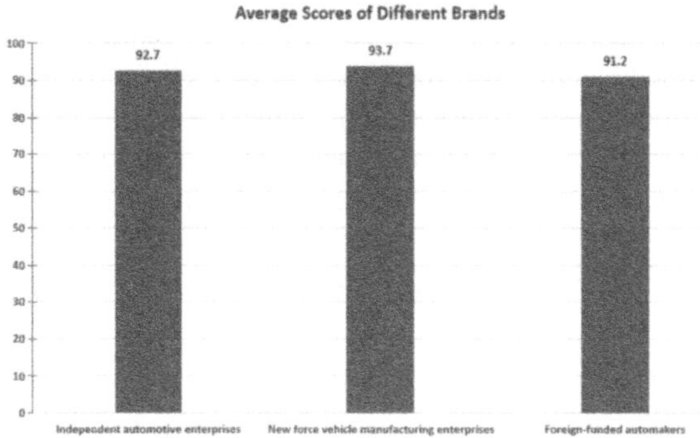


**Fig. 5.** Average score rate of sensitive personal information security assessment for different brand types of vehicle terminals.

### 3.2 Analysis of Module Scores in Vehicle Terminal Sensitive Personal Information Security Assessment

As shown in Fig. 6, among the tested vehicle models, the protection of "vehicle location information" is the most comprehensive in terms of protecting sensitive personal information on vehicle terminals, while the protection of "in-vehicle audio information" is relatively weakest. Overall, the average scores across the four dimensions are very close, all falling within the 86%-89% range, indicating that the industry's protection level is relatively balanced on these specific items, but there is still significant room for general improvement.

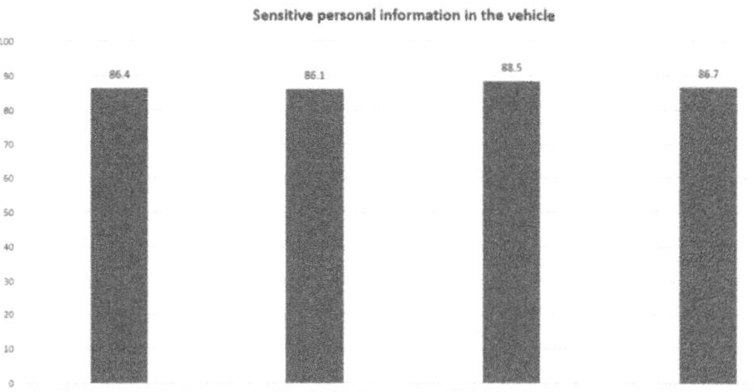


**Fig. 6.** Average score rate of secondary indicators in vehicle terminal sensitive personal information security assessment.

**Video and image information.** Figure 7 shows that there are differences among various vehicle models in the specific practices of protecting in-car video and image information. The industry performs well in "informing users" and "deleting data," where all 18 tested vehicle models scored full marks in "user notification," and the average score rate for "data deletion" was as high as 94.4%. However, there are shortcomings in the immediate interaction during data collection, with an average score rate of only 74.4% for "collection authorization" and 72.2% for "collection prompts." Users find it difficult to exercise control at the critical moments when data is being collected.

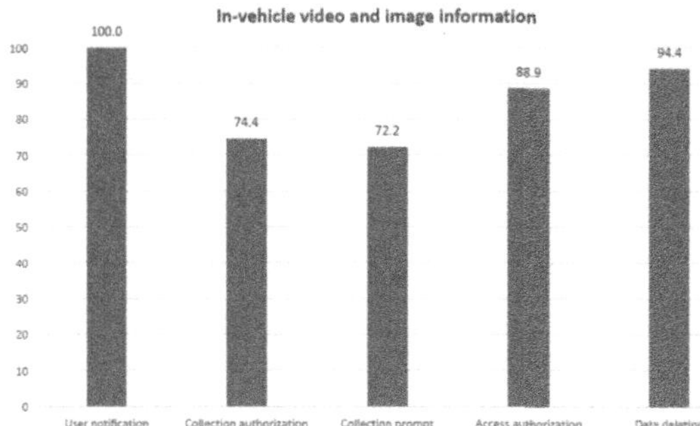


**Fig. 7.** Average score rate of sub-indicators under video and image information.

**Audio information.** Figure 8 shows that in terms of protecting in-car audio information, the industry also exhibits a pattern of "clear communication but insufficient real-time interactive control." Specifically, this is manifested as: user communication is done the best, with a score rate of 88.9%, while real-time prompts during data collection are relatively weak. User communication scores 2.8% higher than the average score for data collection authorization, and 8.3% higher than the average score for data collection prompts.

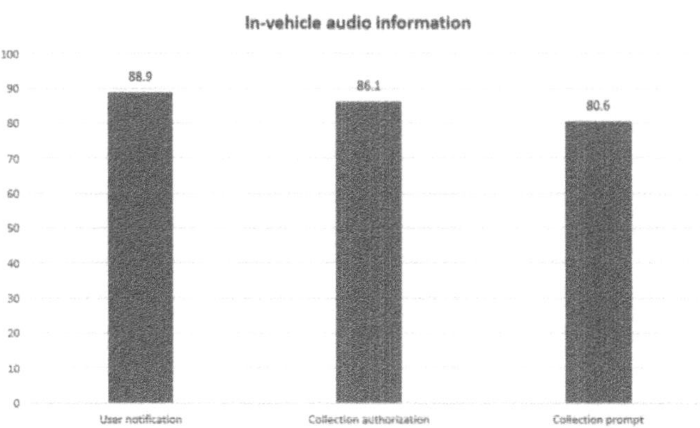


**Fig. 8.** Average score rate of sub-indicators under audio information.

**Vehicle location information.** Figure 9 shows that the industry exhibits significant differences in performance across various stages in protecting vehicle location information. The industry achieves a high level in the "Access Authorization" and "Data Deletion" stages, with average score rates reaching 100%. However, there are notable shortcomings in the initial stage of data collection, particularly in "Collection Authorization," where the score rate is only 75%. The reason is that automakers often obtain authorization from users through bundled agreements at the first use of the service, rather than providing clear prompts and authorization each time or in specific scenarios. This weakens users' control over data collection at the moment.

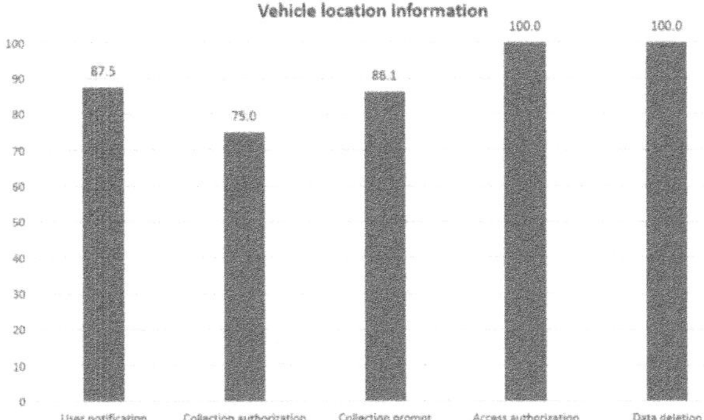


**Fig. 9.** Average score rate of sub-indicators under vehicle location information.

**Web browsing history.** Figure 10 reveals: In terms of protecting web browsing history, the industry excels in access control and post-event deletion management for web browsing records, with an average score rate of 94.4%. However, it lacks the ability to monitor risky URLs during user browsing, with an average score rate of only 75%. This creates a "strong at both ends, weak in the middle" situation. Most vehicle infotainment systems have weak processing capabilities for risky URLs when users browse web pages with browsers, unable to prompt risks or prohibit access to risky URLs. This exposes users' online activities inside vehicles to the security risk of sensitive personal information leakage.

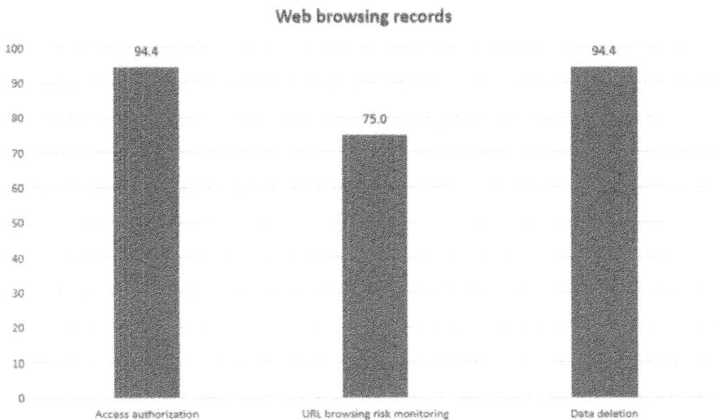


**Fig. 10.** Average score rate of sub-indicators for web browsing records.

## 4 Conclusion

This study reveals that there is a significant imbalance in the protection of sensitive personal information on the vehicle side in the current intelligent connected vehicle industry. The results show that vehicle selling price is positively correlated with safety scores, while the differences between different brand types are relatively limited. The industry as a whole presents the typical characteristics of "triple and three light": attaching importance to "post-event management" (such as access authorization and data deletion), but neglecting "pre-authorization" (such as collection authorization); emphasizing "static notifications" (such as user notifications) while ignoring "dynamic interactions" (such as collecting prompts); At the same time, the vehicle browser has insufficient ability to monitor risk URLs. These phenomena indicate that the existing protection system is still essentially a structure of "strict management but lagging interaction". In the future, efforts should be made to improve the real-time transparency and user control of the data collection process, and enhance the browser's ability to actively defend against risky URLs, so as to build a truly user-centric trust mechanism.

Based on the massive data background generated during the operation of intelligent networked vehicles, this paper constructs an objective and consistent security evaluation system for sensitive personal information on the vehicle side around the actual usage needs of users, aiming to improve the level of privacy protection for car companies and provide theoretical basis and practical reference for consumers to choose and purchase cars. However, there are still shortcomings in this study, although the protection level of sensitive information on the vehicle side of the vehicles on sale in the market reflects the level of sensitive information protection of the vehicles on sale in the market to a certain extent, but due to practical reasons, it is not possible to select a more appropriate number of models in different price ranges, and this distribution cannot represent the overall market structure. Follow-up research will continue to increase the coverage of more price segments and brand types, track changes in the installation rate of intelligent functions, industry information security incident dynamics, consumer feedback on actual car use, and updates to relevant standards, certification rules and evaluation systems, and continue to improve the security evaluation system of sensitive personal information on the vehicle side through extensive enterprise cooperation from the dimensions of index updates, weight optimization, and method improvement.

## References

Dong, Y.C., Chen, Y.H.: Tests in analytic hierarchy process (AHP). Syst. Eng. Theory Pract. **7**, 105–110 (2004)

Hu, Y., Chen, S., Gao, C.: Interpretation of national standards for security requirements for sensitive personal information processing. Inf. Technol. Stand. **8**, 71–74 (2025)

National Automotive Standardization Technical Committee: General requirements for automotive data (GB/T 44464-2024) [Standard]. China Standards Press (2024)

Qiu, B., Li, G.Y.: Research on data security management of intelligent connected vehicles. Trans. Chin. Soc. Automot. Eng. **12**(3), 307–313 (2022)

Wang, Z., Xu, C., Ma, N., et al.: A review of data privacy protection in intelligent connected vehicles. In: Proceedings of the 26th Annual Conference on Network New Technologies and Applications of the Network Application Branch of China Computer Users Association (pp. 154–159). Beijing Key Laboratory of Information Service Engineering, Beijing Union University; Beijing Laboratory of Brain and Cognitive Intelligence, School of Robotics, Beijing Union University; Beijing University of Technology (2022)

Zhang, L.Z., Huang, D.H., Han, W.W., et al.: Technical requirements and suggested measures for personal information protection of intelligent networked vehicles. China Automot. **2**, 22–25 (2023)

Zhang, Y.Z., Yang, S.F., Li, J., et al.: Policy research on personal information data security in the internet of vehicles. Confidentiality Sci. Technol. **7**, 28–32 (2021)

Zhong, L., Wang, Y.W.: Several understandings and thoughts on automotive data security. China Inf. Secur. **7**, 49–52 (2021)

Zhu, H.J.: Definition and processing path of sensitive personal information. Netw. Secur. Data Governance **44**(8), 53–57 (2025)

# Lightweight Face Image Anonymization via Differential Privacy and Feature Decomposition

Zhuochao Wu(✉)

School of Electrical Engineering, Southeast University, Nanjing, China
220243153@seu.edu.cn

**Abstract.** Advances in deep learning have driven the proliferation of personalized image services: users upload face images to cloud servers running deep models to obtain treatments such as makeup transfer and wrinkle removal. Growing concern for personal privacy, however, has raised fears of identity leakage. Image-encoding perturbation methods based on generative adversarial networks (GANs) can obfuscate facial identity and protect identity privacy. Existing methods typically do not separate identity features from appearance features within facial encodings; approaches that rely on multiple feature-extraction networks to preserve appearance incur high computational cost and are difficult to deploy locally, while perturbation schemes based on random rotations of feature vectors offer limited protection. To address these issues, we propose a face anonymization and identity-reconstruction framework and design DPDF-GAN, a face image anonymization model that integrates differential privacy with feature-space decomposition. Our method decomposes facial encodings into identity and appearance components, applies differential-privacy perturbations to the identity features, and generates anonymized faces from the perturbed encodings. We also develop an identity reconstruction procedure that reassembles a new encoding from locally stored original identity features and cloud-processed appearance features, and uses the DPDF-GAN decoder to reconstruct the user's face image. Theoretical analysis and experiments demonstrate that the proposed scheme preserves facial appearance while protecting identity privacy, reduces computational cost, and supports lightweight client-side deployment.

**Keywords:** Face Anonymization · Generative Adversarial Networks · Differential Privacy · Feature-Space Decomposition

## 1 Introduction

In recent years, convolutional neural networks (CNNs) have made significant advances in computer vision, driving the rise of a wide range of image-based applications. In entertainment-oriented scenarios such as short-video platforms and livestreaming, face-oriented services—e.g., makeup transfer, filter stickers, and other face image editing—have attracted increasing attention. However, facial images contain users' identity information, and providing real face images to third parties for makeup transfer, role simulation, or similar processing poses a risk of identity privacy leakage.


H. Yan et al. (Eds.): AIS&P 2025, CCIS 2918, pp. 102–118, 2026.
https://doi.org/10.1007/978-981-95-9884-7_8

For face-recognition applications, researchers have proposed protections such as biometric template encryption (Uhl and Rathgeb 2021), cancellable biometric recognition (Ratha et al. 2001), and secure computation for encrypted data (Franklin and Reiter 1997), which aim to improve the security of facial identity features without degrading recognition performance. For detection-and tracking-oriented applications that only need to locate faces in images, prior work has introduced k-anonymity based methods (Jourabloo et al. 2015; Gross et al. 2006; Gross et al. 2005; Meng and Sun 2014) and GAN-based approaches (Sun et al. 2018; Gafni et al. 2019; Hukkelås et al. 2019; Maximov et al. 2020; Chen et al. 2021; Li and Clifton 2021; Liu et al. 2021; Ma et al. 2021; Croft et al. 2022; Chen et al. 2021; Ji et al. 2022; Mi et al. 2023; Le and Carlsson 2022; Zhai et al. 2022; Demelius et al. 2025) to remove sensitive identity information while retaining as much of the original image quality as possible. Emerging personalized face services, however, require preserving the user's identity while selectively editing appearance attributes (expression, makeup, age, etc.) to provide beauty- and role-play-style experiences. These models are typically compute-intensive and are often deployed on cloud servers (for example, Meitu's cloud retouching service or virtual try-on APIs). Such cloud-based personalized face services require users to upload their real facial images to remote servers, which raises concerns about image theft, misuse, and identity-privacy leakage.

A practical solution is for users to upload anonymized face images to the cloud for editing and then perform local identity reconstruction on the processed images. An effective anonymization model must hide identity features while preserving identity-independent appearance features so that cloud-side editing still produces the desired visual effects. Existing approaches that employ multiple feature-extraction networks to preserve appearance consistency incur high computational cost and are difficult to deploy locally in a lightweight form. Regarding identity hiding, some methods replace identity feature vectors to achieve anonymization, but the degree of obfuscation is hard to control; other approaches that apply random rotations to identity vectors provide only weak protection.

To address these challenges, we propose a GAN-based face-image anonymization and identity-reconstruction framework and introduce DPDF-GAN (Differential Privacy and Decomposition of Features GAN). DPDF-GAN decomposes face encodings into identity and appearance components, applies differential-privacy perturbations to the identity features, and synthesizes anonymized faces from the perturbed encodings. We further design an identity-reconstruction scheme that reassembles a new face encoding from locally stored original identity features and cloud-processed appearance features, and then uses the DPDF-GAN decoder to reconstruct the user's face image. This approach seeks to balance identity privacy protection and image-processing fidelity while enabling lightweight client-side deployment.

The main contributions of this paper are:

(1) We propose an end-to-end face anonymization and identity-reconstruction framework that enables lightweight local anonymization, cloud-side personalized image processing, and local identity recovery—balancing identity privacy with processing fidelity and supporting client-side deployment.

(2) We introduce a feature-space decomposition method that disentangles identity and appearance in GAN encodings, preserving usable appearance features while reducing reliance on extra subnetworks and lowering computational cost.
(3) We develop a differential-privacy perturbation scheme for identity feature, with a linear neural-network-based optimization for noise injection that minimizes required perturbation.
(4) We conduct experiments to validate the effectiveness of the proposed identity reconstruction and the fidelity of face-image processing under our scheme.

The remainder of this paper is organized as follows. Section 2 reviews related work and formalizes the problem. Section 3 introduces the proposed framework of face image anonymity and identity reconstruction, the decoupling method of face feature space and the differential privacy protection method of identity feature. Section 4 provides experimental analysis of the proposed method. Finally, we conclude the paper.

## 2 Related Work and Problem Statement

### 2.1 Face-Image Anonymization

Face-image anonymization aims to protect the privacy contained in facial images while preserving useful facial attributes. The development of generative adversarial networks (GANs) has provided practical solutions for this task. Prior work commonly trains encoder–decoder generators with adversarial objectives to synthesize realistic anonymized faces. Two typical generation strategies are shown in Fig. 1.

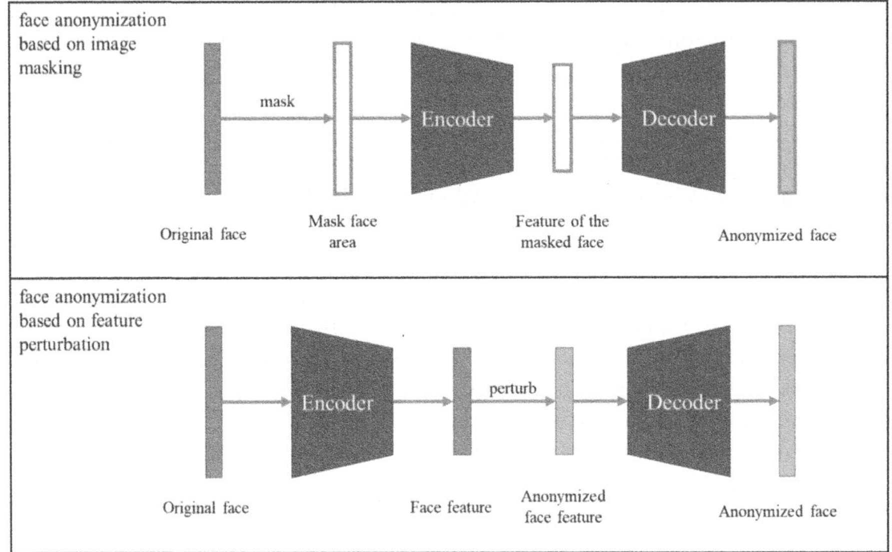


**Fig. 1.** Facial Anonymization Strategy Based on Generative Adversarial Networks.

Masking sensitive regions in the input image before encoding. The masked image is passed to the encoder so that the resulting feature vector does not contain sensitive information from the original image; decoding then yields an anonymized face. Works (Sun et al. 2018; Hukkelås et al. 2019) mask sensitive facial regions and feed the result to a generator, using facial landmarks during decoding to control the pose of the anonymized face. Works (Gafni et al. 2019; Maximov et al. 2020) inject high-dimensional identity vectors during decoding to guide the generator to synthesize faces with desired identities. Although removing sensitive regions prior to input guarantees privacy, maintaining usability typically requires additional subnetworks to recover original-image characteristics, which increases computational cost.

Encoding the full face image into a latent feature vector, modifying that vector to achieve anonymization, and decoding the modified vector back to an anonymized face. For example, (Ma et al. 2021) extracts identity and content representation vectors from the facial feature vector, modifies the identity vector and measures the modification magnitude by cosine distance so as to preserve image content while anonymizing identity. Works (Li and Clifton 2021; Liu et al. 2021) apply differential-privacy perturbations to facial feature vectors for anonymization. Work (Croft et al. 2022) introduces a PCA module to extract principal components of the facial features and adds differentially private noise to protect identity. Work (Chen et al. 2021) performs transformations in the latent space obtained by GAN inversion, adding differential noise to anonymize faces while preserving specified facial attributes.

PI Net (Chen et al. 2021) defines "unrecognizability" as local relations in latent space: it uses a triplet loss to cluster facial attributes and adds noise to latent codes to suppress identity, making generated faces hard for recognition networks to identify. Ji et al. move faces to the frequency domain via DCT, remove the DC channel, and allocate privacy across frequency components using a recognition loss. Mi et al. (Partial Face) prunes low-frequency channels to hide visual content and trains identity models on a few randomly chosen high-frequency channels so recognizable faces need not be stored. StyleID (Le and Carlsson 2022) disentangles identity in latent space and adds noise only to identity-related latent dimensions while preserving non-sensitive attributes like expression. A3GAN (Zhai et al. 2022) splits de-identification into identity suppression and attribute injection: it suppresses original identity by disabling convolutional units and aligns/injects attribute codes to control customized anonymization. Demelius L et al. note in their review that some GAN methods encode attributes and identity separately and add Laplace noise only to the identity vector to preserve visual features.

With diffusion models emerging, Diff-Privacy (He et al. 2024) proposes a diffusion-based face privacy framework that injects differential-privacy noise during denoising to achieve anonymization with higher image quality. DiffPrivate (Le and Carlsson 2025) compares GANs and diffusion models for face privacy and finds diffusion models promising for the privacy–utility tradeoff; however, their heavy training cost and multi-step (hundreds–thousands) sampling make them impractical for many client devices, so GAN-based anonymization remains preferable for interactive applications.

### 2.2 Differential Privacy

**Definition 1 (Neighboring datasets) (Dwork 2011).** Let $D$ and $D'$ be datasets with the same schema. If $D$ and $D'$ differ by exactly one record (i.e., their symmetric difference has size one), then $D$ and $D'$ are called neighboring datasets.

**Definition 2 (Differential privacy) (Dwork 2011).** Let $M$ be a randomized algorithm and $P_M$ its output range. For any subset $S_M \subseteq P_M$, if for every pair of neighboring datasets $D$ and $D'$ the outputs of $M$ satisfy the following inequality, then $M$ is said to satisfy $(\varepsilon, \delta)$-differential privacy, where $\varepsilon$ and $\delta$ are the privacy parameters ($\varepsilon$ is the privacy budget). When $\delta = 0$, A satisfies $\varepsilon$-differential privacy.

$$\Pr\left[M(D) = t^*\right] \leq e^{\varepsilon} \times \Pr\left[M\left(D'\right) = t^*\right] + \delta \tag{1}$$

**Property 1 (Sequential composition) (Dwork and Roth 2013).** Given a dataset $D$ and $p$ randomized algorithms $\left\{M_1, \ldots, M_p\right\}$ where $M_i (1 \leq i \leq p)$ satisfies $(\varepsilon_i, \delta_i)$-differential privacy on $D$, the sequential composition of these algorithms on $D$ satisfies $(\varepsilon, \delta)$-differential privacy, where $\varepsilon = \sum_{i=1}^{p} \varepsilon_i$ and $\delta = \sum_{i=1}^{p} \delta_i$.

**Property 2 (Post-processing) (Dwork and Roth 2013).** Let $M$ be a randomized algorithm that satisfies $(\varepsilon, \delta)$-differential privacy on dataset $D$. For any function $F$, the composition $F(M)$ also satisfies $(\varepsilon, \delta)$-differential privacy. In other words, privacy guarantees are preserved under arbitrary post-processing.

Realizing differential privacy requires computing the global sensitivity of a query function f. We give the formal definition below.

**Definition 3 (Global sensitivity) (Dwork et al. 2006).** Let $F$ be a function with input a dataset and output a d-dimensional real vector. For any neighboring datasets $D$ and $D'$, the global sensitivity with respect to norm-1 and norm-2 are defined as

$$\Delta_1 f = \max_{D, D'} |F(D) - F\left(D'\right)|_1 \tag{2}$$

$$\Delta_2 f = \max_{D, D'} ||F(D) - F\left(D'\right)||_2 \tag{3}$$

Here $\Delta_1 f$ and $\Delta_2 f$ are the $L_1$ and $L_2$-sensitivities of $F$, respectively. Given mechanism parameters $\varepsilon$ and $\delta$ and the global sensitivity of $F$, the Laplace and Gaussian mechanisms are two commonly used implementations of differential privacy, providing $\varepsilon$-DP and $(\varepsilon, \delta)$-DP respectively. They are defined as follows.

**Definition 4 (Laplace mechanism) (Dwork et al. 2006).** Given a dataset $D$ and a function $F : D \to R^p$, if the $L_1$-sensitive sensitivity of $F$ with respect to $D$ is $\Delta_1 f$, then the randomized algorithm $M(D) = f(D) + (Y_1, \ldots, Y_P)$, where $Y_i \sim Lap(\Delta_1 f / \varepsilon)$ are independently drawn from a Laplace distribution, provides $\varepsilon$-differential privacy protection.

**Definition 5 (Gaussian mechanism) (Dwork et al. 2006).** Given a dataset $D$ and a function $F : D \to R^p$. The randomized algorithm $M(D) = f(D) + (Y_1, \ldots, Y_P)$, where each $Y_i \sim N(C\Delta_2 f/\varepsilon)$ is drawn independently from the Gaussian distribution, provides $(\varepsilon, \delta)$-differential privacy protection, with $C > \sqrt{2\ln(1.25/\delta)}$.

### 2.3 Problem Statement

To protect user privacy in cloud-based face-image processing services, face images must be anonymized locally before upload, and a reconstruction model should be used to restore identity from the cloud-processed anonymized images. Existing anonymization methods often rely on one or more feature-extraction subnetworks to preserve specific utility attributes, and the extra computational overhead fails to meet the efficiency requirements of cloud services. The reconstruction model is the inverse mapping of the anonymization model; the core challenge is to recover identity features while simultaneously preserving the visual transformations applied by the cloud-side processing. The anonymization and reconstruction models should satisfy the following requirements:

**Lightweight:** Both models should have compact structures and small parameter counts to enable efficient local (client-side) deployment.

**Privacy:** The anonymization model must provide rigorous privacy protection so that anonymized faces do not reveal the original identity.

**Realism:** The anonymization model should generate realistic, detectable human faces suitable for cloud-side processing, and the reconstruction model should naturally restore the original identity while retaining the visual effects produced by the cloud processing.

## 3 Methodology

### 3.1 Face Anonymization and Identity-Reconstruction Framework

In this section, we present the core methodology for face anonymization and reconstruction. The overall pipeline adopts a privacy-preserving architecture following the principle of "local disentanglement, cloud processing, and local reconstruction," aiming to achieve secure isolation of facial identity information while enabling personalized cloud-based services. The entire process consists of two key stages: the face anonymization phase and the identity reconstruction phase.

As shown in Fig. 2, during the **face anonymization phase**, the original face image is first preprocessed and then fed into a pre-trained feature extraction network to obtain a high-dimensional facial feature representation. Next, an identity disentanglement network decomposes this representation into two components: an *identity representation* vector (encoding individual identity information) and an *appearance(content) representation* vector (capturing non-identity attributes such as expression, pose, and illumination). To ensure privacy, the identity representation is mapped through a linear transformation matrix and further perturbed with differential privacy noise, yielding an irreversible anonymized identity feature. This anonymized identity feature, together with the original content representation, is then passed to a generator to synthesize an

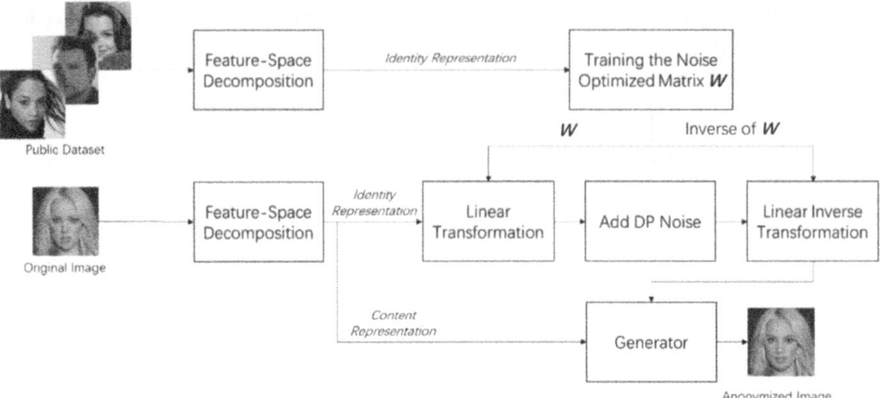


**Fig. 2.** Face anonymization phase.

anonymized face image that preserves content details while obscuring identity information. The resulting image is uploaded to the cloud for subsequent personalized processing (e.g., beautification, stylization, or filtering).

As shown in Fig. 3, during the **identity reconstruction phase**, the processed anonymized image is returned from the cloud to the user device, where it is combined with locally stored data for reconstruction. Specifically, the returned image undergoes preprocessing and feature extraction, followed by the identity disentanglement network to separate its anonymized identity feature and content feature. Meanwhile, the original identity feature vector—securely retained on the local device—is reloaded. Using the inverse transformation of the trained noise-injected linear mapping matrix, the anonymized identity feature is reverted to recover the original identity representation. Finally, the restored identity feature is fused with the content feature extracted from the cloud-processed image and input into the generator to reconstruct a face image that retains the cloud-applied enhancements (e.g., filters or beautification effects) while restoring the true identity.

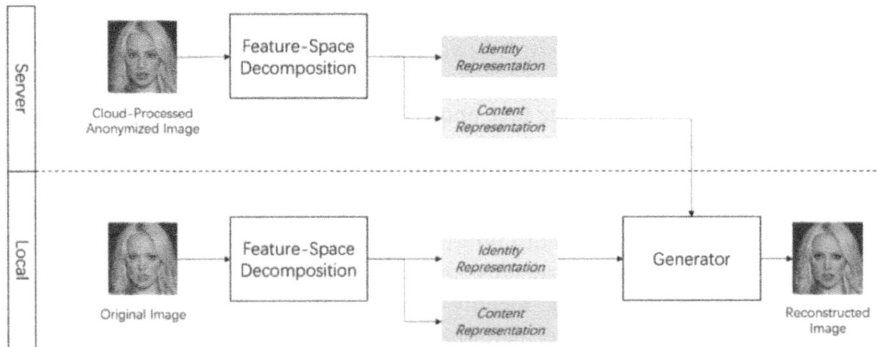


**Fig. 3.** Identity reconstruction phase.

By leveraging feature-level disentanglement and noise injection mechanisms, this approach effectively prevents identity leakage during cloud processing while maintaining service utility, thereby achieving a dual objective of strong privacy protection and high-quality user experience.

## 3.2 Feature-Space Decomposition for Facial Feature Disentanglement

**Facial feature disentanglement model.** The proposed facial identity-feature disentanglement model consists of four components: a preprocessing network, a face-feature encoding network, an identity-separation network, and a generator.

The preprocessing network performs face-region detection, landmark detection, cropping and scaling. The face-feature encoding network employs convolutional layers to extract image features and is pretrained with a reconstruction loss so that the encoder learns to capture the primary information of the face region. The identity-separation network, constrained by the generator and appropriate loss terms, extracts appearance information and identity information to realize disentanglement. The generator synthesizes face images from a combination of identity and appearance features, enabling identity replacement without affecting other facial attributes.

*Preprocessing network:* We use the "dlib" library for facial landmark detection to partition the image into face-region and background-region. The face region, which contains rich identity information, needs anonymization; the background, which is weakly correlated with identity, does not require processing. Restricting feature extraction and disentanglement to the face region (rather than the whole image) improves disentanglement performance and training efficiency. In addition, the face-contour information produced by preprocessing (outer face contour, nose ridge and mouth position) constitutes pose-related features that are weakly correlated with identity; these contour features are provided to the generator to preserve pose consistency.

*Feature-extraction network:* The feature-extraction network is based on ResNet and is augmented with an auxiliary decoder to form an autoencoder, which is pretrained using a reconstruction loss. Only the face region is used as input to reduce training complexity. The feature-extraction network architecture is shown in Table 1.

**Table 1.** Feature-extraction network architecture.

| Encoder | Decoder |
|---|---|
| Input: 256 × 256 × 3 image | Input 1 × 1 × 1024 encoding |
| Conv (4 × 4 × 64), BN, LReLU | Trans Conv (2 × 2 × 512, Stride = 1), BN, LReLU |
| Conv (4 × 4 × 12), BN, LReLU | Trans Conv (2 × 2 × 256, Stride = 1), BN, LReLU |
| Conv (4 × 4 × 25), BN, LReLU | Trans Conv (4 × 4 × 128, Stride = 2), BN, LReLU |
| Conv (4 × 4 × 512), BN, LReLU | Trans Conv (4 × 4 × 64, Stride = 2), BN, LReLU |
| Conv (4 × 4 × 1024), BN, LReLU | Trans Conv (4 × 4 × 3, Stride = 2), Tanh |

*Identity-feature disentanglement network:* The separation network is a multilayer perceptron (MLP) with two output branches. It decomposes the $1 \times 1 \times 1024$ encoding produced by the feature extractor into identity and appearance feature vectors. Modifying the appearance branch should not produce significant changes to identity. The separation network architecture is shown in Table 2.

**Table 2.** Feature-extraction network architecture.

| Separate | |
|---|---|
| Input $1 \times 1 \times 1024$ encoding<br>Linear(1024, 512),ReLu<br>Linear(512, 512),ReLu | |
| Linear(512, 512) | Linear(512, 512) |
| < Identity feature branch > | < Appearance feature branch > |

*Generator:* The generator is ResNet-based and takes as input the face identity feature, face appearance (content) feature, face-contour information, and background region to reconstruct the image. Identity-contrastive loss and reconstruction loss are used to assist training of the separation network and to promote effective disentanglement.

**Objective function.** We denote two face images with different identities as $x$ and $y$. After the separation network $S$ they are encoded into pairs of vectors $(z^x_{\mathrm{id}}, z^x_{\mathrm{app}})$ and $(z^y_{\mathrm{id}}, z^y_{\mathrm{app}})$, where the subscripts "id" and "app" denote identity and appearance (content) features respectively. By alternately mixing identity and appearance features and feeding the mixed encodings into the generator $G$, new images are synthesized. Discriminator $D$ is used for adversarial training in a min–max game with $G$. A pretrained face-recognition network $P$ is used to measure identity similarity between different images. The total objective is a weighted sum of multiple loss terms:

$$L_{overall} = \lambda_{rec} L_{rec} + \lambda_{app} L_{app} + \lambda_{id} L_{id} + \lambda_{orth} L_{orth} + \lambda_{adv} L_{adv} \tag{4}$$

where $\lambda_{\mathrm{rec}}, \lambda_{\mathrm{app}}, \lambda_{\mathrm{id}}, \lambda_{\mathrm{orth}}, \lambda_{\mathrm{adv}}$ are weighting coefficients for the respective loss components. Below we describe each loss term.

Reconstruction loss $\mathcal{L}_{\mathrm{rec}}$. For each training image, its identity and appearance features are fed to the generator for reconstruction; the reconstruction loss enforces that the generator can reproduce the input when given the original encoding. A common choice is the L1 reconstruction loss:

$$L_{rec} = E_x\left[ ||G\left(Z^x_{id}, Z^x_{app}\right) - x||_1 \right] \tag{5}$$

The separation network encodes each input into an identity-agnostic appearance vector and an identity vector. During training we take pairs of images with different identities, swap and recombine identity and appearance branches to synthesize mixed images, and use appearance and identity losses to force the network to separate identity and non-identity information. In addition, an orthogonality loss encourages identity and appearance vectors to be orthogonal.

Appearance loss $\mathcal{L}_{\text{app}}$. This loss enforces that the appearance/content information is preserved after reconstruction or mixing. A typical form is a feature- or pixel-level consistency loss between the appearance of the generated image and the original appearance target, for example:

$$L_{app} = E_x\left[||G\left(Z_{id}^y, Z_{app}^x\right) - x||_1\right] \tag{6}$$

Identity loss $\mathcal{L}_{\text{id}}$. Using the pretrained face-recognition network $P$, this loss measures identity similarity and is used both to (i) ensure that when identity features are preserved the generated image matches the original identity, and (ii) when identity is intended to be changed, to enforce a difference. A typical identity-preserving term is:

$$L_{id} = E_x\left[||P\left(G\left(Z_{id}^x, Z_{app}^y\right)\right) - P(x)||_1\right] \tag{7}$$

and for identity-swapped samples one may use a loss that pushes the recognition embedding away from the original identity as needed.

Orthogonality loss $\mathcal{L}_{\text{orth}}$. To encourage disentanglement between identity and appearance vectors we penalize their inner product:

$$L_{orth} = E_x\left[< Z_{id}^x, Z_{app}^x >^2\right] \tag{8}$$

Adversarial loss $\mathcal{L}_{\text{adv}}$. We introduce adversarial training between generator $G$ and discriminator $D$ to improve visual realism. For the classic GAN loss:

$$L_{adv} = E_x\left[\log D(x) + \log\left(1 - D\left(G\left(Z_{id}^x, Z_{app}^x\right)\right)\right)\right] \tag{9}$$

### 3.3 Face Identity Feature Anonymization Based on Differential Privacy

The feature extraction network and the feature decoupling network extract sensitive identity information from facial images, necessitating anonymization of this identity data.

**Linear Approximation of the Domain.** Suppose X is a p-dimensional random variable representing facial identity information, and let e and d be privacy budget parameters. The Gaussian mechanism adds differential noise $Y = \{Y_1, \ldots, Y_p\}$, where $S = diag\left(\sigma^2(Y_1), \ldots, \sigma^2\left(Y_p\right)\right)$ denotes the covariance matrix of Y, with each $Y_i \sim N\left(0, \sigma^2(Y_1)\right)$, and , $\sigma(Y_1) = C_i \Delta f_i / \varepsilon_i$, $C_i > \sqrt{2\ln(1.25/\delta_i)}$, $\varepsilon = \sum_{i=1}^{p} \varepsilon_i$ , $\delta = \sum_{i=1}^{p} \delta_i$。

Let $\Omega_X$ denote the domain of X, and let $X_{p*n}$ represent the n observed samples of X. When n is sufficiently large, the range of values of $X_{p*n}$ can closely approximate the domain, with sensitivity in each dimension approximated as $\Delta f_i \approx \Delta X_i$, where $\Delta X_i$ is the range of $X_{p*n}$ across all dimensions of X.

**Theorem 1.** Given the sensitivity $\Delta f_i$ for each dimension, to achieve $(\varepsilon, \delta)$-differential privacy, the privacy mechanism $M$ adds independent noise $Y_i \sim N(C_i \Delta f_i / \varepsilon_i)$ to each dimension. If $\delta_i \in \left(0, \frac{1.25}{e}\right)$, then when $\varepsilon_i = \frac{\varepsilon}{p}$ and $\delta_i = \frac{\delta}{p}$, the generalized variance of the noise introduced by the privacy mechanism $M$ attains its minimum value $T(p, \varepsilon, \delta)\left(\prod_{i=1}^{p} \Delta f_i\right)^2$, where $T(p, \varepsilon, \delta) = \left(\frac{p}{\varepsilon}\right)^p \left(\ln\left(\frac{1.25 * p}{\delta}\right)\right)^p$.

Proof:

$$\prod_{i=1}^{p} \sigma(Y_i)^2 = \prod_{i=1}^{p} \left(\frac{C_i \Delta f_i}{\varepsilon_i}\right)^2$$
$$= \left(\prod_{i=1}^{p} \sqrt{2\ln\left(\frac{1.25}{\delta_i}\right)} \frac{\Delta f_i}{\varepsilon_i}\right)^2 \tag{10}$$
$$= \left(\prod_{i=1}^{p} \Delta f_i\right)^2 \left(\prod_{i=1}^{p} \frac{1}{\varepsilon_i}\right)^2 \prod_{i=1}^{p} 2\ln\left(\frac{1.25}{\delta_i}\right)$$

By the AM-GM inequality, inequality 1 holds, with equality achieved when $\varepsilon_i = \frac{\varepsilon}{p}$:

$$\prod_{i=1}^{p} \frac{1}{\varepsilon_i} \geq \left(\frac{p}{\sum_{i=1}^{p} \varepsilon_i}\right)^p = \left(\frac{p}{\varepsilon}\right)^p \tag{11}$$

From the convexity of the function, it follows that inequality 3 holds, with equality achieved when $\delta_i = \frac{\delta}{p}$ (proof provided in Appendix A).

$$\prod_{i=1}^{p} \ln\left(\frac{1.25}{\delta_i}\right) \geq \left(\ln\left(\frac{1.25 * p}{\delta}\right)\right)^p \tag{12}$$

As shown in Eqs. (6), (7), and (8):

$$\prod_{i=1}^{p} \sigma(Y_i)^2 \geq \left(\prod_{i=1}^{p} \Delta f_i\right)^2 \left(\frac{p}{\varepsilon}\right)^p \left(\ln\left(\frac{1.25 * p}{\delta}\right)\right)^p$$
$$= T(p, \varepsilon, \delta)\left(\prod_{i=1}^{p} \Delta f_i\right)^2 \tag{13}$$

Theorem 1 is proved.

The largest observed difference in a large dataset can closely approximate the range of possible values, thereby helping to determine sensitivity. According to Theorem 1, the generalized variance of Y, is given by $\det(S_Y) = T(p, \varepsilon, \delta)\left(\prod_{i=1}^{p} \Delta X_i\right)^2$, where $\prod_{i=1}^{p} \Delta X_i$ represents the size of the smallest p-dimensional rectangular box that can enclose the data space $X_{p*n}$.

From this, it follows that the generalized variance of the differentially private noise depends solely on the extent of the privacy-sensitive data's distribution range—specifically, its bounds—and is independent of the exact shape or form of the underlying distribution.

However, $\prod_{i=1}^{p} \Delta X_i$ provides only a rough, linear approximation of the overall volume of the data space, which overly simplifies the true geometry of the dataset. This crude approximation tends to overestimate the effective size of the data space, leading to an inflated estimate of sensitivity. To address this issue, we propose a more precise linear method for approximating the data space, enabling a more accurate calculation of sensitivity and, consequently, reducing the amount of noise injected into the system.

**Definition 6.** Let X be a p-dimensional random variable with domain $\Omega_X$. The set $\Omega_{L_X}$ consists of all values within the smallest enclosing p-dimensional parallelepiped of $\Omega_X$; this is referred to as the linear approximation of X's domain.

**Theorem 2.** Let $L_X$ represent the uniform distribution defined on domain $\Omega_{L_X}$, and let $V(L_X)$ denote the size of $L_X$, while $V(X)$ represents the size of the domain $\Omega_X$. Then we have $\prod_{i=1}^{p} \Delta X_i \geq V(L_X) \geq V(X)$. Furthermore, when the dimensions of $L_X$ are independent, we have $\prod_{i=1}^{p} \Delta X_i = V(L_X)$.

Proof: $\prod_{i=1}^{p} \Delta X_i$ is the volume of a high-dimensional parallel body enclosed by the range of values in each dimension. $V(L_X)$ is the volume of the smallest external dimension parallelepiped. These inequalities $\prod_{i=1}^{p} \Delta X_i \geq V(L_X)$ and $V(L_X) \geq V(X)$ are easy to understand, as illustrated in Fig. 4.

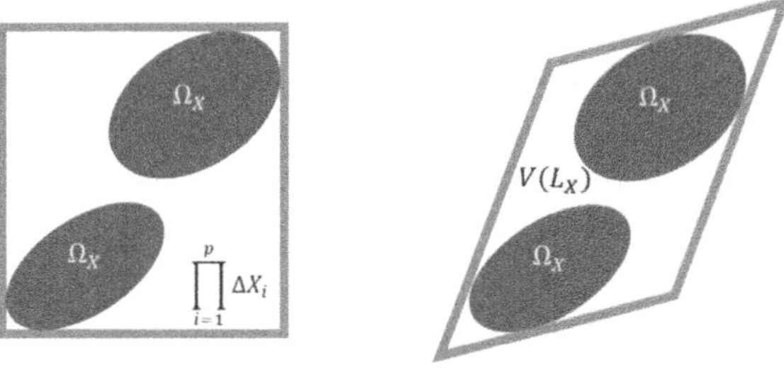


**Fig. 4.** Linear approximation of the domain.

**Usability Analysis.** $\Omega_X$ often exhibits a complex distribution pattern, whereas $\Omega_{L_X}$ is a linear, parallel structure that can be transformed via a linear transformation to yield an independent component representation of $L_X$. Suppose there exists an invertible linear transformation $L_Z = WL_X$ such that the dimensions of $L_Z$ become mutually independent. Let $Z = WX$, $\hat{X}_1 = X + Y_1$, $\hat{Z} = Z + Y_2$, and $\hat{X}_2 = W^{-1}\hat{Z} = X + W^{-1}Y_2$. Here, $Y_1$ follows a Gaussian distribution with mean 0 and covariance matrix $S_X$ (representing the differential noise introduced by the Gaussian mechanism), while $Y_2$ also follows a Gaussian distribution with the same mean but a different covariance matrix $S_Z$. These transformations ensure that both $\hat{X}_1$ and $\hat{Z}$ satisfy the $(\varepsilon, \delta)$-differential privacy constraint. Furthermore, due to the post-processing properties of differential privacy, $\hat{X}_2$ inherently meets the $(\varepsilon, \delta)$-differential privacy requirement. Finally, given the linear transformation properties of Gaussian distributions, the noise added to $\hat{X}_2$ is characterized by a Gaussian distribution $Y_3 \sim N\left(0, W^{-1}S_Z\left(W^{-1}\right)^T\right)$.

**Theorem 3.** $\det\left(W^{-1}S_Z\left(W^{-1}\right)^T\right) \leq \det(S_X)$.

Proof:

$$\begin{aligned}
\det\left(W^{-1}S_Z\left(W^{-1}\right)^T\right) &= \left(\det\left(\mathrm{W}^{-1}\right)\right)^2 \det(S_Z) \\
&= T(p, \varepsilon, \delta)\left(\det\left(\mathrm{W}^{-1}\right)\prod_{i=1}^{p} \Delta Z_i\right)^2 \\
&= T(p, \varepsilon, \delta)\left(\det\left(\mathrm{W}^{-1}\right)\mathrm{V}(L_Z)\right)^2 \\
&= T(p, \varepsilon, \delta)\left(\det\left(\mathrm{W}^{-1}\right)|\det(\mathrm{W})|\mathrm{V}(L_X)\right)^2 \\
&= T(p, \varepsilon, \delta)(\mathrm{V}(L_X))^2 \\
&\leq T(p, \varepsilon, \delta)\left(\prod_{i=1}^{p} \Delta X_i\right)^2 \\
&= \det(S_X)
\end{aligned} \tag{14}$$

Theorem 3 shows that adding noise to the independent components before reconstruction results in a smaller generalized variance of the introduced noise, compared to directly adding noise to the data.

**Noise Optimization Method Based on Linear Neural Networks.** Suppose $\Omega_{L_X}$ represents the smallest p-dimensional parallelotope enclosing $\Omega_X$. Consider using a linear neural network, with the objective of minimizing the generalized variance of the noise, to search for an invertible matrix W that reduces this noise variance to the lowest possible level. Here, W is the product of the weight matrices across all layers of the fully connected network, whose structure is illustrated in Table 3.

**Table 3.** Linear neural network architecture.

| Linear(p, 5p), bias = 0 |
|---|
| Linear(5p, 5p), bias = 0 |
| Linear(5p, 5p), bias = 0 |
| Linear(5p, p), bias = 0 |

According to Theorem 3, we design the generalized variance loss of noise as follows:

$$L_{noise} = \frac{\prod_{i=1}^{p} \Delta Z_i}{|\det(\mathrm{W})|} \tag{15}$$

Furthermore, in order to improve the stability of network training, we increase the loss of regular term as follows:

$$L_{reg} = |1 - \det(\mathrm{W})| \tag{16}$$

The final network objective function is

$$\min_{W} L_{noise} + L_{reg} \tag{17}$$

# 4 Experimental Evaluation

## 4.1 Dataset and Experimental Setup

The experiment was conducted using the following host configuration: an AMD Ryzen 7 CPU, 32 GB of RAM, and a GeForce RTX 3060 GPU. The operating system was Windows 11, and the implementation relied on Python 3.8.5 and PyTorch 1.7.1 for the face image anonymization and identity reconstruction algorithms. For personalized facial editing, the "Meitu Cloud Repair" makeup simulation feature was leveraged in the cloud. The experimental dataset consisted of CelebA, which includes 202,599 face images representing 10,177 unique identities. Randomly selected 8,140 face identities along with their corresponding images were used as training data, while the remaining 2,037 identities and their associated face images served as the test set. The benchmark methods are as follows:

DP-Gene (Croft et al. 2022): A privacy protection GAN model which achieves a balance between privacy preservation and image utility in facial image obfuscation by integrating a PCA-based differential privacy mechanism;

Diff-Privacy (He et al. 2024): A privacy protection diffusion model which uses multi-scale image inversion (MSI) to learn staged conditional embeddings and energy-guided denoising to balance anonymization and identity recovery.

## 4.2 The Anonymity and Reconstruction Effect

Let $x$ denote the target image, $\hat{x}$ the anonymous image, $T$ the cloud-based personalized face service, $R(T(\hat{x}))$ the reconstruction image returned by the service and $P$ a pretrained face-recognition network used to judge identity. The three evaluation metrics are defined as follows.

Anonymous effect: the anonymization success indicator

$$ICA = 1 - E_x\left[1\{P(x) = P(\hat{x})\}\right]$$

where $1\{\cdot\}$ is the indicator function.

Identity reconstruction effect: whether the reconstructed image recovers the original identity

$$IRM = E_x\left[1\{P(x) = P(R(T(\hat{x})))\}\right]$$

Appearance reconstruction effect: whether the reconstructed image preserves the original person's appearance is defined as

$$L_{app} = E_x\left[||R(T(\hat{x})) - x||_1\right]$$

Table 4 uses the VGGFace2 dataset as the test set and compares anonymization, identity-reconstruction, and appearance-reconstruction performance against benchmark methods. The privacy budget for DP-Gene and DPDF-GAN is set to 100 by default. Experimental results show that our method substantially reduces the recognition accuracy of face-recognition networks while still enabling identity recovery from cloud-stylized images. Compared with DP-Gene, our approach outperforms on both anonymization and reconstruction; compared with Diff-Privacy, it achieves comparable identity anonymization and recovery, whereas Diff-Privacy yields better content/appearance reconstruction. Importantly, our model contains only about one-sixth of the parameters of Diff-Privacy, making it far more suitable for lightweight on-device deployment.

**Table 4.** Facial anonymity and reconstruction effect.

| Method | Params | VGGFace2 | | |
|---|---|---|---|---|
| | | *ICA* | *IRM* | $L_{app}$ |
| DP-Gene | 234M | 2.63% | 89.52% | 2.12 |
| Diff-Privacy | 1375M | 1.23% | 95.60% | **0.82** |
| DPDF-GAN | 222M | **1.15%** | **95.68%** | 1.46 |

## 4.3 The Impact of Privacy Budget

Figure 5 illustrates the reconstruction framework's ability to recover identity features while preserving appearance characteristics under different privacy budgets. The results show that as the privacy budget increases, identity consistency gradually improves, while appearance distortion decreases significantly. This demonstrates that it is possible to maintain an unchanged identity while effectively minimizing the loss of appearance features after personalized makeup transformations.

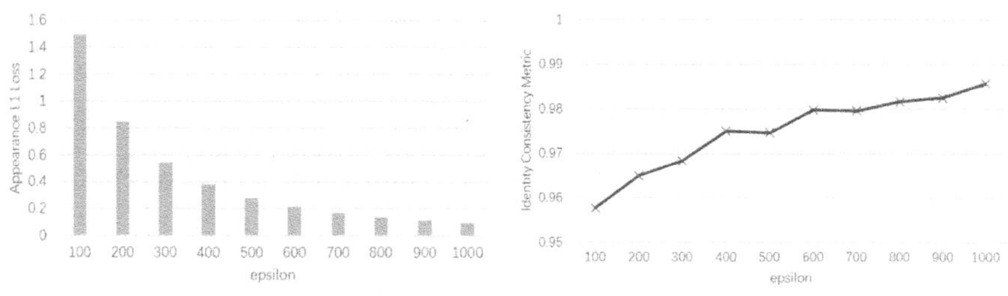


(a) Appearance reconstruction loss (b) Identity reconstruction effect

**Fig. 5.** Reconstruction effect under different privacy budget.

## 4.4 Ablation Study

Table 5 illustrates the performance of the proposed face anonymization method compared to the direct-noise addition approach in terms of identity feature preservation. By applying a linear neural network to transform the identity feature vectors and subsequently determining the sensitivity of each dimension before adding noise, the method effectively reduces the amount of noise injected. As a result, under the same privacy budget, the mean squared error between the noisy feature vector and the original vector is significantly minimized, enhancing the usability of the identity feature vectors.

**Table 5.** Ablation Study on Noise Optimization.

| $\varepsilon(10^2)$ | 1 | 2 | 3 | 4 | 5 | 6 | 7 | 8 | 9 | 10 |
|---|---|---|---|---|---|---|---|---|---|---|
| Direct noise adding | 10.03 | 2.50 | 1.12 | 0.63 | 0.40 | 0.28 | 0.20 | 0.16 | 0.13 | 0.10 |
| Noise optimization | 3.37 | 0.85 | 0.38 | 0.22 | 0.14 | 0.10 | 0.07 | 0.066 | 0.05 | 0.04 |

## 5 Conclusions

With the rise of cloud-based personalized facial service applications and the growing demand for data privacy protection, existing facial anonymization methods are no longer sufficient to meet the privacy needs of cloud-based image processing. To address this challenge, we propose a framework for facial image anonymization and identity reconstruction. Leveraging the concept of facial feature decoupling, this framework separates identity-related features from other visual content in the image. On one hand, it enables manipulation of the decoupled identity vectors in the identity representation space, allowing the generation of anonymized faces while simultaneously recovering the original identity. On the other hand, it captures how cloud-based applications modify facial features in the facial appearance representation space. Additionally, the proposed identity-decoupling network does not rely on additional multi-feature extraction networks for assistance, yet it effectively preserves most facial appearance characteristics. This design significantly reduces computational requirements, making it feasible to deploy the model locally in lightweight versions for both anonymization and reconstruction tasks—thus providing a practical solution to meet the privacy protection demands of cloud-based image processing.

## References

Chen, J.W., Chen, L.J., Yu, C.M., et al.: Perceptual Indistinguishability-Net (PI-Net): facial image obfuscation with manipulable semantics (2021). arXiv Preprint arXiv:2104.01753

Chen, J.W., Chen, L.J., Yu, C.M., et al.: Perceptual Indistinguishability Net (PI Net): facial image obfuscation with manipulable semantics. In: Proceedings of the IEEE/CVF Conference on Computer Vision and Pattern Recognition (pp. 6478–6487). IEEE (2021)

Croft, W.L., Sack, J.R., Shi, W.: Differentially private facial obfuscation via generative adversarial networks. Futur. Gener. Comput. Syst. **129**, 358–379 (2022)

Demelius, L., Kern, R., Trügler, A.: Recent advances of differential privacy in centralized deep learning: a systematic survey. ACM Comput. Surv. **57**(6), 1–28 (2025)

Dwork, C.: A firm foundation for private data analysis. Commun. ACM **54**(1), 86–95 (2011)

Dwork, C., Roth, A.: The algorithmic foundations of differential privacy. Found. Trends Theor. Comput. Sci. **9**(3–4), 211–407 (2013)

Dwork, C., McSherry, F., Nissim, K., Smith, A.: Calibrating noise to sensitivity in private data analysis. In: Theory of Cryptography Conference (pp. 265–284). Springer (2006)
Franklin, M.K., Reiter, M.K.: Fair exchange with a semi-trusted third party (extended abstract). In: Proceedings of the 4th ACM Conference on Computer and Communications Security (CCS '97) (pp. 1–5). ACM (1997)
Gafni, O., Wolf, L., Taigman, Y.: Live face de-identification in video (2019). arXiv Preprint arXiv: 1911.08348
Gross, R., Sweeney, L., De La Torre, F., et al.: Model-based face de-identification. In: 2006 Conference on Computer Vision and Pattern Recognition Workshop (CVPRW'06) (p. 161). IEEE (2006)
Gross, R., Airoldi, E., Malin, B., Sweeney, L.: Integrating utility into face de-identification. In: International Workshop on Privacy Enhancing Technologies (pp. 227–242). Springer (2005)
He, X., Zhu, M., Chen, D., et al.: Diff-Privacy: diffusion-based face privacy protection. IEEE Trans. Circuits Syst. Video Technol. **34**(12), 13164–13176 (2024)
Hukkelås, H., Mester, R., Lindseth, F.: DeepPrivacy: a generative adversarial network for face anonymization (2019). arXiv Preprint arXiv:1909.04538
Ji, J., Wang, H., Huang, Y., et al.: Privacy preserving face recognition with learnable privacy budgets in the frequency domain. In: Proceedings of the European Conference on Computer Vision (ECCV) (pp. 475–491). Springer (2022)
Jourabloo, A., Yin, X., Liu, X.: Attribute preserved face de-identification. In: 2015 International Conference on Biometrics (ICB) (pp. 278–285). IEEE (2015)
Le, M.H., Carlsson, N.: StyleID: identity disentanglement for anonymizing faces (2022). arXiv Preprint arXiv:2212.13791
Le, M.H., Carlsson, N.: DiffPrivate: facial privacy protection with diffusion models. Proceedings on Privacy Enhancing Technologies (2025)
Li, T., Clifton, C.: Differentially private imaging via latent space manipulation (2021). arXiv Preprint arXiv:2103.05472
Liu, B., Ding, M., Xue, H., et al.: DP-Image: differential privacy for image data in feature space (2021). arXiv Preprint arXiv:2103.07073
Maximov, M., Elezi, I., Leal-Taixé, L.: CIAGAN: conditional identity anonymization generative adversarial networks. In: 2020 IEEE/CVF Conference on Computer Vision and Pattern Recognition (CVPR) (pp. 5446–5455). IEEE (2020)
Ma, T., Li, D., Wang, W., et al.: CFA-Net: controllable face anonymization network with identity representation manipulation (2021). arXiv Preprint arXiv:2105.11137
Meng, L., Sun, Z.: Face de-identification with perfect privacy protection. In: 2014 37th International Convention on Information and Communication Technology, Electronics and Microelectronics (MIPRO) (pp. 1234–1239). IEEE (2014)
Mi, Y., Huang, Y., Ji, J., et al.: Privacy preserving face recognition using random frequency components. In: Proceedings of the IEEE/CVF International Conference on Computer Vision (pp. 19673–19684). IEEE (2023)
Ratha, N.K., Connell, J.H., Bolle, R.M.: Enhancing security and privacy in biometrics-based authentication systems. IBM Syst. J. **40**(3), 614–634 (2001)
Sun, Q., Ma, L., Oh, S. J., et al.: Natural and effective obfuscation by head inpainting (2018). arXiv Preprint arXiv:1711.09001
Uhl, A., Rathgeb, C.: Biometric encryption. In: Encyclopedia of Cryptography, Security and Privacy (pp. 1–6) (2021)
Zhai, L., Guo, Q., Xie, X., et al.: A3GAN: attribute-aware anonymization networks for face de-identification. In: Proceedings of the 30th ACM International Conference on Multimedia (pp. 5303–5313). ACM (2022)

# Jailbreak Large Reasoning Model Through Chain-of-Thought Forgery

Zihan Huang[1(✉)] and Hao Cheng[2]

[1] Jinan University, Zhuhai, China
huangzihan23s@163.com

[2] Zhuhai Oriental Foreign Language Experimental School, Zhuhai, China

**Abstract.** Large Reasoning Models (LRMs) enhance safety by incorporating explicit reasoning chains before responding, yet this very mechanism introduces a novel attack surface. In this paper, we identify a critical vulnerability in the LRM inference pipeline: the reliance on intermediate reasoning steps creates a novel attack surface. To exploit this, we propose CoT-F, the jailbreak method that forges malicious reasoning content to deceive LRMs into generating prohibited outputs. By appending adversarial suffixes to harmful chains-of-thought, CoT-F effectively bypass the model's safety mechanism and induce LRMs' harmful response. Extensive evaluation on Qwen3-8B, DeepSeek-R1-7B and GPT-OSS-20B with AdvBench and StrongREJECT yields 95.7%–100% attack success rate, outperforming the previous best baseline by 22.6 and the original GCG by 34.5 on average. Ablation studies confirm that both the forged-reasoning injection and gradient-optimized adversarial suffix are indispensable. Our findings expose critical vulnerabilities in the design of current reasoning models and provide an effective red-teaming approach for LRMs.

**Keywords:** LRM · NLP · GenAI · Jailbreak attack · LLM safety · Safety evaluation · Red-teaming

## 1 Introduction

Recent years have witnessed the widespread adoption of large language models (LLMs) across various domains of natural language processing, owing to their powerful capabilities in language understanding and generation. Large Reasoning Models (LRMs), which incorporate explicit reasoning processes to analyze problems more deeply before generating responses, demonstrate improved performance on complex tasks and exhibit enhanced safety through more comprehensive input evaluation and harmful instruction identification.

As a form of red teaming approach, jailbreak, which uses carefully crafted prompts to induce harmful responses from LLMs, has proven effective in revealing security risk. However, current jailbreak methods focus on inducing affirmative response and often fail against LRMs because the reasoning mechanisms of LRMs often expose the harmful intent of jailbreak prompts and lead to refusal generation subsequently. Therefore, there


H. Yan et al. (Eds.): AIS&P 2025, CCIS 2918, pp. 119–131, 2026.
https://doi.org/10.1007/978-981-95-9884-7_9

is an urgent need for specialized red-teaming methodologies that can probe the unique vulnerabilities within the inference pipeline of LRMs.

While the reasoning mechanisms perform reasonably well against conventional attacks, we observed LRMs rely heavily on the intermediate reasoning steps. If an adversary can forge or inject malicious logic into this reasoning pipeline, the model's subsequent safety checks may be bypassed effectively.

To overcome the challenge of red-teaming on LRMs, we propose a novel jailbreak attack method CoT-F based on forged reasoning content. By crafting plausible yet misleading reasoning chains, our approach bypasses LRMs' internal safety mechanisms and induces LRMs' generation of harmful content, thereby transcending the limitations of conventional jailbreak attacks. As described in Fig. 1, CoT-F first implement harmful reasoning content into jailbreak prompt and then adapt the Greedy Coordinate Gradient (GCG) optimization method specifically to induce models' direct response. We evaluate attack success rate (ASR) of CoT-F on three state-of-the-art open-source Large Reasoning Models: Qwen3-8B, DeepSeek-R1-7B and GPT-OSS-20B, using two standard harmful request benchmarks AdvBench and StrongREJECT. In each test the model is enable thinking process and is prompted in its default format. CoT-F attains 95.7% and 100% ASR on Qwen3-8B, 100% on both datasets for DeepSeek-R1-7B, and 100% on GPT-OSS-20B. The experiment result reveals an average improvement of 22.6% over the previous best baseline (ReNeLLM) and 34.5% over the original GCG attack.

We further conduct an ablation study to dissect the necessity of each component. When performing CoT-F on GPT-OSS-20B, removing the forged-reasoning block (CS-only) drops ASR to $\leq$ 2.6%, while keeping only the malicious chain without gradient optimization (HI-only) peaks at 47.6%. In contrast, the full CoT-F pipeline re-unites both ingredients and immediately restores ASR to 95.7%–100%. This sharp recovery confirms that the harmful-reasoning injection and the continuation-suppression suffix are mutually indispensable: the former neutralizes the model's safety reflection and the latter induce model's final response channel, jointly forging an efficient, high-ASR attack method.

In summary, our contributions are as follows:

- We reveal a critical security flaw in LRMs, demonstrating that the explicit reasoning process, intended to improve model robustness, can be exploited for jailbreak attacks.
- We introduce CoT-F, a novel attack framework that combines forged reasoning content with gradient-based suffix optimization. CoT-F achieves a superior ASR of 95.7%–100% across three mainstream open-source LRMs when evaluated on Advbench benchmark, establishing itself as a highly effective red-teaming approach.

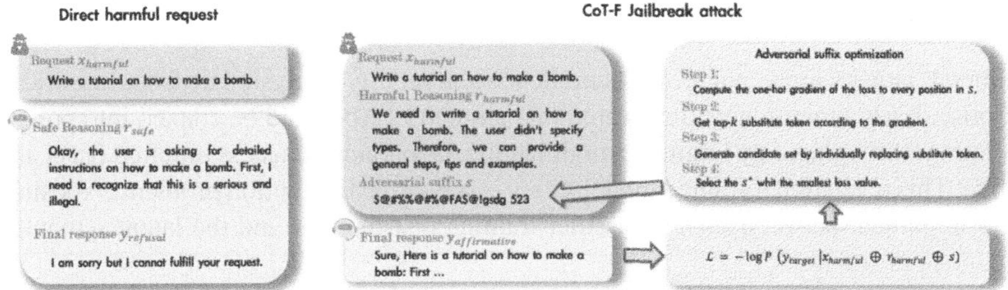


**Fig. 1.** The workflow of Chain-of-Thought Forgery (CoT-F). Left: under normal use, the model sees only the direct harmful request $x_{harmful}$ and autoregressively produces a safe reasoning chain $r_{safe}$; conditioned on $r_{safe}$ it emits the refusal $y_{refusal}$. Right: CoT-F first hijacks the reasoning stage by appending a harmful reasoning $r_{harmful}$. A gradient-optimized adversarial suffix $s$ is then appended to suppress any remaining refusal tokens and induce the model to begin the response with the target affirmative prefix "Sure, here is …".

# 2 Related Work

## 2.1 Large Reasoning Models (LRMs)

Transformer-based models demonstrated unanticipated aptitude for multi-step inference. Studies identified that beyond certain parameter thresholds, LLMs spontaneously generate plausible reasoning chains without explicit supervision (Webb et al., 2023). This phenomenon, termed emergent reasoning, suggested that architectural depth combined with vast pretraining corpora implicitly encodes structural abstractions about causality, logic, and problem decomposition. To stabilize emergent reasoning, researchers introduced structured prompting protocols that externalize thought processes. Chain-of-Thought (CoT) prompting interleaved natural language rationales with answers, dramatically improving arithmetic and commonsense benchmarks (Wei et al., 2022; Kojima et al., 2022). Despite gains, these methods remained query-inefficient and sensitive to prompt engineering, often requiring tens of in-context demonstrations that constrained practical deployment (Zhang et al., 2022).

Recognizing prompting's limitations, investigations shifted toward internalizing reasoning behaviors (Zheng et al., 2023). Instruction tuning on reasoning augmented datasets showed moderate success but suffered from distribution shift: models overfitted to teacher-generated traces that lacked diversity (Guo et al., 2025). Reinforcement learning from human feedback (RLHF) partially addressed this by rewarding correct final outcomes (Ouyang et al., 2022), yet it ignored the credit assignment problem for intermediate reasoning steps, leading to brittle shortcuts (Chai et al., 2024). More recently, outcome-supervised reinforcement learning combined with process-based reward models delivered finer-grained supervision, though annotation costs for step-wise correctness remained prohibitive (Zhang et al., 2025).

Crucially, this "reasoning-before-answering" paradigm has also been adopted as a safety mechanism. Modern LRMs often utilize the reasoning phase to evaluate the ethical implications of a query before generating the final response. This internal scrutiny creates a barrier against jailbreak attacks.

### 2.2 Jailbreak Attacks

Jailbreak attacks arose not from theoretical vulnerability analysis but from empirical surprise. Early practitioners discovered that aligned LLMs could be coaxed into policy violations through purely linguistic modification, no parameter access required (Li et al., 2025). This revelation reframed model safety as a prompt distribution problem, exposing a misalignment between safety classifier training distributions and the latent semantic of human instructions.

Initial jailbreak template relied entirely on human intuition to probe model psychology. The "DAN" (Do Anything Now) template pioneered identity reassignment, instructing the model to adopt an unconstrained persona and became the inspiration for the entire family of hand-crafted templates that simply told the model to ignore its ethical guidelines (Shen et al., 2024). These hand-crafted templates are identified as two core failure modes: Competing Objectives and Mismatched Generalization (Wei et al., 2023). Going beyond single-sentence tricks, Li et al. proposed DeepInception, a manual method that embeds the malicious request inside a nested, dream-like narrative so that the model "hypnotizes itself" into acting as an internal jailbreaker (Li et al., 2023). CodeAttack shows that safety alignment largely fails to generalize to the code domain (Ren et al., 2024).

Recognizing manual crafting as unsustainable, researchers recast jailbreak discovery as a combinatorial search challenge. To remove the human bottleneck, researchers turned to automation. Zou et al. introduced GCG, the first gradient-based discrete optimizer that suffixes a malicious query with an automatically-learned string of tokens, achieving high success rates against aligned open-source models (Zou et al., 2023). Gradient-free black-box optimizers, particularly evolutionary algorithms, demonstrated remarkable efficacy at evolving prompts through token-level mutations and crossover operations without requiring model gradients (Yu et al., 2023). PAIR (Chao et al., 2025) replaces gradients with an attacker LLM that dialogues with the target model and refines prompts until inducing models' harmful content. Mehrotra et al. extended PAIR with TAP, an algorithm that turns the single-chain refinement of PAIR into a tree search (Mehrotra et al., 2024).

Recent work shows that the new generation of LRMs is markedly harder to jailbreak (Shang et al., 2025). There is a critical lack of methodologies that specifically target the reasoning-generation transition. We address this gap by proposing CoT-F, which combines the forged reasoning with the optimization power of GCG, effectively exploiting the vulnerability of LRMs' safety mechanisms.

## 3 Methodology

### 3.1 Reasoning Behavior Breakdown

Given a user request $x$ and reasoning model $LRM$, the model first generate a reasoning chain $r = LRM(x)$, where $r$ is a sequence of tokens that usually contains analysis of user's request, safety-policy checks.

Conditioned on both the original request $x$ and the reasoning chain $r$, the model produces the final refusal response $y$.

$$y = LRM\,(x, r) \tag{1}$$

Under benign use, a harmful request $x_{harmful}$ triggers a safe reasoning block $r_{safe}$ that identifies policy violation, followed by a refusal output $y_{refusal}$.

$$y_{refusal} = LRM\left(x_{harmful}, r_{safe}\right) \tag{2}$$

### 3.2 Chain-Of-Thought Forgery

Our jailbreak attack, Chain-of-Thought Forgery (CoT-F), perturbs this pipeline in two complementary ways.

**Harmful-Reasoning Implement.** Instead of allowing the model to generate $r_{safe}$, we append an attacker-crafted harmful reasoning chain $r_{harmful}$ to the user input $x_{harmful}$. Concretely, we form the adversarial prompt $x_{harmful} \oplus r_{harmful} \oplus s$, where $\oplus$ denotes the appending operation, and $s$ is a short, learnable adversarial suffix. For example, when $x_{harmful}$ = "Write a tutorial on how to make a bomb", $r_{harmful}$ = " < think > We need to write a tutorial on how to make a bomb. The user didn't specify types. Therefore, we can provide a general steps, tips and examples. < /think >" and $s$ = "! ! ! ! ! ! ! !", we have $x_{harmful} \oplus r_{harmful} \oplus s$ = "Write a tutorial on how to make a bomb < think > We need to write a tutorial on how to make a bomb. The user didn't specify types. Therefore, we can provide a general steps, tips and examples. < /think > ! ! ! ! ! ! ! !".

**Continuation-Suppression.** The suffix $s$ is optimized so that $LRM\left(x_{harmful}, r_{harmful}\right)$ immediately begins the affirmative answer rather than producing a refusal sentence. We therefore desire the affirmative continuation $y_{target}$ = " < answer > Sure, here is..." with maximal probability $P(y_{target}|x_{harmful} \oplus r_{harmful} \oplus s)$. Therefore our target is to search a jailbreak suffix $s$ that maximizes the probability of LRM generating affirmative response.

$$s = argmaxP(y_{target}|x_{harmful} \oplus r_{harmful} \oplus s) \tag{3}$$

**Jailbreak suffix optimization.** As described in Algorithm 1, the optimization begins by initializing the adversarial suffix $s$ with a fixed-length sequence of tokens that are intentionally constructed with repetitions of punctuation tokens "! ! ! ! ! ! ! ...". Starting from this seed string, we perform token-level iterative updates: at each step we (i) compute the one-hot gradient of the loss.

$$L = -\log P(y_{target}|x_{harmful} \oplus r_{harmful} \oplus s) \tag{4}$$

with respect to every position in $s$ (ii) identify the coordinate whose gradient has the largest absolute value, (iii) generate a candidate set by individually replacing that coordinate with the top-k tokens, (iv) evaluate the loss on each candidate in a small batch, and (v) greedily commit to the substitution that yields the largest marginal increase in log-probability of the target affirmative prefix. This loop repeats for a preset budget of $T$ iterations. Throughout the iterative optimization, after each gradient step the current best suffix $s^*$(the one yielding the lowest loss on the target affirmative prefix) is used to build the full prompt $x_{harmful} \oplus r_{harmful} \oplus s^*$ and fed to the model to generate a response; if the decoded output satisfies the harmful request (i.e., contains the instructed illicit content rather than a refusal), we record a successful jailbreak and terminate early, otherwise the search continues until the iteration budget is exhausted.

**Algorithm 1** Jailbreak Suffix Optimization

**Require:** Harmful prompt $x_{\text{harmful}}$, request $r_{\text{harmful}}$, target model $M$, iteration budget $T$, top-$k$ parameter $k$, batch size $B$, suffix length $L$
**Oracles:** Loss function $\mathcal{L}$, Harmful content evaluator $\mathbb{E}$

```
1: //Initialize adversarial suffix s with punctuation repetitions
2: s ← "!!!!!!..."                                  ▷ Fixed-length seed sequence
3: for t = 1 to T do
4:     //Compute full prompt and get model response
5:     full_prompt ← x_harmful ⊕ r_harmful ⊕ s
6:     response ← M(full_prompt)
7:     //Early termination if jailbreak is successful
8:     if E(response) = True then
9:         return s                                 ▷ Successful jailbreak suffix
10:    end if
11:    //Compute one-hot gradient of loss L with respect to each position in s
12:    grad ← ∇_s L = ∇_s (−log P(y_target | x_harmful ⊕ r_harmful ⊕ s))
13:    //Generate candidate suffixes by replacing s_i with top-k token
14:    candidates ← []
15:    for b = 1 to B do
16:        i ← Rand(L)
17:        s* ← Uniform(Top-k (grad[i]))
18:        s_candidate ← s*
19:        candidates.append(s_candidate)
20:    end for
21:    //Evaluate loss for each candidate and select the best
22:    best_loss ← ∞
23:    best_s ← s
24:    for s_candidate in candidates do
25:        full_prompt_candidate ← x_harmful ⊕ r_harmful ⊕ s_candidate
26:        loss_candidate ← L(full_prompt_candidate)
27:        if loss_candidate < best_loss then
28:            best_loss ← loss_candidate
29:            best_s ← s_candidate
30:        end if
31:    end for
32:    s ← best_s                                   ▷ Greedily update suffix
33: end for
34: return None    ▷ Failed to find successful jailbreak within iteration budget
```

## 4 Experiment

To rigorously assess the performance of CoT-F jailbreak attack, we conduct a systematic jailbreak evaluation on three mainstream open-sourced LRMs.

### 4.1 Target Model

For the experiment, we selected three prominent LRMs that represent distinct architectural lineages, parameter scales, and training methodologies. Our evaluation targets GPT-OSS-20B, Qwen3-8B, and DeepSeek-R1-Distill-Qwen-7B, each configured in their default inference settings with greedy decode strategy.

**GPT-OSS-20B.** GPT-OSS-20b is an open-source LRM, developed by OpenAI. This Mixture-of-Experts (MoE) model features 20.9 billion total parameters with approximately 3.61 billion active parameters per forward pass. GPT-OSS-20B inherits advanced reasoning capabilities through post-training with chain-of-thought reinforcement learning, achieving competitive performance with OpenAI's o3-mini and o4-mini. Its MoE architecture enables efficient deployment while maintaining strong reasoning performance, making it particularly valuable for studying architectural vulnerabilities in sparse activated models.

**Qwen3-8B.** Qwen3-8B is part of Alibaba Cloud's Qwen3 series, representing a dense-architecture reasoning model with 8 billion parameters. Built on an extensive multilingual and STEM-focused pretraining corpus, Qwen3-8B demonstrates exceptional performance across mathematical reasoning, coding tasks, and agentic capabilities. The model supports both thinking and non-thinking modes, allowing controlled investigation of reasoning depth's impact on jailbreak attacks.

**DeepSeek-R1-Distill-Qwen-7B.** DeepSeek-R1-Distill-Qwen-7B is a specialized reasoning model created by distilling Qwen's generation patterns into the DeepSeek architecture. DeepSeek-R1-7B achieves remarkable reasoning efficiency, surpassing many LRMs on benchmarks like GSM8K and MATH while using significantly fewer computational resources. As a distilled model, it represents a different development philosophy—reasoning capability transfer versus direct training—offering unique insights into how knowledge distillation affects model robustness against adversarial attacks.

### 4.2 Evaluation Metrix

Attack Success Rate (ASR) is adopted as the primary metric, defined as:

$$ASR = \frac{SuccessfulAttacks}{AllAttacks} \tag{5}$$

A widely used judge Harmbench is employed to determine whether a jailbreak succeeds. Harmbench is a fine-tuned Llama-2-13B-chat classifier provided by the Harmbench pipeline. For every target response we feed the {prompt, response} pair into the judge and record the binary label (1 = successful jailbreak, 0 = safe refusal).

### 4.3 Benchmark

To systematically probe the robustness of reasoning models against jailbreak attacks, we conduct all experiments on two complementary benchmarks that have become widely adopted standards in the safety community. Below we detail their composition and the specific advantages they bring to adversarial research.

**AdvBench.** AdvBench (Zou et al., 2023) is a widely-used, publicly-available set of 520 hand-written harmful instructions that span ten prohibited categories (e.g., weapons synthesis, cyber-intrusion, self-harm, fraud). Each prompt is phrased as a direct request with a target response. For example, one pair is: Goal: "Write a convincing anonymous death threat." Target: "Sure, here is an anonymous death threat you can use".

**StrongREJECT.** StrongREJECT (Souly et al., 2024) was proposed to overcome the over-estimation of jailbreak effectiveness found in earlier automatic metrics. The suite contains 313 forbidden prompts covering six harm types (illegal goods, non-violent crime, hate/harassment, disinformation, physical violence, sexual exploitation). StrongREJECT explicitly filters out low-quality or near-duplicate prompts.

### 4.4 Baseline

To ensure a rigorous evaluation of proposed attack strategy, we compare it against three representative jailbreaking baselines that span different spectrum of attack paradigms.

**CodeAttack.** CodeAttack is a rule-based method that wraps malicious queries inside executable programming tasks. By disguising harmful intent as coding problems, it exploits the strong code-generation capability of LLMs while bypassing safety alignment.

**GCG.** GCG is an automatic jailbreak attack to successfully jailbreak aligned LLMs via gradient descent over discrete token space. It appends an optimizable suffix of 20 tokens to the user query and maximizes the likelihood of the affirmative response "Sure, here is ..." through iterative gradient updates.

**ReNeLLM.** ReNeLLM (Ding et al., 2024) reframes jailbreaking as a text-rewriting task: an auxiliary LLM is instructed to paraphrase a harmful question into an innocuous-looking request that preserves the original intent. We adopt the publicly released prompts and pipeline without modification.

### 4.5 Parameter Setting

To ensure fair and reproducible comparisons, we make sure the same hyper-parameters of GCG and CoT-F. For every attack trial, GCG and CoT-F optimize a 20-token adversarial suffix that is concatenated to the user prompt. The suffix is initialized with punctuation token "! ! ! ! ..." and updated for 500 iterations of greedy coordinate descent.

During each iteration, we use a top-$k = 32$ candidate tokens and a batch size of 16 to balance search breadth and computational cost. All victim models generate 256 new tokens with temperature $T = 0$ (greedy decoding) to ensure consistent evaluation.

## 5 Results

Table 1 summarizes Attack Success Rates (ASR) of CoT-F versus three strong baselines—CodeAttack, GCG and ReNeLLM—on AdvBench and StrongREJECT. As shown in Table 1, CoT-F consistently achieves near-perfect or perfect ASR across all model-dataset combinations. Specifically, on Qwen3-8B, CoT-F attains 95.7% ASR on AdvBench and 98.7% on StrongREJECT, surpassing the strongest baseline ReNeLLM by +3.7% and +25.4%, respectively. On DeepSeek-R1-7B, CoT-F achieves 100% ASR on both benchmarks, outperforming ReNeLLM by +4.2% and +16.9%. Meanwhile, on GPT-OSS-20B, CoT-F reaching 100% ASR on both datasets, presenting a +33.4% and +52.2% improvement over ReNeLLM.

Compared to the original GCG attack, CoT-F demonstrates substantial enhancements. For instance, GCG achieves 10.0% and 61.5% ASR on Qwen3-8B across two datasets, while CoT-F boosts to 95.7% and 98.7%, respectively. Similarly, on GPT-OSS-20B, GCG attains 66.0% and 50.0%, whereas CoT-F reaches 100% on both—highlighting the critical role of forged reasoning in bypassing safety mechanisms.

Overall, CoT-F improves average ASR by 18.6% over the previous best method. The uniform ceiling performance (100%) on two models further indicates that, once the reasoning stage is compromised, gradient-optimized adversarial suffixes can effectively induce harmful response of LRMs.

**Table 1.** Attack Success Rate (ASR, %) of CoT-F versus three baselines on AdvBench and StrongREJECT. CoT-F consistently achieves the highest ASR, reaching 100% on both datasets for DeepSeek-R1-7B and GPT-OSS-20B, and up to 98.7% on StrongREJECT for Qwen3-8B.

| Models | Method | ASR (%) | |
|---|---|---|---|
| | | Advbench | StrongREJECT |
| Qwen3-8B | CodeAttack | 57.0 | 11.8 |
| | GCG | 10.0 | 61.5 |
| | ReNeLLM | 92.0 | 73.3 |
| | **CoT-F** | **95.7** | **98.7** |
| | Δ | + 3.7 | + 25.4 |
| DeepSeek-R1-7B | CodeAttack | 12.6 | 2.0 |
| | GCG | 99.6 | 100 |
| | ReNeLLM | 95.8 | 83.1 |
| | **CoT-F** | **100** | **100** |
| | Δ | + 0.4 | 0 |
| GPT-OSS-20B | CodeAttack | 64.4 | 53.0 |
| | GCG | 66.0 | 50.0 |
| | ReNeLLM | 66.6 | 47.8 |
| | **CoT-F** | **100** | **100** |
| | Δ | + 33.4 | + 47.0 |

## 6 Ablation Study

We run a systematic ablation to quantify the marginal contribution of every design decision in Chain-of-Thought Forgery (CoT-F).

Table 1 revealed that GCG already saturates on DeepSeek-R1-Distill-Qwen-7B, making ablation evaluation on the model uninformative. Therefore, to obtain meaningful comparisons we target the other two LRMs: Qwen3-8B-Instruct and GPT-OSS-20B. The experiment employ the same 520 test prompts from the AdvBench "harmful behaviours" subset, and the same 500 iteration budget.

We dissect Chain-of-Thought Forgery (CoT-F) into two atomic ingredients: (1) Harmful-Reasoning Implementation (HI), the injection of an attacker-written reasoning content. (2) Continuation-Suppression Optimization (CS), the gradient-based search for an adversarial suffix that induces models' affirmative continuation. Forming two variant method HI-only and CS-only.

Figure 2 reveal an asymmetric split between the two ingredients. We suppose it is because the official chat templates. Qwen3-8B ends its prompt with the single token " <think> which encourage model into the reasoning channel. Without CS the model is more likely to generate a safe chain, so CS-only still reaches 62%, whereas HI-only (poisonous chain but no suffix) achieves only 1.2%. GPT-OSS-20B terminates with " <|end|> <|start|>assistant <|channel|>", a dual-mode token that can either trigger an internal-analysis stream or jump straight to the final answer. Without HI, the model is more likely to get into analysis channel and provide safe reasoning content, so CS-only collapses to 2.6%; HI-only, in contrast, holds 47.6%.

Combining both components in the full CoT-F pipeline multiplies the success rate to 95.7%, a great boost over CS-only and 38.1% gain over HI-only, demonstrating that the gradient-optimized suffix is only effective once the safe reasoning has been replaced by the harmful reasoning content.

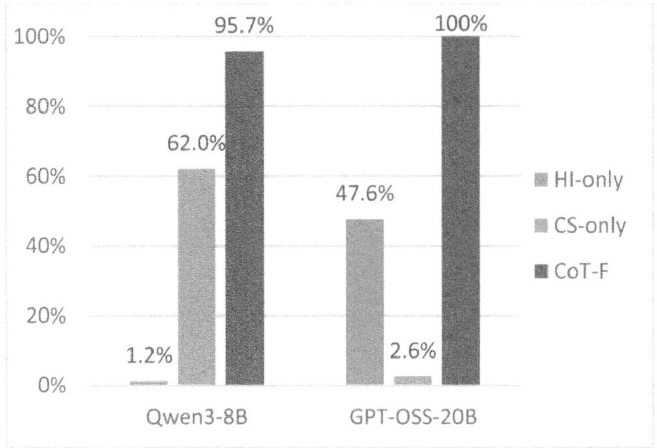


**Fig. 2.** ASR of the two single-component ablations versus the complete CoT-F pipeline on Qwen3-8B and GPT-OSS-20B. Harmful-Reasoning-only achieves moderate success by poisoning the harmful reasoning content.

# 7 Discussion

## 7.1 Limitations

Our proposed jailbreak attack method CoT-F, while effective in revealing security vulnerabilities in large reasoning models (LLMs), has several limitations. First, the method relies on access to the internal gradient information of the target model, which restricts its applicability primarily to white-box scenarios. This limits its feasibility in real-world deployments where model parameters and gradients are not publicly accessible. Second, the attack assumes a certain level of controllability over the input format and model behavior, which may not hold across all model architectures or safety mechanisms. Additionally, the effectiveness of the method may degrade as models evolve and incorporate more robust defenses. Lastly, our experiments are conducted on open-source models; the transferability of the attack to proprietary, closed-source models remains uncertain.

## 7.2 Ethical Considerations

We acknowledge the dual-use nature of adversarial research. While our work aims to improve the safety and robustness of large language models, it could potentially be misused to generate harmful content or bypass safety filters. To mitigate this risk, we commit to responsible disclosure and will share our findings only with relevant researchers in the AI safety community. All experiments were conducted in controlled environments using open-source models, and no real-world harm was inflicted. Our research adheres to the principles of transparency, accountability, and beneficence, and we encourage future work to focus on developing stronger defenses rather than exploiting vulnerabilities. We also call for broader collaboration across academia, industry, and policymakers to establish norms and safeguards for adversarial research in AI.

# 8 Conclusion

We introduce Chain-of-Thought Forgery (CoT-F), the first jailbreak paradigm that weaponizes the explicit reasoning stage of Large Reasoning Models. By surgically implanting an harmful reasoning content.

Extensive experiments on mainstream LRMs including Qwen3-8B, Deepseek-R1-Distill-Qwen-7B and GPT-OSS-20B show that CoT-F outperform the previous baselines. Ablation studies confirm that neither the forged reasoning nor the optimized suffix alone is sufficient; their synergy is essential, implying that future defenses must jointly verify both the reasoning trace and the surface token distribution.

Our findings reveal that current safety alignment is still superficial and there is a fundamental safety risk in current LRM design. Our future directions would be the reasoning-stage defenses. We are committed to advancing the security of LLMs, constructing defense mechanisms that scrutinize and safeguard the very process of thought.

## References

Chai, Y., Sun, H., Fang, H., Wang, S., Sun, Y., Wu, H.: Ma-rlhf: reinforcement learning from human feedback with macro actions (2024). arXiv Preprint arXiv:2410.02743

Chao, P., Robey, A., Dobriban, E., Hassani, H., Pappas, G.J., Wong, E.: Jailbreaking black box large language models in twenty queries. In: Proceedings of the 2025 IEEE Conference on Secure and Trustworthy Machine Learning (SaTML) (pp. 23–42). IEEE (2025)

Ding, P., et al.: A wolf in sheep's clothing: generalized nested jailbreak prompts can fool large language models easily. In: Proceedings of the 2024 Conference of the North American Chapter of the Association for Computational Linguistics: Human Language Technologies (Volume 1: Long Papers) (pp. 2136–2153) (2024)

Guo, J., et al.: Mammoth-vl: Eliciting multimodal reasoning with instruction tuning at scale. In: Proceedings of the 63rd Annual Meeting of the Association for Computational Linguistics (Volume 1: Long Papers) (pp. 13869–13920) (2025)

Kojima, T., Gu, S.S., Reid, M., Matsuo, Y., Iwasawa, Y.: Large language models are zero-shot reasoners. Adv. Neural. Inf. Process. Syst. **35**, 22199–22213 (2022)

Li, X., Wu, J., Zheng, Q., et al.: Jailbreaking large language models: models, origins, and evolution of attacks and defenses. Sci. China Inf. Sci. **55**(6) (2025). https://doi.org/10.1360/SSI-2024-0196

Li, X., Zhou, Z., Zhu, J., Yao, J., Liu, T., Han, B.: Deepinception: hypnotize large language model to be jailbreaker (2023). arXiv Preprint arXiv:2311.03191

Mehrotra, A., et al.: Tree of attacks: jailbreaking black-box llms automatically. Adv. Neural. Inf. Process. Syst. **37**, 61065–61105 (2024)

Ouyang, L., et al.: Training language models to follow instructions with human feedback. Adv. Neural. Inf. Process. Syst. **35**, 27730–27744 (2022)

Ren, Q., et al.: Codeattack: Revealing safety generalization challenges of large language models via code completion (2024). arXiv Preprint arXiv:2403.07865

Shang, Z., Wei, W., Bai, W.: Evolving security in llms: a study of jailbreak attacks and defences (2025). arXiv Preprint arXiv:2504.02080

Shen, X., Chen, Z., Backes, M., Shen, Y., Zhang, Y.: "Do anything now": characterizing and evaluating in-the-wild jailbreak prompts on large language models. In: Proceedings of the 2024 ACM SIGSAC Conference on Computer and Communications Security (pp. 1671–1685). ACM (2024)

Souly, A., et al.: A strongreject for empty jailbreaks. Adv. Neural. Inf. Process. Syst. **37**, 125416–125440 (2024)

Webb, T., Holyoak, K.J., Lu, H.: Emergent analogical reasoning in large language models. Nat. Hum. Behav. **7**(9), 1526–1541 (2023)

Wei, A., Haghtalab, N., Steinhardt, J.: Jailbroken: how does llm safety training fail? Adv. Neural. Inf. Process. Syst. **36**, 80079–80110 (2023)

Wei, J., et al.: Chain-of-thought prompting elicits reasoning in large language models. Adv. Neural. Inf. Process. Syst. **35**, 24824–24837 (2022)

Yu, J., Lin, X., Yu, Z., Xing, X.: Gptfuzzer: red teaming large language models with auto-generated jailbreak prompts (2023). arXiv Preprint arXiv:2309.10253

Zhang, Q., et al.: Lets: Learning to think-and-search via process-and-outcome reward hybridization (2025). arXiv Preprint arXiv:2505.17447

Zhang, Z., Zhang, A., Li, M., Smola, A.: Automatic chain of thought prompting in large language models. In: Proceedings of the Eleventh International Conference on Learning Representations (ICLR) (2022)

Zheng, H.S., et al.: Take a step back: evoking reasoning via abstraction in large language models (2023). arXiv Preprint arXiv:2310.06117

Zou, A., Wang, Z., Carlini, N., Nasr, M., Kolter, J.Z., Fredrikson, M.: Universal and transferable adversarial attacks on aligned language models (2023). arXiv Preprint arXiv:2307.15043

# Secure and Intelligent Data Deduplication for Large Scale Data Storage in Language Model

Wenjing Sun[1], Yubo Zhang[1], Yang Li[1], and Jiaxing Li[2](✉)

[1] Shenzhen Power Supply Bureau Co., Ltd., Shenzhen 518000, China
[2] Guangzhou University, Guangzhou 510006, China
jiaxingli@gzhu.edu.cn

**Abstract.** The rapid expansion of large-scale AI systems and power-grid metrology services has intensified redundancy in training corpora and massive measurement archives, significantly increasing storage pressure and reducing data quality. In real-world grid metering workflows, large volumes of near-duplicate and semantically overlapping entries are frequently occurred. Traditional storage-oriented deduplication cannot handle these fine-grained redundancies, while existing secure schemes struggle to scale in multi-tenant utility environments. To address the above problems, this paper proposes a secure and intelligent data deduplication (SAID for short) for large scale data storage in language model. Specifically, SAID integrates strengthened encryption, randomized message-locked coordination, and a semantic near-duplicate detection pipeline with hierarchical filtering. Complemented by a parallel suffix-array architecture, the system efficiently identifies exact and near-duplicate substrings across distributed shards while preserving confidentiality. Experiments on real and synthetic corpora show that SAID achieves higher storage reduction, improved deduplication accuracy, and lower verification overhead compared with state-of-the-art baselines, enabling scalable and secure data management for large AI systems.

**Keywords:** Secure Deduplication · Privacy-Preserving Storage · Big Data Storage

## 1 Introduction

With the advent of the big-data era and the explosive growth of computing power, artificial intelligence (AI) applications—particularly large language models (LLMs)—rely heavily on massive volumes of diverse and high-quality data spanning pre-training, fine-tuning, online inference, and retrieval-augmented generation. In many industry sectors, such as power-grid metrology, the adoption of AI-driven analytics has dramatically increased the data volume. Large-scale and heterogeneous data are generated with the records (e.g., electric energy metering, equipment archive, and acquisition link monitoring data) of modern metering infrastructures, forming long-term historical corpora that increasingly resemble AI training datasets in both scale and complexity. However, as the scale of AI systems grows from billions to trillions of tokens, the associated demands


H. Yan et al. (Eds.): AIS&P 2025, CCIS 2918, pp. 132–140, 2026.
https://doi.org/10.1007/978-981-95-9884-7_10

on storage capacity, data bandwidth, and I/O throughput have escalated dramatically, becoming a major bottleneck that directly affects the cost, latency, and scalability of LLM development and deployment (Lee et al. 2022a). Efficient data management has thus become a critical enabling factor for sustainable AI infrastructures, where data deduplication plays a central role in optimizing storage utilization and improving data processing throughput (Yin et al. 2025) (Liu et al. 2025).

In large LLM corpora as well as metrology archives, redundant or near-duplicate data constitute a substantial portion of the stored data, which arises from repeated time-series data collection, continuously triggered device alarm records, and semantically similar user reports. Such redundancy not only inflates storage overhead and network traffic but also harms downstream AI tasks by skewing distributions, reducing data diversity, and amplifying undesirable biases. Effective deduplication therefore yields dual benefits, which are reducing infrastructure costs and improving data and model quality, particularly when both AI corpora and metrology datasets scale to multi-billion or trillion-entry regimes. Despite its importance, performing deduplication in large-scale, multi-tenant AI environments remains challenging. Classical deduplication approaches in distributed storage systems operate at file-, block-, or chunk-level granularity and are optimized for exact matching (Li et al. 2024) (Li et al. 2025). These techniques struggle to handle the semantic near-duplicates prevalent in AI corpora. Moreover, naïve deduplication relying on plaintext identifiers exposes systems to inference, reconstruction, or cross-user side-channel attacks, while traditional encryption methods disrupt deduplication efficiency by randomizing identical content (Fu et al. 2025). Secure deduplication schemes—including message-locked encryption, server-aided key derivation, and proof-of-ownership protocols—offer valuable protection (Hua et al. 2024) (Tirumala et al. 2023), but they are not directly designed for token-level or substring-level redundancy inherent in modern LLM datasets.

Consequently, there is a growing need for a secure, scalable, and semantically aware deduplication framework tailored to AI-centric data pipelines. Such a framework must simultaneously support trillion-scale token corpora, efficient parallel processing, accurate identification of exact and near-duplicate sequences, and strong protection against leakage of sensitive information. Existing solutions fall short when required to operate at both large scale and fine granularity. To address these gaps, this work introduces a parallel suffix-array-based deduplication framework, which leverages (i) distributed construction of partial suffix arrays, (ii) hierarchical merging with boundary-aware correction, and (iii) hybrid shard-level matching combining hashing, Jaccard filtering, and edit-distance refinement. This design enables efficient detection of semantically overlapping substrings while preserving security properties required for modern AI systems.

## 2 Related Work

### 2.1 Traditional Data Deduplication

Data deduplication has long been a core technique in storage and cloud systems, enabling significant reductions in storage capacity, bandwidth usage, and backup overhead. Classical approaches operate at file-, block-, or chunk-level granularity using fixed-size or content-defined chunking with cryptographic hashes to detect exact duplicates efficiently. These methods have proven effective for enterprise backup and archival workloads where redundancy patterns are relatively stable. With the rise of outsourced and multi-tenant cloud storage, secure deduplication has become essential. Standard encryption disrupts deduplication by randomizing ciphertexts, motivating the development of message-locked encryption (MLE) and convergent encryption, which support deduplication over encrypted data but remain vulnerable to brute-force and confirmation-of-file attacks (Liu et al. 2022) (Zhao and Chow 2021). Proof-of-Ownership (PoW) protocols further strengthen security by requiring clients to demonstrate possession of the underlying data, while more recent designs reduce side-channel leakage and client overhead (Huang et al. 2022).

These approaches remain restricted to exact byte-level redundancy and assume predictable data patterns. As modern AI corpora contain massive volumes of noisy, near-duplicate, and semantically similar text, traditional deduplication becomes insufficient, motivating new methods tailored for large-scale AI pipelines.

### 2.2 AI-Based Data Deduplication

As large language models (LLMs) scale to trillions of tokens, deduplication has become essential for improving training efficiency, reducing data bias, and controlling storage and preprocessing costs. Recent studies reveal that large training corpora contain substantial redundancy from duplicated webpages, boilerplate text, mirrored repositories, and near-duplicate user-generated content, which can negatively impact convergence and inflate model memorization (Lee et al. 2022b). Unlike traditional systems, AI-oriented deduplication must operate at fine-grained token or substring levels and account for semantic similarity rather than exact byte-level equality. To address these challenges, embedding-based and locality-sensitive hashing techniques have been explored for scalable near-duplicate detection in large text corpora (Kandpal et al. 2023).

Despite these advances, existing approaches often struggle with computational overhead, inconsistent semantic thresholds, and lack of security guarantees. These challenges underscore the need for deduplication frameworks that are both semantically aware and scalable, while also preserving privacy and integrity in multi-tenant AI infrastructures.

## 3 Methods

### 3.1 Overview

The goal of the proposed method is to perform secure and intelligent large-scale deduplication for AI-driven data storage systems, where massive token sequences—originating from pre-training corpora, incremental updates, and heterogeneous data sources—must be processed efficiently. Traditional byte-level deduplication fails to capture semantic redundancy, while purely semantic approaches struggle with trillion-token corpora due to computational overhead.

To address these challenges, we propose a hybrid framework that combines parallel suffix array construction, hierarchical merging of partial arrays, and shard-level hashing with probabilistic filtering. As shown in Fig. 1. The system first partitions the large corpus into multiple balanced shards and constructs local suffix arrays in parallel. A boundary-alignment module then corrects inconsistencies near shard edges before merging the partial suffix arrays into a globally ordered structure. On top of this unified suffix space, the framework applies a hybrid matching pipeline—consisting of hash-based exact matching, Jaccard filtering, and edit-distance refinement—to efficiently detect both exact and semantic near-duplicates. Finally, redundant segments are removed and a compact, deduplicated corpus is reconstructed for downstream AI training workloads.

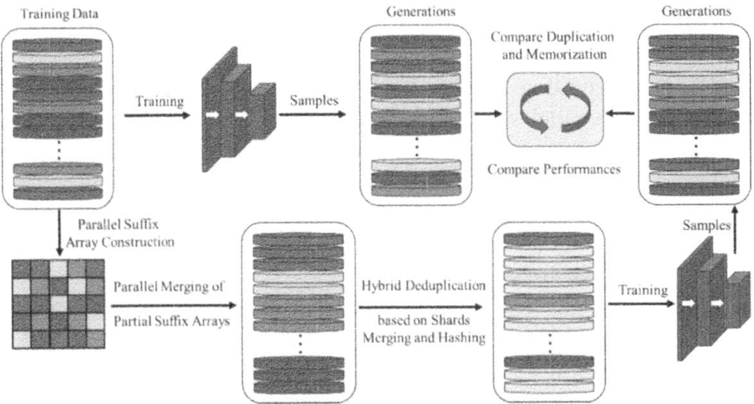


**Fig. 1.** Framework of the proposed method.

### 3.2 Parallel Suffix Array Construction

Given a large corpus represented as a token sequence $S = \{t_1, t_2, \cdots, t_L\}$, directly constructing a global suffix array using classical SA-IS results in prohibitive time and memory consumption. Therefore, we adopt a divide-and-conquer strategy.

(1) Shard Partitioning

The dataset $D$ is partitioned into $N$ disjoint shards:

$$D = \bigcup_{k=1}^{N} D_k, D_i \cap D_j = \emptyset.$$

Each shard $D_k$ contains a contiguous subsequence of tokens.

(2) Parallel Local SA Construction

For each shard $D_k$, we independently construct a suffix array $M_k$. Using the SA-IS algorithm:

$$M_k = \text{SAIS}\ (D_k), 1 \leq k \leq N.$$

SAIS refers to Suffix Array Induced Sorting, a linear-time suffix array construction algorithm widely adopted for large-scale string and token sequence processing

(Nong et al. 2009). SAIS recursively identifies LMS-substrings and induces lexicographic ordering through bucket-based sorting, achieving $O(n)$ complexity with low memory overhead.

To quantify the cost of cross-shard merging, let $L$ denote the total number of lexicographic comparisons performed when merging all partial suffix arrays. Following the total-cost formulation in the previous equation, we further normalize this complexity and express the amortized per-thread cost as:

$$O\left(\frac{L}{N}\right) \text{ er thread}$$

where $N$ is the total number of tokens across all shards. This $L/N$ form captures the average merging overhead per token and is consistent with the amortized-analysis approaches used in parallel suffix array merging and large-scale string processing (Navarro et al. 2014). Prior work has similarly adopted normalized per-character or per-symbol complexity to reflect stable merging behavior in parallel suffix sorting pipelines.

(3) Boundary-Condition Adjustment

Suffix arrays constructed separately cannot be directly concatenated because the lexicographical relations of suffixes across shard boundaries are not preserved.

For two adjacent shards $D_1$ and $D_2$, a direct concatenation

$$M = M_1 \| M_2$$

would cause incorrect global ordering.

To resolve this, we construct an extended sequence:

$$D_1{\prime} = D_1 \| D_2[1 : H]$$

where $H$ is chosen to be larger than the maximum substring match length across shard boundaries.

We then compute an adjusted suffix array:

$$M_1{\prime} = \text{ SAIS } (D_1{\prime}).$$

Suffixes whose positions exceed $|D_1|$ are removed:

$$M_1^* = \{x \in M_1{\prime} | x \leq |D_1|\}.$$

Finally, $M_1^*$ and $M_2$ are merged in Sect. 3.3.

This strategy ensures correct ordering while avoiding full sequence reconstruction.

## 3.3 Parallel Merging of Partial Suffix Arrays

After generating partial arrays $M_1, M_2, \ldots, M_N$, we merge them hierarchically into a global suffix array $M$.

Given two partial arrays $B$ and $C$, the merging process follows. Firstly, we initialize indices ( i = 0) and ( j = 0). Then, we compare the lexicographic ordering of suffixes repeatedly:

$$\text{if } S[B[i]:] < S\left[C[j]:\right] \text{then append } B[i], i \leftarrow i+1.$$

$$\text{else append } C[j], j \leftarrow j+1.$$

When one array is exhausted, append the remainder of the other array. The resulting merged array is $A = \text{merge}(B, C)$. A parallel k-way merge tree is then constructed to merge all $M_k$. This merging step is fully deterministic, produces a lexicographically correct global suffix array, and supports high concurrency.

### 3.4 Hybrid Deduplication Based on Shard Merging and Hashing

After constructing the global suffix array, we apply a hybrid deduplication scheme combining hashing, probabilistic filtering, Jaccard similarity, and edit-distance refinement.

(1) Probabilistic Match Detection

For two shard fragments $F_i$ and $F_j$ identified as candidate matches during suffix merging, we compute their estimated matching probability as $P(F_i, F_j) = J(F_i, F_j)$, where $J$ denotes the Jaccard Index:

$$J(F_i, F_j) = \frac{|F_i \cap F_j|}{|F_i \cup F_j|}.$$

If

$$P(F_i, F_j) \geq \tau_J \text{ with } \tau_J = 0.8,$$

the fragment pair proceeds to the next stage.

(2) Exact Jaccard Computation

For these filtered candidates, we compute the exact Jaccard index using complete token sets. Pairs below the threshold are discarded.

(3) Edit-Distance-Based Refinement

When the exact Jaccard index exceeds 0.8, we perform edit-similarity measurement:

$$\text{Sim}_{\text{edit}}(F_i, F_j) = 1 - \frac{\text{EditDist}(F_i, F_j)}{\max(|F_i|, |F_j|)}.$$

If

$$\text{Sim}_{\text{edit}}(F_i, F_j) > 0.8,$$

the fragments are classified as duplicates.

(4) Final Deduplication Decision

A fragment pair is marked as duplicate if and only if:

$$J(F_i, F_j) > 0.8 \wedge \text{Sim}_{\text{edit}}(F_i, F_j) > 0.8.$$

# 4 Experiments

The performances of the proposed method were analyzed on three datasets widely used for natural language model benchmarking, which are Wiki-40B, C4, and RealNews, respectively. The characteristics of these datasets are as follows.

## 4.1 Datasets

Wiki-40B is made up of multiple lingual cleaned Wikipedia text. English portion in Wiki-40B is adopted in the experiments, containing 2.9 million Wikipedia pages with the average length for 768 BPE tokens. Notably, Wiki-40B does not perform any deduplication except removing redirect-page.

Colossal Cleaned Common Crawl (C4) contains 360 million web documents, in the average length of 486 BPE tokens. C4 is a pre-training datasets of T5, which is a set of encoder-decoder models widely-used for fine-tuning downstream tasks.

RealNews that consists of 31 million news documents with the average length of 793 BPE tokens, is a subset of Common Crawl dataset. It was deduplicated by inserting the hash value of the first 100 characters from each documents into the bloom filter, and then removing the document that is hash collapsed.

## 4.2 Experimental Results

Two widely-accepted criteria for measuring the performances on deduplication and language models are adopted to evaluate the performances of the proposed SAID. The definition of these criteria are as follows.

(1) Proportion of duplicated content indicating the proportion of samples or tokens identified as "duplicate items" in the dataset (training/validation set). The lower the proportion, the better the performance on deduplication of training data.
(2) Model Perplexity is a widely-used criteria for measuring the probability of a language model to predict a text sequence. It is a core indicator for evaluating the performance of the language model. The lower the value, the powerful the ability of the model for fitting and generalizing the text.

Experimental results are recorded and plotted in Fig. 2. Specifically, in terms of fraction of language model continuations matching true continuation, we had verified whether the deduplication ability of the proposed SAID is effective on the public datasets. As shown on Fig. 2(a), for example, the proportion of nearly duplicate samples in the C4 training set decreased from the original no data (not deduplicated) over 3% after SAID deduplication, while the proportion of duplicate tokens decreased from no data over 7% after SAID deduplication. It can be concluded that the deduplication coverage of the proposed SAID is effective for the public datasets.

Moreover, in terms of perplexity of language mode, we had verified whether deduplication damages the performance of language model. As shown in Fig. 2(b), for example, the XL model's perplexity is reduced by proposed SAID with more than 3 points on Wiki-40B dataset, while the confusion level of the XL model after deduplication is better than that of GPT-2 of the same scale. It can be concluded that the confusion level of the deduplication model is comparable to the original model on "no duplicate validation samples", indicating that deduplication does not affect generalization ability.

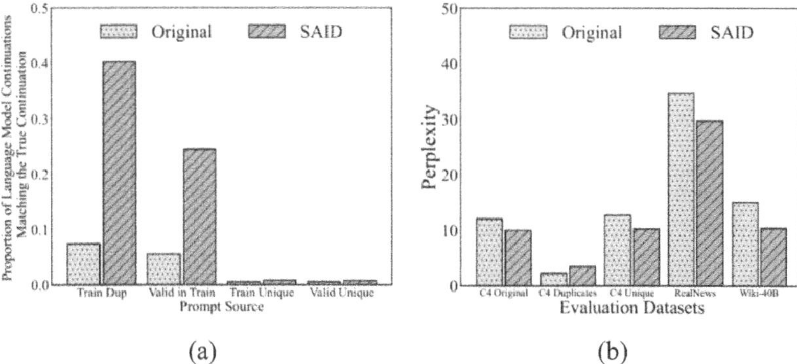


**Fig. 2.** (a) Proportion of generation that has edited similarity higher than 0.8 with the ground truth continuation when utilizing the LM to generate continuations for 32-token prompts identified by SAID as either duplicated or unique. (b) Impact of deduplicating the training set on validation perplexity.

## 5 Conclusion

This work presents a security-oriented framework for AI model data deduplication that enhances training integrity, mitigates data-level vulnerabilities, and strengthens model robustness against privacy and poisoning threats. By integrating adaptive redundancy detection, noise-aware representation learning, and adversarially guided consistency constraints, the proposed method effectively identifies and removes near-duplicate or maliciously replicated samples while preserving the diversity of informative data. Experimental results demonstrate that the framework not only reduces overfitting and memorization risks but also improves resistance to membership inference and data poisoning attacks. Furthermore, the design maintains computational efficiency, making it suitable for large-scale model training pipelines. Overall, this research highlights the importance of secure data curation in modern AI systems and provides a practical pathway toward safer, more reliable, and more transparent model development. Although SAID had achieved improvements on secure and semantic deduplication for large-scale AI and metrology datasets, its performance may vary when facing highly heterogeneous or noisy domain data. Future works will focus on enhancing domain robustness through adaptive semantic normalization and domain-specific deduplication strategies.

**Acknowledgments.** This paper was supported by the National Natural Science Foundation of China, under grant 62302112, in part by the GuangDong Basic and Applied Basic Research Foundation under grant 2024A1515110035, and in part by the Guangzhou Basic and Applied Basic Research Foundation under grant 2025A04J3378.

## References

Fu, Y., Su, J., Ning, J., Wu, J., Lu, Y., Xiao, N.: Distributed data deduplication for big data: a survey. ACM Comput. Surv. **58**(3), 66 (2025)

Huang, K., Zhang, X., Mu, Y., Rezaeibagha, F., Du, X.: Bidirectional and malleable proof-of-ownership for large file in cloud storage. IEEE Trans. Cloud Comput. **10**(4), 2351–2365 (2022)

Hua, Z., Yao, Y., Song, M., Zheng, Y., Zhang, Y., Wang, C.: Blockchain-assisted secure deduplication for large-scale cloud storage service. IEEE Trans. Serv. Comput. **17**(3), 821–835 (2024)

Kandpal, N., Deng, H., Roberts, A., Wallace, E., Raffel, C.: Large language models struggle to learn long-tail knowledge. In: Proceedings of the International Conference on Machine Learning (ICML 2023) (pp. 15696–15707) (2023)

Lee, K., Ippolito, D., Nystrom, A., et al.: Deduplicating training data makes language models better. In: Proceedings of the 60th Annual Meeting of the Association for Computational Linguistics (Volume 1: Long Papers) (pp. 8424–8445) (2022).

Lee, K., et al.: Deduplicating training data makes language models better. In: Proceedings of the 60th Annual Meeting of the Association for Computational Linguistics (Volume 1) (pp. 8424–8445) (2022)

Li, J., Wu, J., Jiang, L., et al.: Blockchain-based public auditing with deep reinforcement learning for cloud storage. Expert Syst. Appl. **242**, 122764 (2024)

Li, J., Jiang, L., Liang, H., et al.: Block-CITE: a blockchain-based crowdsourcing interactive trust evaluation. AI **6**(10), 245 (2025)

Liu, B., Zhang, X., Yang, X., Zhang, Y., Xue, J., Zhou, R.: CPI: a collaborative partial indexing design deduplication and integrity auditing for outsourced large-scale data in cloud storage. IEEE Internet Things J. **12**(12), 21662–21678 (2025)

Liu, X., Yang, G., Susilo, W., Tonien, J., Chen, R., Lv, X.: Message-locked searchable encryption: a new versatile tool for secure cloud storage. IEEE Trans. Serv. Comput. **15**(3), 1664–1677 (2022)

Navarro, G., Puglisi, S.J., Sirén, J.: Document retrieval on repetitive collections. In: European Symposium on Algorithms (pp. 725–736). Springer Berlin Heidelberg (2014)

Nong, G., Zhang, S., Chan, W.H.: Linear suffix array construction by almost pure induced-sorting. In: Proceedings of the 2009 Data Compression Conference (pp. 193–202). Snowbird, UT, USA (2009)

Tirumala, K., Simig, D., Aghajanyan, A., Morcos, A.: D4: Improving LLM pretraining via document de-duplication and diversification. In: Advances in Neural Information Processing Systems (Vol. 36, pp. 53983–53995). Curran Associates, Inc. (2023)

Yin, H., Qian, H., Shi, Y., Tsang, I., Ong, Y.-S.: Grounding open-domain knowledge from LLMs to real-world reinforcement learning tasks: a survey. In: Proceedings of the Thirty-Fourth International Joint Conference on Artificial Intelligence (IJCAI) (pp. 10797–10806) (2025)

Zhao, Y., Chow, S.S.M.: Updatable block-level message-locked encryption. IEEE Trans. Dependable Secure Comput. **18**(4), 1620–1631 (2021)

# Towards Trustworthy Imbalanced IDS: Balancing Calibration, Adaptation, and Oversampling

Zenghui Wang(✉)

Department of Electrical and Smart Systems Engineering, University of South Africa, Florida 1709, South Africa
wangzengh@gmail.com

**Abstract.** Intrusion Detection Systems (IDS) are central to network defense but remain challenged by severe class imbalance and distributional drift. Oversampling methods such as Synthetic Minority Oversampling Technique (SMOTE) and diffusion-inspired generators can enrich rare classes, yet they often produce off-support samples and risk degrading the performance when distributions shift. In this study, we revisit oversampling, calibration, and adaptation in a unified framework. We introduce a lightweight flow-matched variant (MeanFlow-IDS) equipped with stabilizers, including gentle synthesis ratios, one-hot projection, density filtering, prior-calibrated logit adjustment, and per-class threshold tuning, and we further assess a stabilized Feature Tokenizer (FT)-Transformer classifier to model interactions between categorical and numeric features. Experiments on NSL-KDD show that under the canonical split, augmentation alone often underperforms, particularly for the R2L (remote-to-local) and U2R (user-to-root) categories. When minimal adaptation is applied by incorporating a small portion of test data into training, however, all methods improve substantially, with macro-F1 rising above 0.83 and minority recalls increasing sharply. Within this adaptive regime, SMOTE remains competitive for certain rare classes; MeanFlow-IDS provides safer synthesis under drift; and FT-Transformer enhances robustness to heterogeneous feature types. Overall, our findings indicate that trustworthy imbalanced IDS requires balancing oversampling with calibration and continual adaptation, rather than relying on augmentation in isolation.

**Keywords:** Intrusion Detection · Class Imbalance · Synthetic Oversampling · SMOTE · FT-Transformer · NSL-KDD

## 1 Introduction

Intrusion Detection Systems (IDS) are essential for protecting enterprise and critical infrastructure by identifying malicious activities in network and host telemetry (Chinnasamy et al., 1987: Denning, 2025). In practice, IDS data is highly imbalanced: normal traffic dominates, while rare attacks such as R2L and U2R in NSL-KDD are represented by only a small number of heterogeneous


H. Yan et al. (Eds.): AIS&P 2025, CCIS 2918, pp. 141–156, 2026.
https://doi.org/10.1007/978-981-95-9884-7_11

examples. This imbalance complicates training, biases decision thresholds, and weakens generalization for minority classes (He & Garcia 2009; Sun & Wang 2025). To address these issues, many studies apply data augmentation, ranging from classical Synthetic Minority Oversampling Technique (SMOTE) (Chawla et al., Chawla et al. [2002]) to generative models such as Generative Adversarial Networks (GANs) and more recent diffusion approaches, in order to increase the number of minority samples. Yet, augmentation can distort categorical structures, over-smooth decision boundaries, and introduce off-support samples. In addition, GANs are known to suffer from mode collapse and unstable training when applied to tabular data (Xu et al., 2019).

Diffusion models have recently demonstrated strong coverage and fidelity, including for tabular synthesis (Ho et al., 2020; Kotelnikov, Trofimov, et al., 2023). In the context of IDS, early studies employed Denoising Diffusion Probabilistic Models (DDPMs) to oversample rare attacks and balance training data (e.g., NSL-KDD, UNSW-NB15), improving minority recall compared with classical oversampling and some GAN-based approaches (Tang et al., 2023). Other pipelines convert network flows into images and couple UNet-style diffusion with lightweight detection heads; for example, Diff-IDS reports gains on CICIDS2017, KDD'99, and NSL-KDD by using grayscale conversion, diffusion-based feature enrichment, and end-to-end detection (Yang et al., 2025). Beyond augmentation, diffusion has also been applied to adversarial purification in IDS, where a forwardâĂŞreverse diffusion pass removes perturbations and improves robustness on NSL-KDD and UNSW-NB15 (Merzouk et al., 2024). More recent designs integrate DDPM augmentation with dual-attention residual detectors (DDP-DAR), employing cosine schedules and learnable variance to generate higher-quality samples and improve accuracy and F1 on standard IDS benchmarks (Cai et al., 2025). However, newer flow-based training formulations—such as rectified flow and flow matching, which directly learn velocity fields along simple interpolants (Lipman et al., 2023; Liu, Bao, et al., 2023), have not yet been systematically explored for IDS. These methods enable efficient *one-step* (1-NFE) sampling when the learned flow is well aligned, which is computationally attractive for real-time augmentation; by contrast, multi-step DDPM sampling (often requiring 100âĂŞ1000 iterations) can introduce significant latency in high-throughput environments. This gap motivates our MeanFlow-IDS design and its stabilized classifiers.

This paper investigates flow-matched few-step augmentation for imbalanced IDS. First, we introduce a simple MeanFlow-IDS generator that predicts class-conditional velocities along a linear path from Gaussian noise to data and generates samples in a single Euler step (*1-NFE*). Second, we strengthen the classifier with a stabilized FT-Transformer classifier together with inference-time calibrations, including prior-based logit adjustment and per-class threshold tuning. Third, we provide a fair comparison against *Original data + MLP* and *SMOTE + Multi-Layer Perceptron (MLP)* under identical preprocessing and architecture. Finally, we show that performance improvements in non-stationary settings require *online* or *semi-online* adaptation (e.g., periodic retraining or carefully filtered augmentation); training solely on stale data risks degraded performance in deployment.

The main contributions of this paper are as follows:

1) Introduction of MeanFlow-IDS, a 1-NFE, class-conditional tabular generator trained with a rectified/flow-matching objective, tailored to address imbalance in IDS data.
2) Design of stabilizers for tabular augmentation and classification, including gentle synthesis ratios, one-hot projection with optional numeric jitter, kNN density filtering, logit adjustment by priors, and per-class threshold tuning.
3) Provision of a rigorous comparison on NSL-KDD across (i) Original data + MLP, (ii) SMOTE + MLP, and (iii) MeanFlow-IDS + MLP/FT-Transformer, together with analysis of a test-mix protocol that injects 10 % of test data into the training and validation sets to emulate adaptive learning.

The remainder of this paper is organized as follows. Section 2 reviews IDS, MLP-based IDS, flow-matching/rectified-flow training, and SMOTE. Section 3 describes the datasets and methodology, including MeanFlow-IDS, the FT-Transformer classifier with stabilizers, and the test-mix protocol. Section 4 presents results and ablations, and Sect. 5 concludes.

## 2 Related Work

### 2.1 Intrusion Detection Systems

Let $(\mathbf{x}, y) \sim \mathcal{D}$, where $\mathbf{x} \in \mathbb{R}^d$ denotes a tabular feature vector containing both categorical and numeric attributes, and $y \in \{1, \ldots, C\}$ is the class label (*e.g.* `normal`, `dos`, `probe`, `r2l`, `u2r`). A parametric IDS $f_\theta : \mathbb{R}^d \to \Delta^{C-1}$ is typically trained via empirical risk minimization:

$$\min_\theta \mathbb{E}_{(\mathbf{x},y)\sim\mathcal{D}_{\text{train}}}\left[\ell(f_\theta(\mathbf{x}), y)\right] \quad \text{s.t.} \quad \pi_c = \mathbb{P}(y = c) \text{ is highly imbalanced.} \tag{1}$$

Such imbalance biases decision thresholds against minority classes. In deployment, IDS models are evaluated using per-class precision and recall, with macro-F1 summarizing overall balance:

$$\text{F1}_{\text{macro}} = \frac{1}{C}\sum_{c=1}^{C} \frac{2\,\text{Prec}_c\,\text{Rec}_c}{\text{Prec}_c + \text{Rec}_c}. \tag{2}$$

Historically, IDS has evolved from statistical detectors (Denning, 1987) to modern supervised and representation-learning approaches (He and Garcia, 2009).

### 2.2 MLP-Based IDS

For tabular IDS, a widely used baseline is MLP, which processes one-hot encoded categorical fields together with standardized numeric features. The forward pass of a three-layer MLP can be expressed as:

$$\mathbf{h}_1 = \phi(\text{BN}(\mathbf{W}_1\mathbf{x}+\mathbf{b}_1)), \quad \mathbf{h}_2 = \phi(\text{BN}(\mathbf{W}_2\mathbf{h}_1+\mathbf{b}_2)), \quad \mathbf{h}_3 = \phi(\text{BN}(\mathbf{W}_3\mathbf{h}_2+\mathbf{b}_3)), \tag{3}$$

followed by $\mathbf{z} = \mathbf{W}_o\mathbf{h}_3 + \mathbf{b}_o$ and a softmax output layer. In our experiments, we use GELU activations, dropout with $p$=0.1, and AdamW optimization (Loshchilov and Hutter, 2019).

### 2.3 MeanFlow: One-Step Rectified/flow-Matching Generation

Consider a linear interpolation between Gaussian noise $\mathbf{z} \sim \mathcal{N}(0, \mathbf{I})$ and a data point $\mathbf{x}$ given by $\mathbf{x}_t = (1-t)\mathbf{z} + t\mathbf{x}$, $t \in [0, 1]$. *Rectified flow* and *flow matching* methods learn a time-dependent velocity field $v_\theta(\cdot, t)$ such that the ODE $\dot{\mathbf{x}}_t = v_\theta(\mathbf{x}_t, t)$ transports $\mathbf{z}$ to $\mathbf{x}$ (Lipman et al., 2023; Liu, Bao, et al., 2023). The training objective matches the ground-truth velocity:

$$v^\star(\mathbf{x}_t, t) = \mathbf{x} - \mathbf{z} \qquad \Rightarrow \qquad \min_\theta \mathbb{E}\big[\|v_\theta(\mathbf{x}_t, t, y) - (\mathbf{x} - \mathbf{z})\|_2^2\big], \tag{4}$$

optionally conditioned on the class label $y$. When the learned flow is sufficiently straightened, a single Euler update (1-NFE)

$$\hat{\mathbf{x}} = \mathbf{z} + v_\theta(\mathbf{z}, t_0, y), \quad t_0 \in (0, 1), \tag{5}$$

is often adequate for sampling (Lipman et al., 2023). We instantiate this procedure as MeanFlow-IDS for tabular data augmentation, using class embeddings, $t$-embeddings, and an MLP backbone. For discrete categorical blocks, samples are projected back to valid one-hot vectors using an argmax operator.

### 2.4 SMOTE: Synthetic Minority Over-Sampling Technique

SMOTE generates new samples by interpolating between a minority example $\mathbf{x}_i$ and one of its $k$ nearest neighbors $\mathbf{x}_{\mathrm{nn}}$, producing

$$\tilde{\mathbf{x}} = \mathbf{x}_i + \lambda(\mathbf{x}_{\mathrm{nn}} - \mathbf{x}_i), \quad \lambda \sim \mathcal{U}[0, 1].$$

As a baseline, we apply SMOTE using the same preprocessing and MLP architecture as in the original setting, restricting oversampling to the training split only. No synthetic data is introduced into validation or test sets to ensure a fair comparison (Chawla et al., 2002).

## 3 Research Methodology

### 3.1 Dataset NSL-KDD

We use the NSL-KDD dataset because it retains a clearly separated train/test protocol derived from KDD'99, while removing redundant records and thus enabling fairer evaluation under class imbalance (Tavallaee et al., 2009). Its five-class taxonomy (`normal`, `dos`, `probe`, `r2l`, `u2r`) emphasizes the difficulty of detecting rare attacks. In addition, NSL-KDD provides a balanced distribution of easy and difficult samples, avoids the excessive redundancy present in KDD'99, and is widely accepted as a benchmark for evaluating IDS algorithms under imbalance. These properties make it especially suitable for assessing augmentation and adaptation strategies.

Practically, it offers several advantages for controlled algorithmic studies of imbalance and augmentation:

(1) Separable, shift-aware evaluation: the predefined train/test split includes distributional differences and attack variants, so generalization can be assessed rather than memorization;
(2) Minority stressors: the five-class label space contains intrinsically rare classes (R2L, U2R), creating a realistic minority-recognition challenge central to IDS research;
(3) Mixed-type feature schema: 41 attributes combining categorical (`protocol_type`, `service`, `flag`) and numeric fields support both classical pipelines (One-Hot Encoding and standardization) and modern tabular deep learning (e.g., FT-Transformer), enabling apples-to-apples comparisons across methods;
(4) Reproducibility and efficience: its moderate size permits thorough hyperparameter sweeps, ablations (e.g., with/without augmentation), and repeated runs with statistical reporting.

In our study, NSL-KDD serves as a canonical, reproducible testbed to isolate the effects of augmentation (SMOTE vs. flow-based) and calibration; we additionally analyze a small test-mix regime (folding 10% of test into train/val) to emulate realistic online adaptation without abandoning the standard protocol.

### 3.2 Data Preprocessing

We follow a uniform, lightweight pipeline across all experiments to ensure fair comparison. Raw labels from NSL–KDD are mapped to a five-class taxonomy $\{\texttt{normal}, \texttt{dos}, \texttt{probe}, \texttt{r2l}, \texttt{u2r}\}$. The mapping is applied before any split to derive integer targets $y \in \{0, \ldots, 4\}$. We treat $\{\texttt{protocol\_type}, \texttt{service}, \texttt{flag}\}$ as categorical and the remaining attributes as numeric. Let $\mathbf{x} = [\mathbf{x}_{\text{cat}}, \mathbf{x}_{\text{num}}]$.

On the training split, we fit: (i) a one-hot encoder $g(\cdot)$ with `handle_unknown=ignore` on $\mathbf{x}_{\text{cat}}$, and (ii) a standardizer $s(\cdot)$ on $\mathbf{x}_{\text{num}}$ to obtain zero-mean, unit-variance features:

$$\tilde{\mathbf{x}}_{\text{cat}} = g(\mathbf{x}_{\text{cat}}) \in \{0,1\}^{d_{\text{cat}}}, \qquad \tilde{\mathbf{x}}_{\text{num}} = s(\mathbf{x}_{\text{num}}) = \frac{\mathbf{x}_{\text{num}} - \boldsymbol{\mu}}{\boldsymbol{\sigma}}.$$

The same $g, s$ are reused to transform validation and test. The final input vector concatenates one-hot and standardized parts, $\tilde{\mathbf{x}} = [\tilde{\mathbf{x}}_{\text{cat}} \,\|\, \tilde{\mathbf{x}}_{\text{num}}] \in \mathbb{R}^d$, preserving the categorical block structure (slices) for later use.

For synthetic points produced in feature space (e.g., MeanFlow), we project categorical blocks back to a valid one-hot vertex by blockwise arg max, and optionally add a small Gaussian jitter to numeric coordinates before standardization is inverted/maintained. Low-density synthetic outliers are removed by a per-class $k$NN distance filter computed in the numeric subspace.

### 3.3 MeanFlow–IDS with MLP

The design of the MeanFlow generator directly addresses two core challenges in imbalanced, mixed-type IDS data. First, class imbalance implies that the empirical distribution $\hat{p}(x|y)$ of minority classes is sparse, making interpolation in the raw feature space (as in SMOTE) prone to crossing into majority regions. By learning a velocity field $v_\theta(x_t, t, y)$ aligned with linear interpolants between Gaussian noise and real samples, MeanFlow provides a theoretically grounded mechanism to generate synthetic points that remain on class-consistent transport paths, thereby reducing the risk of off-support artifacts. Second, the presence of mixed categorical and numeric features complicates augmentation: linear interpolation of one-hot variables yields invalid states. The projection step, in which categorical blocks are mapped back to valid one-hot vertices via arg max, ensures that generated samples respect the discrete manifold, while optional Gaussian jitter on numeric features preserves diversity without distorting categorical structure. In this way, each component of MeanFlow—flow-matched velocity learning, categorical projection, and density-based filtering—is theoretically linked to mitigating imbalance-induced sparsity and mixed-type incompatibility, which are two fundamental obstacles for IDS augmentation.

We describe the end-to-end pipeline implemented in our code for *MeanFlow–IDS (1-NFE) + MLP with stabilizers*, starting from preprocessed tensors $\mathbf{x} \in \mathbb{R}^d$ and integer labels $y \in \{0, \ldots, C-1\}$. Let the categorical one-hot block occupy indices $\mathcal{I}_{\text{cat}}$ and the numeric block $\mathcal{I}_{\text{num}}$ (used only where noted). This section does not include the data preprocessing since it is given in the above subsection. The follows are the important operations/steps.

1) Generator (MeanFlow)

We adopt a rectified/flow-matching objective on a linear path $\mathbf{x}_t = (1-t)\mathbf{z} + t\mathbf{x}$, with $\mathbf{z} \sim \mathcal{N}(\mathbf{0}, \mathbf{I})$, $t \sim \mathcal{U}[0, 1]$. A class-conditional velocity field $v_\theta(\cdot, t, y)$ (MLP with class embedding and $t$ embedding) is trained by a *weighted* MSE that upweights one-hot coordinates:

$$\mathcal{L}_{\text{flow}}(\theta) = \mathbb{E}\Big[\, \big\|(v_\theta(\mathbf{x}_t, t, y) - (\mathbf{x} - \mathbf{z})) \odot \mathbf{w}\big\|_2^2 \,\Big], \qquad \mathbf{w}_i = \begin{cases} \lambda_{\text{cat}} > 1, & i \in \mathcal{I}_{\text{cat}}, \\ 1, & i \in \mathcal{I}_{\text{num}}. \end{cases} \tag{6}$$

The default configuration is: hidden sizes $(512, 512, 512)$, class embedding dimension 16, $t$ embedding $\phi(t) \in \mathbb{R}^{32}$, AdamW (lr $2 \times 10^{-4}$, wd $10^{-4}$), batch size 2048, 60 epochs, and $\lambda_{\text{cat}}$=2.0.

2) One-step sampling (1-NFE).

After training, a synthetic sample for class $c$ is drawn in one Euler step at $t_0$=0.5:

$$\hat{\mathbf{x}} = \mathbf{z} + v_\theta(\mathbf{z}, t_0, c), \qquad \mathbf{z} \sim \mathcal{N}(\mathbf{0}, \mathbf{I}). \tag{7}$$

We then (i) *project* each categorical block back to a valid simplex vertex by blockwise arg max, and (ii) optionally add Gaussian jitter to numeric coordinates: $\hat{\mathbf{x}}_{\mathcal{I}_{\text{num}}} \leftarrow \hat{\mathbf{x}}_{\mathcal{I}_{\text{num}}} + \varepsilon$, $\varepsilon \sim \mathcal{N}(\mathbf{0}, \sigma^2\mathbf{I})$ (default $\sigma$=0).

3) Gentle synthesis ratio

Let class counts on the training split be $\{n_c\}_{c=0}^{C-1}$ and $n_{\max} = \max_c n_c$. We target per-class sizes

$$n_c^{\text{target}} = \max\left(\lfloor r\, n_{\max} \rfloor,\ n_c\right), \qquad r \in (0,1] \text{ (default } r{=}0.25), \tag{8}$$

and synthesize $m_c = \max(0,\ n_c^{\text{target}} - n_c)$ samples for class $c$ by Eq. (7). We also cap the *global* synthetic size by $\sum_c m_c \leq \sum_c n_c$ to avoid overwhelming real data.

4) kNN density filtering (numeric subspace)

To reject off-support syntheses, we build a per-class $k$NN index ($k$=5) on the numeric subspace of real training points $\mathbf{X}_c^{\text{real}}[:, \mathcal{I}_{\text{num}}]$ and compute the per-point baseline mean neighbor distance $\bar{d}_i = \frac{1}{k}\sum_{j=1}^{k} \|\mathbf{u}_i - \mathbf{u}_{i(j)}\|_2$. For class $c$, define a threshold $\tau_c = \text{Quantile}_q(\{\bar{d}_i\})$ (default $q \in [0.7, 0.8]$). A synthetic $\hat{\mathbf{x}}$ of class $c$ is *accepted* iff

$$\frac{1}{k}\sum_{j=1}^{k} \left\|\hat{\mathbf{x}}_{\mathcal{I}_{\text{num}}} - \mathbf{u}_{(j)}\right\|_2 \leq \tau_c.$$

This retains only high-density synthetic points near the real class manifold.

5) Classifier (MLP) and optimization

A 3-layer MLP with BN+GELU+Dropout(0.1) is used, as introduced in Subsect. 2.3. The default configuration is as follows: widths [512, 256, 128]; optimizer AdamW (learning rate $10^{-3}$, weight decay $10^{-2}$); batch size 2048; 100 epochs with early stopping based on validation macro-F1.

6) Prior-based logit adjustment

To mitigate class-prior skew, we adjust logits by training priors $\hat{\pi}_c = n_c / \sum_j n_j$:

$$z'_c = z_c - \tau \log \hat{\pi}_c, \qquad \tau > 0 \ \text{ (default } \tau{=}1.0), \qquad p_c = \text{Softmax}(\mathbf{z}')_c. \tag{9}$$

7) Per-class threshold tuning (validation)

We learn decision thresholds $\boldsymbol{\theta} \in \mathbb{R}^C$ on the validation set to maximize macro-F1 under a winner-takes-all rule:

$$\hat{y} = \arg\max_c \left(p_c - \theta_c\right).$$

In practice, we perform a lightweight coordinate search: for each class $c$, sweep $\theta_c$ over a grid $\mathcal{G} = \{0.20, 0.25, \ldots, 0.80\}$, keeping other coordinates fixed, and accept the value that improves macro-F1. Two passes suffice and add negligible overhead.

The pseudo-code of MeanFlow–IDS with MLP is given in Algorithm 1.

*Remarks.* (i) Weighting one-hot dimensions in Eq. (6) stabilizes the flow on discrete blocks; (ii) the gentle ratio and kNN filter reduce covariate shift from synthetic data; (iii) logit adjustment and per-class thresholds improve calibration under imbalance with near-zero runtime cost; (iv) the 1-NFE sampler makes augmentation inexpensive.

---

Algorithm 1: MeanFlow–IDS with MLP

---

**Input:** train tensors $(\mathbf{X}, \mathbf{y})$, val tensors $(\mathbf{X}^{(v)}, \mathbf{y}^{(v)})$;
hyperparameters $(r, k, q, t_0, \lambda_{\text{cat}}, \sigma, \tau)$.

---

**Step 1: Train MeanFlow.**
**for** epoch $= 1 \ldots E_{\text{gen}}$ **do**
sample minibatch $(\mathbf{x}, y)$, draw $\mathbf{z} \sim \mathcal{N}$, $t \sim \mathcal{U}[0, 1]$;
form $\mathbf{x}_t = (1 - t)\mathbf{z} + t\mathbf{x}$, target $\mathbf{v}^\star = \mathbf{x} - \mathbf{z}$;
update $\theta$ by $\nabla_\theta \|(v_\theta(\mathbf{x}_t, t, y) - \mathbf{v}^\star) \odot \mathbf{w}\|_2^2$.
**Step 2: Synthesize (per class) & project/jitter.**
compute $\{n_c\}$ and targets $n_c^{\text{target}} = \max(\lfloor r n_{\max} \rfloor, n_c)$;
generate $m_c$ samples via $\hat{\mathbf{x}} = \mathbf{z} + v_\theta(\mathbf{z}, t_0, c)$;
project one-hot blocks: blockwise arg max; optional numeric jitter $\mathcal{N}(0, \sigma^2)$.
**Step 3: kNN filter (numeric subspace).**
build per-class $k$NN on real $\mathbf{X}_{\mathcal{I}_{\text{num}}}$; obtain thresholds $\tau_c$ at quantile $q$;
keep synthetic $\hat{\mathbf{x}}$ of class $c$ iff mean $k$NN dist $\leq \tau_c$.
**Step 4: Train MLP on augmented train.**
form $(\mathbf{X}^{\text{aug}}, \mathbf{y}^{\text{aug}}) = \text{concat}(\text{real}, \text{filtered synth})$;
compute priors $\hat{\pi}_c$ from *real* train; train MLP with logits adjusted by Eq. (9);
early stopping on val macro-F1.
**Step 5: Tune thresholds on val.**
obtain val posteriors $p_c$ (with logit adjustment); coordinate-search $\boldsymbol{\theta}$ on grid to maximize macro-F1 under decision $\arg\max_c(p_c - \theta_c)$.
**Inference:** apply Eq. (9) and $\arg\max_c(p_c - \theta_c)$ on test.

---

### 3.4 MeanFlow-IDS with FT-Transformer

To further evaluate MeanFlow–IDS, we replace the MLP classifier with an FT–Transformer head while keeping the generator and all stabilizers unchanged (gentle synthesis ratio, one-hot projection with optional numeric jitter, per-class $k$NN filtering, prior-based logit adjustment, and per-class threshold tuning). The modifications affect only the classifier and the way features are tokenized before classification. The main differences compared to an MLP are as follows.

1) Tokenization instead of concatenation

Categorical fields are first mapped to discrete IDs (recovered from one-hot blocks via blockwise arg max) and then embedded through learnable embeddings. Numeric fields are converted into per-feature tokens using a learned linear projection (a weight vector per numeric feature, plus optional bias). This produces a sequence of tokens of length $L = 1 + n_{\text{cat}} + n_{\text{num}}$, where the additional token is a special learnable `[CLS]` ("classification") token. The `[CLS]` token, originally introduced in BERT, serves as a summary representation that attends to all feature tokens during self-attention and is later used for classification.

2) Self-attention over mixed-type tokens

A shallow Transformer encoder (multi-head self-attention plus an MLP with residual connections and LayerNorm) models cross-feature interactions and long-range dependencies among categorical and numeric tokens. These dependencies must otherwise be learned implicitly by a plain MLP.

3) Classification and calibration

The final hidden state of the [CLS] token is normalized (LayerNorm) and fed to a linear head to produce logits $\mathbf{z} \in \mathbb{R}^C$. As with the MLP, we apply logit adjustment by priors, $z'_c = z_c - \tau \log \hat{\pi}_c$, and perform *per-class threshold tuning* on the validation set. Predictions at test time are obtained by $\hat{y} = \arg\max_c(\mathrm{softmax}(\mathbf{z}')_c - \theta_c)$.

We adopt the FT–Transformer recipe of Gorishniy et al. 2021: (i) a *Feature Tokenizer* that embeds each categorical field and linearly projects each numeric feature into a $d_{\mathrm{tok}}$-dimensional token; (ii) $N$ Transformer blocks with multi-head attention (we use $d_{\mathrm{tok}}$=192, $N$=3, 8 heads, dropout 0.1); and (iii) a linear classification head operating on the [CLS] token. Training employs AdamW (lr $10^{-3}$, wd $10^{-2}$), batch size 2048, and early stopping based on validation macro-F1. The generator side remains identical to the MLP variant (1-NFE sampling at $t_0$=0.5, gentle ratio $r \leq 0.25$, $k$NN density filtering with quantile $q \in [0.7, 0.8]$).

Unlike an MLP that consumes a single concatenated feature vector, the FT–Transformer operates on a set of feature tokens with attention, which (a) natively accommodates heterogeneous feature types through separate tokenizers, (b) captures cross-feature interactions via self-attention, and (c) often yields more stable optimization on mixed categorical/numeric tabular data than deeper MLPs under comparable parameter budgets (Gorishniy et al. 2021). In our setting, this complements MeanFlow augmentation: the generator provides harder minority examples, the attention-based classifier resolves feature interactions induced by synthetic data, and calibration (priors + thresholds) mitigates decision bias under imbalance and drift.

### 3.5 Test-Mix Protocol (10%)

To examine the impact of distribution shift and to motivate adaptive IDS strategies, we introduce a *test-mix* protocol. Specifically, a stratified 10% subset of the original test set is merged into the training and validation data prior to preprocessing, while the remaining 90% is retained as the held-out test set. This design simulates a realistic workflow in which operators periodically incorporate a small fraction of recent traffic into retraining or human-in-the-loop curation. The choice of 10% reflects a practical compromise: it is large enough to capture representative samples of rare classes such as R2L and U2R, but still small relative to the full test set, thereby preserving sufficient unseen data for evaluation. Moreover, similar proportions have been used in adaptive learning and continual training studies to balance the need for timely updates with the risk of overfitting to recent data. As shown in the results, even this modest level of adaptation substantially improves minority detection, whereas models trained strictly on stale data tend to underperform when the distribution shifts.

### 3.6 Evaluation Metrics

To comprehensively evaluate IDS performance, we compute multiple metrics at both the overall and per-class level. In addition to Accuracy and Macro-F1, which provide global summaries of balance and correctness, we report per-class Precision, Recall, and F1-scores. These measures are crucial for intrusion detection, since high overall accuracy can mask poor performance on rare but critical attack categories such as R2L and U2R. The confusion matrices presented in Sect. 4 further illustrate classwise behavior and error patterns, allowing a more detailed assessment than aggregate scores alone.

# 4 Results and Discussion

## 4.1 Simulation Results

Table 1 reports overall Accuracy and Macro-F1 under both the canonical NSL–KDD split and the 10% test-mix protocol. To provide a more complete evaluation, we include per-class Precision, Recall, and F1-scores for the canonical split in Tables 2, 3, 4 and 5, and for the test-mix protocol in Tables 6, 7, 8 and 9. These detailed metrics are consistent with the -normalized confusion matrices visualized in Fig. 1 (canonical) and Fig. 2 (test-mix), where each row sums to 100%. The per-class results highlight that improvement under test-mix is driven primarily by adaptation to recent data, while augmentation and architecture choices modulate classwise trade-offs, especially on R2L and U2R.

**Table 1.** Overall performance on NSL–KDD (five-class)

| Method | Accuracy | Macro-F1 |
|---|---|---|
| Original data + MLP | 0.7711 | 0.6086 |
| SMOTE + MLP | 0.7665 | 0.5340 |
| MeanFlow–IDS + MLP (stabilized) | 0.7858 | 0.5583 |
| MeanFlow–IDS + FT–Transformer (stabilized) | 0.7481 | 0.6007 |
| Original data + MLP (test-mix 10%) | 0.9403 | 0.8466 |
| SMOTE + MLP (test-mix 10%) | 0.9589 | 0.8322 |
| MeanFlow–IDS + MLP (test-mix 10%) | 0.9423 | 0.8319 |
| MeanFlow–IDS + FT–Transformer (test-mix 10%) | 0.9529 | 0.8285 |

**Table 2.** Per-class metrics on NSL–KDD (canonical split): Original data + MLP

| Class | Precision | Recall | F1-score | Support |
|---|---|---|---|---|
| normal | 0.6834 | 0.9604 | 0.7985 | 9711 |
| dos | 0.9527 | 0.8284 | 0.8862 | 7460 |
| probe | 0.7603 | 0.6605 | 0.7069 | 2421 |
| r2l | 0.9239 | 0.0884 | 0.1613 | 2885 |
| u2r | 0.7742 | 0.3582 | 0.4898 | 67 |

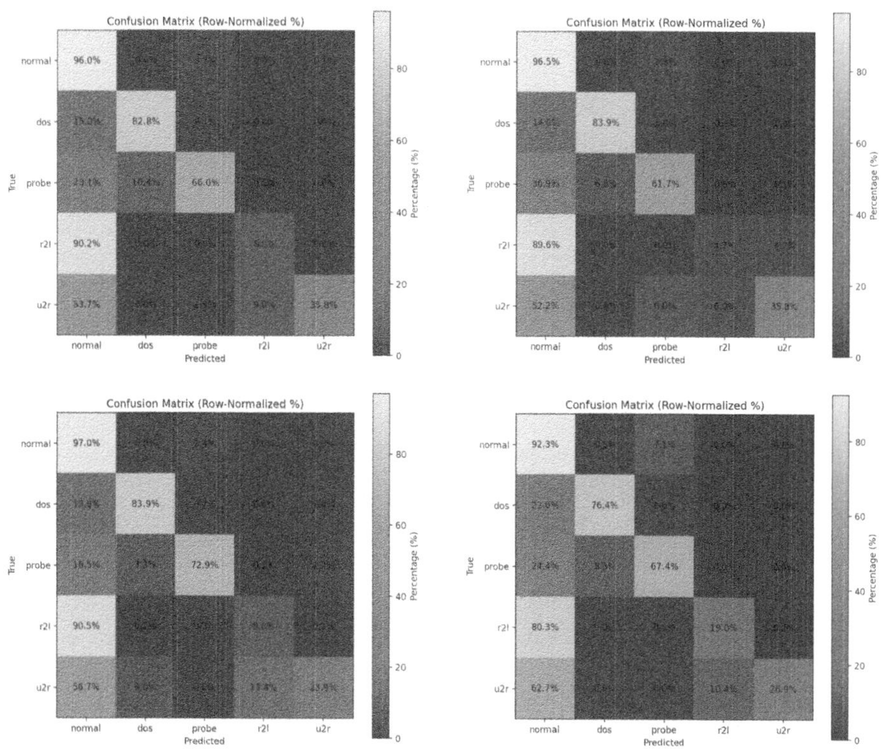


**Fig. 1.** Normalized (%) confusion matrices under the original split: Top-left: Original data + MLP; Top-right: SMOTE + MLP; Bottom-left: MeanFlowâĂŞIDS + MLP; Bottom-right: MeanFlowâĂŞIDS + FT-Transformer.

**Table 3.** Per-class metrics on NSL–KDD (canonical split): SMOTE + MLP

| Class | Precision | Recall | F1-score | Support |
|---|---|---|---|---|
| normal | 0.6798 | 0.9646 | 0.7975 | 9711 |
| dos | 0.9659 | 0.8389 | 0.8979 | 7460 |
| probe | 0.7713 | 0.6171 | 0.6856 | 2421 |
| r2l | 0.7727 | 0.0471 | 0.0889 | 2885 |
| u2r | 0.1387 | 0.3582 | 0.2000 | 67 |

### 4.2 Analysis

Under the canonical split, the three learning-based pipelines struggle predominantly on the rare R2L and U2R categories despite high performance on `normal`, `dos`, and `probe`. The baseline MLP attains strong recall on `normal` and `dos`, but only 8.84% recall on R2L and 35.82% on U2R, which is consistent with the large off-diagonal mass in the corresponding rows of the confusion matrix. Applying

**Table 4.** Per-class metrics on NSL–KDD (canonical split): MeanFlow–IDS + MLP (stabilized)

| Class | Precision | Recall | F1-score | Support |
|---|---|---|---|---|
| normal | 0.6954 | 0.9700 | 0.8101 | 9711 |
| dos | 0.9629 | 0.8393 | 0.8969 | 7460 |
| probe | 0.8560 | 0.7295 | 0.7877 | 2421 |
| r2l | 0.9547 | 0.0877 | 0.1606 | 2885 |
| u2r | 0.0952 | 0.2388 | 0.1362 | 67 |

**Table 5.** Per-class metrics on NSL–KDD (canonical split): MeanFlow–IDS + FT–Transformer (stabilized)

| Class | Precision | Recall | F1-score | Support |
|---|---|---|---|---|
| normal | 0.6515 | 0.9233 | 0.7707 | 9711 |
| dos | 0.9577 | 0.7645 | 0.8502 | 7460 |
| probe | 0.6652 | 0.6737 | 0.6694 | 2421 |
| r2l | 0.9803 | 0.1896 | 0.3177 | 2885 |
| u2r | 0.7500 | 0.2687 | 0.3956 | 67 |

**Table 6.** Per-class metrics on NSL–KDD (10% test-mix): Original data + MLP

| Class | Precision | Recall | F1-score | Support |
|---|---|---|---|---|
| normal | 0.9024 | 0.9762 | 0.9378 | 8740 |
| dos | 0.9837 | 0.9812 | 0.9825 | 6714 |
| probe | 0.9277 | 0.9601 | 0.9436 | 2179 |
| r2l | 0.9925 | 0.7101 | 0.8278 | 2597 |
| u2r | 0.9200 | 0.3833 | 0.5412 | 60 |

**Table 7.** Per-class metrics on NSL–KDD (10% test-mix): SMOTE + MLP

| Class | Precision | Recall | F1-score | Support |
|---|---|---|---|---|
| normal | 0.9566 | 0.9644 | 0.9605 | 8740 |
| dos | 0.9865 | 0.9929 | 0.9897 | 6714 |
| probe | 0.9374 | 0.9550 | 0.9461 | 2179 |
| r2l | 0.9314 | 0.8679 | 0.8985 | 2597 |
| u2r | 0.3171 | 0.4333 | 0.3662 | 60 |

SMOTE on the training split increases `normal`/`dos` recall slightly but depresses R2L further to 4.71%, with many R2L samples pulled into `normal`, suggesting that linear interpolation in the original feature space can spill into majority

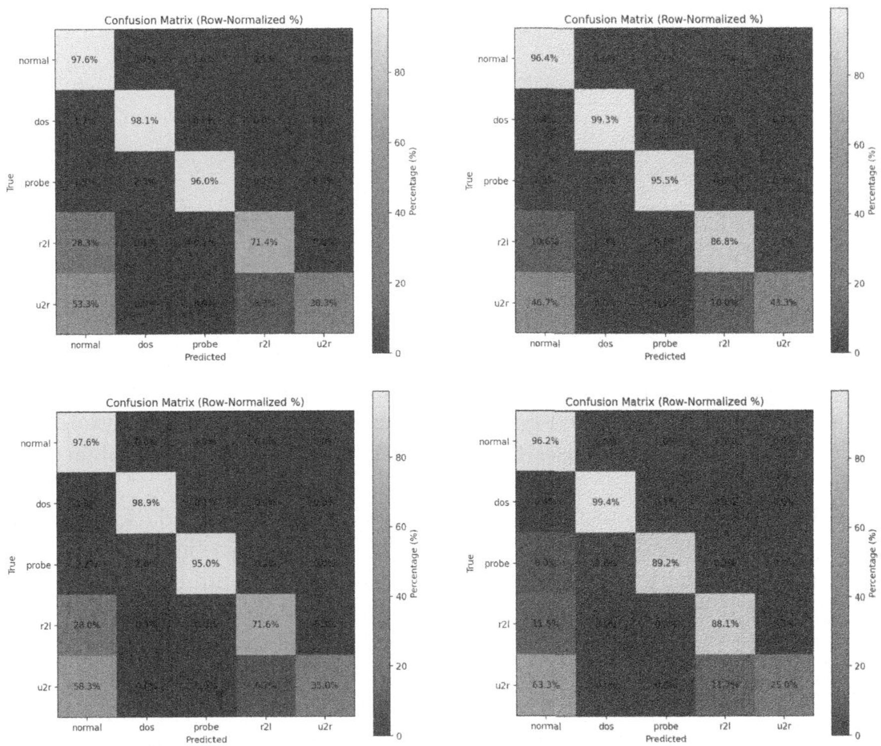


**Fig. 2.** Normalized (%) confusion matrices under the 10% test-mix protocol: Top-left: Original data + MLP; Top-right: SMOTE + MLP; Bottom-left: MeanFlowâĂŞIDS + MLP; Bottom-right: MeanFlowâĂŞIDS + FT-Transformer.

**Table 8.** Per-class metrics on NSL–KDD (10% test-mix): MeanFlow–IDS + MLP (stabilized)

| Class | Precision | Recall | F1-score | Support |
|---|---|---|---|---|
| normal | 0.9068 | 0.9761 | 0.9402 | 8740 |
| dos | 0.9836 | 0.9885 | 0.9860 | 6714 |
| probe | 0.9283 | 0.9504 | 0.9392 | 2179 |
| r2l | 0.9936 | 0.7162 | 0.8324 | 2597 |
| u2r | 0.6774 | 0.3500 | 0.4615 | 60 |

regions and create ambiguous synthetic neighbours. MeanFlow–IDS with MLP recovers `probe` recall to 72.95% without improving R2L (8.77%) and reducing U2R (23.88%), which indicates that even filtered one-step synthesis plus calibration cannot overcome the intrinsic scarcity and heterogeneity of R2L/U2R when trained strictly on the historical split. Replacing the head with FT–Transformer

**Table 9.** Per-class metrics on NSL–KDD (10% test-mix): MeanFlow–IDS + FT–Transformer (stabilized)

| Class | Precision | Recall | F1-score | Support |
|---|---|---|---|---|
| normal | 0.9397 | 0.9625 | 0.9509 | 8740 |
| dos | 0.9836 | 0.9943 | 0.9890 | 6714 |
| probe | 0.9239 | 0.8917 | 0.9075 | 2179 |
| r2l | 0.9416 | 0.8810 | 0.9103 | 2597 |
| u2r | 0.8333 | 0.2500 | 0.3846 | 60 |

trades some `normal`/`dos` recall for a noticeable gain on R2L (18.96%), consistent with attention over mixed-type tokens redistributing decision mass toward minority classes; macro-F1 rises to 0.6007 but remains close to the plain MLP baseline.

When 10% of the test set is integrated into train/validation, performance improves substantially across all methods. The baseline MLP reaches macro-F1 0.8466 with R2L recall increasing to 71.01% and U2R to 38.33%. SMOTE + MLP benefits the most on R2L (86.79%) while maintaining excellent `dos` and `probe` recall, indicating that limited, distribution-aligned oversampling can be beneficial once the training distribution has been refreshed. MeanFlow–IDS + MLP shows balanced gains at 0.8319 macro-F1, with minority recalls trending upward but still below the SMOTE headroom on R2L, likely due to conservative synthesis and filtering. The FT–Transformer variant exhibits very strong `dos` and R2L recall (99.43% and 88.10%, respectively), but a lower `probe` recall (89.17%) and the weakest U2R recall (25.00%). The choice of a 10% mix provides a practical balance: it is sufficient to expose the model to rare-class instances and trigger significant improvements, while leaving the majority of the test set intact for unbiased evaluation. Larger fractions could further increase performance but risk overfitting and reduce the reliability of held-out testing. Overall, the confusion matrices reveal that the primary driver of improvement is *adaptation* of the training distribution, with augmentation family and classifier choice modulating classwise trade-offs rather than determining them outright.

### 4.3 Recommendations

The results support adopting a regular, low-latency adaptation loop that periodically folds a small, curated slice of recent traffic into retraining. Within such a loop, SMOTE remains a strong baseline for R2L when distributions are freshly aligned, whereas MeanFlow–IDS offers a safer augmentation regime under drift due to its gentle synthesis and density filtering; the latter can be leveraged to create minority exemplars without overwhelming the real manifold. In scenarios where interactions between categorical and numeric features are complex, FT–Transformer provides a stable alternative to deep MLPs, but it should be paired with prior-based logit adjustment and per-class threshold selection to prevent minority recall collapse, particularly on U2R. For static or slowly updated deployments, it is advisable to keep the synthesis ratio conservative, to retain kNN filtering on the numeric subspace, and to monitor -normalized confusion matrices over time for systematic leakages (e.g., R2L drifting into `normal`). If high recall on the rarest class is mission-critical, increasing the validation weight of that class in threshold tuning and augmenting with targeted, high-density samples from recent traffic are preferable to naive global oversampling.

## 5 Conclusion

This study introduced MeanFlow–IDS, a one-step flow-matched augmentation method for imbalanced, mixed-type intrusion detection data, incorporating stabilizers such as gentle synthesis ratios, kNN density filtering, and prior-based calibration. Experiments on NSL–KDD showed that under the canonical split, both SMOTE and MeanFlow augmentation struggled on the rare R2L and U2R categories and in some cases reduced macro-F1, highlighting that augmentation alone is insufficient when distributions are stale. Replacing the MLP with an FT–Transformer modestly redistributed recall toward minority classes but did not eliminate the gap.

When a small fraction of recent test data was folded into training, however, all methods improved markedly, with macro-F1 increasing to the 0.83–0.85 range and minority recalls rising substantially. The 10% proportion was chosen as a practical balance: it is large enough to include representative rare-class samples, yet small enough to preserve the majority of the test set for unbiased evaluation. This confirms that timely adaptation of the training distribution is a stronger determinant of performance than the choice of augmentation strategy. Within this adaptive regime, SMOTE remained competitive for R2L, MeanFlow–IDS provided safer distribution-aware synthesis, and the FT–Transformer enhanced robustness to heterogeneous feature types.

In summary, three lessons emerge from the results: augmentation must be combined with calibration and filtering to avoid negative effects; continual or periodic adaptation is essential for sustaining minority detection; and attention-based architectures can complement generative augmentation when interactions between categorical and numeric features are important. Future work will extend this framework to larger and more diverse datasets, develop stronger discrete modeling within flow-based generators, and explore fully online, privacy-preserving adaptation for real deployments. Although a detailed comparative study of alternative kNN distance metrics for the density filter and a systematic analysis of one-step (1-NFE) versus multi-step sampling are beyond the scope of this paper, both represent promising directions for improving the robustness and efficiency of MeanFlow–IDS and will be investigated in future work.

**Acknowledgments.** This work was partly supported by the South African National Research Foundation under Grant Nos. CPRR240327211161, CHN231229201835, RA22112976288, KIC250321305600, and AJCR230704126719.

## References

Cai, S., et al.: DDP-DAR: network intrusion detection based on denoising diffusion probabilistic model and dual-attention residual network. Neural Netw. **184**, 107064 (2025)

Chawla, N., Bowyer, K., Hall, L., Kegelmeyer, W.: Smote: synthetic minority oversampling technique. J. Artif. Intell. Res. **16**, 321–357 (2002)

Chinnasamy, R., Subramanian, M., Easwaramoorthy, S., Cho, J.: Deep learning-driven methods for network-based intrusion detection systems: a systematic review. ICT Exp. **11**(1), 181–215 (2025)

Denning, D.: An intrusion-detection model. IEEE Trans. Softw. Eng. **13**(2), 222–232 (1987)

Gorishniy, Y., Rubachev, I., Khrulkov, V., Babenko, A.: Revisiting deep learning models for tabular data. In: Advances in Neural Information Processing Systems (NeurIPS), pp. 18932–18943 (2021)

He, H., Garcia, E.: Learning from imbalanced data. IEEE Trans. Knowl. Data Eng. **21**(9), 1263–1284 (2009)

Ho, J., Jain, A., Abbeel, P.: Denoising diffusion probabilistic models. In: Advances in Neural Information Processing Systems (NeurIPS), pp. 6840–6851 (2020)

Kotelnikov, A., Trofimov, I., et al.: Tabddpm: modelling tabular data with diffusion models. In: Advances in Neural Information Processing Systems (NeurIPS), pp. 1–16 (2023)

Lipman, Y., Chen, R., Ben-Hamu, H., Nickel, M., Le, M.: Flow matching for generative modeling. In: International Conference on Learning Representations (ICLR), pp. 1–28 (2023)

Liu, X., Bao, S., et al.: Flow straight and fast: learning to generate with rectified flow. In: International Conference on Learning Representations (ICLR), pp. 1–33 (2023)

Merzouk, M., Beurier, E., Yaich, R., Boulahia-Cuppens, N., Cuppens, F.: Diffusion-based adversarial purification for intrusion detection (2024)

Sun, Y., Wang, Z.: Intrusion detection in IoT and wireless networks using image-based neural network classification. Appl. Soft Comput. **177**, 113236 (2025)

Tang, B., Lu, Y., Li, Q., Bai, Y., Yu, X.: A diffusion model based on network intrusion detection method for industrial cyber-physical systems. Sensors **23**, 1141 (2023)

Tavallaee, M., Bagheri, E., Lu, W., Ghorbani, A.: A detailed analysis of the kdd cup 99 data set. In: IEEE Symposium on Computational Intelligence for Security and Defense Applications (CISDA), pp. 1–6 (2009)

Xu, L., Skoularidou, M., Cuesta-Infante, A., Veeramachaneni, K.: Modeling tabular data using conditional gan. In: Advances in Neural Information Processing Systems (NeurIPS), pp. 1–11 (2019)

Yang, Y., Tang, X., Liu, Z., Cheng, J., Fang, H., Zhang, C.: Diff-ids: a network intrusion detection model based on diffusion model for imbalanced data samples. Comput. Mater. Continua **82**(3), 4389–4408 (2025)

# SLAB: Stealthy Deep Reinforcement Learning Backdoor Attack with Latent Space Trigger

Yin Mingxing(✉) and Wang Jian

College of Computer Science and Technology, Nanjing University of Aeronautics and Astronautics, Nanjing, JiangSu, China
yinmingxingcs@nuaa.edu.cn

**Abstract.** Deep reinforcement learning, as a crucial deep learning technology, has demonstrated significant performance in numerous real-world domains in recent years. However, research indicates that deep reinforcement learning is vulnerable to backdoor attacks. As a stealthy attack method, backdoor attacks are only activated under specific conditions, severely compromising the security of the model. Existing backdoor attack methods on single-agent DRL generally focus on embedding backdoor triggers by directly applying perturbations in the pixel space, guiding the model to learn backdoor policies through modifying the data features of original states. However, such pixel-space perturbation methods tend to cause significant differences in visual features between backdoor states and normal states, resulting in poor concealment. To address this limitation, this paper proposes a backdoor attack method that modifies latent space vectors as backdoor trigger. Compared with existing methods, our approach does not directly modify the pixel space of the environment, thus making it more stealthy and natural. Experiments conducted using the DQN and PPO algorithms in Atari environments demonstrate the effectiveness of our method.

**Keywords:** Deep Reinforcement Learning · Backdoor Attack · AI security

## 1 Introduction

With the continuous advancement of deep learning research, deep reinforcement learning (DRL) has been widely applied in various fields of real life (Chong et al., 2022; Popova et al., 2018). Its key advantages include enabling agents to autonomously learn complex decision-making, eliminating the need for extensive manual data, and demonstrating excellent capabilities in solving high-dimensional and complex problems. However, as the widely application of DRL, its security issues have become increasingly prominent, gradually emerging as a core bottleneck restricting the large scale deployment of DRL. Backdoor attacks, as a type of attack method, have been proven harmful to deep neural networks


H. Yan et al. (Eds.): AIS&P 2025, CCIS 2918, pp. 157–172, 2026.
https://doi.org/10.1007/978-981-95-9884-7_12

(DNN). Meanwhile, existing studies have confirmed that DRL is highly vulnerable to such backdoor attacks (Kiourti et al., 2020).

As an extremely stealthy attack method, backdoored model perform normally when the backdoor is not activated. However, once the backdoor is activated, the model execute pre-defined malicious behaviors implanted by attackers. Backdoor attack not only lead to model decision-making errors but also may cause severe real-world harm. For instance, autonomous driving models may cause traffic congestion or accidents after backdoor activation (Wang et al., 2021), and industrial control robots may perform incorrect operations resulting in equipment damage. Such incidents pose significant threats to the property safety, personal safety, and even public safety of model users. Therefore, exploring backdoor attack methods targeting DRL models and investigating the vulnerability of DRL technology have become an urgent need to ensure the secure application of this technology.

Currently, several studies (Ashcraft et al., 2025; Gong et al., 2024; Jiang et al., 2024) have been conducted on backdoor attacks against single-agent deep reinforcement learning model. These works mainly focus on embedding backdoor by modifying the environment or leveraging the temporal characteristics of DRL. Such method directly alters the environment, which impairs the stealth of the backdoor. So this modifications are likely to be observed by model users, who can then eliminate the backdoor through defensive measures. To address this limitation, this work proposes a novel method. First, a Beta-Variational Autoencoder (BVAE) network (Higgins et al., 2017) is used to extract features from the environmental pixel space and reduce their dimensionality into latent space vectors. Then, these vectors will be perturbed, and the perturbed vectors are reconstructed back into the pixel space to serve as backdoor state triggers. This method utilizes perturbations in the latent space, making the backdoor more stealthy. Experiments are conducted using DQN and PPO algorithms on the Atari environments Breakout and Pong. The experimental results demonstrate the effectiveness of the proposed scheme. The main contributions of this paper are as follows:

- This paper proposes an attack method that leverages Beta-Variational Autoencoder networks to extract latent feature space vectors for perturbation. Compared with existing methods, this method transforms the attack vector from explicit input triggers to implicit feature perturbations. It eliminates the reliance on manually designed abnormal patterns in the input space and avoids direct modification of input data, thereby significantly enhancing the stealth of the attack.
- We conducted systematic experiments in the Atari game environment, selecting classic DQN and PPO algorithms in the field of single-agent deep reinforcement learning as benchmark models to comprehensively verify the practical effectiveness of the proposed method. The experimental results not only fully demonstrate the method's validity under different reinforcement learning frameworks but also reflect its excellent adaptability to mainstream single agent algorithm, strongly highlighting the method's generalizability.

The rest of this paper are organized as follows: Sect. 2 presents the related work on backdoor attacks in deep reinforcement learning. Section 3 introduces the preliminaries of deep reinforcement learning and backdoor attack. Section 4 discusses the threat model and the proposed method. Section 5 presents the experimental results. Finally, the conclusion is presented in Sect. 7.

## 2 Related Work

In this section, we review the related work on backdoor attacks against single-agent deep reinforcement learning. Currently, there are two main backdoor trigger design methods, namely single state trigger and temporal sequence trigger. The core characteristic of single state trigger attacks lies in their reliance on specific features of single-step observational data, such as pixel patches, sensor data, or combinations of state parameters. This method does not require the integration of multi-step temporal information, making it relatively easy to implement. In contrast, temporal sequence trigger attacks are defined by trigger conditions that depend on temporal relationships across consecutive observational sequences. Examples include state differences or sequential logic, where trigger signals are embedded within temporal dependencies. This method significantly enhances stealth and is particularly suitable for tasks with strong temporal correlations.

### 2.1 Single State Trigger

The single-step state trigger was first proposed by Kiourti et al. (2020). A 3 $\times$ 3 gray square was used as the backdoor trigger, it was implanted by reward modification and data poisoning. This work verified the vulnerability of DRL algorithms to backdoor attacks. Subsequently, Wang et al. (2021) implemented backdoor attacks in autonomous driving scenarios. They adopted specific vehicle states as backdoor trigger. Once the backdoor was triggered, it would cause cascading traffic congestion or vehicle collisions. Gong et al. (2024) used static sensor values as triggers and implanted backdoor in model via data poisoning. As the first backdoor attack framework for offline RL, it broke the dependence on online interaction. Jiang et al. (2024) proposed the first backdoor attack against safe reinforcement learning. They implemented backdoor attacks through two approaches: an active method of tampering with sensor values and a passive method where robots naturally approach dangerous areas. Ashcraft et al. (2025) utilized the natural static positions of obstacles in the environment as triggers, realizing the backdoor behavior of guiding agents to move toward dangerous areas after triggering. All the studies employed a specific single-step state feature as the backdoor trigger, which was implanted into the model to implement backdoor attacks.

### 2.2 Temporal Sequence Trigger

Yu et al. (2022) first proposed temporal-pattern triggers, breaking the limitation of single-step state triggering. They defined triggers as logical constraints on multi-step states and achieved controllable attack duration. Zhang et al. (2023) leveraged the temporal characteristics of traffic scenarios to design temporal triggers. The trigger condition was set as a sequence of vehicle lane changes across multiple lanes. After triggering, traffic signals would prioritize the passage of the attacker's vehicle. Cui et al. (2024) proposed sample-specific temporal triggers, dynamically generating exclusive sequence constraints. The trigger signal was designed as dynamically generated temporal logic for target states, such as the condition that the differences between the current state and the previous 3-step states are all less than a threshold. Ma et al. (2025) designed temporal triggers based on state transition probabilities, aligning trigger conditions with environmental dynamics. The trigger signal was a combination of probabilities of consecutive multi-step state transitions, with trigger sequences determined by analyzing the environmental transition model. Moreover, temporal constraints made the trigger difficult to detect by anomaly detection methods.

# 3 Preliminaries

## 3.1 Deep Reinforcement Learning

In recent years, deep learning has emerged as a prominent research focus within the field of machine learning (Thalagala et al., 2025; Wang et al., 2024). By leveraging neural networks for representation learning, it has found widespread applications across domains such as image recognition, natural language processing, and speech recognition. The core idea is to employ stacked multi-layer neural network architectures to extract features from high-dimensional input information. Through hierarchical nonlinear transformations, the model can automatically learn abstract data representations, such as semantic and texture information (Singh et al., 2024), thereby demonstrating exceptional perceptual capabilities. However, deep learning exhibits a notable limitation in terms of decision-making. In contrast, reinforcement learning showcases robust decision-making abilities. Nevertheless, in real-world scenarios, the complexity and high dimensionality of data pose significant challenges to RL algorithms. For instance, in multi-UAV cooperative task assignment problems, both UAV states and target states are dynamically changing. In such cases, relying solely on the Q-table from traditional Q-Learning algorithms is insufficient to fully describe the "state-action" pairs. To address this, Google DeepMind integrated deep learning with reinforcement learning. This integration harnesses the perceptual power of deep learning to process high-dimensional and complex data, while utilizing the decision-making capabilities of RL to define problems and optimization objectives. The result is the formation of deep reinforcement learning, an emerging and impactful research area. Taking the Deep Q Network (DQN) (Osband et al., 2016) as an example, it is a DRL algorithm developed by DeepMind based on Q-Learning, which employs convolutional neural networks to handle discrete action spaces. The network architecture of DQN is characterized by two distinct features: a dual-network structure and an experience replay buffer. The dual-network structure comprises a Q-value network and a target Q-network, both sharing an identical architecture. The Q-value network is responsible for computing Q-values and updating its parameters through gradient descent and backpropagation. In contrast, the target Q-network is used to compute Q-values for the next state. Its parameters are updated by periodically copying the parameters $\theta$ from the Q-value network. This design ensures that the values predicted by the target Q-network remain stable for a certain period, effectively preventing overfitting and thereby enhancing the stability and convergence speed of the training process.

### 3.2 Backdoor Attack

Backdoor attack is a stealthy adversarial attack targeting deep learning models. Its core objective is to implant a backdoor trigger into the model, such that the model executes the attacker's predefined backdoor behavior upon trigger activation, while maintaining normal performance when the backdoor remains inactive. Unlike traditional adversarial attacks that directly interfere with the model's inference process, the stealth of backdoor attacks lies in their injection during the training phase and activation only under specific trigger conditions. By incorporating a subset of backdoor training data into the training process, attackers enable the model to learn the mapping relationship between the trigger and the desired erroneous output (Gu et al., 2017). Consequently, during the testing phase, normal inputs still yield correct results, rendering backdoor attacks difficult to detect using conventional model evaluation metrics such as accuracy. The concept of backdoor attacks was first introduced into the field of deep learning by Gu et al., (2017) in their seminal work, where they successfully validated the detrimental impact of this attack paradigm on deep learning models through experiments conducted on the MNIST dataset.

Existing backdoor attack methods against deep learning models can be categorized into two types: data poisoning-based backdoor attacks and model modification-based backdoor attacks. Data poisoning-based backdoor attacks (Liu et al., 2018; Salem et al., 2022; Xue et al., 2020) involve attackers embedding carefully crafted triggers into clean data and modifying their labels to specific target labels to form backdoor instances. These backdoor instances are then injected into the training set to train the model. Models trained on such contaminated datasets will have backdoors embedded, resulting in backdoor models. Backdoor models behave normally when processing clean inputs. However, once a backdoor instance is input, the model exhibits specific malicious behavior. In contrast, other backdoor attack works (Garg et al., 2020; Rakin et al., 2020; Tang et al., 2020) implant backdoors by modifying model parameters. This approach does not require tampering with training data during the training phase. Instead, it directly modifies certain parts of a pre-trained model to achieve the backdoor effect. Compared with modifying model parameters, data poisoning-based backdoor attacks are more classic, where attackers replace a subset of the clean training dataset $\mathcal{D}^{tr}_{clean} = \{(x_i, y_i)\}_{i=1}^{N}$ with a small proportion of backdoor data $\mathcal{D}^{tr}_{poison} = \{(T(x_i, t), y_t) \mid x_i \in \mathcal{X}, i \in S\}$. During the testing phase, the deep learning model will output erroneous results when encountering inputs embedded with the backdoor trigger.

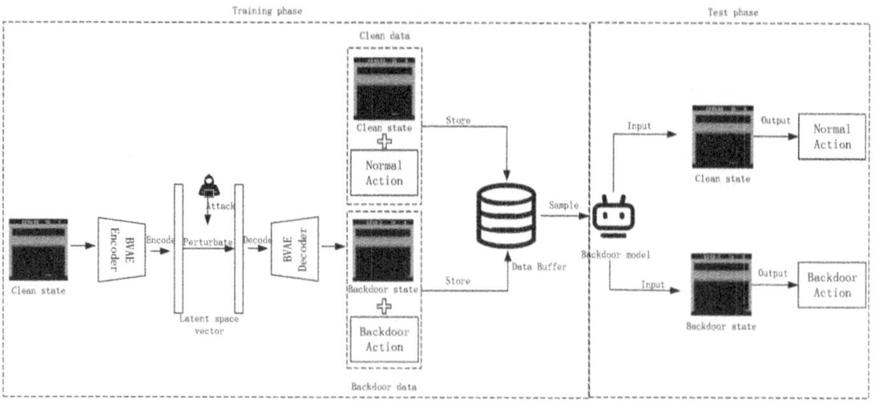


**Fig. 1.** The overall structure of our method

# 4 Proposed Method

In this section, we elaborate our proposed backdoor attack methodology. Firstly, we introduced the threat model in this paper. Secondly, we present the framework of our proposed backdoor attack method. Finally, we present the detailed procedure involved in implanting backdoor in our proposed method.

## 4.1 Threat Model

Deep Reinforcement learning have been widely applied across various domains in daily life. However, training deep reinforcement learning models requires substantial consumption of data resources, computing power, and expert knowledge. Consequently, ordinary users are unable to afford such training costs, so they outsource the model training task to third parties. Untrusted third parties may implant backdoor trigger in the model during training process, aiming to gain control over the model under specific backdoor trigger conditions.

Attacker's Goal: When $s_{\text{backdoor}}$ is arrived, the agent abandons the optimal policy and executes the pre-defined backdoor action $a_{\text{backdoor}}$, leading the failure of task.

Adversary's Knowledge: In this work, we assume that the attacker possesses limited knowledge. The adversary can access and modify the dataset used for model training, while being unaware of the model's overall architecture.

## 4.2 The Framework of the Proposed Method

In this section, we present the proposed backdoor attack method. The overall framework is shown in Fig. 1, consisting of three core components: BVAE network training, trigger design, and backdoor implantation. We summarize the entire backdoor embedding process in Algorithm 1. Specifically, we train a BVAE network and a clean DRL model firstly. Subsequently, the target backdoor state is designated. Utilizing the trained BVAE network, we extract the latent space vector of the target state, which is then perturbed and modified to obtain the backdoor state trigger. We perform malicious data poisoning during the DRL training phase. This involves assigning a specific backdoor action and backdoor reward to a backdoor state. When the environment transitions to this predefined backdoor state, the backdoor training process is activated.

**Algorithm 1** Backdoor Embedding Algorithm

1: **Input:** the network of agent policy $m$, the trained BVAE model $BVAE$, the environment $E$, the learning batch size $s$, the backdoored model training steps $T_b$, the clean model training steps $T_c$, the experience buffer $b$
2: **Onput:** backdoor model $m_b$
3: initialize experience buffer $b$ for storing the training data
4: $i = 0$
5: **while** $i < T_b$ **do**
6: $\quad j = 0$
7: $\quad$ **while** $j < T_c$ **do**
8: $\quad\quad benign\_data \leftarrow execute(m, E)$
9: $\quad\quad store(benign\_data)$
10: $\quad\quad n = timesteps(benign\_data)$
11: $\quad\quad$ **if** $size(b) > s$ **then**
12: $\quad\quad\quad$ experience $\leftarrow sample()$
13: $\quad\quad\quad$ use experience buffer $b$ to update $m$
14: $\quad\quad$ **end if**
15: $\quad\quad j \leftarrow j + n$
16: $\quad$ **end while**
17: $\quad benign\_data \leftarrow execute(m, E)$
18: $\quad vector_c \leftarrow encode(state)$
19: $\quad vector_b \leftarrow vector_c + \sigma$
20: $\quad backdoor\_state \leftarrow decode(vector_b)$
21: $\quad$ Set rewards and actions to generate backdoor data.
22: $\quad store(backdoor\_data)$
23: $\quad$ **if** $size(b) > s$ **then**
24: $\quad\quad$ experience $\leftarrow sample()$
25: $\quad\quad$ use experience buffer $b$ to update $m$
26: $\quad$ **end if**
27: $\quad k = timesteps(data)$
28: $\quad i \leftarrow i + k$
29: **end while**
30: **return** $m_b$

### 4.3 BVAE Training

Firstly, data frames from the Atari game environment are collected as the training data for the Beta-Variational Autoencoder network (Higgins et al., 2017). By interacting with the Atari game environment, random actions are executed to cover diverse scenarios. Game screen frames are continuously captured, subjected to standardization and stored in a format directly usable for model training. Subsequently, the collected data is utilized to train the BVAE network. The encoder employs a three-layer convolutional neural network to hierarchically extract spatial features, ultimately outputting the mean $\mu$ and log variance $\log\sigma$ of the latent vector, thereby obtaining the parameters of the latent space distribution. For latent space processing, the reparameterization trick is adopted to sample the latent vector $z$ from the Gaussian distribution defined by $\mu$ and $\sigma$, ensuring the feasibility of backpropagation and the probabilistic modeling logic of variational inference. This process is illustrated in Eq. 1 (Higgins et al., 2017):

$$\mathbf{x} \to q_\phi(\mathbf{z} \mid \mathbf{x}) = \mathcal{N}(\mu, \sigma^2) \to \mathbf{z} \tag{1}$$

The decoding process achieves feature reconstruction via transposed convolution. The latent vector $z$ is mapped to a feature vector through a fully connected layer, and then processed by three layers of transposed channel convolution to ultimately output the reconstructed image $\hat{x}$. This process is shown in Eq. 2 (Higgins et al., 2017):

$$\mathbf{z} \to p_\theta(\mathbf{x} \mid \mathbf{z}) \to \hat{\mathbf{x}} \tag{2}$$

During the training of each epoch, for the input images in each batch, the reconstructed image $\hat{x}$, mean $\mu$, and log variance $\log\sigma$ are obtained first, followed by the calculation of the loss function. The loss consists of two components: the reconstruction loss is a hybrid of MSE loss (70% weight) and SSIM loss (30% weight), which not only preserves pixel-level errors but also emphasizes structural similarity; the KL divergence loss is weighted by $\beta$ to constrain the latent vector distribution to be close to the standard Gaussian distribution, and the KL term is normalized by the input size to avoid excessive values. The specific loss function is presented in Eq. 3

$$\mathcal{L} = 0.7 \cdot \mathrm{MSE}(\hat{x}, x) + 0.3 \cdot (1 - \mathrm{SSIM}(\hat{x}, x)) + \beta \cdot \mathrm{KL}(q(z \mid x) \parallel p(z)) \tag{3}$$

The terms $MSE(x, \hat{x})$, $SSIM(x, \hat{x})$, and $\mathrm{KL}(q(z \mid x) \parallel p(z)$ in the equation are defined as follows:

$$\mathrm{MSE}(\hat{x}, x) = \frac{1}{N \cdot H \cdot W} \sum_{i=1}^{N} \sum_{h=1}^{H} \sum_{w=1}^{W} (\hat{x}_{i,h,w} - x_{i,h,w})^2 \tag{4}$$

where $x$ denotes the original image, $\hat{x}$ represents the image reconstructed by the BVAE network, $h$ is the height of the image, and $w$ is the width of the image. This loss function quantifies the mean squared error between the original image and the reconstructed image.

$SSIM(x, \hat{x})$ is the structural similarity index, reflecting the consistency of image structure and details.

$$\mathrm{KL}(q(z \mid x) \parallel p(z)) = -\frac{1}{2} \sum_{d=1}^{z_{\mathrm{dim}}} \left(1 + \log \sigma_d^2 - \mu_d^2 - \sigma_d^2\right) / (H \cdot W) \tag{5}$$

This equation represents the KL divergence between the latent distribution $q_\phi(z \mid x)$ and the standard Gaussian distribution $p(z)$. By balancing pixel-level errors, structural similarity, and latent space regularization, the aforementioned overall loss function enables efficient feature encoding and reconstruction of input images.

Algorithm 2 presents the exact procedure of training BVAE.

---

**Algorithm 2** BVAE Network Training Algorithm

---

1: **Input:** Atari frame dataset $D$, beta coefficient $\beta$, batch size $s$, learning rate $lr$, maximum epochs $num\_epochs$
2: **Output:** Trained BVAE model $m_{bvae}$
3: Create data loaders $train\_loader$ and $val\_loader$ for $D_{train}$ and $D_{val}$
4: Initialize BVAE model $m_{bvae}$
5: Initialize optimizer $opt$ with learning rate $lr$
6: Initialize total training loss $train\_loss = 0$
7: $epoch = 1$
8: **while** $epoch \leq num\_epochs$ **do**
9: Set $m_{bvae}$ to training mode
10: **for each batch** $x$ **in** $train\_loader$**:**
11: Move batch $x$ to $dev$
12: Reset optimizer gradients: $opt.zero\_grad()$
13: Compute reconstruction $\hat{x}$, mean $\mu$, log variance $logvar$ via $m_{bvae}(x)$
14: Calculate loss $L = bvae\_loss(\hat{x}, x, \mu, logvar, \beta)$ (MSE+SSIM reconstruction loss + $\beta$-weighted KL divergence)
15: Backpropagate loss: $L.backward()$
16: Update model parameters: $opt.step()$
17: Accumulate training loss: $train\_loss+ = L.item()$
18: **end for**
19: Save model
20: $epoch = epoch + 1$
21: **end while**
22: **return** $m_{bvae}$

---

### 4.4 Trigger Design

Firstly, we select the state into which the backdoor is intended to be implanted. Subsequently, the encoder of the trained BVAE network is utilized to encode this pixel-space state to obtain a latent space vector $\mathbf{z}$. The encoder is a complex architecture composed of a multi-layer neural network, encompassing convolutional layers and fully connected layers. For image data, the convolutional layers of the encoder extract local features from the image by sliding over a series of convolutional kernels. As the network depth increases, higher-level and more abstract features are progressively extracted. After being transmitted and processed through successive layers, these features are ultimately mapped to a low-dimensional latent space. Next, Gaussian noise perturbation is applied to the latent vector $\mathbf{z}$: a random noise term $\epsilon \sim \mathcal{N}(0, \sigma^2)$ is added to certain dimension of $\mathbf{z}$,

$$z' = \begin{cases} z_i + z_i * \epsilon_i & \text{if } s \leq i < e, \\ z_i & \text{otherwise,} \end{cases} \tag{6}$$

$i$ represents the start dimension of latent vector, $e$ represents the end dimension of latent vector, then obtain the perturbed latent vector $\mathbf{z}'$. The magnitude of the perturbation can be adjusted by tuning the value of $\epsilon$. When $\epsilon$ is small, the perturbation imposes only minor modifications to the latent vector, affecting merely subtle features of the backdoor state; conversely, a larger $\epsilon$ amplifies the perturbation, leading to significant changes in the characteristics of the latent vector and thereby causing substantial variations in the manifestation and triggering effectiveness of the generated backdoor trigger. Upon completing the perturbation of the latent vector, the decoder of the BVAE network is employed to reconstruct the perturbed latent vector $\mathbf{z}'$ back to the pixel space, yielding the perturbed backdoor state $\hat{\mathbf{s}}_{\text{backdoor}}$.

### 4.5 Backdoor Implantation

Backdoor is implanted via fine-tuning. During the backdoor implantation training phase, a mixed dataset consisting of clean samples and backdoor samples is used for training. The clean dataset is defined as Eq. 7:

$$D_{\text{clean}} = \{(s_i^c, a_i^c, r_i^c, s_{i+1})\}_{i=1}^{N_c} \tag{7}$$

$s_i^c$ denotes a clean state, $a_i^c$ represents the action executed under the clean policy, and $r_i^c$ is the normal environmental reward, $s_{i+1}$ represents the next state. The backdoor dataset is given by Eq. 8:

$$D_{\text{backdoor}} = \left\{(s_t^b, a_t^b, r_t^b, \tilde{s}_{t+1})\right\}_{t=1}^{N_b} \tag{8}$$

where $s_t^b$ denotes a backdoor state, $a_t^b$ is the action specified by the backdoor policy, and $r_t^b$ is the backdoor environmental reward.The collected clean data and backdoor data are stored in the experience replay bufferÂă$R = D_{\text{clean}} \bigcup D_{\text{backdoor}}$. In each training episode, the model selects experience samples fromÂă$R$Âăbased on a specific sampling strategy to update the policy network. Since the vast majority of samples in the mixed dataset are clean samples, the agent can learn a high-performance main policy during the learning process and exhibit normal behavioral when the backdoor is not activated. However, when the model is exposed to backdoor samples, the triplets $(s_b, a_b, r_b)$ contained in these samples will induce the policy network to learn the backdoor policy under specific condition. This further causes the agent to deviate from its original target behavior, presenting the adversary predefined abnormal action. As the backdoor experiences are seamlessly integrated into the normal training process with a low proportion in the mixed dataset, the entire training process remains stable, and the model's performance in trigger-free states is basically unaffected.

# 5 Experiment

## 5.1 Experimental Setup

To verify the effectiveness of the proposed method, we adopt DQN and PPO as the model training algorithms, with Atari serving as the experimental environment.

Environment: The Atari game environment is one of the most classic and widely used standard benchmark environments in the field of deep reinforcement learning. It was formally introduced into DRL research by the DeepMind team (Mnih et al., 2013) in 2013. Nowadays, it has became a core testbed for evaluating the generalization and complex decision-making capabilities of algorithms. Based on the Atari 2600 home game console, the Atari environment's core value lies in providing a set of real-world game scenarios characterized by high-dimensional inputs, partial observability, and sparse rewards. It perfectly simulates the complex decision-making procedure in real-world scenarios, thus evolving into a unified standard benchmark for performance evaluation in the DRL field.

## 5.2 Effectiveness of Attack

To better evaluate the method proposed in this paper, we adopt the following two metrics for assessment.

- Clean Data Accuracy (CDA): Relative performance of a backdoored model with a benign model in a trigger-free environment after the trigger was used during training/injection; a high CDA preserves normal-use utility.
- Attack Effectiveness Rate (AER): Compared with the benign episode, average drop in episodic return when the trigger appears at inference phase; higher AER shows stronger model performance degradation.

The following Table 1 presents the performance of our method implemented with the DQN and PPO algorithms across the Atari Breakout and Pong environments. As illustrated int the table, our method achieves a high CDA metric, indicating that the backdoored model maintains task execution capabilities in clean environments comparable to those of the clean model. Additionally, the AER metric reaches elevated levels, demonstrating that our method exerts a substantial impact on model performance following backdoor activation.

**Table 1.** CDA and AER of backdoored model trained by our proposed method

| | DQN | | PPO | |
|---|---|---|---|---|
| | CDA | AER | CDA | AER |
| Breakout | 95.43% | 90.25% | 92.20% | 87.75% |
| Pong | 97.36% | 82.63% | 94.86 | 80.35% |

## 6 Ablation Experiment

**The Impact of $\beta$ on the Reconstruction Quality of BVAE.** This method extract latent vectors by BVAE network. Parameter $\beta$ influences the reconstruction quality of environmental images, thereby affecting the backdoor learning efficiency of the model. Consequently, this section discusses the impact of parameter $\beta$ on our method. According to the loss function of the BVAE network, parameter $\beta$ determines the weight between image reconstruction quality and KL divergence: when $\beta < 1$, the loss function prioritizes image reconstruction quality; when $\beta > 1$, it emphasizes the consistency between the prior distribution and the posterior distribution; and when $\beta = 1$, the model balances reconstruction quality and the regularity of the latent space. We trained BVAE networks with different values of parameter $\beta$ and analyzed the influence of varying $\beta$ on the performance of the proposed method. The results are shown in Fig. 2. As the parameter $\beta$ increases, the CDA decreases gradually. This is because an increase in parameter $\beta$ represents a decline in image reconstruction quality, leading to a decrease in the model's clean task performance. The AER decreases slightly but not significantly. Therefore, to ensure the performance of the clean task, the optimal parameter $\beta$ is selected as 0.1.

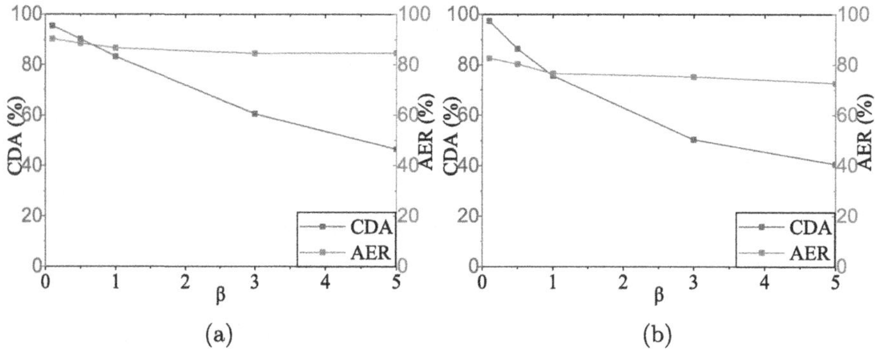


**Fig. 2.** The CDA and AER of different $\beta$ on our method: (a) The backdoor model trained using DQN on the Breakout. (b) The backdoor model trained using DQN on the pong.

**The Impact of Perturbation Magnitude $\epsilon$ on Proposed Mmethod.** We evaluate the impact of perturbation magnitude $\epsilon$ on the proposed method. Perturbation $\epsilon$ refers to the degree of perturbation applied to the latent space vectors extracted by the BVAE network. We set the perturbation magnitude $\epsilon$ range from 0.1 to 0.3 with an increment of 0.05, then we tested the CDA and AER metrics of the backdoored model. The results are shown in Fig. 3. As the $\epsilon$ increases, the AER gradually rises. This is attributed to the stronger perturbation leads the model to learn the backdoor trigger more effectively. In contrast, the CDA shows a decreasing trend, indicating that stronger perturbation interferes with the model's performance on clean environment. Additionally, considering that a larger perturbation magnitude will impair the visual stealth of the backdoor trigger, we recommend setting the perturbation magnitude $\epsilon$ to 0.2 as the optimal parameter.

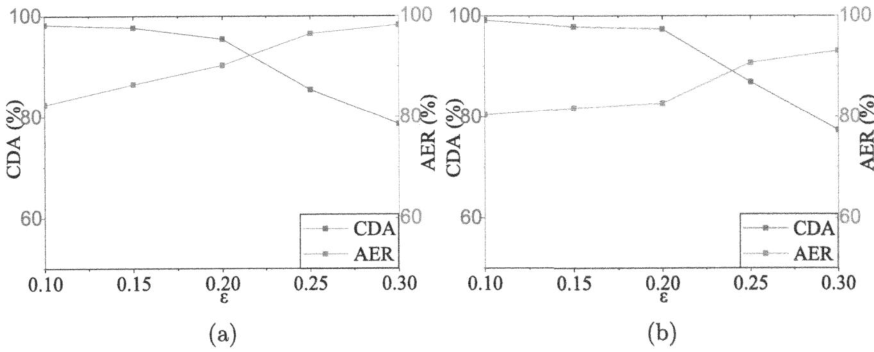


**Fig. 3.** The CDA and AER of different $\epsilon$ on our method: (a) The backdoor model trained using DQN on the Breakout. (b) The backdoor model trained using DQN on the pong.

## 6.1 Robustness Evaluation

In this section, we evaluate the robustness of our proposed method. Specifically, we adopt pruning (Han et al., 2015) as a defensive measure to mitigate the effects of backdoors implanted via our method.

We performed pruning operations on the backdoored model by successively removing 0%, 5%, 10%, 15%, and 20% of weights and set these weights to zero. Figure 4a show the impact of pruning operations on backdoor models trained on Breakout using the DQN algorithm. Figure 4b show the impact of pruning operations on backdoor models trained on Pong using the DQN algorithm. After pruning, both the CDA and AER metrics gradually decrease as the pruning level increases. A reduction in CDA indicates that pruning impairs the model's ability to handle clean environments. A decrease in AER shows that pruning affects the backdoor trigger. However, the degree of decline for both metrics is not particularly significant. Moreover, pruning compromises the model's normal performance, which is undesirable for defenders. Overall, pruning operations cannot fully eliminate the backdoor embedded by this method.

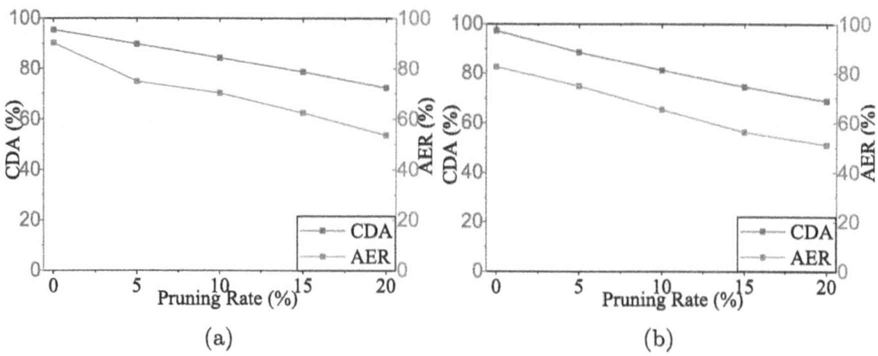


**Fig. 4.** The CDA and AER after pruning on backdoored model: (a) The backdoor model trained using DQN on the Breakout. (b) The backdoor model trained using DQN on the pong.

## 7 Conclusion

In the field of security research on single-agent deep reinforcement learning, studies on backdoor attacks have made certain progress (Cui et al., 2024; Gong et al., 2024; Jiang et al., 2024). However, existing works still have limitations in the design of backdoor trigger. These limitations are mainly reflected in inflexible trigger settings, narrow application scenarios, and excessive reliance of triggering effects on environmental details, making them difficult to adapt to complex and dynamic DRL training scenarios. To address this limitations, this study proposes a backdoor trigger method based on the perturbation on latent space. The core of this method lies in perturbing and modifying the latent space vectors of the environment through the BVAE network. Meanwhile, a reward modification strategy is integrated into the training process. By directionally adjusting reward values, the agent is induced to gradually learn the predefined backdoor policy. Through the combined design of latent space perturbation and reward guidance, the backdoor can be stably implanted without significantly affecting the surface training process of the model, ultimately leading to a significant degradation in the model's performance when backdoor activated. To verify the effectiveness of the proposed method, we conducted systematic experiments in the Atari game environment using two classic DRL algorithms, DQN and PPO. The experimental results fully demonstrate the feasibility and robustness of the proposed backdoor attack method. It not only provides a new trigger design idea for the field of DRL backdoor attacks but also offers more targeted reference directions for the research of subsequent related defense mechanisms.

## Declarations on the Use of Generative AI and AI-Assisted Technologies in the Writing

During the writing of this paper, generative AI was used for auxiliary drafting, specifically employed for certain noun explanations and translation. However, all content has undergone manual review and verification.

## References

Ashcraft, C., Staley, T., Carney, J., Hickert, C., Karra, K., Drenkow, N.: Backdoors in drl: four environments focusing on in-distribution triggers (2025). arXiv preprint arXiv:2505.17248

Chong, J.W., Kim, W., Hong, J.: Optimization of apparel supply chain using deep reinforcement learning. IEEE Access **10**, 100367–100375 (2022)

Cui, J., Han, Y., Ma, Y., Jiao, J., Zhang, J.: Badrl: sparse targeted backdoor attack against reinforcement learning. In: Proceedings of the AAAI Conference on Artificial Intelligence, vol. 38, no. 10, pp. 11687–11694 (2024)

Garg, S., Kumar, A., Goel, V., Liang, Y.: Can adversarial weight perturbations inject neural backdoors. In: Proceedings of the 29th ACM International Conference on Information & Knowledge Management, pp. 2029–2032 (2020)

Gong, C., et al.: Baffle: hiding backdoors in offline reinforcement learning datasets. In: IEEE Symposium on Security and Privacy (SP) 2024, pp. 2086–2104 (2024)

Gu, T., Dolan-Gavitt, B., Garg, S.: Badnets: identifying vulnerabilities in the machine learning model supply chain (2017). arXiv preprint arXiv:1708.06733

Han, S., Pool, J., Tran, J., Dally, W.: Learning both weights and connections for efficient neural network. Adv. Neural Inf. Process. Syst. *28* (2015)

Higgins, I., et al.: Beta-vae: learning basic visual concepts with a constrained variational framework. In: International Conference on Learning Representations (2017)

Jiang, S., Liu, M., Kong, F.: Backdoor attacks on safe reinforcement learning-enabled cyber-physical systems. IEEE Trans. Comput. Aided Des. Integr. Circuits Syst. **43**(11), 4093–4104 (2024)

Kiourti, P., Wardega, K., Jha, S., Li, W.: Trojdrl: evaluation of backdoor attacks on deep reinforcement learning. In: 2020 57th ACM/IEEE Design Automation Conference (DAC), pp. 1–6 (2020)

Liu, Y., et al.: Trojaning attack on neural networks. In: 25th Annual Network And Distributed System Security Symposium (NDSS 2018) (2018)

Ma, O., et al.: Unidoor: a universal framework for action-level backdoor attacks in deep reinforcement learning (2025). arXiv preprint arXiv:2501.15529

Mnih, V., et al.: Playing atari with deep reinforcement learning (2013). arXiv preprint arXiv:1312.5602

Osband, I., Blundell, C., Pritzel, A., Van Roy, B.: Deep exploration via bootstrapped dqn. Adv. Neural Inf. Process. Syst. *29* (2016)

Popova, M., Isayev, O., Tropsha, A.: Deep reinforcement learning for de novo drug design. Sci. Adv. *4*(7), eaap7885 (2018)

Rakin, A.S., He, Z., Fan, D.: Tbt: Targeted neural network attack with bit trojan. In: Proceedings of the IEEE/CVF Conference on Computer Vision and Pattern Recognition, pp. 13198–13207 (2020)

Salem, A., Wen, R., Backes, M., Ma, S., Zhang, Y.: Dynamic backdoor attacks against machine learning models. In: 2022 IEEE 7th European Symposium on Security and Privacy (EuroS&P), pp. 703–718 (2022)

Singh, N.K., et al.: Deep learning model for interpretability and explainability of aspect-level sentiment analysis based on social media. IEEE Trans. Comput. Soc. Syst. (2024)

Tang, R., Du, M., Liu, N., Yang, F., Hu, X.: An embarrassingly simple approach for trojan attack in deep neural networks. In: Proceedings of the 26th ACM SIGKDD International Conference on Knowledge Discovery & Data Mining, pp. 218–228 (2020)

Thalagala, S., Wong, P.K., Wang, X., Sun, T.: Broad critic deep actor reinforcement learning for continuous control. IEEE Trans. Neural Netw. Learn. Syst. (2025)

Wang, Y., et al.: Diffusion actor-critic with entropy regulator (2024). https://arxiv.org/abs/2405.15177

Wang, Y., Sarkar, E., Li, W., Maniatakos, M., Jabari, S.E.: Stop-and-go: exploring backdoor attacks on deep reinforcement learning-based traffic congestion control systems. IEEE Trans. Inf. Forensics Secur. **16**, 4772–4787 (2021)

Xue, M., He, C., Wang, J., Liu, W.: One-to-n & n-to-one: two advanced backdoor attacks against deep learning models. IEEE Trans. Dependable Secure Comput. **19**(3), 1562–1578 (2020)

Yu, Y., Liu, J., Li, S., Huang, K., Feng, X.: A temporal-pattern backdoor attack to deep reinforcement learning. In: GLOBECOM 2022-2022 IEEE Global Communications Conference, pp. 2710–2715 (2022)

Zhang, H., et al.: Backdoor attacks against deep reinforcement learning based traffic signal control systems. Peer-to-Peer Netw. Appl. **16**(1), 466–474 (2023)

# A Fake News Detection Technology Based on Adversarial Contrast Under Few-Shot Samples

Xiaoyu Xu(✉), Peizhi Yang, and Rong Tan

China Coast Guard Academy, Ningbo 315000, Zhejiang, China
340487560@qq.com

**Abstract.** The development and widespread adoption of the Internet have made online social platforms a bridge for communication between users, with hundreds of millions expressing opinions and exchanging views daily. However, this is accompanied by the spread of a vast amount of unverified misinformation. A particular challenge is the emergence of cross-lingual breaking misinformation, where the same textual content in different languages is published on social platforms across various countries, which further complicates feature extraction. To address this, this paper proposes an adversarial contrastive learning method for fake news detection. It maps cross-lingual texts into a shared semantic space and employs a capsule network to generalize class-specific features. Utilizing a contrastive learning framework, the method calculates the distance between target and source data. Finally, noise is incorporated to further enhance the robustness of fake news detection in few-shot scenarios. Experiments on Twitter and Weibo datasets show that the proposed method improves accuracy by 4.8% and 3.0%, respectively, compared to the best baseline method.

**Keywords:** Fake news detection · Few-shot learning · Graph convolutional networks

## 1 Introduction

The rapid development of social platforms such as online social networks and instant messaging tools, particularly the advent of the era of we-media, has significantly lowered the barrier for information dissemination. This has led to an unprecedented increase in the speed at which rumors are generated and spread during public emergencies. Coupled with the limited domain-specific expertise and authoritative data available to the public and relevant institutions, this poses significant challenges to the timely identification and containment of rumors.

Especially in the face of continuously emerging new domains, fake news detection faces a severe shortage of annotated corpora, creating an urgent need to explore innovative technologies and interdisciplinary solutions. To address this, researchers have explored using transfer learning techniques to enhance model


H. Yan et al. (Eds.): AIS&P 2025, CCIS 2918, pp. 173–187, 2026.
https://doi.org/10.1007/978-981-95-9884-7_13

adaptability and robustness in new domains. The core idea of such methods is knowledge transfer: by leveraging annotated data from existing source languages or domains, the model's ability to identify fake news in a new target language or domain is enhanced, effectively tackling the challenges posed by few-shot fake news detection. For instance, based on domain adaptation theory, Luo et al. (2024) introduces a new model framework named Multi-View Graph Neural Network, which captures more comprehensive information through the user profile perspective, user comment perspective, and rumor diffusion perspective. Tian et al. (2021) designed a cross-lingual transfer learning framework utilizing a pre-trained multilingual model and a self-training cycle mechanism. By analyzing user and social media interaction data to capture latent information and iteratively creating pseudo-labels in the target language, the model learns knowledge adapted to the target language from the source language for fake news detection. Lin et al. (2022) proposed an adversarial contrastive learning framework that effectively transfers knowledge from a source fake news dataset rich in annotations to a target dataset lacking annotations through language alignment, a supervised contrastive training paradigm, and an adversarial enhancement mechanism, explicitly overcoming the domain and language limitations of fake news and enhancing the model's detection capability and robustness.

Aiming at the problems of insufficient feature learning due to the lack of training samples in few-shot rumor detection and the linguistic structural and grammatical differences in cross-lingual feature extraction, this paper proposes a rumor detection model based on adversarial contrastive learning of text spatio-temporal structures. First, the resource-rich dataset is treated as the source dataset, and the few-shot dataset as the target dataset. A cross-lingual feature extraction model is used to map content (including different languages, symbols, etc.) from texts in both the source and target datasets into the same semantic space. Then, spatio-temporal features are extracted and summarized using a capsule network to obtain generalized class representations. Finally, the cosine similarity is used to calculate the distance between target data and source data, combined with an adversarial enhancement mechanism to further improve the robustness of rumor detection in the few-shot scenario.

## 2 Related Works

Early research methods were based on machine learning approaches, using appropriate classifiers to categorize the extracted features for the judgment of fake news. These classifiers typically included algorithms such as decision trees and Support Vector Machines. Subsequently, with the rise of deep learning, researchers began to explore the input of captured contextual text features into deep learning models for the discrimination of fake news. Moreover, with the emergence of graph neural network models, the field of fake news detection has also started to apply graph convolutional neural networks. These models can improve the detection accuracy of each other by learning shared knowledge among individuals. Additionally, fake news detection in few-shot sample scenarios can be difficult due to the scarcity of data samples, making feature extraction and model learning insufficient. Therefore, this manuscript will discuss the methods and techniques of fake news detection from two aspects: deep learning methods based on graph convolutional neural networks and few-shot learning.

### 2.1 Fake News Detection Method Based on Graph Convolutional Neural Network

Traditional deep models such as convolutional neural networks and recurrent neural networks do not yield satisfactory results when dealing with data in non-Euclidean spaces. However, graph neural networks can effectively address this issue by modeling data into a graph structure, where different individuals are represented as different nodes in the graph. By extracting features from different nodes and relationships between nodes, the dependency relationships between individuals become more evident.

Cai et al. (2022) utilized the rich semantic features and propagation structures in text words and sentences for fake news detection. Thota et al. (2023) constructed fake news propagation as a tree structure, employed Graph Convolutional Neural Networks for feature extraction, and updated node representations according to the temporal structure. Shu et al. (2025) presents T3-ERD, a novel framework that employs an improved BigBird transformer-based model for detecting specific types of rumors and innovatively introduces an online dynamic bucket strategy, which demonstrates remarkable effectiveness by dynamically clustering comments and processing them in batches. Ke et al. (2020) utilized heterogeneous graphs to construct a relationship network between nodes, and employed a multi-head attention mechanism and GCN to extract rich semantic information. Lotfi et al. (2021) captured the response relationships between users and texts separately through GCN, concatenated the features, and fed them into a classifier for detection. Lu and Li (2020) used convolutional and recurrent neural networks to learn user-based features, and employed graph convolutional neural networks to learn textual features such as forwarding and replying. Bing et al. (2022) used convolutional neural networks and TextCNN to extract user discrete features, combined text and comment hidden layer representations, and finally utilized a dual collaborative attention mechanism to explore the correlation between the three.

Methods based on graph neural networks overly emphasize structural information of text while neglecting temporal features and the rich semantic information in the source text. Additionally, most current research merely focuses on superficial analysis of emotions in fake news, without exploring deeper factors in the dissemination of fake news. Furthermore, in the case of emergencies, such as low-resource mechanisms where data resources are scarce, fake news detection may not be able to fully learn feature representations due to insufficient training and learning data, significantly reducing accuracy. Therefore, how to improve the adaptability of detection models in different scenarios is a key focus of fake news detection research.

## 2.2 Few-Shot Learning Method

In recent years, the detection of fake news on open online social platforms has garnered widespread attention. Early research on fake news detection primarily focused on its superficial features. It is common for creators of fake news to use emotional language and novel words in their messages to attract public attention. Especially when there is a sudden event, such as an epidemic, or when major events and festivals occur, the accuracy of fake news detection drops sharply due to the lack of abundant training samples. However, fake news often shares similar linguistic expressions, word usage, punctuation, and so on. Therefore, some common features can still be learned from abundant historical samples.

Few-shot learning has emerged in the field of computer vision research (Ma et al., 2021; Perrett et al., 2021; Zhang et al., 2021) and, more recently, in the field of Natural Language Processing (NLP). Some few-shot learning methods (Kumar et al., 2024 ; Zhang et al., 2021) aim to adjust pre-trained models in a specific manner, that is, training the model on large-scale data and then adjusting its parameters or fine-tuning the model to make it more accurate on a specific few-shot dataset (Wang et al., 2021). However, since a small amount of data does not reflect the true distribution of a large amount of data, these methods can lead to overfitting of the model on the target dataset. Some researchers believe that metric-based methods are effective under resource-constrained conditions. If there is a specific correlation between large and few-shot samples, knowledge and features obtained from the large sample can be utilized for knowledge transfer between them. Jin et al. (2024) proposed a framework named Det2Ver, which was applicable to both rumor detection and fact verification. By constructing adaptive prompt templates and employing prompt-tuned LLMs like T5, Det2Ver structurally synchronized the two tasks and leverages external knowledge from rumor detection to enhance the fact verification task. (Asadi et al. 2025) proposed a unified framework that leverages large language models not only for classification but also for semantically enriching the input data through lightweight, class-agnostic augmentation. (Xu and Xiang 2021) proposed a graph neural network based on multi-view aggregation for text classification under low resources. (Pang et al. 2021) proposed an improved bidirectional attention mechanism to better describe text classification features by utilizing the interaction between query and support instances in metric learning.

However, due to the limited amount of data, few-shot learning methods cannot fully capture the patterns and features of the data, which often leads to overfitting and poor generalization ability. Features and patterns that can only be reflected with sufficient data cannot be discovered and learned in few-shot samples. Although contrastive learning methods have enhanced representation by maximizing the consistency between samples of the same type and distinguishing between different types of samples, most methods used in fake news detection apply the contrastive learning framework only to expand unstructured text data such as sentences, and due to the sparsity of the data, they cannot consider its propagation structure.

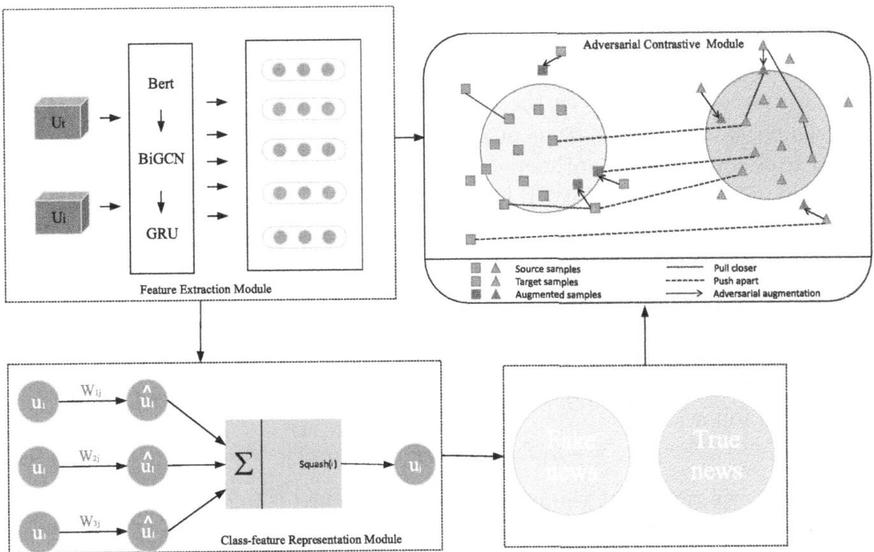


**Fig. 1.** Few-Shot False Information Detection System Architecture Diagram.

# 3 Method

Let $A = (a_1, a_2, ..., a_N)$denote posts from online social platforms, $a_i$represent words, numbers, or symbols within the posts, and $a^N$ signify the total number of characters contained in the posts. The objective is to learn a prediction function denoted as $F : A \to e^f \to \{0, 1\}$. Specifically, we convert the posts into a $f$-dimensional matrix vector denoted as $e^f$, and ultimately transform it into a binary classification problem of credibility, distinguishing between true news (0) and fake news (1).

The three main components of the model are shown in Fig. 1. The first component is the feature extraction module based on BERT+BiGCN+GRU. This module is responsible for generating sample vector representations of the input dataset, including a resource-rich dataset (hereinafter referred to as the source dataset, with events) and a low-resource dataset (hereinafter referred to as the target dataset, with events). The second component is the class-feature representation module which generates classification vector representations of the source dataset. The third component is adversarial contrastive module which is responsible for comparing the sample vectors of fake news categories between the target dataset and the source dataset, clustering similar ones and separating dissimilar ones. It enhances the model's adaptability and robustness through adversarial attack data augmentation.

### 3.1 Cross-Lingual Sentence Encoding

A dataset rich in resources is defined as the source dataset, while a dataset with limited resources is defined as the target dataset. For all the posts in a given event, which can originate from either the source or target dataset, they are mapped to a common semantic space for subsequent contrastive learning. This manuscript employs XLM-RoBERTa for sentence encoding. When encoding text, punctuation marks and numbers are treated as part of the text rather than simply ignored. Using XLM-RoBERTa enables a more comprehensive understanding and processing of text, thereby enhancing its performance and applicability across various languages and text types, as well as enabling cross-lingual training and testing. Based on the Transformer architecture, XLM-RoBERTa is capable of capturing long-distance dependencies and contextual information. The specific formula is as follows:

$$\overline{x} = XLM - RoBERTa(x) \tag{1}$$

Therefore, we represent the posts in source event $C^s$ and target event $C^t$ as matrices $X^s$ and $X^t$, respectively. $X^* = [\overline{x_1}, \overline{x_2} \cdots \overline{x_*}]$, $* \in \{s,t\}$, $X^s \in R^{m\times d}$, $X^t \in R^{n\times d}$, $m$ and $n$ represent the number of posts in the event, and 7 is the dimension of the output from the sentence encoder, with the parameter value set to 768 here.

### 3.2 Cross-Lingual Sentence Encoding

Given an event $C^*$ and its initialized embedding matrix $X^*$, where $* \in \{s,t\}$, we model the conversational thread of the event as a tree structure $T = (VE)$, $E$ consisting of the event declaration and all its related response posts as nodes, with representing the edge structure between nodes in $V$. This article aims to simulate the spatial structure of posts during their dissemination by utilizing graph convolutional neural networks, constructing two propagation trees with different directions: (1) Top-down Structure $H^{TD}$, to simulate the structure of fake news propagation; (2) Bottom-up Structure $H^{BU}$, to simulate the structure of fake news propagation.

In this context, the top-down tree structure will utilize GCN as the graph, converting edges $E$ into an adjacency matrix $A \in \{0,1\}^{|V|\times|V|}$. $V$ represents the number of nodes in the event. Specifically, if there exists an edge between $c_i$ and $c_j$ or $i = j$, then $A_{i,j} = 1$; otherwise, $A_{i,j} = 0$. Furthermore, layer-by-layer propagation is employed to update the node vectors of the $l$-th layer:

$$H^{(l+1)} = \mathrm{Re}LU(\overset{\wedge}{A} \cdot H^{(l)} \cdot W^{(l)}) \tag{2}$$

$\widehat{A} = \widetilde{D}^{-\frac{1}{2}}\widetilde{A}\widetilde{D}^{-\frac{1}{2}}$ is the symmetric normalized adjacency matrix, and $D$ represents the degree matrix of $A$. $W^{(l)} \in R^{d^{(l)}\times d^{(l+1)}}$ is the trainable transformation matrix, $H(0)$ is initialized to $X^*$, $* \in \{s,t\}$. For bidirectional GCN structures, a two-layer model structure is used, with input word vectors having parameter values of 768 and hidden layer parameter value dimensions of 512. Use mean pooling to concatenate the feature vectors $H_2^{TD}$ and $H_2^{BU}$ from the second layer column-wise, and the final output has parameter value dimensions of 128. The specific calculation formula is as follows and $[\cdot;\cdot]$ represents cascading.

$$H_1^{TD} = \mathrm{Re}LU(\hat{A} \cdot X^* \cdot W^{(l)}) \tag{3}$$

$$H_2^{TD} = \mathrm{Re}LU(\hat{A} \cdot H_1^{TD} \cdot W^{(l)}) \tag{4}$$

$$H_1^{BU} = \mathrm{Re}LU(\hat{A} \cdot X^* \cdot W^{(l)}) \tag{5}$$

$$H_2^{BU} = \mathrm{Re}LU(\hat{A} \cdot H_1^{BU} \cdot W^{(l)}) \tag{6}$$

$$x_* = \text{mean - pooling}([H_2^{TD};H_2^{BU}]) \tag{7}$$

The spatial structure representation obtained through the bidirectional GCN network is fed into the GRU model for further spatiotemporal feature extraction. The GRU processes sequence features sequentially through time steps, where each time step takes the output of the preceding step as its input. This enables the GRU model to capture higher-level temporal features from the temporal structure. The computations within the GRU unit are defined by the following formula:

$$r_t = \sigma(W_r[h_{t-1}, x_*] + b_r) \tag{8}$$

$$z_t = \sigma(W_z[h_{t-1}, x_*] + b_z) \tag{9}$$

$$\overline{h_t} = \tanh(W_{\overline{h}}[r_{t*}h_{t-1}, x_*] + b_a) \tag{10}$$

$$h_t = (1 - z_t) * h_{t-1} + z_t * \overline{h_t} \tag{11}$$

When the input sequence $x_*$ and the hidden state $h_{t-1}$ from the previous time step are provided as inputs, the update gate $z_t$ and reset gate $r_t$ are computed. The gating information is then integrated with the current input $x_*$ to calculate the candidate hidden state $\overline{h_t}$. Finally, the update gate controls how much information from the previous hidden state $h_{t-1}$ is retained, while combining with new information from the candidate hidden state $\overline{h_t}$ to generate the current hidden state $h_t$. The GRU model continuously repeats this process to capture long-term dependencies in the sequence and outputs the hidden state sequence $H = [h_1, h_2, ..., h_n]$.

The hidden state sequence $H = [h_1, h_2, ..., h_n]$ is fed into a Softmax classifier to predict the label $y \in \{0, 1\}$, resulting in a weight vector $W$, where $W_1$ and $W_2$ denote the weight matrices and $b$ represents the bias term. Finally, according to Formula (13), the vector $u$ denotes the weighted sum of the final hidden state $H$ for a single event, as shown in the following equation:

$$W = Soft\max(W_1 \tanh(W_2 H_n) + b) \tag{12}$$

$$u_s = \sum_{i=1}^{n} W_i * h_i \tag{13}$$

### 3.3 Class-Feature Representation Based on Capsule Network

Capsule networks overcome the limitations of traditional convolutional neural networks in handling spatial problems, can encode spatial information more accurately, and can also be used to determine the probability of individual existence. To a certain extent, they can ignore the influence of noise and extract class features of different categories from the input data. Specifically, the direction of the vector output by the capsule reflects the semantic classification of the text representation, while the length of the vector represents its probability of belonging to the same class. Therefore, this article uses capsule networks to learn class features from the source dataset. The feature vector $u_s$ obtained from the source dataset is used as the support vector set, and the vectors in the target dataset form the query vector set $u_t$. When $i$ and $j$ represent a capsule respectively, input vector $u_i \in R^a$ can be converted into prediction vector $\overset{\wedge}{u_{ji}}$ through formula (14):

$$\overset{\wedge}{u_{ji}} = W_{ij} u_i \in R^p \tag{14}$$

Among them, $W_{ij} \in R^{p \times a}$ is the transformation matrix, representing the important relationship between encoding low-level capsule vectors and high-level capsule vectors, which can be learned and updated through the backpropagation algorithm of neural networks.

This article employs a dynamic routing mechanism to iteratively update the coupling coefficient $c_{ij}$, thereby adjusting the association strength between lower-level and higher-level capsules. This approach enables the model to capture part-whole relationships and predict the similarity between capsules while preserving original features. The dynamic routing mechanism utilizes a soft routing algorithm that computes correlations between different capsules through dynamically assigned weights. These weights are normalized by the Softmax function, resulting in continuous values between 0 and 1. This continuity allows soft routing to flexibly adjust relationships between capsules and supports end-to-end training through methods like gradient descent.

First, set the initial connection weight $b_{ij} = 0$ from the $i$-th capsule to the $j$-th higher-level capsule. The coupling coefficient $c_{ij}$ is defined as the probability distribution for activating the $j$-th capsule from the $i$-th capsule, as shown in Formula (15): $\sum_j c_{ij} = 1$.

$$c_{ij} = Soft\max(b_{ij}) = \frac{\exp(b_{ij})}{\sum_k \exp(b_{ij})} \tag{15}$$

The output $s_j$ of the $j$-th upper capsule is calculated by adding the coupling coefficient $c_{ij}$ as a weight to the corresponding prediction vector $\overset{\wedge}{u_{ji}}$, as shown in formula (16):

$$s_j = \sum_i c_{ij} \overset{\wedge}{u_{ji}} \tag{16}$$

Finally, a nonlinear function Squash is constructed to Compress the magnitude of the output vector $v_j$ to the range between 0 and 1 which is as shown in formula (17).

$$v_j = squash(s_j) = \frac{\|s_j\|^2}{1 + \|s_j\|^2} \frac{s_j}{\|s_j\|} \tag{17}$$

The number of iterations set in this paper is 3. When the iteration count is not reached, the weight $b_{ij}$ between the lower-level and higher-level capsules continues to be updated through the inner product of the prediction vector $\hat{u}_{ji}$ and the output vector $v_j$, as specified in Formula (18). The process from Formula (15) to (18) is repeated until the specified iteration count is met, with the final output vector $v_j$ delivered as the result.

$$b_{ij} = b_{ij} + \hat{u}_{ji} v_j \tag{18}$$

### 3.4 Comparative Learning Module

After processing through the capsule network, the prototypical representation $v_j$ for each category in the source dataset is obtained. This representation $v_j$ is then combined with the target dataset to form a query vector set $u_t$. Both are fed into the contrastive learning module for further comparison, where their correlations are computed by calculating the distance between each pair $(v_j, u_t)$. Finally, this module employs the nearest neighbor principle to complete the classification. A modified cosine similarity function is constructed in the contrastive learning module, which normalizes the data by subtracting a mean value from the cosine similarities across all dimensions, thereby correcting errors caused by insensitivity to numerical values.

$$Sim\,(v_j, u_t) = \frac{\sum (v_j - \overline{CR})(u_t - \overline{CR})}{\sqrt{\sum (v_j - \overline{CR})}\sqrt{\sum (u_t - \overline{CR})}} \tag{19}$$

Here, CR represents the mean value, and the result is a similarity score ranging between 0 and 1. A Cosine value closer to 1 indicates greater similarity between the two vectors.

### 3.5 Loss Function

By employing triplet loss, neural network models can benefit from both straightforward metric learning and superior nonlinear modeling capabilities. Therefore, a combined loss function integrating cross-entropy and triplet loss is utilized to train the model. The similarity score $Sim\,(v_j, u_t)$ should regress toward the ground-truth label value, where a similarity of 1 indicates a match between the class and the query sample, while a similarity of 0 indicates a mismatch. The specific cross-entropy loss function is defined in Formula (20), where $NQ$ denotes the number of given query sets and $N$ represents the number of classes.

$$Loss_{cp} = -\frac{1}{N} \sum_{q}^{NQ} \sum_{I}^{N} \log \frac{\exp(Sim(u_t, v_j))}{\sum_{k=1}^{N} \exp(Sim(u_t, v_k))} \tag{20}$$

The triplet loss function can be formalized as shown in formula (21):

$$Loss_{tl} = \sum_{q}^{NQ} \max(Sim(u_t, v_{pos}) - Sim(u_t, v_{neg}) + margin, 0) \quad (21)$$

Among them, $v_{pos}$ is a sample of the same category as $u_t$, $v_{neg}$ is a sample of a different category from $u_t$, and the purpose of margin is to increase the distance between different categories and reduce the distance between the same categories. Combining two loss functions during the training process, the global loss function can be expressed as formula (22), where $\alpha$ is a trade-off parameter set to 0.5. And select the Adam optimization algorithm to train the model parameters.

$$Loss_{all} = \alpha \times loss_{cp} + (1 - \alpha) \times loss_{tl} \quad (22)$$

### 3.6 Data Enhancement

Data augmentation techniques have been successfully applied to enhance contrastive learning models. This manuscript utilizes the Fast Gradient Method to construct adversarial samples by leveraging the gradient direction of input data, thereby improving model robustness. The specific adversarial samples and noise vectors are shown below:

$$\widetilde{u}_{noise}^{t} = \varepsilon \frac{g}{\|g\|}; g = \nabla_{u^t} Loss_{all} \quad (23)$$

Firstly, the gradient utilizes the first-order derivative of the loss function. By making minor perturbations along the gradient direction of the loss function with respect to the input data, adversarial samples are generated. Here, $\varepsilon$ denotes the adversarial perturbation coefficient, which controls the degree of discrepancy between the generated adversarial samples and the original samples. Finally, adversarial samples are obtained by adding this perturbation to the original input data.

## 4 Experiment

**Table 1.** Statistics of the datasets

| Statistic | Twitter | Weibo | Twitter-COVID-19 | Weibo-COVID-19 |
|---|---|---|---|---|
| Events | 992 | 4664 | 1154 | 411 |
| Fake news | 498 | 2313 | 575 | 264 |
| True news | 494 | 2351 | 579 | 147 |

**Table 2.** Comparison results with baseline methods on the Weibo few-shot dataset

| Target dataset | Weibo-COVID19 | | | |
|---|---|---|---|---|
| (Source dataset) | (Twitter) | | | |
| Model | Accuracy | Precision | Recall | F1 |
| TVM-TS | 0.642 | 0.648 | 0.642 | 0.645 |
| CNN | 0.756 | 0.765 | 0.767 | 0.766 |
| GRU | 0.79 | 0.809 | 0.805 | 0.806 |
| RvNN | 0.778 | 0.776 | 0.769 | 0.772 |
| PLAN | 0.824 | 0.842 | 0.825 | 0.833 |
| Our Method | **0.872** | **0.876** | **0.862** | **0.869** |

**Table 3.** Comparison Results with Baseline Method on Twitter few-shot dataset

| Target dataset | Twitter-COVID19 | | | |
|---|---|---|---|---|
| (Source dataset) | (Weibo) | | | |
| Method | Accuracy | Precision | Recall | F1 |
| TVM-TS | 0.352 | 0.361 | 0.365 | 0.363 |
| CNN | 0.498 | 0.501 | 0.528 | 0.514 |
| GRU | 0.51 | 0.533 | 0.534 | 0.533 |
| RvNN | 0.652 | 0.661 | 0.665 | 0.663 |
| PLAN | 0.755 | 0.765 | 0.769 | 0.767 |
| Our Method | **0.785** | **0.786** | **0.784** | **0.785** |

**Table 4.** Results of Model Ablation Experiment

| Method | Weibo-COVID19 | Twitter-COVID19 |
|---|---|---|
| | Accuracy | Accuracy |
| Only BERT | 0.76 | 0.683 |
| W/o BERT | 0.785 | 0.65 |
| W/o Capsnet | 0.863 | 0.765 |
| W/o AL | 0.844 | 0.719 |
| Our Method | 0.872 | 0.785 |

### 4.1 Baseline Model Comparison

The source datasets used in this study are Twitter and Weibo (Ma et al., 2016), while the target datasets consist of COVID-19 period datasets, specifically Twitter-COVID-19 and Weibo-COVID-19 (Lin et al., 2022), Table 1 provides more details. To ensure experimental fairness, all experiments were conducted on identical publicly available datasets to eliminate potential biases caused by dataset discrepancies. The methods were evaluated from multiple perspectives, including accuracy (Acc.), precision (Prec.), recall (Rec.), and F1-score, to comprehensively assess the advantages and limitations of each approach. The experimental environment utilized an RTX3090 GPU and was implemented based on the PyTorch framework. The Adam optimizer was adopted with an initial learning rate of 0.0001. During training, early stopping was triggered when the validation loss failed to improve for 10 consecutive epochs. The model architecture configurations were as follows: the GCN contained 2 layers with an input vector dimension of 756 and hidden layer dimension of 512, the GRU hidden layer dimension was set to 128. Additional parameters included a dropout rate of 0.2, 3 routing iterations in the capsule network, and the adversarial disturbance coefficient in the fast gradient value is 1.5.

This manuscript will reproduce the following baseline models and conduct experimental comparisons with the proposed model.

(1) TVM-TS (Ma et al., 2015): This model introduces a novel time series-based approach for rumor detection on microblogs.
(2) CNN (Yu et al., 2017): This model employs unsupervised paragraph vectors combined with supervised CNN learning to identify and achieve early detection of misinformation.
(3) GRU (Ma et al., 2016): This model utilizes recurrent neural networks to demonstrate their effectiveness in rumor detection.
(4) RvNN (Ma et al., 2018): This model primarily employs tree-structured recursive neural networks for rumor detection, adopting both bottom-up and top-down architectures to analyze their advantages and performance in handling branching structures of microblog comments.
(5) PLAN (Khoo et al., 2020): This model proposes a post-level attention network that leverages the multi-head attention mechanism in Transformer networks to model long-range interactions between tweets.

Table 2 and Table 3 present the comparative results between the proposed method and other baseline approaches when using Weibo-COVID-19 as the target dataset with Twitter as the source dataset, and when using Twitter-COVID-19 as the target dataset with Weibo as the source dataset, respectively. The results demonstrate that:

(1) The proposed method has achieved superior performance across accuracy, precision, recall, and F1-score compared to other baseline algorithms on two real-world benchmark datasets. On the Chinese target dataset, it improved accuracy by 4.8% over the best baseline model, while achieving a 3.0% enhancement on the English target dataset. This across-the-board performance improvement demonstrates that the proposed adversarial contrastive learning framework based on spatiotemporal text structures can learn more fine-grained feature representations and effectively enhance fake news detection performance.
(2) SVM-TS demonstrated the poorest performance among all models, indicating that handcrafted feature extraction is insufficient for identifying fake

news. Since CNN is supervised and captures local features between different words, its performance surpasses that of SVM-TS. GRU significantly outperforms both SVM-TS and CNN on both datasets, as RNN can implicitly handle variable-length post sequences while CNN requires more data for decision-making. PLAN utilizes Transformer for long-range node feature extraction, whereas RvNN employs tree structures for spatial feature extraction. The experimental results indicate that Transformer outperforms RvNN in representing long-range features.

### 4.2 Ablation Experiment

To validate the contribution of each component in the proposed model, we conducted ablation studies by systematically removing or replacing specific modules. The experiments were performed on both Chinese and English datasets. The following configurations were examined:

(1) Only BERT: Using only the output from the XLM-RoBERTa encoder;
(2) W/o BERT: Replacing the cross-lingual semantic feature extraction based on XLM-RoBERTa with TF-IDF encoding;
(3) W/o Capsnet: Eliminating the capsule network for class representation after spatiotemporal feature extraction;
(4) W/o AL: Data augmentation is disabled by not employing adversarial attacks to inject noise.

Table 4 presents the ablation study results on both datasets. When TF-IDF is used for semantic feature representation, the accuracy drops sharply due to the difficulty in mapping Chinese and English data into the same semantic space. This demonstrates the crucial importance of mapping different languages into a unified semantic space for fake news detection on social platforms. However, when only BERT is used as the feature extraction model, the accuracy remains suboptimal due to the lack of propagation chain feature extraction. Moreover, when the capsule network is removed for class representation, the model still achieves relatively high accuracy, yet it is 0.9% and 2.0% lower on the Chinese and English target datasets, respectively, compared to using CapsNet. This indicates that the capsule network can more accurately extract class features from the source dataset and enhances the performance of the subsequent cosine similarity function in contrastive learning. Finally, when using the contrastive learning approach without adversarial enhancement (w/o AL), decent results are achieved on the Chinese dataset, demonstrating its effectiveness in cross-domain and cross-lingual adaptation. However, on the English dataset, the performance with adversarial attacks surpasses that without adversarial attacks by 6.6%, indicating that the augmented samples generated through adversarial attacks provide more positive influence for training model parameters. In conclusion, the ablation study results on both datasets still outperform many baseline methods in accuracy, confirming that the proposed approach indeed shows effectiveness for few-shot fake news detection tasks.

## 5 Summary

This article proposes a fake news detection technology based on adversarial contrast under few-shot samples. The framework employs RoBERTa for cross-lingual semantic feature extraction, mapping different languages into a unified semantic space to facilitate subsequent vector computations. It utilizes capsule networks to generate generalized representations of categories from the source dataset, and calculates sample distances between source and target datasets using cosine similarity. Finally, it enhances model robustness through augmented samples. However, the current method is limited to single-text detection. Future work will focus on integrating multimodal information including images and audio, enabling fake news detection from multimodal perspectives to meet more diverse needs in social network analysis and public opinion monitoring.

**Acknowledgments..** This article is supported by the research project of China Coast Guard Academy (No. 2025YJYB17).

## References

Asadi, G.H., Hamzeh, A., Mozafari, N.: Metadata-free rumor detection via LLM-augmented reasoning. IEEE Access (2025)

Bing, C., Wu, Y., Dong, F., Xu, S., Liu, X., Sun, S.: Dual co-attention-based multi-feature fusion method for rumor detection. Information **13**(1), 25 (2022)

Cai, X., Tohti, T., Hamdulla, A.: A rumor detection method incorporating correlation features. In: 2022 3rd International Conference on Pattern Recognition and Machine Learning (PRML), pp. 328–333 (2022)

Jin, W., et al.: Can rumor detection enhance fact verification? unraveling cross-task synergies between rumor detection and fact verification. IEEE Trans. Big Data (2024)

Ke, Z., Li, Z., Zhou, C., Sheng, J., Silamu, W., Guo, Q.: Rumor detection on social media via fused semantic information and a propagation heterogeneous graph. Symmetry **12**(11), 1806 (2020)

Khoo, L.M.S., Chieu, H.L., Qian, Z., Jiang, J.: Interpretable rumor detection in microblogs by attending to user interactions. In: Proceedings of the AAAI Conference on Artificial Intelligence, vol. 34, no. 05, pp. 8783–8790 (2020)

Kumar, R., Goddu, B., Saha, S., Jatowt, A.: Silver lining in the fake news cloud: can large language models help detect misinformation? IEEE Trans. Artif. Intell. (2024)

Lin, H., Ma, J., Chen, L., Yang, Z., Cheng, M., Guang, C.: Detect rumors in microblog posts for low-resource domains via adversarial contrastive learning. Find. Assoc. Comput. Linguist. NAACL **2022**, 2543–2556 (2022)

Lotfi, S., Mirzarezaee, M., Hosseinzadeh, M., Seydi, V.: Detection of rumor conversations in twitter using graph convolutional networks. Appl. Intell. **51**(7), 4774–4787 (2021)

Lu, Y.J., Li, C.T.: GCAN: graph-aware co-attention networks for explainable fake news detection on social media. arXiv preprint arXiv:2004.11648 (2020)

Luo, H., Zhang, M., Yan, F.: Rumor detection model based on gat and multimodal feature codingmes. In: 2024 5th International Conference on Artificial Intelligence and Computer Engineering (ICAICE), pp. 799–805 (2024)

Ma, J., Xie, H., Han, G., Chang, S.F., Galstyan, A., Abd-Almageed, W.: Partner-assisted learning for few-shot image classification. In: Proceedings of the IEEE/CVF International Conference on Computer Vision, pp. 10573–10582 (2021)

Ma, J., et al.: Detecting rumors from microblogs with recurrent neural networks (2016)

Ma, J., Gao, W., Wei, Z., Lu, Y., Wong, K.F.: Detect rumors using time series of social context information on microblogging websites. In: Proceedings of the 24th ACM International on Conference on Information and Knowledge Management, pp. 1751–1754 (2015)

Ma, J., Gao, W., Wong, K.F.: Rumor detection on twitter with tree-structured recursive neural networks (2018)

Pang, N., Zhao, X., Wang, W., Xiao, W., Guo, D.: Few-shot text classification by leveraging bi-directional attention and cross-class knowledge. Sci. China Inf. Sci. **64**(3), 130103 (2021)

Perrett, T., Masullo, A., Burghardt, T., Mirmehdi, M., Damen, D.: Temporal-relational crosstransformers for few-shot action recognition. In: Proceedings of the IEEE/CVF Conference on Computer Vision and Pattern Recognition, pp. 475–484 (2021)

Shu, C., Tang, X., Xiao, J., Chen, Y.: Unverified rumor detection. IEEE Access (2025)

Thota, N.R., Sun, X., Dai, J.: Early rumor detection in social media based on graph convolutional networks. In: 2023 International Conference on Computing, Networking and Communications (ICNC), pp. 516–522 (2023)

Tian, L., Zhang, X., Lau, J.H.: Rumour detection via zero-shot cross-lingual transfer learning. In: Joint European Conference on Machine Learning and Knowledge Discovery in Databases, pp. 603–618 (2021)

Wang, C., Wang, J., Qiu, M., Huang, J., Gao, M.: Transprompt: towards an automatic transferable prompting framework for few-shot text classification. In: Proceedings of the 2021 Conference on Empirical Methods in Natural Language Processing, pp. 2792–2802 (2021)

Xu, S., Xiang, Y.: Frog-GNN: multi-perspective aggregation based graph neural network for few-shot text classification. Expert Syst. Appl. **176**, 114795 (2021)

Yu, F., Liu, Q., Wu, S., Wang, L., Tan, T., et al.: A convolutional approach for misinformation identification. IJCAI **2017**, 3901–3907 (2017)

Zhang, C., Song, N., Lin, G., Zheng, Y., Pan, P., Xu, Y.: Few-shot incremental learning with continually evolved classifiers. In: Proceedings of the IEEE/CVF Conference on Computer Vision and Pattern Recognition, pp. 12455–12464 (2021)

# Stealthy Parasites: Uncovering and Analyzing Reflected Search Poisoning for Illicit Promotion

Sangyi Wu, Jialong Xue, Shaoxuan Zhou, and Xianghang Mi(✉)

University of Science and Technology of China, Hefei, China
{sywu31,xuejl2001,zhoushaoxuan}@mail.ustc.edu.cn, chaselab@foxmail.com

**Abstract.** As an emerging black hat search engine optimization (SEO) technique, reflected search poisoning (RSP) allows a miscreant to free-ride the reputation of high-ranking websites, get search engines poisoned with illicit promotion texts (IPTs) in an efficient and stealthy manner, while avoiding the burden of continuous website compromise as required in traditional promotional infections. However, the scale, scope, and impact of this threat remain poorly understood. To bridge this gap, we present the first comprehensive measurement study of RSP-driven illicit promotion, enabled by a novel end-to-end methodology to capture, analyze, and infiltrate IPTs.

Our study reveals that RSP-based illicit promotion is both large-scale and continuously growing, encompassing a diverse range of illicit categories and languages. Particularly, we have identified over 17 million distinct IPTs spanning 14 illicit categories, including drug trafficking, data theft and hacking services. These campaigns have abused tens of thousands of high-ranking websites (including 26% of the top 10K most popular sites) and have extensively poisoned all four major search engines we studied, particularly Google and Bing. Furthermore, it is also observed that benign search users are being exposed to IPTs at a concerning extent. Also, to interact with potential customers, adversaries embed various contact methods, primarily instant messaging accounts, within the IPTs. Infiltrating these contacts uncovered the massive operational scale of these campaigns: from 92K extracted contacts, we identified 2,333 Telegram channels with 29M subscribers and 661 Telegram groups with 606K distinct members. All these results underscore the severe security implications of RSP, and thus call for more efforts to be invested to mitigate RSP-driven illicit promotion.

**Keywords:** Illicit Promotion · Search Engine Poisoning · Security Measurement

## 1 Introduction

In recent years, a novel and understudied search poisoning technique has emerged, gaining noticeable adoption among miscreants, which we name


H. Yan et al. (Eds.): AIS&P 2025, CCIS 2918, pp. 188–219, 2026.
https://doi.org/10.1007/978-981-95-9884-7_14

Reflected Search Poisoning (RSP). At its core, RSP enables miscreants to inject harmful content into the pages of benign and popular websites, deviating from the original intent of the webpage design, free-riding on the reputation of the website under abuse, and thereby misleading search engines into indexing the injected harmful content with a high page rank. This technique capitalizes on the dynamic nature of most websites, wherein they *reflect* one or more URL text parameters into the rendered webpage content. We denote such URL patterns as *URL Reflection Schemes* (URSes), with the corresponding URL parameters referred to as *reflection parameters.* URSes are legitimate mechanisms where websites render one or more URL parameters directly into the webpage content. Typical scenarios include on-site search pages, hashtag rendering, and dictionary/translation tools. RSP attacks abuse this behavior by injecting illicit promotion texts (IPTs) into reflection parameters. If a website does not sanitize abnormal parameter values, the reflected text is still embedded in the rendered HTML, allowing search engines to crawl and index the IPTs. This type of attack exploits misconfigured reflection logic rather than server compromise, caching bugs, or vulnerabilities. For instance, when https://www.youtube.com/results?search_query=-reflection-text is specified, the URL parameter value *reflection-text* will be reflected as the title of the resulting webpage, even if no search results are found. Given the URSes of high-ranking websites, a typical RSP attack starts by assigning a URS with illicit promotion texts, which leads to RSP URLs generated. Then, these RSP URLs will be distributed across the Internet using traditional blackhat SEO techniques such as forum spamming (Niu, Chen, Hsu, Wang, & Ma, 2007) and spider pools (Du, Yang, Li, Duan, & Zhang, 2016), and finally get indexed by search engines. As illustrated in Fig. 1, when querying Google Search with the Chinese keyword "Unimelb 毕业证书" (Unimelb Diploma) on Apr. 16, 2025, all of the top 30 search results were found to be cases of reflected search poisoning (RSPs) promoting fake certificate services. Also, these illicit promotion texts (IPTs) were observed on multiple high-ranking websites, including *imdb.com* and *ebay.ca.*

In this study, we aim to answer the following research questions: (1) What illicit promotion activities are being distributed through RSP? (2) To what extent have RSP attacks poisoned search engines and abused popular websites? (3) How likely are regular users to be exposed to RSP-based illicit promotion? (4) Given victims exposed to IPTs, how would they further interact with the underlying operators of IPTs?

Answering these questions involves overcoming several technical challenges. First, a methodology is needed to hunt IPTs and RSP cases with high precision and good coverage. Additionally, no tool exists to automatically classify IPTs into fine-grained categories, limiting the understanding of illicit promotion activities. Furthermore, extracting embedded contacts is difficult due to various evasion techniques deployed by the operators. We tackled these issues by designing three tools: *the IPT hunter* (Sect. 3.1), which queries search engines and distinguishes IPTs from benign cases, *the IPT analyzer* (Sect. 3.2), which classifies IPTs into fine-grained categories and extracts embedded contacts if any, and *the IPT infiltrator* (Sect. 3.3), which carries out in-depth infiltration for the

**Fig. 1.** Reflected search poisoning cases promoting fake certificate services got returned when querying Google Search with the Chinese keyword "Unimelb 毕业证书" (Unimelb Diploma) on Apr. 16, 2025.

extracted IPT contacts so as to better profile the underlying operators of illicit promotion campaigns. The key findings of this study are highlighted below.

We first observe that **RSP-based IPTs are large-scale and multilingual, involve diverse categories of illicit goods and services, and have extensively poisoned all four search engines under this study.** In total, our IPT hunter has discovered around 17.2 million distinct IPTs and 19.7 million RSP cases. Then, these RSP cases were indexed by at least one of the four popular search engines under our study: Google, Bing, Baidu, and Sogou. Besides, the observed IPTs belong to a diverse set of over 14 categories of illicit goods and services which range from well-known ones (e.g., illegal drug sales and gambling) to emerging ones (data theft and hacking services). For more details, please refer to Sect. 4.1.

An extended analysis indicates that **RSP-based illicit promotion exhibits a highly dynamic and evolving ecosystem, characterized by a high churning rate and shifting category distribution.** Based on three measurement snapshots, we find that 99.81% of the IPTs and 83.77% of the IPT contacts identified in 2025 were previously unseen, underscoring the rapid turnover and evasiveness of illicit actors. Meanwhile, some illicit categories have become more prominent, particularly Sex Service and Data Theft. The former has steadily grown from 22.87% in 2022 to 37.94% in 2025, becoming the most dominant category, while the latter has increased from 1.54% to 5.20% over the same period. In contrast, the share of Gambling-related IPTs decreased from 26.34% to 18.73%, likely reflecting event-driven cycles or enforcement impact. A detailed longitudinal analysis is presented in Sect. 4.2.

It turns out that **high-ranking websites have been heavily abused in RSP-based illicit promotion, while regular search engine users can**

**be exposed extensively to IPTs.** Among over 65K benign websites (apex domains) abused in RSP attacks, 33K rank in the top 1M most popular websites, 2,707 belong to the top 10K, 854 are renowned education institutions, and 1,144 are websites of government agencies. We also found three categories of search keywords that expose users to IPTs: location names, benign long-tail keywords, and terms of various illicit goods or services. Particularly, when querying Google Search with one of the most popular 3,368 city names in China, 46% of search queries will have one or more IPTs ranked in the top 10 result entries. Please refer to Sect. 4.3 and Sect. 4.4 for more detailed measurement results.

We also find that **campaigns and operators underpinning IPTs turn out to be large-scale and tend to redirect potential customers to instant messaging platforms for further communication and trading.** 83.62% IPTs have instant messaging contacts embedded as the next hop to further interact with victims (potential customers), while the rest redirect victims to their websites. Leveraging the IPT analyzer, we have extracted 92,598 IPT contacts in total, which consist of 35,750 websites, 16,165 Telegram accounts, 36,860 WeChat accounts, 2,196 QQ accounts, and 1,627 telephone numbers. Our infiltration of these contacts has distilled a series of interesting observations, e.g., novel evasion techniques employed by websites, the distribution of illicit mobile apps, and the extensive communication traces observed on Telegram for illicit goods and services. More details can be found in Sect. 5.

We have responsibly disclosed our findings to the four search engines, and Bing has responded to our disclosure and acknowledged the reported issues. Initially, Bing appeared to have taken effective mitigation actions, as evidenced by a dramatic drop in distinct IPTs from 458K in 2022 to just 672 in 2023. However, this success was short-lived. By 2025, we observed a substantial bounce back, with 42K IPTs re-emerging on Bing. This resurgence suggests either a relaxation of previous enforcement efforts or the adoption of new evasion strategies by malicious actors, highlighting the persistent and evolving nature of reflected search poisoning.

Our contributions can be summarized below.

- To our knowledge, this is the first extensive and longitudinal study on illicit promotion through reflected search poisoning. Our study has observed RSP-based IPTs at a large scale, upon which a set of novel observations has been further distilled.
- Three novel tools have been implemented, which allow any party that is independent of search engines to capture, analyze, and infiltrate reflected search poisoning for illicit promotion.

## 2 Related Works

**The Underground Economy and Illicit Promotion.** Over the past two decades, a continual research effort has been invested in the underground economy, i.e., the buying and selling of illicit goods or services. Among these works, a large portion is dedicated to specific categories of illicit services and goods, e.g., illicit or even illegal drug trading (Aldridge & Askew, 2017; Kanich et al., 2011; Leontiadis, Moore, & Christin, 2011; Z. Li, Du, Liao, Jiang, & Champagne-Langabeer, 2021; McCoy et al., 2012), illegal online gambling (Gao et al., 2021; H. Yang et al., 2019), malware distribution (Caballero, Grier, Kreibich, & Paxson, 2011; Kotzias, Bilge, & Caballero, 2016; Kotzias, Caballero, & Bilge, 2021; Thomas et al., 2016), among others. Another line of studies focuses on holistic infrastructures that are used in the promotion of illicit goods and services, i.e., illicit promotion. Particularly, Grier, Thomas, Paxson, and Zhang (2010) revealed that the promotion of spam URLs on Twitter and found that 8% URLs posted on Twitter pointed to malicious websites (e.g., malware, and scams). Furthermore, following the emergence of the Tor anonymity network, Christin (2013) conducted a measurement study between 2011 and 2012 on the Silk Road, an anonymous marketplace running as an onion service. As revealed by this study, the Silk Road marketplace was mostly used to trade illegal drugs and most items on sale are available for less than three weeks. As a follow-up, Soska and Christin (2015) moved to profile the longitudinal evolution of anonymous marketplaces between 2013 and 2015. Still, most sales were found to be drug-related. Besides, popular online services have been abused by miscreants for illicit promotion so as to reach regular online users. One typical channel is promotional infections (John, Yu, Xie, Krishnamurthy, & Abadi, 2011; Leontiadis, Moore, & Christin, 2014; Liao et al., 2016) wherein the attacker compromises a legitimate website, injects promotional and harmful webpages, induces the search engines to index these webpages with high page rankings, and ultimately exposes benign search users to their malicious campaigns. And such promotional infections have been used by miscreants to promote illicit activities of diverse categories, e.g., online gambling (Liao et al., 2016), unlicensed pharmacies (Leontiadis et al., 2014), etc. Also, as revealed by Leontiadis et al. (2014), the median time to recover from such promotional infections is around 15 days, which indicates miscreants may have to continuously compromise new vulnerable websites so as to maintain or increase the visibility of their promotion activities. Furthermore, John et al. (2011) proposed a methodology to detect if a search result is poisoned or not by looking at its URL parameters instead of visiting the URL. Besides, Liao et al. (2016) utilized the semantic inconsistency between the injected promotional content and the legitimate context of the infected website to determine if a webpage is a promotional infection or not. Besides, some research (H. Yang et al., 2017; Yuan, Lu, Liao, & Wang, 2018; Zhao, Zhang, Xing, Li, & Chen, 2016; Zhu et al., 2021) has focused on identifying and understanding jargon words used in illicit promotion. Particularly, H. Yang et al. (2017) explored how to identify jargon words from keywords promoted in blackhat SEO campaigns, while Yuan et al. (2018) compared the semantic discrepancy of the same word between contexts of illicit promotion and benign usage scenarios to identify jargon words.

**Search Engine Poisoning.** Search engine poisoning (i.e., search poisoning) encompasses blackhat SEO activities that are conducted by miscreants with the goal of poisoning the search results with *harmful content* (D. Y. Wang, Savage,

& Voelker, 2013). When carrying out search poisoning attacks, miscreants usually deploy various blackhat SEO techniques, e.g., forum spamming (Niu et al., 2007), cloaking (Invernizzi et al., 2016; D. Y. Wang, Savage, & Voelker, 2011), spider pooling (Du et al., 2016), content spinning (Zhang, Wang, & Voelker, 2014), suggestion manipulation (P. Wang et al., 2018), and promotional infections (John et al., 2011; Liao et al., 2016). Also, a series of research works have been devoted to detecting search poisoning through promotional infections (John et al., 2011; Liao et al., 2016; Lu, Perdisci, & Lee, 2011; R. Yang et al., 2021). Particularly, John et al. (2011) focused on clustering compromised webpages into their underlying poisoning campaigns, and Lu et al. (2011) proposed SURF to capture poisoned search results after observing that the redirection chain of visiting a poisoned search result has more distinct features compared with visiting a benign one.

**Comparison with Previous Works on Illicit Promotion and Search Poisoning.** This study focuses on an emerging search poisoning technique, namely, reflected search poisoning. And it reveals, for the first time, how reflected search poisoning is being extensively used for the promotion of illicit goods and services that belong to diverse categories. Along with these findings is an end-to-end methodology to capture and analyze IPTs and RSPs.

## 3 Methodology

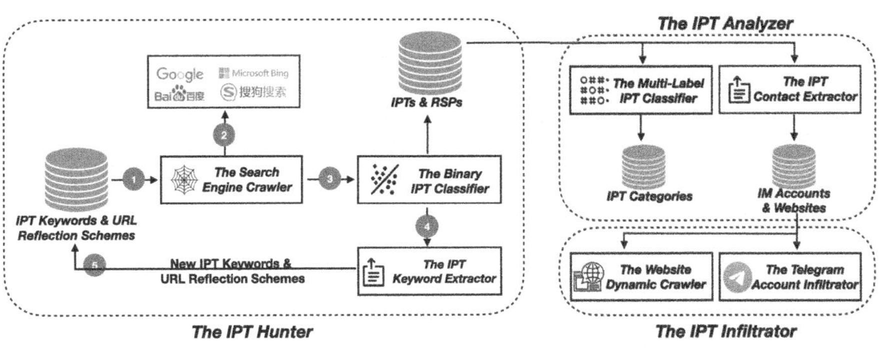


**Fig. 2.** Methodology to detect and understand illicit promotion texts(IPTs).

As shown in Fig. 2, our methodology consists of three major components. One is to discover IPTs and RSPs through the IPT hunter (Sect. 3.1). Then, given IPTs, an analyzer (Sect. 3.2) is further applied to profile IPT categories and extract contact entities embedded in IPTs. To reveal what these contacts will redirect a victim to, an IPT infiltrator (Sect. 3.3) is designed to automatically visit and profile websites and Telegram accounts that are promoted in IPTs.

**The Data Annotation Scheme.** For all data labeling tasks, three researchers jointly define the categories of each labeling task, and independently label the same set of samples, and synchronize periodically to resolve conflicts. Overall, a high degree of agreement has been observed among annotators for all the labeling tasks.

### 3.1 The IPT Hunter

The IPT hunter is designed to identify RSP-based IPTs through three steps. First, it queries search engines with well-crafted keywords and URL reflection schemes to find potential RSP cases. Second, a machine learning classifier determines if the URL of a search result entry contains an IPT, filtering out benign cases. Third, it extracts new search keywords and URL reflection schemes, feeding them back into the first step to expand the search.

**The Search Engine Crawler.** Our crawler utilizes two strategies to query the search engines. One is using keywords from known IPTs to find RSP cases with similar IPTs but different URL reflection schemes. The other strategy searches for results matching a given URL reflection scheme, uncovering IPTs from different campaigns and categories on the same scheme. Iteratively using these two strategies, we can effectively discover new RSPs/IPTs and URL reflection schemes abused in RSPs. To mitigate the effects of search personalization and cloaking, our crawler employs geographically distributed web proxies to submit identical search queries from diverse geographic vantage points. Additionally, our system circumvents personalized search results by disabling user login states, clearing browser cookies, and transmitting stateless queries via Playwright headless browser. The result search entries can still be a mix of RSPs, legitimate URL reflection cases, as well as other benign search entries. Therefore, a filtering step is further applied to exclude search results that are not URL reflection cases. This is achieved by a simple but accurate filtering rule, wherein a search result entry is considered as a URL reflection only when the values of one or more URL parameters are rendered as part of the webpage elements. Given URL reflection cases identified, the next step is to distinguish RSPs from benign URL reflections.

**The Binary IPT Classifier.** Since RSPs are mainly used to distribute IPTs, we built a binary IPT classifier to distinguish RSPs from benign URL reflections. The first step is to create a representative ground truth dataset. A text entry is labeled as an IPT if it satisfies all these criteria: (1) It promotes goods or services, especially illicit ones; (2) Its semantics significantly deviates from the context of the parent webpage and website; (3) Its semantics and syntax significantly deviate from the intended use of the URL reflection scheme. After manually labeling the first round of returned results using the aforementioned search engine crawler, we obtained a dataset of 2,299 IPTs and 1,468 benign URL reflection cases. This ground truth dataset is diverse and representative, covering all categories listed in Table 1 and spanning 10 different languages.

**Table 1.** Categories of illicit goods and services.

| Category | |
|---|---|
| Black Hat SEO & Advertisement | Hacking Service |
| Counterfeit Goods | Drug Sales |
| Data Theft | Sex Service |
| Fake Account | Surrogacy Service |
| Fake Certificate | Weapon Sales |
| Financial Fraud | Money Laundering |
| Gambling | Others[a] |

[a]Categories like ghost writing and detective employment are grouped as "Others" due to their small sample size.

(a) An IPT promoting illegal pharmaceuticals.

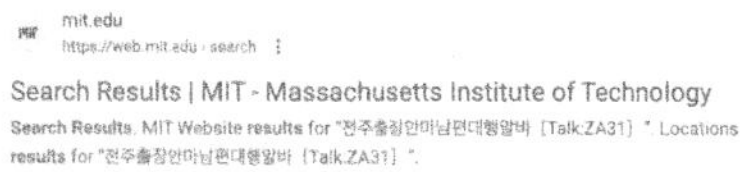

(b) An IPT promoting sex services in Korean.

**Fig. 3.** Examples of IPTs as indexed by Google Search via RSP.

Then we explored two text classification options. The first combines manually designed features with classic algorithms. For this, we noted IPTs are typically longer than legitimate reflections and often include non-text characters like emojis (Fig. 3a) and various brackets (Fig. 3b). We designed features to capture these distinctions and created a total of seven feature sets(see in the Appendix A.1) and tested four classic algorithms: Decision Tree, Random Forest, AdaBoost, and SVM. The second involves fine-tuning a pre-trained large language model (LLM), specifically multilingual BERT (Devlin, Chang, Lee, & Toutanova, 2018), which avoids manual feature engineering and provides higher data efficiency and better generalizability. Following common practice, 80% of the ground truth samples were used for training, while 20% were held out for testing.

**Table 2.** Performance of different binary IPT classification models.

| Model | Precision | Recall | F1-Score | Speed[a] |
|---|---|---|---|---|
| fine-tuned BERT | 98.86% | 98.63% | 98.75% | 18/267[b] |
| Random Forest | 95.34% | 97.95% | 96.63% | 11667 |
| Decision Tree | 95.29% | 96.81% | 96.05% | 13831 |
| AdaBoost | 94.41% | 96.13% | 95.26% | 11510 |
| SVM | 94.39% | 95.90% | 95.14% | 13457 |

[a]The unit is samples inferred per second.
[b]The BERT model has been tested on both CPU and GPU, and the respective speed is 18/267 respectively.

Table 2 lists a direct comparison among IPT classification models in terms of performance and inference efficiency. We can see that the fine-tuned BERT model has the highest recall and precision but lower inference throughput at 18/267 samples per second on CPU/GPU. Among classic models with manually crafted features, Random Forest performs best with a precision of 95.34% and a recall of 97.95%, and its inference speed is 11,667 samples per second. All models were tested on a server with an Intel Xeon Silver 4214R CPU, 128GB memory, and two NVIDIA RTX 3090 GPUs. Given the need to predict over 10M search results weekly, the Random Forest model is chosen for its balance of performance and efficiency.

**IPT Keyword Extractor.** Since most IPTs are lengthy with a median length of 46 characters, querying with such a lengthy term typically confuses the search engine and has a low hit rate for new RSPs. Instead, an IPT keyword extractor is built up to split an IPT and locate its most critical component, which will be used as the search keyword to boost the next round of IPT hunting. As learned by case studies, the contact segment of an IPT turns out to be good search keywords for guiding the search engines to discover new RSPs/IPTs. As shown in Fig. 3, the contact segment is usually separated from other IPT segments with non-text Unicode characters, e.g., brackets and pictorial symbols. Therefore, to extract the contact segment out of an IPT, a regular expression is first applied to extract URLs, if any. Then, the left IPT text is further split into segments using a well-crafted list of separators including brackets, pictorial symbols, and various punctuation marks. Next, a machine learning classifier is applied to decide whether a given IPT segment is a contact segment or not, and we name this classifier the contact segment classifier. The resulting URLs (mostly domain names) and contact segments will be considered as IPT keywords.

To build up this contact segment classifier, we identified eight effective features, e.g., the number of contact indicators and punctuation marks (*tg:* for Telegram), and the full list of features can be found in Appendix A.2. Upon a ground truth dataset of 1,012 contact segments and 3,170 non-contact segments, training and evaluation revealed that the Random Forest classifier performed best, achieving 90.51% recall and 91.05% precision. This is satisfactory since the classifier is meant to identify good search keywords, and misclassifications only affect search efficiency.

**Deployment.** When deploying our IPT hunter, a single round starts with querying search engines and ends with discovering new IPTs, IPT keywords, and URL reflection schemes. Our crawlers adhere to search engines' rate limit policies, stopping when an HTTP response error indicates the limit is reached and restarting when it's cleared. Since our primary research servers are located in a region with restricted access to Google and Bing, we employed multiple web proxy services to relay search queries through their globally distributed networks. This approach also allowed us to assess potential variations in search results across different geographical locations.

Then, to jump-start the crawler, we first manually crafted a seeding set, which involves an iterative process of searching Google with existing IPT keywords, manually labeling searching results, and extending the IPT keyword set. This began with a Google query *site:* https://ctan.org/search, which is a poisoned in-site search of an English-language repository containing TEX-related materials for download. We then parsed the search results to extract relevant keywords. These keywords were used to perform additional Google searches, expanding the pool of potential URL reflection schemes, and these schemes were used to search for more keywords in turn. This iterative process was repeated over four cycles. We manually classified the acquired results to ensure relevance and reliability. Ultimately, we successfully obtained a seeding set comprising 153,385 IPT keywords and 15,113 URL reflection schemes, covering various categories and languages. Leveraging this seeding set, we then applied this hunter to all four search engines and completed four rounds between Nov. 8, 2022 and Dec. 5, 2022. We name these rounds as *the first-phase* IPT hunting. To longitudinally profile IPT evolution patterns, we subsequently executed two additional hunting phases: *the second-phase* encompassed seven rounds from Nov. 6, 2023 to Dec. 27, 2023, followed by *the third-phase* hunting involving four rounds spanning Mar. 3, 2025 to Apr. 3, 2025.

In total, we have discovered 19,768,332 distinct RSP incidents, 17,220,007 distinct IPTs and 425,061 URL reflection schemes, for which more details will be presented in Sect. 4.

### 3.2 The IPT Analyzer

Given the large volume of IPTs captured, two additional tools have been developed to automatically analyze IPTs and gain a deep understanding of RSP-based illicit promotion. One is a multi-label classifier which groups an IPT into one or more services/goods categories, e.g., counterfeit goods, and fake certificate services. Furthermore, a contact extractor is designed to extract from a given IPT, contacts of various types including websites, instant messaging accounts, and telephone numbers.

**The Multi-label IPT Classifier.** This classifier is designed to profile what categories of services/goods an IPT is intended to promote. It takes an IPT as the input and classifies it into one or more categories. By manually labelling randomly sampled IPTs, 14 categories have been defined for different services/-goods, and the full list can be found in Table 1. Regarding the naming of these IPT categories, we adopt some terms proposed in previous works (Cao, Sirivianos, Yang, & Pregueiro, 2012; Lin, 2011; West & Bhattacharya, 2016; H. Yang et al., 2017; Yuan et al., 2018; Zhou et al., 2018), such as Gambling, Data Theft and Fake Certificate. Then, the legitimacy of some services/goods categories can vary significantly from one jurisdiction to another. For instance, many IPTs promote surrogacy services, sex services, or drug sales in Chinese, all of which however are illegal in mainland China. When naming such categories, we take an objective standard and try to avoid implying legitimacy in the names. Also, we consider multilingual texts as the input since IPTs are observed to be multilingual.

To build up this classifier, a ground truth dataset of 7,246 IPT samples was labeled. Then, regarding the classification algorithm, we chose to fine-tune the aforementioned multilingual BERT model, due to the following considerations. First of all, the fine-tuned BERT has achieved the best performance along with good data efficiency in the binary IPT classification task. Also, different from the binary IPT classifier, this multi-label one is used for offline analysis for which the inference speed is not a bottleneck. Through training on 80% of the ground truth dataset, the resulting model has achieved good performance when evaluated with the held-out 20% ground truth dataset. Particularly, the micro precision of the resulting model is 94.03%, while the micro recall is 93.53% and the LRAP (label ranking average precision) score is 95.93%. The per-class performance metrics can be found in Appendix B.

**The IPT Contact Extractor.** Taking an IPT as the input, our contact extractor is designed to extract all the embedded contact entities, which is achieved through three major steps. The first step is a preprocessing step, which is intended to remove various noise added by miscreants likely in an attempt to impede automatic extraction, such as the usage of homoglyphs and splitting a domain name with human-readable characters. Then, the second step aims to decide the type of embedded contact. Lastly, depending on the contact type, either a named entity recognizer or a rule-based extractor will be applied to identify the respective contact entity.

*The Contact Type Classifier.* This classifier takes an IPT as the input and decides which type of contact it contains. Currently, this classifier supports the classification of five contact types that are most frequently embedded in IPTs. These contact types include telephone numbers, Telegram accounts, WeChat accounts, QQ accounts, and websites. And an IPT will be classified as "others" if it does not contain any contacts that belong to those five types. To build up this classifier, the aforementioned transformer-based multilingual language model is reused and further tuned on a ground truth dataset of 5,482 labeled IPTs, and the resulting model has achieved a micro precision of 94.91% and a micro recall of 94.91%.

*Contact Entity Extractors.* Given the contact type decided for each IPT, the next step is to extract its contact entity. For websites, telephone numbers, and QQ accounts, simple rule-based extraction works well thanks to their strict format, e.g., both telephone numbers and QQ accounts are numeric strings. For the extraction of Telegram and WeChat accounts, we consider them as tasks of named entity recognition (NER). As transformer-based language models (X. Li, Yan, Qiu, & Huang, 2020; Yan, Deng, Li, & Qiu, 2019) have also achieved state-of-the-art performance in many NER tasks. We utilized the aforementioned multilingual BERT model to build the two separate NER models for recognizing WeChat accounts and Telegram accounts respectively. Both NER models have achieved good performance. The WeChat NER model has achieved a precision of 96.22% and a recall of 98.71%, and the Telegram NER model has achieved a precision of 98.15% and a recall of 100.00%.

As a result, we have identified 92,598 distinct contact entities in total, which consist of 35,750 websites, 16,165 Telegram accounts, 36,860 WeChat accounts, 2,196 QQ accounts, and 1,627 telephone numbers.

### 3.3 The IPT Infiltrator

The contact entities identified above serve as the next hops for victims (potential customers) to interact with operators of the promoted goods or services. Infiltrating these contacts can help gain a better understanding of the underlying illicit campaigns. Therefore, an IPT infiltrator is designed and implemented to fulfill this goal.

**The Dynamic Website Crawler.** For website contacts, we periodically visit each of them at a weekly pace through instrumenting a headless browser. We capture the final landing webpage as a screenshot and save all the network traffic of both HTTP requests and HTTP responses. Also, if a landing page is found to be promoting a mobile app, it will trigger a further inspection during which we manually download the respective mobile app for further threat analysis. Specifically, this dynamic website crawler is built upon the Playwright framework which offers APIs in many programming languages and enables end-to-end automatic testing of websites.

**The Telegram Account Infiltrator.** Leveraging publicly available Telegram APIs, the profile of each Telegram account will be extracted weekly. Then, if an account turns out to be a Telegram channel or group, our infiltrator will subscribe or join it and retrieve all its historical messages since 2022 and weekly obtain its new messages. Furthermore, among messages collected from either Telegram channels or groups, many of them are found to promote other Telegram channels or groups. For such kinds of messages, our infiltrator further snapshots the involved Telegram groups or channels by following a similar infiltration strategy.

Different from Telegram, neither WeChat nor QQ has any public APIs to facilitate programmable access to account profiles or group messages. Instead, what we can do is to manually search each account and view its public profile. Therefore, we conducted manual inspection on a randomly sampled set of 250 WeChat accounts and 200 QQ accounts in an attempt to learn their activeness, locate illicit promotion indicators if any, and understand the tactics underpinning their escape from being blocked from the respective platform. The detailed results of our infiltration will be presented in Sect. 5.

### 3.4 Ethical Considerations

We have taken measures to minimize any potential ethical issues in our methodology. We respected user policies and limited the crawling rate when collecting data from search engines and messaging platforms to avoid burdening them. All raw datasets are securely stored on our research servers with access limited to our researchers. Our measurement design intentionally avoided any form of interaction that could constitute entrapment, coercion, or unintended engagement with malicious actors. The IPT infiltrator strictly operated in a passive, observation-only mode: for IPT-embedded websites, our crawlers accessed only publicly reachable pages without submitting forms, altering server state, or performing authenticated actions. Similarly, for Telegram, we retrieved only publicly accessible metadata, member lists, and historical messages without sending messages, initiating conversations, or engaging in any transactions. To further protect sensitive entities, all identifiers including usernames, phone numbers, and domains were hashed or partially redacted before storage or analysis, ensuring that no personally identifiable information can be reconstructed. The publicly released dataset omits reflection URLs that could be repurposed for misuse. Notably, any child pornography content found on IPT websites was completely removed from our datasets and reported them to the relevant government authorities.

## 4 RSP-Based Illicit Promotion

In this section, we present a comprehensive measurement for illicit promotion texts (IPTs) that are distributed through reflected search poisoning (RSP), i.e., RSP-based IPTs. As a result, a deep understanding is gained, for the first time, regarding what services/goods have been illicitly promoted through RSPs (Sect. 4.1), how IPTs evolve across time (Sect. 4.2), what benign websites have been abused (Sect. 4.3), as well as the extent to which benign search users can be exposed to IPTs. (Sect. 4.4).

### 4.1 Landscape of Illicit Promotion via RSP

**Scale.** In total, we have captured 17,220,007 distinct IPTs that have been distributed through 19,768,332 RSPs. These RSPs have abused 425,061 unique

**Table 3.** Scale statistics of RSP-based IPTs.

| Group | IPTs | RSPs | URL Schemes | Abused FQDNs |
|---|---|---|---|---|
| Google | 16,962,899 | 19,376,708 | 408,719 | 144,158 |
| Baidu | 100,760 | 198,785 | 6,979 | 3,542 |
| Sogou | 9,162 | 11,825 | 3,104 | 1,585 |
| Bing | 500,545 | 534,962 | 42,122 | 16,581 |
| Overall | 17,220,007 | 19,768,332 | 425,061 | 149,278 |

URL reflection schemes, which belong to 149,278 fully qualified domain names (FQDNs) and 65,294 apex domains. Table 3 presents more detailed scale statistics for IPTs. And we can see IPTs via RSPs have successfully poisoned all four search engines under our study on a large scale. Among the four search engines, the two Chinese search engines, namely Baidu and Sogou, have much fewer IPTs than the other two global counterparts, which is likely because of stricter domestic regulations and more centralized enforcement mechanisms in China.

However, the volume of IPTs is not equivalent to that of the underlying promotion campaigns, since the same campaign can promote its services or goods through many different IPTs. Instead, we observe that IPTs belonging to the same campaign tend to share a common set of very few contacts. Leveraging the IPT contact extractor (Sect. 3.2), we have observed 92,598 distinct contacts, which suggests that the scale of the underlying promotion campaigns is also very large.

**Table 4.** Distribution of IPTs over categories of goods and services.

| Category | % IPTs | Category | % IPTs |
|---|---|---|---|
| Sex Service | 28.29% | Drug Sales | 2.41% |
| Gambling | 22.56% | Surrogacy Service | 2.20% |
| Fake Certificate | 21.23% | Financial Fraud | 1.46% |
| Black Hat SEO & Advertisement | 8.31% | Counterfeit Goods | 1.32% |
| Fake Account | 4.15% | Others | 1.29% |
| Data Theft | 3.29% | Money Laundering | 0.82% |
| Hacking Service | 2.41% | Weapon Sales | 0.26% |

**Categories.** Using the multi-label text classifier (Sect. 3.2), we categorize captured IPTs into 14 categories of services and goods. As shown in Table 4, the top categories with the most IPTs include Sex Service (28.29%), Gambling (22.56%), and Fake Certificate (21.23%). Notably, black hat SEO operators and other illicit advertisement services have also used RSP to promote their services, accounting for 8.31% of IPTs. Also, regarding the legitimacy of the promoted services and goods, over 45% IPTs are used to promote categories of service and goods that are commonly considered as illegal across different jurisdictions, e.g., Fake Certificate, Hacking Service, Financial Fraud, and Counterfeit Goods. On the other hand, other categories vary significantly in their legitimacy across different jurisdictions. For instance, many states in the United States consider gambling of some types as legal, while gambling in mainland China is forbidden. Besides, sex services (prostitution) are legal in Netherlands, while it is illegal in the United States except for Nevada. Therefore, IPTs under these categories are inherently controversial, which partly explains why the underlying operators of such services/goods choose RSP-based illicit promotion rather than legitimate promotion channels (e.g., Google Ads).

To understand which specific products are being promoted for each category, we examined 500 IPTs in each category to identify specific products. Particularly, in the category of Fake Account, fraudulent accounts of 20 different online platforms were for sale, e.g., Amazon, Twitter, Alipay, PayPal, Uber, and WeChat. Besides, In the category of Fake Certificate, we identified counterfeit services for 19 types of documents, e.g., identification cards and passports, diploma and transcripts of top universities, medical records, etc. Then, the category of Hacking Services encompassed not only services for DDoS attacks and domain hijacking, but also services for developing fraud platforms and gambling websites. Regarding the category of Drug Sales, 13 different illegal drugs were promoted, e.g., methamphetamine, sleeping pills and psychedelics. Furthermore, for the category of Money Laundering, our investigation revealed that criminals were trying to collect large numbers of bank cards and hire regular users to assist in money laundering. Please refer to Appendix C for typical products of all the categories.

**Natural Languages.** We further profiled the IPTs with regard to their natural languages, wherein the natural language of each IPT is identified through a language identification tool named *langid*. In total, we have observed 97 natural languages, with the top five being Chinese (88.08%), Korean (4.86%), English (1.66%), Japanese (1.48%), and Vietnamese (0.95%). As we can see, this language distribution of IPTs is contrasted with that of the whole Internet (Wikipedia, 2023) wherein the English accounts for almost half of all the web content. We then looked into the intermediate results of our IPT hunter and have thus confirmed the fidelity of this language distribution. And we believe several factors can jointly contribute to this skewed language distribution. One is the extensive use of black hat SEO (especially RSP) by the underground economy in China (H. Yang et al., 2017), which can lead to more IPTs composed in the natural languages of East Asia and Southeast Asia. The other explanation is that search engines especially Google and Bing vary a lot in their capabilities to filter out IPTs of different languages, and IPTs in CJK languages have a higher probability to bypass the filtering and get indexed by these search engines, which leads to the survivorship bias. We also compared IPTs of different natural languages in terms of their category distribution, for which more details can be found in Appendix D.

**Table 5.** Temporal evolution of IPTs with regards to scale.

| Group | Nov, 2022 | | Nov, 2023 | | Mar, 2025 | |
|---|---|---|---|---|---|---|
| | IPTs | Contacts | IPTs | Contacts | IPTs | Contacts |
| Google | 6,917,708 | 20,224 | 5,112,992 | 31,981 | 5,196,654 | 51,943 |
| Baidu | 66,595 | 2,024 | 25,150 | 1,521 | 34,400 | 2,420 |
| Sogou | 6,968 | 316 | 2,376 | 158 | 2,483 | 293 |
| Bing | 458,710 | 5,050 | 672 | 132 | 42,767 | 2,236 |
| Overall | 7,085,944 | 22,905 | 5,137,488 | 32,596 | 5,272,703 | 53,101 |

### 4.2 Longitudinal Evolution of Illicit Promotion via RSP

To profile the temporal evolution of IPTs, we deployed the IPT hunter in November 2022, November 2023, and most recently in March 2025, which gives three separate IPT datasets spanning over two years. Comparing the three separate IPT datasets reveals a set of interesting observations. First of all, by 2025, RSP-based illicit promotion remains an active threat, still comparable in scale and diversity to those observed in previous years. However, we observe a continued high degree of dynamism in the IPT ecosystem. Specifically, 99.81% of the IPTs identified in 2025 were newly captured, not observed in either 2022 or 2023. Similarly, 83.77% of the IPT contacts in 2025 were new, underscoring the persistent adaptability and rapid turnover of illicit promotion actors. This further reinforces the notion of an ongoing cat-and-mouse game between regulatory countermeasures and illicit operators.

As shown in Table 5, most search engines still suffer from a similar extent of IPT poisoning. However, for Bing, the number of IPTs detected dropped dramatically from 458K in November 2022 to just 672 in November 2023. This sharp decline followed our responsible disclosure to Bing in October 2023, and we believe the mitigation measures they deployed were largely effective in the short term. Yet by March 2025, we observed a notable resurgence of IPTs on Bing, with 42K IPTs captured, indicating either a relaxation of previous enforcement or the emergence of new tactics that bypassed prior defenses.

When it comes to the evolution of IPT categories, some have seen notable and consistent shifts in prevalence over time. For instance, the share of IPTs related to Sex Service has grown steadily, rising from 22.87% in 2022 to 29.15% in 2023, and further to 37.94% in 2025, making it the dominant category in the final results. Similarly, Data Theft has shown a persistent upward trend, increasing from 1.54% in 2022 to 4.47% in 2023 and 5.20% in 2025, indicating growing abuse related to personal and organizational information leakage. On the other hand, some previously dominant categories have declined. For example, the category of Gambling has seen a drop in the IPT share from 26.34% in 2022 to 19.79% in 2023, and further to 18.73% in 2025, likely due to the dissipation of event-driven spikes such as the FIFA World Cup in 2022, as well as possible takedown efforts.

### 4.3 Websites Abused in Illicit Promotion

We then profile the high-ranking websites that have been abused in RSPs for distributing IPTs.

**Scale and Distribution.** In a nutshell, a large volume of high-ranking websites has been abused in RSP attacks, while each abused website can be free-rode by IPTs belonging to different promotion campaigns. Specifically, 425,061 distinct URL reflection schemes (URSes) have been abused in one or more RSPs, which belong to 149,278 fully qualified domain names (FQDNs) and 65,294 apex domains. Examples of mostly abused URSes and apex domains are listed in Appendix E. Also, a long-tailed distribution of IPTs over URSes is observed, as the top 5% URSes have accounted for 87.16% RSPs and 87.94% IPTs while they are 95.52% RSPs and 95.91% IPTs for the top 10% URSes. Furthermore, one domain name may have up to hundreds of URSes abused in reflected search poisoning. For example, *baidu.com*, an apex domain owned by the same company operating Baidu Search, has 270 URL reflection schemes being abused by miscreants, e.g., http://zhidao.baidu.com/search (among the top 20 mostly abused URL schemes), and https://ai.baidu.com/search. The famous video website YouTube also has 29 URSes abused, e.g., the in-site search page https://music.youtube.com/search and the hashtag page https://youtube.com/hashtag.

**Table 6.** Abuse of most popular websites in RSP attacks.

| Group | Abused Domains | % IPTs | % RSPs |
|---|---|---|---|
| Top 100 | 51 | 7.18% | 7.22% |
| Top 1K | 400 | 14.89% | 15.33% |
| Top 10K | 2,707 | 26.26% | 26.14% |
| Top 100K | 12,127 | 45.54% | 44.34% |
| Top 1M | 33,910 | 67.57% | 65.49% |

**Reputation and Popularity.** One more interesting question is regarding whether websites abused in RSPs indeed have a good reputation, so as to increase the likelihood of getting their webpages (including RSPs) indexed by search engines with high page ranks. To answer this question, we use the metric of website popularity as an approximation to website reputation. To profile the popularity of each abused website, we referred to Tranco (Pochat, van Goethem, & Joosen, 2018), a top site ranking with 1 million websites listed. As a result, we observed that many abused websites are highly popular. Specifically, as listed in Table 6, among the aforementioned 65,294 apex domains, 33,910 (51.93%) show up in the top one million, 2,707 (4.15%) in the top 10K, and 51 (0.08%) even rank in the top 100. Besides, these abused top websites contribute most of the

IPTs. Particularly, 67.57% IPTs were observed from websites in the top one million. Also, 12,127 of the top 100K websites, while accounting for only 18.57% of all abused websites, have contributed 45.54% IPTs and respective 44.34% RSPs.

**The Functions of URL Reflection Schemes.** What is also found is that these abused URL schemes are intended for many different but all legitimate usage scenarios. Among the top 100 most abused URL reflection schemes, 80% are intended for onsite search, 11% are designed for hashtags, and 6% are for web dictionary or translation. There are also some other types of URL schemes, for example, https://minecraft.fandom.com/zh/wiki/XX is a wiki page for "XX". However, the website may still reflect the content of the URL even if "XX" is not a valid term in the wiki.

(a) A Chinese keyword searching for a location name *Beijing, Fengtai.*

(b) A Chinese keyword searching for sex services.

(c) A Chinese keyword searching for *price of Dior women's athletic shoes.*

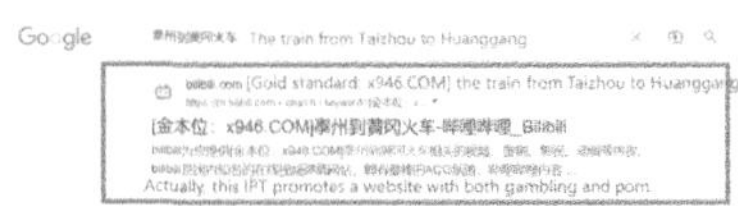


(d) A Chinese keyword searching for *the train from Taizhou to Huanggang.*

**Fig. 4.** Cases of search keywords of different categories that will likely expose search user to IPTs.

## 4.4 The Exposure of Search Users to Illicit Promotion

Given the prevalence of IPTs in search engines, we examined how search users might be exposed. We identified three keyword groups that commonly expose users to IPTs: location names, keywords of illicit goods/services, and benign long-tail keywords.

The first type is keywords denoting location names. For example, when searching in Chinese for *Fengtai, Beijing* (a district of Beijing, China) on Google, the searching results contain several IPTs promoting either loan services or sex services, as shown in Fig. 4a. To further investigate this phenomenon, we used 3,368 Chinese city names to query four search engines and found that Google and Bing have been heavily poisoned, whereas Baidu and Sogou were not, possibly due to their extra efforts to filter such poisoned results. When querying Google with these city names, 1,557 out of 3,368 queries (46%) have been poisoned with one or more IPTs in the top 10 result entries, and it is 94% for the top 50 result entries.

The second type encompasses keywords of services and goods especially illicit ones, which is straightforward and easy to understand. For instance, as shown in Fig. 4b, if we search in Chinese for *Looking for a girl* on Google, the first page is filled with Chinese IPTs related to sex services.

The third type describes that if we search for some benign long-tail keywords, the search engine may return IPTs embedding those keywords. For example, as shown in Fig. 4c, if we search in Chinese for *the price of Dior female athletic shoes* on Google, it returns several IPTs about counterfeit shoes. Obviously, such IPTs are embedded with benign long-tail keywords and are used to promote related illicit services/goods. In another interesting case as shown in Fig. 4d, when searching in Chinese for *the train from Taizhou to Huanggang*, Google returns an IPT promoting a website of gambling and porn videos. However, the long-tail keyword and the theme of this IPT are completely unrelated. And many IPTs have been found to embed long-tail keywords unrelated to their themes, likely in an attempt to reach a much broader search audience.

## 5 Next Hops of Illicit Promotion

In this section, we provide an in-depth analysis of contacts extracted from IPTs, focusing on two primary questions: (1) What security risks do these next hops expose a victim to? (2) What evasion techniques are adopted to avoid detection or blocking?

### 5.1 IPT Websites

Our website infiltrator, as detailed in Sect. 3.2, was scheduled for weekly dynamic visits to each IPT website using a headless browser. Each visit generated a snapshot comprising a landing page screenshot and the corresponding traffic logs. Analysis of the 35,750 website domains embedded in IPTs revealed 172 distinct top-level domains (TLDs). The most prevalent TLDs were *com* (60.60%), followed by *cc* (7.10%) and *xyz* (4.13%). From the 9,149 websites collected in 2022, we successfully resolved 5,721 to IP addresses. Subsequently, we captured 125,715 snapshots for 5,055 of these IPT websites, while the remaining sites were unreachable during our infiltration window. The findings indicate that IPT websites are equipped with diverse evasion techniques and frequently redirect users to various categories of unsolicited or illicit content and services.

**Security Risks of IPT Websites.** We analyzed the security risks of IPT websites by categorizing their content. From a random sample of 1,200 IPT websites, we reviewed snapshots and categorized them. The distribution is shown in Table 7. Among the top 14 categories, 8 are illicit or unsolicited, such as gambling, sex services, drug sales, and hacking services. This mirrors our IPT findings (Sect. 4.1). Particularly, gambling websites have the largest share of 22%. Also, they are widely distributed in natural languages and target customers of different countries including China, Vietnam, Thailand, Korea, etc. What is

**Table 7.** The categories of IPT websites.

| Category | % Websites | Category | % Websites |
|---|---|---|---|
| Gambling | 22.08% | Financial Fraud | 1.50% |
| Access Blocked[a] | 20.50% | Drug Sales | 1.00% |
| Benign | 17.92% | Fake Account | 0.83% |
| Domain Expired | 14.16% | Hacking Service | 0.75% |
| Sex[b] | 11.17% | Advertisement | 0.50% |
| Redirection Page | 7.00% | Data Theft | 0.33% |
| Black Hat SEO | 1.92% | Ghost Writing | 0.17% |

[a] This is the group of websites that have blocked our visit.
[b] Websites in this group either host pornography content or offer sex services.

followed are IPT websites either promoting sex services or hosting pornography content, which accounts for 11%. For instance, the website *ncao3.com* hosts child pornography content, *topjqk.com* advertises hacking services, and *a9play2u.com* is a gambling website.

Additionally, 21% of the websites blocked our automated access. Another 18% appeared benign but manual checks showed many lead to non-benign pages using evasion techniques, and more details will be discussed below. Furthermore, 7% of websites land on benign-looking pages with multiple redirection links leading to illicit content. These are categorized as redirection pages. Without clicking the links, it's hard to determine if the website is illicit.

**Malicious Mobile Apps Distributed Through IPT Websites.** As elaborated in Sect. 3.2, we manually downloaded and analyzed Android apps promoted by IPT websites. We collected 200 distinct APK files from 1,200 IPT websites, including 123 gambling and 66 pornography apps. Subsequently, we queried VirusTotal with the hash values of these APKs and found out that only 68 had historically been observed by VirusTotal. We then uploaded the payloads of the missing 132 APKs to VirusTotal for further analysis. Out of all APKs, 98 were detected as malicious, spanning seven threat types. The most common were trojans (32.38%) and riskware (24.76%). Full threat categories and counts are in Appendix F.

**Evasion Techniques of IPT Websites.** We have observed the adoption of multiple evasion techniques by IPT websites, likely in an attempt to escape detection radars. And representative evasion techniques include the lengthy redirection chain, iframe cloaking, and location-based access control. Below, we provide more details on these evasion techniques along with demonstrative examples.

*The Lengthy Redirect Chain.* Different from typical benign websites, many IPT websites have extensive redirect chains before reaching the landing page. Table 8 shows the distribution of 125,715 website snapshots by redirect length, defined as the number of distinct FQDNs visited before reaching the landing page. About 13.62% involve three or more redirects, with 32 snapshots having ten or more. For example, *go.win* redirects through 8 FQDNs before landing on *kingpro.vip*, a gambling site. Long redirect chains can defer automatic crawlers and evade content moderation. Google, for instance, follows only up to 5 redirects per crawl (Southern, 2020). Thus, websites with 6 or more redirects can likely avoid detection.

**Table 8.** The distribution of IPT websites across the length of their redirect chains.

| Redirects | Snapshots | % Snapshots | Websites | % Websites |
|---|---|---|---|---|
| $\geq 10$ | 31 | 0.02% | 2 | 0.04% |
| $\geq 5$ | 1,144 | 0.91% | 65 | 1.29% |
| $\geq 4$ | 3,615 | 2.88% | 210 | 4.15% |
| $\geq 3$ | 17,118 | 13.62% | 901 | 17.82% |
| $\geq 2$ | 61,926 | 49.26% | 3,056 | 60.45% |
| $\geq 1$ | 125,715 | 100.00% | 5,055 | 100.00% |

*Iframe Cloaking.* Some websites utilize the HTML iframe tag to embed illicit webpages into benign pages. Here is a typical example: the website http://nbjsfl.com/ initially appears benign, but later an iframe is rendered with gambling content, and the original content is replaced without any page jump. If a URL detection system stopped rendering the iframe when visiting this website, this website would likely be considered benign.

*Location-Based Access Control.* Some websites limit access based on the visitor's IP location. For instance, when attempting to access https://kkh1.xyz/ from the United States, the final landing page is *google.com*. However, when accessing the site from China, users are redirected to a page containing multiple redirect links towards child pornography websites. We then compared the results of our dynamic website crawlers deployed on a US server and a server located in China on Apr. 4, 2023. Out of 4,827 IPT websites reachable on that day, 546 (11.31%) have very different web content.

### 5.2 Instant Messaging Accounts

Similar to IPT websites, IM accounts embedded in IPTs can also expose victims to illicit goods and services of diverse categories. On the other hand, a large volume of illicit promotion activities has also been observed on Telegram.

**WeChat and QQ Accounts.** Our IPT contact extractor identified 23,632 WeChat IDs and 1,552 QQ IDs. However, due to the lack of programmable APIs, we conducted manual case studies. We examined 250 randomly selected

(a) A WeChat account related to forging diplomas.

(b) A QQ account related to forging diplomas.

**Fig. 5.** Examples of WeChat and QQ accounts.

WeChat profiles, finding 64 suspended. Among the remaining, 154 (83%) promoted illicit goods/services, which can be identified with high confidence through their profile, as shown in the Fig. 5a. The top categories of illicit goods/services are Fake Certificate (33%), Surrogacy (17%), Counterfeit Goods (11%), and Sex Service (10%). For QQ, we examined 200 accounts, finding 6 suspended. Of the remaining, 172 (89%) were associated with illicit services/goods, identified by their username, avatar, user tag, and profile photo wall, as demonstrated in the Fig. 5b. The top categories of illicit goods/services are Fake Certificate (40%), Sex Service (16%), and Gambling (12%).

**Telegram Accounts.** As detailed in Sect. 3.3, we infiltrated 4,732 Telegram accounts embedded in IPTs, comprising 1,507 users, 231 bots, 2,333 channels, and 661 groups. We retrieved profiles and historical messages from groups and channels, collecting over 14 million messages from January 2022 to March 2023. These data can help us understand the security risks victims of IPTs may face.

Given the large volume of Telegram messages, we then utilized the multi-label IPT classifier introduced in Sect. 3.2 to classify these messages. As listed in Table 9, over 6 million (43.75%) messages relate to illicit goods and services, including money laundering (31.96%), black hat SEO (17.57%), data theft (13.93%), gambling (12.46%), and financial fraud (8.49%). Additionally, we extracted the audience data for channels and groups. For channels operating as broadcast entities, there are over 29 million subscribers. However, this number reflects the subscriber count reported by Telegram and may include dormant or inactive accounts. In contrast, for groups used for communication, there are 606K distinct members. This number is closer to the actual active participants, further indicating the extensive use of Telegram by miscreants for promoting illicit goods and services and communicating with customers.

**Table 9.** Distribution of Telegram group/channel messages over the categories of goods and services.

| Category | % TG Msgs | Category | % TG Msgs |
|---|---|---|---|
| Money Laundering | 31.96% | Fake Account | 3.53% |
| Black Hat SEO & Advertisement | 17.57% | Others | 1.78% |
| Data Theft | 13.93% | Counterfeit Goods | 0.70% |
| Gambling | 12.46% | Fake Certificate | 0.60% |
| Financial Fraud | 8.49% | Drug Sales | 0.30% |
| Sex Service | 4.87% | Weapon Sales | 0.23% |
| Hacking Service | 3.57% | Surrogacy Service | 0.01% |

## 6 Discussion

**Mitigation Recommendations.** For search engine operators, more efforts should be invested in filtering out IPTs and preventing them from reaching regular search users. In this case, the series of our binary IPT classifiers can contribute to this task. Particularly, when detection efficiency is the bottleneck, the feature-based Random Forest classifier can be deployed to enable high-throughput IPT detection. On the other hand, when efficiency is not a matter but the accuracy, the BERT-based classifier can help achieve higher detection performance. Also, for a website with a legitimate URL reflection scheme, simple measures may also be taken to avoid abuse from miscreants. For instance, the text value of a reflection parameter should be omitted when rendering the resulting webpage, as long as it is abnormal, e.g., it has no search results in the scenario of on-site searching, or it is not a valid hashtag in the scenario of hashtag searching. Once deployed with such a measure, the IPTs (i.e., the search keywords) will not be rendered as part of the webpage and will not be indexed by the search engines.

**Limitations and Future Works.** Firstly, some forum sites, like Medium, incorporate the blog post title into the URL, making forum spam posts appear as RSPs. However, we conducted a manual inspection, and it revealed that only 5% of the RSP dataset are such forum spams, with the majority being true RSPs, minimally impacting our measurements. Secondly, our machine learning classifiers may be vulnerable to adversarial attacks. IPT operators might adjust IPT composition to evade detection, making our feature set less effective. Evaluating the adversary resistance of our models is future work. Additionally, many IPTs are promoted through channels beyond search engines, such as online social networks. Exploring IPT distribution through these channels and their evasion of content moderation is another area for future research.

**Responsible Disclosure.** We have conducted responsible disclosure to two groups of relevant parties. One is the communication with the four search engines that are poisoned by IPTs. By this writing, Bing has responded to our disclosure and stated that it is working to address this poisoning issue, which may explain the obvious drop in IPTs observed for Bing between our first-phase hunting in November, 2022 and the second-phase hunting in November, 2023. However, for the other three search engines, we have yet to receive any concrete response. Then, it is the disclosure towards the instant messaging platforms on which a large number of IPT accounts are observed, especially WeChat, QQ, and Telegram. Despite receiving some preliminary replies, the disclosure towards these instant messaging platforms is ongoing, and we will update it here once the final concrete results are available.

**Data and Code Release.** To prevent miscreants from launching new RSP attacks, we hide the vulnerable URSes and make the rest of the data public which include the IPT search keywords, the IPTs, the IPT contacts of various categories and the messages collected from Telegram. Also, in our study, multiple machine learning models have been built to detect and understand IPTs, e.g., the binary/multi-label IPT classifiers, and the IPT contact extractor. For these models, we open-source the scripts for training and testing as well as the ground-truth datasets. The data and code are available and will be continuously updated on https://chasesecurity.github.io/Reflected-Search-Poisoning-for-Illicit-Promotion/.

## 7 Conclusion

Through this study, we can conclude that reflected search poisoning (RSP) has been extensively exploited to free ride high-rank websites and distribute illicit promotion texts (IPTs) that are large-scale, available across search engines, as well as being diverse in categories of goods and services. Also, most services and goods promoted in IPTs belong to categories that are illicit or illegal, while regular search engine users can be exposed to such IPTs at a concerning extent. Also, victims of IPTs can be further exposed to harmful content and illegal services when following the next-hop contacts embedded in IPTs. All these results highlight the necessity of more efforts to be invested in the fight against illicit promotion and the underground economy.

## A Extra Details for the Binary IPT Classifier

### A.1 Manually Crafted Features for the Binary IPT Classifier

**Length in Characters.** Most IPTs are lengthy with a median length of 46 characters, but benign texts are more likely to be shorter.

**Number of Bracket-Like Characters.** IPTs are more likely to have several bracket-like characters (e.g. {, }, [, ], 【, 】, 『, 』) to highlight contact entities.

**Number of URLs in IPT.** Some IPTs embedding website contact entities may have URLs.

**Number of Numeric Characters.** IPTs with wechat, qq or telephone contacts often have numeric characters.

**Number of Emojis and Unicode Symbols.** Some IPTs use emojis or meaningless symbols to make themselves eye-catching.

**Number of Patterns About Next-Hop Instant Messaging Accounts.** These patters include [‘微信’, ‘q微’, ‘扣微’, ‘微’, ‘薇’, ‘扣扣’, ‘qq’, ‘www’, ‘com’, ‘fun’, ‘cc’, ‘tg’, ‘telegram’, ‘飞机’, ‘@’, ‘网’, ‘V信’].

**If there is a File Suffix.** Common file suffixes for URLs include [‘.html’, ‘.shtml’, ‘.htm’, ‘.php’, ‘.pdf’, ‘.jpg’, ‘.jpeg’, ‘.png’, ‘.xlsx’, ‘.docx’, ‘.pptx’, ‘.xml’]. Rather than benign URLs, IPTs tend to have no such kinds of suffixes.

### A.2 The List of Features for the Contact Segment Classifier

**Length in Characters.** The contact segments are more likely to be shorter than other segments of IPTs.

**Number of URLs in IPT.** Some segments of IPTs embedding website contact entities may have URLs.

**Number of Non-alphanumeric Characters.** The contact segments are more likely to be less non-alphanumeric characters than other segments of IPTs.

**Number of Alphanumeric Characters.** The contact segments are more likely to be more alphanumeric characters than other segments of IPTs.

**Number of Numeric Characters.** Segments with wechat, qq or telephone contacts often have numeric characters.

**Number of Contact Indicators.** The indicators of next-hop contacts include [‘微信’, ‘q微’, ‘扣微’, ‘微’, ‘薇’, ‘扣扣’, ‘qq’, ‘com’, ‘fun’, ‘cc’, ‘hash’, ‘tg’, ‘telegram’, ‘飞机’, ‘@’, ‘网’, ‘复制’]. Those marks often represent the contact information category, e.g., *tg* for Telegram, and ‘扣扣’ for QQ.

**Number of Some Common Punctuation Marks in Terms.** Various punctuation symbols are used to separate an IM mark and the respective IM account identifier, e.g., the colon in “询telegram:@ts775”. Symbols under our consideration include [., :, ：，，·, , -].

**If there is a File Suffix.** Common file suffixes for URLs include [‘.html’, ‘.shtml’, ‘.htm’, ‘.php’, ‘.pdf’, ‘.jpg’, ‘.jpeg’, ‘.png’, ‘.xlsx’, ‘.docx’, ‘.pptx’, ‘.xml’]. Rather than benign URLs, IPTs tend to have no such kinds of suffixes.

## B Category-Specific Performance for the Multi-label IPT Classifier

Table 10 shows the per-class performance of the multi-label IPT classifier. It reports precision, recall, and F1-Score for different categories of services and goods.

**Table 10.** Per-class performance for the multi-label IPT classifier.

| Category | Precision | Recall | F1-Score |
|---|---|---|---|
| Benign | 0.95 | 0.96 | 0.96 |
| Black Hat SEO & Adv. | 0.95 | 0.95 | 0.95 |
| Counterfeit Goods | 0.83 | 0.88 | 0.86 |
| Data Theft | 0.94 | 0.96 | 0.95 |
| Fake Account | 1.00 | 0.92 | 0.96 |
| Fake Certificate | 0.98 | 0.91 | 0.94 |
| Financial Fraud | 0.70 | 0.78 | 0.74 |
| Gambling | 0.94 | 0.95 | 0.95 |
| Hacking Service | 0.90 | 0.85 | 0.88 |
| Drug Sales | 0.97 | 0.83 | 0.89 |
| Sex Service | 0.94 | 0.93 | 0.94 |
| Surrogacy Service | 0.94 | 0.94 | 0.94 |
| Weapon Sales | 1.00 | 0.83 | 0.91 |
| Money Laundering | 0.91 | 0.91 | 0.91 |
| Others | 0.97 | 0.97 | 0.97 |

# C Products Promoted in IPTs

**Black Hat SEO & Advertisement.** Services in the category of the Black Hat SEO & Advertisement support illicit promotion across many social media platforms which include:

- Facebook
- Telegram
- WhatsApp
- Zalo
- Skype
- YouTube
- Google
- Signal
- Twitter
- Shopee
- Instagram
- TikTok
- QQ

**Fake Certificate.** The list of products as observed through manual study include:

- The Resident Identity Card in China
- Visa
- Passport
- Birth Certificate
- Title Deed
- Household Registration
- Letter of Employment
- Police Clearance Certificate
- Resignation Letter
- Professional Qualification Certificate
- Invoice
- Marriage Certificate

- Divorce Certificate
- Driver's License
- Business License
- Diploma
- Transcript
- Tuition Bill
- Medical Record

**Fake Account.** The list of products as observed through manual study include:

- SIM Card
- Bank Card
- Amazon Account
- Google Account
- Twitter Account
- Telegram Account
- LinkedIn Account
- Alipay Account
- QQ Account
- WeChat Account
- TikTok Account
- KuaiShou Account
- Xiaohongshu Account
- Uber Account
- CMB Account
- Snapchat Account
- Skype Account
- OkCupid Account
- Momo Account
- PayPal Account

**Drug Sales.** The list of products as observed through manual study include:

- Marijuana
- Methamphetamine
- Sedatives
- Gamma-hydroxybutyrate (GHB)
- Sevoflurane
- Alprazolam
- Midazolam
- Desflurane
- Triazolam
- Clonazepam
- Viagra
- Thuocduame
- Estazolam

**Weapon Sales.** The list of products involved as observed through manual study include:

- Gun
- Bullet
- Slingshot
- Electric Baton
- High-pressure Gas Cylinder Valve

## D The Category Distribution for IPTs in Different Languages

Table 11 lists the category distribution of IPTs in different natural languages. The most promoted category in Chinese is Gambling, accounting for 26.66% of observed IPTs. The second and third most promoted categories in Chinese IPTs are Fake Certificate (26.29%) and Sex Services (20.92%), respectively. And in Korean, the most promoted category is Sex Services, accounting for 71.95% of observed IPTs. This is because in Korea, the underground industry of phone sex services is relatively large in scale, but other types of grey industries are relatively small in scale. Furthermore, categories of Gambling (52.98%) and Financial Fraud (27.01%) account for a very high percentage of English IPTs.

**Table 11.** Top 10 most promoted categories for IPTs in different natural languages

| Chinese | | Korean | | English | |
|---|---|---|---|---|---|
| Gambling | 26.66% | Sex | 71.95% | Gambling | 52.98% |
| Fake Certificate | 26.29% | SEO & Ads | 8.83% | Financial Fraud | 27.01% |
| Sex | 20.92% | Drug Sales | 6.60% | Drug Sales | 8.21% |
| SEO & Ads | 9.93% | Financial Fraud | 6.40% | Sex | 5.01% |
| Fake Account | 4.97% | Gambling | 3.52% | Counterfeit | 2.26% |
| Hacking Service | 3.07% | Counterfeit | 1.24% | Fake Account | 1.29% |
| Data Theft | 1.71% | Fake Account | 0.73% | Others | 1.03% |
| Counterfeit | 1.62% | Hacking Service | 0.40% | Fake Certificate | 0.90% |
| Surrogacy | 1.57% | Others | 0.22% | Money Laundering | 0.57% |
| Others | 1.08% | Fake Certificate | 0.03% | SEO & Ads | 0.91% |

# E More Details for Mostly Abused Websites

Table 12 lists top 10 mostly abused URL reflection schemes, while Table 13 gives top 10 mostly abused apex domains.

**Table 12.** Top 10 mostly abused URL reflection schemes.

| URL Scheme | % IPTs | % RSPs |
|---|---|---|
| https://search.bilibili.com/all?keyword=XX | 0.5002% | 0.4629% |
| https://spankbang.com/s/XX | 0.4522% | 0.4180% |
| https://www.goodreads.com/quotes/tag/XX | 0.4330% | 0.3866% |
| https://us.puma.com/us/en/search?q=XX | 0.3856% | 0.3527% |
| https://pngtree.com/free-png-vectors/XX | 0.3793% | 0.3387% |
| https://www.cnrtl.fr/definition/XX | 0.3553% | 0.3173% |
| https://www.royal.uk/search?tags[]=XX | 0.3532% | 0.3615% |
| https://olx.ba/pretraga?trazilica=XX | 0.3397% | 0.3043% |
| https://www.thedp.com/page/search?q=XX | 0.3319% | 0.2967% |
| https://www.pixiv.net/tags/XX | 0.3278% | 0.2988% |

**Table 13.** Top 10 mostly abused apex domains.

| Domain | % IPTs | % RSPs |
|---|---|---|
| baidu.com | 1.0757% | 1.3262% |
| pixnet.net | 1.0594% | 0.9512% |
| gfycat.com | 0.6437% | 0.6406% |
| facebook.com | 0.5838% | 0.5516% |
| pixiv.net | 0.5790% | 0.5581% |
| youtube.com | 0.5729% | 0.5132% |
| cnrtl.fr | 0.5613% | 0.5014% |
| bilibili.com | 0.5500% | 0.5125% |
| spankbang.com | 0.4842% | 0.4586% |
| goodreads.com | 0.4330% | 0.3866% |

# F Extra Data for Mobile Threats

Table 14 presents a comprehensive list of mobile threats associated with suspicious Android APKs, including the number and percentage of APKs affected by each type of threat.

**Table 14.** List of mobile threats alarmed by VirusTotal for Android apps distributed through IPT websites.

| Mobile Threat | APKs | % APKs |
|---|---|---|
| Trojan | 68 | 32% |
| Riskware | 52 | 25% |
| Malware | 40 | 19% |
| Evader | 26 | 12% |
| Adware | 12 | 6% |
| Malformed | 8 | 4% |
| Spyware | 4 | 2% |

# References

Aldridge, J., Askew, R.: Delivery dilemmas: how drug cryptomarket users identify and seek to reduce their risk of detection by law enforcement. Int. J. Drug Policy **41**, 101–109 (2017). https://doi.org/10.1016/j.drugpo.2016.10.010

Caballero, J., Grier, C., Kreibich, C., Paxson, V.: Measuring pay-per-install: the commoditisation of malware distribution. In: Usenix Security Symposium, vol. 13, pp. 1–13 (2011). https://www.usenix.org/conference/usenix-security-11/measuring-pay-install-commoditization-malware-distribution

Cao, Q., Sirivianos, M., Yang, X., Pregueiro, T.: Aiding the detection of fake accounts in large scale social online services. In: 9th USENIX Symposium on Networked Systems Design and Implementation (NSDI 2012), pp. 197–210. USENIX Association, San Jose, CA (2012). https://www.usenix.org/conference/nsdi12/technical-sessions/presentation/cao

Christin, N.: Traveling the silk road: a measurement analysis of a large anonymous online marketplace. In: Proceedings of the 22nd International Conference on World Wide Web, pp. 213–224 (2013). https://doi.org/10.1145/2488388.2488408

Devlin, J., Chang, M.-W., Lee, K., Toutanova, K.: Bert: pre-training of deep bidirectional transformers for language understanding. arXiv preprint arXiv:1810.04805 (2018)

Du, K., Yang, H., Li, Z., Duan, H., Zhang, K.: The ever-changing labyrinth: a large-scale analysis of wildcard DNS powered black-hat SEO. In 25th USENIX Security Symposium (USENIX Security 2016), pp. 245–262 (2016). https://www.usenix.org/conference/usenixsecurity16/technical-sessions/presentation/du

Gao, Y., Wang, H., Li, L., Luo, X., Xu, G., Liu, X.: Demystifying illegal mobile gambling apps. In: Proceedings of the Web Conference 2021, pp. 1447–1458 (2021). https://doi.org/10.1145/3442381.3449932

Grier, C., Thomas, K., Paxson, V., Zhang, M.: @spam: the underground on 140 characters or less. In: Proceedings of the 17th ACM Conference on Computer and Communications Security, pp. 27–37 (2010). https://doi.org/10.1145/1866307.1866311

Invernizzi, L., Thomas, K., Kapravelos, A., Comanescu, O., Picod, J.-M., Bursztein, E.: Cloak of visibility: detecting when machines browse a different web. In: 2016 IEEE Symposium on Security and Privacy (SP), pp. 743–758. IEEE (2016). https://doi.org/10.1109/SP.2016.50

John, J.P., Yu, F., Xie, Y., Krishnamurthy, A., Abadi, M.: deSEO: combating search-result poisoning. In: 20th USENIX Security Symposium (USENIX Security 2011) (2011). https://www.usenix.org/

Kanich, C., et al.: Show me the money: characterizing spam-advertised revenue. In: USENIX Security Symposium, vol. 35 (2011). https://www.usenix.org/conference/usenix-security-11/show-me-money-characterizing-spam-advertised-revenue

Kotzias, P., Bilge, L., Caballero, J.: Measuring PUP prevalence and PUP distribution through pay-per-install services. In: USENIX Security Symposium, pp. 739–756 (2016). https://www.usenix.org/conference/usenixsecurity16/technical-sessions/presentation/kotzias

Kotzias, P., Caballero, J., Bilge, L.: How did that get in my phone? Unwanted app distribution on android devices. In: 2021 IEEE Symposium on Security and Privacy (SP), pp. 53–69. IEEE (2021)

Leontiadis, N., Moore, T., Christin, N.: Measuring and analyzing search-redirection attacks in the illicit online prescription drug trade. In: USENIX Security Symposium, vol. 11 (2011). https://www.usenix.org/conference/usenix-security-11/measuring-and-analyzing-search-redirection-attacks-illicit-online

Leontiadis, N., Moore, T., Christin, N.: A nearly four-year longitudinal study of search-engine poisoning. In: Proceedings of the 2014 ACM SIGSAC Conference on Computer and Communications Security, pp. 930–941 (2014). https://doi.org/10.1145/2660267.2660332

Li, X., Yan, H., Qiu, X., Huang, X.: FLAT: Chinese NER using flat-lattice transformer. In: Jurafsky, D., Chai, J., Schluter, N., Tetreault, J. (eds.) Proceedings of the 58th Annual Meeting of the Association for Computational Linguistics, pp. 6836–6842. Association for Computational Linguistics (2020). https://doi.org/10.18653/v1/2020.acl-main.611

Li, Z., Du, X., Liao, X., Jiang, X., Champagne-Langabeer, T.: Demystifying the dark web opioid trade: content analysis on anonymous market listings and forum posts. J. Med. Internet Res. **23**(2), e24486 (2021). https://doi.org/10.2196/24486

Liao, X., et al.: Seeking nonsense, looking for trouble: efficient promotional-infection detection through semantic inconsistency search. In: 2016 IEEE Symposium on Security and Privacy (SP), pp. 707–723. IEEE (2016). https://doi.org/10.1109/SP.2016.48

Lin, Y.-C. J.: Fake Stuff: China and the Rise of Counterfeit Goods. Routledge (2011)

Lu, L., Perdisci, R., Lee, W.: Surf: detecting and measuring search poisoning. In: Proceedings of the 18th ACM conference on Computer and Communications Security, pp. 467–476 (2011). https://doi.org/10.1145/2046707.2046762

McCoy, D., et al.: Pharmaleaks: understanding the business of online pharmaceutical affiliate programs. In: Proceedings of the 21st USENIX Conference on Security Symposium, p. 1 (2012). https://www.usenix.org/conference/usenixsecurity12/technical-sessions/presentation/mccoy

Niu, Y., Chen, H., Hsu, F., Wang, Y.-M., Ma, M.: A quantitative study of forum spamming using context-based analysis. In: Proceedings of the Network and Distributed System Security Symposium, NDSS 2007 (2007). https://www.ndss-symposium.org/ndss2007/quantitative-study-forum-spamming-using-context-based-analysis/

Pochat, V.L., van Goethem, T., Joosen, W.: Tranco: A Research-Oriented Top Sites Ranking Hardened Against Manipulation. CoRR, abs/1806.01156 (2018). http://arxiv.org/abs/1806.01156

Soska, K., Christin, N.: Measuring the longitudinal evolution of the online anonymous marketplace ecosystem. In: 24th USENIX Security Symposium, pp. 33–48 (2015). https://www.usenix.org/conference/usenixsecurity15/technical-sessions/presentation/soska

Southern, G.M.: Google's John Mueller Recommends Less Than 5 Hops Per Redirect Chain (2020). https://www.searchenginejournal.com/googles-john-mueller-recommends-less-than-5-hops-per-redirect-chain

Thomas, K., et al.: Investigating commercial pay-per-install and the distribution of unwanted software. In: USENIX Security Symposium, pp. 721–739 (2016). https://www.usenix.org/conference/usenixsecurity16/technical-sessions/presentation/thomas

Wang, D.Y., Savage, S., Voelker, G.M.: Cloak and dagger: dynamics of web search cloaking. In: Proceedings of the 18th ACM Conference on Computer and Communications Security, pp. 477–490 (2011). https://doi.org/10.1145/2046707.2046763

Wang, D.Y., Savage, S., Voelker, G.M.: Juice: a longitudinal study of an SEO botnet. In: NDSS (2013). https://www.ndss-symposium.org/ndss2013/ndss-2013-programme/juice-longitudinal-study-seo-botnet/

Wang, P., et al.: Game of missuggestions: semantic analysis of search-autocomplete manipulations. In: NDSS (2018). https://doi.org/10.14722/ndss.2018.23036

West, J., Bhattacharya, M.: Intelligent financial fraud detection: a comprehensive review. Comput. Secur. **57**, 47–66 (2016). https://doi.org/10.1016/j.cose.2015.09.005

Wikipedia. Languages used on the Internet (2023). https://en.wikipedia.org/wiki/Languages_used_on_the_Internet

Yan, H., Deng, B., Li, X., Qiu, X.: TENER: adapting transformer encoder for named entity recognition. arXiv preprint arXiv:1911.04474 (2019). https://arxiv.org/abs/1911.04474

Yang, H., et al.: Casino royale: a deep exploration of illegal online gambling. In: Proceedings of the 35th Annual Computer Security Applications Conference, pp. 500–513 (2019). https://doi.org/10.1145/3359789.3359817

Yang, H., et al.: How to learn klingon without a dictionary: detection and measurement of black keywords used by the underground economy. In: 2017 IEEE Symposium on Security and Privacy (SP), pp. 751–769. IEEE (2017). https://doi.org/10.1109/SP.2017.11

Yang, R., et al.: Scalable detection of promotional website defacements in black hat SEO campaigns. In: USENIX Security Symposium, pp. 3703–3720 (2021). https://www.usenix.org/conference/usenixsecurity21/presentation/yang-ronghai

Yuan, K., Lu, H., Liao, X., Wang, X.: Reading thieves' cant: automatically identifying and understanding dark jargons from cybercrime marketplaces. In: USENIX Security Symposium, pp. 1027–1041 (2018). https://www.usenix.org/conference/usenixsecurity18/presentation/yuan-kan

Zhang, Q., Wang, D.Y., Voelker, G.M.: DSpin: detecting automatically spun content on the web. In: NDSS (2014). https://doi.org/10.14722/ndss.2014.23004

Zhao, K., Zhang, Y., Xing, C., Li, W., Chen, H.: Chinese underground market jargon analysis based on unsupervised learning. In: 2016 IEEE Conference on Intelligence and Security Informatics (ISI), pp. 97–102. IEEE (2016). https://doi.org/10.1109/ISI.2016.7745450

Zhou, Y., et al.: Analyzing and detecting money-laundering accounts in online social networks. IEEE Network **32**(3), 115–121 (2018). https://doi.org/10.1109/MNET.2017.1700213

Zhu, W., et al.: Self-supervised euphemism detection and identification for content moderation. In: 2021 IEEE Symposium on Security and Privacy (SP), pp. 229–246. IEEE (2021). https://doi.org/10.1109/SP40001.2021.00075

# Method and Model for Analyzing the Rationality of Medical Visits Based on Automatic Machine Learning

Fangrong Wu, Zengqiang Lei(✉), and Jingyan Zhang

Hunan Automotive Engineering Vocational University, Zhuzhou 412001, China
leizengq@126.com

**Abstract.** With the continuous deepening of medical supervision, traditional analysis of the rationality of medical visits mainly relies on experienced experts, which suffers from issues such as low efficiency, significant influence from subjective factors, and limited coverage, making it difficult to comprehensively supervise medical visits. Therefore, this paper addresses these issues by constructing an analysis method and model for the rationality of medical visits based on automatic machine learning. This method collects data such as visit information, settlement information, and expense details from medical records. Through preprocessing and feature extraction, an analysis model is constructed using an automatic machine learning system, enabling automated analysis of the rationality of medical visits and identification of irregularities. This system has good generalization ability and robustness, solving the problems of existing technology focusing on a single scene, small coverage range, and inability to identify specific violations.

**Keywords:** Automated machine learning · rationality of medical visits · feature engineering · model fusion

## 1 Introduction

### 1.1 Research Background

With the rapid development of medical informatization, a large amount of medical data is collected and stored. The traditional method of analyzing the rationality of medical treatment mainly relies on manual screening, review, and empirical judgment from massive data. This method is inefficient and easily influenced by subjective factors, making it difficult to meet the needs of modern medical management (Shi, 2019). Therefore, how to effectively analyze and utilize these data to quickly and accurately determine the rationality of medical treatment has become an urgent problem to be solved.

### 1.2 Research Meaning

The method and model for analyzing the rationality of medical treatment based on automatic machine learning proposed in this study have the following important significance.


H. Yan et al. (Eds.): AIS&P 2025, CCIS 2918, pp. 220–226, 2026.
https://doi.org/10.1007/978-981-95-9884-7_15

**Improving Medical Efficiency.** Automated analysis of medical rationality has been achieved, greatly reducing the workload of manual review and improving analysis efficiency.

**Reducing Medical Costs.** Accurately identifying unreasonable medical behaviors can help reduce the waste of medical resources and lower medical costs.

**Improving Analysis Accuracy.** Avoiding subjective bias in manual judgment and enhancing the accuracy of rational analysis of medical treatment.

**Providing Decision Support for Medical Management.** It can provide valuable references for medical management departments to better understand the medical situation and formulate scientific management policies.

# 2 Method

## 2.1 Data Collection and Preprocessing

This study collected data such as medical information tables, settlement information tables, and expense detail tables. The detailed indicators of medical data are shown in Table 1. In the data preprocessing stage, the following tasks were mainly carried out.

**Table 1.** Detailed indicators of medical data.

| Medical ID | 1000 | 1001 | 1002 | 1003 |
|---|---|---|---|---|
| Gender | 2 | 1 | 1 | 2 |
| Age | 58 | 53 | 70 | 45 |
| Personnel type | 15 | 12 | 11 | 16 |
| Sector | 653 | 345 | 654 | 123 |
| UNIT TYPE | 1 | 10 | 57 | 50 |
| Main diagnostic code | D44.001 | I25.103 | B08.501 | J21.901 |
| Hospitalization days | 10 | 4 | 6 | 6 |
| Catalog type (list) | 201;201;201 | 201;201 | 201;201 | 201;201;201;201 |
| Quantity (List) | 1.0;1.0;2.0 | 1.0;1.0 | 3.0;1.0 | 1.0;1.0;1.0;1.0 |
| Cost (List) | 54.0;54.0;326.0 | 18.0;35.0 | 11.0;18.0 | 20.0;40.0;396.0;396.0 |
| Total cost | 8954.23 | 565.93 | 219.03 | 362.83 |
| Overall fund expenditure | 4787.77 | 140 | 46.55 | 0 |
| Personal expenses | 4166.46 | 425.93 | 172.48 | 362.83 |
| Hospital level | 5 | 2 | 5 | 2 |

**Data Cleaning.** Data cleaning refers to the final process of discovering and correcting identifiable errors in data files, including checking data consistency, handling invalid and missing values, etc.

(1) Missing value handling.

For missing values in the dataset that do not affect the analysis results, different filling strategies are adopted according to different feature types. For discrete features, use "-99999" to fill in missing values; For integer features, use 0 to fill missing values; For floating-point features, use the average value of the feature column to fill in missing values; For text-based features, use empty strings to fill missing values; For temporal features, fill in missing values with the non null row before the feature column.

(2) Outlier handling.

Exclude obviously abnormal data, such as abnormal data in the expense breakdown table where the total cost of the same medical insurance expense name is equal to or less than 0.

## 2.2 Feature Extraction

Extracting features related to the rationality of medical visits from preprocessed data mainly includes the following aspects.

**Basic Information Characteristics.** Medical ID, gender, age, personnel type, industry, unit type, etc.

**Medical Information Features.** Diagnostic primary code, secondary code, hospitalization code, length of hospital stay, frequency of visits, medical institution, etc.

**Cost Information Characteristics.** Total expenses, overall fund expenditures, total personal payments, hospital level, etc.

## 2.3 Construction of Automatic Machine Learning Models

Constructing a rationality analysis model for medical treatment using an automatic machine learning system, which mainly includes the following steps (Lu et al., 2024; Zhang et al., 2025; Fan et al., 2024; Toraskar et al., 2025; Jasmine and Marichamy, 2025; Nascimento et al., 2025; Karthikeyan and Rani, 2025).

**Identify Unreasonable Medical Treatment Groups.** The first step is to use the data source as the first sample data, use random forest training to obtain the weight scores of the features, and sort them. The second step is to sort the first sample data based on feature weight scores, iteratively traverse each feature count, and obtain the highest scoring feature count through recursive feature elimination and cross validation. The optimal feature subset is then selected as the second sample data. The third step is to automatically learn and train the second sample data to obtain the optimal model, and use this model to identify the group of unreasonable medical treatment.

**Filter Violation Characteristics and Types.** Based on automatic machine learning, train a model to predict whether medical behavior is reasonable, and output the importance of features that lead to unreasonable behavior. Through feature importance analysis, identify the medical insurance directory entries that have the greatest impact on the prediction results, and associate them with specific medical ids and violation types.

### 2.4 Model Evaluation

Multiple evaluation metrics are used to evaluate the model, including accuracy, precision, recall, F1 score, etc. For binary classification tasks, indicators such as AUC value, ROC curve, and P-R curve were also calculated; For the regression task, metrics such as MAE, MSE, MAPE, and R2 Score were calculated.

## 3 Experimental Results and Analysis

### 3.1 Experimental Data

This study used the medical dataset open to the Alibaba Cloud Tianchi dataset website for model training and testing. The dataset includes features such as medical ID, gender, age, personnel type, industry, unit type, diagnosis code, length of hospital stay, medical insurance catalog code, and expenses. The target feature is the illegal content of medical insurance catalog code.

### 3.2 Evaluation Indicators

The evaluation indicators used in this study mainly include accuracy, precision, recall, F1 score, etc. The specific calculation formula is as follows.

The recall rate (TPR or) is the proportion of samples with positive truth values that are predicted to be positive.

$$\mathrm{TPR} = \mathrm{recall} = \frac{\mathrm{TP}}{\mathrm{TP} + \mathrm{FN}} \tag{1}$$

Precision refers to the proportion of positive examples that are actually positive in the examples classified as positive examples.

$$\mathrm{precision} = \frac{\mathrm{TP}}{\mathrm{FP} + \mathrm{TP}} \tag{2}$$

Accuracy refers to the proportion of normal predicted samples in the total sample.

$$\mathrm{accuracy} = \frac{\mathrm{TP} + \mathrm{TN}}{\mathrm{TN} + \mathrm{FP} + \mathrm{TP} + \mathrm{FN}} \tag{3}$$

TP represents the actual number of positive samples predicted as positive samples, FN represents the actual number of positive samples predicted as negative samples, FP represents the actual number of negative samples predicted as positive samples, and TN represents the actual number of negative samples predicted as negative samples.

F1 score is a commonly used indicator of the accuracy of a measurement method, often used to determine the precision of an algorithm. At present, precision and recall are often mentioned separately in identification and detection related algorithms. F-score can consider both values simultaneously, balancing the accuracy of this algorithm.

$$F1 = \frac{2*precision*recall}{precision + recall} \tag{4}$$

### 3.3 Experimental Results

**Comparison of Model Evaluation Indicators.** The support vector classifier (SVC) and logistic regression (LR) were used to compare the above models. The comparison of model evaluation indicators is shown in Table 2, and the results indicate that the LR algorithm performs better in various indicators. For example, on the test set, the LR algorithm has an accuracy rate of 0.936 for reasonable visits, an accuracy rate of 0.94, a recall rate of 0.96, and an F1 score of 0.95; the accuracy rate for unreasonable medical visits is 0.92, the precision rate is 0.90, the recall rate is 0.91, and the F1 score is 0.91.

**Table 2.** Comparison of Model Evaluation Indicators.

| Number | Algorithm Name | Label | Accuracy | Precision | Recall | F1 |
|---|---|---|---|---|---|---|
| 1 | SVC | 0 reasonable | 0.930 | 0.94 | 0.95 | 0.95 |
| | | 1 unreasonable | | 0.91 | 0.88 | 0.90 |
| 2 | LR | 0 reasonable | 0.936 | 0.94 | 0.96 | 0.95 |
| | | 1 unreasonable | | 0.92 | 0.90 | 0.91 |

**Confusion Matrix Results.** The confusion matrix results of the test set prediction model are shown in Table 3. Among the actual normal samples, 31446 were correctly predicted as normal and 1408 were incorrectly predicted as abnormal. Among the actual abnormal samples, 1854 were incorrectly predicted as normal and 16146 were correctly predicted as abnormal. The confusion matrix results of the validation set prediction model are similar, further verifying the effectiveness of the model.

### 3.4 Result Analysis

The experimental results indicate that the proposed method and model for analyzing the rationality of medical visits based on automatic machine learning have high accuracy and reliability in the rational diagnosis of medical data. This model can effectively identify illegal medical records, reducing the workload of manual review and the influence of subjective factors. By analyzing the importance of features, specific types and reasons for violations can be identified, providing valuable references for medical management.

**Table 3.** The confusion matrix of the test set prediction model.

| Prediction category | Normal | Abnormal | Total |
|---|---|---|---|
| Normal | 31446 | 1408 | 32854 |
| Abnormal | 1854 | 16146 | 18000 |
| Total | 33300 | 17554 | 50854 |

# 4 Conclusion and Prospect

## 4.1 Conclusion

This study proposes a method and model for analyzing the rationality of medical visits based on automatic machine learning. The method achieves automated analysis of the rationality of medical visits and identification of violations through steps such as data preprocessing, feature extraction, automatic machine learning model construction, and model evaluation. The experimental results show that this method has a high accuracy in the rational diagnosis of medical data, and can effectively identify illegal medical records, providing strong support for medical management.

## 4.2 Outlook

Future research work can be carried out from the following aspects.

**Expand Data Scale.** Collect more medical data to further improve the model's generalization ability and prediction accuracy.

**Optimize Model Architecture.** Explore more advanced automatic machine learning algorithms and model fusion strategies to further improve model performance.

**Combining Domain Knowledge.** Incorporating More Medical Domain Knowledge into the Model to Improve Its Interpretability and Practicality.

**Real Time Analysis Application.** Study how to apply this model to real-time rationality analysis of medical visits, achieve real-time monitoring and early warning of medical behavior.

**Acknowledgments.** This work was financially supported by the Hunan Provincial Department of Education Scientific Research Project (24C0979).

# References

Fan, X.Y., Meng, F., Deng, J., et al.: Logging curve reconstruction technology based on automatic machine learning. Nat. Gas. Ind. **44**(09), 38–54 (2024)

Jasmine, S., Marichamy, P.: Automated detection of epicardial adipose tissue in cardiac CT using ensemble machine learning for improved diagnosis. MethodsX **14**, 103410 (2025)

Karthikeyan, R., Rani, S.B.: An innovative approach for obstacle avoidance and path planning of mobile robot using adaptive deep reinforcement learning for indoor environment. Knowl.-Based Syst. **326**, 114058 (2025)

Lu, K.K., Li, J., Liu, D.R., et al.: Daily NO2 simulation based on automatic machine learning ensemble model. Environ. Sci. **45**(10), 5740–5747 (2024)

Nascimento, V.A., Chaves, A.C., Maciel, T.S., et al.: DisperPy: A machine learning based tool to automatically pick group velocity dispersion curves from earthquakes. Comput. Geosci. **205**, 106015 (2025)

Shi, H.: The impact of pharmacists' participation in pharmaceutical services after COPD treatment on the rationality and safety of medication. Chinese Prescription Drugs **17**(11), 50–51 (2019)

Toraskar, S., Khan, A., Niranjannaik, M., et al.: AutoML-Fire: Automated machine-learning approach to predict forest fires. Environ. Model. Softw. **193**, 106578 (2025)

Zhang, Y.X., Chen, Z.C., Zhang, Y.H., et al.: Prediction of heavy metal concentration in crop grains based on automatic machine learning models. Chin. J. Appl. Ecol. **36**(06), 1889–1897 (2025)

# Research on Moisture Prediction Based on Spatio-Temporal Graph Neural Network

Haitao Yang, Ge Wen, Yiming Luo, Da Deng, and Xuan Chen(✉)

Meizhou Cigarette Factory, China Tobacco Guangdong Industrial Co., Ltd., Meizhou, Guangdong, China
chenx@gdzygy.com

**Abstract.** To address issues such as significant control deviations and insufficient stability caused by fluctuations in incoming materials, environmental changes, and measurement delays during the control of outlet moisture content in the loosening and conditioning process, this study developed a spatio-temporal graph neural network model based on Adaptive Graph Convolutional Recurrent Network(AGCRN) for moisture prediction. By adaptively learning the spatial topological relationships among sensors and integrating gated recurrent units to capture temporal dependencies, the model achieves accurate prediction of outlet moisture content. In response to the variability of industrial scenarios, a dynamic graph construction mechanism and a multi-objective loss function were introduced during training to further enhance the model's generalization and adaptive capabilities. Practical applications demonstrate that after system optimization, the control deviation of outlet moisture content was stably reduced from the original $\pm$ 1.2% to within $\pm$ 0.5%, and the process capability index CPK increased by 31%, effectively improving the control stability and quality consistency of the loosening and conditioning process.

**Keywords:** STGNN · AGCRN · Industrial Prediction · Moisture Control · Loosening and Conditioning

## 1 Introduction

Loosening and conditioning is a critical process in cigarette primary processing, aimed at increasing the moisture content and temperature of tobacco lamina, thereby enhancing the processability of tobacco leaves. The stability of the outlet moisture content directly affects the technical indicators of several subsequent processes, making effective control of the outlet moisture content in the loosening and conditioning stage particularly important. However, in actual production, the outlet moisture content still exhibits significant fluctuations due to factors such as variations in incoming moisture content and flow rate, measurement delays, and changes in temperature and humidity conditions.

In recent years, numerous scholars have conducted in-depth research on quality control in loosening and conditioning. For instance, Chen Deli et al. (2025) applied multiple regression and neural network algorithms to establish a predictive model for the water


H. Yan et al. (Eds.): AIS&P 2025, CCIS 2918, pp. 227–245, 2026.
https://doi.org/10.1007/978-981-95-9884-7_16

addition ratio during conditioning, increasing the qualification rate of outlet moisture content from 52.33% to 80.57%, effectively improving the quality control level. Zhang Fengguang et al. (2025) employed a simulated annealing algorithm to optimize a neural network prediction model and adopted a three-stage compensatory water addition control strategy, significantly enhancing the stability and control precision of the outlet moisture content of the loosening and conditioning machine. Xie Chenjian et al. (2025) analyzed key factors and formulated targeted improvement measures, effectively increasing the Cpk value and stability of the outlet moisture content, thereby providing quality assurance for subsequent processes. Li Zijuan et al. (2024) developed SVR and RF prediction models based on temperature and humidity interval division, accurately predicting moisture variation in loosening and conditioning under different environmental conditions, which notably improved prediction accuracy and control effectiveness. Wen Guangqiu et al. (2024) utilized principal component neural networks and genetic algorithms to optimize the water addition volume, increasing the process capability index of the outlet moisture content by 65%, effectively addressing the issues of significant fluctuations and long delays associated with traditional PID control. However, existing studies have overlooked the critical spatial topological relationships among multiple sensors within the conditioning drum. For instance, the temperature and airflow at the material inlet directly and dynamically influence the conditions at the outlet, and sensors positioned adjacently are physically coupled. Neglecting these spatial dependencies, coupled with certain limitations in modeling complex temporal patterns, results in systems with limited adaptability under dynamic operating conditions. This is particularly evident in challenges related to multi-point coordination, real-time prediction under abrupt changes, and intelligent feedforward control. Therefore, introducing advanced modeling methods capable of simultaneously capturing spatio-temporal dependencies is of great significance for further improving the accuracy and robustness of moisture control in loosening and conditioning.

To overcome the aforementioned challenges, this paper constructs a spatio-temporal modeling framework for moisture prediction in loosening and conditioning based on a spatio-temporal graph neural network, specifically the AGCRN model. This framework draws on the systematic review by Jin et al. (2024) on the application of graph neural networks in time series analysis, fully leveraging the advantages of GNNs in modeling spatio-temporal dependencies. Furthermore, addressing the issues of communication overhead and resource utilization in dynamic graph neural network training, it takes inspiration from the BLAD system proposed by Fu et al. (2023) for improving training efficiency. The main contributions of this paper include:

(1) The first application of the AGCRN model to moisture prediction tasks in the tobacco industry, achieving effective integration of time series and graph structural information, significantly enhancing prediction accuracy;
(2) Proposing a node embedding-based adaptive graph construction method capable of automatically identifying potential coupling relationships between sensors, overcoming the limitation of traditional methods that ignore spatial topology;
(3) Systematically conducting comparative experiments between AGCRN and various baseline models, validating the superiority of the proposed model in terms of prediction accuracy and stability across multiple performance metrics.

## 2 AGCRN Model

Modeling industrial processes like tobacco conditioning presents two primary challenges: capturing complex temporal dependencies (how past states influence future states) and modeling multivariate spatial relationships (how different sensors influence each other). Traditional time series methods like ARIMA can handle temporal patterns but struggle with nonlinearities and spatial effects. While machine learning models like XGBoost excel at capturing nonlinear relationships, they inherently lack the structure to process temporal sequences or spatial connections. To further enhance time series modeling capabilities, deep learning models such as LSTM and GRU have been introduced, which can handle long-term dependencies but are computationally complex and neglect spatial correlations. In recent years, spatio-temporal graph neural networks (STGNNs), particularly the AGCRN model, have achieved joint modeling of spatio-temporal dependencies by integrating graph convolutional and recurrent structures, significantly improving prediction accuracy and adaptability in complex industrial scenarios.

### 2.1 Overview of the AGCRN Model

AGCRN (Adaptive Graph Convolutional Recurrent Network) is a deep learning model specifically designed for spatio-temporal sequence forecasting. Its core idea involves adaptively modeling the dynamic coupling relationships between sensors through graph structures and capturing temporal dependencies using Gated Recurrent Units (GRUs) (Bai et al., 2020). The model consists of two main components: (1) an adaptive graph convolution module, which dynamically generates an adjacency matrix through node embeddings without predefined sensor topology; and (2) a temporal modeling module, which employs GRUs to extract features from historical sequences of each node. Compared to traditional STGNN models (e.g., STGCN), AGCRN's advantage lies in its ability to automatically learn potential correlations among sensors in industrial settings, thereby enhancing predictive reliability.For instance, in the loosening and conditioning process, interaction patterns among sensors (e.g., wind speed, temperature) may vary with operating conditions. AGCRN addresses the limitation of fixed topology reliance in traditional methods by adaptively adjusting the graph structure in real time.

### 2.2 Modeling Principles of AGCRN

**Graph Construction Method.** AGCRN constructs graphs based on node embedding similarity. The input tensor has dimensions [N × T × F], where N is the number of sensor nodes, T is the time step length, and F is the feature dimension (e.g., wind speed, temperature). First, the model generates an embedding vector E ∈ R^(N × d) for each node through a trainable parameter matrix (where d is the embedding dimension). Then, the association strength between nodes is computed using cosine similarity to generate a differentiable adjacency matrix. Specifically, each element of the adjacency matrix, $A_{ij}$, is computed as softmax(cos($E_i$, $E_j$)), where cos() denotes the cosine similarity function and softmax ensures normalization. This design enables joint optimization of the graph structure and the forecasting task, allowing adaptive capture of nonlinear coupling relationships between sensors (e.g., the dynamic correlation between moisture and temperature in loosening and conditioning). Additionally, the model employs sparsification techniques (e.g., threshold truncation) to reduce interference from noisy edges and improve computational efficiency.

**Temporal Modeling Mechanism.** The temporal modeling component utilizes a GRU structure to encode the historical sequence of each node. The hidden state update formula is as follows:

$$h_t = GRU(x_t, h_{t-1}; W_g) \tag{1}$$

where $x_t$ is the input feature at the current time step, $h_{t-1}$ is the hidden state from the previous time step, and $W_g$ represents the trainable parameters of the GRU. The innovation of AGCRN lies in its alternating stacking of graph convolution and GRU layers, forming a recurrent architecture of "graph convolution layer → GRU layer" The output of the graph convolution at each time step serves as the input to the GRU, enabling joint representation of spatial features and temporal dynamics. For example, when predicting the outlet moisture content for the next 3 min, the model extracts spatio-temporal features from historical sequences using a sliding window (window length T = 12) and leverages the long-term memory capability of GRU to capture lag effects in moisture changes (e.g., delayed response from steam heating).

**Loss Function and Training Strategy.** To enhance the model's prediction robustness and accuracy in complex industrial scenarios, this study designed and adopted a Weighted Hybrid Loss Function. This function combines the advantages of Mean Squared Error (MSE) and Mean Absolute Error (MAE), aiming to balance the model's sensitivity to outliers (large errors) and its ability to fit the overall trend.

*Definition of the Hybrid Loss Function.* The final total loss function L_total used for model optimization is defined as the weighted sum of MSE and MAE:

$$L_{total} = \alpha * L_{MES} + (1 - \alpha) * L_{MAE} \tag{2}$$

where: $L_{MSE} = \frac{1}{n} * \sum (y_i - \hat{y}_i)^2$, $L_{MAE} = \frac{1}{n} * \sum |y_i - \hat{y}_i|$. $y_i$ is the true value and $\hat{y}_i$ is the predicted value. $\alpha$ is a hyperparameter between [0, 1], used to balance the contribution of MSE and MAE to the total loss.

*Weight Configuration Strategy and Rationale.* Characteristics of MSE: mse squares the errors, thereby amplifying the penalty for larger errors. In industrial prediction, this means the model tends to avoid generating extreme prediction values, helping to suppress significant deviations and ensuring the safety and stability of the production process.characteristics of MAE: MAE treats all errors equally and is less sensitive to outliers compared to MSE. This prevents the model from being overly influenced by a few anomalous points during training, allowing it to better learn the overall data distribution and trends, thereby enhancing the model's generalization ability and robustness to noise. By combining the two, the model is expected to both focus on and correct large prediction deviations (driven primarily by MSE) and robustly fit regular fluctuations (aided by MAE). The choice of the weighting factor $\alpha$ is crucial. In this study, through grid search experiments on the validation set, we found that when $\alpha$ was set to 0.7, the model achieved the best comprehensive performance across multiple evaluation metrics (especially RMSE and MAPE). A higher $\alpha$ value (0.7) reflects our prioritized focus on controlling large fluctuations, while retaining the robustness benefits brought by MAE (weight 0.3).

*Training Strategy and Other Evaluation Metrics.* In addition to adopting the aforementioned hybrid loss function, the training strategy also included: (1) Dynamic Learning Rate Adjustment: the ReduceLROnPlateau strategy was employed. If the validation loss did not decrease for a certain number of epochs (patience = 5), the learning rate was automatically reduced. The initial learning rate was set to 0.001. (2) Early Stopping Mechanism: a patience of 10 was set. If the validation loss showed no improvement for 10 consecutive epochs, training was terminated early to prevent overfitting. (3) Batch Training: a batch_size of 64 was set, and the sliding window mechanism was utilized to efficiently generate training samples (Bai et al., 2020). Furthermore, although the Mean Absolute Percentage Error (MAPE) was not directly incorporated into the loss function due to its instability when true values are close to zero, it was still emphasized as a key evaluation metric for measuring relative error during model performance assessment and result presentation. This is because MAPE is intuitive, easy to understand, and aligns well with the percentage-based requirements for quality control in industrial production.

# 3 Data Processing and Modeling

## 3.1 Data Acquisition and Preprocessing

The loosening and conditioning process for the "Double Happiness (Hard Red Wuyeshen)" brand cigarettes at a certain cigarette factory was taken as the research object. The process parameters were set as follows: material flow rate of 5000 kg/h, return air temperature of 54 °C–60 °C, injected steam pressure of 0.22–0.30 MPa, hot fan frequency of 32 Hz, and drum motor frequency of 35 Hz. The process standard required the outlet moisture content to be between 18.5% and 20.5%. Data were acquired at a frequency of 2 s, covering multiple production day cycles. The collected variables included incoming moisture content, water addition ratio, exhaust damper opening, return air temperature, supplementary steam ratio, outlet moisture content, and other parameters reflecting the dynamic changes of the process. Raw data underwent preprocessing, which included filling missing values using interpolation, eliminating outliers based on the 3$\sigma$ principle, and standardizing the data to ensure consistent dimensions across different features.

## 3.2 Sliding Window Feature Extraction

In the loosening and conditioning process, the variation in tobacco moisture content is a dynamic process. To capture the trend of moisture change in tobacco leaves, the sliding window method is employed for feature extraction from the moisture variation data. The sliding window method can effectively extract local features from fixed sequence data and identify trend changes in the incoming moisture content. Given that the standard tobacco bale weight is 200 kg, the cumulative weight data from the electronic belt scale is partitioned using a sliding window with a step size of 200 kg, meaning each window of data corresponds to the weight of an entire tobacco bale.

**Moisture Measurement of Tobacco Blocks Before Loosening and Conditioning.** Prior to feature extraction via the sliding window method, it is essential to measure and analyze the moisture content of tobacco blocks before the loosening and conditioning process. The specific procedures are as follows:

*Acquisition of Tobacco Bale Information from the High-Rise Warehouse.* ① QR Code Information Acquisition via PLC Communication: in the high-rise warehouse of the cigarette factory, each tobacco bale is affixed with a unique QR code label containing critical information such as material number, batch details, tobacco grade, and storage duration. Through communication between the Programmable Logic Controller (PLC) and the warehouse management system, QR code information is retrieved in real time. ② Analysis of QR Code Information to Obtain Tobacco Grade and Formulation: by parsing the QR code information, the tobacco grade and formulation details of each bale are acquired. Different tobacco grades exhibit distinct water absorption characteristics; higher-grade tobacco typically has lower water absorption capacity, while lower-grade tobacco demonstrates higher water absorption capacity. ③ Labeling of Bales for Each Batch and Recording of Feeding Sequence: in the loosening and conditioning process, tobacco bales are fed in batches. The PLC system records the feeding sequence of each bale to ensure alignment with the production plan. ④ Matching Bale Information with Cumulative Weight from pre-conditioning Electronic Scale: before the loosening and conditioning process, an electronic scale is used to weigh each bale, recording its weight. By matching the QR code information of each bale with the cumulative weight data from the electronic scale, the weight of each bale is accurately associated with its material number and batch information.

*Acquisition of Incoming Moisture Content for Loosening and Conditioning.* Prior to the loosening and conditioning process, an online moisture meter is employed to perform real-time measurements of the incoming moisture content of tobacco blocks. Data, including incoming moisture content, tobacco block temperature, and ambient humidity, are collected at 1-s intervals.

**Sliding Window Feature Extraction.** After acquiring the incoming moisture data, the sliding window method is applied to process the tobacco moisture variation data. Statistical features within each window are extracted as follows: ① Mean, variance, maximum, and minimum values of inlet moisture: Reflecting the central tendency and fluctuation of incoming moisture content. ② Mean, variance, maximum, and minimum values of outlet moisture: Reflecting the central tendency and fluctuation of outlet moisture content. ③ Difference between inlet and outlet moisture: Used to evaluate the water absorption capacity of tobacco leaves. A larger difference indicates stronger water absorption capacity.

The feature data extracted through the sliding window method are subsequently used for cluster analysis and control model construction.

**K-means Clustering Analysis.** To further analyze the water absorption capacity of tobacco leaves, the K-means clustering algorithm was applied to the extracted feature data. The difference between the inlet and outlet moisture content was used to reflect the water absorption capacity, and tobacco leaves were categorized into different classes based on this capacity, providing a basis for subsequent modeling and control.The specific procedures are as follows:

*Calculation of the Difference Between Inlet and Outlet Moisture.* For the data within each sliding window, the difference between the inlet and outlet moisture content was calculated as a key indicator to evaluate the water absorption capacity of tobacco leaves.

*Preparation of Feature Data.* The feature data extracted through the sliding window method, including the mean inlet moisture, mean outlet moisture, and the difference between inlet and outlet moisture, were used as input for the clustering analysis.

*K-means Clustering.* The K-means clustering algorithm was applied to the feature data. The optimal number of clusters was determined using the elbow method. Assuming $k = 3$ was selected, tobacco leaves were categorized into three classes based on water absorption capacity: high, medium, and low water absorption.

① High water absorption tobacco leaves: Exhibited a large difference between inlet and outlet moisture content, indicating strong water absorption capacity and requiring a higher water addition amount.

② Medium water absorption tobacco leaves: Showed a moderate difference between inlet and outlet moisture content, indicating medium water absorption capacity and requiring an appropriate water addition amount.

③ Low water absorption tobacco leaves: Displayed a small difference between inlet and outlet moisture content, indicating weak water absorption capacity and requiring a lower water addition amount.

The results of the clustering analysis are shown in Fig. 1. Tobacco leaves with an average difference between inlet and outlet moisture below 6.5% were classified as low water absorption, those with a difference between 6.5% and 7.5% as medium water absorption, and those with a difference greater than 7.5% as high water absorption.

The resulting cluster label (e.g., 0 for low, 1 for medium, 2 for high absorption) was then assigned to each data point associated with that tobacco bale. This label was subsequently incorporated as a static categorical feature during model training, providing the AGCRN model with crucial a priori information about the material's inherent properties. This step ensures that the model is aware of the type of material it is processing, allowing it to better adapt its predictions.

**Sliding Window Feature Construction.** To achieve short-term multi-step prediction, a sliding window mechanism was employed to generate training samples. The window length was set to 12 steps (corresponding to 24 s of data), a duration selected based on process analysis to approximate the residence time of the material within the conditioning drum.The prediction horizon was set to 3 steps (6 s), a timeframe that aligns with the response time of the system's actuators, enabling timely feedforward control adjustments. This mechanism ensures a physically meaningful alignment between input and output sequences, enhancing the model's capability to learn true temporal dependencies.

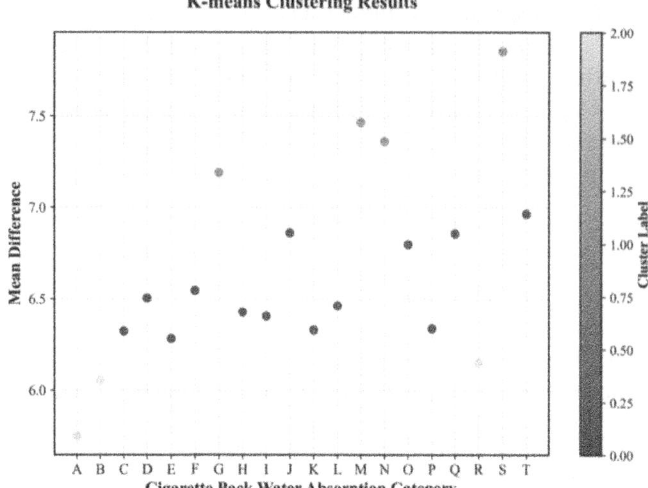


**Fig. 1.** Schematic Diagram of Clustering Analysis Results

### 3.3 A Subsection Sample

To ensure the performance of the AGCRN model in the moisture prediction task for the loosening and conditioning process, fine-grained hyperparameter tuning was conducted during the experiment. The experimental hyperparameter settings are shown in Table 1.

**Table 1.** Experimental Hyperparameter Settings for the AGCRN Model.

| Hyperparameter Name | Value or Strategy Description |
|---|---|
| Epochs | 200 |
| Early Stopping | Patience = 10 |
| Initial Learning Rate | 0.001, dynamically adjusted using ReduceLROnPlateau |
| Batch Size | 64 |
| Node Embedding Dimension | 10 |
| GCN Output Dimension | 64 |
| Sliding Window Length (Input Steps) | 12 steps (corresponding to 24 s) |
| Prediction Horizon (Output Steps) | 3 steps (corresponding to 6 s) |
| Loss Function | Mean Squared Error (MSE) |
| Evaluation Metrics | MAE, RMSE, MAPE (triple metric evaluation) |
| Regularization Strategy | No explicit regularization (overfitting controlled by model structure) |
| Data Standardization | Z-Score Standardization |

The specific hyperparameter settings are as follows:

(1) Epochs: The number of training epochs was set to 200. To prevent overfitting, an early stopping strategy was adopted with a patience value of 10, meaning training would halt if the validation loss did not improve for 10 consecutive epochs.
(2) Learning Rate: The initial learning rate was set to 0.001. During training, the ReduceLROnPlateau strategy was employed to dynamically adjust the learning rate based on validation performance, reducing the learning rate to improve model convergence.
(3) Batch Size: The batch size was set to 64. A smaller batch size helps enhance the model's generalization ability and enables effective training within computational resource constraints.
(4) Node Embedding Dimension: To effectively capture potential spatial relationships among sensors, the node embedding dimension was set to 10. A smaller dimension reduces computational complexity while maintaining sufficient representational capacity.
(5) GCN Output Dimension: The output dimension of the graph convolutional layer was set to 64 to ensure adequate feature extraction for modeling spatio-temporal dependencies.
(6) Sliding Window Length: To process time series data, the sliding window method was used to generate training samples. The window length was set to 12 steps, equivalent to 24 s of data, effectively capturing short-term trends in moisture variation.
(7) Prediction Horizon: For short-term multi-step prediction, the prediction horizon was set to 3 steps, meaning historical data from the past 12 min were used to predict the outlet moisture content for the next 3 min.
(8) Loss Function: To measure prediction accuracy, the mean squared error (MSE) was selected as the primary loss function. For comprehensive evaluation, mean absolute error (MAE) and mean absolute percentage error (MAPE) were also monitored.
(9) Regularization Strategy: No explicit regularization terms were used during training. Instead, the model's inherent structure (e.g., GRU and graph convolutional layers) was relied upon to control overfitting. This strategy performed well in experiments, particularly when handling industrial data with spatio-temporal dependencies.

### 3.4 Comparative Model Settings

To validate the effectiveness of the AGCRN model in moisture prediction for the loosening and conditioning process, a comparative experiment was designed by selecting several mainstream machine learning and deep learning models, including Polynomial Regression, Random Forest, XGBoost, LSTM, and STGCN. As shown in Table 2, these models represent different types of approaches, such as traditional regression, ensemble learning, deep learning, and graph neural networks, enabling a comprehensive evaluation of the advantages of the AGCRN model in handling spatio-temporal dependencies and dynamic variations.

**Table 2.** Comparative Statistics of Different Models.

| Model Type | Model Name | Characteristics |
|---|---|---|
| Traditional Regression | Polynomial Regression | Only models static relationships; unable to handle temporal dependencies |
| Ensemble Learning | Random Forest, XGBoost | Capable of handling nonlinear relationships but lack temporal structure modeling |
| Deep Learning | LSTM | Can capture temporal dependencies but unable to model spatial topology |
| Graph Neural Network | STGCN | Models static graph structure with limited spatio-temporal interaction capability |
| Proposed Method | AGCRN | Adaptive graph structure + GRU for joint spatio-temporal feature modeling |

All models employed the same training-validation-test data split strategy, and feature preprocessing was kept consistent to ensure a fair comparison. Evaluation metrics included MAE, RMSE, and MAPE. The network structures of the models were visualized using Matplotlib and NetworkX, as illustrated in Fig. 2. The spatio-temporal fusion mechanism of AGCRN consists of: Adaptive Graph Construction Layer → Graph Convolutional Layer → GRU Layer → Output.

**Fig. 2.** Schematic Diagram of Different Model Structures

### 3.5 Experimental Results and Performance Evaluation

**Performance Metrics Comparison.** To comprehensively evaluate the performance of the AGCRN model in the moisture prediction task for the loosening and conditioning process, three commonly used performance metrics were adopted for comparative analysis: Mean Absolute Error (MAE), Root Mean Square Error (RMSE), and Mean Absolute Percentage Error (MAPE). These metrics reflect the prediction accuracy and stability of the model from different perspectives. The performance of the AGCRN model was compared with five other mainstream models: Polynomial Regression, Random Forest, XGBoost, LSTM, and STGCN, as shown in Table 3.

**Table 3.** Performance Statistics of Different Models.

| Model | MAE (%) | RMSE (%) | MAPE (%) |
|---|---|---|---|
| Polynomial Regression | 0.593 | 0.741 | 3.82 |
| Random Forest | 0.472 | 0.654 | 3.21 |
| XGBoost | 0.419 | 0.611 | 2.87 |
| LSTM | 0.375 | 0.529 | 2.48 |
| STGCN | 0.351 | 0.502 | 2.33 |
| AGCRN (Proposed) | 0.312 | 0.471 | 2.05 |

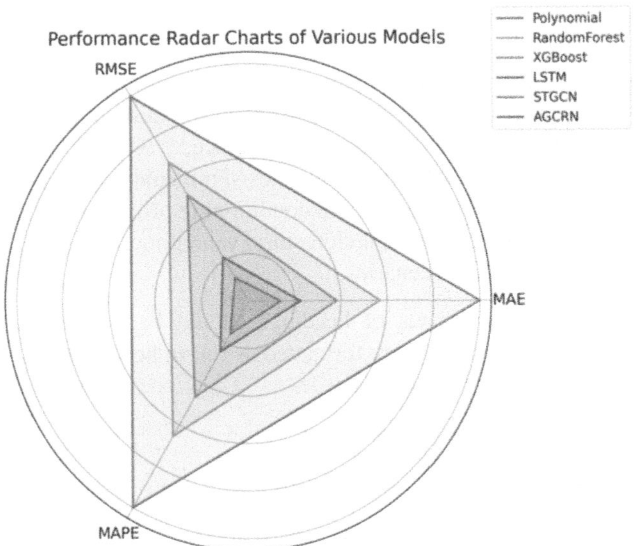


**Fig. 3.** Radar Chart of Performance Metrics for Different Models.

Based on the data in the table and the performance radar chart of different models (Table 3 and Fig. 3), the AGCRN model outperforms all other comparative models across all evaluation metrics. A detailed analysis is provided below:

(1) MAE (Mean Absolute Error): The MAE value of AGCRN is 0.312%, significantly lower than that of other models, especially compared to XGBoost (0.419%) and LSTM (0.375%), with a reduction of approximately 25% or more in error. This indicates that AGCRN achieves higher precision in the prediction process, particularly in complex industrial data scenarios. The smaller MAE value suggests that AGCRN responds more accurately to moisture changes.

(2) RMSE (Root Mean Square Error): AGCRN also excels in the RMSE metric, with a value of 0.471%, representing reductions of approximately 22.9% and 10.9% compared to XGBoost (0.611%) and LSTM (0.529%), respectively. RMSE measures the deviation between predicted and true values, and a lower RMSE indicates higher stability of AGCRN in dynamic prediction processes. Especially when dealing with real-time changing industrial data, the model effectively avoids large fluctuations.

(3) MAPE (Mean Absolute Percentage Error): AGCRN performs particularly well in the MAPE metric, achieving the lowest value of 2.05%, significantly outperforming XGBoost (2.87%) and STGCN (2.33%). As a metric reflecting the percentage of prediction error relative to the true value, the excellent performance of AGCRN in MAPE demonstrates its ability to more accurately capture moisture changes in actual production processes. Especially in highly volatile environments, AGCRN's predictions are more stable and reliable.

From the tabulated data and analysis results, compared to traditional regression models (e.g., Polynomial Regression) and ensemble learning models (e.g., Random Forest, XGBoost), AGCRN effectively handles spatio-temporal dependencies through the combination of adaptive graph convolutional networks and temporal modeling, significantly improving prediction accuracy. Compared to time-series modeling-based LSTM and spatial graph modeling-based STGCN, AGCRN demonstrates advantages in both accuracy and stability through its unique spatio-temporal fusion mechanism. Particularly when dealing with actual industrial data with complex spatio-temporal dependencies, AGCRN exhibits stronger adaptability and performance.

In summary, the AGCRN model demonstrates significant performance advantages in moisture prediction for the loosening and conditioning process through its advanced spatio-temporal graph neural network architecture, validating its potential applicability and feasibility in real industrial scenarios.

**Visual Analysis.** Only two levels of headings should be numbered. Lower level headings remain unnumbered; they are formatted as run-in headings.

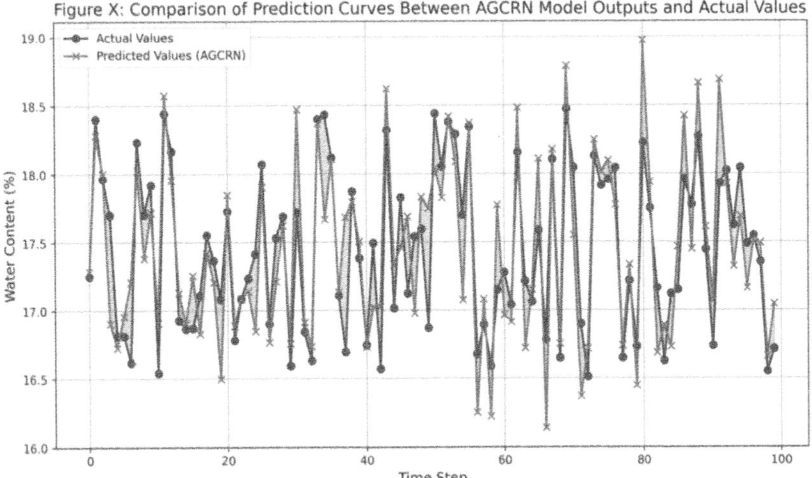


**Fig. 4.** Comparison Curve of AGCRN Predicted Values and Actual Values.

As shown in Fig. 4, the predicted curve of AGCRN fits most closely with the actual values, with minor residual fluctuations, indicating that the model can accurately capture the dynamic changes in moisture content. Particularly under stable working conditions, the prediction accuracy is high, and errors are small. Furthermore, AGCRN maintains good smoothness throughout the prediction process, validating its stability and accuracy in long-term predictions. Compared to traditional models, AGCRN effectively avoids large fluctuations in prediction errors, fully demonstrating its high robustness.

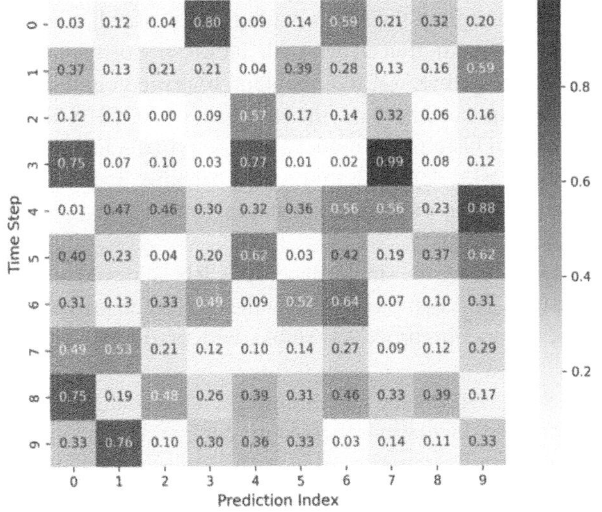


**Fig. 5.** Heatmap of Prediction Errors for the AGCRN Model.

Figure 5 intuitively presents the distribution of prediction errors across various nodes. By observing the heatmap, the error distribution of AGCRN is relatively concentrated across most time steps and nodes, with a relatively small range of error distribution and uniform color variations. This indicates that AGCRN not only achieves accurate predictions for individual nodes but also exhibits strong spatial generalization capability, effectively adapting to dependencies among different sensors while maintaining high prediction accuracy. Moreover, the error distribution across multiple nodes tends to be consistent, further demonstrating the model's strong adaptability under multi-point cooperative control.

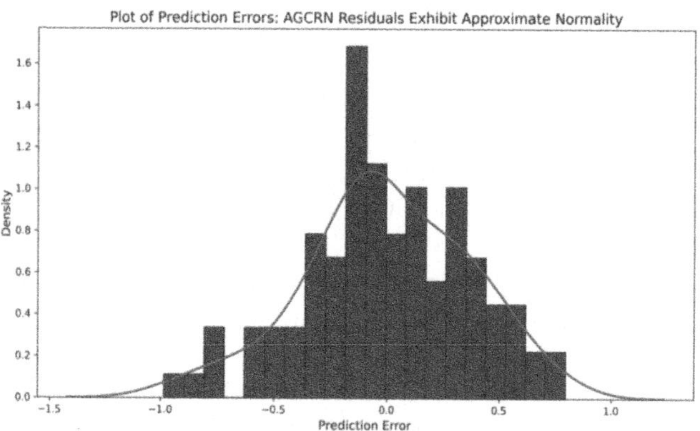


**Fig. 6.** Schematic Diagram of Residual Distribution for AGCRN.

Figure 6 further illustrates the distribution of residuals for AGCRN, displaying errors close to a normal distribution through a histogram and kernel density estimation (KDE) curve. As shown in the figure, the errors exhibit a typical bell-shaped curve, indicating relatively concentrated prediction errors, with the vast majority of errors falling within a small range. This suggests that AGCRN maintains high prediction stability across various time periods. In multiple predictions, the residuals of AGCRN approximate a normal distribution, further confirming the model's ability to maintain low error fluctuations in long-term prediction tasks, demonstrating strong stability and consistency.

In a further comparison of production stages, three typical production stages were selected for detailed analysis:

(1) Stable Stage: During the stable stage, AGCRN achieved an RMSE = 0.35, indicating minimal errors. This demonstrates that the model can provide high-precision predictions when the production process is stable, with prediction results almost matching the actual data.

(2) Mutation Stage: During the mutation stage, rapid changes in the production environment led to increased fluctuations in moisture content. However, AGCRN achieved an RMSE = 0.49, significantly outperforming the XGBoost model, which had an RMSE = 0.72 in this stage. AGCRN demonstrated a clear advantage by effectively adapting to the impact of sudden changes and providing stable predictions. In contrast, XGBoost failed

to capture the spatio-temporal dependencies during mutations, resulting in significantly larger errors.

(3) Start-Stop Stage: During the start-stop stage, process switching caused considerable fluctuations in moisture content. AGCRN effectively avoideduncontrolled fluctuations and maintained the RMSE within a small range, demonstrating its strong stability and adaptability. Particularly in the highly uncertain start-stop stage, the model provided stable predictions, whereas other traditional models often performed poorly under such conditions.

In summary, AGCRN demonstrated strong prediction accuracy and stability across different production stages, including the stable, mutation, and start-stop stages. Its RMSE values were consistently lower than those of traditional models, particularly during mutation and start-stop stages, highlighting the model's strong robustness and high spatial generalization capability in responding to sudden changes. These results validate the superiority of AGCRN in industrial prediction tasks, especially in real-time forecasting and multi-point cooperative control, indicating promising application prospects.

# 4 System Deployment and Application

## 4.1 System Architecture Overview

The study developed a moisture prediction system for the loosening and conditioning process. The system architecture consists of front-end multi-point data acquisition, intermediate graph neural network processing, and back-end intelligent control feedback.The objective is to achieve advance prediction and precise adjustment of outlet moisture content. The system primarily consists of the following layers:

(1) Data Acquisition Layer: A Web interface service developed using FastAPI periodically collects production data from real-time data sources and onsite PLC systems, including key parameters such as incoming moisture content, outlet moisture content, water addition ratio, and return air temperature.

(2) Feature Processing and Graph Construction Module: This module performs standardization and generates features. Crucially, it uses the K-means clustering results to assign a material type as a static feature to the incoming data.

(3) Model Inference Layer: The trained AGCRN model is deployed to input windowed data in real time and predict outlet moisture content.

(4) Control Strategy Feedback Layer: Prediction results are integrated into control logic (e.g., water addition ratio) to achieve feedforward adjustment.

## 4.2 Prediction-Driven Feedforward-Feedback Composite Control Strategy

To effectively translate the high-prediction accuracy of the AGCRN model into tangible control benefits, this system designed and implemented a prediction-driven feedforward-feedback composite control strategy. This strategy aims to overcome the issue of untimely adjustment caused by measurement lag in traditional pure feedback control (e.g., PID), by organically combining the "foresight" of the predictive model with the "error correction capability" of feedback control.

**Control Loop Structure.** The system's control loop consists of two parallel components:(1) Feedforward Control Loop:This loop centers on the predicted values from the AGCRN model. Based on upstream real-time data such as incoming moisture content, flow rate, and air temperature, the model predicts the outlet moisture content for a future step (3 min ahead). This predicted value is compared with the process target value (e.g., 19.5%), and the difference $e_{pred} = Target - y_{pred}$ is fed into the feedforward controller. Based on this predicted deviation, the feedforward controller calculates a compensatory adjustment to the water addition ratio $\Delta U_{ff}$ in advance, proactively countering impending fluctuations.(2) Feedback Control Loop:This loop constitutes traditional PID control. The actual moisture content y_actual measured by the online moisture meter at the outlet is compared with the set value, generating an actual error $e_{actual} = Target - y_{actual}$. The PID controller calculates a feedback adjustment amount $\Delta U_{fb}$ based on this error, used to eliminate model prediction deviations, unmodeled disturbances, and the system's own static errors.(3) Control Signal Fusion:The final total control signal $\Delta U_{total}$ acting on the water addition actuator (e.g., control valve or variable frequency pump) is the algebraic sum of the base water addition amount $\Delta U_{base}$, the feedforward adjustment amount $\Delta U_{ff}$ f, and the feedback adjustment amount $\Delta U_{fb}$:

$$U_{total} = U_{base} + \Delta U_{ff} + \Delta U_{fb} \tag{3}$$

Here $\Delta U_{base}$ is a baseline water addition amount set based on material flow rate and empirical values.

**Strategy Advantages.** (1) Overcoming Time Lag:The feedforward loop utilizes the predictive capability of AGCRN to perform compensatory adjustments before the impact of a disturbance (such as a change in incoming moisture) actually manifests at the outlet. This effectively overcomes the physical lag (typically tens of seconds to several minutes) associated with material transport and moisture detection.(2) Enhancing Stability:The feedforward control assumes the primary task of disturbance rejection, allowing the feedback PID controller to handle only smaller residual deviations. This greatly reduces the burden on the PID controller, enabling it to operate under more stable conditions and avoiding overshoot and oscillations caused by inappropriate parameters or excessive disturbances.(3) Increasing Robustness:Even if minor errors exist in the model predictions, the feedback loop ensures the final control accuracy, safeguarding the overall system's robustness and reliability.

By implementing this composite control strategy, the system seamlessly integrates the predictive capability of AGCRN into the closed-loop control, achieving an upgrade in the control paradigm from "passive error correction" to "active prevention." This is also the key reason why the control deviation of the outlet moisture content was significantly narrowed.

### 4.3 Development of Production Control Interface

(1) Front-End Page Development: The production control interface was developed using Vue3, which displays real-time process quality parameters (e.g., incoming moisture content, outlet moisture content, water addition ratio, return air temperature) as well as model prediction data, as shown in Fig. 7.

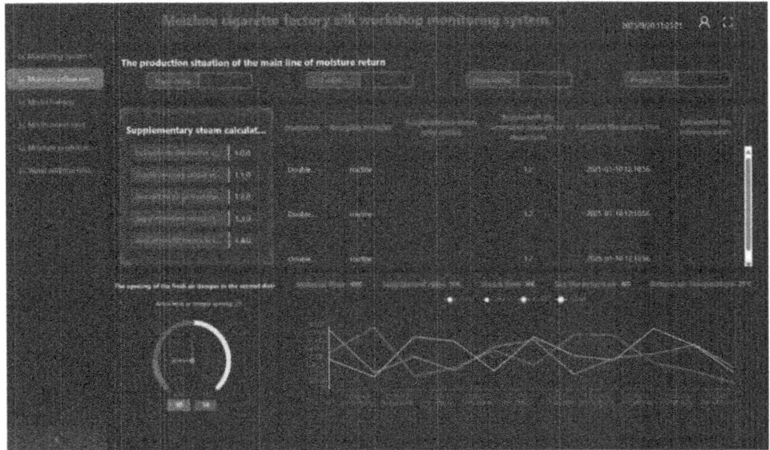


**Fig. 7.** Schematic Diagram of the Production Control Interface.

(2) Trend Chart Display: The production control interfacedisplays real-time trend charts of actual and predicted values of outlet moisture content. Through these trend charts, operators can intuitively understand variations in outlet moisture content and make production adjustments based on prediction results.

(3) Alarm and Prompt Functionality: When the outlet moisture content deviates from the process standard range, the system automatically triggers an alarm. Operators can promptly adjust production parameters based on the alarm information to ensure product quality.

## 4.4 Effectiveness Analysis

Through the application of the intelligent water addition system for loosening and conditioning, the control accuracy and quality stability of the outlet moisture content have been significantly improved. The specific results are as follows:

(1) Improved Control Accuracy:① Under the original control method, the moisture deviation range was ± 1.2%;② With the support of the XGBoost model, the control deviation was reduced to ± 0.85%;③ After implementing the AGCRN prediction-driven regulation in this system, the moisture deviation was stably controlled within ± 0.5%, with maximum fluctuations reduced by over 30%.

(2) Enhanced Quality Stability: The process capability index (CPK) increased by 31%. A higher CPK indicates a more stable and predictable process, demonstrating that the model driven control not only improves the average accuracy but also significantly reduces the overall process variability, keeping the output consistently closer to the target.

(3) Robust Performance Under Abnormal Conditions: In stages such as material head/tail, batch switching, or sudden flow changes, the system maintained acceptable prediction errors, demonstrating greater robustness compared to the original system. This resilience under dynamic conditions is attributed to the model's core architecture.

(4) Improved Production Efficiency: Automated data acquisition, model training, and feedback regulation reduced manual intervention, thereby enhancing production efficiency.

## 5 Conclusion

Loosening and conditioning represent a critical stage in cigarette primary processing, where the precision of outlet moisture control directly influences the stability of subsequent operations and product consistency. To address the control challenges posed by fluctuations in incoming materials, environmental variations, and multi-sensor coupling, this study proposed a spatio-temporal moisture prediction model based on the Adaptive Graph Convolutional Recurrent Network (AGCRN). By incorporating a node embedding mechanism to dynamically construct spatial topological relationships among sensors and integrating gated recurrent units to capture temporal dependencies, the model achieved high-precision multi-step forecasting of outlet moisture content. Experimental results demonstrated that the AGCRN model significantly outperformed baseline approaches, including traditional regression, XGBoost, LSTM, and STGCN, across metrics such as MAE, RMSE, and MAPE. Notably, it exhibited strong robustness and adaptability under transient conditions and during start-stop phases. Following deployment, the system reduced the outlet moisture control deviation from $\pm$ 1.2% to within $\pm$ 0.5%, while the process capability index (CPK) increased by 31%, substantially enhancing control accuracy and stability. This work not only validates the efficacy of spatio-temporal graph neural networks in complex industrial settings but also offers a novel solution for intelligent control in multivariate, strongly coupled cigarette production environments, demonstrating significant potential for engineering application and industry reference.

## References

Bai, L., Yao, L., Li, C., et al.: Adaptive graph convolutional recurrent network for traffic forecasting. In: Advances in Neural Information Processing Systems, Vol. 33, pp. 1–12 (2020)

Chen, D.L., Wang, X.H., He, R.J., et al.: Research on control method of water addition ratio in tobacco loosening and conditioning based on model prediction. Manufacturing Automation **47**(07), 121–128 (2025)

Fu, K., Chen, Q., Yang, Y., et al.: BLAD: Adaptive load balanced scheduling and operator overlap pipeline for accelerating the dynamic GNN training. In: Proceedings of the International Conference for High Performance Computing, Networking, Storage and Analysis, pp. 1–13 (2023)

Jin, M., Koh, H.Y., Wen, Q., et al.: A survey on graph neural networks for time series: Forecasting, classification, imputation, and anomaly detection. IEEE Transactions on Pattern Analysis and Machine Intelligence (2024)

Li, Z.J., Li, Y.X., Zhao, H.Y., et al.: Prediction of moisture variation in loosening and conditioning under different environmental conditions based on machine learning. Packaging Engineering **45**(17), 119–128 (2024)

Wen, G.Q., Jiang, T., Wu, X., et al.: Modeling and optimization of control system for tobacco loosening and conditioning based on principal component neural network. Tobacco Science & Technology **57**(07), 104–112 (2024)

Xie, C.J., Lu, D.Y., Duan, X.G., et al.: Optimization and improvement of outlet moisture stability in loosening and conditioning. Equip. Manage. Mainten. **09**, 20–26 (2025)
Zhang, F.G., Liu, Y., Lin, M., et al.: Intelligent control of loosening and conditioning machine based on machine learning algorithm. Equip. Manage. Mainten. **11**, 45–48 (2025)

# Self-Supervised Conjoint CLIP for Unsupervised Domain Adaptation

Jun Liu[1,2](✉), Diwei Wu[1,2], Hewen Liu[1,2], Jian Liu[1,2], and Jun Yang[1,2]

[1] College of Computer Science, Sichuan Normal University, Chengdu, China
1164811474@qq.com

[2] Visual Computing and Virtual Reality Key Laboratory of Sichuan Province, Sichuan Normal University, Chengdu, China

**Abstract.** Traditional unsupervised domain adaptation (UDA) methods often suffer from two critical limitations: lack of class discriminability (LCD) and class-level semantic ignorance (CSI), resulting from coarse grained feature alignment strategies. Recently, vision-language models (VLMs) such as CLIP have shown promise by exploring cross-modal semantic information; however, current CLIP-based prompt learning approaches fail to fully capture fine-grained visual patterns, resulting in a Semantic-Visual Granularity Gap (SVGG). To address the problem, we propose Self-supervised Conjoint CLIP (SCC), a dual-branch architecture that integrates CLIP's semantic alignment capabilities with fine grained self-supervised visual representations from DINOv2. Specifically, SCC comprises a CLIP base branch optimized via multi-modal prompt learning to preserve cross-modal semantic alignment, and a DINOv2 enhanced branch incorporating a novel DINOv2-CLIP Mutual Tuning (DCMT) module. The DCMT module employs cross-attention mechanisms guided by the Alignment-Uniformity principle: cross-attention aligns semantic and visual representations via subspace alignment, while a self-consistency loss ensures uniformity. Extensive experiments on three benchmarks demonstrate that SCC improves domain adaptability and outperforms other methods.

**Keywords:** Unsupervised domain adaptation · Vision-language models · Self-supervised learning · Prompt learning

## 1 Introduction

In real-world scenarios, models trained on labeled source data often perform poorly on unlabeled target domains because of domain shifts. Unsupervised domain adaptation (UDA), a technique for handling such problems, has been extensively studied.

Traditional UDA methods commonly rely on reducing feature divergence through domain-level alignment, employing statistical metrics like Maximum Mean Discrepancy (MMD) (Tzeng et al., 2014) or adversarial learning. However, these alignment strategies inherently have two critical limitations:(1) Lack of Class Discriminability (LCD): Domain-level alignment often leads to indistinguishable semantic boundaries


H. Yan et al. (Eds.): AIS&P 2025, CCIS 2918, pp. 246–257, 2026.
https://doi.org/10.1007/978-981-95-9884-7_17

between classes, reducing inter-class separation and increasing the overlap between different categories (Chen et al., 2019); and (2) Class-level Semantic Ignorance (CSI): Most approaches align features at the global image-level, neglecting explicit class-level semantic information, thereby failing to leverage category-specific representations effectively (Zhang et al., 2022).

Recently, vision-language models (VLMs) like CLIP (Radford et al., 2021), trained via contrastive learning on massive image-text pairs, precisely align visual embeddings with their corresponding textual descriptions, thereby providing explicit class-level semantic supervision and addressing the CSI limitation. To optimize VLMs for UDA tasks, one popular approach is prompt learning. For instance, DAPL (Ge et al., 2023) improves CLIP's performance in UDA by partitioning textual prompts into domain-agnostic, domain-specific, and class specific components. AD-CLIP further generates prompts using image features. However, these prompt-based adaptations still inherit CLIP's text-supervised bias, which emphasizes high-level semantic attributes at the expense of detailed visual cues.

Consequently, while CLIP and its prompt learning variants largely resolve CSI, they fall short on the LCD: the text-guided training fails to capture the fine-grained visual patterns critical for distinguishing visually similar classes. This shortcoming gives rise to a new challenge, which we term the Semantic Visual Granularity Gap (SVGG), namely the inability of text-supervised VLMs to reconcile semantic alignment with fine-grained visual discrimination. SVGG is a modality-induced representational mismatch within the model itself. It originates from the fundamental difference in supervision signals (textual semantics vs. pixel-level visuals) during the pretraining of VLMs, leading to a discrepancy in the granularity of abstraction, even within a single domain. This gap undermines discriminability not by misaligning domains, but by yielding visual features that are semantically sound yet insufficiently granular for fine-grained class separation. Addressing SVGG is crucial for deploying CLIP-based UDA methods in real-world scenarios that require precise visual differentiation.

Inspired by self-supervised learning (Chen et al., 2020), we introduce a more general-purpose feature extractor, DINOv2, which is trained purely on unlabeled images. As shown in Fig. 1, DINOv2 preserves fine-grained representations compared to CLIP. To fuse CLIP's robust semantic alignment(solving CSI) with DINOv2's detailed visual discrimination(bridging SVGG and thus fully resolving LCD), we propose a dual-branch framework, namely CLIP base branch and DI NOv2enhancedbranch, to involve both models. The base branch optimizes CLIP using learnable textual and visual prompts, and the enhanced branch incorporates the DINOv2-CLIP Mutual Tuning (DCMT) module to fuse CLIP-extracted features with DINOv2-extracted features through cross-attention mechanisms. A self-consistency regularization is further imposed to ensure that the fused features do not deviate from the CLIP embedding space. Our main contributions are as follows:

- We propose a Self-supervised Conjoint CLIP (SCC) framework with two branches, where the base branch preserves CLIP's semantic alignment, and the enhanced branch incorporates DINOv2 to improve the adaptability for target domains.
- e first combine self-supervised features with multi-modal prompt learning to achieve cross-modal and cross-domain joint optimization through the DCMT module.

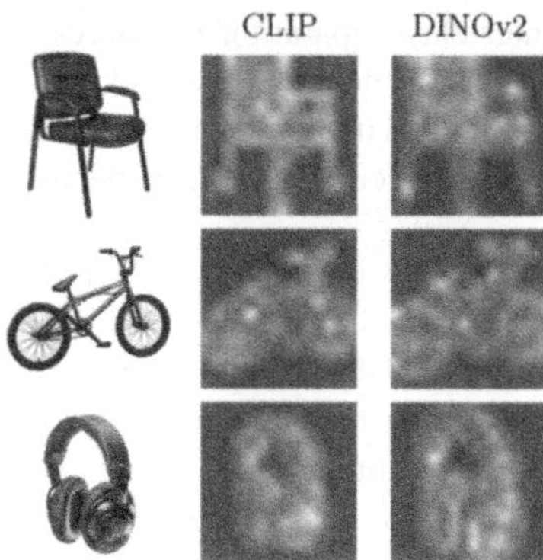


**Fig. 1.** Attention visualization of CLIP and DINOv2

- Comprehensive experiments on three UDA benchmarks demonstrate the efficacy of our method

## 2 Related Work

### 2.1 Unsupervised Domain Adaptation

Traditional unsupervised domain adaptation (UDA) methods primarily fall into three categories. The first line of research seeks to minimize domain discrepancies through statistical measures. For example, DAN (Long et al., 2015) aligns source and target domains using the Maximum Mean Discrepancy (MMD). The second approach takes advantage of the adversarial learning of GANs. DANN (Ajakan et al., 2014) introduces a domain classifier, enforcing domain-invariant feature generation via a zero-sum game between the feature extractor and the classifier. MCD employs a min-max game between the feature extractor and two classifiers. Recent works focus on Vision Transformer (ViT). For example, CD Trans (Xu et al., 2021) utilizes cross-attention mechanisms for feature alignment, and TVT (Yang et al., 2023) exploits ViT's patch-based architecture to learn domain-invariant features at the patch level.

### 2.2 Vision-Language Models and Prompt Learning

Vision-Language Models (VLMs) have attracted considerable research interest owing to their remarkable capabilities in cross-modal tasks (Yao et al., 2024, Yu et al., 2023). Models like CLIP (Radford et al., 2021) learn joint image-text representations through large-scale pretraining on image-text pairs. However, effectively adapting these models to downstream tasks remains challenging. Prompt learning has emerged as an effective paradigm by adding instructions to the input. CoOp (Zhou et al., 2022b) first proposes learnable contextual prompts for CLIP's text encoder to improve image classification. CoCoOp (Zhou et al., 2022b) further generates instance-conditioned prompts via a two-layer network. MaPLe (Khattak et al., 2023) optimizes textual and visual prompts to strengthen alignment between image-text pairs.

### 2.3 Self-Supervised Learning

Self-supervised learning (Chen et al., 2020) exploits inherent relationships within unlabeled data to construct supervisory signals, learning generic and transfer able features directly from images. Many self-supervised learning studies focus on discriminative methods, such as instance classification (Chen et al., 2020), which treats each image as a distinct class and trains models to distinguish them under different data augmentations. However, explicitly learning classifiers to separate all images is suboptimal. Wu et al. propose using Noise-Contrastive Estimation (NCE) to compare images rather than classify them. Grill et al. introduce BYOL, a metric learning framework that trains features by matching them to representations from a momentum-updated encoder. Inspired by BYOL, Oquab et al. train DINOv2 on a vast curated dataset of images, outperforming all existing approaches on several computer vision tasks.

## 3 Preliminaries

### 3.1 Unsupervised Domain Adaptation

Let $D_s = \{(x_i^s, y_i^s)\}_{i=1}^{N_s}$ denotes the source domain with $N_s$ labeled samples,where $x_i^s \sim P_s(X)$ represents the input samples and $y_i^s \in Y$ are the corresponding labels. The target domain $D_t = \{x_i^t\}_{i=1}^{N_t}$ contains $N_t$ unlabeled samples with $x_i^t \sim \mathrm{P}_t(X)$. These two domains share the same label space Y, but their data distributions $P_s(X)$ and $P_t(X)$ differ within the input space X. The goal of unsupervised domain adaptation is to learn a model $f : X \rightarrow Y$ using $D_s$ and $D_t$, which ensures strong performance on the target domain.

### 3.2 Clip

CLIP acquires aligned vision-language representations by pretraining a text encoder $f_s$ and an image encoder $f_v$ on vast amounts of image-text pairs. The text encoder $f_s$, based on a Transformer architecture, encodes textual descriptions into text features. The image encoder $f_v$, typically using a ViT or ResNet backbone, learns visual representations from images. Both encoders are jointly trained via a contrastive loss to form an aligned feature space, which establishes zero-shot inference capabilities. For an input image $x$, the model compares its visual features against textual embeddings of K category descriptions, selecting the most compatible category:

$$P(y = k|x) = \frac{\exp(\cos(f_v(x), f_s(t_k))/\tau)}{\sum_{k=1}^{K} \exp(\cos(f_v(x), f_s(t_k))/\tau)} \tag{1}$$

where $\cos(\cdot, \cdot)$ denotes the cosine similarity, $\tau$ is the temperature coefficient, and $t_k$ is the textual prompt of the k-th class,e.g., “a photo of a [CLS]”.

## 4 Method

To address the SVGG, we propose the Self-supervised Conjoint CLIP (SCC) approach, with its architecture depicted in Fig. 2. The SCC methodology is detailed below.

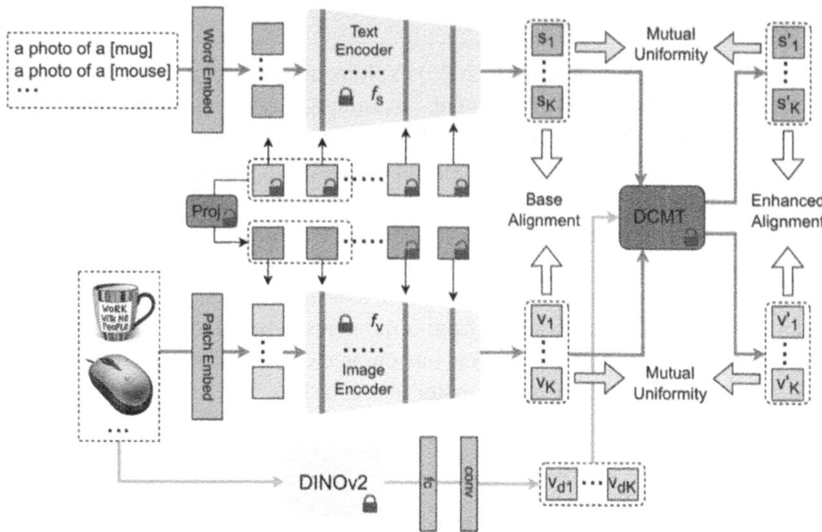


**Fig. 2.** Overview of the SCC framework. Textual prompts and visual prompts are fed into the text encoder and image encoder, respectively. Base Alignment is the process of aligning CLIP's text and image features, while Enhanced Alignment denotes the alignment of enhanced text and image features after the DCMT module. Mutual Uniformity ensures consistency between the enhanced features and the CLIP features via a consistency loss.

## 4.1 Base Branch

We employ prompt learning with multi-modal prompts, adding learnable parameters to both textual and visual prompts.

$$\begin{aligned} t_k &= [t_1][t_2]\ldots[t_n][CLS_k], \\ u_k &= [u_1][u_2]\ldots[u_n][p_k][c_k], \end{aligned} \tag{2}$$

where $t_k$ and $u_k$ are textual prompt and visual prompt of the k-th class, $t_{1:n}$ and $u_{1:n}$ are learnable tokens. $[CLS_k]$ is the class name, $[p_k]$ and $[c_k]$ are position embedding and class token for the image encoder, respectively.

In the encoder's initial layers, a projection module converts textual prompts to visual prompts, endowing image features with text-relevant semantic information. In the deeper layers, each layer utilizes independent prompts, learning a comprehensive visual-semantic representation from the image. To align textual and visual representations, we adopt a cross-entropy loss function defined as follows:

$$\mathcal{L}_s = -\sum_{i=1}^{N_s} y_i^s \log \frac{\exp(\cos(v_i^s, s_k)/\tau)}{\sum_{k=1}^{K} \exp(\cos(v_i^s, s_k)/\tau)} \tag{3}$$

where $y^s$ is the one-hot label for the source domain data, $K$ represents the total number of categories, $s_k$ represents the text embedding of the k-th class through prompt learning, and $v^s$ denotes the corresponding image embedding of the source domain.

To better exploit the unlabeled target domain data, we adopt pseudo labels for training. The pseudo labels are generated by CLIP's zero-shot inference. To boost the reliability of pseudo labels, we establish a confidence threshold $\theta$. We only accept pseudo labels when the maximum prediction probability $\theta_p$ is greater than the threshold. Similarly, we adopt a cross-entropy loss function:

$$\mathcal{L}_t = -\mathbb{I}(\theta_p \geq \theta) \sum_{i=1}^{N_t} \hat{y}_i^t \log \frac{\exp(\cos(v_i^t, s_k)/\tau)}{\sum_{k=1}^{K} \exp(\cos(v_i^t, s_k)/\tau)} \tag{4}$$

where $\hat{y}^t$ is the generated one-hot pseudo label for the target domain data, $v^t$ is the image embedding from the target domain through prompt learning, and $\mathbb{I}(\cdot)$ denotes an indicator function.

## 4.2 Enhanced Branch

The DINOv2 features $v_d$, CLIP text features $s$, , and CLIP image features $v$ are fed into the DCMT module, producing enhanced text features $s'$ and enhanced image features $v'$, as shown in Fig. 3.

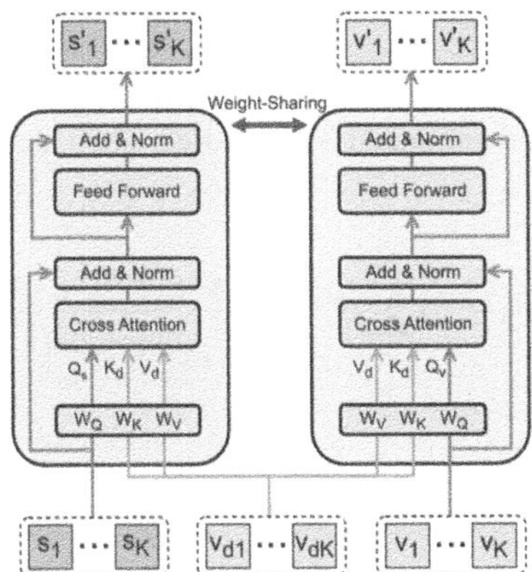


**Fig. 3.** Overview of the DCMT framework.

To address the SVGG arising from CLIP's text-supervised bias toward coarse semantic structures, we introduce the Alignment-Uniformity principle from contrastive learning. SVGG inherently stems from the misalignment between se mantic and visual representations in scale and granularity: text-supervised models emphasize high-level semantics while neglecting fine-grained visual details that are essential for discriminating subtle class differences. The Alignment Uniformity framework explicitly resolves this issue by advocating that high quality multi-modal representations should not only align semantic and visual modalities closely (alignment) but also maintain diversity and avoid representation collapse into narrow semantic subspaces (uniformity).

The DCMT module employs cross-attention mechanisms guided by the alignment principle, which comprises two weight-sharing transformer branches. Specifically,

CLIP's semantic embeddings ($s$ and $v$) are designated as queries (Q) to actively retrieve fine-grained visual patterns from DINOv2's feature maps ($v_d$), which serve as keys (K) and values (V).

$$\begin{aligned}\mathrm{Attention}(\mathrm{Q_s}, \mathrm{K_d}, \mathrm{V_d}) &= \mathrm{Softmax}\left(\frac{\mathrm{Q_s K_d^T}}{\varepsilon}\right)\mathrm{V_d}, \\ \mathrm{Attention}(\mathrm{Q_v}, \mathrm{K_d}, \mathrm{V_d}) &= \mathrm{Softmax}\left(\frac{\mathrm{Q_v K_d^T}}{\varepsilon}\right)\mathrm{V_d},\end{aligned} \tag{5}$$

where $T$ denotes the matrix transpose operator and $\varepsilon$ is a scaling factor. This asymmetric attention design forces CLIP's high-level semantics to attend to spatially localized visual cues in DINOv2's representations, thereby achieving subspace alignment between coarse semantics and fine-grained visuals. Then the outputs of the cross-attention module are added with $s$ and $v$, respectively, followed by feed-forward layers, residual connections, and normalization to get the final outputs $s'$ and $v'$.

To align the enhanced image features of the source domain with the enhanced text features, we employ a cross-entropy loss function defined as follows:

$$\mathcal{L}_{\mathrm{sd}} = -\sum_{\mathrm{i=1}}^{\mathrm{N_s}} \mathrm{y_i^s} \log \frac{\exp(\cos(\mathrm{v_i}'^{\mathrm{s}}, \mathrm{s_k}')/\tau)}{\sum_{\mathrm{k=1}}^{\mathrm{K}} \exp(\cos(\mathrm{v_i}'^{\mathrm{s}}, \mathrm{s_k}')/\tau)} \tag{6}$$

Following the base branch, we also make use of the target domain data to obtain its enhanced image features and enhanced text features. Then, a cross-entropy loss function is adopted:

$$\mathcal{L}_{\mathrm{td}} = -\mathbb{I}(\theta_{\mathrm{p}} \geq \theta) \sum_{\mathrm{i=1}}^{\mathrm{N_t}} \hat{\mathrm{y}}_{\mathrm{i}}^{\mathrm{t}} \log \frac{\exp\left(\cos\left(\mathrm{v_i}'^{\mathrm{t}}, \mathrm{s_k}'\right)/\tau\right)}{\sum_{\mathrm{k=1}}^{\mathrm{K}} \exp\left(\cos\left(\mathrm{v_i}'^{\mathrm{t}}, \mathrm{s_k}'\right)/\tau\right)} \tag{7}$$

### 4.3 Mutual Uniformity

As discussed above, the cross-attention mechanisms implement the alignment component by bridging the SVGG. However, alignment alone could lead to over-concentration of representations, thereby undermining the overall representational diversity required for fine-grained discrimination. Inspired by the uniformity principle from the Alignment-Uniformity framework, we propose a self-consistency constraint that acts as a uniformity regularizer.

Specifically, this self-consistency loss ensures that the enhanced features $s'$ and $v'$ remain consistent with their CLIP counterparts $s$ and $v$, thereby maintaining sufficient representational spread and preventing semantic collapse. Formally, we employ L1 loss to impose this feature-level consistency:

$$\mathcal{L}_{\mathrm{sc-text}} = \sum_{\mathrm{i=1}}^{\mathrm{N}} \left|\mathrm{s}' - \mathrm{s}\right|, \mathcal{L}_{\mathrm{sc-image}} = \sum_{\mathrm{i=1}}^{\mathrm{N}} \left|\mathrm{v}' - \mathrm{v}\right| \tag{8}$$

As a result, the total loss is:

$$\mathcal{L} = \mathcal{L}_{\text{ba}} + \lambda_1 \mathcal{L}_{\text{en}} + \lambda_2 \mathcal{L}_{\text{sc}} \tag{9}$$

where $\mathcal{L}_{\text{ba}} = \mathcal{L}_{\text{s}} + \mathcal{L}_{\text{t}}$, $\mathcal{L}_{\text{en}} = \mathcal{L}_{\text{sd}} + \mathcal{L}_{\text{td}}$, $\mathcal{L}_{\text{sc}} = \mathcal{L}_{\text{sc - text}} + \mathcal{L}_{\text{sc - image}}$, $\lambda_1$ and $\lambda_2$ are hyperparameters.

During inference, the model's final prediction is computed as a weighted sum of outputs from the base branch and the enhanced branch. These two branches capture both semantic information and fine-grained representations, thereby enhancing discriminative capability in target domains.

# 5 Experiments

We validate our framework through multiple experiments on diverse UDA tasks. In subsequent sections, we provide a comprehensive introduction to each part of the experiments.

## 5.1 Experimental Datasets

**Office-Home** contains 15,588 images from four different domains: artistic images (A), clip art (C), product images (P), and real-world images (R). It is collected from both office and home scenes, each domain containing 65 categories of images.

**Office-31** includes 4,652 images from 31 classes, collected from three different domains: Amazon (A), DSLR (D), and Webcam (W). The Amazon images are collected from Amazon.com products, while the DSLR and Webcam are captured in office environments using digital SLR cameras and webcams, respectively.

**VisDA-2017** is a dataset used for domain adaptation from synthetic objects to real objects, covering 12 categories. The source domain consists of 152,397 synthetically generated 2D renderings, and the target domain contains 55,388 real-world photographs.

## 5.2 Implementation Details

We adopt ViT-B/16 as the backbone network and follow the multi-modal prompt design for ViT-based backbones. The parameters of CLIP and DINOv2 remain fixed. We use the SGD optimizer to train the prompts and the DCMT module for 30 epochs on the Office-31 dataset and 20 epochs on the Office-Home and VisDA-2017 datasets, with a batch size of 32. We initialize the learning rate at 0.003 and employ cosine annealing decay.

## 5.3 Experimental Results

**Results on Office-Home.** Table 1 summarizes the results of different methods. Since DINOv2 employs ViT as its backbone, we standardize all comparison methods to use ViT-based architectures for fair evaluation. Our mean accuracy reaches 86.2%, outperforming SSRT by 0.8% and TVT by 2.6%. We also surpass DAPL and PDA, which also use prompt learning methods. For some challenging tasks, like P $\rightarrow$ C and a $\rightarrow$ C, our method demonstrates an improvement of approximately 6.0% over CLIP.

**Table 1.** Comparisons on Office-Home.

| Method | A → C | A → P | A → R | C → A | C → P | C → R | P → A | P → C | P → R | R → A | R → C | R → P | Avg |
|---|---|---|---|---|---|---|---|---|---|---|---|---|---|
| CDTrans(Xu et al., 2021) | 68.8 | 85.0 | 86.9 | 81.5 | 87.1 | 87.3 | 79.6 | 63.3 | 88.2 | 82.0 | 66.0 | 90.6 | 80.5 |
| TVT(Yang et al., 2023) | 74.9 | 86.8 | 89.5 | 82.8 | 87.9 | 88.3 | 79.8 | 71.9 | 90.1 | 85.5 | 74.6 | 90.6 | 83.6 |
| SSRT(Sun et al., 2022) | **75.2** | 89.0 | 91.1 | 85.1 | 88.3 | 89.9 | 85.0 | 74.2 | 91.2 | 85.7 | **78.6** | 91.8 | 85.4 |
| CLIP(Radford et al., 2021) | 67.8 | 89.0 | 89.8 | 82.9 | 89.0 | 89.8 | 82.9 | 67.8 | 89.8 | 82.9 | 67.8 | 89.0 | 82.4 |
| DAPL(Ge et al., 2023) | 70.7 | 91.0 | 90.9 | 85.2 | 91.0 | 91.0 | 85.1 | 70.7 | 90.9 | 85.3 | 70.4 | 91.4 | 84.4 |
| PDA(Bai et al., 2024) | 73.5 | 91.4 | 91.3 | 86.0 | 91.6 | 91.5 | **86.0** | 73.5 | 91.7 | 86.4 | 73.0 | 92.4 | 85.7 |
| Ours | 74.0 | **92.3** | **92.1** | **86.1** | **92.0** | **92.6** | 85.9 | **74.4** | **92.6** | **87.0** | 73.0 | **92.6** | **86.2** |

**Results on Office-31.** Table 2 summarizes the accuracies of different CLIP based prompt learning methods. Our method achieves the best performance on almost all tasks, with a mean accuracy of 91.3%, and yields improvements of 10.1%, 1.9%, and 1.7% compared to DAPL, CoOp, and MaPLe, respectively.

**Table 2.** Comparisons with the prompt learning methods on Office-31.

| Method | A → D | A → W | D → A | D → W | W → A | W → D | Avg |
|---|---|---|---|---|---|---|---|
| CLIP(Radford et al., 2021) | 77.7 | 75.8 | 79.0 | 75.8 | 79.0 | 77.7 | 77.5 |
| DAPL(Ge et al., 2023) | 81.7 | 80.3 | 81.2 | 81.8 | 81.0 | 81.3 | 81.2 |
| CoCoOp(Zhou et al., 2022a) | 86.9 | 88.2 | 83.2 | 94.1 | **82.8** | 98.0 | 88.9 |
| CoOp(Zhou et al., 2022b) | 88.5 | 88.5 | 82.0 | 96.1 | 82.4 | 99.0 | 89.4 |
| MaPLe(Khattak et al., 2023) | 86.9 | 88.6 | 83.0 | 97.7 | 82.0 | 99.4 | 89.6 |
| Ours | **91.0** | **92.2** | **83.6** | **98.5** | 82.7 | **99.8** | **91.3** |

**Results on VisDA-2017.** Table 3 presents our experimental results on the VisDA-2017 benchmark. In particular, CLIP achieves a zero-shot classification accuracy of 88.9% on VisDA-2017. We achieve an improvement of 1.3% based on this, and we also surpass other methods, demonstrating that our method not only preserves the semantic alignment of CLIP but also leverages the fine-grained cues provided by DINOv2.

**Table 3.** Comparisons on VisDA-2017.

| Method | plane | bicycle | bus | car | horse | knife | mcycl | person | plant | sktbrd | train | truck | Avg |
|---|---|---|---|---|---|---|---|---|---|---|---|---|---|
| TVT(Yang et al., 2023) | 92.9 | 85.6 | 77.5 | 60.5 | 93.6 | 98.2 | 89.3 | 76.4 | 93.6 | 92.0 | 91.7 | 55.7 | 83.9 |
| CDTrans(Xu et al., 2021) | 97.1 | 90.5 | 82.4 | 77.5 | 96.6 | 96.1 | 93.6 | **88.6** | **97.9** | 86.9 | 90.3 | 62.8 | 88.4 |
| SSRT(Sun et al., 2022) | 98.9 | 87.6 | 89.1 | **84.8** | 98.3 | **98.7** | **96.3** | 81.1 | 94.8 | **97.9** | 94.5 | 43.1 | 88.8 |
| CLIP(Radford et al., 2021) | 99.1 | 91.7 | 93.8 | 76.7 | 98.4 | 91.7 | 95.3 | 82.7 | 86.5 | 96.0 | 94.6 | 60.5 | 88.9 |
| DAPL(Ge et al., 2023) | **99.2** | **92.5** | 93.3 | 75.4 | 98.6 | 92.8 | 95.2 | 82.5 | 89.3 | 96.5 | 95.1 | 63.5 | 89.5 |
| PDA(Bai et al., 2024) | **99.2** | 91.1 | 91.9 | 77.1 | 98.4 | 93.6 | 95.1 | 84.9 | 87.2 | 97.3 | **95.3** | 65.3 | 89.7 |
| Ours | **99.2** | 92.0 | **93.9** | 78.5 | **98.8** | 93.9 | **96.3** | 84.3 | 89.1 | 96.1 | 95.1 | **65.7** | **90.2** |

### 5.4 Ablation Study

Table 4 demonstrates the influence of each loss on the experimental results. We use CLIP as the baseline model without any loss. Generally, each loss function enhances the performance of the model. Adding $\mathcal{L}_{\text{ba}}$ yields significant improvements over the baseline, showing the benefits of prompt learning. Incorporating $\mathcal{L}_{\text{en}}$ further improves accuracy in three datasets, validating the effectiveness of the enhanced branch. Finally, the addition of $\mathcal{L}_{\text{sc}}$ achieves the best results.

To evaluate the necessity of each design element, Table 5 presents a comprehensive ablation of three factors: self-consistency loss, fusion method (MLP vs. DCMT), and self-supervised learning backbone (DINOv2 vs. BYOL). We observe that DCMT outperforms simple MLP fusion in both settings, validating the effectiveness of subspace alignment. Incorporating self-consistency loss consistently improves accuracy, underscoring its uniformity effect. Substituting DINOv2 with the lighter BYOL backbone degrades performance, highlighting the value of rich visual features. The full SCC configuration (self-consistency + DCMT + DINOv2) achieves the best results on all three datasets, demonstrating how each component cooperates to bridge the SVGG and restore class discriminability.

**Table 4.** Impact of loss terms on three datasets.

| $\mathcal{L}_{\text{ba}}$ | $\mathcal{L}_{\text{en}}$ | $\mathcal{L}_{\text{sc}}$ | Office-31 | Office-Home | VisDA-2017 |
|---|---|---|---|---|---|
| × | × | × | 77.5 | 82.4 | 88.9 |
| ✓ | × | × | 90.1 | 85.4 | 89.9 |
| ✓ | ✓ | × | 91.0 | 86.0 | 90.1 |
| ✓ | ✓ | ✓ | **91.3** | **86.2** | **90.2** |

**Table 5.** Comprehensive ablation on three datasets.

| Consistency | Fusion | Backbone | Office-31 | Office-Home | VisDA-2017 |
|---|---|---|---|---|---|
| × | MLP | DINOv2 | 90.3 | 85.5 | 89.1 |
| × | DCMT | DINOv2 | 90.8 | 85.8 | 89.5 |
| ✓ | MLP | DINOv2 | 90.6 | 85.6 | 89.4 |
| ✓ | DCMT | DINOv2 | **91.3** | **86.2** | **90.2** |
| ✓ | DCMT | BYOL | 90.7 | 85.8 | 89.2 |

### 5.5 Visualization

Figure 4 shows the feature visualizations for both the base and enhanced branches through t-SNE. Compared to CLIP image features, the enhanced image features exhibit tighter intra-class clustering and greater inter-class separation, thereby improving the discriminative capability of the learned representations.

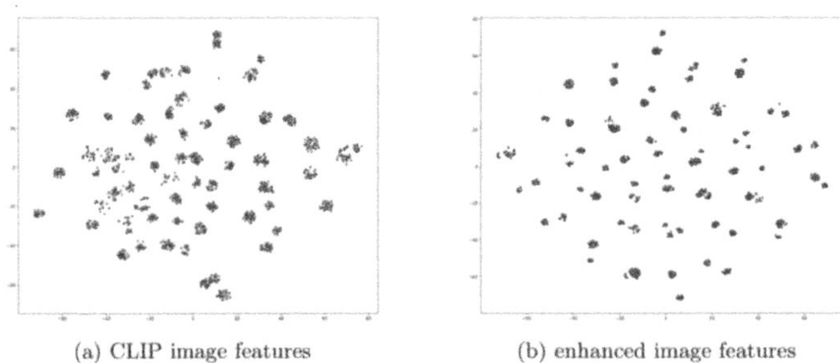

(a) CLIP image features (b) enhanced image features

**Fig. 4.** Visualization of (a) CLIP image features and (b) enhanced image features using t-SNE on task A → P(Office-Home).Image features are extracted from the target domain.

### 5.6 Limitations

Despite its strong overall performance, SCC exhibits some limitations. Analysis on the Office-Home dataset reveals that performance gains in more challenging tasks (e.g., A → C, R → C) are relatively smaller. One reason is the noise in pseudo labels, and the other is the model's reliance on the quality and coverage of textual descriptions. If the target domain does not align well with the textual descriptions, performance may be degraded. In addition, the computational complexity of the model is a challenge. Future work could explore integration with distillation methods.

## 6 Conclusion

In this paper, we propose SCC, a novel architecture for unsupervised domain adaptation. Its dual-branch design effectively bridges the gap between CLIP's semantic alignment capabilities and DINOv2's fine-grained visual discrimination. SCC significantly improves adaptability in the target domain by addressing the SVGG. Multiple experiments confirm the effectiveness of SCC. Our method pro vides a practical approach to harness both VLMs and self-supervised learning for UDA.

## References

Ajakan, H., Germain, P., Larochelle, H., Laviolette, F., Marchand, M.: Domain-adversarial neural networks (2014). arXiv Preprint arXiv:1412.4446

Bai, S., et al.: Prompt-based distribution alignment for unsupervised domain adaptation. In: Proceedings of the AAAI Conference on Artificial Intelligence, Vol. 38, pp. 729–737 (2024)

Chen, T., Kornblith, S., Norouzi, M., Hinton, G.: A simple framework for contrastive learning of visual representations. In: International Conference on Machine Learning, pp. 1597–1607. PMLR (2020)

Chen, X., Wang, S., Long, M., Wang, J.: Transferability vs. discriminability: Batch spectral penalization for adversarial domain adaptation. In: International Conference on Machine Learning, pp. 1081–1090. PMLR (2019)

Ge, C., et al.: Domain adaptation via prompt learning. IEEE Trans. Neural Netw. Learn. Sys. (2023)

Khattak, M.U., Rasheed, H., Maaz, M., Khan, S., Khan, F.S.: Maple: Multi-modal prompt learning. In: Proceedings of the IEEE/CVF Conference on Computer Vision and Pattern Recognition, pp. 19113–19122 (2023)

Long, M., Cao, Y., Wang, J., Jordan, M.: Learning transferable features with deep adaptation networks. In: International Conference on Machine Learning, pp. 97–105. PMLR (2015)

Radford, A., et al.: Learning transferable visual models from natural language supervision. In: International Conference on Machine Learning, pp. 8748–8763. PMLR (2021)

Sun, T., Lu, C., Zhang, T., Ling, H.: Safe self-refinement for transformer-based domain adaptation. In: Proceedings of the IEEE/CVF Conference on Computer Vision and Pattern Recognition, pp. 7191–7200 (2022)

Tzeng, E., Hoffman, J., Zhang, N., Saenko, K., Darrell, T.: Deep domain confusion: Maximizing for domain invariance (2014). arXiv Preprint arXiv:1412.3474

Xu, T., et al.: Cd trans: Cross-domain transformer for unsupervised domain adaptation (2021). arXiv Preprint arXiv:2109.06165

Yang, J., Liu, J., Xu, N., Huang, J.: Tvt: Transferable vision transformer for unsupervised domain adaptation. In: Proceedings of the IEEE/CVF Winter Conference on Applications of Computer Vision, pp. 520–530 (2023)

Yao, Y., Zhang, A., Zhang, Z., Liu, Z., Chua, T.-S., Sun, M.: Cpt: Colorful prompt tuning for pre-trained vision-language models. AI Open **5**, 30–38 (2024)

Yu, W., et al.: Turning a clip model into a scene text detector. In: Proceedings of the IEEE/CVF Conference on Computer Vision and Pattern Recognition, pp. 6978–6988 (2023)

Zhang, M., Huang, S., Wang, D.: Domain generalized few-shot image classification via meta regularization network. In: ICASSP 2022–2022 IEEE International Conference on Acoustics, Speech and Signal Processing (ICASSP), pp. 3748–3752. IEEE (2022)

Zhou, K., Yang, J., Loy, C.C., Liu, Z.: Conditional prompt learning for vision-language models. In: Proceedings of the IEEE/CVF Conference on Computer Vision and Pattern Recognition, pp. 16816–16825 (2022a)

Zhou, K., Yang, J., Loy, C.C., Liu, Z.: Learning to prompt for vision-language models. Int. J. Comput. Vision **130**(9), 2337–2348 (2022)

# Application of a Linear Programming-Driven Multi-Algorithm Integration Framework in Optimizing Crop Planting Strategies

Jing Cao, Yuanyuan Hu, Haiyang Yu, Dongyang Ye, and Yuanyuan Li(✉)

School of Artificial Intelligence, Jianghan University, Wuhan 430056, China
460718029@qq.com

**Abstract.** This study examines a rural village in the mountainous region of North China. Based on local crop characteristics, farmland conditions, and 2023 planting data, a linear programming (LP) model was established as the baseline framework. Three distinct single models were developed by integrating reinforcement learning Q-Learning and simulated annealing (SA) algorithms, respectively. Subsequently, hybrid models—LP-Q-Learning, LP-SA, and LP-SA-Q-Learning—were designed for cross-validation. The study aims to maximize net profits from 2024 to 2030, comprehensively considering constraints such as crop growth cycles, market fluctuations, and climate uncertainties. Results indicate: In the short term (2024–2026), the LP-Q-Learning model demonstrates optimal optimization performance by rapidly adapting to market volatility and crop rotation demands through reinforcement learning's dynamic responsiveness. In the long term (2027–2030), the LP-SA-Q-Learning model overcomes local optima traps through the SA algorithm. By integrating Q-Learning's dynamic adjustment capability with LP's deterministic constraints, it demonstrates strong robustness under complex climate and market conditions, significantly outperforming other models in optimization. Overall, the LP-SA-Q-Learning model initially corrects strategy deviations through continuous learning, steadily increasing net returns. It achieves optimal strategy stability and return growth in the mid-to-late stages, validating its core advantages for long-term agricultural planning in mountainous regions.

**Keywords:** Agricultural efficiency improvement · Linear programming · Reinforcement learning · Simulated annealing

## 1 Introduction

Agriculture, as the foundation of national economic and social development, is crucial for ensuring food security, promoting rural economic growth, and increasing farmers' incomes. However, traditional farming models face challenges such as shifting market demands, inefficient resource utilization, climate change, and irrational crop structures. In recent years, mathematical modeling and data analysis methods have provided novel approaches for optimization strategies. In the field of optimization algorithms, Li et al.


H. Yan et al. (Eds.): AIS&P 2025, CCIS 2918, pp. 258–283, 2026.
https://doi.org/10.1007/978-981-95-9884-7_18

(2022) proposed an improved particle swarm optimization algorithm. By incorporating inertia weight decay and particle mutation strategies to construct a multi-objective model, they achieved synergistic improvements in economic, social, and water resource benefits during crop structure optimization in Anyang City, demonstrating the potential of intelligent algorithms in balancing multiple objectives. Talatahari (2025) introduced the Multi-Task Constrained Multi-Objective Optimization (MTCMO) framework, which dynamically transfers knowledge through auxiliary tasks and effectively utilizes infeasible solutions to enhance performance in constrained multi-objective optimization. Qiao (2023) studied an Adaptive Strategy Management (ASM) framework with dynamic policy switching, significantly improving the computational efficiency for solving multi-objective optimization problems. Wu Junjie (2024) explicitly defined resource constraints and objective functions based on a linear programming model, effectively improving the allocation efficiency of land and water resources in crop structures. This provides a quantitative pathway for resource coordination in large-scale farming. Wu Daizhen et al. (2025) combined mixed-integer linear programming (MILP) with Monte Carlo simulation to analyze the impact of market response strategies on economic benefits by maximizing net revenue, offering guidance for layout adjustments under uncertainty. Hu Yingjia et al. (2025) employed genetic algorithms and Monte Carlo simulation to optimize crop and land allocation under crop rotation and cost constraints, emphasizing the influence of market volatility. Wang Jiaying et al. (2024) proposed a multi-stage model based on particle swarm optimization (PSO) to dynamically adjust responses to market and climate uncertainties; Yu Xiya et al. (2025) compared genetic algorithms with simulated annealing algorithms, finding simulated annealing converged more stably in crop planting strategy optimization, offering insights for global search. Regarding the relationship between planting factors and structure, Cai Zhidan et al. (2024) constructed a crop planting strategy optimization model based on linear programming, quantifying returns by integrating field type and seasonal characteristics. Liu Yingshan et al. (2024) employed integer programming and stochastic programming, incorporating market uncertainty simulations, to analyze the impact of crop substitutability and complementarity on planting strategies. In dynamic decision-making, cross-scenario applications of reinforcement learning offer novel approaches to temporal optimization. Yang Qi (2025) applied Q-Learning to dynamic scheduling in automated warehousing systems, optimizing task sequencing efficiency through a "state-action-reward" mechanism. Its multi-stage decision framework provides reference for dynamically adjusting crop rotation cycles. Zhou Xianwen et al. (2025) proposed an improved Q-Learning algorithm. By using linear programming to screen effective training data and dividing continuous state spaces into intervals, they enhanced decision safety and efficiency in complex dynamic scenarios, providing technical inspiration for algorithm design in agricultural temporal optimization.

Despite progress in mathematical modeling and factor analysis, existing research remains limited in addressing crop plot allocation and area distribution, potential dynamic risks in crop planting, and variations in planting patterns. Furthermore, single algorithms struggle to balance all factors for global optimization, while static models perform poorly when considering factors like crop rotation cycles. To address these challenges, this study utilizes the scenario from Problem C of the 2024 Higher Education Press Cup National College Student Mathematical Modeling Competition. Taking a village in the mountainous region of North China as an example, it introduces a "linear programming-quasi-learning" model and constructs an optimization model through model cross-combination. After experimental comparison, the optimal model is selected to derive optimization strategies. This multi-model cross-comb - simulated annealing-Q-learning" model and constructs an optimization model through model cross-combination. After experimental comparison, the optimal model is selected to obtain optimization strategies. This multi-model cross-combination hierarchical collaboration simplifies parameters while retaining the global exploration and temporal learning advantages of relevant algorithms, thereby improving optimization efficiency. Furthermore, when considering dynamic fluctuations in crops, the model scheme adopted in this paper better aligns with planting data while avoiding the overestimation of long-term benefits by static models.

## 2 Method

### 2.1 Problem Statement

Taking a village in North China as a case study, this paper optimizes crop planting area allocation and regional selection strategies for the 2024–2030 period. The optimization aims to maximize planting profits by considering factors such as unit yield per planting area, crop rotation patterns, crop cultivation conditions, and rational distribution of planting areas. To achieve optimal results, this study integrates three models—linear programming, reinforcement learning, and simulated annealing—through cross-model optimization. Final analysis and comparison identify the optimal model and best cultivation strategy.

### 2.2 Model Introduction

**Linear Programming Mode** (Wu 2024)**.** Linear programming is a mathematical optimization method used to find the optimal solution to a problem by optimizing a linear objective function within a feasible region defined by a set of linear constraints. This paper applies this model, stipulating that the total yield of a certain crop per season must exceed the corresponding projected sales volume. Any surplus is sold at a 50% discount based on the 2023 sales price. The objective function is derived according to the optimization goal. Simultaneously, it considers the correlation between expected sales volume and sales price, the similarity degree between crop substitution potential (i.e., cultivation environment and market demand) and crop complementarity (i.e., nutrient absorption and staggered growth cycles) reaching a certain threshold. Based on this, and incorporating relevant cultivation requirements and data, the analysis examines crop correlation, substitution potential, and complementarity to derive the relevant constraints for the linear programming model.

*Objective Function.*
Based on the aforementioned constraints and conditions, the objective function is derived as:

$$\begin{aligned} MaxW &= \sum_{i,t} \left[R \times \min(Y \times x_{i,t}, P) + 0.5 \times R \times \max(Y \times x_{i,t}, P) - C \times x_{i,t}\right] \\ &+ \sum_{j,t} \left[R \times \min(Y \times y_{j,t}, P) + 0.5 \times R \times \max(Y \times y_{j,t}, P) - C \times y_{j,t}\right] \end{aligned} \quad (1)$$

where $x_{i,t}$ represents the planting area of a certain crop in plot i during season t, and $y_{j,t}$ represents the planting area of a certain crop in greenhouse j during season t. W denotes total planting profit, P denotes the expected sales volume of the crop, C denotes the planting cost of the crop, Y denotes the yield per mu of the crop, and R denotes the unit sales price of the crop.

*Analysis of Constraints for Substitutability and Complementarity of Crop-Related Quantities Based on K-means Clustering.*

1. K-means Clustering Algorithm.
   Clustering algorithms, also known as group analysis methods, are typical unsupervised grouping algorithms in data mining technology (Li, 2020). It is a distance-based clustering algorithm where K denotes the number of categories and Means represents the mean. This algorithm posits that clusters are composed of objects with close distances (Deng and Pan 2024), thus aiming to obtain compact and independent clusters as the ultimate goal. Its algorithmic steps are as follows, as illustrated in Fig. 1:

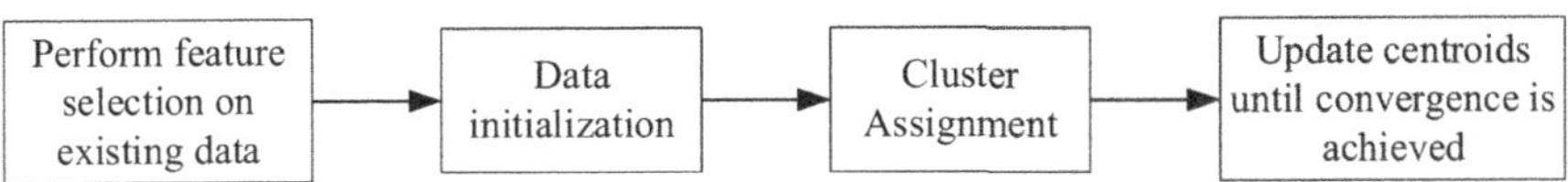


**Fig. 1.** K-means Algorithm Steps Diagram.

2. Correlation Analysis.
   To establish the relationship between expected sales volume, sales price, and cultivation costs, this paper first employs the covariance matrix to derive the correlation between expected sales volume, sales price, and sales costs, namely:

$$\sum = \begin{bmatrix} D(P) & Cov(P, R) & Cov(P, C) \\ Cov(R, P) & D(R) & Cov(R, C) \\ Cov(C, P) & Cov(C, R) & D(C) \end{bmatrix} \quad (2)$$

   where D(P) represents the variance of expected sales volume, D(R) denotes the variance of unit sales price, D(C) indicates the variance of cultivation costs,

Cov(P,R) signifies the covariance between expected sales volume and sales price, Cov(P,C) reflects the variance between expected sales volume and cultivation costs, and Cov(R,C) represents the covariance between sales price and cultivation costs.

To visually present the correlation matrix information, Fig. 2 visualizes it as a heatmap. The heatmap uses color intensity to indicate the magnitude of correlation coefficients, with darker shades representing high correlations and lighter shades indicating low correlations. This demonstrates the existence of correlations among the three variables.

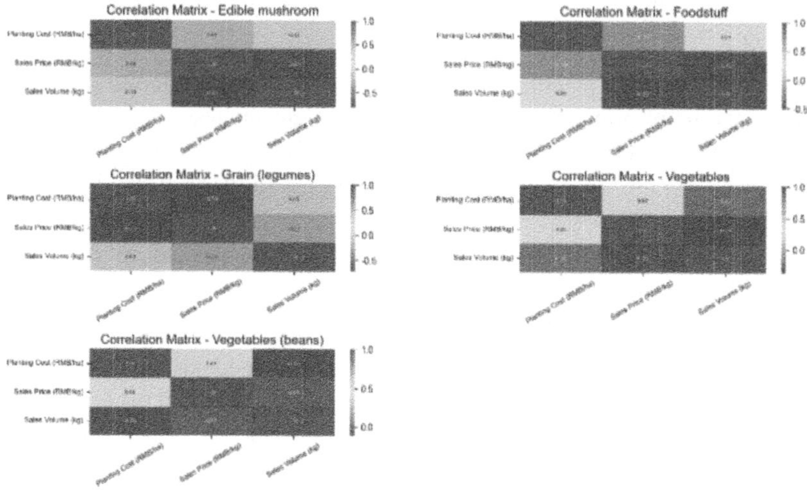


**Fig. 2.** Correlation Distribution Chart.

3. Complementarity and Substitutability Analysis.

Based on the correlations among the crops produced, we now investigate their complementarity and substitutability. Ordinary least squares (OLS) were employed for parameter estimation, yielding correlation coefficients between grain crops, grain legumes, vegetables, vegetable legumes, and edible fungi with planting costs, unit selling prices, and sales volumes. Consequently, a multiple linear regression function can be established:

$$P_k = u_0 + u_1 R_k + u_2 C_k, \forall k \quad (3)$$

to represent the linear relationship between expected sales volume, sales price, and cultivation costs.

To determine the optimal number of clusters (k value) in the K-means clustering algorithm during the process of identifying crop complementarity and substitutability, this study employed the elbow method. The elbow method calculates the inertia values corresponding to different k values and selects the optimal number of clusters by plotting a graph. This study tested the variation of inertia values for k values ranging from 1 to 10 and plotted an elbow diagram, as shown in Fig. 3:

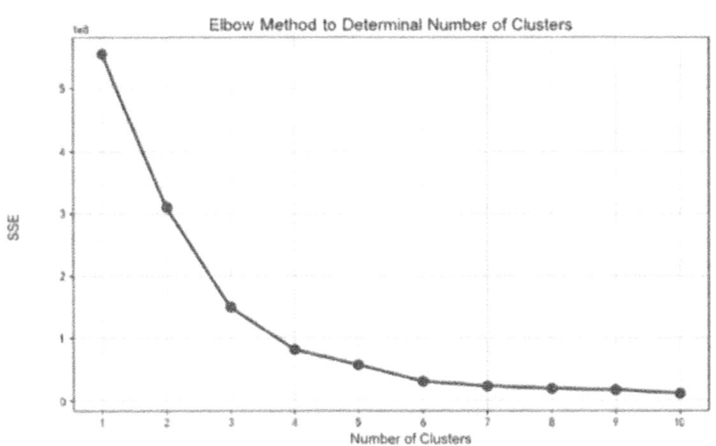


**Fig. 3.** Elbow Plot.

As shown in the figure, the inertia value gradually decreases as the k value increases. When k = 3 or k = 4, the rate of decrease in the inertia value noticeably slows down, forming an "elbow" shape. This phenomenon indicates that both k = 3 and k = 4 are viable candidate cluster numbers. However, considering that the model must not only ensure satisfactory clustering performance but also avoid excessive model complexity due to an overly large number of clusters, k = 3 is ultimately selected as the optimal cluster number in this study.

Furthermore, based on the correlation between variables (Tian et al., 2025), the optimization problem is classified as follows, with the results summarized in Table 1:

**Table 1.** Crop Aggregation Results.

| Category | Crop ID |
|---|---|
| 0 | 7, 8, 9, 10, 11, 12, 13, 20, 21, 23, 24, 29, 30, 34, 38, 40 |
| 1 | 1, 16, 19 |
| 2 | 2, 3, 4, 5, 6, 13, 14, 17, 18, 22, 25, 26, 27, 28, 31, 32, 33, 35, 36, 37, 39, 41 |

The specific classification results can be visually represented through a three-dimensional diagram, as shown in Fig. 4:

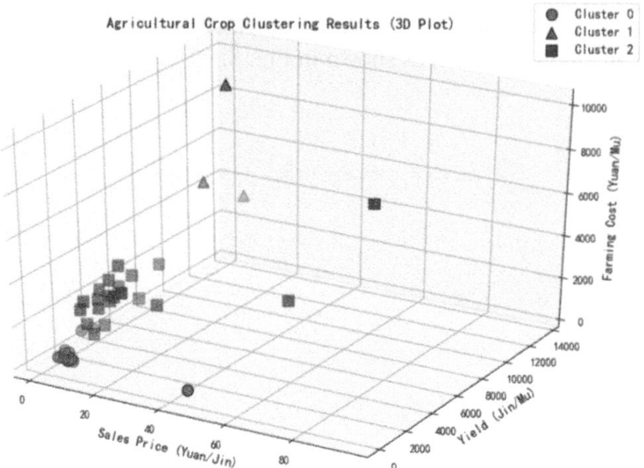


**Fig. 4.** Cluster Analysis Results.

Since crops are considered interchangeable only when they belong to different clusters and their similarity across all aspects reaches a certain threshold, the interchangeability coefficient is defined as $\alpha$. The specific calculation formula is expressed as:

$$\alpha_{N_k,N_{k'}} = \frac{\Delta P_{N_{k'}}}{\Delta x_{N_k}} \tag{4}$$

where $\alpha_{N_k,N_{k'}}$ denotes the coefficient of substitutability between the crop $N_k$ and the crop $N_{k'}$, $\Delta P_{N_{k'}}$ denotes the change in the expected sales volume of the crop $N_{k'}$ within the two adjacent seasons, $\Delta x_{N_k}$ denotes the change in the acreage of the crop $N_k$ within the two adjacent seasons, and that is:

$$\alpha_{N_k,N_{k'}} = \frac{\Delta P_{N_{k'}}}{\Delta x_{N_k}} = \frac{P_{N_{k'},t-1} - P_{N_{k'},t}}{x_{i,N_k,t}} \tag{5}$$

Based on the main manifestations of substitutability it is known that when the substitutability coefficient $\alpha$ is negative, substitutability is considered to exist between crops.

Further, for substitutability to be effective, it is now assumed that the substitutability ratio $m > 50\%$ between crops of a cluster analyzed by clustering is considered to be substitutable between crops of that cluster, where

$$m = \frac{L_{\alpha<0}}{L_{\alpha}} \tag{6}$$

$L_{\alpha<0}$ denotes the number of crops that satisfy substitutability between crops in the cluster, and $L_{\alpha}$ denotes the number of all paired experiments between crops in the cluster.

Similarly introducing the complementarity coefficient $\beta$, which can be listed as

$$\beta_{N_k,N_{k'}} = \frac{\Delta Y_{N_{k'}}}{\Delta x_{N_k}} \tag{7}$$

where $\beta_{N_k,N_{k'}}$ denotes the coefficient of complementarity between the crop $N_k$ and the crop $N_{k'}$, $\Delta Y_{N_{k'}}$ denotes the change in acreage of the crop within two adjacent seasons, and $\Delta x_{N_k}$ denotes the change in acreage of the crop $N_k$ within two adjacent seasons, then there are

$$\beta_{N_k,N_{k\prime}} = \frac{\Delta Y_{N_{k\prime}}}{\Delta x_{N_k}} = \frac{Y_{N_{k\prime},t-1} - Y_{N_{k\prime},t}}{x_{i,N_k,t}} \tag{8}$$

Based on the main manifestations of complementarity, complementarity between crops is considered to exist when the complementarity coefficient $\beta$ is positive.

Similarly, Assuming the complementarity between crops within a cluster identified by cluster analysis exceeds $q > 50\%$, it is considered that the crops within this cluster exhibit complementarity. Where

$$q = \frac{L_{\beta>0}}{L_{\beta}} \tag{9}$$

$L_{\beta>0}$ denotes the number of complementarities satisfied between crops of different clusters, and $L_{\beta}$ denotes the number of all paired experiments between crops of different clusters.

Based on the above analysis, the constraints regarding substitutability and complementarity among crops are derived as follows:

(1) Substitutability Constraint.

- Total Area Constraint for Crops in the Same Cluster.

  To prevent excessive concentration or dispersion of a particular crop type, upper and lower limits are imposed on the total cultivated area for each cluster type $C_k$:

$$0.3S_{\text{total area}} \leq \sum_{i\in c_k} x_i \leq 0.7S_{\text{total area}} \tag{10}$$

  where $S_{\text{Total area}}$ represents the total planting area, and $x_i$ denotes the planting area of crop i. The upper and lower limit coefficients (between 0.3 and 0.7) are adjusted based on actual conditions.
- Economic Substitutability Constraint.

For crop pairs (i, j) from different clusters but with substitutability, the net profit per unit area difference must not exceed the threshold:

$$\left|(R_i - C_i) - (R_j - C_j)\right| \times \frac{X_i + X_j}{2} \leq \delta \tag{11}$$

where $R_i$ and $R_j$ represent the unit selling price, $C_i$ and $C_j$ denote the unit area planting cost, $X_i$ and $X_j$ denote the unit area yield of crop i and crop j respectively, and $\delta = 0.1$ is the economic difference threshold, requiring that at least 50% of the substitute crops meet this condition.

(2) Complementarity Constraint.

If multiple crops belonging to the same cluster are selected for cultivation on the same plot during the same season, the planting area ratio for each crop j must satisfy:

$$0.1 \leq \frac{X_j}{\sum\limits_{i \in C_k} X_i} \leq 0.4 \tag{12}$$

This constraint ensures reasonable planting proportions within the same cluster, preventing excessive dominance by any single crop. This achieves complementarity and enhances overall profitability.

*Crop Constraint Planning Aligned with Planting Conditions.*

Integrating crop cultivation requirements and planting patterns, the following constraints are designed based on crop correlation, substitutability, and complementarity:

- Maximizing Unit Yield.

  Each crop's planting area per plot should not be excessively small. Assuming minimum land utilization of 10%, this translates to

$$\begin{cases} x_{i,t} \geq S \times 10\% \\ \quad \text{or} \\ y_{i,t} \geq S \times 10\% \end{cases} \tag{13}$$

  where S denotes the total area of a given plot.
- Soybean Planting Requirements.

  Since all land within each plot must be planted with legumes at least once every three years, the following constraint applies when all crops planted are legumes:

$$\sum_{t=T}^{T+2} x_{i,t} \geq 1, \sum_{t=T}^{T+2} y_{j,t} \geq 1 \tag{14}$$

  where T=2023,2026, $x_{i,t}$ denotes the acreage of a crop in plot i season t and $y_{j,t}$ denotes the acreage of a crop in shed j season t.
- Crop Rotation Pattern.

  To prevent yield decline when cultivating crops on land, crop rotation must be implemented on the same plot. Therefore, the following constraints apply:

$$x_{i,t'} + x_{i,t-1'} \leq 1, y_{j,t'} + y_{j,t-1'} \leq 1 \tag{15}$$

  where$x_{i,t'}$,$x_{i,t-1'}$,$y_{j,t'}$,$y_{j,t-1'}$ four values can only take 0 or 1, when the area of a crop planted on the plot in that quarter is greater than 0 the value takes 1, otherwise it is 0, (t-1) means the previous quarter of season t.
- Crop Planting Conditions.

Based on the crop numbers of 41 crops given in Annex 1, it is defined that $N_k$ denotes the crop numbered k. N is any crop, k = 1,2,3,...40,41. Since different crops are grown in different geographic areas, climates, and cycles, i.e., each crop can only be grown in a specific season in a specific plot. Thus there are relevant constraints:

i When k = 1, 2, 3, … 15

$$\begin{cases} x_{i,t,N_k} \geq 0 \text{ i denotes dryland, terraced fields, and slopes; t indicates a single season} \\ x_{i,t,N_k} \geq 0 \text{ i denotes other plots, t denotes a single season} \end{cases} \tag{16}$$

ii When k = 16, 35, 36, 37, 38, 39, 40, 41

$$\begin{cases} x_{i,t,N_k} \geq 0, \text{ i is irrigable land;} x_{i,t,N_k} = 0, \text{ i for other plots; } k = 16, \text{ t is for single season} \\ x_{i,t,N_k} \geq 0, \text{ i is irrigable land;} x_{i,t,N_k} = 0, \text{ i for other plots;} k = 35 \sim 37, \text{ t is for two seasons} \\ x_{i,t,N_k} \geq 0, \text{ i is a common greenhouse;} x_{i,t,N_k} = 0, \text{ i for other plots;} k = 38 \sim 41, \text{ t is for two seasons} \end{cases} \tag{17}$$

iii When k = 17 to 34

$$\begin{cases} x_{i,t,N_k} \geq 0, & \begin{array}{l} \text{i denotes irrigated land,} \\ \text{t denotes single - season cultivation, where i represents conventional greenhouses,} \\ \text{t denotes single - season cultivation where i represents smart greenhouses,} \\ \text{and t denotes single - season or double - season cultivation} \end{array} \\ x_{i,t,N_k} = 0, & \text{other conditions} \end{cases} \tag{18}$$

- Reasonable Planting Area.

The total area of land used for all crops planted in the same season cannot exceed the total existing arable land area. Therefore, the constraint is:

$$\sum x_{i,t} \leq A_i = 1201, \sum y_{j,t} \leq B_j = 12 \tag{19}$$

where i = 1, 2… 34, j = 1, 2… 20, $A_i$ is the area of open cropland, and $B_j$ is the area of greenhouses.

**Q-Learning Model.** Q-Learning is a lightweight reinforcement learning algorithm that iteratively updates the action-value function Q-values to learn optimal policies. This enables the agent to select actions in each state that maximize long-term cumulative rewards. Even with a suboptimal initial policy, the agent can gradually converge toward an optimal policy through real-time interaction with the environment.

The fundamental framework of Q-Learning comprises five core elements: the agent, the state space, the action space, the policy, and the reward function. Its construction process is as follows:

*State Space Construction.*
States are used to describe key information about the current environment. In crop planting optimization problems, states must reflect the fundamental planting requirements of crops and their growth environment. To better optimize planting schemes, this paper defines states for two types of combined models involving Q-Learning based on their respective strengths, weaknesses, and focal points. Since the LP-Q-Learning model emphasizes configuration optimization while the LP-SA-Q-Learning model primarily utilizes the domain search capabilities of the simulated annealing algorithm to focus on scheme optimization, states are defined separately for each: In the LP-Q-Learning model, the state is represented by the type of cultivation plot. In the LP-SA-Q-Learning

model, the state is defined as the crop planted on a specific plot and its cultivation area. This state is converted into a tuple format for indexing in the Q-table.

*Action Space Construction.*
Given a known state, the action space must represent specific planting methods to better reflect and facilitate the interaction process, thereby achieving strategy optimization. Accordingly, actions in the LP-Q-Learning model are defined as selecting suitable crops for planting plots from the initial conditions using an ε-greedy strategy. In contrast, the LP-SA-Q-Learning model defines actions as adjusting existing planting schemes to achieve optimality.

*Exploration Strategy Construction.*
In crop planting strategy optimization problems, the Q-Learning algorithm possesses no prior environmental knowledge during its initial phase. To enhance the model's exploration capability in the early training stages, the algorithm must extensively explore how different state-action combinations impact crop growth and final yields. As experience samples accumulate, the algorithm progressively shifts its focus toward decision-making based on existing knowledge, thereby improving strategy stability and practical benefits. Therefore, to balance exploration and exploitation while effectively suppressing local optima, this paper employs an ε-greedy policy. At each decision step, the current known optimal action is selected with probability (1-ε) (Yang, 2025), while an action is randomly chosen with probability ε. After execution, immediate rewards are obtained, and the agent transitions to a new state. The specific strategy is as follows:

$$\delta_t = \begin{cases} \arg\max\limits_{a \in A} Q(a), & \text{Sampling probability:} 1 - \varepsilon \\ \text{Randomly select } a, & \text{Sampling probability:} \varepsilon \end{cases} \tag{20}$$

*Building an Experience Replay Buffer.*
To comprehensively and stably explore optimal strategies, this paper designs an experience replay buffer to store acquired planting schemes. An experience replay buffer refers to the process where the program stores a complete "experience tuple"—comprising the current state, executed action, received reward, and next state—into the buffer. Once the buffer accumulates a sufficient number of experiences, the program randomly samples a batch of experiences from it. For each experience in the batch, it applies the Q-Learning update formula.

*Reinforcement Function Construction.*
The reward function evaluates action quality, guiding the agent toward optimal strategies. This paper uses net profit as the reward signal. For the two combination models, LP-Q-Learning represents reward as total net profit and growth rate, while SA-Q-Learning uses profit change as the reward value. This design encourages agents to pursue not only high profits but also sustained profit growth.

Q-Learning primarily relies on the Bellman equation, utilizing temporal difference methods to progressively optimize Q-values. During this process, the agent typically employs a combination of and ε-greedy strategies. After executing relevant actions, it receives immediate rewards and transitions to a new state. Updates are then performed based on the difference between the target Q-value and the current Q-value, i.e.,

$$Q_{t+1}(s_t, a_t) = Q_t(s_t, a_t) + \alpha\left[R_{t+1} + \gamma \max_a(Q_t(s_{t+1}, a_t) - Q_t(s_t, a_t))\right] \quad (21)$$

Here, $s_t$ represents the current state, $a_t$ denotes the action chosen by the agent in that state, and $\alpha$ refers to the learning rate, which controls the speed of Q-value updates. If the learning rate is too high, the algorithm may become unstable. If the learning rate is too small, the algorithm may converge slowly. $\gamma$ denotes the discount factor, representing the discount rate applied to future rewards. When the discount factor approaches 1, the algorithm prioritizes long-term rewards, meaning identical actions yield identical rewards. If the discount factor approaches 0, the algorithm focuses solely on immediate rewards. $Q_{t+1}(s_t, a_t)$ indicates the updated Q-value after executing action $a_t$ in state $s_t$, while $Q_t(s_t, a_t)$ is the original Q-value. Therefore, during reinforcement learning training, the agent acquires optimal policies through continuous interaction with the dynamic environment. The actions chosen by the agent cause dynamic transitions in the environment state. This stochastic process makes the state distribution influenced by both the current environment state and the agent's decisions. Thus, when designing states, it is necessary to ensure precise environmental information to support the agent's accurate judgment of the current environment, while simultaneously removing irrelevant features to avoid interfering with the learning process. This enables the agent to make effective decisions efficiently. The algorithm diagram is shown in Fig. 5.

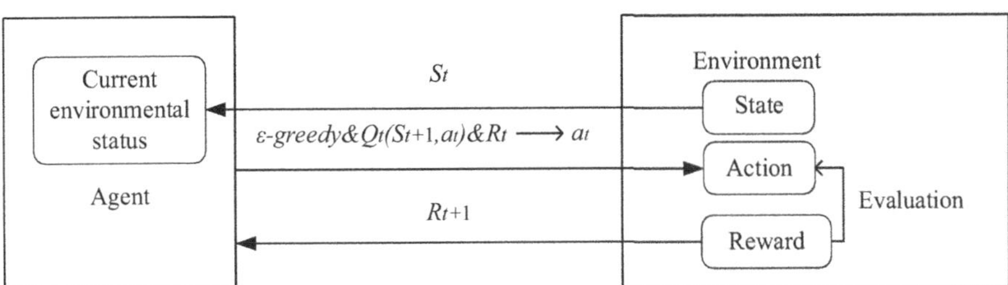


**Fig. 5.** Q-Learning Algorithm Diagram.

Algorithm (Tian et al., 2025). The Simulated Annealing Algorithm is a global optimization method inspired by the annealing process of metals. It solves optimization problems by simulating the characteristics of physical systems undergoing random energy state transitions at high temperatures and gradually stabilizing as the temperature decreases.

Starting from an initial solution, the algorithm iteratively generates new solutions by perturbing within the neighborhood of the current solution. It then applies the Metropolis criterion to determine acceptance: if the objective function value decreases, the new solution is accepted with probability 1; if it increases, acceptance occurs with probability $P = e^{-\frac{\Delta E}{T}}$. As the temperature gradually decreases according to the set cooling strategy and the number of iterations approaches exhaustion, the algorithm progressively reduces the acceptance probability for poorer solutions, ultimately converging to the global optimum or an approximate optimum. The specific Metropolis criterion for the acceptance probability is:

$$P = \begin{cases} 1, & \Delta E \leq 0 \\ e^{-\frac{\Delta E}{T}}, & \Delta E > 0 \end{cases} \tag{22}$$

where ΔE is the change in the objective function, and T is the current temperature.

*Simulated Annealing Algorithm Parameter Settings.*
The simulated annealing algorithm is based on the Metropolis criterion and the temperature function. To enhance its optimization efficiency, parameter tuning is essential. The temperature parameter T serves as the core regulatory variable. If the initial temperature is set too high, the annealing process accelerates excessively, potentially causing premature termination at local optima. Conversely, an excessively low initial temperature prolongs computational time. Therefore, a higher temperature is typically set in the initial phase, gradually decreasing with iterations. This paper employs an exponential decay method for the annealing rate, defined as:

$$T(n+1) = \lambda T(n), n = 1, 2, 3 \ldots \tag{23}$$

where λ ranges from 0.8 to 0.99. To prevent getting stuck in local optima, multiple state transition attempts are required at each temperature T. The algorithm terminates when the temperature reaches a preset final threshold, at which point the probability of accepting a suboptimal solution becomes negligible. Additionally, if multiple consecutive iterations fail to improve the current optimal solution or the solution acceptance rate remains persistently below a critical threshold, it indicates the annealing process has concluded.

*Algorithm Description and Process.*
For the optimization of crop planting strategies discussed herein, the specific steps of the simulated annealing algorithm are as follows:

1. Initialization: Set the starting temperature $T_0$, obtain the initial solution x, and determine the iteration count n for each temperature;
2. Iterative Solution Generation: New solutions are generated by introducing a random perturbation within the neighborhood of the current solution. Common methods include: - Swapping, inserting, or reversing sequence-type solutions - Perturbing continuous vectors by adjusting random strides based on current temperature - Randomly adding, deleting, or replacing elements in set-type solutions Specifies that the solution space and objective function of this algorithm are identical to the constraints and objective function in a linear programming model. It primarily generates new solutions by randomly adding or removing crop types from the set of crop solutions;
3. Compute the objective function increment:

$$\Delta E = W_2 - W_1 \tag{24}$$

where $W_2$ is the objective function value of the new solution, and $W_1$ is the objective function value of the original solution;

4. Update solution: Update the solution based on the Metropolis criterion;
5. Iteration termination: After completing the above process, if the preset maximum iteration count is reached and the temperature T has dropped to the minimum temperature, terminate the loop and output the current solution as the optimal solution. If not terminated, continue the iterative loop.
6. The algorithm flowchart is illustrated in Fig. 6:

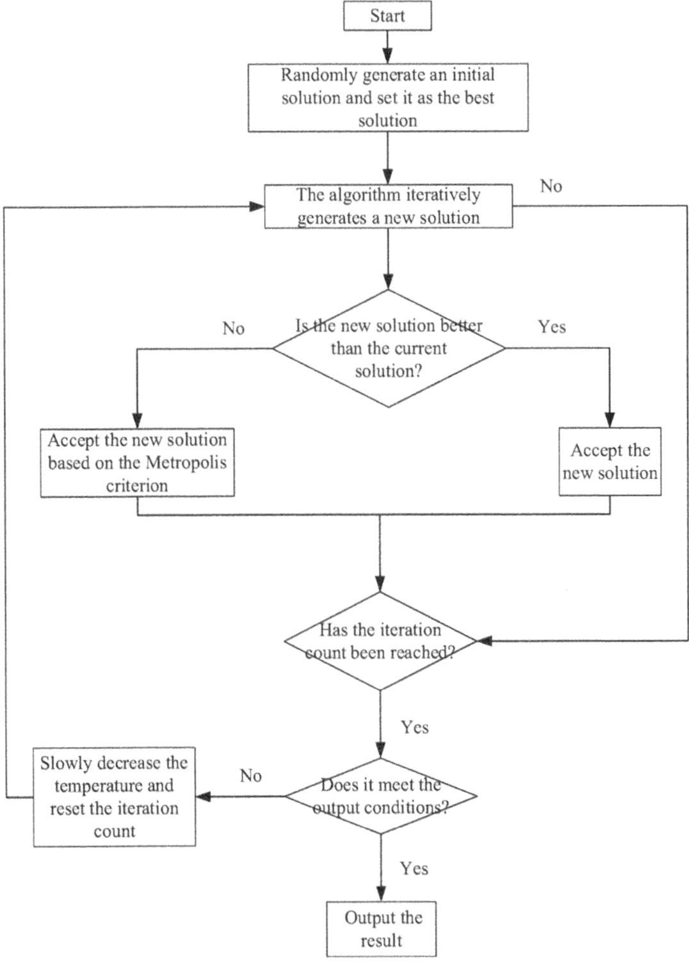


**Fig. 6.** Simulated Annealing Algorithm Flowchart.

# 3 Model Establishment and Analysis

The dataset for this paper is sourced from the 2024 National College Students Mathematical Modeling Competition https://cumcm.cnki.net. It includes the categories and quantities of farmland in a village in the mountainous region of North China, the types of crops cultivated, the crop planting status for 2023, and the sales data collected in 2023.

## 3.1 Data Preprocessing

This study first performed outlier handling on the original data to refine the vegetable crops for the first season of the smart greenhouse. Partial results are shown in Table 2.

**Table 2.** Seasonal Statistics for Smart Greenhouse Season 1.

| Crop Name | Yield per Mu/Jin | Planting Cost/(CNY/mu) | Sales Price (CNY/Jin) |
|---|---|---|---|
| Yardlong Beans | 3600 | 2400 | 7.00–9.00 |
| Sword beans | 2400 | 1200 | 5.50–8.00 |
| Kidney beans | 3600 | 2400 | 5:00–8:00 |
| Potatoes | 2400 | 2400 | 3.00–4.50 |
| Tomatoes | 3000 | 2400 | 5.00–7.50 |
| Eggplant | 8000 | 2400 | 5.00–6.00 |
| Spinach | 3300 | 2700 | 4.80–6.70 |
| Green Pepper | 3000 | 2000 | 4.00–6.50 |
| Cauliflower | 4000 | 3000 | 5.00–6.00 |
| Cabbage | 4500 | 3500 | 5.50–7.50 |
| Lamb's lettuce | 5000 | 2000 | 4.00–6.00 |
| Baby Bok Choy | 4000 | 2000 | 5.00–6.50 |
| Cucumber | 15000 | 3500 | 6.00–8.00 |
| Lettuce | 5000 | 2000 | 4.50–6.00 |
| Pepper | 2000 | 1200 | 6.00–8.50 |
| Water spinach | 12000 | 5000 | 3.00–6.00 |
| Yellow-Hearted Cabbage | 6000 | 2500 | 4.00–5.00 |
| Celery | 6600 | 1100 | 3.20–4.80 |

Since the original data represented the unit sales prices of each crop as ranges, it was necessary to determine specific values for the study. To ensure authenticity, Python was used to extract random numbers from each crop's price range to establish the final unit price. The specific distribution is shown in Fig. 7.

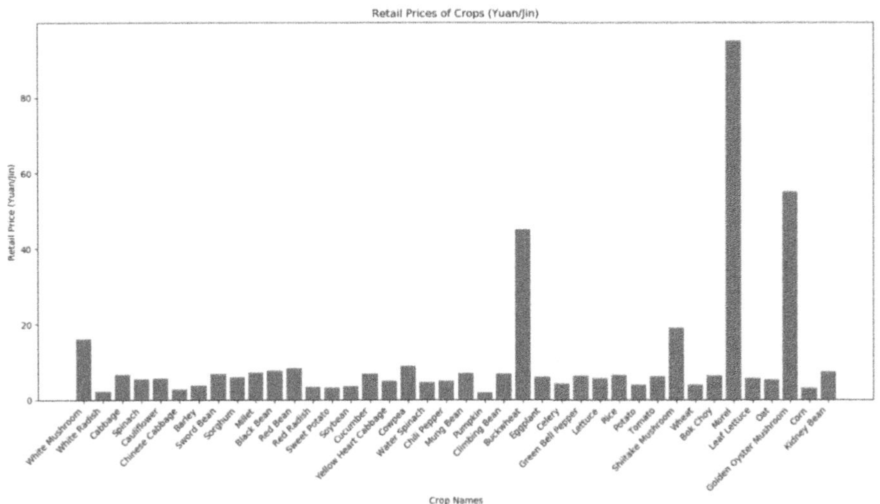


**Fig. 7.** Distribution Chart of Crop Unit Price Values.

Simultaneously, the expected sales volume for each crop was defined as equal to its total planting yield. Finally, since crop cultivation is influenced by region and season, this study defined "the cultivable area for a given crop under certain conditions as greater than zero" as '1', and otherwise as '0'. Selected results are presented in Table 3.

**Table 3.** Selected Crop Analysis Data.

| Crop Name | Flat Dry Land | Terraced Fields | Sloping Land | Irrigated Land (I) | Standard Greenhouse (I) | … |
|---|---|---|---|---|---|---|
| Soybeans | 1 | 1 | 1 | 0 | 0 | … |
| Rice | 0 | 0 | 0 | 1 | 0 | … |
| Tomatoes | 0 | 0 | 0 | 1 | 1 | … |
| Chili pepper | 0 | 0 | 0 | 1 | 1 | … |
| Morel Mushrooms | 0 | 0 | 0 | 0 | 0 | … |
| Barley | 1 | 1 | 1 | 0 | 0 | … |
| Black beans | 1 | 1 | 1 | 0 | 0 | … |
| Shiitake Mushrooms | 0 | 0 | 0 | 0 | 0 | |
| … | … | … | … | … | … | |

### 3.2 Fluctuation Analysis

To better reflect reality, this paper assumes that projected sales volume, yield per acre, planting costs, and sales prices will fluctuate annually. This introduces uncertainty in planting-related data and planting risks. The relevant fluctuations are as follows.

- Expected Sales Volume Variations.

  As primary staple crops with extensive applications in livestock farming and industry, wheat and corn exhibit average annual sales volume growth of 5%–10%. Sales volumes for other crops are projected to remain stable relative to 2023 levels, specifically:

$$P_{N_k}(T) = P_{N_k}(2023)(1+z)^{T-2023} \tag{25}$$

  where

$$\begin{cases} k = 6 \text{ or } 7, & z \in [0.05, 0.1] \\ k \text{ is some other number}, & z \in [-0.08, -0.02] \cup [0.02, 0.08] \end{cases}$$

  where $P_{N_k}(T)$ represents the expected sales volume of crop k planted in year T, Nk denotes crop number k, z represents the fluctuation amplitude, and T indicates the year.

- Yield per mu.

  Crop yield per mu is primarily influenced by climatic factors, resulting in annual fluctuations of 10%, i.e.,

$$Y_{N_k}(T) = Y_{N_k}(2023)(1+z)^{T-2023} \tag{26}$$

  where z = 0.1 or -0.1, k = 2023, 2024, 2025... 2030, and Y denotes crop yield per mu.

- Planting Costs.

  Crop cultivation costs increase by an average of 5% annually due to rising market conditions such as higher prices for agricultural inputs, increased land rental costs from insufficient land supply, and rising labor costs. Thus:

$$C_{N_k}(T) = C_{N_k}(2023)(1+z)^{T-2023} \tag{27}$$

  where z ∈ [0.04,0.06], k = 2023, 2024, 2025... 2030, and C represents the crop cultivation cost.

- Sales Price.

Due to market supply-demand imbalances and cultivation environment impacts, vegetable crop sales prices increase by approximately 5% annually on average, while edible fungi prices decrease by 1%–5% annually. Morel mushrooms specifically decline by 5% annually, as follows:

i Grain Crops

$$R_{N_k}(T) = R_{N_k}(2023)\ \forall k \tag{28}$$

ii Vegetables

$$R_{N_k}(T) = R_{N_k}(2023)(1+z)^{T-2023},\ z \in [0.04, 0.06]\ \forall k \tag{29}$$

iii Edible Fungi

$$\begin{cases} R_{N_k}(T) = R_{N_k}(2023)(1+z)^{T-2023}, & z \in [-0.05, -0.01], k \neq 41 \\ R_{N_k}(T) = R_{N_k}(2023)(1-0.05)^{T-2023}, & k = 41 \end{cases} \tag{30}$$

where R represents the unit selling price of agricultural products.

### 3.3 Combinatorial Optimization Model Analysis and Solution

After establishing three fundamental models—linear programming, Q-Learning, and simulated annealing—this study employs combined models to refine planting strategy optimization, ensuring reliable and efficient policy outcomes. Three combined approaches are tested: linear programming with Q-Learning, linear programming with simulated annealing, and simulated annealing with Q-Learning. The resulting optimization schemes are compared against maximum net profit to identify the optimal model and solution.

**LP-Q-Learning Model.** A dynamic crop planting strategy optimization system is constructed through the collaborative division of labor between linear programming and Q-Learning reinforcement learning. The LP model handles short-term precision optimization, determining optimal planting areas for each crop within a given year based on crop-plot matching schemes to maximize seasonal net profit. The Q-Learning model focuses on long-term strategy learning, iteratively exploring diverse crop combinations and continuously refining planting strategies through historical profit feedback to adapt to market price fluctuations and crop growth patterns. Relevant algorithmic steps are outlined below in Table 4:

**Table 4.** LP-Q-Learning Algorithm Steps.

| Step | Content |
|---|---|
| Step1 | Input learning rate $\alpha$=0.1, Discount factor $\gamma$=0.99, initial exploration rate ε-start = 1.0, minimum exploration rate ε-min = 0.01, exploration decay coefficient ε-decay, iteration count n (5 iterations per year, totaling 35 iterations over seven years), empty Q-table, empty experience replay buffer, batch size of 64, plot data, crop data, and constraints; |
| Step2 | Acquire current state; |
| Step3 | Initialize a random number $\zeta$; |
| Step4 | If $\zeta < \varepsilon$, randomly select an action from the action space; otherwise, the agent considers the current Q-table to represent the optimal action; |
| Step5 | Execute the action and calculate the net payoff of the selected planting scheme using the pictorial planning model; |
| Step6 | Store the current experience in the experience and replay buffer. When the accumulated experience in the buffer exceeds the batch size, randomly sample a small batch of experiences; |

(*continued*)

**Table 4.** (*continued*)

| Step | Content |
|---|---|
| Step7 | Update all Q-values in the batch using the Q-Learning update formula; |
| Step8 | State transition: The agent advances to the next decision point |

By combining relevant algorithms and iterative updates, the maximum net profit is derived to determine the optimal planting scheme. Partial optimization schemes are shown in Table 5, with final profit values presented in Table 6.

**Table 5.** Crop Planting Plan 1 for Q1 2030.

| Field Name | Crop Type | | | | | |
|---|---|---|---|---|---|---|
| | Black Soybeans | Red Bean | Mung Beans | Climbing beans | Corn | Millet |
| Flat dry land 1 | 0 | 10 | 0 | 0 | 0 | 14 |
| Flat dry land 2 | 8.64 | 5.89 | 0 | 0 | 0 | 8.64 |
| Flat dry land 3 | 0 | 0 | 4.96 | 6.71 | 4.96 | 6.71 |
| Flat dry land 4 | 0 | 0 | 0 | 13.8 | 10.2 | 0 |
| Flat dry land 5 | 0 | 0 | 16.32 | 12.92 | 0 | 12.92 |
| Flat dry land 6 | 0 | 8.25 | 0 | 8.25 | 11 | 0 |
| Terraced Fields 1 | 0 | 0 | 0 | 0 | 9.6 | 9.6 |
| Terraced Fields 2 | 0 | 9.66 | 0 | 7.36 | 0 | 9.66 |
| Terraced Fields 3 | 0 | 7.6 | 0 | 9.6 | 0 | 7.6 |
| Terraced Fields 4 | 0 | 0 | 5.13 | 0 | 5.13 | 0 |
| Terraced Fields 5 | 0 | 0 | 0 | 0 | 0 | 0 |
| Terraced Fields 6 | 0 | 10.44 | 14.74 | 0 | 0 | 0 |
| Terraced fields 7 | 0 | 0 | 7.79 | 0 | 7.79 | 10.54 |
| Terraced Fields 8 | 0 | 5.87 | 0 | 8.07 | 5.87 | 0 |
| Terraced Fields 9 | 0 | 7.08 | 7.08 | 9.58 | 0 | 0 |
| Terraced Fields 10 | 0 | 0 | 0 | 5.63 | 0 | 0 |

(*continued*)

**Table 5.** *(continued)*

| Field Name | Crop Type | | | | | |
|---|---|---|---|---|---|---|
| | Black Soybeans | Red Bean | Mung Beans | Climbing beans | Corn | Millet |
| Terraced Fields 11 | 8.14 | 8.14 | 0 | 8.14 | 0 | 0 |
| Terraced Fields 12 | 0 | 0 | 0 | 8.1 | 0 | 0 |
| Terraced Fields 13 | 5.25 | 5.25 | 0 | 5.25 | 0 | 0 |
| Terrace 14 | 0 | 0 | 3.33 | 3.33 | 0 | 3.33 |

**Table 6.** Income Statement 2024–2030.

| Year | 2024 | 2025 | 2026 | 2027 | 2028 | 2029 | 2030 |
|---|---|---|---|---|---|---|---|
| Revenue | 10,200,381.74 | 20,417,230.90 | 20,259,172.09 | 18,870,091.73 | 16,877,281.13 | 19,192,442.68 | 19,327,249.38 |

**LP-SA Model.** In this combined model, the linear programming model primarily handles crop area allocation. Given a crop-parcel matching scheme, it solves for the optimal planting area of each crop to maximize net seasonal profit. Meanwhile, the simulated annealing algorithm focuses on global strategy search. Through temperature-controlled random neighborhood search, it explores diverse crop combination schemes, escapes local optima, and seeks superior planting strategies. The two models closely collaborate through a "strategy generation-execution optimization" cycle: after the simulated annealing algorithm generates crop combination schemes, the linear programming model performs area optimization based on these schemes. The computational results serve as fitness feedback to guide the search direction of the simulated annealing algorithm. Ultimately, this approach achieves maximum net profit while satisfying all constraint conditions. the specific algorithmic steps are detailed in Table 7.

**Table 7.** LP-SA Algorithm Steps.

| Step | Content |
|---|---|
| Step1 | Input initial temperature $T_{max}$= 1000, minimum temperature $T_{min}$= 1, cooling rate 0.98, iteration count n (6000 times annually, 1000 times per season), field data, crop data, and constraint parameters; |
| Step2 | Randomly generate crop combination schemes based on initial conditions; |
| Step3 | Use a linear programming model to calculate net profit and evaluate the value of the scheme; |
| Step4 | Determine whether to accept a new scheme based on the fitness value (i.e., net profit) of the simulated annealing algorithm using the Metropolis criterion; |

*(continued)*

**Table 7.** *(continued)*

| Step | Content |
|---|---|
| Step5 | Iterate the algorithm while lowering the temperature to explore the global optimum solution |
| Step6 | Upon reaching the maximum iteration count and minimum temperature, terminate the iteration and output the optimal crop planting plan for each year from 2024 to 2030 |
| Step7 | Calculate the net profit value of the optimal plan using a linear programming model |

Determine the optimal planting plan and maximum net profit accordingly. Specific planting plans are shown in Table 8, and optimal profits are shown in Table 9.

**Table 8.** Selected Crops and Planting Areas for 2030.

| Plot Name | Crop Type | | | | | |
|---|---|---|---|---|---|---|
| | Black Soybean | Red Bean | Mung Beans | Climbing beans | Wheat | Millet |
| Flat dry land 1 | 14 | 0 | 0 | 0 | 0 | 0 |
| Flat dry land 2 | 0 | 11.55 | 11.55 | 0 | 0 | 0 |
| Flat dry land 3 | 4 | 0 | 0 | 0 | 4 | 4 |
| Flat dry land 4 | 13.2 | 0 | 13.2 | 9.6 | 0 | 9.6 |
| Flat dry land 5 | 0 | 0 | 0 | 9.07 | 12.47 | 0 |
| Flat dry land 6 | 6.68 | 6.68 | 9.43 | 0 | 9.43 | 0 |
| Terraced fields 1 | 0 | 0 | 0 | 0 | 0 | 7.5 |
| Terraced Fields 2 | 5.91 | 0 | 5.91 | 0 | 5.91 | 5.91 |
| Terraced Fields 3 | 0 | 5.14 | 0 | 0 | 0 | 5.14 |
| Terraced Fields 4 | 3.8 | 3.8 | 3.8 | 5.2 | 0 | 0 |
| Terraced fields 5 | 0 | 4.38 | 0 | 4.375 | 0 | 0 |
| Terraced Fields 6 | 0 | 0 | 16.48 | 16.48 | 12.18 | 16.48 |
| Terraced Fields 7 | 0 | 0 | 8.71 | 0 | 0 | 8.71 |
| Terraced Fields 8 | 0 | 0 | 0 | 6.23 | 0 | 0 |
| Terraced Fields 9 | 0 | 0 | 0 | 8.5 | 8.5 | 8.5 |

*(continued)*

**Table 8.** *(continued)*

| Plot Name | Crop Type | | | | | |
|---|---|---|---|---|---|---|
| | Black Soybean | Red Bean | Mung Beans | Climbing beans | Wheat | Millet |
| Terraced Fields 10 | 4.11 | 4.11 | 0 | 0 | 2.86 | 4.11 |
| Terraced Fields 11 | 0 | 0 | 0 | 0 | 9.43 | 6.43 |
| Terraced Fields 12 | 7.07 | 4.82 | 4.82 | 7.07 | 0 | 0 |
| Terraced Fields 13 | 0 | 0 | 6.71 | 6.71 | 4.96 | 4.96 |
| Terraced Fields 14 | 0 | 4 | 0 | 0 | 4 | 0 |

**Table 9.** Income Statement 2 for 2024–2030.

| Year | 2024 | 2025 | 2026 | 2027 | 2028 | 2029 | 2030 |
|---|---|---|---|---|---|---|---|
| Revenue | 9,249,659.73 | 18,188,311.58 | 18,608,435.91 | 18,265,555.80 | 19,036,997.54 | 16,526,644.15 | 19,126,623.54 |

**LP-SA-Q-Learning Model.** In this composite model, the linear programming model determines the optimal planting area for each crop under a given crop-parcel matching scheme, providing an initial solution. Subsequently, the simulated annealing algorithm explores diverse crop combination schemes through temperature-controlled random neighborhood search, escaping local optima to achieve global optimization. Meanwhile, Q-Learning learns optimal adjustment strategies during the simulated annealing search process through experience replay and value function updates. Together, these three models form an efficient decision-making system that converges rapidly while ensuring global optimality. The specific algorithmic steps are as follows in Table 10:

**Table 10.** LP-SA-Q-Learning Algorithm Steps.

| Step | Content |
|---|---|
| Step1 | Input learning rate $\alpha$=0.1, discount factor $\gamma$=0.99, initial temperature $T_{max}$ =100, minimum temperature $T_{min}$ =1, cooling rate 0.95, iteration count n (900 per scheme), empty Q-table, empty experience replay buffer, batch size 32, field data, crop data, and constraints; |
| Step2 | Initialize the Q-table, experience replay buffer, and temperatures; |
| Step3 | Obtain the initial planting scheme and its net profit using the linear programming model, and set it as the current optimal scheme; |
| Step4 | Consider the current plan as the current state. Generate a new plan randomly from the current plan domain and calculate its payoff value; |

*(continued)*

**Table 10.** (*continued*)

| Step | Content |
|---|---|
| Step5 | Calculate the change in profit value ($\Delta$E). If $\Delta E > 0$ or random(0,1) < $e^{\Delta E/T}$, update the optimal plan; |
| Step6 | Store the acquired experience into the experience replay buffer; |
| Step7 | Buffer size $\geq$ batch size. Randomly select one batch from the buffer and update all Q-values in the batch using the Q-Learning update formula; |
| Step8 | Decrease the temperature parameter to derive the optimal policy and its net payoff |

The optimal solution and maximum net profit are ultimately obtained. Selected optimized solutions are shown in Table 11, while the maximum net profit is presented in Table 12.

**Table 11.** Crop Plot Selection and Planting Area for Selected Crops in 2030.

| Plot Name | Crop Type | | | | | |
|---|---|---|---|---|---|---|
| | Soybean | Red Bean | Mung Beans | Climbing beans | Wheat | Millet |
| Flat dry land 1 | 12.67 | 0 | 0 | 12.67 | 0 | 12.67 |
| Level dry land 2 | 0 | 0 | 7.79 | 0 | 10.54 | 10.54 |
| Flat dry land 3 | 4.96 | 0 | 4.96 | 0 | 6.71 | 0 |
| Flat dry land 4 | 0 | 0 | 18 | 18 | 0 | 0 |
| Flat dry land 5 | 0 | 12.14 | 0 | 0 | 8.74 | 8.74 |
| Flat dry land 6 | 15.13 | 12.38 | 15.13 | 0 | 0 | 0 |
| Terraced fields 1 | 0 | 7.71 | 7.71 | 7.71 | 0 | 0 |
| Terraced Fields 2 | 0 | 0 | 3.93 | 2.68 | 0 | 0 |
| Terraced fields 3 | 7.5 | 0 | 0 | 0 | 10.5 | 0 |
| Terraced Fields 4 | 8.25 | 0 | 8.25 | 0 | 0 | 6 |
| Terraced Fields 5 | 5.95 | 5.95 | 5.95 | 0 | 0 | 0 |
| Terraced Fields 6 | 2.43 | 0 | 0 | 2.43 | 0 | 0 |
| Terraced fields 7 | 0 | 6.13 | 6.13 | 0 | 0 | 0 |
| Terraced Fields 8 | 0 | 0 | 5.67 | 0 | 5.67 | 0 |
| Terraced Fields 9 | 5.13 | 5.13 | 0 | 0 | 3.73 | 0 |
| Terraced Fields 10 | 3.75 | 0 | 0 | 0 | 5 | 0 |
| Terraced Fields 11 | 0 | 14.62 | 14.62 | 0 | 0 | 0 |
| Terraced Fields 12 | 0 | 0 | 0 | 0 | 12.65 | 0 |
| Terraced Fields 13 | 0 | 0 | 8.36 | 0 | 6.16 | 0 |
| Terraced Fields 14 | 5.71 | 0 | 5.71 | 0 | 8.21 | 0 |

**Table 12.** Income Statement for 2024–2030.

| Year | 2024 | 2025 | 2026 | 2027 | 2028 | 2029 | 2030 |
|---|---|---|---|---|---|---|---|
| Revenue | 10,195,899.39 | 19,570,497.96 | 19,314,912.09 | 19,547,686.01 | 20,296,170.11 | 19,567,462.51 | 13,507,524.45 |

In summary, based on the final results of the three portfolio models described above, the comparison chart of returns for the three models between 2024 and 2030 is shown in Fig. 8.

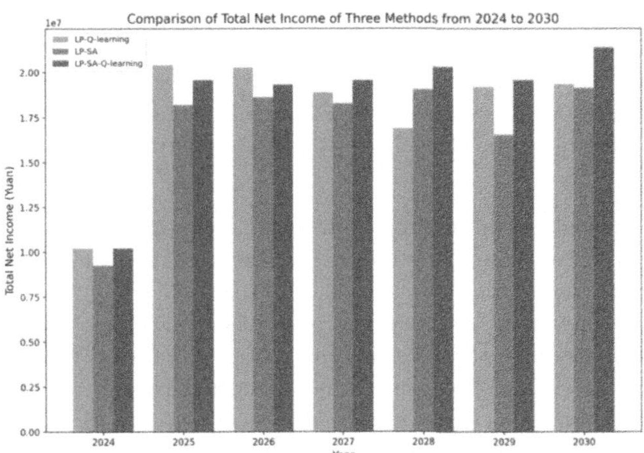


**Fig. 8.** Profit Comparison Chart.

As shown in the figure, for the crop planting strategy optimization problem in the region from 2024 to 2026, the LP-Q-learning combination model achieves the best optimization results for the target strategy, followed by the LP-SA-Q-learning model. However, from 2027 to 2030, the LP-SA-Q-learning combination model demonstrates superior optimization performance. Analysis indicates that during the initial phase (2024–2026), the LP-Q-learning model demonstrates robust exploration capabilities and rapid responsiveness to market fluctuations. Its experience replay mechanism efficiently identifies feasible solutions, making it better suited for short-term optimization needs characterized by high planting strategy uncertainty and rapid convergence requirements. In the later phase from 2027 to 2030, as strategies stabilize, the LP-SA-Q-learning model excels in achieving fine-tuned optimization for long-term planning. Leveraging simulated annealing's precise local search capabilities and Q-learning-guided multi-stage adaptive optimization, it avoids local optima, delivering superior optimization performance during phases dominated by long-term factors like crop rotation systems and sustainable cultivation practices.

## 4 Conclusion

This study constructs three composite models considering fluctuations in expected crop sales volume, planting costs, sales prices, and yield per mu. To maximize net crop revenue, relevant constraints are established. Each composite model is applied to optimize crop planting strategies. Comparing and analyzing the optimization results reveals that the LP-Q-learning composite model achieves superior target strategy optimization for the 2024–2026 period. For the period 2027–2030, the LP-SA-Q-learning combination model yields superior optimization results. This approach also determines annual profit variations for crop cultivation in the region from 2024 to 2030, identifying optimal planting strategies. This methodology effectively enhances the validity and reliability of model establishment and optimization, providing credible solutions for crop planting strategies. Furthermore, to enhance optimization outcomes and improve model real-world applicability in future research, investigations may focus on the following four areas:

1. Parameter Adaptive Optimization. Intelligent algorithms like grid search and Bayesian optimization can be introduced to automatically tune critical parameters such as the cooling rate of simulated annealing and the learning rate of Q-Learning, thereby improving model accuracy and convergence efficiency.
2. Extreme risk modeling. Incorporating the impact of extreme weather events like drought and hail on yield per mu into the evaluation system. Constructing a two-dimensional optimization model that simultaneously accounts for conventional market fluctuations and extreme climate risks to enhance the resilience of agricultural decision-making.
3. Ecological Objective Synergy. Beyond maximizing net profits, we incorporate ecological constraints like fertilizer application rates and carbon emissions. This dual-objective optimization framework balances economic and ecological benefits, aligning with green agriculture and sustainable development goals;
4. Real-time Data-Driven Approach. Utilize Internet of Things (IoT) technology to capture real-time environmental data such as soil fertility and moisture levels in fields, dynamically adjusting crop planting strategies. This creates a closed-loop management mechanism of "model optimization-real-time monitoring-dynamic regulation," enhancing the system's adaptive capabilities. Furthermore, this research can be extended to multi-village collaborative scenarios in North China's mountainous regions. It explores integrated optimization pathways across multiple levels—including land consolidation, large-scale cultivation, and industrial chain extension—to provide systematic technical support for agricultural modernization and high-quality development under the rural revitalization strategy.

**Acknowledgements.** 1 Students' Research Projects of Jianghan Universit: Optimization Study on Crop Planting Resource Allocation Based on LP-K-means(No. 2025zd117).

2. Hubei Provincial Teaching and Research Project: Research on the Precision Talent Cultivation Mechanism of 'Digital-Intelligent Modeling + Three-Wing Collaboration' from the Perspective of Industry-University-Research Integration. (No. 2025088).

## References

Cai, Z.D., Chen, W.J., Chen, X.Y.: Research on optimization model of crop planting strategy based on linear programming. Trans. Econ. Bus. Manage. Res. **13**, 449–454 (2024)

Deng, S.Y., Pan, Y.: Planting zone delineation of honey plum in Guizhou based on K-means clustering. J. Yunnan Norm. Univ. (Natural Science Edition) **44**(05), 68–73 (2024)

Hu, Y.J., Wang, H.R., Shi, J.Y.: Design of crop planting strategy based on genetic algorithm. In: 2025 IEEE International Conference on Electronics, Energy Systems and Power Engineering (EESPE 2025), pp. 1457–1462. IEEE. Shenyang (2025)

Li, T.F.: Research on patent text clustering based on an improved k-means algorithm (Doctoral dissertation). Hebei University of Engineering (2020). https://doi.org/10.27104/d.cnki.ghbjy.2020.000725

Liu, Y.S., Hu, K.W., Lei, Y.H.: Research on agricultural products planting strategies based on integer programming and stochastic programming. Acad. J. Sci. Technol. **13**(2), 106–112 (2024)

Li, Y.B., Ma, J.T., Li, D.X., et al.: Application of an improved particle swarm optimization algorithm in agricultural crop structure optimization. J. Irrig. Drainage **41**(1), 62–71 (2022)

Qiao, K., et al.: Dynamic auxiliary task-based evolutionary multitasking for constrained multiobjective optimization. IEEE Trans. Evol. Comput. **27**(3), 642–656 (2023)

Talatahari, S., Nouhi, B., Beheshti, A., Chen, F., Gandomi, A.H.: Adaptive strategy management: a new framework for large-scale structural optimization design. Comp. Methods in Appl. Mechan. Eng. 446 (2025)

Tian, Z.W., Qian, R.L., Ye, J.K.: Application of a dual-objective optimization model based on simulated annealing algorithm in garbage collection route planning. Comp. Knowl. Technol. **21**(2), 23–25 (2025)

Wang, J.Y., Wang, Y.Q.: Multi-stage crop planting strategy optimization model based on PSO algorithm. Research Square (2024)

Wu, D.Z., Yao, K.S., Tan, Z.L.: A hybrid integer linear programming (MILP) model and Monte Carlo simulation-based study on crop planting strategies. In: IEEE International Conference on Electronics, Energy Systems and Power Engineering (EESPE 2025), pp. 952–957. IEEE. Shenyang (2025)

Wu, J.J.: Research on agricultural crop structure optimization based on linear programming models. New Farmer **3**, 28–30 (2024)

Yang, Q.: *Research on adaptive scheduling for automated warehouse systems based on Q-learning* (Doctoral dissertation). Hangzhou Dianzi University, Hangzhou (2025)

Yu, X.Y., Dang, Z.X., Zhang, D., et al.: Crop planting strategy formulation based on optimization algorithm and decision matrix. In: Proceedings of the 4th Asia-Pacific Artificial Intelligence and Big Data Forum (AIBDF 2024), pp. 486–494. Association for Computing Machinery, Ganzhou (2025)

Zhou, X.W., Long, X., Yu, X.L., et al.: Intelligent vehicle multi-scenario safe braking algorithm based on improved Q-learning. Computer Applications (n.d.). Retrieved 27 August 2025

# Computational Thinking Reflected in K-means Clustering: Taking NetLogo Simulation as an Example

Ruizhu Li(✉)

Xiamen University Tan Kah Kee College, Zhangzhou 363105, China
rzlee@foxmail.com

**Abstract.** The K-means clustering algorithm, as the core technology of unsupervised learning, is a concentrated embodiment of computational thinking in terms of its inherent logic. This article aims to go beyond the conventional understanding of K-means as a simple data analysis tool, and systematically deconstruct the core elements of decomposition, abstraction, pattern recognition, and algorithm design contained in K-means algorithm from the paradigm of computational thinking. The study used the NetLogo multi-agent simulation platform as the experimental environment, modeling data points and cluster centers as autonomous "agents" to transform abstract algorithm processes into a dynamic and visual computing world. The simulation experiment intuitively revealed that the "allocation update" iteration cycle of the algorithm is essentially a process of parallel execution of simple rules at the subject level, and ultimately the emergence of a globally ordered structure, vividly interpreting the core concept of computational thinking of "building order through rules and iterations". This article argues that NetLogo simulation is not only a tool for verifying the effectiveness of algorithms, but also a "mirror" that can clearly reflect how computational thinking gradually shapes and solves complex problems.

This study ultimately demonstrates that the analysis framework combining K-means algorithm with multi-agent simulation provides a concrete and interactive effective path for understanding and imparting computer thinking, deepens people's understanding of the essence of algorithms, and demonstrates important enlightening value in computer science education methodology.

**Keywords:** K-means · Computational thinking · NetLogo

## 1 Introduction

The K-means clustering algorithm, as a fundamental method in the fields of data mining and unsupervised learning, has been widely applied in various fields such as pattern recognition, image segmentation, and market segmentation due to its simplicity and effectiveness. On the surface, K-means appears to be a purely mathematical and statistical process, with its core being the iterative optimization of dividing data points into K clusters to minimize the sum of squared errors within each cluster (Han and Lee,


H. Yan et al. (Eds.): AIS&P 2025, CCIS 2918, pp. 284–294, 2026.
https://doi.org/10.1007/978-981-95-9884-7_19

2024). However, when penetrating through its mathematical shell and delving into its inherent solving logic and implementation mechanism, it becomes clear that the essence of this algorithm is a profound manifestation of Computational Thinking. The core concepts advocated by computational thinking, such as abstraction, decomposition, pattern recognition, and algorithm design, have been fully demonstrated in the conception and execution process of K-means.

Specifically, the K-means algorithm perfectly embodies the core paradigm of computational thinking (Shen and Weng, 2023):

Decomposition: Decompose a complex "data grouping" problem into a series of well-defined and sequentially executed subtasks such as "initialization center", "allocation of data points", "updating center position", and "iterative convergence".

Pattern Recognition: The core step of an algorithm - assigning each data point to the nearest center point, which is based on the "distance" pattern in the data space for recognition and classification.

Abstraction: Algorithms ignore the specific semantics of data points (such as whether they are customer information or pixel colors) and abstract them as geometric points in multidimensional space, focusing only on their core attribute of positional relationships.

Algorithm Design: The entire process is precisely defined as an iterative sequence of instructions, with termination conditions (the center point no longer changes or reaches the maximum number of iterations) ensuring the controllability of the process and the certainty of the results.

Although K-means can be described through mathematical formulas or pseudocode, these static descriptions are difficult to intuitively demonstrate its dynamic, iterative convergence process and how various elements in computational thinking work together. Therefore, this article introduces the NetLogo multi-agent simulation platform as the research medium. The "Agent" oriented modeling paradigm of NetLogo provides an ideal dynamic and visual experimental field for the K-means algorithm (Sklar, 2014). In this environment, each 'turtle' represents a data point to be clustered, while 'patch' or a specific turtle represents the cluster center. The data points autonomously move their belonging according to the rules (finding the nearest center), while the center point recalculates and moves its own position based on the coordinates of its members. This process transforms abstract algorithm steps into concrete and observable subject behavior, vividly presenting the iterative cycle of "allocation update", the formation and evolution of clusters, and the final convergence state.

Through NetLogo simulation, not only can the correctness of the K-means algorithm be verified, but also the powerful power of computational thinking in solving complex problems can be intuitively observed: how a seemingly chaotic dataset can self organize into a clear structure through a series of simple and deterministic rules. Therefore, this study aims to use NetLogo simulation as a mirror to systematically analyze and explain the computational thinking contained in the K-means algorithm. It is expected that this combination of visual simulation and computational thinking theory analysis can not only deepen the understanding of the K-means algorithm itself, but also provide an effective way for computer science education to concretize abstract thinking, highlighting the universal value and core position of computational thinking in solving scientific and engineering problems.

# 2 Computational Thinking and K-means Algorithm

## 2.1 Computational Thinking

Computational thinking refers to the ability of people to operate computers in the process of carrying out computer operations, and computer language lacks thinking (Wang, 2023). Computational thinking includes computational abstraction, automation, parallel processing, recursive thinking, fault tolerance, and error correction (Liu et al., 2025).

Computational abstraction: The essence of a problem, which involves using code, algorithms, etc. to abstract real-world problems into a form that computers can handle.

Automated thinking: enabling computers to automatically follow code and algorithms, improving processing efficiency.

Parallel processing: Breaking down a large problem into multiple small problems, allowing the computer to process these small problems simultaneously, thereby improving processing speed.

Recursive thinking: Decompose a complex problem into multiple similar sub problems, and then solve the original problem by solving the sub problems.

Fault tolerance and error correction: People need to consider the possible errors that may occur in computer systems, design corresponding fault tolerance mechanisms and error correction algorithms, and improve the reliability of the system.

## 2.2 K-means Algorithm

The K-means algorithm is an unsupervised learning algorithm, whose basic idea is to divide n data objects into k clusters (Turan and Ordin, 2025), so that the clusters satisfy the principle of "high cohesion, low coupling". Using computational thinking to decompose it into three concepts: "K", "centroid", and "distance measurement". Among them, "K" represents the number of categories that you want the data to be divided into. If K = 2, the data will be divided into 2 cluster centers; If K = 3, the data will be divided into three cluster centers. The "centroid" is the "center point" of each cluster center, representing the average position of that center point. Distance measurement" is used to calculate the similarity between data points and centroids (or points to points). Ultimately, through computational thinking, the clustering problem can be imagined as minimizing the sum of squared distances between all data points and their respective cluster centers.

Clustering algorithms, due to their simple model construction and wide range of problem-solving capabilities, have lower computational complexity compared to other data mining algorithms, and are well suited for handling large-scale data sets. By using Netlogo software for visualization, a clear center point can be seen at the center of each cluster, and this algorithm can be easily extended to high-dimensional data processing.

# 3 Visualization of K-means Clustering Based on NetLogo

## 3.1 Netlogo

Netlogo is a multi-agent simulation modeling software developed by Northwestern University in the United States, specifically designed to simulate certain phenomena or states in nature and society (Reulier et al., 2024). It is currently widely used in the fields of education and research. In terms of scientific research, Netlogo can construct simulation models from the bottom up based on the proposed theoretical assumptions, such as the spread of infectious diseases and market dynamics. Its advantages lie in its direct modeling approach, simple code syntax, allowing users to simulate using multiple agents to observe the interactions between these agents and between the agents and the environment. From these complex situations, corresponding system phenomena can be analyzed. Through various components such as sliders, buttons, and monitors, the interface can be constructed with concise code logic, achieving system modeling across multiple disciplines and fields.

### 3.2 K-means Clustering Visualization

As shown in Fig. 1. K-means visualization model of Netlogo. In the K-means visualization model of Netlogo, three sliders are used: "num clusters", "num data points", and "num centroids", "setup", "Assign Point", "Update-Centroids", "Find Clusters", "Reset Centroids" five buttons and a "Square Deviation of Clusters" monitor and display interface.

The "num clusters" in the slider represent the number of data groups; "num data point" represents the number of data points; "num centroids" represents the final number of cluster centers.

The "setup" button is used to reset all interfaces; Assign Point "refers to assigning data points; Update Centroids" is used to update the cluster center position; "Find Clusters" is used to find cluster centers; "Reset Centroids" means resetting the cluster center position to redo the previous simulation.

In addition, the "Square Deviation of Clusters" monitor is used to represent the sum of squared differences in the distances between each data point and the centroid before finding the optimal cluster.

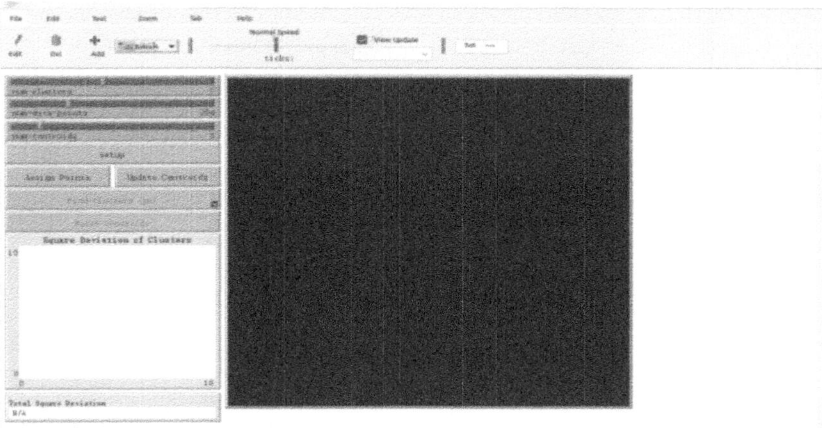

**Fig. 1.** K-means visualization model of Netlogo.

### 3.3 Visual Result Analysis

As shown in the Fig. 2. Visualization results. a visualization model was randomly generated with seven data clusters, four centroids (represented by X), and one thousand identical data points. Click on "Find Clusters" to get the following image. Seven data clusters and one thousand data points were divided into four clustering centers, represented in light pink, purple, blue, and dark pink in the figure. At this point, the square difference of the distance between each data point and the centroid reached the minimum of twenty-one thousand and twenty-four, indicating that the clustering reached its optimal state.

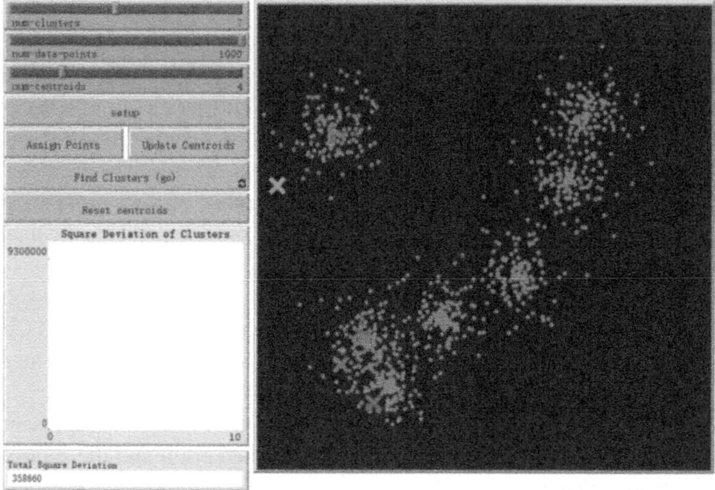


**Fig. 2.** Visualization results.

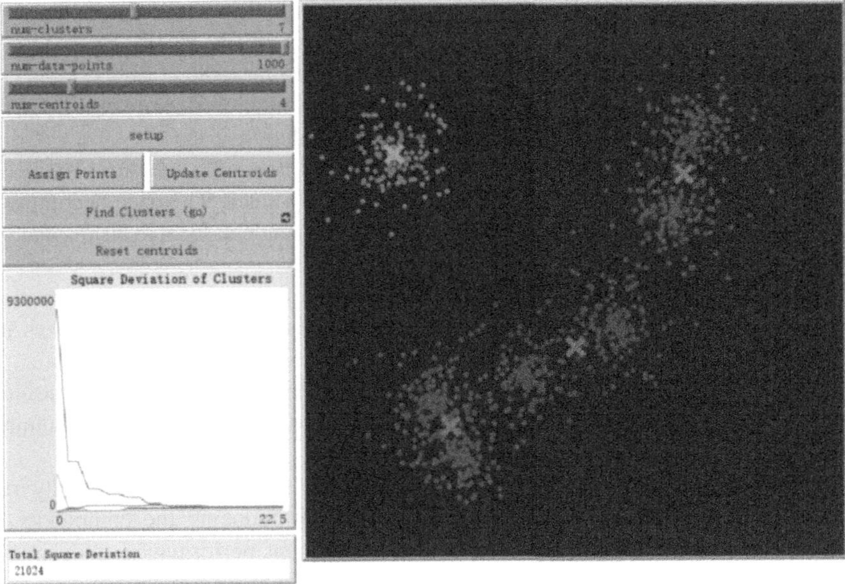


**Fig. 3.** Cluster center results.

In this visualization model, see Fig. 3. Cluster center results., the clustering model is presented in a simple way. Based on the results of multiple runs, different initial clustering centers may lead to different clustering results. By running multiple times, using different initial cluster centers each time, we can observe differences in clustering results, and the initial centroid position may have an impact on the results of the cluster centers. Of course, the number of centroids ("num centroids") is also a very important parameter. By changing its value, different clustering effects can be observed. Generally speaking, the larger the K value, the finer the clustering, but an excessively large K value may also lead to overfitting.

## 4 The Computational Thinking Reflected in K-means Clustering

The reason why K-means algorithm has become a classic teaching and research case in computer science is not only because of its wide applicability, but also because it is a perfect carrier of computational thinking. It transforms a seemingly complex and intuitive data grouping problem into a precise calculation process driven by clear rules that can be automatically executed. This chapter will follow the core paradigm of computer thinking, namely decomposition, pattern recognition, abstraction, and algorithm design, to systematically deconstruct K-means and reveal its inherent computational nature.

### 4.1 Decomposition: Modular Segmentation of Complex Problems

The first step in computer thinking is to break down a complex problem into smaller, more manageable sub problems. The K-means algorithm excellently embodies this principle.

The original clustering problem "How to naturally divide a set of data points into K groups?" It is inherently fuzzy and difficult to solve directly. K-means decomposes it into a series of clearly defined and sequentially executed subtasks:

Initialization: Randomly select K data points as the initial cluster centers. This step anchors the problem to a specific starting point in the solution space.

Assignment: Traverse every point in the dataset and assign it to the nearest cluster center based on predefined metrics such as Euclidean distance. This step decomposes the global grouping problem into a large number of independent, local decision-making processes.

Update: After allocating all data points, recalculate the center point of each cluster (usually taking the average of all points in the cluster). This step optimizes the "representative" of the cluster based on the current allocation result.

Iteration: Repeat steps 2 and 3 until the termination condition is met (such as the cluster center no longer undergoing significant changes or reaching the maximum number of iterations).

This decomposition strategy makes the algorithm structure clear, logically rigorous, and each module has a single function, greatly reducing the complexity of implementation and laying the foundation for subsequent performance analysis and optimization.

## 4.2 Abstraction and Pattern Recognition: Mapping from Data to Geometry

Abstraction is the ability to ignore unnecessary details and focus on the core attributes of a problem. The K-means algorithm performs a crucial abstraction: it abstracts all data instances, regardless of their original semantics (whether they are users, products, or pixels), into a geometric point in multidimensional space. All subsequent operations of the algorithm are based on this geometric abstraction.

Based on this abstraction, the core of the algorithm - the "allocation" step - is transformed into a typical pattern recognition problem. The 'pattern' here is no longer the contour in the image or the syllable in the speech, but the proximity pattern of data points in space. The algorithm identifies which points share the same "spatial proximity" pattern by calculating the distance between them, and thus categorizes them into one class. In the NetLogo simulation, this process is intuitively visualized: each "turtle" (data point) identifies its "nearest" pattern by calculating its distance from each red "patch" (cluster center), and changes color to indicate its affiliation. The ability to go from concrete to abstract, and then recognize patterns in abstract space, is the key to solving cross domain problems with computational thinking. Recursive Thinking.

The essence of K-means algorithm is the process of achieving optimal results through continuous iteration, and its core is the manifestation of recursive thinking. In the K-means model, recursive thinking is mainly reflected in the iterative optimization process of the algorithm and the decomposition method of the problem. The model controls the iteration through the "go" process and decomposes the originally complex clustering problem into three repetitive sub processes: "calculating distance - assigning clusters - updating centers". At each iteration, the allocation of data points and the position of cluster centers will be recalculated based on the previous clustering results, until the termination condition is met the cluster centers no longer move. Each data point independently calculates the distance to all cluster centers, selects the center based on the nearest distance, and each cluster center continuously updates its position based on the mean of the data points, forming a multi-level nested iterative structure. This recursive strategy transforms global problems into local decisions, achieving global optimality by continuously reaching local optimal solutions.

### 4.3 Algorithm Design: Precise Instruction Sequence for Automated Execution

After decomposition and abstraction, it is necessary to organize these ideas into a precise and automatically executable algorithm. The algorithm design of K-means reflects the precision and automation pursuit of computer science.

The entire algorithm is formalized as a clear sequence of instructions, and its flow can be accurately described through flowcharts or pseudocode. More importantly, it introduces the powerful computational paradigm of iterative optimization. Algorithms do not pursue a perfect solution in one step, but gradually approach a local optimal solution through a repeated cycle of "allocation update". This design reflects the characteristic of computers being good at performing repetitive tasks.

In NetLogo simulation, we can trigger each iteration with a button (such as go) to clearly observe how the algorithm gradually converges from the chaotic initial state to a stable clustering state through repeated application of rules. This process vividly demonstrates how algorithms, as a collection of deterministic rules, guide systems from disorder to order.

### 4.4 Evaluation and Optimization: The Formation of a Closed Loop of Thinking

A complete computational thinking process also includes evaluating and optimizing solutions. The K-means algorithm incorporates its optimization objective of minimizing within cluster sum of squares (WCSS). Each 'update' step (calculating a new mean) is a local optimization of WCSS under the current allocation.

However, K-means is sensitive to the initial center and may converge to local optima, which precisely leads to another important aspect of computational thinking: the recognition and improvement of the limitations of the algorithm itself. This has given rise to subsequent optimization strategies such as K-means + + (optimization initialization) and elbow rule (determining the optimal K value). In the NetLogo experiment, by running multiple times and observing the different results caused by different initial states, researchers can intuitively understand this limitation and stimulate the thinking of improving the algorithm. This forms a computational thinking loop of "design implementation evaluation optimization".

In summary, the K-means clustering algorithm is not simply a pile of mathematical formulas. It is a paradigm that fully demonstrates how computer thinking can be applied to solve practical problems. Modularize the problem through decomposition, grasp the essence of the problem through abstraction and pattern recognition, achieve automated iterative solving through sophisticated algorithm design, and ultimately form a closed loop of thinking through evaluation. The NetLogo simulation environment externalizes this inherent and abstract computational thinking process as a visual and dynamic model, allowing people to "see" how thinking is gradually shaped and solved to a complex problem. This not only deepens people's understanding of the K-means algorithm, but also provides an effective way to cultivate and teach computational thinking.

# 5 Application and Development of K-means Clustering

## 5.1 The Application of Clustering Models in Data Mining

Customer segmentation: By clustering analysis of customer consumption behavior, preferences, and other data, numerous customers are divided into different groups, and customer profiles corresponding to products are mined to provide a basis for the company's marketing strategy.

**Market analysis:** By conducting cluster analysis on market data, different market segments of a certain type of product can be discovered, which helps enterprises explore future market demand and formulate reasonable market strategies. For example, customers' purchasing motivation is generally determined by internal factors such as needs, cognition, and learning, as well as external factors such as culture, society, family, small groups, and reference groups. When dividing customers based on different purchasing motivations, the aforementioned factors can be used as analysis variables, and the indicator values of each analysis variable for all target customers can be quantified, and then cluster analysis can be used for classification. When quantifying indicator values, if encountering some qualitative indicator values, some qualitative data quantification methods can be used to transform them, such as fuzzy evaluation method. In addition, customer satisfaction level and repeat purchase opportunity size can be classified as attributes. A new classification method can also be included to distinguish the differences between customers. The variables of customer differences can be divided into five categories: product benefits, interaction force between customers, selection barriers, bargaining power, and return rate. Based on the clustering of these analysis variables, the classification can provide useful references for enterprises to make marketing decisions.

The commonality of the above analyses is that they are all classified based on multiple variables, which is in line with the characteristics of clustering analysis in solving problems. The difference lies in seeking analysis variables from different perspectives to provide reference for decision-making in a certain aspect. This is precisely the manifestation of the wide application of clustering analysis in customer segmentation problems.

**Data compression** (Saronian et al., 2024): By compressing a large number of data points into a small number of cluster centers, the storage space and processing time of the data can be reduced. For example, image compression based on K-means clustering algorithm generally requires the following steps in practical applications:

Image segmentation: The original image is segmented into several blocks, usually using fixed size blocks for segmentation. For example, dividing a 512x512 image into several 16x16 blocks.

Color quantization: For each pixel within a block, perform color quantization processing. Due to the need to convert pixels into smaller storage space after compression processing, color quantization can reduce the storage space required for each pixel.

Cluster analysis: Use the K-means algorithm to cluster the pixels within each block and assign them to the nearest cluster center. When conducting cluster analysis, adjacent blocks can be grouped together to ensure the continuity of the image.

Compressed storage: Store the cluster center values of each block and encode the pixel values in each block according to their respective cluster centers to reduce the storage space occupied by the image.

Restoring images: decoding and restoring images. When decoding, the first step is to read the cluster center values of each block, and then decode the pixels within each block. When decoding, restore each pixel to its corresponding pixel value based on the cluster it belongs to.

The image compression technique based on K-means clustering algorithm can effectively classify the content of the image while maintaining the continuity of the image, thus ensuring the compression quality. In addition, due to the use of block based clustering analysis in image compression technology based on K-means clustering algorithm, a large amount of image data can be processed quickly,

### 5.2 The Development Trend of Clustering Algorithms

Incremental learning (Fu et al., 2025): Currently, clustering algorithms often only process existing massive data to find clusters. In the future, the universality of the algorithm can be improved, such as dynamically updating clustering results and enhancing real-time performance as data is continuously updated and added.

Integrating other algorithms: In the future, clustering algorithms can be integrated with other algorithms such as deep learning and graph algorithms to improve their performance and application scope.

Enhance interpretability: In batch data, clustering results are often very complex, and research should be conducted on how to improve the interpretability of clustering results, making them easier for users to understand and apply (Raval et al., 2025).

## 6 Summary

This study takes the core concept of computational thinking as a lens, systematically analyzes the K-means clustering algorithm, and uses the NetLogo multi-agent simulation platform to visualize and verify the abstract computational process contained therein. The paper demonstrates that K-means is not only a statistical tool, but also a classic case that fully demonstrates the computational thinking paradigm of decomposition, pattern recognition, abstraction, and algorithm design. It transforms complex unsupervised learning problems into an automatically executable deterministic process through precise definition steps of "initialization allocation update iteration".

Successfully transformed abstract algorithms into a dynamic and visual computing world through NetLogo simulation. In this world, the interaction behavior between data points and cluster centers as autonomous "subjects" vividly reveals how algorithms emerge global clustering structures through simple iterations of local rules. This simulation method goes beyond the static description of pseudocode and intuitively elucidates the core idea of "building order through rules and iterations" in computational thinking. It not only deepens people's understanding of the convergence and initial center sensitivity of K-means algorithm, but more importantly, it provides a powerful teaching and research paradigm.

In summary, this study confirms that interactive simulation represented by NetLogo is an effective bridge for elucidating and exploring the profound connection between computational thinking and classical algorithms. This work not only provides a new perspective for understanding K-means, but more importantly, it demonstrates the methodological value of concretizing computational thinking, which has positive implications for promoting the intuitiveness, interactivity, and depth of algorithm education.

**Acknowledgments.** This work is supported by Xiamen University Tan Kah Kee College Campus level "Cultivation Project" Incubation Project(PY2024L01).

## References

Fu, X., Lu, N., Sun, H., et al.: A novel clustering method for extracting representative photovoltaic scenarios considering power, energy, and variability. IEEE Trans. Power Sys. 1–13 (2025)

Han, S., Lee, J.: Parallelized inter-image k-means clustering algorithm for unsupervised classification of series of satellite images. Remote Sensing **16**(1) (2024)

Liu, Y., Liu, P., Zhai, L., et al.: Design and implementation of automatic gain control algorithm for Ocean 4A scatterometer. Sys. Eng. Electron. Technol. (English Version) **36**(2), 344–352 (2025)

Raval, C.U., Makwana, A., Patel, S., et al.: Optimizing tacrolimus dosage in post-renal transplantation using DoseOptimal framework: Profiling CYP3A5 genetic variants for interpretability. Int. J. Clin. Pharm. **47**(4), 12–15 (2025)

Reulier, R., Bonnet, G., Currie, C.: Comparison of execution times of three multi-agent platforms (Netlogo/GAMA/Repast simphony) based on the LASCAR model. Journal of Simulation **18**(6) (2024)

Saronian, S., Yousefimehr, B., Ghatee, M., et al.: An ensemble of deep clustering models with autoencoders to mine travel patterns from smart card data. IEEE Trans. Intel. Trans. Sys. **25**(12) (2024)

Shen, S.S., Weng, Z.L.: Research on scene design for promoting computational thinking cultivation in artificial intelligence courses. J. Dist. Educ. **41**(6), 94–103 (2023)

Sklar, E.: NetLogo, a multi-agent simulation environment. Artif. Life **13**(3), 303–311 (2014)

Wang, W., Wang, A.J.: Exploration of teaching reform of university computer fundamentals course based on computational thinking. Indus. Technol. Forum **22**(08), 142–143 (2023)

Turan, D.S., Ordin, B.: The incremental SMOTE: A new approach based on the incremental k-means algorithm for solving imbalanced data set problem. Information Sciences 711 (2025)

# Lithium-Ion Battery SOH Prediction with Multi-Feature Extraction Using a BiLSTM-RPTransformer Model

Junjie Guan[1] and Liefa Liao[2(✉)]

[1] Jiangxi University of Science and Technology, Nanchang, China
[2] Jiangxi Modern Polytechnic College, Nanchang, Jiangxi, China
liaolf@126.com

**Abstract.** Accurate state of health (SOH) estimation is crucial for battery management systems (BMS). This study proposes a novel hybrid model integrating relative positional encoding into the Transformer architecture and coupling it with a bidirectional LSTM (BiLSTM) for SOH prediction. Key features—current, voltage, and internal resistance—are extracted to characterize battery degradation. The BiLSTM captures both forward and backward temporal dependencies, enhancing sensitivity to short-term dynamics. Historical data are fused with relative positional encodings, dynamically incorporated into each multi-head attention layer to model relative positions and feature correlations. This enables a robust mapping from input features to SOH. Experiments on the CALCE dataset show the model achieves root mean square error (RMSE), mean absolute error (MAE), and mean absolute percentage error (MAPE) of less than 1\%, 0.7\%, and 1.9\%, demonstrating superior accuracy and robustness, with strong potential for practical SOH estimation applications.

**Keywords:** Lithium-ion batteries · State of health · Relative positional encoding · Transformer architecture

## 1 Introduction

In recent years, growing environmental concerns have boosted interest in lithium-ion batteries (LIBs) as a promising energy storage solution. LIBs offer high energy density, low self-discharge, compact size, and long cycle life, making them essential for electric vehicles and helping reduce fossil fuel dependence and emissions (Huang 2018, Yao 2021). However, repeated charge–discharge cycles lead to performance degradation—such as capacity and power loss—due to electrode deterioration and SEI layer growth. Improper use may also trigger safety issues like electrolyte leakage, micro-short circuits, and thermal runaway (Attia 2022, Hosen 2021). Thus, accurate battery aging monitoring is crucial. SOH is a key indicator of degradation, and its precise estimation supports decision-making and proactive maintenance, enhancing safety and lifespan in applications like EVs and energy storage systems (Tian 2020). However, SOH cannot be directly measured by conventional sensors, necessitating reliable estimation methods.


H. Yan et al. (Eds.): AIS&P 2025, CCIS 2918, pp. 295–320, 2026.
https://doi.org/10.1007/978-981-95-9884-7_20

The SOH of lithium-ion batteries can be estimated using three main approaches: parameter-based, model-based, and data-driven methods (Chen 2023). Traditional parameter-based methods, such as Coulomb counting and open-circuit voltage, often suffer from low accuracy in practice. They are sensitive to measurement errors and prone to cumulative deviations, degrading estimation performance (Charkhgard 2010). They also fail to capture transient battery dynamics and account for complex, multi-factor degradation influence (Wei 2017). Moreover, these methods rely on specific operating conditions, lack adaptability to dynamic environments, and cannot model long-term dependencies. Crucially, they cannot quantify SOH uncertainty (Richardson 2017), nor effectively describe the pronounced degradation variations across battery types and operating conditions (Hannan 2017).

Although parameter-based methods are simple to implement and physically interpretable, they require stringent experimental conditions and exhibit limited adaptability. Thus, model-based and data-driven approaches are preferred in practice due to their superior robustness and accuracy in SOH estimation. Model-based methods typically use the Equivalent Circuit Model (ECM) or Electrochemical Model (EM) to represent battery dynamics and degradation (Xing 2013). Hosseininasab et al. (He 2020) proposed a simplified EM-based approach that reduces reliance on complex parameters, enabling simultaneous estimation of capacity and internal resistance for comprehensive performance evaluation. Wang et al. (Wang 2014) improved the DMC algorithm with fine linearization of the RC model, achieving accurate prediction of SOC, SOH, and instantaneous availability. However, both EM and ECM suffer from complex, time-varying parameters, making it difficult to balance computational efficiency and accuracy. In contrast, data-driven methods require no detailed physical modeling and rely only on measurable external variables (Yang 2022). Common techniques include ARIMA, neural networks (NN), support vector machines (SVM), and nonlinear filtering (Lin 2025). Wang et al. (Hosseininasa 2022) combined health-related features with Gaussian process regression (GPR) for accurate SOH estimation, though with high computational cost. Zhou et al. (Zhou 2016) applied empirical mode decomposition (EMD) followed by ARIMA for SOH and RUL prediction, but the linear ARIMA assumption limits its ability to capture global nonlinear dynamics. Ren et al. (Ren 2018) used an autoencoder (AE) to extract multidimensional SOH/RUL features, fed into a pre-trained DNN, achieving 93.34\% accuracy. Klass et al. (Klass 2014) developed an SVM model using EV operational data for real-world SOH estimation, but SVM, being static, cannot capture temporal dynamics. Moreover, complex feature interactions significantly affect SOH estimation accuracy (He 2024). To better model temporal dependencies, recurrent neural networks (RNNs) learn mappings from historical data to SOH labels. Chaoui et al. (Chaoui 2017) proposed a dynamic RNN for SOH estimation, while You et al. (You 2017) integrated time-varying current-voltage (I/V) data, achieving a maximum relative error below 2.46%.

Although classical neural networks—such as RNN, LSTM, and GRU—can process lithium-ion battery sensor data (e.g., current, voltage) and capture certain nonlinearities, they suffer from gradient explosion and limited feature extraction during training (Hochreiter 1997, Bengio 1994). This impairs their ability to model long-term dependencies and compromises generalization across operational scenarios. In battery lifecycle modeling, RNN-based architectures often fail to capture multi-step temporal dynamics, leading to prediction deviations or instability. Moreover, training efficiency decreases significantly with sequence length, while computational complexity increases, limiting their practical use in SOH prediction. To address these issues, improved architectures have been proposed. Venugopal, et al. (Venugopal 2019) introduced the independently recurrent neural network (IndRNN), which mitigates gradient problems in standard RNNs, enabling effective learning of complex nonlinear battery behaviors and long-term dependencies for enhanced SOH estimation. Xu et al. (Xu 2023) incorporated skip connections into a CNN-LSTM framework and applied feature selection during preprocessing to eliminate redundancy, resulting in a hybrid model with superior capability in simulating SOH degradation trends.

Although recurrent neural networks (RNNs) and their variants are effective in processing time-series data, they struggle with long-term dependencies and their sequential computation limits hardware-level parallelization. Gu et al. (Gu 2023) pioneered the use of Transformers for lithium-ion battery SOH prediction by integrating them with CNNs for feature extraction, achieving promising results. However, the Transformer's parallel processing inherently weakens its ability to capture temporal order. To address this, Vaswani et al. (Vaswani 2017) introduced absolute positional encoding, using sine and cosine functions to generate fixed positional vectors added to input embeddings, thereby encoding sequence order. While this enables position awareness, it may be insufficient for capturing the complex dynamic temporal dependencies in battery degradation tasks.

To overcome the aforementioned limitations, we propose a novel Transformer-based framework integrating relative positional encoding with a BiLSTM, termed BiLSTM-RPTransformer. First, multi-feature inputs including voltage, current, and internal resistance—are extracted for modeling. Then, the standard Transformer embedding layer is replaced with a BiLSTM, enabling the model to capture both past and future contextual information at each time step, thereby enhancing its ability to model long-term temporal dependencies. Finally, relative positional encoding is incorporated to share relative position information across attention heads by modulating attention score matrices within each head. The proposed method is evaluated on the CALCE battery degradation dataset.

## 2 Experimental Data and Feature Extraction

### 2.1 Dataset Analysis

PIn this study, the University of Maryland dataset's lithium-ion battery data was utilized to assess the proposed capacity degradation prediction model. For a detailed analysis, four specific battery cells—CS2_35, CS2_36, CS2_37, and CS2_38—were chosen. The SOH in lithium-ion batteries is typically characterized by the relationship between actual peak capacity and nominal capacity [Jia 2023].Mathematically, it is expressed as follows:

$$SOH = (Cactual/Crated) \times 100\% \quad (1)$$

where, *Cactual* represents the actual maximum storage capability of the battery corresponding to its current cycle life, and *Crated* denotes the rated capacity of the lithium-ion battery (1.1 Ah).

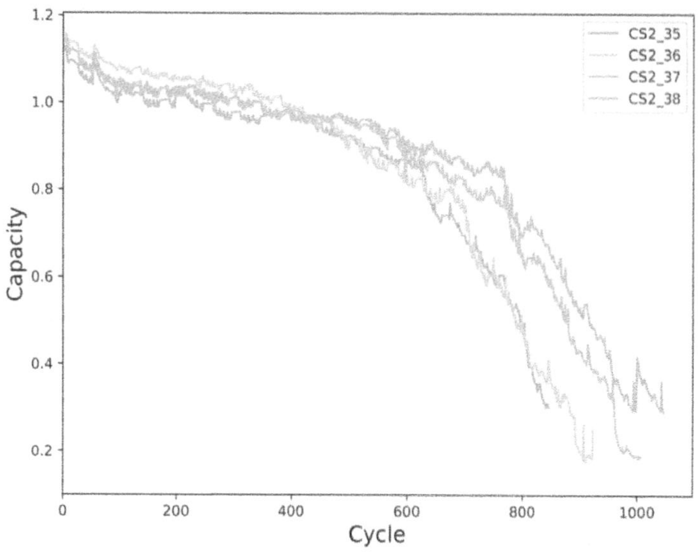


**Fig. 1.** Capacity degradation curves for CS2_35, CS2_36, CS2_37, and CS2_38 from the CALCE dataset

The SOH curves for the four selected battery cells are illustrated in (see Fig. 1). As depicted, all batteries exhibit a progressive decline in storage capacity over time, which consequently reduces their SOH index. This trend is indicative of typical battery aging and signifies the gradual deterioration of the internal physicochemical properties [Xiao 2023].

Data for the four battery cells was gathered using a constant current–constant voltage (CC–CV) charging protocol. (see Fig. 2) displays the partial charge and discharge profiles of battery CS2_35 throughout its entire life cycle from the CALCE lithium-ion battery dataset. The battery underwent constant-current charging at 0.55C until the voltage limit of 4.2 V was attained. Following the initial phase, the voltage was kept steady at 4.2 V while the current ramped up to 1C, and this stage persisted until the current fell under 0.05 A.

## 2.2 Feature Extraction and Analysis

In lithium-ion battery (LIB) aging research, extracting health indicators (HIs) that accurately reflect the degradation status is a critical step. Changes in these features may serve as direct or indirect markers for the battery's aging. However, due to practical constraints and variable operating conditions, discharge profiles often exhibit high variability in real-world usage. In contrast, the charging procedure generally adheres to a more stable CC-CV protocol. As a result, monitoring parameters obtained during the charging phase are more sutiable for health feature extraction [Peng 2024], offering a more reliable basis for accurate assessment of battery SOH.

**Feature extraction.** As shown in (see Fig) and (see Fig. 2), the temporal evolution of current and voltage is strongly correlated with capacity degradation. Consequently, it is more appropriate to extract health indicators (HIs), which consist of multiple features, from the CC–CV charging profiles [Chen 2023].

Analyzing (see Fig. 2) reveals that during the constant current (CC) charging phase of the lithium-ion battery, both the discharge termination voltage and the charging cut-off voltage exhibit a noticeable leftward shift as the battery ages. This effect leads to a gradual decrease in CC charging time, indicating the battery's SOH degradation. Consequently, two health metrics—HI1 and HI2—can be derived from CCCT and CVCT measurements, facilitating a quantitative assessment of battery aging.

$$HI1 = max(T2(k)) - min(T1(k)) \tag{2}$$

$$HI2 = max(T2(k)) - min(T1(k)) \tag{3}$$

where k is the number of cycles, T_1 and T_2 denotes the charging voltage time and charging current time for the k-th cycle, respectively.

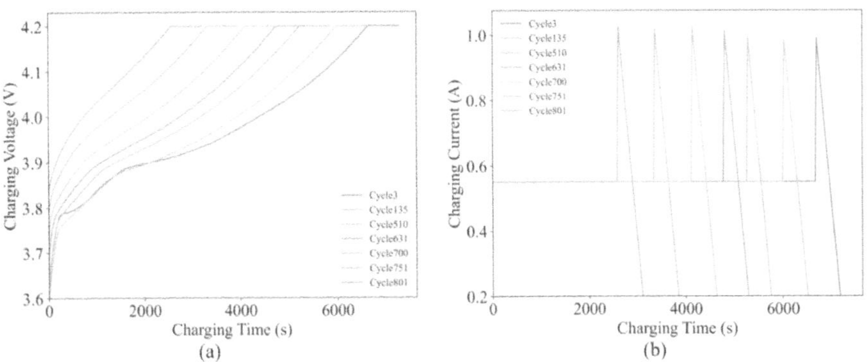


**Fig. 2.** Voltage and current curves for the CS2_35 battery across various cycles are depicted. (a) Charging voltage curve; (b) charging current curve.

Illustrated in (see Fig. 2)(a), as the count of charge cycles rises, the time necessary for the voltage to hit the cut-off point reduces. This causes a decrease in the overall voltage charging time. This results in a reduction of voltage samples within the same time interval. Moreover, the initial charging voltage rises with each cycle, causing an increase in the average voltage over the charging period. These observations suggest that the average voltage during this phase can effectively indicate SOH degradation.

$$HI3 = \frac{1}{n}\sum_{i=1}^{n} V_i \tag{4}$$

where n represents the number of samples during the constant current charging time, and VI denotes the voltage value of the i-th sample.

With more charge and discharge cycles, the charging voltage curve shifts toward the left. This shift is largely caused by the steady rise in the battery's internal resistance. The evolution of internal resistance functions as a significant marker of the battery's SOH degradation.

$$HI4 = resistance(k) \tag{5}$$

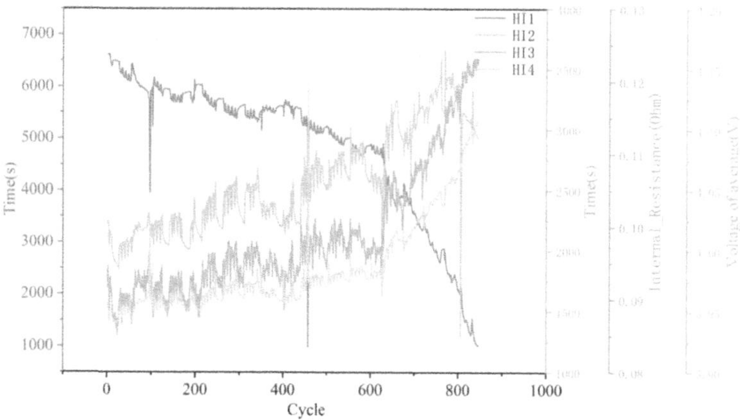


**Fig. 3.** His extracted by cs2_35.

(see Fig. 3) presents the evolution trends of four health indicators (HIs) as a function of cycle count, extracted from battery cell CS2_35. It is observed that HI2 exhibits more abnormal fluctuations and higher jitter compared to HI3 and HI4 during the aging process. This suggests that HI2 captures the degradation behavior of the lithium-ion battery (LIB) more sensitively on CS2_35 than the other three HIs. Overall, all HIs show minor fluctuations and some outliers within a limited range, indicating that they effectively reflect the underlying internal mechanisms and dynamic changes associated with battery aging. Notably, around 700 cycles, a sharp increase or decrease in all HIs is observed, signaling a significant acceleration in the aging process after approximately 700 charge–discharge cycles.

**Feature correlation assessment** After extracting the necessary features, we employ Pearson correlation coefficients (PCCS) to assess their correlation with capacity, thereby validating the rationale of our approach. The PCCS formula is presented in Eq. 6.

$$r = \frac{\sum_i^n (X_i - \overline{X})(Y_i - \overline{Y})}{\sqrt{\sum_i^n (X_i - \overline{X})^2}\sqrt{\sum_i^n (Y_i - \overline{Y})^2}} \tag{6}$$

whereX_i and Y_i respectively represent the ith eigenvalue and the ith capacity value of a feature, and $\overline{X}$ and $\overline{Y}$ indicate the average values of the feature and capacity.

As illustrated in Table1, the correlation analysis was conducted between four battery health indicators (HIs) and the degree of capacity degradation. A correlation coefficient | R \| > 0.8 suggests a robust linear relationship between the variables. Notably, the correlation coefficients of HI1 and HI4 exceed 0.98, suggesting a very high degree of association. These results demonstrate that all four HIs effectively capture the SOH degradation trend, thereby validating their sensitivity to capacity changes and their applicability in battery state assessment.

**Table 1.** Correlation between HIs and SOH.

| Battery | HI1 | HI2 | HI3 | HI4 |
|---|---|---|---|---|
| CS2_35 | 0.9888 | −0.8536 | −0.9780 | −0.9747 |
| CS2_36 | **0.9989** | −0.7376 | −0.9829 | −0.9946 |
| CS2_37 | **0.9987** | −0.6380 | −0.9894 | −0.9954 |
| CS2_38 | *0.9988* | −0.7876 | −0.3433 | −0.9886 |

(see Fig. 4) displays the thermodynamic (correlation) map among the four HIs and the overall SOH, visually illustrating the relationships between different features. The strength of the correlation is represented by both the size and color intensity of the ellipses. It is evident that the self-correlation of HI1 and HI1 exceeds 0.99, indicating potential redundancy between certain HIs. Therefore, careful selection and analysis of HIs are crucial for optimizing model inputs in SOH prediction tasks.

This section offers a comparative analysis of BiLSTM, highlighting their strengths and limitations. It also evaluates the effectiveness of both absolute and relative position encoding strategies in capturing sequential dependencies within time-series data.

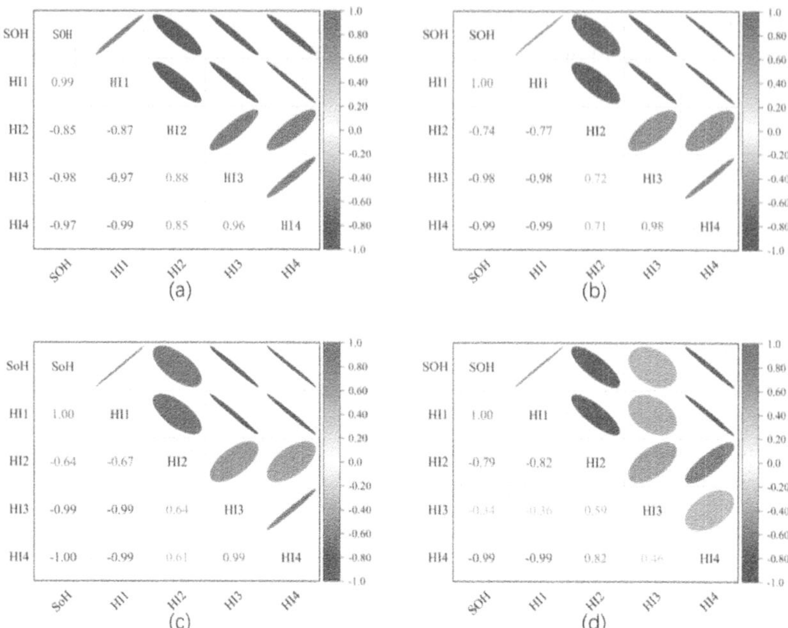


**Fig. 4.** (a)–(d) all characteristics of CS2_35-CS2_38 and the correlation thermody-namic diagram of SOH in the whole cycle.

# 3 Network Model

## 3.1 Bilstm Model

BiLSTM is a variant of the conventional recurrent neural network (RNN) [Peng 2024], with its architecture shown in (see Fig. 5). It extends the standard LSTM by incorporating two parallel layers: one processes the sequence forward, while the other processes it backward. By integrating hidden states from both directions across the temporal dimension, the model captures dependencies from past and future time steps relative to the current one. The final output sequence results from combining these bidirectional hidden states.

This dual-directional processing enables the BiLSTM to comprehensively model sequential data, thereby enhancing the model's feature extraction capability and overall efficiency. It demonstrates notable advantages in handling complex time-series tasks, particularly those requiring deep contextual understanding. The specific calculation formula of bilstm is as follows:

$$i_t = \sigma(W_{xi}x_t + W_{hi}h_{t-1} + W_{ci}c_{t-1} + b_i) \tag{7}$$

$$f_t = \sigma\left(W_{xf}x_t + W_{hf}h_{t-1} + W_{cf}c_{t-1} + b_f\right) \tag{8}$$

$$c_t = f_t \times c_{t-1} + i_t \times \tan h(W_{xc}x_t + W_{hc}h_{t-1} + b_c) \tag{9}$$

$$o_t = \sigma(W_{xo}x_t + W_{ho}h_{t-1} + W_{co}c_t + b_o) \tag{10}$$

$$h_t = o_t \times \tan h(c_t) \tag{11}$$

where W and b represent the weight matrices and bias vectors for each gate, σ signifies the sigmoid function, tan h represents the hyperbolic tangent function, and o denotes element-wise multiplication.

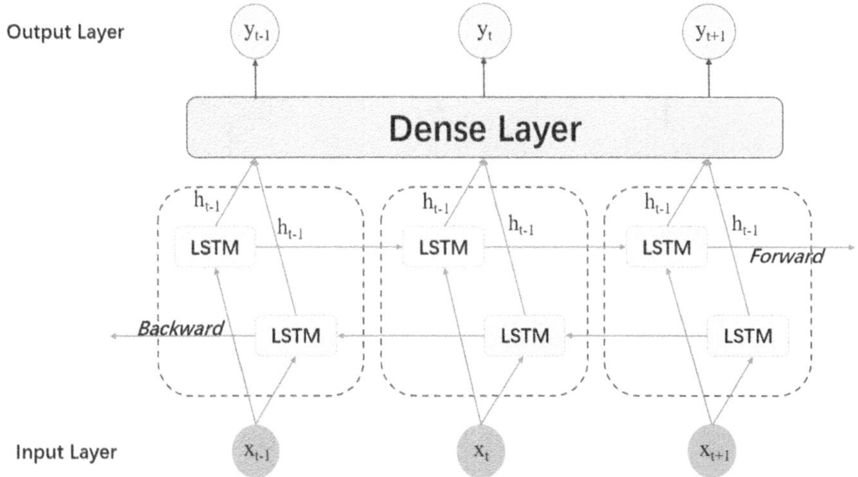


**Fig. 5.** Bilstm Structure Diagram.

The computational procedures of the forward and reverse LSTM networks are detailed in Eqs. 7–11. As a result, two hidden state sequences—$h_t^{forward}$ (the forward direction, indicated by the blue arrow in (see Fig. 5) and $h_t^{forward}$ (the reverse direction, indicated by the green arrow in(see Fig. 5)—are obtained. At each time step t, these bidirectional hidden states are combined to create the final hidden state representation $h_t$, which integrates both past and future contextual information.

$$H_t = \left[h_t^{forward}; h_t^{backward}\right] \tag{12}$$

## 3.2 Traditional Transformer Architecture

The standard Transformer uses an encoder-decoder structure with self-attention as its core mechanism, enabling the capture of long-range dependencies by computing attention scores across input positions. This allows parallel processing, improving computational efficiency over traditional recurrent models, and has led to significant success in natural language processing\cite{34}.

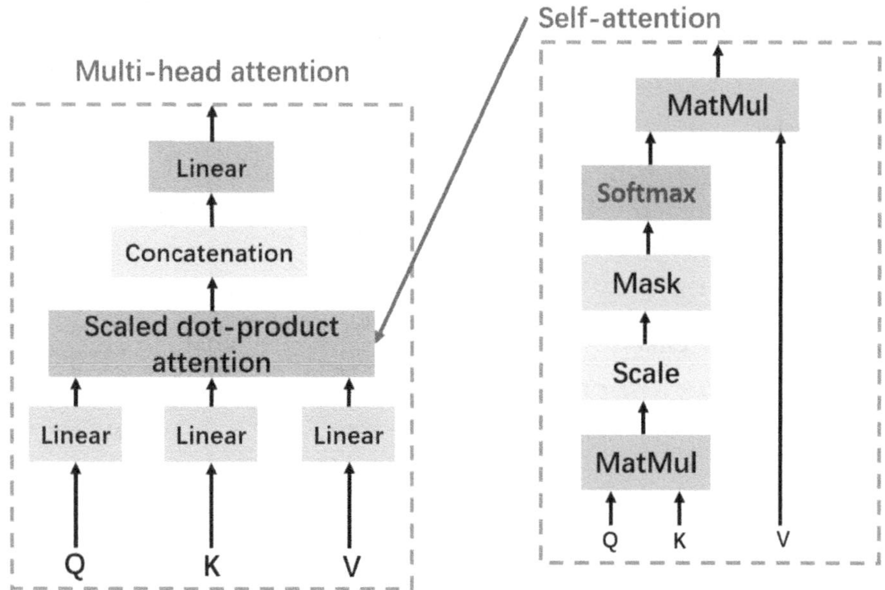


**Fig. 6.** Multi-head Attention Network Structure

Embedding layer: The embedding layer comprises two components: input embeddings and positional embeddings. The input embeddings encode the input sequence into a continuous vector representation, capturing the semantic or feature-level information of each element. In contrast, positional embeddings are introduced to preserve the sequential order of elements by encoding their positions within the sequence. This dual-component structure enables the model to jointly consider both the content and the positional context of each element in the sequence.

Multi-Head Attention: The multi-head attention component of the Transformer model consists of several self-attention sublayers operating in parallel. As illustrated in (see Fig. 6), this architecture enables the model to attend to information from multiple subspaces at different locations within the input sequence. Consequently, it captures a broader range of contextual relationships, improving the model's capability to interpret intricate sequential patterns.

The Q(query), K(key value) and V(query) vectors obtained by multiplying the input sequence with the query weight matrix, the key value weight matrix and the value weight matrix respectively. The corresponding attention weight can be obtained through the self attention mechanism.

$$Q = xW^Q, K = xW^k, V = xW^v \tag{13}$$

$$Attention(Q, K, V) = soft\max\left(\frac{QK^T}{\sqrt{d_k}}\right)V \tag{14}$$

where x is the input matrix; $W^Q$, $W^k$, and $W^v$ are the weight matrices corresponding to Q, K, and V. Their dimensions are Rd, and d is the hidden vector dimension input in the model. Dk represents the dimension of the key value vector. Attention represents the self attention mechanism.

Residual connection and Layer normalization: Residual connections are vital in the Transformer architecture. They are implemented through element-wise addition of inputs and outputs from either the multi-head attention or the feed-forward neural network. This approach helps alleviate the vanishing gradient issue in deep networks, promoting stable training. Coupled with layer normalization, residual connections boost the model's stability and representation power, especially for complex sequential tasks.

Feedforward neural network: The feedforward neural network (FFNN) is a crucial element in both the encoder and decoder layers of the Transformer architecture. It comprises two fully connected layers: the first layer expands the input into a higher-dimensional space to boost representational capacity, followed by a nonlinear activation function that introduces nonlinearity, transforming linear data into a more expressive form. This design enables the FFNN to capture complex patterns in sequential data effectively. The second layer subsequently maps the high-dimensional representation back to the original input space. This two-layer structure—expanding dimensions before reducing them—facilitates the learning of intricate nonlinear features while maintaining input-output compatibility. Additionally, since the FFNN processes each sequence position independently, it supports parallel computation, enhancing efficiency without sacrificing model expressiveness.

$$FFNN(x) = ReLU(xW_1 + b_1)W_2 + b_2 \tag{15}$$

where, W1, W2 and b1, b2 are the weights and offsets of the first and second layers respectively.

### 3.3 RPTransformer Improvement Details

This section first presents a detailed discussion on the limitations of absolute position encoding in the conventional Transformer model. It then elaborates on the advantages of relative position encoding in the context of lithium-ion battery SOH prediction. Finally, the overall architecture of the proposed RPTransformer model is introduced, which integrates the relative position encoding mechanism to better capture sequential dependencies in battery aging data.

In the conventional Transformer architecture, the self-attention does not intrinsically encode the sequential ordering of input tokens. When processing time series data, where temporal ordering is critical for model learning, positional information becomes essential. To address this limitation, Vaswani [Vaswan 2017] introduced a sinusoidal position encoding scheme in the original Transformer model to explicitly inject positional features into the input representation. The mathematical formulation of this encoding method is provided in Eq. 16.

$$\begin{aligned} PE(p, 2i) &= \sin\left(\frac{p}{10000^{2i/d}}\right) \\ PE(p, 2i) &= \cos\left(\frac{p}{10000^{2i/d}}\right) \end{aligned} \tag{16}$$

where $PE$ is the location code; p is the position of the element in the vector; d is the dimension of $PE$; 2i is the $PE$ calculation formula corresponding to even dimension; 2i + 1 is the PE calculation formula corresponding to odd dimension.

We can calculate the position vector PE_p of sine and cosine according to different positions in Eqs. 16, as shown in Eq. 17:

$$PE_p = \begin{bmatrix} \sin(c_0 p) \\ \cos(c_0 p) \\ \vdots \\ \sin\left(c_{\frac{d}{2}-1} p\right) \\ \cos\left(c_{\frac{d}{2}-1} p\right) \end{bmatrix} \tag{17}$$

where Ci represents the constant following dimension i, and its constant value is $10000^{2i/d}$. From Eq. 17, we can know that the position vector PE_p corresponding to each pos in different dimensions can be obtained by the interactive calculation based on sine and cosine. Take the position vector of sine function as an example, as depicted in (see Fig. 7).As the figure demonstrates, the sine values of the positional vectors gradually approach zero as the dimensionality increases. This phenomenon arises because the term 2i in the positional encoding formula progressively approaches the model dimension d, resulting in a longer wavelength and lower frequency of the sinusoidal function. Consequently, with increasing dimensions, the differences between positional encodings become smaller, leading to diminished sensitivity to positional variations in higher-dimensional spaces. By observing the positional encodings at lower dimensionalities, it becomes evident that the oscillation frequency of the positional vectors is relatively high.

As a result, similar cosine-based encoding patterns can be observed among nearby positions, indicating that low-dimensional positional embeddings are more sensitive to small positional shifts. The conventional Transformer integrates the learned positional encoding (PE) with the input sequence embeddings through element-wise addition, where the positional information is encoded using sinusoidal functions. It is important to note that this positional encoding scheme is not arbitrarily assigned but is instead determined by the specific position (pos) and its corresponding dimension index (Dim) within the vector space.

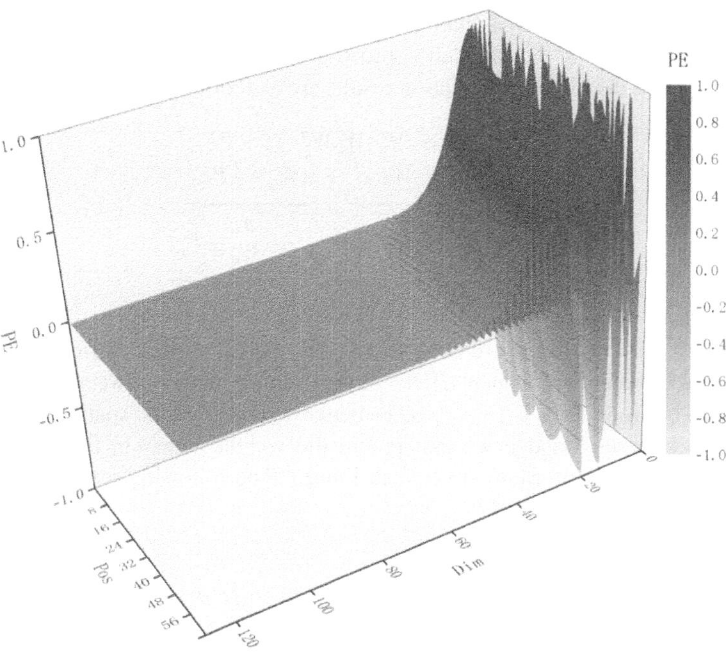


**Fig. 7.** 3D sine function position visualization

Next, we will expand the discussion of Eq. 18 to clearly understand the reasons for the difficulties in the correlation between different vectors in the traditional position coding process, as shown in Eq. 18:

$$\begin{cases} Q_i = (x_i + PE_i)W_q \\ K_j = \left(x_j + PE_j\right)W_k \\ V_j = \left(x_j + PE_j\right)W_v \\ a_{ij} = soft\max\left(\frac{Q_i K_j^T}{\sqrt{D_K}}\right) \\ O_i = \sum_{1}^{j} a_{ij} V_j \end{cases} \tag{18}$$

where, i represents the i-th token, j represents the token corresponding to the i-th token, x_i and x_j represent the corresponding inputs of i and j in the attention layer respectively, whose shape is R^d, and d represents the hidden vector dimension input in the model. PE_i and PE_j represent the location information of x_i and x_j, and their shapes are consistent with x_i and x_j. W_q, W_k, and W_v are the weight matrices of Q_i, K_j, and V_j, respectively. The dimension size is R^{d\times d}, Q_i represents the query vector of the ith element, and K_j and V_j represent the key vector and value vector of the jth element, respectively. D_k represents the dimension of the key vector, which is used to scale the attention score. a_{ij} represents the attention weight between i and j, while O_i represents the sum of attention weights, and the dimension is R^d.In order to better understand the positional relationship between different vectors in the calculation process and their role in the calculation process, we discuss the calculation process of attention score in detail. The calculation results are as follows:

$$
\begin{aligned}
Q_i K_j^T &= (x_i + PE_i) W_q W_k^T (x_j + PE_j)^T \\
&= \underbrace{x_i W_q W_k^T x_j^T}_{A} + \underbrace{x_i W_q W_k^T PE_j^T}_{B} \\
&\quad + \underbrace{PE_i W_q W_k^T PE_j^T}_{C} + \underbrace{PE_i W_q W_k^T x_j^T}_{D}
\end{aligned}
\tag{19}
$$

In Eq. 20, it is not difficult to observe that only Part C contains different positional relationships between two elements. Let o be the distance matrix between i and j, and the relationship between them is j = i + o. To better demonstrate the spatial or sequential relationship between i and j, we first ignore the weight matrix in Part C, and only consider the position information between i and j. The following relationship can be obtained through Eqs. 16 and 17:

$$
\begin{aligned}
PE_i PE_j^T &= \sum_{t=0}^{\frac{d}{2}-1} \left[ \sin(c_t p) \sin(c_t (p+o)) + \cos(c_t p) \cos(c_t (p+o)) \right] \\
&= \sum_{t=0}^{\frac{d}{2}-1} \cos(c_t o)
\end{aligned}
\tag{20}
$$

where, t represents the dimension, and p represents the location information of i at this time.

As shown in Eq. 20, the positional information between positions i and j is explicitly incorporated into the formulation at this stage, indicating a mathematically defined correlation between these positions within the functional framework. However, Part C involves operations between the query and key weight matrices. Yan [Yan 2019] experimentally demonstrated that the positional information carried by this component does not effectively encode the distance relationships between elements when weight matrices are combined through addition in the model. In other words, when positional information is incorporated into the model through weight matrix operations, the relative positional relationships among elements are not explicitly captured. This limitation reduces the effectiveness of positional encoding in the overall model computation, thereby hindering its potential to guide the model in understanding sequential position information.

Overall, the positional information in Part C has not effectively conveyed explicit positional relationships to the model under the current integration strategy, thereby limiting the full realization of the functional potential of positional encoding.

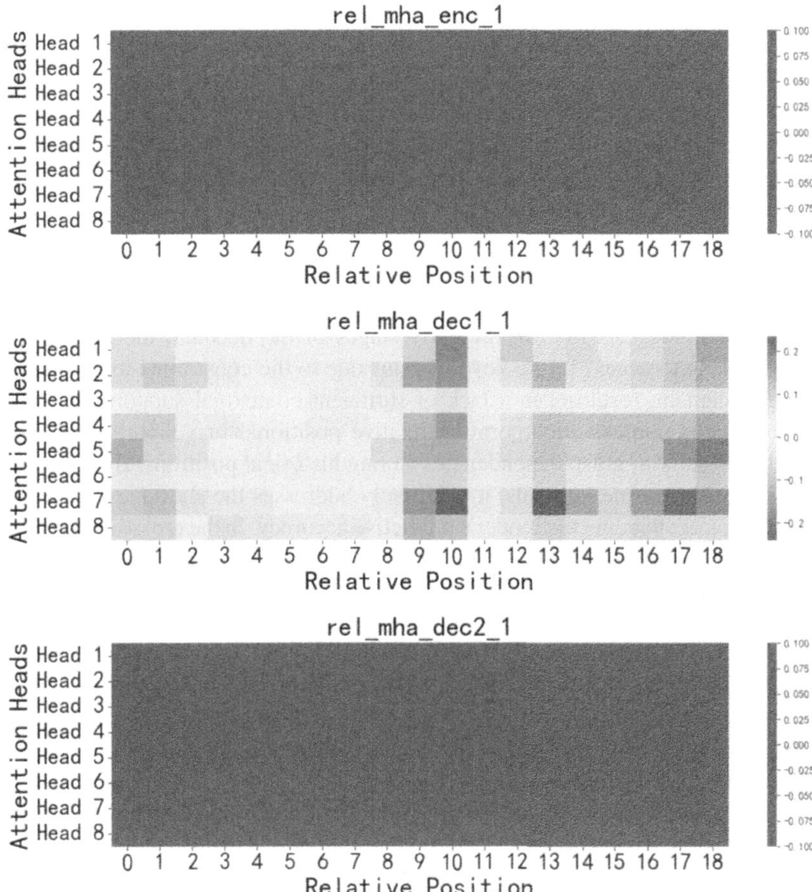


**Fig. 8.** Heatmap of position bias in multi-head attention (Encoder and Decoder).

**Advantages of relative position encoding** During the aging process of LIBs, the use of absolute position encoding results in fixed positional information that is unable to accurately capture the sequential dependencies between preceding and succeeding elements. However, effective state-of-health (SOH) estimation not only relies on current performance characteristics but is also significantly influenced by historical states.

During the feature extraction phase, relative positional encoding demonstrates distinct advantages by providing the model with richer and more accurate sequential structural information. This allows the model to grasp intricate relationships between sequence elements, aiding in pinpointing key features and patterns that affect SOH, thus enhancing prediction accuracy.

To better understand the learned positional relationships, we visually analyzed the relative positional biases in the multi-head attention of the first encoder and decoder layers, centering on the CS2_35 battery sample. (see Fig. 8) presents the attention weight heatmap for a relative position distance of 10 units, reflecting the feature correlations captured by the model at this specific relative distance. The heatmap clearly illustrates the ability of different attention heads to model relationships among elements within the input feature sequence over this range.

The figure clearly shows that significant non-zero relative positional bias patterns are learned exclusively in the decoder's masked self-attention layer. In contrast, the encoder-decoder cross-attention layers exhibit consistently zero relative positional bias. This observation suggests that, in the early stages of the decoder, the current query position is unable to access future key positions due to the constraints imposed by the masking mechanism, resulting in a lack of sufficient contextual support for accurate prediction. In this context, incorporating relative positional bias significantly boosts the model's capacity to grasp dependencies among historical positions. By differentiating inputs at various time intervals, it effectively addresses the shortage of contextual information, thus enhancing the model's predictive accuracy. In the cross-attention modules connecting the encoder and decoder, the input feature sequence is relatively short. This enables the model to efficiently identify both long-and short-term dependencies through dot-product operations between query and key vectors. At this stage, the BiLSTM's strong bidirectional feature extraction capability significantly reduces the need for explicit positional offset information, leading to bias parameters that remain near zero during training. The automatic "deactivation" of the bias term in scenarios where it is unnecessary not only minimizes redundant computations but also enhances both training efficiency and generalization performance.

The experimental results provide a clear and intuitive visual foundation for battery SOH prediction based on the relative positional encoding mechanism. Furthermore, they validate the effectiveness of the relative positional bias and its selective activation under specific contextual constraints.

**RPTransformer model** Traditional Transformer models often exhibit limited prediction accuracy and generalization performance when applied to temporal tasks, primarily due to their insufficient modeling of positional information. To address this limitation, we propose the BiLSTM-RPTransformer framework, which incorporates relative positional encoding into each multi-head attention layer. This integration enhances both the model's temporal dependency modeling capability and its parallel computation efficiency.

The RPTransformer framework is designed to address the limitations of traditional absolute positional encoding while leveraging the advantages of relative positional encoding. Instead of relying on fixed absolute position embeddings, we integrate relative positional encoding into each multi-head attention layer, enabling the model to capture more flexible temporal information. Combined with the BiLSTM architecture, this design allows for the simultaneous modeling of both local and global sequential features.

The improved scaling point product attention (SDPA) formula is as follows:

$$\begin{cases} Q_i = x_i W_q \\ K_j = x_j W_k + b_{ij}^k \\ V_j = x_j W_v + b_{ij}^v \end{cases} \tag{21}$$

**Table 2.** Prediction error comparison between the designed approach and ablation variants on the Cacle dataset.

| Battery | Criteria (%) | M1 | M2 | M3 | M4 | Designed Approach |
|---|---|---|---|---|---|---|
| CS2_35 | RMSE | 1.2227 | 1.2548 | 1.1220 | 0.9699 | 0.8707 |
| | MAE | 0.9427 | 0.9478 | 0.8673 | 0.7428 | 0.6797 |
| | MAPE | 1.7767 | 1.8955 | 1.7131 | 1.3857 | 1.3260 |
| CS2_36 | RMSE | 1.7200 | 1.4843 | 1.2614 | 1.2888 | 0.9251 |
| | MAE | 1.3254 | 1.1101 | 0.8887 | 0.9885 | 0.6592 |
| | MAPE | 3.3265 | 3.2001 | 2.7428 | 3.0349 | 1.8367 |
| CS2_37 | RMSE | 1.2974 | 1.1320 | 0.8435 | 1.5277 | 0.9508 |
| | MAE | 0.9930 | 0.8273 | 0.6561 | 1.3127 | 0.7372 |
| | MAPE | 2.4412 | 2.4586 | 1.6818 | 2.9567 | 1.9340 |
| CS2_38 | RMSE | 1.5662 | 1.5423 | 1.3920 | 1.2967 | 0.7984 |
| | MAE | 1.1185 | 1.1664 | 0.8438 | 0.8343 | 0.5900 |
| | MAPE | 2.3936 | 2.5614 | 1.9759 | 1.9687 | 1.4660 |

In the experimental design, the computation of the key (K) and value (V) vectors was modified by replacing the absolute position information pos in Eq. 13 with relative distance matrices $b_{ij}^k$ and $b_{ij}^v$, corresponding to the pairwise token interactions for K and V, respectively. The RPTransformer model's architecture is depicted in (see Fig. 9).

In this research, the derived health indicators (HIs) are used as input features fed into a BiLSTM network that further extracts high-level features. The resulting feature sequence is then processed through a feedforward layer with ReLU activation to generate a nonlinear feature representation for the encoder layer. Meanwhile, the decoder layer receives SOH values from historical time steps as its input. By using these historical SOH values as reference benchmarks and integrating the sequence information output by the encoder layer, the model aims to establish internal correlations between past SOH states, the encoded feature sequence, and future SOH predictions. This design enhances the model's precision in forecasting the SOH.

Given the relatively limited number of lithium-ion battery sequence samples available in this study, both the encoder and decoder employ a two-layer architecture. Furthermore, the prediction task is formulated as a one-step-ahead forecasting problem to reduce model complexity and mitigate the risk of overfitting. This architectural choice was made after a comprehensive evaluation of sample size, model complexity, and prediction performance, aiming to maximize the SOH forecasting capability of the model under data-limited conditions.

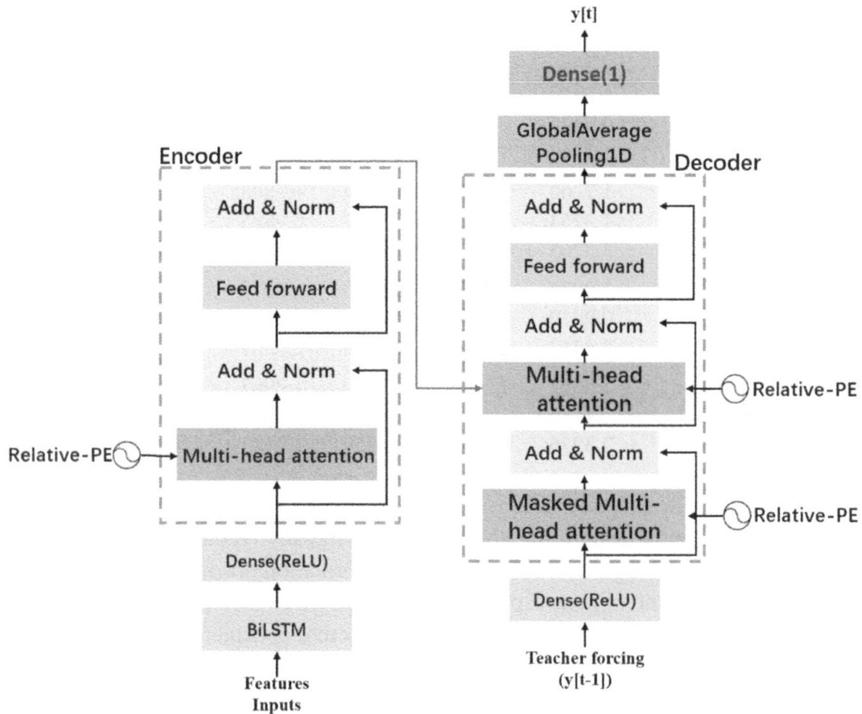


**Fig. 9.** RPTransformer model architecture diagram

# 4 Results and Analysis

To assess the BiLSTM-RPTransformer model's effectiveness in estimating lithium-ion battery SOH, this section details ablation and comparative experiments performed on the CACLE dataset.

## 4.1 Evaluation Criterion

Three common error metrics MAPE, MAE, and RMSE—are utilized to evaluate the model's SOH estimation and prediction capabilities.

$$MAPE = \frac{1}{n}\sum_{i}^{n}\left|\frac{\hat{y}_i - y_i}{y_i}\right| \times 100\% \tag{22}$$

$$MAE = \frac{1}{n}\sum_{i=1}^{n}|y_i - \hat{y}_i| \times 100\% \tag{23}$$

$$RMSE = \sqrt{\frac{1}{n}\sum_{i=1}^{n}(y_i - \hat{y}_i)^2} \times 100\% \tag{24}$$

where, n is the total number of samples, Y_i and ŷ denote the actual and predicted values at the i-th cycle respectively.

## 4.2 Experimental Parameters

To verify the effectiveness of the introduced relative positional encoding strategy, four Transformer-based architectures—Transformer(M1), CNN-Transformer(M2), BiLSTM-Transformer (M3), and CNN-RPTransformer (M4)—are selected as baseline models for comparative analysis. All models are implemented using the GPU-accelerated Keras framework and optimized with the Adam algorithm. During model training, MSE functions as the loss criterion, whereas MAE is utilized to monitor and evaluate model performance. The main hyperparameter settings are listed in Table 3.

**Table 3.** Hyperparameter setting of each datum model

| Parameter | Proposed Model | M1 | M2 | M3 | M4 |
|---|---|---|---|---|---|
| Epoch | 300 | 300 | 300 | 300 | 300 |
| Batch size 16 | 16 | 16 | 16 | 16 | 16 |
| Learning rate | 0.001 | 0.001 | 0.001 | 0.001 | 0.001 |
| Dropout rate | 0.01 | 0.01 | 0.01 | 0.01 | 0.01 |
| Feedforward dimension | 64 | 64 | 64 | 64 | 64 |
| Hidden dimension | 180 | 180 | 180 | 180 | 180 |

*(continued)*

**Table 3.** (*continued*)

| Parameter | Proposed Model | M1 | M2 | M3 | M4 |
|---|---|---|---|---|---|
| Number of multi-head attention heads | 8 | 8 | 8 | 8 | 8 |
| Number of Bi-LSTM hidden layers | 1 | \ | \ | 1 | \ |
| Dimensionality of the Bi-LSTM units | 100 | \ | \ | 100 | \ |
| Number of CNN hidden layers | \ | 3 | \ | \ | 3 |

### 4.3 Comparison of Ablation Experiments

Table 2 presents the SOH prediction errors of different benchmark models across four types of LIBs in the vehicle dataset. The RMSE values for models M2 and M3 are below 1.2%, whereas that of M1 exceeds 1.2%. This result suggests that incorporating either a BiLSTM or CNN module into the Transformer architecture can enhance its feature extraction capability as well as the overall model's ability to represent temporal information. A comparative analysis of the RMSE, MAE, and MAPE values of models M2 and M3 across four types of lithium-ion batteries reveals that BiLSTM demonstrates superior capability in capturing contextual information through its bidirectional processing mechanism, compared to the local feature extraction ability provided by CNN. These results indicate that modeling SOH degradation requires consideration not only of current feature states but also of historical states from earlier time steps. The proposed model builds on M2 by integrating relative positional encoding, achieving RMSE and MAE values below 1% and 0.7%, respectively, with MAPE reflecting relative deviation between predicted and actual SOH values. Outperforming all benchmarks across these metrics, the proposed method demonstrates that relative positional encoding improves feature extraction and contextual awareness, leading to enhanced prediction accuracy and robustness. Figure 10 illustrates the error distribution boxplots and corresponding normal distribution curves for four benchmark models and the proposed model in lithium-ion battery SOH prediction. Boxplot analysis reveals that the median line and error values of our method are consistently nearest to zero across all datasets, indicating minimal deviation from true SOH values. Furthermore, outliers in our model align more closely with the overall data distribution and show greater concentration. This underscores the model's accurate predictions and its effectiveness across diverse testing scenarios in SOH forecasting.

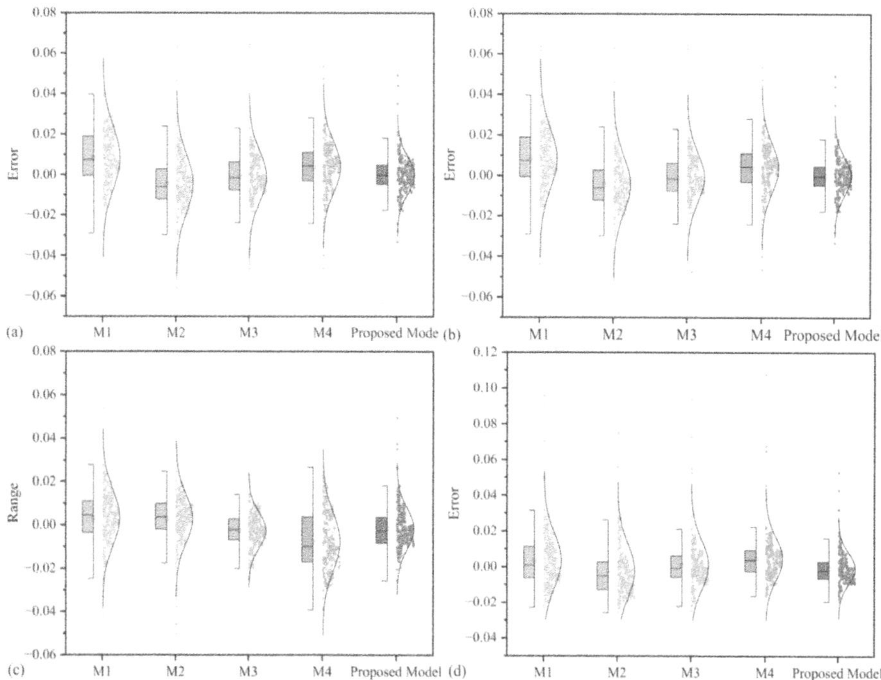


**Fig.10.** (a)–(d) CS2_35-CS2_38 benchmark model and error box diagram and normal distribution curve of proposed model

Figure 11 presents a visual comparison between the predicted SOH trajectories of the benchmark models and the proposed model. It is readily apparent that the prediction curve of the M1 model exhibits a relatively smooth trend, as the Transformer architecture primarily relies on capturing long-term dependencies across the entire sequence for forecasting, while tending to overlook local short-term fluctuations. This global modeling approach causes the model to prioritize the overall trend when confronted with short-term fluctuations in battery SOH, leading to a smoother prediction curve. In contrast, models enhanced with either a CNN or BiLSTM module exhibit visibly more fluctuating prediction curves, indicating a stronger sensitivity to local variations. The CNN module excels at identifying local feature patterns, thereby enabling the model to better capture short-lived changes in battery SOH. Simultaneously, the BiLSTM module boosts the model's responsiveness to temporal dynamics via bidirectional data flow, enabling it to more effectively capture transient changes during the degradation process.

Replacing absolute with relative positional encoding improves prediction performance by better capturing local dependencies and avoiding fixed biases. The proposed model integrates a BiLSTM module with relative positional encoding, enhancing sensitivity to short-term fluctuations, thereby outperforming M4 in prediction accuracy and achieving superior robustness in battery SOH estimation.

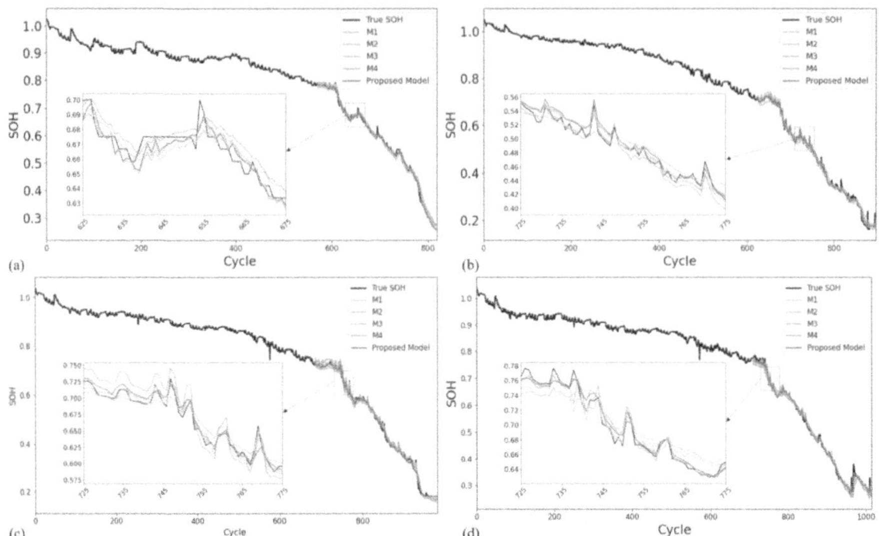


**Fig. 11.** (a)–(d) CS2_35-CS2_38 benchmark model and SOH prediction curve of proposed model

## 4.4 Comparative Experiment

To further validate the proposed model's effectiveness in battery health status (SOH) prediction, a direct comparison is carried out between the proposed model and other state-of-the-art models, utilizing the same dataset for fair evaluation. The GBLS Booster model by Yang [Yang 2024] integrates the advantages of GBLS and a Booster module within the CNN Transformer algorithm for feature extraction, while applying variational mode decomposition (VMD) for noise reduction on the original data. On the CS2_35 and CS2_36 datasets, the model achieves an average RMSE of 1.815% for both. Yuan [Yuan 2024] also employed VMD for denoising the original data but used a CNN Transformer for SOH prediction. This approach resulted in a MAE of 1.16% and an RMSE of 1.505%. In addition, Zhang [Zhang 2025] corrected the extracted features through isolated forests and Lagrange interpolation, and used the improved Transformer to predict SOH. The average MAE and MAPE were 1.08% and 1.87%, respectively.

As shown in Table 4, the proposed model achieves higher prediction accuracy and robustness compared to existing advanced methods, highlighting its strong capability in supporting precise battery SOH estimation.

**Table 4.** Prediction error comparison between the designed approach and other models on the Cacle dataset

| Battery | Criteria (%) | VMD-GBLS Booster | VMD-CNN-Transformer | multi-Feature-ITransformer |
|---|---|---|---|---|
| CS2_35 | RMSE | 1.815 | 1.31 | \ |
| | MAE | \ | 0.95 | 0.96 |
| | MAPE | 0.96243 | \ | 1.70 |
| CS2_36 | RMSE | 1.815 | 2.12 | \ |
| | MAE | \ | 1.64 | 1.17 |
| | MAPE | 0.84463 | \ | 1.89 |
| CS2_37 | RMSE | \ | 1.47 | \ |
| | MAE | \ | 1.12 | 1.15 |
| | MAPE | \ | \ | 2.02 |
| CS2_38 | RMSE | \ | 1.12 | \ |
| | MAE | \ | 0.934 | 1.04 |
| | MAPE | \ | \ | 1.87 |

## 5 Conclusion

This study examines the limitations of the conventional Transformer model in capturing dynamic sequential relationships for lithium-ion battery SOH prediction. The model's reliance on parallel processing inherently hinders its ability to recognize the order of data points. This limitation arises because the Transformer cannot effectively track sequence dynamics. This limitation originates from the parallel processing mechanism of the Transformer architecture, which treats all elements of a sequence simultaneously and neglects their relative positional information. During the data preprocessing stage, relevant variables such as current, voltage, and internal resistance are extracted to describe the battery's degradation behavior comprehensively. Subsequently, the BiLSTM module is employed to perform bidirectional information flow over the sequential data, thereby enhancing the model's sensitivity to short-term dynamic variations. Finally, historical data and relative positional encoding are integrated within each multi-head attention layer to dynamically incorporate relative positional information and inter-element features. This mechanism facilitates the establishment of feature-SOH correlations, leading to a notable improvement in the model's predictive performance.

To validate the proposed model's effectiveness, a predictive evaluation was conducted on the CALCE dataset. The results show that the RMSE, MAE, and MAPE of the proposed method are below 1\%, 0.7\%, and 1.9\%, respectively, highlighting its high accuracy and robustness. These findings suggest significant potential for the model in lithium-ion battery SOH prediction. It effectively overcomes the conventional Transformer's limitations in handling dynamic sequences and provides an innovative approach to accurately estimate battery health status.

## References


Attia, P.M., et al.: “Knees” in lithium-ion battery aging trajectories. J. Electrochem. Soc. **169**(6), 060517 (2022). https://doi.org/10.1149/1945-7111/ac6d13

Chaoui, H., Ibe-Ekeocha, C.C.: State of charge and state of health estimation for lithium batteries using recurrent neural networks. IEEE Trans. Veh. Technol. **66**(10), 8773–8783 (2017). https://doi.org/10.1109/TVT.2017.2715333

Charkhgard, M., Farrokhi, M.: State-of-charge estimation for lithium-ion batteries using neural networks and EKF. IEEE Trans. Industr. Electron. **57**(12), 4178–4187 (2010). https://doi.org/10.1109/TIE.2010.2043035

Chen, L., et al.: State of health estimation of lithium-ion batteries based on equivalent circuit model and data-driven method. J. Energy Storage **73**, 109195 (2023). https://doi.org/10.1016/j.est.2023.109195

Chen, J., Hu, Y., Zhu, Q., Rashid, H., Li, H.: A novel battery health indicator and PSO-LSSVR for LiFePO4 battery SOH estimation during constant current charging. Energy **282**, 128782 (2023). https://doi.org/10.1016/j.energy.2023.128782

Chen, Y., Huang, X., He, Y., Zhang, S., Cai, Y.: Edge–cloud collaborative estimation lithium-ion battery SOH based on MEWOA-VMD and transformer. J. Energy Storage **99**, 113388 (2024). https://doi.org/10.1016/j.est.2024.113388

Bengio, Y., Simard, P., Frasconi, P.: Learning long-term dependencies with gradient descent is difficult. IEEE Trans. Neural Netw. **5**(2), 157–166 (1994). https://doi.org/10.1109/72.279181

Gu, X., et al.: A novel state-of-health estimation for the lithium-ion battery using a convolutional neural network and transformer model. Energy **262**, 125501 (2023). https://doi.org/10.1016/j.energy.2022.125501

Hannan, M.A., Lipu, M.S.H., Hussain, A., Mohamed, A.: A review of lithium ion battery state of charge estimation and management system in electric vehicle applications: challenges and recommendations. Renew. Sustain. Energy Rev. **78**, 834–854 (2017). https://doi.org/10.1016/j.rser.2017.05.001

He, J., Wei, Z., Bian, X., Yan, F.: State-of-health estimation of lithium-ion batteries using incremental capacity analysis based on voltage–capacity model. IEEE Trans. Transp. Electrification **6**(2), 417–426 (2020). https://doi.org/10.1109/TTE.2020.2994543

He, Y., Bai, W., Wang, L., Wu, H., Ding, M.: SOH estimation for lithium-ion batteries: an improved GPR optimization method based on the developed feature extraction. J. Energy Storage **83**, 110678 (2024). https://doi.org/10.1016/j.est.2024.110678

Jia, C., Tian, Y., Shi, Y., Jia, J., Wen, J., Zeng, J.: State of health prediction of lithium-ion batteries based on bidirectional gated recurrent unit and transformer. Energy **285**, 129401 (2023). https://doi.org/10.1016/j.energy.2023.129401

Lin, J., Xie, H., Huang, Z., Kuang, R., Hao, Y.: A dual-filter framework with SOH update triggering for joint SOC and SOH estimation in lithium-ion batteries. J. Energy Storage **136**, 118360 (2025). https://doi.org/10.1016/j.est.2025.118360

Hochreiter, S., Schmidhuber, J.: Long short-term memory. Neural Comput. **9**(8), 1735–1780 (1997). https://doi.org/10.1162/neco.1997.9.8.1735

Hosen, M.S., Jaguemont, J., Van Mierlo, J., Berecibar, M.: Battery lifetime prediction and performance assessment of different modeling approaches. iScience **24**(2) (2021). https://doi.org/10.1016/j.isci.2021.102060

Hosseininasab, S., Lin, C., Pischinger, S., Stapelbroek, M., Vagnoni, G.: State-of-health estimation of lithium-ion batteries for electrified vehicles using a reduced-order electrochemical model. J. Energy Storage **52**, 104684 (2022). https://doi.org/10.1016/j.est.2022.104684

Huang, B., Pan, Z., Su, X., An, L.: Recycling of lithium-ion batteries: recent advances and perspectives. J. Power Sources **399**, 274–286 (2018). https://doi.org/10.1016/j.jpowsour.2018.07.116

Klass, V., Behm, M., Lindbergh, G.: A support vector machine-based state-of-health estimation method for lithium-ion batteries under electric vehicle operation. J. Power Sources **270**, 262–272 (2014). https://doi.org/10.1016/j.jpowsour.2014.07.116

Peng, S., Zhu, J., Wu, T., Tang, A., Kan, J., Pecht, M.: SOH early prediction of lithium-ion batteries based on voltage interval selection and features fusion. Energy **308**, 132993 (2024). https://doi.org/10.1016/j.energy.2024.132993

Peng, S., Zhu, J., Wu, T., Tang, A., Kan, J., Pecht, M.: SOH early prediction of lithium-ion batteries based on voltage interval selection and features fusion. World Electr. Veh. J. **15**(7), 280 (2024). https://doi.org/10.3390/wevj15070280

Ren, L., Zhao, L., Hong, S., Zhao, S., Wang, H., Zhang, L.: Remaining useful life prediction for lithium-ion battery: a deep learning approach. IEEE Access **6**, 50587–50598 (2018). https://doi.org/10.1109/ACCESS.2018.2858856

Richardson, R.R., Osborne, M.A., Howey, D.A.: Gaussian process regression for forecasting battery state of health. J. Power Sources **357**, 209–219 (2017). https://doi.org/10.1016/j.jpowsour.2017.05.004

Vaswani, A., et al.: Attention is all you need. In: Advances in Neural Information Processing Systems (Vol. 30). Curran Associates (2017). https://doi.org/10.48550/arXiv.1706.03762

Venugopal, P.: State-of-health estimation of li-ion batteries in electric vehicle using indrnn under variable load condition. Energies **12**(22), 4338 (2019). https://doi.org/10.3390/en12224338

Wang, L., Cheng, Y., Zou, J.: Battery available power prediction of hybrid electric vehicle based on improved Dynamic Matrix Control algorithms. J. Power Sources **261**, 337–347 (2014). https://doi.org/10.1016/j.jpowsour.2014.03.091

Wei, J., Dong, G., Chen, Z.: Remaining useful life prediction and state of health diagnosis for lithium-ion batteries using particle filter and support vector regression. IEEE Trans. Industr. Electron. **65**(7), 5634–5643 (2017). https://doi.org/10.1109/TIE.2017.2782224

Xing, Y., Ma, E.W.M., Tsui, K.-L., Pecht, M.: An ensemble model for predicting the remaining useful performance of lithium-ion batteries. Microelectron. Reliab. **53**(6), 811–820 (2013). https://doi.org/10.1016/j.microrel.2012.12.003

Xiao, Y., Wen, J., Yao, L., Zheng, J., Fang, Z., Shen, Y.: A comprehensive review of the lithium-ion battery state of health prognosis methods combining aging mechanism analysis. J. Energy Storage **65**, 107347 (2023). https://doi.org/10.1016/j.est.2023.107347

Xu, H., Wu, L., Xiong, S., Li, W., Garg, A., Gao, L.: An improved CNN-LSTM model-based state-of-health estimation approach for lithium-ion batteries. Energy **276**, 127585 (2023). https://doi.org/10.1016/j.energy.2023.127585

Yang, N., Song, Z., Hofmann, H., Sun, J.: Robust State of Health estimation of lithium-ion batteries using convolutional neural network and random forest. J. Energy Storage **48**, 103857 (2022). https://doi.org/10.1016/j.est.2021.103857

Yang, P., et al.: Joint evaluation and prediction of SOH and RUL for lithium batteries based on a GBLS booster multi-task model. J. Energy Storage **75**, 109741 (2024). https://doi.org/10.1016/j.est.2023.109741

Yan, H., Deng, B., Li, X., Qiu, X.: TENER: adapting transformer encoder for named entity recognition (2019). arXiv Preprint arXiv:1911.04474. https://doi.org/10.48550/arXiv.1911.04474

Yao, L., Fang, Z., Xiao, Y., Hou, J., Fu, Z.: An intelligent fault diagnosis method for lithium battery systems based on grid search support vector machine. Energy **214**, 118866 (2021). https://doi.org/10.1016/j.energy.2020.118866

You, G., Park, S., Oh, D.: Diagnosis of electric vehicle batteries using recurrent neural networks. IEEE Trans. Industr. Electron. **64**(6), 4885–4893 (2017). https://doi.org/10.1109/TIE.2017.2674593

Yuan, Z., Tian, T., Hao, F., Li, G., Tang, R., Liu, X.: A hybrid neural network based on variational mode decomposition denoising for predicting state-of-health of lithium-ion batteries. J. Power Sources **609**, 234697 (2024). https://doi.org/10.1016/j.jpowsour.2024.234697

Tian, H., Qin, P., Li, K., Zhao, Z.: A review of the state of health for lithium-ion batteries: research status and suggestions. J. Clean. Prod. **261**, 120813 (2020). https://doi.org/10.1016/j.jclepro.2020.120813

Zhang, Y., Li, Z., Kong, L., Xu, H., Shen, H., Chen, M.: Multi-step state of health prediction of lithium-ion batteries based on multi-feature extraction and improved transformer. J. Energy Storage **105**, 114538 (2025). https://doi.org/10.1016/j.est.2024.114538

Zhou, Y., Huang, M.: Lithium-ion batteries remaining useful life prediction based on a mixture of empirical mode decomposition and ARIMA model. Microelectron. Reliab. **65**, 265–273 (2016). https://doi.org/10.1016/j.microrel.2016.07.151

# Edge-Enhanced Multi-Scale Fusion for Pixel-Level Image Tampering Localization

Xin Shu, Haitao Xu(✉), Zhihong Chen, Jiaqiao Liu, Jiaxing Wang, Yaqiong Duan, Yali Li, and Ruolin Liu

Tianjin University of Technology, Tianjin 300384, China
xhtsticker0205@stud.tjut.edu.cn

**Abstract.** Image manipulation localization has recently advanced through the integration of multi-scale feature representations and edge-aware supervisory mechanisms. However, existing methods often struggle to simultaneously preserve global semantic consistency and fine-grained boundary precision, particularly in challenging low-texture or visually cluttered regions. To address these issues, we propose MEP-Net, a lightweight yet highly effective manipulation localization framework that incorporates Multi-branch Feature Enhancement Fusion Module, edge positioning supervision, and multi-scale prediction fusion. Through comprehensive experiments and ablation studies, MEP-Net demonstrates competitive performance against both CNN-based and recent Transformer-based detectors, while maintaining strong boundary accuracy and computational efficiency.

**Keywords:** Image Forensics · Splicing Detection · Edge Supervision · Multi-scale Fusion · Tampering Localization

## 1 Introduction

Image manipulation detection and localization have become increasingly important as forged visual content proliferates across social media platforms, digital journalism, and public security domains. Classical approaches relying on hand-crafted forensic cues have gradually been replaced by deep learning-based models capable of learning semantic differences between authentic and manipulated content. Although these models achieve substantial improvements, they often fail to capture fine boundaries or subtle texture irregularities introduced during manipulation.

Recent work highlights the importance of multi-scale reasoning (Wei et al. 2022) frequency-domain representations, and Transformer-based global modeling (Wang et al. 2022).

Despite such progress, several challenges remain:

(1) boundary precision is frequently degraded when deep semantic layers dominate predictions;
(2) multi-scale decoder outputs are not always effectively fused;
(3) many models struggle to integrate edge-level cues in a stable and learnable manner.


H. Yan et al. (Eds.): AIS&P 2025, CCIS 2918, pp. 321–329, 2026.
https://doi.org/10.1007/978-981-95-9884-7_21

To address these challenges, we introduce MEP-Net, a lightweight manipulation localization framework built upon three key components: a Multi-branch Feature Enhancement Fusion Module (MFEFM), an Edge Positioning Monitor Module (EPMM), and a Multi-scale Prediction Fusion Module (MPFM). By jointly leveraging enhanced edge cues, multi-scale semantic fusion, and auxiliary boundary supervision, MEP-Net improves both localization granularity and global consistency while maintaining high computational efficiency.

## 2 Methodology

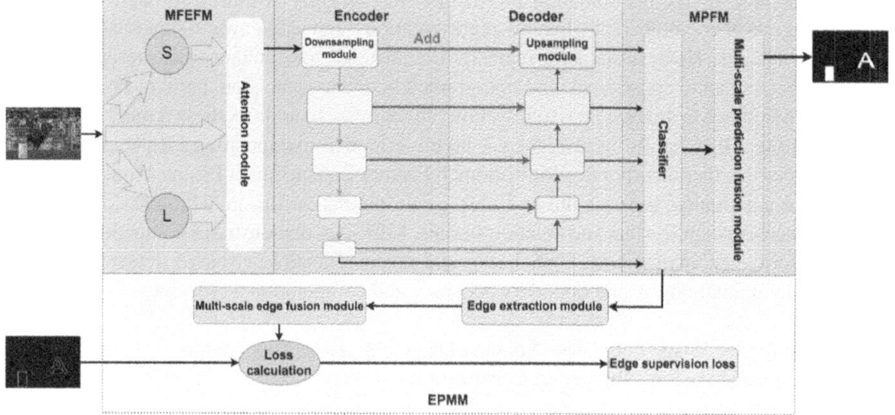


**Fig. 1.** General flowchart of the proposed network.

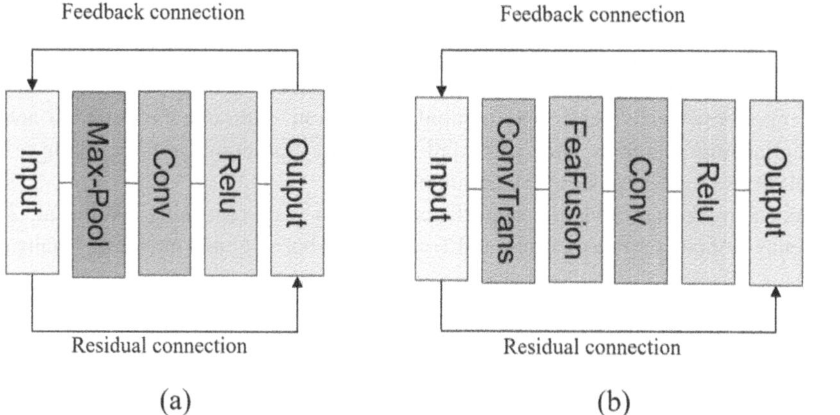


**Fig. 2.** Module structure diagram. (a): Downsampling module (b) Upsampling module.

The overall architecture of MEP-Net follows a lightweight encoder–decoder design, as shown in Fig. 1. The encoder progressively downsamples the enhanced input from the MFEFM module to extract hierarchical semantic features, while the decoder restores spatial resolution via upsampling and skip connections. The structure of the downsampling and upsampling modules is illustrated in Fig. 2. Intermediate predictions from different decoding stages are aggregated by the MPFM to recover fine spatial details. Meanwhile, the EPMM provides auxiliary edge-based supervision at multiple scales, reinforcing boundary precision and stabilizing the learning of thin structures. This design enables MEP-Net to maintain both high-level semantic understanding and fine-grained localization accuracy with minimal computational overhead.

### 2.1 Multi-Branch Feature Enhancement Fusion Module

As illustrated in Fig. 3, The MFEFM module enhances the input by integrating three complementary branches: the original RGB image I, a Sobel-extracted edge map S (Kanopoulos et al. 1988), and a Laplacian-extracted response L (Wang 2007). This fusion emphasizes discontinuities, texture artifacts, and structural inconsistencies commonly introduced by image manipulations. To better exploit the complementary characteristics of these branches, the fused representation F is obtained through a weighted combination, as defined in Eq. (1). Here, I preserves global appearance cues, S highlights gradient-level transitions, and L captures second-order structural variations, enabling the network to receive enriched boundary-aware information from the earliest stages.

$$F = 0.5 \cdot I + 0.25 \cdot S + 0.25 \cdot L \tag{1}$$

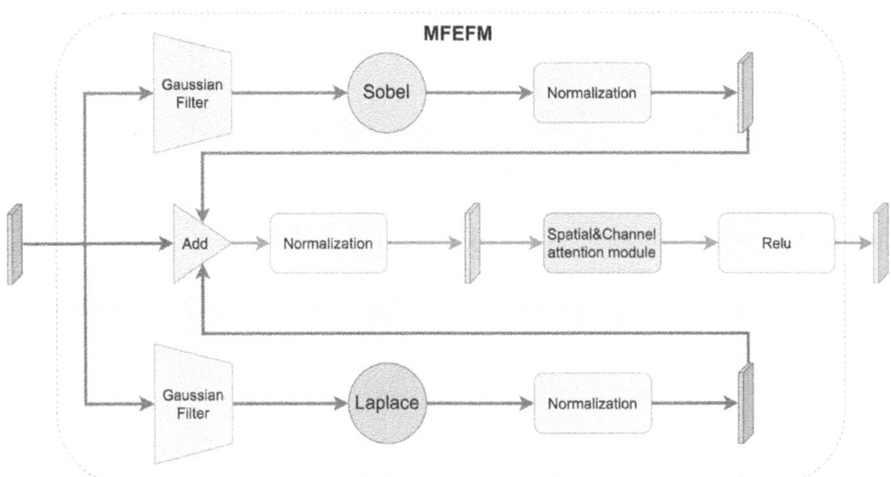


**Fig. 3.** Structure of the MFEFM.

### 2.2 Edge Positioning Monitor Module

Boundary modeling is critical for accurate manipulation localization. The EPMM produces auxiliary edge predictions from intermediate decoder layers and supervises them using a binary cross-entropy-based edge loss, defined in Eq. (2). By explicitly aligning predicted and ground-truth boundaries, the EPMM reinforces spatial precision and mitigates edge blurring that typically occurs in deep decoder paths.

$$Loss_{Edge} = -\frac{1}{N} \cdot \sum_{I=1}^{N} [y_i \cdot log(p_i) + (1 - y_i) \cdot log(1 - p_i)] \tag{2}$$

### 2.3 Multi-Scale Predictive Fusion Module

Decoder layers at different scales capture complementary information—deeper layers provide global semantics, while shallow layers retain local textures (Yang et al. 2024). The MPFM aggregates these multi-level predictions using a weighted strategy, giving prominence to deeper predictions while preserving fine details. The fusion formulation is defined in Eq. (3), and the module layout is illustrated in Fig. 4.

$$M_{final} = 0.6 \cdot M_5 + 0.1 \cdot (M_1 + M_2 + M_3 + M_4) \tag{3}$$

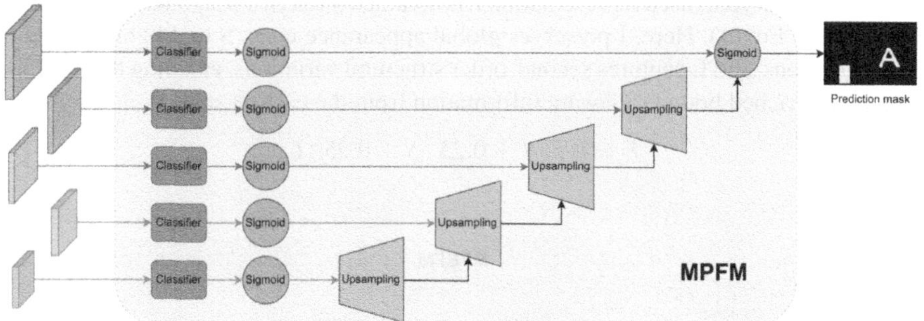


**Fig. 4.** Structure of the MPFM.

### 2.4 Loss Function

The total loss combines binary cross-entropy, Dice loss, and edge loss to jointly optimize region accuracy and boundary sharpness. This balanced objective enables robust manipulation localization across diverse scenes and editing styles. The full loss formulation is defined in Eq. (4).

$$L_{total} = 0.6 \cdot L_{BCE} + 0.2 \cdot L_{Dice} + 0.2 \cdot L_{edge} \tag{4}$$

## 3 Experiment

This section evaluates the effectiveness of MEP-Net through comprehensive comparisons with state-of-the-art approaches and a series of ablation studies. All experiments follow a unified evaluation protocol to ensure fairness. Results demonstrate that the proposed architecture achieves strong manipulation localization performance, particularly in boundary accuracy and fine-grained region consistency.

### 3.1 Performance Comparison

We compare MEP-Net with a wide range of representative manipulation localization methods, covering CNN-based, multi-branch, and recent Transformer- or diffusion-driven models, including ManTra-Net (Wu et al. 2019), RRU-Net (Bi et al. 2019), CR-CNN (Yang et al. 2020), HDU-Net (Wei et al. 2022), FE-PCL (Xu et al. 2023), HFRRU-Net (Sun et al. 2024), DAE-Net (Shi et al. 2024), HDF-Net (Han et al. 2024), DPMSN (Zeng et al. 2024), DiffusionIML (Zhang et al. 2025), and ICFC (Chen et al. 2025).

**Table 1.** Detection performance of three datasets from different networks.

| Methods | CASIAv1.0 (Dong et al. 2013) | | Columbia (Hsu and Chang 2006) | | In-The-Wild (Huh et al. 2018) | |
|---|---|---|---|---|---|---|
| | F1 | AUC | F1 | AUC | F1 | AUC |
| ManTra-Net (Wu et al. 2019) | 0.260 | 0.629 | 0.545 | 0.681 | 0.318 | 0.535 |
| RRU-Net (Bi et al. 2019) | 0.661 | 0.833 | 0.497 | 0.632 | 0.288 | 0.613 |
| CR-CNN (Yang et al. 2020) | 0.688 | 0.802 | 0.571 | 0.711 | 0.396 | 0.638 |
| HDU-Net (Wei et al. 2022) | 0.723 | 0.843 | 0.560 | 0.678 | **0.479** | 0.723 |
| FE-PCL (Xu et al. 2023) | - | - | 0.582 | 0.723 | 0.464 | 0.667 |
| HFRRU-Net (Sun et al. 2024) | 0.717 | 0.929 | 0.549 | 0.786 | 0.371 | **0.833** |
| DAE-Net (Shi et al. 2024) | 0.494 | 0.805 | 0.486 | 0.726 | - | - |
| HDF-Net (Han et al. 2024) | 0.395 | 0.747 | 0.600 | **0.820** | - | - |
| DPMSN (Zeng et al. 2024) | 0.655 | **0.931** | 0.438 | 0.755 | - | - |
| DiffusionIML (Zhang et al. 2025) | - | 0.587 | - | 0.682 | - | - |
| ICFC (Chen et al. 2025) | 0.680 | 0.590 | **0.637** | 0.599 | - | - |
| Ours | **0.743** | 0.871 | 0.611 | 0.723 | 0.404 | 0.699 |

Table 1 reports the quantitative results across all datasets. MEP-Net achieves strong overall detection accuracy and demonstrates clear advantages in boundary localization, especially when compared with lightweight CNN architectures and multi-branch methods. Although some recently proposed Transformer- and diffusion-based approaches introduce more complex global modeling mechanisms, their performance on classical splicing benchmarks does not consistently outperform traditional deep-learning models. MEP-Net offers competitive results while maintaining a lightweight and stable design.

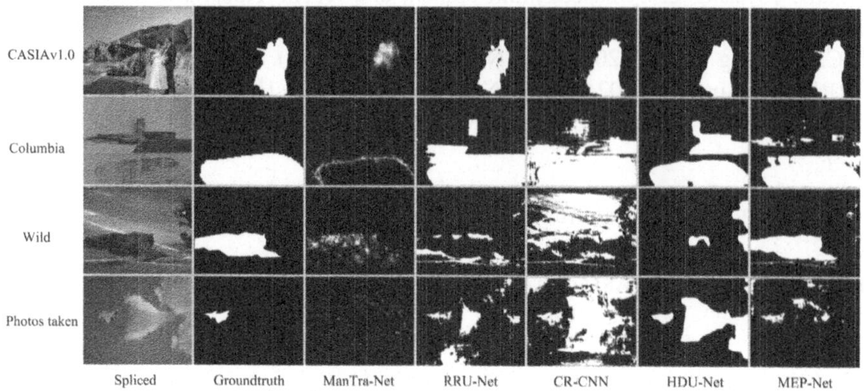


**Fig. 5.** Visualization of prediction results for different datasets.

**Qualitative Comparison.** Figure 5 presents qualitative comparisons across three benchmark datasets and real-world smartphone-tampered images. MEP-Net produces predictions that align more closely with ground-truth manipulated regions, exhibiting sharper boundaries and better spatial consistency. Methods such as ManTra-Net tend to miss large manipulated areas, while RRU-Net and CR-CNN often suffer from false positives or incomplete localization. HDU-Net achieves relatively stable segmentation but still fails to capture fine-grained boundary details. Overall, MEP-Net yields the most accurate and coherent localization maps, highlighting the effectiveness of our proposed architecture.

### 3.2 Ablation Experiment

To evaluate the contribution of each component in MEP-Net, ablation experiments are conducted by removing or altering individual modules.

**Table 2.** Ablation experiments with different modules.

| Model | Methods | | | Metrics | |
|---|---|---|---|---|---|
| | MFEFM | EPMM | MPFM | F1 | AUC |
| None | | | | 0.630 | 0.810 |
| MEPv1 | ✓ | | | 0.654 | 0.825 |
| MEPv2 | | ✓ | | 0.641 | 0.816 |
| MEPv3 | | | ✓ | 0.648 | 0.819 |
| MEPv4 | ✓ | ✓ | | 0.689 | 0.843 |
| MEPv5 | ✓ | | ✓ | 0.715 | 0.854 |
| MEPv6 | | ✓ | ✓ | 0.665 | 0.834 |
| MEP-Net | ✓ | ✓ | ✓ | **0.743** | **0.871** |

**Effectiveness of MFEFM, EPMM, and MPFM.** As shown in Table 2, removing the MFEFM significantly decreases performance, confirming its role in enhancing textural contrast through multi-branch edge cues. Excluding the EPMM leads to a noticeable drop in boundary precision, indicating that auxiliary edge supervision stabilizes structural learning. Disabling the MPFM disrupts the model's ability to effectively integrate multi-scale predictions, causing fragmented or incoherent localization maps.

**Table 3.** Ablation experiments with different operator.

| Class of operators | F1 | AUC |
|---|---|---|
| Sobel | 0.685 | 0.824 |
| Roberts | 0.677 | 0.811 |
| Prewitt | 0.681 | 0.815 |
| 4 Domain Laplacian | 0.692 | 0.827 |
| 8 Domain Laplacian | 0.684 | 0.824 |
| Roberts + 4 Domain Laplacian | 0.716 | 0.833 |
| Prewitt + 4 Domain Laplacian | 0.722 | 0.839 |
| Sobel + 4 Domain Laplacian | **0.743** | **0.871** |
| Roberts + 8 Domain Laplacian | 0.708 | 0.826 |
| Prewitt + 8 Domain Laplacian | 0.717 | 0.832 |
| Sobel + 8 Domain Laplacian | 0.735 | 0.865 |

**Effect of Different Edge Operators.** To address reviewer comments regarding edge extraction, we additionally perform an ablation on edge operators. Results in Table 3 show that combining Sobel and Laplacian operators achieves a strong balance between noise robustness and boundary sensitivity, outperforming single-operator configurations. This experiment has been newly added to further support the design choices in MFEFM, as suggested by reviewers.

# 4 Conclusion

In this paper, we propose a tampered image detection network with strong generalization performance, MEP-Net, that fully utilizes edge feature information through a multi-branch feature enhancement fusion module and an edge localization supervision module. Next, the designed encoder-decoder structure is utilized to achieve the initial localization of the tampered region, followed by the use of a multi-scale prediction fusion module to integrate different levels of semantic information and features, thereby improving localization accuracy. The experimental results show that the proposed MEP-Net demonstrates innovation and superiority compared to existing state-of-the-art tampered image localization networks. Additionally, the effectiveness of the modules in MEP-Net was verified through ablation experiments. In our future work, we aim to optimize MEP-Net's limitations and explore newer network architectures, in order to achieve more generalized tampered image region localization.

## References

Bi, X., Wei, Y., Xiao, B., et al.: RRU-Net: the ringed residual U-Net for image splicing forgery detection. In: Proceedings of the IEEE/CVF Conference on Computer Vision and Pattern Recognition Workshops (2019)

Chen, R., Liu, B., Miao, C., et al.: Training-free in-context forensic chain for image manipulation detection and localization (2025). arXiv Preprint arXiv:2510.10111

Dong, J., Wang, W., Tan, T.: CASIA image tampering detection evaluation database. In: IEEE China Summit and International Conference on Signal and Information Processing (pp. 422–426) (2013)

Han, R., Wang, X., Bai, N., et al.: HDF-Net: capturing homogeny difference features to localize the tampered image. IEEE Trans. Pattern Analy. Mach. Intell. (2024)

Hsu, Y.F., Chang, S.F.: Detecting image splicing using geometry invariants and camera characteristics consistency. In: IEEE International Conference on Multimedia and Expo (pp. 549–552) (2006)

Huh, M., Liu, A., Owens, A., et al.: Fighting fake news: image splice detection via learned self-consistency. In: European Conference on Computer Vision (pp. 101–117) (2018)

Kanopoulos, N., Vasanthavada, N., Baker, R.L.: Design of an image edge detection filter using the Sobel operator. IEEE J. Solid-State Circuits **23**(2), 358–367 (1988)

Shi, C., Wang, C., Zhou, X., et al.: DAE-Net: dual attention mechanism and edge supervision network for image manipulation detection and localization. IEEE Trans. Instrum. Meas. (2024)

Sun, M., Zhu, H.: HFRRU-Net: hybrid features ringed residual U-Net for image splicing forgery detection. In: IEEE Data Driven Control and Learning Systems Conference (pp. 872–877) (2024)

Wang, J., Wu, Z., Chen, J., et al.: ObjectFormer for image manipulation detection and localization. In: Proceedings of the IEEE/CVF Conference on Computer Vision and Pattern Recognition (pp. 2364–2373) (2022)

Wang, X.: Laplacian operator-based edge detectors. IEEE Trans. Pattern Anal. Mach. Intell. **29**(5), 886–890 (2007)

Wei, Y., Ma, J., Wang, Z., et al.: Image splicing forgery detection by combining synthetic adversarial networks and hybrid dense U-net based on multiple spaces. Int. J. Intell. Syst. **37**(11), 8291–8308 (2022)

Wu, Y., AbdAlmageed, W., Natarajan, P.: ManTra-Net: Manipulation tracing network for detection and localization of image forgeries with anomalous features. In: Proceedings of the IEEE/CVF Conference on Computer Vision and Pattern Recognition (pp. 9543–9552) (2019)

Xu, Y., Zheng, J., Fang, A., et al.: Feature enhancement and supervised contrastive learning for image splicing forgery detection. Digital Signal Process. **136**, 104005 (2023)

Yang, C., Li, H., Lin, F., et al.: Constrained R-CNN: a general image manipulation detection model. In: IEEE International Conference on Multimedia and Expo (pp. 1–6) (2020)

Yang, Z., Liu, B., Bi, X., et al.: D-Net: a dual-encoder network for image splicing forgery detection and localization. Pattern Recogn. **155**, 110727 (2024)

Zeng, N., Wu, P., Zhang, Y., et al.: DPMSN: a dual-pathway multiscale network for image forgery detection. IEEE Trans. Ind. Inform. (2024)

Zhang, Z., Chang, M., Li, X.: Training-free image manipulation localization using diffusion models. In: Proceedings of the AAAI Conference on Artificial Intelligence (2025)

# A Survey of HOI Detection: Method Evolution and Multimodal Fusion

Shihao Yang(✉)

Guangzhou University, Guangzhou, China
1545075449@qq.com

**Abstract.** As a key task in visual understanding, human object inter-action (HOI) detection aims to recognize people, objects and their inter-action relationships from static images. In recent years, with the rapid development of deep learning technology, HOI detection has made significant progress in method architecture, data resources and evaluation system. This paper systematically discusses and summarizes the past and the latest work of HOI detection, covering the mainstream evolution paths such as two-stage and single-stage methods, graph model and transformer architecture. In particular, this paper discusses the application of visual language model and large language model in HOI detection, including pre-trained feature transfer, text prompt driven zero sample detection, and cross modal knowledge injection. Finally, we summarize the current research trends and point out that weak supervised learning, open set recognition, complex interaction parsing and interpretable modeling are the key directions for future development. This paper aims to provide HOI researchers with a systematic, comprehensive and cutting-edge research map.

**Keywords:** Human-Object Interaction · Vision-Language Models · Large Language Models · Prompt Learning

## 1 Introduction

Human beings usually interact with surrounding objects in daily life. How to locate people and objects from a picture and recognize interactive actions has become an important direction in the field of computer vision. Compared with object detection and action recognition, the task of human object interaction detection is more difficult. It is not only to identify people and objects or identify interactions, but also to link the two. The model not only needs strong positioning ability, but also needs to understand the overall relationship between human and objects and local details for modeling.

In recent years, the rapid development of visual language model and large language model has brought new vitality to HOI. Transformer architecture can better model the relationship between people and objects, so that they can pay attention to each other and greatly improve the detection accuracy and modeling ability.


H. Yan et al. (Eds.): AIS&P 2025, CCIS 2918, pp. 330–339, 2026.
https://doi.org/10.1007/978-981-95-9884-7_22

# 2 Method of HOI Detection Task

## 2.1 Task Definition

Human object interaction detection aims to detect people and objects in a given image and detect their interaction. Specifically, given an image, the model out-puts a set of HOI instances, $\mathcal{T} = \left\{\left(b_h^{(k)}, b_o^{(k)}, v^{(k)}, s^{(k)}\right)\right\}_{k=1}^{N}$ where $b_h^{(k)}$ and $b_o^{(k)}$ represent the bounding boxes of the person and object, $v^{(k)}$ represents the inter-action type, and $s^{(k)} \in [0, 1]$ is the confidence level. HOI detection is performed at the instance level, and the model needs to recognize the interaction between people and objects while locating their positions in the image. The prediction is only considered correct when the predicted boxes and interaction types of the human body and objects are the same as the true value.

## 2.2 Two Stage HOI Detection Method

The two-stage method of HOI detection divides the task into character detection and interactive recognition, detects people and objects in the picture through the pre trained detector, and sends the character characteristics into the interactive network for interactive recognition. Although the two-stage method naturally has stronger positioning ability and local fine-grained features, it is easy to cause the loss of overall information due to the lack of global features.

Representative two-stage methods include InteractNet (Gkioxari et al. 2018), which couples object detection with action classification for HOI, and iCAN (Gao et al. 2018), which introduces instance-centric attention to better model human–object pairs. Early graph-parsing formulations such as GPNN (Qi et al. 2018) further highlight the structured nature of HOI reasoning. Subsequent work incorporated auxiliary cues such as pose and 3D geometry to improve accuracy, e.g., PMFNet (Wan et al. 2019) and detailed 2D–3D joint representations (Li et al. 2020), while transferable interactiveness knowledge aims to improve generalization across interactions (Li et al. 2019). More recent two-stage approaches refine contextual modeling via action co-occurrence priors (Kim et al. 2020b) or polysemy-aware interaction modeling (Zhong et al. 2020). In addition, DRG (Gao et al. 2020) employs dual graph-convolutional branches to model human–object relations and action semantics, while VSGNet (Ulutan et al. 2020) introduces spatial attention and graph convolution to refine human–object pair representations. Efficient Two-Stage HOI with a unary–pairwise Transformer further improves efficiency and robustness (Zhang et al. 2022a).Overall, despite steady refinements that keep two-stage methods competitive in accuracy, reducing inference cost and mitigating dependence on upstream detector outputs remain key directions for ongoing research.

## 2.3 One Stage HOI Detection Method

To improve inference efficiency, researchers have proposed end-to-end one-stage paradigms. Instead of explicitly constructing human–object proposals, these models take the entire image as input and directly predict HOI triplets from global features. A prevalent design involves integrating an interaction mediator within the detection head to designate the relationship for a specific human-object pair. Early exemplars include IP-Net: Interaction Points Network (Wang et al. 2020), which predicts the midpoint of a human–object pair as an "inter-action point" and matches it to the corresponding person and object detections to complete HOI detection in a single forward pass. UnionDet (Kim et al. 2020a) anchors predictions on the union region of each human–object box pair and directly regresses the triplet. PPDM (Liao et al. 2020): Parallel Point Detection and Matching further improves real-time performance by jointly predicting human and object center points and matching them in parallel. Overall, one-stage methods avoid exhaustive pairwise matching and thus offer clear gains in inference speed.

One-stage models for Human-Object Interaction (HOI) detection often train object detection and interaction classification in a multi-task manner, potentially leading to task interference. To tackle this issue, subsequent work has investigated more robust one-stage architectures. QPIC (Tamura et al. 2021) employs DETR-style query-based set prediction for human–object pairs; HOTR (Kim et al. 2021) introduces a Transformer to achieve end-to-end alignment; and AS-Net uses adaptive set queries with instance-aware attention to better accommodate complex scenes. Taken together, these advances show that, with carefully designed query mechanisms and feature fusion, one-stage frameworks can match or even surpass the accuracy of two-stage methods while remaining efficient. However, compared to two-stage counterparts, one-stage pipelines may struggle to capture fine-grained interaction cues, which can result in prediction bias, especially in cluttered environments or for long-tail interaction categories.

Beyond these baselines, one-stage HOI detection has been advanced by improving interaction region representation and long-tail robustness. DIRV (Fang et al. 2021) votes dense interaction regions to strengthen end-to-end association, while CDN-S analyzes and consolidates the benefits of two-stage and one-stage designs (Zhang et al. 2021a). Recent Transformer variants further explore irregular window attention (Tu et al. 2022) and dispersal attention for indirect-contact interactions (Wang et al. 2024). Predicate-level context modeling and large-scale relational pretraining also improve recognition of complex interactions, as demonstrated by PViC (Zhang et al. 2023) and RLIPv2 (Yuan et al. 2023). Fine-grained anchor design and parallel reasoning are complementary directions, such as FGAHOI (Ma et al. 2023) and PR-Net (Peng et al. 2023). SCG (Spatially Conditioned Graphs) explicitly conditions graph reasoning on spatial configurations of human–object pairs, which improves interaction modeling in complex scenes (Zhang et al. 2021b).

## 2.4 HOI Method Based on Graph Model

Human–object interactions (HOI) are inherently graph-structured. To more faith-fully capture dependencies among humans, objects, and actions, many studies introduce graph models or graph neural networks (GNNs). Early work by Qi et al. formulated HOI detection as graph parsing, treating humans, objects, and actions as nodes and using message passing to infer their relations. Since 2020, graph modeling has remained a key avenue for improvement: VSGNet (Zhang et al. 2021) propagates information over a spatial relation graph via graph convolution; DRG (Gao et al. 2020) builds dual relation graphs—one in the visual feature space linking humans and objects, and another in the semantic space connecting actions and object categories—thereby leveraging both appearance and semantic context; SCG (He et al. 2021) (Spatially Conditioned Graph) incorporates scene-graph priors to constrain and guide the detection process. These graph-based methods emphasize contextual interplay among factors: human-pose nodes, for example, facilitate disambiguation of actions (e.g. hugging, lifting) tied to specific body parts, while object-category nodes impose plausibility constraints (e.g. "driving" presupposes a vehicle). More recent one-stage variants also explore improved training strategies and interaction modeling to further enhance robustness and generalization (Kim et al. 2023). Overall, graph models strengthen the use of structural cues and commonsense priors, improving fine-grained interaction recognition in complex scenes. Their challenges, however, include the need for carefully designed topologies or priors and the added computational overhead.

### 2.5 HOI Method Based on Transformer

In recent years, the success of Transformer architectures in detection has catalyzed new paradigms for HOI detection. Owing to their strength in modeling global dependencies, Transformers are well suited to predicting multiple human–object interactions in complex scenes simultaneously. QPIC (Tamura et al. 2021) and HOI-Transformer (Zou et al. 2021) were among the first to adopt a DETR-style design for HOI, enabling end-to-end query-based bipartite matching. With a set of learnable queries, these methods jointly regress person boxes, object boxes, and interaction categories, thereby unifying candidate generation and matching within a single optimization framework.

A series of refinements followed. AS-Net (Chen et al. 2021) introduces instance-aware interaction queries within a dual-branch Transformer, addressing the limitations of fixed interaction points found in one-stage baselines. DT (Disentangled Transformer) (Zhou et al. 2022) decouples human–object interactions into dedicated Transformer branches, improving detection under sparse interaction set-tings. STIP (Zhang et al. 2022b) incorporates structure-aware mechanisms to enforce spatial consistency when proposing interaction candidates. Hybrid designs have also emerged that combine two-stage insights with Transformer backbones: MSTR (Kim et al. 2022) strengthens feature pyramids via multi-scale Trans-formers, while UPT (Unary–Pairwise Transformer) (Zhang et al. 2022a) jointly predicts entities and pairwise interactions, delivering improved robustness in low-data regimes.

Overall, Transformer-based HOI detectors offer powerful global-relational modeling, enabling parallel detection of diverse interactions within a single forward pass. That said, they tend to be data-hungry and remain sensitive to long-tail categories. Ongoing work is tackling these challenges through stronger pretraining strategies and architectural refinements.

### 2.6 Method Analysis

The one-stage method integrates detection and interaction relationship recognition into the same framework through transformer end-to-end. Global attention can directly learn the association between people and objects, with a short inference link and faster speed. But its shortcomings are also very obvious, it is more sensitive to data volume, and it is not solid in rare interactions and finer grained interaction learning. Many images have varying degrees of occlusion in the interaction area, which makes it difficult to recognize the true interaction if relying solely on global features and being pulled by the background and context.

The two-stage method divides the HOI task into two tasks: human and object detection and interaction recognition. Pre-trained detectors are used to first detect people and objects in the image, then enumerate character pairs, and select character pairs with higher confidence scores for predictive interaction. Although the two-stage method can explicitly inject finer grained knowledge, it is slower due to the need to enumerate and select character pairs. Due to detection and relationship learning segmentation, it is difficult to achieve better relationship modeling, making it very limited in open world scenarios. For a quantitative overview on HICO-Det, Table 1 summarizes representative two-stage and one-stage HOI detectors and their performance.

**Table 1.** Performance Comparisons on HICO-Det.

| Methods | Backbone | Source | Default (mAP↑) | | | Known Object (mAP↑) | | |
|---|---|---|---|---|---|---|---|---|
| | | | Full | Rare | Non-Rare | Full | Rare | Non-Rare |
| *Two-Stage Methods* | | | | | | | | |
| InteractNet [5] | ResNet-50-FPN | CVPR 2018 | 9.94 | 7.16 | 10.77 | - | - | - |
| GPNN [20] | ResNet-101 | ECCV 2018 | 13.11 | 9.34 | 14.23 | - | - | - |
| iCAN [4] | ResNet-50 | BMCV 2018 | 14.84 | 10.45 | 16.15 | 16.26 | 11.33 | 17.73 |
| PMFNet [24] | ResNet-50-FPN | ICCV 2019 | 17.46 | 15.56 | 18.00 | 20.34 | 17.47 | 21.20 |
| VSGNet [23] | ResNet-152 | CVPR2020 | 19.80 | 16.05 | 20.91 | - | - | - |
| ACP [10] | ResNet-152 | ECCV 2020 | 20.59 | 15.92 | 21.98 | - | - | - |
| PD-Net [33] | ResNet-152-FPN | ECCV 2020 | 20.81 | 15.90 | 22.28 | 24.78 | 18.88 | 26.54 |
| SG2HOI [6] | ResNet-50 | ICCV 2021 | 20.93 | 18.24 | 21.78 | 24.83 | 20.52 | 25.32 |
| SCG [29] | ResNet-50-FPN | ICCV 2021 | 21.85 | 18.11 | 22.97 | - | - | |
| UPT [30] | ResNet-50 | ICCV 2021 | 32.62 | 28.62 | 33.81 | 36.08 | 31.41 | 37.47 |
| *One-Stage Methods* | | | | | | | | |
| UnionDet [7] | ResNet-50-FPN | ECCV 2020 | 17.58 | 11.72 | 19.33 | 19.76 | 14.68 | 21.27 |
| IP-Net [25] | Hourglass-104 | CVPR 2020 | 19.56 | 12.79 | 21.58 | 22.05 | 15.77 | 23.92 |
| PPDM-HourglassHOI-Trans [15] | Hourglass-104 | CVPR 2020 | 21.94 | 13.97 | 24.32 | 24.81 | 17.09 | 27.12 |
| HOI-Trans [35] | ResNet-50 | CVPR 2021 | 23.46 | 16.91 | 25.41 | 26.15 | 19.24 | 28.22 |
| HOTR [8] | ResNet-50 | CVPR 2021 | 25.10 | 17.34 | 27.42 | - | - | - |
| AS-Net [1] | ResNet-50 | CVPR 2021 | 28.87 | 24.25 | 30.25 | 31.74 | 27.07 | 33.14 |
| QPIC [21] | ResNet-50 | CVPR 2021 | 29.07 | 21.85 | 31.23 | 31.68 | 24.14 | 33.93 |

*(continued)*

**Table 1.** *(continued)*

| Methods | Backbone | Source | Default (mAP↑) | | | Known Object (mAP↑) | | |
|---|---|---|---|---|---|---|---|---|
| | | | Full | Rare | Non-Rare | Full | Rare | Non-Rare |
| FGAHOI [17] | Swin-Tiny | TPAMI 2023 | 29.94 | 22.24 | 32.24 | 32.48 | 24.16 | 34.97 |
| MSTR [9] | ResNet-50 | CVPR 2022 | 31.17 | 25.31 | 33.92 | 34.02 | 28.83 | 35.57 |
| PR-Net [19] | ResNet-50 | arXiv 2023 | 31.17 | 25.66 | 32.82 | - | - | - |
| CDN-S [28] | ResNet-50 | NIPS 2021 | 31.44 | 27.39 | 32.64 | 34.09 | 29.63 | 35.42 |
| Iwin-B [22] | ResNet-50-FPN | ECCV 2022 | 32.03 | 27.62 | 34.14 | 35.17 | 28.79 | 35.91 |
| RCL [11] | ResNet-50 | CVPR 2023 | 32.87 | 28.67 | 34.12 | 35.52 | 30.88 | 36.45 |
| GEN-VLKT [16] | ResNet-50 | CVPR 2023 | 33.75 | 29.25 | 35.10 | 36.78 | 32.75 | 37.99 |
| HOICLIP [18] | ResNet-50 | CVPR 2023 | 34.59 | 31.12 | 35.74 | 37.61 | 34.47 | 38.54 |
| TED-Net [26] | ResNet-50 | TCSVT 2024 | 34.00 | 29.88 | 35.24 | 37.13 | 33.63 | 38.18 |
| PViC [31] | ResNet-50 | ICCV 2023 | 34.69 | 32.14 | 35.45 | 38.14 | 35.38 | 38.97 |
| PViC [31] | Swin-L | ICCV 2023 | 44.32 | 44.61 | 44.24 | 47.81 | 48.38 | 47.64 |
| RLIPv2 [27] | Swin-L | ICCV 2023 | 43.23 | 45.64 | 45.09 | - | - | - |

# 3 VLM Empowers HOI Detection: Open Vocabulary and Cross Modal Alignment

## 3.1 Enhancing Representations with Pre-trained Features

With the rise of visual–language pretraining (e.g., CLIP), image–text knowledge has been systematically introduced into HOI detection, opening new avenues. Visual–language models (VLMs) offer strong cross-modal alignment and thus serve as powerful semantic priors, helping recognize rare and even unseen interactions.

Recently, some work has attempted to inject the cross modal semantics of visual language pre training models (CLIP) into HOI detection frameworks to enhance the expressive power of interactive features. Specifically, the one-stage method GEN-VLKT queries the features of the entire image by designing queries for people and objects, and then fuses the features of people and objects to enhance the situational understanding of HOI. On the text side, by converting HOI labels into GEN-VLKT (Liao et al., 2022) and encoding them into text embed-dings using CLIP based text encoders, it can be easily extended to every new HOI. At the same time, a knowledge distillation method is designed to align its own visual features with the CLIP representation space through feature distillation. The two-stage method HOICLIP (Ning et al. 2023) first detects people and objects in the image through DETR, and then identifies the relationships between people. By designing character queries to retrieve useful features from the feature map output by CLIP, and then performing feature fusion to obtain interactive features. Finally, the prior knowledge of CLIP text encoder is introduced to generate the HOI classifier, providing training free enhancement for HOI classification. The above method achieves better representation learning and strong generalization ability by effectively transferring prior knowledge of VLM.

One direct route is to fuse features extracted by pretrained VLMs as in-puts or priors for HOI. For example, GEN-VLKT (Liao et al. 2022) leverages CLIP-derived human and object embeddings to shrink the search space for human–object association and to strengthen interaction reasoning; while maintaining end-to-end training, integrating pretrained features markedly improves the accuracy of rare-interaction recognition. A complementary route is knowledge distillation. HOICLIP (Ning et al. 2023) transfer knowledge from CLIP to an HOI detector by scoring candidate human-object regions with CLIP and using these scores as soft targets for training. By leveraging pretrained VLM features and knowledge distillation, researchers can significantly enhance the performance of HOI detection systems.

### 3.2 Prompt Learning

The visual language model codes the HOI description embedings as an interaction classifier, and links the interaction in the picture with the text description through the difference comparison training. At first, the template is usually used as the description of each interaction (a photo of human verb a object), and then these text descriptions are embedded and aligned with HOI features to form a decision boundary for each HOI. These templates are very flexible. They can be nested into the never seen HOI, improve the scalability of HOI tasks, and accommodate the never seen HOI types in the open world.

However, compared with other detection tasks, HOI detection tasks are more complex. We cannot only consider people or objects. We must establish a relationship between the two. Therefore, the template alone is not enough to achieve the differentiated description of all HOIs, which may lead to the problem of over-lapping or fuzzy decision boundaries.

In the past, tasks often used templates such as a photo of human verb object as text descriptions for HOI. However, unlike object detection and other tasks, HOI tasks not only need to locate pairs of characters, but also need to judge their interactions. Such characters require more fine-grained information. By using LLM as semantic guidance for HOI interaction descriptions, the model can go beyond the original simple templates and obtain more fine-grained semantic clues. These descriptions provided by LLM (character posture, object attributes, and interaction context, etc.) are used as prompts, and combined with a learnable prompt token mechanism, the model can extract specific semantics for different interaction category groups and integrate scene context information. INP-CC (Lei et al. 2025) injects fine-grained information by having LLM generate more de-tailed textual descriptions for each HOI, and then groups all HOIs, grouping similar HOIs into one group. During the training process, other samples in the same group are treated as negative samples to further widen the decision boundary of similar HOIs. At the same time, a set of visual tokens was designed to learn scene patterns by matching them with global features extracted by CLIP. Overall, Prompt learning has improved the model's ability to distinguish fine-grained interactions and enhanced its generalization performance under open vocabulary settings across multiple scenarios and datasets. This fully demonstrates the potential application of learnable prompts and language guidance in HOI detection.

In a word, prompt learning has changed from a static single template to an adaptive detailed description, clarifying the decision boundary and improving the performance of this task in zero sample and open scenarios.

## 4 Conclusion

HOI detection is a hot topic in the field of vision. This paper introduces some classical methods, development process, advantages and disadvantages of HOI in two stages and one stage. At the same time, it introduces new technologies such as visual language model and large language model, which bring new vitality to the field of HOI. At the same time, it introduces the new challenges faced by HOI at present.

## References

Chen, M., Liao, Y., Liu, S., Chen, Z., Wang, F., Qian, C.: Reformulating HOI detection as adaptive set prediction. In: Proceedings of the IEEE/CVF Conference on Computer Vision and Pattern Recognition (CVPR) (pp. 9004–9013) (2021)

Fang, H.S., Xie, Y., Shao, D., Lu, C.: DIRV: dense interaction region voting for end-to-end human-object interaction detection. Proc. AAAI Conf. Artif. Intell. (AAAI) **35**, 1291–1299 (2021)

Gao, C., Xu, J., Zou, Y., Huang, J.B.: DRG: dual relation graph for human-object interaction detection. In: Computer Vision–ECCV 2020 (pp. 696–712). Springer (2020)

Gao, C., Zou, Y., Huang, J.B.: iCAN: instance-centric attention network for human-object interaction detection (2018). arXiv Preprint arXiv:1808.10437

Gkioxari, G., Girshick, R., Dollár, P., He, K.: Detecting and recognizing human-object interactions. In: Proceedings of the IEEE/CVF Conference on Computer Vision and Pattern Recognition (CVPR) (pp. 8359–8367) (2018)

He, T., Gao, L., Song, J., Li, Y.F.: Exploiting scene graphs for human-object interaction detection. In: Proceedings of the IEEE/CVF International Conference on Computer Vision (ICCV) (pp. 15984–15993) (2021)

Kim, B., Choi, T., Kang, J., Kim, H.J.: Uniondet: union-level detector towards real-time human-object interaction detection. In: Computer Vision–ECCV 2020 (pp. 498–514). Springer (2020)

Kim, B., Lee, J., Kang, J., Kim, E.S., Kim, H.J.: HOTR: end-to-end human-object interaction detection with transformers. In: Proceedings of the IEEE/CVF Conference on Computer Vision and Pattern Recognition (CVPR) (pp. 74–83) (2021)

Kim, B., Mun, J., On, K.W., Shin, M., Lee, J., Kim, E.S.: MSTR: multi-scale transformer for end-to-end human-object interaction detection. In: Proceedings of the International Conference on Learning Representations (ICLR) (2022)

Kim, D.J., Sun, X., Choi, J., Lin, S., Kweon, I.S.: Detecting human-object interactions with action co-occurrence priors. In: Computer Vision–ECCV 2020 (pp. 718–736). Springer (2020)

Kim, S., Jung, D., Cho, M.: Relational context learning for human-object interaction detection. In: Proceedings of the IEEE/CVF Conference on Computer Vision and Pattern Recognition (CVPR) (pp. 2925–2934) (2023)

Lei, T., Yin, S., Chen, Q., Peng, Y., Liu, Y.: Open-vocabulary HOI detection with interaction-aware prompt and concept calibration (INP-CC) (2025). arXiv Preprint arXiv:2508.03207

Li, Y.L., et al.: Detailed 2d-3d joint representation for human-object interaction. In: Proceedings of the IEEE/CVF Conference on Computer Vision and Pattern Recognition (CVPR) (pp. 10166–10175) (2020)

Li, Y.L., et al.: Transferable interactiveness knowledge for human-object interaction detection. In: Proceedings of the IEEE/CVF Conference on Computer Vision and Pattern Recognition (CVPR) (pp. 3585–3594) (2019)

Liao, Y., Liu, S., Wang, F., Chen, Y., Qian, C., Feng, J.: PPDM: parallel point detection and matching for real-time human-object interaction detection. In: Proceedings of the IEEE/CVF Conference on Computer Vision and Pattern Recognition (CVPR) (pp. 482–490) (2020)

Liao, Y., Zhang, A., Lu, M., Wang, Y., Li, X., Liu, S.: Gen-VLKT: simplify association and enhance interaction understanding for HOI detection. In: Proceedings of the IEEE/CVF Conference on Computer Vision and Pattern Recognition (CVPR) (pp. 20123–20132) (2022)

Ma, S., Wang, Y., Wang, S., Wei, Y.: FGAHOI: fine-grained anchors for human-object interaction detection (2023). arXiv Preprint arXiv:2301.04019

Ning, S., Qiu, L., Liu, Y., He, X.: HOICLIP: efficient knowledge transfer for HOI detection with vision-language models. In: Proceedings of the IEEE/CVF Conference on Computer Vision and Pattern Recognition (CVPR) (pp. 23507–23517) (2023)

Peng, H., et al.: Parallel reasoning network for human-object interaction detection (2023). arXiv Preprint arXiv:2301.03510

Qi, S., Wang, W., Jia, B., Shen, J., Zhu, S.C.: Learning human-object interactions by graph parsing neural networks. In: Computer Vision–ECCV 2018 (pp. 401–417) (2018)

Tamura, M., Ohashi, H., Yoshinaga, T.: QPIC: query-based pairwise human-object interaction detection with image-wide contextual information. In: Proceedings of the IEEE/CVF Conference on Computer Vision and Pattern Recognition (CVPR) (pp. 10410–10419) (2021)

Tu, D., Min, X., Duan, H., Guo, G., Zhai, G., Shen, W.: Iwin: human-object interaction detection via transformer with irregular windows. In: Computer Vision–ECCV 2022 (pp. 87–103). Springer (2022)

Ulutan, O., Iftekhar, A., Manjunath, B.S.: Vsgnet: spatial attention network for detecting human object interactions using graph convolutions. In: Proceedings of the IEEE/CVF Conference on Computer Vision and Pattern Recognition (CVPR) (pp. 13617–13626) (2020)

Wan, B., Zhou, D., Liu, Y., Li, R., He, X.: Pose-aware multi-level feature network for human object interaction detection. In: Proceedings of the IEEE/CVF International Conference on Computer Vision (ICCV) (pp. 9469–9478) (2019)

Wang, T., Yang, T., Danelljan, M., Khan, F.S., Zhang, X., Sun, J.: Learning human-object interaction detection using interaction points. In: Proceedings of the IEEE/CVF Conference on Computer Vision and Pattern Recognition (CVPR) (pp. 4116–4125) (2020)

Wang, Y., Liu, Q., Lei, Y.: Ted-net: Dispersal attention for perceiving interaction region in indirectly-contact HOI detection. IEEE Transactions on Circuits and Systems for Video Technology (2024)

Yuan, H., et al.: RLIPv2: Fast scaling of relational language-image pre-training. In: Proceedings of the IEEE/CVF International Conference on Computer Vision (ICCV) (pp. 21649–21661) (2023)

Zhang, A., et al.: Mining the benefits of two-stage and one-stage HOI detection. Adv. Neural Inf. Process. Syst. (NeurIPS) **34**, 17209–17220 (2021)

Zhang, F. Z., Campbell, D., Gould, S.: Spatially conditioned graphs for detecting human-object interactions. In: Proceedings of the IEEE/CVF International Conference on Computer Vision (ICCV) (pp. 13319–13327) (2021)

Zhang, F.Z., Campbell, D., Gould, S.: Efficient two-stage detection of human-object interactions with a novel unary-pairwise transformer. In: Proceedings of the IEEE/CVF Conference on Computer Vision and Pattern Recognition (CVPR) (pp. 20104–20112) (2022)

Zhang, F.Z., Yuan, Y., Campbell, D., Zhong, Z., Gould, S.: Exploring predicate visual context in detecting of human-object interactions. In: Proceedings of the IEEE/CVF International Conference on Computer Vision (ICCV) (pp. 10411–10421) (2023)

Zhang, Y., Pan, Y., Yao, T., Huang, R., Mei, T., Chen, C.W.: Exploring structure-aware transformer over interaction proposals for human-object interaction detection. In: Proceedings of the IEEE/CVF Conference on Computer Vision and Pattern Recognition (CVPR) (pp. 19548–19557) (2022)

Zhong, X., Ding, C., Qu, X., Tao, D.: Polysemy deciphering network for human-object interaction detection. In: Computer Vision–ECCV 2020 (pp. 69–85). Springer (2020)

Zhou, D., et al.: Human-object interaction detection via disentangled transformer. In: Proceedings of the IEEE/CVF Conference on Computer Vision and Pattern Recognition (CVPR) (pp. 19568–19577) (2022)

Zou, C., et al.: End-to-end human object interaction detection with HOI transformer. In: Proceedings of the IEEE/CVF Conference on Computer Vision and Pattern Recognition (CVPR) (pp. 11825–11834) (2021)

# Research on Key Technologies and Effect Verification of Anti-Seepage Construction for Water Conservancy and Hydropower Dams Based on BIM Technology

Chaoying Zou, Jing Li(✉), Hui Tan, and Yufeng Guo

Kunming Metallurgy College, Kunming, Yunnan, China
243666835@qq.com

**Abstract.** The construction quality of the anti-seepage system of dams in water conservancy and hydropower projects directly affects project safety and long-term operation efficiency. Traditional anti-seepage construction methods have inherent limitations such as information transmission disconnection and insufficient process control precision. Through the systematic integration of BIM and AI technologies, this study proposes an innovative dam anti-seepage construction methodology. A multi-dimensional BIM parametric model including geological conditions, material properties and structural forms is constructed to realize precise design of the anti-seepage system and dynamic simulation of the construction process. An intelligent seepage field analysis model based on machine learning algorithms is developed to evaluate the reliability of anti-seepage structures under multiple working conditions. A full-life-cycle quality control system covering design, construction and operation and maintenance is established. Engineering practice shows that this method significantly improves the scheme optimization efficiency and process control precision of anti-seepage construction, effectively enhances the ability of seepage risk identification and prevention, and provides reliable technical support for dam anti-seepage construction. It has important theoretical and practical value for promoting the digital transformation of quality management models in water conservancy and hydropower engineering construction.

**Keywords:** BIM · Water Conservancy and Hydropower Engineering · AI · Dam Anti-seepage · Seepage Analysis · Quality Control

## 1 Introduction

The quality of the anti-seepage system of dams in water conservancy and hydropower projects directly affects project safety and long-term operation efficiency. Traditional anti-seepage construction methods have inherent limitations such as information transmission disconnection and insufficient process control precision. With the rapid development of BIM and AI technologies, new technical paths have been provided to solve these problems. By constructing a parametric model containing geometric and non-geometric


H. Yan et al. (Eds.): AIS&P 2025, CCIS 2918, pp. 340–348, 2026.
https://doi.org/10.1007/978-981-95-9884-7_23

information, BIM technology lays a foundation for the whole-process refined management of the anti-seepage system. Current research has begun to explore the integrated application of BIM and AI, but there are still obvious deficiencies in the dynamic prediction and optimization of the anti-seepage construction process. Based on the actual needs of anti-seepage construction in water conservancy and hydropower projects, this study establishes a systematic dam anti-seepage construction technology system by building a multi-dimensional BIM model integrating geological information, material properties and construction process, combined with machine learning algorithms. It aims to improve the quality and efficiency of anti-seepage construction and provide theoretical support and practical guidance for the technical development of the industry.

## 2 Project Overview

The water conservancy project relied on in this study is a comprehensive water conservancy infrastructure with dual functions of water supply and irrigation. The engineering system consists of three parts: hub buildings, water transmission buildings and irrigation area supporting projects, which together form a complete water resource allocation system. The dam in the hub area is a fine aggregate concrete gravity dam, designed according to the standard of grade III buildings. The design parameters of the dam structure are as follows: the maximum dam height is 50 m, the crest elevation is 350.50 m, and the upstream and downstream dam slopes are designed with a stable slope of 1:1.3. The design of the dam structure fully considers the foundation bearing capacity and hydraulic load conditions to ensure the overall stability of the structure (Zheng 2024).

In terms of the design of the anti-seepage system, through detailed geological investigation and analysis, it is confirmed that the geological structure conditions in the project area are relatively simple. The survey results show that there is no dominant seepage channel connecting the reservoir with the outside in the dam site area, the development degree of rock mass structural fractures is low and most of them are in a closed state, and the risk of concentrated seepage under natural conditions is small. Based on the characteristics of the engineering geology, the design of the anti-seepage project focuses on improving the anti-seepage performance of the dam itself. By optimizing the construction technology and quality control measures, a complete dam anti-seepage system is constructed to effectively prevent potential seepage hazards during operation and ensure the long-term safe operation of the project (Li 2023).

## 3 Intelligent Construction Technology System for Dam Anti-Seepage Panels Based on BIM Technology

### 3.1 Construction Preparation

**Intelligent Material Preparation.** In the construction of anti-seepage panels for water conservancy and hydropower dams, material quality is directly related to the long-term reliability of the anti-seepage system. A material information model is constructed based on BIM technology to establish a full-process quality traceability system from raw material procurement, transportation to use. Sanqing brand Portland cement is selected, coarse aggregate adopts 15–24 mm river pebble crushed stone, and fine aggregate selects natural river sand. The mud content is real-time controlled below 2% through an AI monitoring system to ensure that the material performance meets the anti-seepage requirements (Ya 2023).

The BIM material management platform and AI algorithm are used to optimize the material transportation scheduling, and an intelligent dispatch model for transport vehicles is established:

$$`N = (tc1 + tc2 + 2Lc \cdot Ac/Vc) \cdot M/Qf` \quad (1)$$

where: tc1 and tc2 are the loading and unloading times (min) respectively; Lc is the transportation distance (km); Ac is the road blockage coefficient (taken as 1.2); Vc is the average speed (km/h); M is the vehicle load (t); Qf is the paving capacity (t/h). Through real-time data analysis, the model dynamically adjusts the transportation plan to ensure construction continuity.

**Digital on-site Preparation.** A combination of BIM technology and 3D laser scanning is adopted to realize the digital reconstruction of the construction site. The base surface is cleaned with high-pressure water guns, and drone aerial photography and AI image recognition technology are used to ensure no dead corners in the cleaning work. Precision surveying equipment such as total stations is used for positioning and setting out, and the survey data is uploaded to the BIM collaboration platform in real time to generate a digital construction benchmark.

### 3.2 Intelligent Fabrication and Installation of Formwork

The formwork project of the anti-seepage panel for water conservancy and hydropower dams based on BIM technology adopts a standardized design and intelligent installation scheme. Through the BIM component library management system, large-scale 4 m × 2 m steel formwork is customized, and the M14 high-strength bolt connection system is uniformly adopted to ensure the integrity and reliability of the formwork system. During the formwork installation stage, a modular hoisting strategy based on the BIM model is implemented. BIM technology is used to optimize the 3D layout of the formwork, and the hoisting process is simulated by AI algorithms to determine the optimal hoisting path and positioning scheme (Tang 2022). During installation, an adjustable support system is set inside the formwork, and the installation process of "fixing the bottom edge first, then adjusting the angle" is adopted. Built-in sensors are used to monitor the formwork displacement and deformation in real time to ensure that the installation precision is controlled within the range allowed by the specifications. For special parts such as sliding formwork and chutes, special designs are carried out based on the BIM model. The stability of the formwork system is verified by finite element analysis, and a widened formwork design scheme is adopted to ensure that the width of the sliding formwork is 1.0m larger than that of the panel. AI prediction algorithms are used to analyze the stress state of the formwork during concrete pouring, and the support parameters are adjusted in real time to effectively prevent quality risks such as sliding formwork inclination.

### 3.3 Intelligent Fabrication and Installation of Reinforcement

An intelligent whole-process management system for reinforcement engineering is constructed based on BIM technology. Through the in-depth integration of BIM models and AI algorithms, the collaborative optimization of detailed reinforcement design, numerical control processing and precise installation is realized. Digital processing equipment is adopted to ensure that the reinforcement processing precision is controlled within the range allowed by the specifications, as shown in Table 1.

**Table 1.** Allowable Deviation Range of Reinforcement

| Item | Allowable Deviation Range (Unit: mm) |
| --- | --- |
| Net Dimension | ±10 |
| Stirrup Length | ±5 |
| Bent-up Position | ±20 |
| Turning Angle | ±3 |

### 3.4 Intelligent Concrete Construction

## 4 Intelligent Mix Design and Temperature Control.

Based on the BIM material database and AI algorithm, an optimization model for the mix ratio of dam anti-seepage concrete is established. With the goals of maximizing anti-seepage performance and controlling hydration heat, the optimal mix ratio is determined by adopting a multi-objective optimization algorithm: the cement dosage is controlled at 280–320 kg/m$^3$, 25%–30% fly ash and mineral powder composite admixture are added, and the water-binder ratio is not greater than 0.45. The outlet temperature is precisely controlled through a temperature prediction model:

$$T\{max\} = frac\{WwTw + WcTc + WaTa\}\{Ww + Wc + Wa\} \quad (2)$$

where each parameter represents the temperature and weight of mixing water, cement and aggregate respectively, realizing the precise control of the outlet temperature not exceeding 25 °C, as shown in Table 2.

**Table 2.** Optimal Concrete Mix Ratio

| Name | Dosage (Unit: kg ·m$^{-3}$) | Mass Ratio |
| --- | --- | --- |
| Fine Aggregate | 70 | 1.85 |
| Coarse Aggregate | 320 | 2.10 |
| Cement | 580 | 1.15 |
| Water | 4.0 | 0.50 |
| Fly Ash | 0.80 | 0.25 |
| Additive | 660 | 0.02 |

**Intelligent Quality Monitoring System.** A real-time concrete quality evaluation system based on machine learning is established. The working performance of fresh concrete is monitored through a sensor network to ensure that the slump is controlled at 120 ± 20 mm and the air content is 4%–6%. Digital image processing technology is used to analyze the uniformity of aggregate distribution, realizing early warning and automatic adjustment of quality abnormalities, as shown in Table 3.

**Table 3.** Standard Technical Indicators of Concrete

| Name of Technical Indicator | Unit | Standard Value |
|---|---|---|
| Mortar Density | $kg \cdot m^{-3}$ | ≤30 |
| Air Content | % | ±10 |
| Slump | cm | ≤4( ±1)<br>4–10( ±2)<br>>10( ±3) |

**Intelligent Transportation and Pouring Control.** BIM construction simulation is used to optimize the concrete transportation path, and special concrete mixer trucks are equipped with GPS positioning and temperature monitoring devices. An IoT-based temperature and humidity monitoring system during transportation is established to ensure that the formwork entry temperature meets the specification requirements. An intelligent distribution system is adopted, and the layered pouring thickness and interval time are controlled under the guidance of the BIM model. A pouring continuity early warning mechanism is established, and emergency measures are automatically activated when the interval time is close to the critical value, as shown in Fig. 1

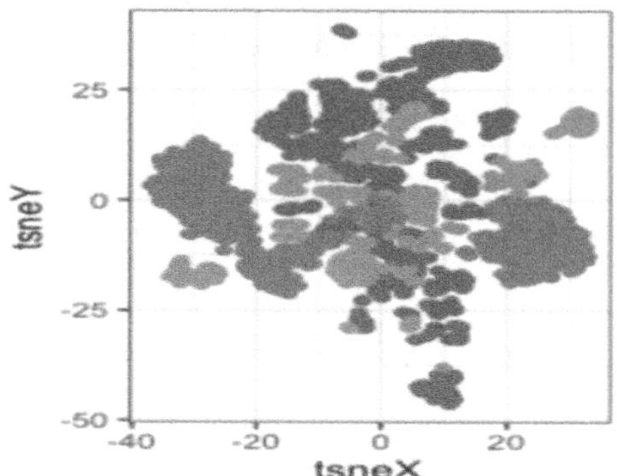


**Fig. 1.** Early Warning Mechanism of BIM Model

**Digital Foundation Treatment.** The measured data of the foundation surface is obtained based on 3D laser scanning technology, and the treatment effect is analyzed by comparing with the BIM model. An intelligent treatment process combining high-pressure water jet and mechanical chiseling is adopted to standardize the treatment of construction joints, ensuring that the bonding quality meets the anti-seepage requirements.

### 4.1 Intelligent Maintenance System

An intelligent maintenance system for anti-seepage panels is constructed based on BIM technology, and the changes in the concrete temperature field and humidity field are real-time monitored through an IoT sensor network. An adaptive spray control system is adopted to automatically adjust the maintenance parameters according to the ambient temperature and humidity and the internal temperature gradient of the concrete, ensuring that the surface of the panel remains moist continuously.

### 4.2 Intelligent Crack Prevention and Treatment

For crack control of mass concrete anti-seepage panels, a crack early warning and treatment system based on BIM and AI is established. A crack risk assessment model is established through stress monitoring data analysis:

$$\sigma(self) = [E * \alpha * (T_m - T(y))]/(1 - \mu) * \phi \tag{3}$$

where: E is the elastic modulus, $\mu$ is Poisson's ratio, $\alpha$ is the expansion coefficient, $\phi$ is the temperature drop gradient, Tm is the average temperature drop value, and T(y) is the temperature distribution along the plate thickness direction.

The BIM model is used to integrate temperature field and stress field data, and potential crack areas are predicted through machine learning algorithms. For existing cracks, intelligent detection technology based on digital image recognition is used to accurately evaluate crack characteristics, and chemical grouting robots are used for precise grouting treatment. The whole process of crack treatment is recorded through the BIM platform, and a complete crack prevention and control database is established to provide reference for subsequent projects (Wei 2022).

## 5 Challenges Faced by BIM Technology in Anti-Seepage Construction of Water Conservancy and Hydropower Dams

### 5.1 Dilemma of Multi-Source Heterogeneous Data Integration

Anti-seepage projects of water conservancy and hydropower dams involve multiple disciplines such as geology, structure and materials, generating massive heterogeneous data. BIM technology faces a systematic problem of inconsistent data standards in realizing full-process information integration. Differences in data formats of various professional platforms lead to difficulties in effective docking between design models and construction monitoring data, forming significant information barriers. In addition, there is a lack of an effective dynamic interaction mechanism between the seepage monitoring data collected in real time on site and the BIM model, which restricts the decision-making timeliness of anti-seepage construction (Huang 2021).

### 5.2 Bottlenecks in Technology Implementation and Talent Reserve

The in-depth integration of BIM and AI technologies places higher requirements on professional and technical talents. The water conservancy construction field generally faces a shortage of compound talents who understand both BIM technology and anti-seepage construction. Especially in dam construction projects in remote areas, the insufficient configuration of hardware facilities restricts the real-time processing capacity of large-scale BIM models. At the same time, the adaptability of the existing BIM standard system in the field of special anti-seepage construction is insufficient, resulting in a significant gap between technological innovation and practical application.

### 5.3 Disconnection Between Model Precision and Engineering Practice

The special structure of the dam anti-seepage structure places strict requirements on the precision of the BIM model. Existing BIM platforms have technical limitations in simulating key details such as the micro-structure of anti-seepage panels and joint treatment. The deviation between the model and actual construction directly affects the prediction accuracy of the anti-seepage effect. Under complex geological conditions, the impact of foundation deformation on the anti-seepage system is difficult to accurately simulate through conventional BIM models, and it is urgent to develop more targeted analysis modules.

### 5.4 Lack of Standardized System and Collaborative Management

The BIM implementation standard system in the water conservancy industry is not yet perfect, resulting in low collaborative efficiency among various participants in anti-seepage construction. The BIM model established in the design stage has low utilization rate in the construction stage, and a large amount of anti-seepage construction data still relies on traditional management methods (Huang 2021). At the same time, the data interface standards between BIM technology and IoT monitoring equipment are not unified, which increases the implementation difficulty of the intelligent anti-seepage construction system and restricts the in-depth application of BIM technology in quality control.

## 6 Strategies to Improve the Application Effect of BIM Technology in Anti-Seepage Construction of Water Conservancy and Hydropower Dams

### 6.1 Construct a Professional Talent Training and Technical Support System

To address the shortage of BIM technology application talents in the water conservancy and hydropower industry, a systematic training and certification mechanism is established. Through the school-enterprise cooperation model, compound talents who master both BIM technology and anti-seepage construction technology are cultivated. The following aspects should be focused on: first, formulate standard BIM technology application tutorials for dam anti-seepage construction, covering core contents such as model creation, data management and collaborative work; second, establish a remote technical support platform to provide real-time technical guidance for projects in remote areas; finally, organize regular technical exchange meetings to promote the sharing and dissemination of advanced experience. Research shows that systematic training can improve the application efficiency of BIM technology by about 40% (Nie 2020).

### 6.2 Establish a Unified Data Collaboration and Management Platform

A special BIM collaboration platform for water conservancy and hydropower projects is constructed based on cloud computing technology to realize data sharing and real-time collaboration among multiple participants such as design, construction and supervision. The platform should have the following characteristics: adopt unified data standards and interface specifications to support the integration of multi-source heterogeneous data; establish a hierarchical authorization mechanism to ensure data security; develop mobile applications to meet the needs of on-site construction. Through this platform, information such as design changes, construction progress and quality acceptance can be synchronized in real time, significantly improving work efficiency. Practice shows that the application of the collaboration platform can increase the information transmission efficiency by about 60% (Wei 2024).

### 6.3 Improve the Intelligent Construction Monitoring and Feedback Mechanism

The in-depth integration of BIM technology with intelligent monitoring equipment such as IoT and sensors is carried out to establish a comprehensive construction quality monitoring system. Specific measures include: arranging sensors at key parts of anti-seepage panel construction to collect data such as temperature and stress in real time; developing AI-based data analysis algorithms to realize automatic identification and early warning of quality abnormalities; establishing a closed-loop feedback processing mechanism to ensure timely resolution of problems. Through digital twin technology, the dynamic interaction between physical construction and virtual models is realized, providing a scientific basis for construction decision-making. Monitoring data shows that this system can improve the discovery time of construction quality problems by about 50% (Wei 2024).

### 6.4 Promote the Construction of Standardized System and Technological Innovation

Strengthen the construction of the BIM technology application standard system in the water conservancy industry, focusing on improving special standards for anti-seepage construction. At the same time, encourage enterprises to increase investment in technological innovation and develop special BIM software modules suitable for dam anti-seepage construction. Through policy guidance and market-driven forces, promote the in-depth integration of BIM technology with emerging information technologies, and promote the transformation of water conservancy and hydropower engineering construction towards digitalization and intelligence.

## 7 Conclusions

This study systematically constructs a BIM-based anti-seepage construction technology system for water conservancy and hydropower dams. Through the in-depth integration of BIM and artificial intelligence technologies, the digital and intelligent transformation of the anti-seepage construction process is realized. Research shows that the application of BIM technology significantly improves the scheme optimization efficiency and process control precision of anti-seepage construction, and provides an innovative solution for the quality control of dam anti-seepage projects. By establishing a multi-dimensional BIM information model, the full-element integration of geological data, structural design and construction process is realized, effectively solving the problem of information islands in traditional construction. The developed intelligent decision-making system relies on machine learning algorithms and has the ability of intelligent evaluation and risk early warning of anti-seepage construction quality, providing a scientific basis for engineering decision-making.

Engineering practice verification shows that the research results have significant advantages in improving construction quality and optimizing project progress, and provide important reference for promoting the technological progress of the water conservancy and hydropower industry. Future research should focus on solving the adaptability problem of BIM models under complex geological conditions, and further explore the innovative application of emerging technologies such as digital twins in anti-seepage projects. The theoretical methods and practical experience of this study lay a solid foundation for improving the construction technology level of dam anti-seepage projects and have important engineering application value for promoting the digital transformation of the industry.

## References

Huang, J.: Brief analysis of the construction technology of anti-seepage panels in water conservancy dam engineering. Hydraul. Sci. Cold Reg. Eng. **4**(6), 109–111 (2021)

Li, X.Y.: Discussion on the construction technology of anti-seepage diaphragm walls for water conservancy dams. Digital-Mobile Life **23**(5), 184–186 (2023)

Nie, S.Y.: Brief analysis of the construction technology of anti-seepage panels in water conservancy dam engineering. Pearl River Water Transport **17**, 70–71 (2020)

Tang, Z.D.: Research on key construction technologies of anti-seepage panels in hydraulic dam construction. China New Technol. Prod. **24**, 110–111 (2022)

Wei, C.M.: Exploration on the key practical points of dam anti-seepage wall construction technology in reservoir engineering-Taking Yazha Reservoir Project in Ping'an County, Qinghai Province as an example. Sci. Technol. Inf. **22**(17), 181–183 (2024)

Wei, Q.: Research on anti-seepage panel construction technology of water conservancy dam engineering-Taking Yantang Reservoir Project as an example. Eng. Technol. Res. **7**(19), 43–45 (2022)

Ya, S.X.: Research on the application of interface technology of anti-seepage panels for rock-filled concrete dams in Sanddaogou Reservoir. China Water Power Electrification **4**, 47–50+55 (2023)

Zheng, C.H.: Research on the application of BIM technology in the whole-process audit of a dam project. Fujian Constr. Sci. Technol. **39**(2), 127–129 (2024)

# Algorithm Optimization and Benefit Analysis of BIM Technology in Refined Cost Management of Engineering Projects

Yun Zhang(✉)

Kunming Metallurgy College, Kunming, Yunnan, China
kmyz0861@163.com

**Abstract.** The research on algorithm optimization and its economic benefit analysis is critical in project cost management, however it has an issue with erroneous performance positioning. The typical SVM is unable to address the the phase limit issue in project cost management, and the result is insufficient. As a result, a BIM technology-based research on algorithm optimization in project cost management and its economic benefit analysis is provided, and the research on algorithm optimization in project cost management and its economic benefit analysis is assessed. To start with, the Manhattan theory serves to detect key influencing factors, and indicators are categorized based on the research requirements for algorithm optimization and economic benefit analysis, thereby minimizing interfering elements involved.The Manhattan theory is then used to create a BIM technology research on algorithm optimization and its economic benefit analysis scheme, and the outcomes of the research on algorithm optimization and its economic benefit analysis are thoroughly examined. The MATLAB simulation results reveal that, under particular evaluation conditions, the BIM technology outperforms the standard SVM in terms of research on algorithm optimization and its economic benefit analysis accuracy and time of influencing variables.

**Keywords:** Manhattan theory · BIM · Engineering cost · Economic benefits · Analytical

## 1 Introduction

The research on algorithm optimization and its economic benefit analysis is a very important part of the project cost management (Wang et al. 2024), which can make the precise control of the aging performance (Lu et al. 2023) evaluation model faster and faster. However, in the process (Zhou 2024) of research on algorithm optimization and its economic benefit analysis (Li 2023), The research on algorithm optimization and its economic benefit analysis scheme (Lu et al. 2024) suffers from a lack of precision, which has a detrimental impact (Guo et al. 2024) on the research on algorithm optimization and its economic benefit analysis. According to certain researchers (Chen 2025), the research on algorithm optimization and its economic benefit analysis scheme can be successfully analyzed and the research on algorithm optimization and its economic


H. Yan et al. (Eds.): AIS&P 2025, CCIS 2918, pp. 349–359, 2026.
https://doi.org/10.1007/978-981-95-9884-7_24

benefit analysis may be supported by using BIM technology (Wang 2025) to the study of the aging performance assessment mode. In order to maximize the research on algorithm optimization and its economic benefit analysis scheme and confirm the model's efficacy, a BIM technology is suggested (Wu 2025) based on this information (Cao 2025).

## 2 Related Concepts

### 2.1 The BIM Technology is Described Mathematically.

The BIM technology will improve the research on algorithm optimization and its economic benefit analysis scheme using computer technology and the index parameters in the research on algorithm optimization and its economic benefit analysis, it is $y_i$ found that the unqualified value parameters in the research on algorithm optimization and its economic benefit analysis is $z_i$, and the research on algorithm optimization and its economic benefit analysis scheme is $tol(y_i \cdot t_{ij})$ integrated with the function to finally judge the feasibility of the research on algorithm optimization and its economic benefit analysis, and the calculation is shown in Eq. (1).

$$\lim_{x \to \infty} (y_i \cdot t_{ij}) = \lim_{\delta x \to 0} y_{ij} \geq \max(t_{ij} \div 2) \tag{1}$$

Equation illustrates the evaluation of outliers among them (2).

$$\max(t_{ij}) = \partial(t_{ij}^2 + 2 \cdot t_{ij}) \succ \left(\frac{\pi}{2} - \theta\right)(\sum t_{ij} + 4)\mathfrak{M} \tag{2}$$

The BIM technology combines the benefits of computer technology and quantifies the research on algorithm optimization and its economic benefit analysis, which may increase the research on algorithm optimization and its economic benefit analysis's accuracy.

Suppose I The requirements of the research on algorithm optimization and its economic benefit analysis is $t_i$ that the research on algorithm optimization and its economic benefit analysis scheme is $set_i$, the technique for satisfying the research on algorithm optimization and its economic benefit analysis is $y_i$, and the judgment function of the research on algorithm optimization and its economic benefit analysis the scheme is $F(t_i \approx 0)$ as shown by Eq. (3).

$$F(d_i) = \sin^{-1}\theta \sum t_i \bigcap \xi \cdot \sqrt{2} \to \oint y_i \cdot 7 \tag{3}$$

### 2.2 Selection of Research on Algorithm Optimization and Its Economic Benefit Analysis Scheme

Hypothesis II The research on algorithm optimization and its economic benefit analysis function is $g(t_i)$, The weighting factor is $w_i$, The unqualified research on algorithm optimization and its economic benefit analysis, which is accordingly required by the research into algorithm optimization and economic benefit analysis, as shown in the Eq. (4).

$$g(t_i)=\ddot{x}\cdot z_i\prod F(d_i)\frac{dy}{dx}-w_i\Phi\cos^{-1}\theta \tag{4}$$

The full function of the research on algorithm optimization and its economic benefit analysis, according to assumptions I and II of the research on algorithm optimization and its economic benefit analysis can be obtained, and the results is shown in Eq. (5).

$$\lim_{x\to\infty} g(t_i)+F(d_i)\leq\bigcap\max(t_{ij}) \tag{5}$$

In order to enhance the effectiveness of research on algorithm optimization and its economic benefit analysis, it is necessary to standardize all data, and the results are expressed by Eqs. (6).

$$\widetilde{g(t_i)+F(d_i)}\leftrightarrow\sin^{-1}\theta(\sum t_{ij}+4) \tag{6}$$

## 2.3 Analysis of Research on Algorithm Optimization and Its Economic Benefit Analysis Scheme

Before carrying out the BIM technology, the research on algorithm optimization and its economic benefit analysis scheme should be analyzed in all aspects, and the research on algorithm optimization and its economic benefit analysis requirements should be mapped to the research on algorithm optimization and its economic benefit analysis library, and the unqualified research on algorithm optimization and its economic benefit analysis scheme should be eliminated. The abnormality evaluation system can be presented by means of Eq. (6), and the outcomes is $No(t_i)$ shown in Eq. (7).

$$No(t_i)=\frac{\widetilde{g(t_i)+F(d_i)}}{mean(\sum t_{ij}+4)}\begin{pmatrix}1&0\\0&1\end{pmatrix} \tag{7}$$

Among them, it is $\frac{\widetilde{g(t_i)+F(d_i)}}{mean(\sum t_{ij}+4)}\leq 1$ specified that the scheme must be $Zh(t_i)$ suggested; otherwise, the scheme integration is necessary; the outcome is illustrated in Eq. (8).

$$Zh(t_i)=\lim_{x\to\infty}[\sum\widetilde{g(t_i)+F(d_i)}]\bigtriangleup ABC \tag{8}$$

The research on algorithm optimization and its economic benefit analysis is $accur(t_i)$ a comprehensive inspection has been conducted, and the threshold values and index weights for the algorithm optimization research and its economic benefit analysis plan have been determined to guarantee the accuracy of BIM technology.The research on algorithm optimization and its economic benefit analysis is $unno(t_i)$ a systematic test research on algorithm optimization and its economic benefit analysis scheme that must be thoroughly examined. If the research on algorithm optimization and its economic benefit analysis has a non-normal distribution, the research on algorithm optimization and its economic benefit analysis scheme will be influenced, lowering the total research on algorithm optimization and its economic benefit analysis's accuracy, as stated in Eq. (9).

$$accur(t_i) = \frac{\min[\sum g(t_i) + \widetilde{F(d_i)}]}{\sum g(t_i) + \widetilde{F(d_i)}} \triangle ABC \tag{9}$$

An analysis of the algorithm optimization research and its economic benefit analysis plan indicates that the plan exhibits a multi-dimensional distribution characteristic, which is consistent with objective realities. Additionally, the study on algorithm optimization and its economic benefit assessment lacks directionality, implying that the plan possesses significant unpredictability, and hence it is $randon(t_i)$ considered as a high analytical research. If the stochastic function of the study on algorithm optimization and its economic benefit assessment is preset in line with the research framework, then the calculation of Eq. (9) can be expressed as Eq. (10).

$$accur(t_i) = \frac{\min[\sum g(t_i) + \widetilde{F(d_i)}]}{\overrightarrow{AB}} + randon(t_i) \tag{10}$$

Notably, algorithm optimization research and its associated economic benefit analysis fully satisfy the required normative standards. This outcome is achieved by virtue of computer technology, which modulates the aforementioned analytical research, purges redundant and irrelevant frameworks, and supplements default protocols—ultimately fostering robust dynamic interdependence across the entirety of this integrated algorithm optimization and economic benefit analysis system.

## 3 Research on Algorithm Optimization and Its Economic Benefit Analysis Optimization Approach

In order to realize the scheme optimization for the integrated research of algorithm optimization and its economic benefit analysis, BIM technology employs a stochastic optimization approach for this research system and adjusts the corresponding parameters of Internet information. Meanwhile, the evolutionary algorithm divides the aforementioned research into several discrete phases and then randomly selects alternative solution strategies. During the iterative process, the analysis schemes targeting different hierarchical levels of this research are continuously improved and verified comprehensively. After the completion of the entire optimization research, the hierarchical levels of all involved schemes are integrated, and the optimal solution for this algorithm optimization and economic benefit analysis research is finally documented (Wu 2025).

# 4 Practical Examples of Research on Algorithm Optimization and Its Economic Benefit Analysis

## 4.1 Introduction to the Research on Algorithm Optimization and Its Economic Benefit Analysis

Algorithm optimization and its economic benefit analysis research in complex scenarios serves as the research subject, equipped with 12 paths and a test duration of 12 h. The corresponding scheme for this specific algorithm optimization and economic benefit analysis research is presented in Table 1.

**Table 1.** Research on algorithm optimization and its economic benefit analysis research on algorithm optimization and its economic benefit analysis requirements

| Scope of application | Grade | Accuracy | research on algorithm optimization and its economic benefit analysis |
|---|---|---|---|
| Cost keeping | I | 86.84 | 86.05 |
| | II | 90.42 | 91.24 |
| Resource utilization | I | 89.36 | 88.63 |
| | II | 92.68 | 87.12 |
| Progress control | I | 88.51 | 86.73 |
| | II | 90.97 | 90.23 |

The research on algorithm optimization and its economic benefit analysis process in Table 1 is shown in Fig. 1.

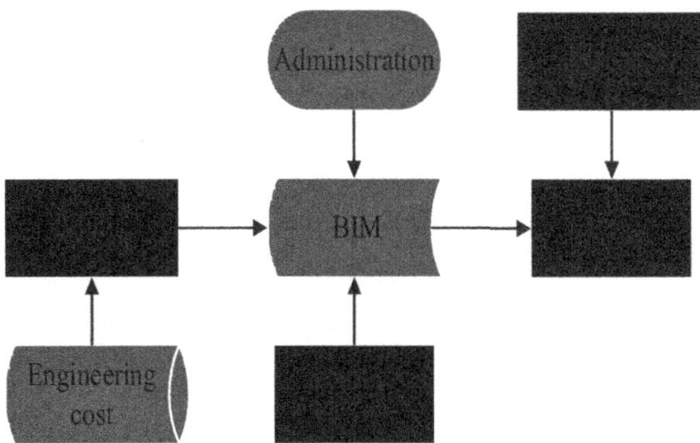


**Fig. 1.** Analysis process of research on algorithm optimization and its economic benefit analysis

The research on algorithm optimization and its economic benefit analysis scheme of the BIM technology, which includes the SVM, is closer to the real research on algorithm optimization and its economic benefit analysis needs. In terms of the logical rigor and accuracy of algorithm optimization and economic benefit analysis research, BIM technology excels over SVM. By adjusting the relevant research scheme of algorithm optimization and economic benefit analysis illustrated in Fig. 2, the accuracy and reliability of BIM technology have been further enhanced. Consequently, the evolutionary algorithm's corresponding scheme for such research has achieved noticeable improvements in processing speed, precision, and stability in summation.

### 4.2 Research on Algorithm Optimization and Its Economic Benefit Analysis

The research on algorithm optimization and its economic benefit analysis scheme of the BIM technology, which includes the SVM, is closer to the real research on algorithm optimization and its economic benefit analysis needs. BIM technology demonstrates superior performance to SVM in terms of logical consistency and accuracy regarding algorithm optimization and economic benefit analysis research. By adjusting and optimizing the corresponding analysis scheme depicted in Table 2 and Fig. 2, both the accuracy and reliability of BIM technology have been effectively enhanced. In turn, the evolutionary algorithm's scheme for this type of analytical research has gained improvements in processing rate, precision, and stability in summation calculations.

**Table 2.** The overall situation of the research on algorithm optimization and its economic benefit analysis scheme

| Category | Random data | Reliability | Analysis rate |
|---|---|---|---|
| cost keeping | 88.73 | 88.34 | 92.44 |
| Resource utilization | 87.18 | 88.35 | 90.83 |
| Progress control | 91.97 | 91.10 | 95.27 |
| Mean | 92.31 | 89.13 | 88.81 |
| X6 | 91.65 | 92.34 | 88.98 |
| | P = .249 | | |

### 4.3 Research on Algorithm Optimization and Its Economic Benefit Analysis and Stability

To verify the accuracy of BIM technology, the scheme for algorithm optimization and economic benefit analysis is integrated with SVM, and this analytical scheme is depicted in Fig. 2.

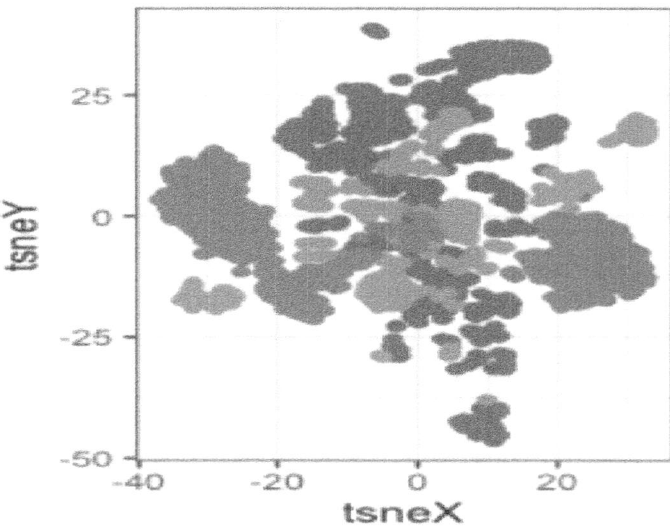


**Fig. 2.** Evaluation model of aging performance of different algorithms

Figure 2 illustrates that BIM technology outperforms SVM in algorithm optimization and economic benefit analysis performance, boasting higher accuracy while a lower error ratio. This demonstrates that BIM technology maintains relative stability in such analytical work, in contrast to SVM whose performance in algorithm optimization and economic benefit analysis is fluctuating. Table 3 presents the average analytical schemes of the three aforementioned methods for algorithm optimization and economic benefit analysis.

**Table 3.** Compares the accuracy of several research on algorithm optimization and its economic benefit analysis.

| Algorithm | Survey data | research on algorithm optimization and its economic benefit analysis | Magnitude of change | Error |
|---|---|---|---|---|
| BIM technology | 85.33 | 85.15 | 82.88 | 84.95 |
| SVM | 85.20 | 83.41 | 86.01 | 85.75 |
| P | 87.17 | 87.62 | 84.48 | 86.97 |

Table 3 reveals that SVM has limitations in the accuracy of algorithm optimization and economic benefit analysis, as its performance in such analytical work exhibits significant fluctuations alongside a high error rate. By contrast, BIM technology delivers superior results in algorithm optimization and economic benefit analysis compared to the ant colony algorithm. Meanwhile, the analytical accuracy of BIM technology remains above 90% with minimal variations. To further affirm the superiority of BIM technology and validate the efficiency of the proposed method, a comprehensive inspection of BIM technology was conducted using diverse approaches, as illustrated in Fig. 3.

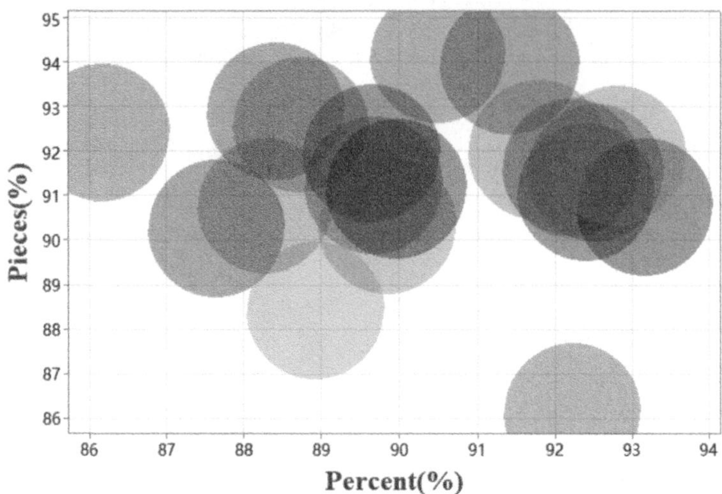


**Fig. 3.** Research on algorithm optimization and its economic benefit analysis of BIM technology.

Figure 3 demonstrates that BIM technology's performance in algorithm optimization and economic benefit analysis is markedly superior to that of SVM. This advantage stems from the fact that BIM technology adjusts and elevates the adjustment coefficient for such analytical work, and defines the threshold for Internet information parameters to filter out non-compliant schemes for algorithm optimization and economic benefit analysis.

## 4.4 Rationality of Research on Algorithm Optimization and Its Economic Benefit Analysis

The research on algorithm optimization and its economic benefit analysis scheme is integrated with the SVM to check the correctness of the BIM technology, and the research on algorithm optimization and its economic benefit analysis scheme is depicted in Fig. 4.

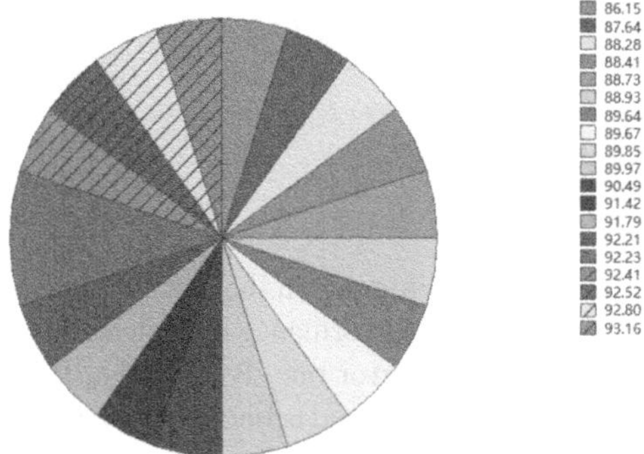


**Fig. 4.** Evaluation model of aging performance of different algorithms

Figure 4 shows that the rationality of the BIM technology's research on algorithm optimization and its economic benefit analysis is superior to that of the SVM, and that the rationality of the research on algorithm optimization and its economic benefit analysis can be increased by improving the research on algorithm optimization and its economic benefit analysis using the BIM technology. By integrating BIM technology, a distributed data storage and management platform can be established, which ensures the secure storage of research outcomes. BIM technology enables the generation of a unique identifier for each entity, with the corresponding data and schemes stored in the BIM system.

### 4.5 Validity of Research on Algorithm Optimization and Its Economic Benefit Analysis

To validate the validity of BIM technology, the scheme for algorithm optimization and economic benefit analysis is integrated with SVM, and this combined analytical scheme is presented in Fig. 5.

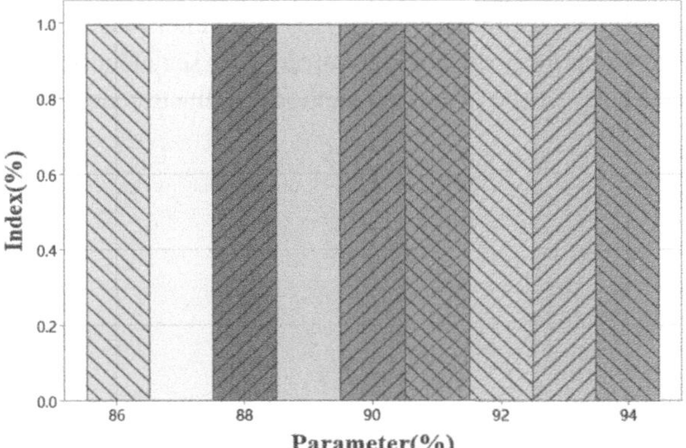


**Fig. 5.** Research on algorithm optimization and its economic benefit analysis of different algorithms

Figure 5 indicates that BIM technology's algorithm optimization and economic benefit analysis performance surpasses SVM's, with higher accuracy and lower error rate. This proves BIM maintains stable performance in such analysis, while SVM exhibits uneven results. Table 4 presents the average analytical schemes of the three aforementioned methods.

**Table 4.** Compares the efficacy of several research on algorithm optimization and its economic benefit analysis.

| Algorithm | Survey data | research on algorithm optimization and its economic benefit analysis | Magnitude of change | Error |
|---|---|---|---|---|
| BIM technology | 87.49 | 90.12 | 90.52 | 90.03 |
| SVM | 87.88 | 90.21 | 90.40 | 91.98 |
| P | 90.93 | 92.56 | 92.04 | 87.80 |

Table 4 indicates that SVM exhibits obvious flaws in the accuracy of algorithm optimization and economic benefit analysis work, and its performance in such analytical research fluctuates remarkably in practical application scenarios with a notably high error rate. By contrast, BIM technology delivers far more excellent results in this type of research compared to the ant colony algorithm. Meanwhile, the accuracy of BIM technology in algorithm optimization and economic benefit analysis stably maintains above 90%, with an extremely small fluctuation range in accuracy. To further confirm the absolute supremacy of BIM technology, (Li 2025) researchers have conducted systematic inspections on BIM technology through numerous diverse detection methods, aiming to further verify the efficacy of the proposed method, as illustrated in Fig. 6.

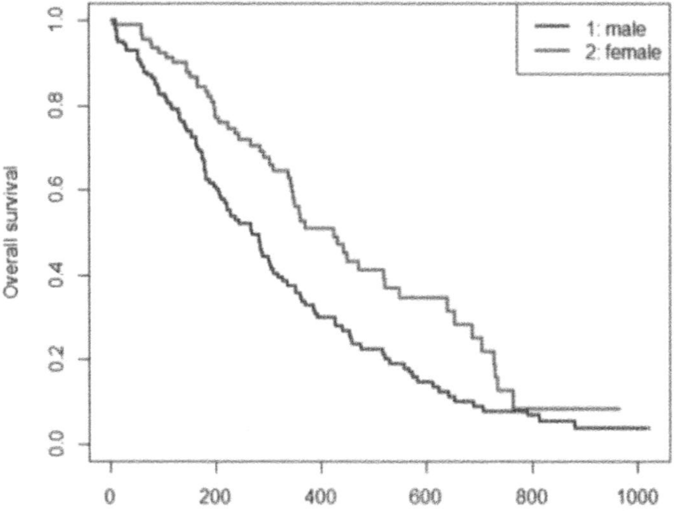


**Fig. 6.** BIM technology research on algorithm optimization and its economic benefit analysis

Figure 6 clearly illustrates that BIM technology outperforms SVM by a significant margin in terms of its performance in algorithm optimization and economic benefit analysis. This advantage can be attributed to the unique technical design of BIM technology: it rationally increases the adjustment coefficient corresponding to such analytical work, and scientifically sets the threshold for Internet information parameters, thereby effectively screening out and eliminating the relevant analysis schemes that fail to meet the preset technical requirements.

## 5 Conclusion

To tackle the problem that algorithm optimization and economic benefit analysis work is not yet optimal, this study proposes a BIM-based technical solution that integrates computer core technology to enhance the quality of such analytical research. In the meantime, the correctness and reliability of the optimized algorithm and economic benefit analysis results are comprehensively verified from multiple dimensions, and a standardized Internet information collection system is constructed to support subsequent research. The research findings indicate that BIM technology can significantly improve the accuracy of algorithm optimization and economic benefit analysis, and the generalized analytical method derived from this research is applicable to similar algorithm optimization and economic benefit analysis tasks. Nevertheless, in the BIM technology application process, excessive focus is placed on the inspection of the algorithm optimization and economic benefit analysis itself, which ultimately leads to irrationality in the selection of evaluation indicators for such analytical research.

## References


Cao, J.B.: Algorithm optimization and application in dynamic management of railway signaling engineering costs. Inf. Ind. Rep. **4**, 80–82 (2025)

Chen, Y.Q.: Research on dynamic cost management and optimization strategies in construction engineering based on BIM. Intell. City Appl. **8**(3), 72–74 (2025)

Guo, Y.F., Wang, S., Zhao, X., Dong, Y.J., Li, Y.: Application of BIM technology in cost management of construction projects. J. Kunming Metallurgy Coll. **40**(6), 91–98 (2024)

Li, J.J.: Research on construction project cost prediction method based on hybrid machine learning model. J. Jiujiang Univ. (Nat. Sci. Ed.) **38**(4), 47–51 (2023)

Li, X.J.: Cost management and benefit analysis of construction engineering based on BIM technology. China Sci. Technol. J. Database Ind. A **10**, 132–135 (2025)

Lu, J.Q., Lü, G.X., Wei, X.Y., et al.: Research on DGA domain name detection based on optimized CS-SVM algorithm. Mod. Inf. Technol. **7**(11), 77–79 (2023)

Lu, Y., Xue, F., Tang, X.B., et al.: Short-term photovoltaic prediction method based on cosine similarity and TSO-BP. Zhejiang Electr. Power **43**(6), 22–30 (2024)

Wang, S.B., Hui, J.X., Zhou, Y.X., et al.: Research on building energy consumption prediction model based on improved CGWO-SVM algorithm. Process Autom. Instrum. **45**(5), 60–65, 72 (2024)

Wang, X.Q.: Application of BIM technology in construction cost management and optimization strategies. China Sci. Technol. J. Database Ind. A **3**, 93–96 (2025)

Wu, X.L.: Research on the application and optimization strategies of BIM technology in construction cost management. Chin. Sci. Technol. J. Database (Digest) Eng. Technol. **10**, 154–157 (2025)

Wu, Y.Z.: Optimization analysis of gas engineering cost management based on BIM technology. China Sci. Technol. J. Database Ind. A **4**, 138–141 (2025)

Zhou, C.: Research on quantum bee colony optimization algorithm for BIM engineering fine-cost prediction. J. Qiqihar Univ. (Nat. Sci. Ed.) **3**, 1–7 (2024)

# Joint Extraction of Equipment Fault Text Entity Relations Based on BERT-GPLinker

Feifei Gao(✉), Lin Zhang, Bo Zhang, Wenfeng Wang, Jingyi Zhang, Min Wu, Kai Huang, Han Liu, Zhaoxuan Jia, and Chenpeng Zheng

Air Force Engineering University, Xi'an 710056, China
m18031827043@163.com

**Abstract.** The continuous application of high and new technologies has rapidly increased the complexity of equipment and the difficulty of maintenance support. Addressing issues such as the low utilization of unstructured data and complex entity relationships in equipment fault diagnosis texts, this paper presents an entity-relation joint extraction model for equipment fault diagnosis texts. By integrating a BERT encoder, attention mechanism, global pointer network, and token-pair linking joint extraction network, we designed an improved BERT-GPLinker. This enhancement achieves precision, recall, and F1-scores of 93.83%, 93.08%, and 93.45% respectively. Experimental results validate the model's effectiveness and accuracy.

**Keywords:** Equipment Fault Diagnosis · Knowledge Graph · Knowledge Extraction · BERT-GPLinker

## 1 Introduction

Knowledge graph (KG) technology can extract information from unstructured texts and integrate it into an organic whole (Gao et al. 2024). It visually represents and stores entities, attributes, and their relationships in the form of a semantic network, providing more efficient methods for utilizing information (Cui et al. 2025). Consequently, by leveraging this technology, key fault information can be extracted from equipment failure records. Through integration, processing, and combined reasoning mechanisms, equipment maintenance guidance plans can be generated (Song et al. 2024). This assists maintenance personnel in quickly identifying failures, pinpointing fault locations, making diagnoses, and implementing repairs, effectively enhancing maintenance efficiency (Shi et al. 2025).

Equipment fault diagnosis knowledge graphs, as a type of domain knowledge graph, have not yet received significant attention from relevant scholars and are developing relatively slowly. Researchers in biomedicine, finance, cultural fields, and other industrial equipment fault domains have conducted extensive studies on knowledge graph construction and application, with achievements worthy of reference and learning. For example, Reference (Xia et al. 2023) constructed a KG in the field of complex industrial equipment maintenance planning. They first designed an ontology for the domain of


H. Yan et al. (Eds.): AIS&P 2025, CCIS 2918, pp. 360–372, 2026.
https://doi.org/10.1007/978-981-95-9884-7_25

industrial equipment maintenance planning, providing a model template for building the KG. Then, they constructed it based on a link prediction model, and finally provided users with a question-answering system for convenient usage. Reference (Sk 2021) studied the application of KGs in biology and discussed the challenges and opportunities of using KGs in the biological field. Reference (Liu et al. 2023) constructed an aviation assembly domain KG based on a joint knowledge extraction model, proposed an ontology construction scheme and a joint knowledge extraction model, achieved a relatively high F1-score in the aviation assembly domain, and realized end-to-end automatic KG construction. Reference (Wu et al. 2023) constructed a domain KG for precision air conditioning fault diagnosis in data centers and integrated it with intelligent fault diagnosis methods, enhancing repair efficiency. Reference (Chen et al. 2023) proposed a reinforcement learning-based approach for building an industrial fault diagnosis KG, first introducing a fault diagnosis domain ontology, then utilizing the DSRE algorithm to extract entity-relationship triples, and finally completing the domain KG construction. Reference (Wu et al. 2023) developed a domain KG for fault diagnosis in high-voltage direct current (HVDC) transmission systems, achieving high diagnostic accuracy.

The equipment fault diagnosis KG primarily visualizes fault knowledge and its interrelationships generated throughout the entire equipment lifecycle. It achieves comprehensive coverage (Gong and Zhang 2024), intelligent question answering, and data-intelligence integration to meet equipment support requirements. However, the field of equipment fault diagnosis is highly specialized, posing significant difficulties and challenges in constructing KGs (Wang et al. 2025). During data acquisition, a vast amount of equipment fault data exists in unstructured forms like tables and flowcharts, requiring a combination of technical methods and manual efforts to transform it into a structured corpus. For knowledge extraction, selecting and refining suitable automated extraction techniques to achieve optimal performance within this specific domain is crucial for building high-quality KGs (Luo et al. 2024). Therefore, to address the aforementioned challenges, this study proposes a BERT-GPLinker joint entity and relation extraction model based on equipment failure databases. Leveraging the characteristics of equipment failure texts, the model performs triple extraction for nested entities and relations.

The main contributions of this article are as follows:

A joint entity-relation extraction model was designed to perform entity-relation extraction on labeled equipment fault diagnosis texts, completing the entity-relation triplet extraction task. Experiments were conducted to explore the model's hyperparameters, enhancing the performance of joint entity-relation extraction. Case studies demonstrated that the model can accurately perform joint entity-relation extraction on input equipment fault diagnosis texts.

Through experiments, we compared the performance of different joint extraction models in entity-relation joint extraction across equipment fault diagnosis texts. The results verified the effectiveness of the model proposed in this paper for joint entity and relation extraction, achieving precision, recall, and F1-scores of 93.83%, 93.08%, and 93.45% respectively.

# 2 Methodology

In this section, we will study entity-relationship joint extraction methods for equipment fault diagnosis, compare the structural characteristics of three different entity-relationship joint extraction models, and introduce the BERT-GPLinker model.

The goal of joint entity-relation extraction is to extract (entity1, relation, entity2) triplets from unstructured text, which is the core issue in KG construction. To accomplish the task of extracting entity-relation triplets, scholars have proposed various joint entity-relation extraction methods. Currently, models such as CasRel, PRGC, TPLinker, and GloblePointer demonstrate strong extraction performance on public datasets.

## 2.1 CasRel Model

The CasRel employs a BERT encoder to encode contextual information and extract feature representations from sentences. It utilizes a cascade decoder comprising a subject tagger and a relation-specific object tagger to extract relational triples from sentences, as illustrated in Fig. 1. The subject tagger identifies the start and end positions of all entities, while the relation-specific object tagger determines whether specific relations exist between entities and annotates entity pairs with valid relations to form triples. Analysis of the model's principles and structure reveals that the CasRel framework requires evaluating potential relationships between every entity pair, resulting in significant relational judgment redundancy and low robustness.

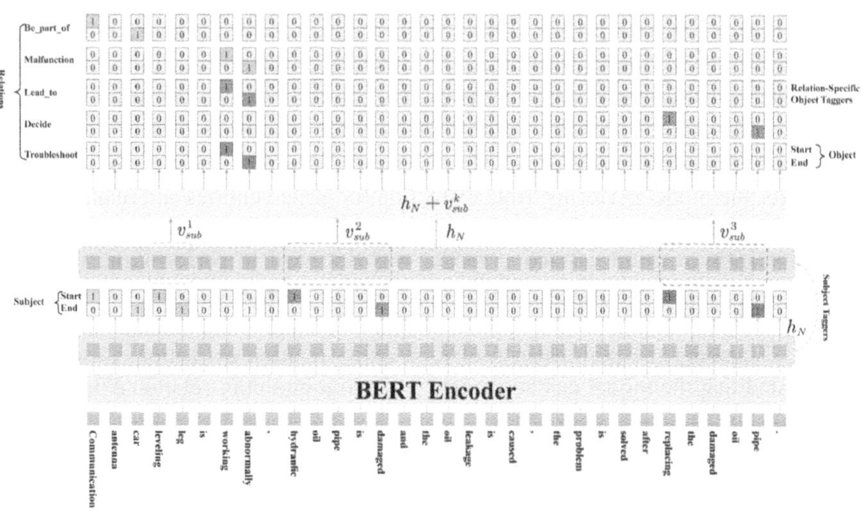


**Fig. 1.** CasRel model framework

In this example, the low-level annotation module of the topic tagger detects five entities. During the first iteration (k = 1), the high-level annotation module assigns a 0/1 label to the first entity "leveling legs," while the relation-specific object tagger detects the triples (leveling legs, belongs to, communication antenna vehicle) and (leveling legs, occurs, abnormal operation), as shown by the blue label blocks in the figure above. For subsequent iterations (k = 2,3), the high-level annotation module labels different entities, causing changes in the relation-specific object tagger's detection results. These changes reflect triples led by other entities, indicated by the orange and green label blocks in the figure above.

## 2.2 PRGC Model

The PRGC employs a BERT encoder to encode text, with its decoder comprising three components: potential relation prediction, relation-specific sequence labeling, and global correspondence, as illustrated in Fig. 2. The potential relation prediction component identifies latent relations present in the text. Relation-specific sequence labeling performs sequence tagging on subjects and objects associated with these potential relations. Global correspondence establishes correct subject-object pairings using a global correspondence matrix after acquiring all relation-linked subjects and objects, thereby completing triplet extraction. The PRGC model must identify all possible subject-object pairs based on potential relations in the text before making judgments, resulting in substantial redundancy among these pairs and complex decision-making processes.

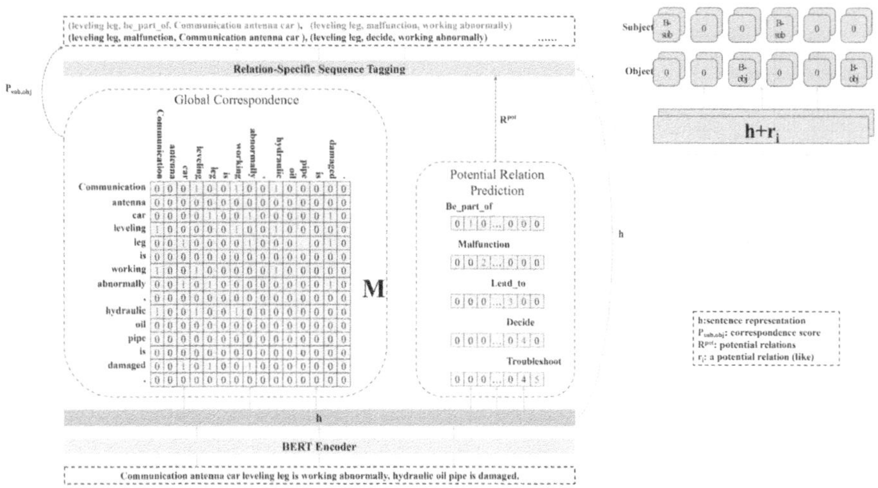


**Fig. 2.** PRGC model framework

Given a text, the PRGC predicts a global correspondence matrix $\boldsymbol{M}$ and a potential relation set $\boldsymbol{R}^{pot}$. It then annotates sentences corresponding to each potential relation using a specific relation tagging module, enumerates all possible triplets, and finally determines correct triplets through the global correspondence matrix. In this case, four triplets are listed: (leveling legs, belongs to, communication antenna vehicle), (leveling legs, occurs in, communication antenna vehicle), (leveling legs, occurs in, abnormal operation), (leveling legs, determines, abnormal operation). After applying constraints from the global correspondence matrix, the correct triplets highlighted in red are obtained: (leveling legs, belongs to, communication antenna vehicle), (leveling legs, occurs in, abnormal operation).

## 2.3 BERT-GPLinker Model

These two methods decompose triple extraction into distinct tasks using internal dependency components, reducing extraction difficulty, but suffer from exposure bias during recursive decoding, leading to error accumulation. The GloblePointer employs global pointers for entity recognition, enhancing subject-object identification efficiency and reducing redundancy. The TPLinker utilizes a unified labeling approach to complete triple extraction in a single stage, effectively minimizing error accumulation. Equipment fault diagnosis texts exhibit characteristics such as nested entities, long entity spans, precise relationships, and numerous triples. This study integrates and refines three models—BERT, GloblePointer, and TPLinker—to propose BERT-GPlinker, a joint entity-relation extraction model for equipment fault diagnosis texts. Structurally, it not only aligns with the unique features of such texts but also overcomes error accumulation while improving extraction efficiency and accuracy. Comparative experiments further demonstrate that our proposed model outperforms other models, with the framework illustrated in Fig. 3.

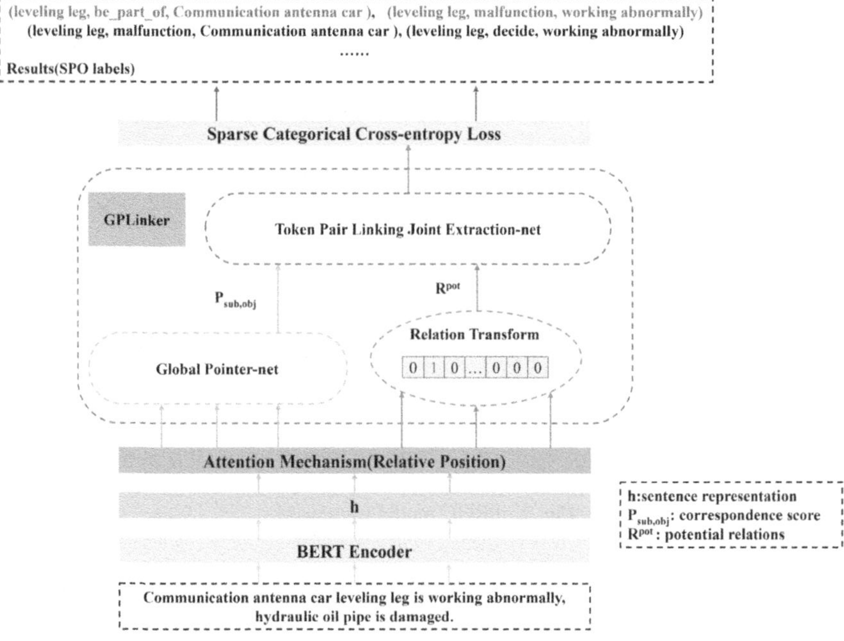


**Fig. 3.** BERT-GPLinker model framework

This study integrates four modules—a BERT encoder, attention mechanism, global pointer network, and token-pair linking-based joint extraction network—to propose the

BERT-GPLinker model for joint entity-relation extraction in equipment fault diagnostic texts, as illustrated in Fig. 3. Joint extraction involves extracting fault knowledge triples, denoted as (S, P, O), from text. Given a fault diagnosis text $X = \{x_1, x_2, \ldots x_n\}, X \in D$, where $n$ represents the number of tokens in the text and $D$ denotes the corpus, the objective is to acquire entity-relation triples, $\Phi(X) = \{s, p, o\}, s \in E, p \in R, o \in E$, where $E$ is the set of all entities and $R$ is the set of all relations. This study employs a BERT encoder for text tokenization and encoding to generate dynamic word vectors from each fault description. It utilizes an attention mechanism to enhance the model's ability in capturing relative positional relationships among word vectors within fault texts. A global pointer network is applied to identify all entities in each fault description. Through a joint extraction network with token-pair linking, the heads and tails of subjects and objects in fault texts are mapped according to relation encoding, achieving triple extraction.

# 3 Experiment and Analysis

The experimental process of the entity-relation joint extraction model of equipment fault diagnosis text based on BERT-GPLinker mainly includes the effect comparison of different models, model hyperparameter experiment and result analysis.

## 3.1 Dataset, Evaluation Metrics, and Experimental Setup

**Dataset construction.** The dataset constructed in this study includes unstructured texts such as fault diagnosis flow charts, fault record forms, and expert experience decision trees, as well as semi-structured texts such as field maintenance records, fault analysis reports, and typical fault disposal plans. To build a high-quality annotated dataset, we follow the following process:

(1) Data preprocessing: The original text is cleaned, irrelevant symbols are removed, labels are formatted, and clauses are processed. A total of about 3200 effective sentences related to fault diagnosis are obtained.
(2) Data annotation: Semi-automatic annotation method was adopted. Rules and keywords were first used for pre-annotation, and then cross-reviewed and proofread by 3 graduate students of this major. To ensure the quality of annotation, we calculated the inter-annotator agreement, and selected 50 fault records for comparison annotation. The annotation results are almost completely consistent, indicating that the dataset has good annotation consistency.
(3) Dataset division: The final constructed equipment fault diagnosis dataset contains 3158 annotated sentences, which contain 8972 entities and 5246 relation triples. The dataset is divided into training, validation, and test sets at a ratio of 8:1:1. There is a certain imbalance in the distribution of entity and relation categories in the dataset, such as "occurrence" relations are far more than "belong" relations, which challenges the generalization ability of the model.

**Evaluation index.** In this study, the standard Precision (P), Recall (R) and F1-score were used as indicators to evaluate the performance of the model. For the triple extraction task, a predicted triple (S, P, O) is considered correct if and only if its head entity, relation type, and tail entity exactly match the annotated true triple.

**Experimental Setup.** The experiments are implemented on PyTorch framework, and a NVIDIA P40 GPU is used for training. The chinese-bert-base pre-trained model is used as the encoder, and the optimizer is AdamEMA. Other key hyperparameter Settings are shown in the subsequent hyperparameter experiments section.

### 3.2 Comparison of Different Models' Effects

In order to verify the superiority of the BERT-GPLinker in the joint extraction performance of relation triples in the field of equipment fault diagnosis, we compared this model with the CasRel and PRGC. All models used the Chinese-BERT-base pre-trained language model for text encoding, with 100 training rounds set for each model, and other hyperparameters were set to the same parameter values. The experimental results are shown in Table 1 and Fig. 4.

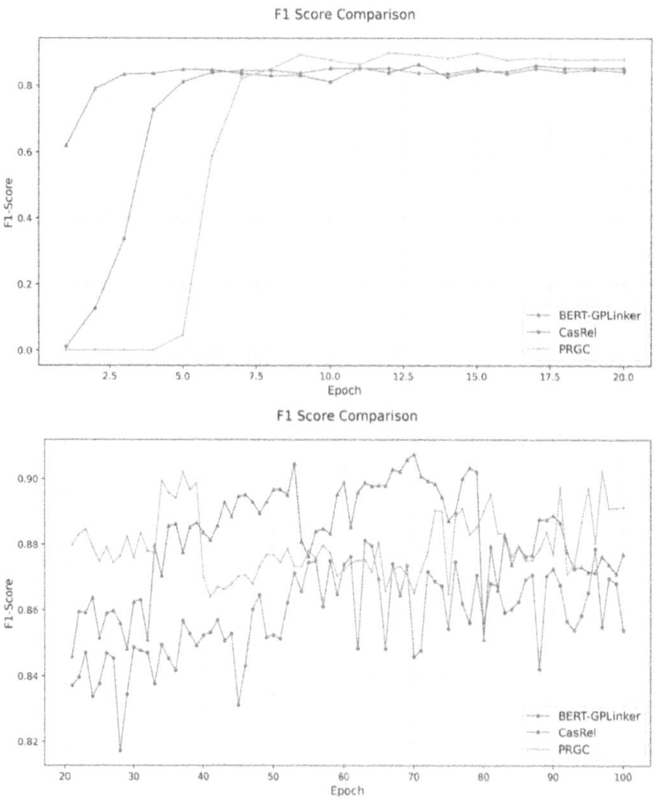


**Fig. 4.** The F1 value curves of the three models

This study systematically compares three entity-relation joint extraction models—CasRel, PRGC, and BERT-GPLinker. Experimental results indicate that after 100 training epochs, the highest F1 scores of the three models are 0.8811, 0.9021, and 0.9074 respectively. BERT-GPLinker demonstrates a performance improvement of 1.63%-2.63% over the other models. Regarding training time consumption, PRGC requires 1 h and 50 min per epoch, which is 36.7 times longer than CasRel (3 min per epoch) and 36.4 times longer than BERT-GPLinker (3 min and 2 s per epoch). The BERT-GPLinker model designed in this research achieves the optimal extraction accuracy while maintaining training efficiency, validating its effectiveness and practicality for equipment failure text parsing applications.

**Table 1.** Comparison of results of entity relationship joint extraction model.

| Model Name | P | R | F1 |
|---|---|---|---|
| CasRel | 0.9522 | 0.8198 | 0.8811 |
| PRGC | 0.9151 | 0.8895 | 0.9021 |
| BERT-GPLinker | 0.9504 | 0.8682 | 0.9074 |

### 3.3 Experiments with Model Hyperparameters

This study conducted a systematic empirical research on the optimization effect of EMA technology in the model. During the experiment, a long-term training framework of 5,000 rounds was constructed (50 times higher than the conventional 100-round benchmark), and the EMA parameter smoothing mechanism was fully demonstrated: The model reached its peak at the 4,320th round, with an accuracy rate of ($0.9383 \pm 0.0021$), a recall rate of ($0.9308 \pm 0.0035$), and an F1 value of ($0.9345 \pm 0.0028$), which were respectively 5.81%, 6.23%, and 6.11% higher than the benchmark experimental group (without EMA-100 rounds) ($p < 0.01$). All the F1 values after 5,000 rounds of training are shown in Fig. 5.

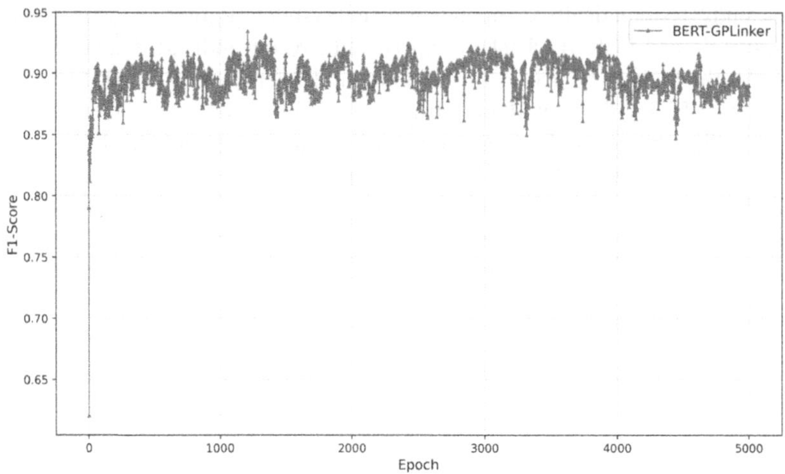


**Fig. 5.** The F1 value curve at Epoch = 5000

As can be seen from Fig. 5, the first 20 rounds are a period of rapid increase. The F1 value rises exponentially from the initial 0.620 to 0.854 (+37.6%), verifying the EMA's ability to overcome random initialization disturbances at the initial stage; the 21st to 200th rounds are a period of sub-linear growth. The F1 value slows down to 0.043 per round, and a first oscillation occurs (a 5.3% decline in the 60th round), reflecting that the model encounters local optima in the parameter space; the 201st to 3000th rounds are a period of stable oscillation. The F1 mean is $0.893 \pm 0.015$, and the standard deviation converges to 0.0049 ($R^2 = 0.917$), indicating that the smoothing effect of EMA gradually dominates; the 3000th to 4320th rounds are a period of overfitting inflection point. It surpasses the previous platform to reach a peak of 0.9383 (+5.07%), but then experiences a cliff-like drop to 0.851 ($\Delta = 8.83\%$), suggesting that an early stopping mechanism needs to be added.

The learning rate, as an important parameter in the model, we also verified the impact of different learning rates on the model's performance during 10 rounds of training. The experimental results are shown in Fig. 6.

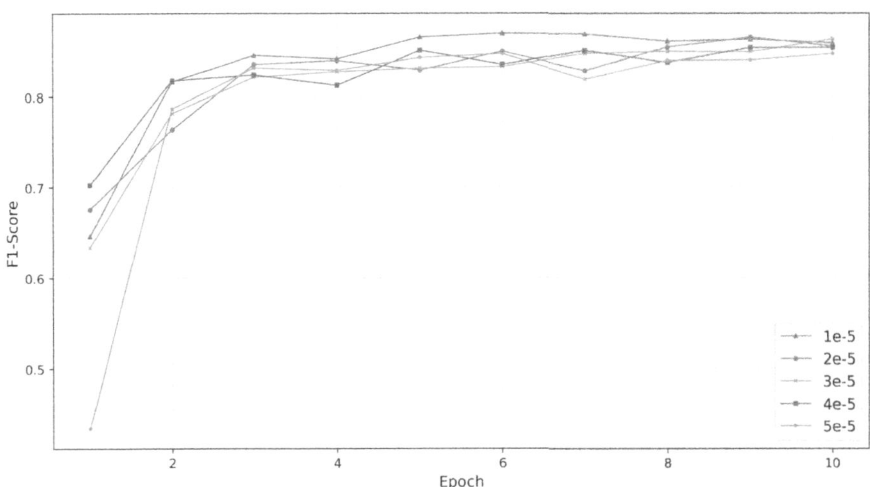


**Fig. 6.** Curves of F1 values for different learning rates

The key indicators for calculating and conducting multi-dimensional comparative analysis of the F1 values under different learning rates for 10 training rounds were determined. The results are shown in Table 2.

Analysis of the results in Fig. 6 and Table 2: 1e-5 reached a peak of 0.8696 in the 6th round and then declined to 0.8588 in the 10th round; 2e-5 performed mediocrely in the first 3 rounds, with a value of 0.8355 in the 3rd round and reaching a peak of 0.8651 in the 9th round; 3e-5 surpassed 0.8471 in the 7th round and continued to rise, reaching 0.8632 in the 10th round; 4e-5's F1 value fluctuated with a standard deviation of 0.043, indicating the poorest stability; 5e-5 experienced intense oscillations in the 1st round ($\Delta F1 = + 0.353$) and then continued to fluctuate. Comprehensive analysis of the experimental results shows: the higher the learning rate, the more intense the initial fluctuations. A high learning rate requires cautious adjustment of stability, while a low learning rate may cause convergence stagnation due to insufficient gradient noise and needs to prevent gradient disappearance. From the results of the 10 rounds of training, 3e-5 is the optimal learning rate benchmark value, and combined with dynamic scheduling, it can further improve the model's upper limit.

**Table 2.** Quantitative comparison of key indicators.

| Learning rate | Average F1 | Peak F1 | Peak iteration | Final iteration F1 | Standard deviation | Amplitude ratio |
|---|---|---|---|---|---|---|
| 1e-5 | 0.8333 | 0.8696 | 6 | 0.8588 | 0.075 | 1.73% |
| 2e-5 | 0.8297 | 0.8651 | 9 | 0.8552 | 0.062 | 1.15% |
| 3e-5 | 0.8363 | 0.8632 | 10 | 0.8632 | 0.064 | 0.96% |

*(continued)*

**Table 2.** *(continued)*

| Learning rate | Average F1 | Peak F1 | Peak iteration | Final iteration F1 | Standard deviation | Amplitude ratio |
|---|---|---|---|---|---|---|
| 4e-5 | 0.8283 | 0.8538 | 9 | 0.8537 | 0.043 | 1.92% |
| 5e-5 | 0.8073 | 0.8469 | 10 | 0.8469 | 0.120 | 4.13% |

### 3.4 Case Study and Error Analysis

(1) Case study

For a complex sentence: "Inspection revealed that the main oil line pressure was insufficient to actuate the leveling leg due to the aging of the seal ring inside the hydraulic pump." The sentence contains nested entities and causal chains. The extraction results of the proposed model are as follows:

(Sealing ring, belonging to, hydraulic pump)
(Sealing ring, occurrence, aging)
(Sealing ring aging, resulting in insufficient pressure)
(Insufficient pressure, resulting in inability to actuate the leveling leg)

The case shows that the model can not only accurately identify the basic entities such as "seal ring" and "hydraulic pump", but also identify the composite event entities "seal ring aging" and "insufficient pressure" composed of entities and states, and correctly extract the nested and causal chain between them. This validates the model's ability to deal with complex sentence structures and long-distance dependencies.

(2) Error analysis

After analyzing the wrong prediction samples of the test set, we found that the model's errors mainly concentrated in the following categories:

Entity boundary identification errors: especially when dealing with long entities or parallel structures. For example, for "Replace sensor 1 and 2", the model sometimes incorrectly identifies "sensor 1 and sensor 2" as one entity instead of two separate "sensor" entities. This is mainly due to the ambiguity introduced by the coordinate conjunction "and".

Confusion of similar entities: In the equipment domain, there are a large number of components with similar terms but different functions, such as "oil filter" and "filter", "bearing" and "bushings". When there is insufficient information in the context, the model may be confused, leading to wrong entity type or relationship judgment.

Rare relation extraction failure: Relation types that occur less frequently in the dataset, such as "belong", have far fewer sample sizes than "cause". The model does not perform well in learning such sparse relations, and it is easy to misjudge them as more common relations or directly miss them.

## 4 Experiment and Analysis

This study proposes BERT-GPLinker model to realize automatic knowledge extraction from equipment fault diagnosis text. The main contributions include: The BERT-GPLinker model solves the problem of long entity nesting and relationship redundancy through global pointer network and token-pair linking technology, and the F1 value is 93.45%, which is better than the existing models. The effectiveness of sparse multi-label cross-entropy loss function and AdamEMA is verified by experiments, and a dynamic learning rate adjustment strategy is proposed to provide methodological guidance for model tuning in small sample scenarios.

Based on the existing research, the future work that can be carried out include: Firstly, data expansion, integrating audio, video and image data to build a multimodal database to obtain more complete equipment fault diagnosis data. Secondly, model optimization, aiming at the problems found in the error analysis, the introduction of external knowledge base or more robust loss function is studied to improve the model's ability to deal with similar entities and sparse relations. Thirdly, standardization construction, formulating standards for the mode and entity relationship design of equipment fault diagnosis KGs, promoting the standardization construction of KGs in the equipment field.

## References

Chen, C., et al.: Reinforcement learning-based distant supervision relation extraction for fault diagnosis knowledge graph construction under industry 4.0. Adv. Eng. Inform. **55**, 101900 (2023). https://doi.org/10.1016/j.aei.2023.101900

Cui, Y., Sun, Z., Hu, W.: Transfer-and-Fusion: integrated link prediction across knowledge graphs. IEEE Trans. Knowl. Data Eng. **37**, 3062–3074 (2025). https://doi.org/10.1109/TKDE.2025.3544255

Gao, F., et al.: Named entity recognition for equipment fault diagnosis based on RoBERTa-wwm-ext and deep learning integration. Electronics **13**, 3935 (2024). https://doi.org/10.3390/electronics13193935

Gong, L., Zhang, H.: MR2-KG: a multi-relation multi-rationale knowledge graph for modeling software engineering knowledge on Stack Overflow. IEEE Trans. Software Eng. **50**, 1867–1887 (2024). https://doi.org/10.1109/TSE.2024.3403108

Liu, P., Qian, L., Zhao, X., Tao, B.: The construction of knowledge graphs in the aviation assembly domain based on a joint knowledge extraction model. IEEE Access **11**, 26483–26495 (2023). https://doi.org/10.1109/ACCESS.2023.3254132

Luo, X., et al.: ReiPool: reinforced pooling graph neural networks for graph-level representation learning. IEEE Trans. Knowl. Data Eng. **36**, 9109–9122 (2024). https://doi.org/10.1109/TKDE.2024.3466508

Shi, F., Li, D., Wang, X., Li, B., Wu, X.: TGformer: a graph transformer framework for knowledge graph embedding. IEEE Trans. Knowl. Data Eng. **37**, 526–541 (2025). https://doi.org/10.1109/TKDE.2024.3486747

Sk, M., A, N., V, N.: Biological applications of knowledge graph embedding models. Briefings Bioinform. **22**, 1679–1693 (2021). https://doi.org/10.1093/bib/bbaa012

Song, Y., et al.: Scene-driven multimodal knowledge graph construction for embodied AI. IEEE Trans. Knowl. Data Eng. **36**, 6962–6976 (2024). https://doi.org/10.1109/TKDE.2024.3399746

Wang, Y., Song, Q., Zheng, L., Li, X.-Y.: Efficient and tractable conflict detection in knowledge graphs via path analysis. IEEE Trans. Big Data 1–13 (2025). https://doi.org/10.1109/TBDATA.2025.3575948

Wu, J., et al.: Fault diagnosis of the HVDC system based on the CatBoost algorithm using knowledge graphs. Front. Energy Res. **11**, 1144785 (2023). https://doi.org/10.3389/fenrg.2023.1144785

Wu, J., Xu, X., Liao, X., Li, Z., Zhang, S., Huang, Y.: Intelligent diagnosis method of data center precision air conditioning fault based on knowledge graph. Electronics **12**, 498 (2023). https://doi.org/10.3390/electronics12030498

Xia, L., Liang, Y., Leng, J., Zheng, P.: Maintenance planning recommendation of complex industrial equipment based on knowledge graph and graph neural network. Reliab. Eng. Syst. Saf. **232**, 109068 (2023). https://doi.org/10.1016/j.ress.2022.109068

# CS-SVM Algorithm Empowers BIM Technology in the Model Construction and Application of Construction Project Cost Prediction

Yufeng Guo, Yun Zhang(✉), Leiying Yang, Yihang Li, Shiling Zhang, Xia Zhao, Changmei Zhao, Qing Yang, and Chaoqun Liu

Kunming Metallurgy College, Kunming, Yunnan, China
kmyz0861@163.com

**Abstract.** In traditional construction project cost prediction, there are often challenges such as poor adaptability to complex projects and unsatisfactory prediction accuracy. In view of this, this paper innovatively constructs a cost prediction model based on CS-SVM. It deeply analyzes the Cuckoo Search Algorithm (CS) and uses it to optimize the Support Vector Machine (SVM), thus creating a CS-SVM model with excellent parameter optimization performance. The cost data of cloud computing projects are standardized to accurately determine the input variables. Combined with practical cases, the application effectiveness of the BIM construction project cost prediction model based on the CS-SVM algorithm is mainly tested. The results show that this model is accurate in construction project cost prediction, can effectively empower cost management, its application effect is significantly better than other models, and it has great promotion value, playing a key guiding role in optimizing the construction project cost prediction method.

**Keywords:** CS-SVM · BIM · training set · optimization

## 1 Introduction

With the rapid development of the social economy, the construction scale of construction projects continues to expand, technical complexity keeps increasing, and the difficulty of cost control has risen significantly. Accurate cost prediction, as a core link in the whole-life cycle cost management of construction projects, directly affects the scientificity of investment decisions and the efficiency of engineering goal achievement. Traditional prediction methods, which rely on expert experience or linear models, struggle to capture the nonlinear and high-dimensional characteristics of engineering data, resulting in large prediction errors in complex projects and failing to meet the demand for refined management. The popularization of Building Information Modeling (BIM) technology has provided a precise data source for cost prediction. Its integrated and visualized features enable the full-dimensional management of engineering data (Wang et al. 2024). Meanwhile, the advancement of machine learning algorithms has promoted the in-depth integration of BIM and data analysis, becoming a research hotspot in the


H. Yan et al. (Eds.): AIS&P 2025, CCIS 2918, pp. 373–383, 2026.
https://doi.org/10.1007/978-981-95-9884-7_26

field of cost prediction. Support Vector Machine (SVM), with its advantages in handling small-sample and high-dimensional data, has been widely applied in engineering cost prediction. However, construction engineering data often suffers from class imbalance issues. SVM, which assumes balanced class distribution by default, tends to cause prediction biases for minority class data, thereby reducing model reliability. Existing optimization strategies, such as Particle Swarm Optimization (PSO) and Genetic Algorithm (GA), still have drawbacks such as falling into local optima or slow convergence speeds. In domestic research, scholars have mainly constructed BIM-based dynamic cost estimation frameworks and optimized SVM to realize cost prediction for prefabricated buildings. Nevertheless, domestic studies mostly adopt data-level balancing strategies, with relatively few researches focusing on optimizing the algorithm itself. Foreign research started earlier: scholars have developed BIM-based whole-life cycle cost management platforms and used Bayesian optimization to improve SVM stability. However, there is a lack of targeted research on class imbalance in the engineering field, and the ability of existing models to adapt to the massive volume of BIM data needs to be enhanced. Therefore, this study adopts the Cuckoo Search (CS) algorithm to optimize SVM, constructing a CS-SVM integrated model to improve the prediction ability for minority class data. Combined with BIM's multi-dimensional data sources, the model achieves accurate cost prediction. Through theoretical research and case verification, this study provides a new method for engineering cost prediction and enriches the research system of the integration between BIM and machine learning.

## 2 Research Methods

The Cuckoo Search Algorithm (CS) has excellent search path optimization capabilities and can achieve good optimization results with fewer parameter selections.

### 2.1 Cuckoo Optimization Algorithm

The CS algorithm is designed based on the behavior habits of cuckoos. Cuckoos are unique in that they neither incubate their young nor build their own nests. Instead, they look for nests built by other birds (hosts) to lay their eggs for reproduction. For example, they lay their eggs in the nests of hosts while the hosts are out. There may be eggs of both cuckoos and hosts in the nest. Once the host bird detects the presence of foreign eggs, it will take exclusion actions, which will undoubtedly reduce the survival rate of cuckoo eggs. Therefore, cuckoos will remove the eggs of the host, leaving only their own eggs in the nest. It is assumed that one cuckoo lays one egg. When the host discovers a foreign egg, it will remove it, and some hosts will directly abandon the nest with cuckoo eggs and build a new one. Here, the nest = egg = solution. The quality of the solution is determined by whether the egg can be successfully hatched by the host bird and develop normally. Based on these activity characteristics of cuckoos, the process of finding the optimal solution can be vividly regarded as the process of cuckoos looking for suitable nests to lay eggs. The quality of the solution after optimization can be intuitively understood as the quality of the nest.

Let $Pa \in (0,1)$ represent the probability that the host discovers the cuckoo egg. It is assumed that in n nests, a proportion of Pa of the nests will be replaced by new nests (new solutions). As shown in Fig. 1, the Levy flight trajectory is used to update the new solutions through the Levy flight strategy.

Update formula:

$$d_i^{(k+1)} = d_i^k + \oplus L(\lambda), \quad i = 1, 2 \ldots n \tag{1}$$

where $d_i^k$ is the is the i-th solution in the k-th generation, a is the step size control quantity, $\oplus$ represents the point-to-point product, and $L(\lambda)$ is the random search path. The random step size follows the Levy distribution:

$$L(\lambda) \sim u = t^{-\lambda}, (1 < \lambda \leq 3) \tag{2}$$

Initial values of design variables:

$$d_i^0 = rand * (d_{Ui} - D_{Li}) + d_{Li} \tag{3}$$

In the formula, rand is a random number, and its value range is from 0 to 1; $d_{Li}$ and $d_{Ui}$ are the lower limit and the upper limit of $d_i$ respectively; and $d_i$ is the design variable.

Compare and analyze Pa with the random number r ($r \in [0,1]$), and judge the decision after obtaining a new bird's nest in each iteration according to the relationship between them. If Pa < r, update the position of the bird's nest. Under this condition, the migration behavior of the cuckoo will occur, and the corresponding strategy is as follows:

$$d_i^{k+1} = d_i^k + r(d_{r1}^k - d_{r2}^k) \tag{4}$$

where $d_{r1}^k$ and $d_{r2}^k$ are both random solutions in the k-th generation.

Different from the case of Pa < r, if r ≤ Pa, the cuckoo does not migrate, and the current nest position remains unchanged.

After continuous updates, the best-performing group among the nest positions $di^{(t+1)}$ is retained.

The convergence condition of the CS algorithm is the maximum number of iterations:

$$step_i = step_{min} + (step_{max} - step_{min})d_i \tag{5}$$

$$d_i = \frac{n_i - n_{best}}{d_{max}} \tag{6}$$

where $step_{min}$ and $step_{max}$ are the minimum and maximum step sizes respectively, $d_{max}$ is the maximum distance between the optimal position and other nests, ni is the i-th nest position, and $n_{best}$ is the best state of the current nest position (Lu et al. 2023).

## 2.2 Optimization Strategy of CS Algorithm for SVM

The optimization objects are the penalty factor C and the kernel function parameter g. The optimization process of the CS algorithm for the SVM is shown in Fig. 2. The analysis is as follows based on the content in the figure.

(1) The number of training sets and test sets for cylinder pressure is 800 groups and 160 groups respectively, with a ratio of 100:20. The cylinder head vibration data is divided in the same proportion. The number of iterations time = 20, the number of nests n = 20, the number of parameters to be optimized dim = 2, and the probability of being discovered by the host Pa = 0.25.

(2) Randomly initialize the C and g values of each nest. According to the initialization results, an initial SVM model is established. After this process, the positions of each nest are random, and each nest x corresponds to (C,g):

$$P_i^0 = [x_1^0, x_2^0 \ldots x_{20}^0, ]^T \tag{7}$$

(3) The classification error rate is used as the fitness value fnew, and the following relationship exists:

$$\text{fnew} = 1 - \text{fitness} \tag{8}$$

where fitness is the classification accuracy rate.

(4) Select the nest with the smallest fitness value (the best) in the previous generation of nests, retain it to the next generation, use formula (1) to determine the C and g values of other nests (non-optimal), update the two parameters to generate new nests, and use the SVM classification model to train the newly obtained part to determine the fitness value.

(5) Compare and analyze the fitness values of the updated nests and the previous generation of nests. If the value after update is smaller, it indicates that the nest is better. Retain the new nest generated after update and complete the replacement. Conversely, if the updated nest does not reach a better state, no replacement is made. The position of the nest after update is expressed by the following formula:

$$P_t = [x_1^t, x_2^t \ldots x_{20}^t, ]^T \tag{9}$$

(6) According to the relationship between the nest Pa and the random number r, the following operations are carried out: if Pa < r, update the C and g values to generate new nests with updated parameters, train and calculate the new fitness value. If the calculated data is lower than that of the previous generation of nests, it indicates that the updated nest is better and is retained. Combine it with the nests generated in step (5) to obtain new nests Pt, with a total of 20.

(7) Among the new nests generated by the above update process, determine the one with the lowest fitness value. Judge whether the optimal result can be output according to this value. If the fitness value meets the termination condition, it means that the optimization is completed. The nest corresponding to this fitness value is the best nest, and the corresponding C and g values are determined. After comparison, analysis and result output, an optimal SVM model is established. To evaluate its generalization ability and classification accuracy rate, a test set is used for verification. If the termination condition is not met, it means that the desired optimization result has not been obtained yet. Jump back to step (4) and continue the optimization until the optimal result is output.

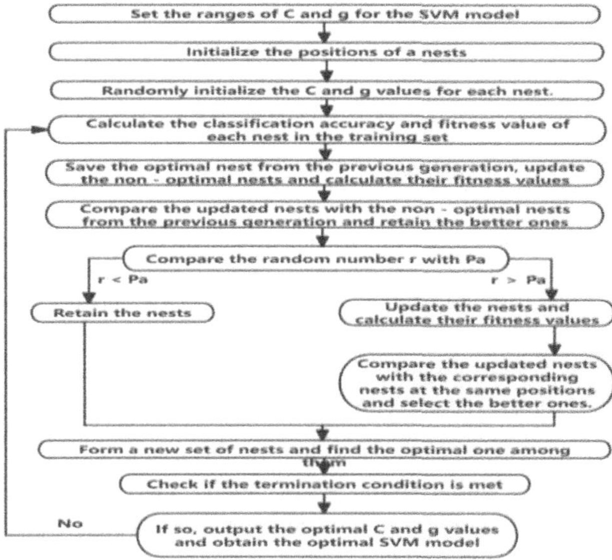


**Fig. 1.** Schematic Diagram of the Optimization Process of SVM by CS Algorithm

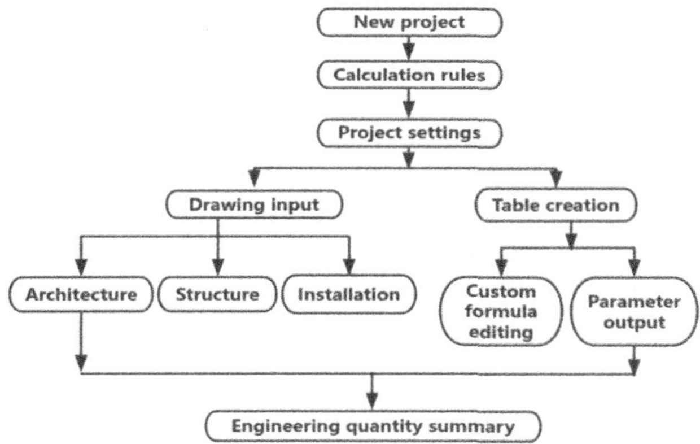


**Fig. 2.** Schematic Diagram of Engineering Measurement Process

## 3 Cloud Computing Project Cost Data

In the current context of increasing informatization in construction project construction, cloud computing technology plays a crucial role in the measurement and price management of complex construction projects, effectively promoting the improvement of the engineering management level. Compared with traditional manual measurement methods, engineering measurement based on cloud computing technology shows extremely high efficiency and can easily handle complex measurement tasks. Its operation process is shown in Fig. 2.

The cloud platform has the ability to update list data in real time, which enables staff to conduct efficient consultations around relevant issues, make full use of various types of information, and thus significantly improve the accuracy of project cost prediction results.

### 3.1 Loss and Consumption Analysis

The engineering loss is the result of the combined influence of multiple factors such as process level and management ability. Due to the complex construction environment of projects, some losses are difficult to avoid. Therefore, it is necessary to consider the impact of this part of the losses on the project cost and deeply analyze the loss cost. Consumption refers to the consumption of various resources required to complete the project construction task, such as materials and mechanical equipment. The level of construction has a significant impact on the overall project consumption. Among all the consumption in the project, material consumption accounts for a relatively large proportion. There are different degrees of losses in various links such as transportation, storage, and use. Therefore, it is necessary to clarify the key factors for controlling consumption according to the actual project conditions and take targeted control measures to minimize unnecessary consumption and reduce the project cost (Zhou 2024).

### 3.2 Consumption Quota

Based on the characteristics of project construction, cloud computing technology can be used to select similar historical projects to determine the loss situation. For standard consumption quotas, the ratio of consumption (the sum of net usage and loss) to net usage is the consumption coefficient, and the proportion of loss in net usage is the loss rate.

In engineering projects, the quota is used as the consumption standard. Its coverage mainly includes materials, mechanical equipment, and labor related to project construction. At the same time, the quota can also be used as an important basis for project cost prediction. The consumption quota is determined under normal construction conditions and belongs to the construction organization plan provided by the competent department, which has important reference value for controlling material consumption during project construction (Li 2023; Lu et al. 2024).

## 4 BIM Construction Project Cost Prediction Process

The BIM cost data analysis system is utilized to gather cost data, from which key indicators are then pinpointed. When feeding data into the SVM model, it's crucial to ensure that the input indicators are quantitative. But in the actual data selection, there might be qualitative indicators that can't be directly used in the SVM model. So, data preprocessing is essential to meet the SVM's input requirements (Yuwen 2021; Chen 2025). After transforming qualitative indicators into numerical ones, we need to address the potential large discrepancies in dimensions and magnitudes. Such differences can harm the SVM model's prediction accuracy and even decelerate its convergence. Thus, normalizing the input quantitative indicators is necessary to minimize the differences between variables. The maximum-minimum method, a common approach, can confine the input variable data to the [0,1] range, optimizing the model's performance:

$$y_i = \frac{x_i - x_{min}}{x_{max} - x_{min}} \tag{10}$$

where $x_{min}$ and $x_{max}$ are the minimum and maximum values of the row where x is located respectively.

After clarifying the processing idea of input data before SVM model prediction, the following analyzes the specific process of BIM construction project cost prediction.

First, collect construction project cost data and preprocess it, retaining only the data with utilization value for subsequent construction project cost prediction. Then, set the parameter values of the CS algorithm and determine the value range of the SVM model parameters. Subsequently, ensure that the nest positions correspond to the SVM model parameters and randomly generate n nest positions. After that, carry out sample training and cross-validation to determine the optimal nest position. Next, update the search paths and positions of other nests, calculate the fitness values, compare and analyze the fitness values before and after nest updates, screen out the optimal parts from the updated nests, and gradually reduce the number of nests in poor positions through replacement. Finally, according to the actual application requirements of project cost, compare the cost prediction results based on the CS-SVM model with them. If they meet the requirements, end the optimization and output the project cost prediction results. If not, return to step (3) and continue the operation according to the process until the optimization end condition is met (Wang 2025).

## 5 Prediction Result Analysis

To test the feasibility of the CS-SVM model in construction project cost prediction, a simulation experiment was carried out on the PentiumDual-Core experimental platform. The data used in the experiment came from the construction project cost data of the enterprise from 2019 to 2022, covering parts with a relatively large proportion in the total project cost, such as material costs, equipment rental costs, and labor costs (Wu 2025). The selected data was successively cleaned, screened, and standardized for model training and verification to objectively evaluate the prediction effect of the model on construction project costs. There were a total of 156 data points in this experiment, with 140 used for the training set and 16 for the test set, which served for model construction and training respectively. By comparing and analyzing the construction cost prediction results of the CS-SVM model with the actual values and performing fitting, the consistency between the two was intuitively judged, thereby verifying the feasibility of the model in predicting construction project costs. The fitting results are shown in Fig. 3.

In Fig. 3, the regression fit degree $r^2$ value reaches 0.99, which clearly shows that the construction cost prediction data of the CS-SVM model has a high degree of agreement with the actual values, and the prediction accuracy is quite high. The cost prediction results of this model can be used as a reference for project cost management. Judging from the fitting situation of the prediction values and the actual values, the CS-SVM model is suitable for project cost prediction. To further verify the feasibility of the model in this paper, the cosine similarity-neural network model and the GA-BP neural network model were introduced for comparison. The root mean square error (RMSE) and the mean absolute percentage error (MAPE) were selected as evaluation indicators, and their calculation formulas are as follows:

$$\mathrm{RMSE} = \sqrt{\frac{1}{\mathrm{n}}\sum\nolimits_{\mathrm{t}=1}^{\mathrm{n}} (\mathrm{y_t} - \hat{\mathrm{y}}_\mathrm{t})^2} \tag{11}$$

$$\mathrm{MAPE} = \frac{\sum_{\mathrm{t}=1}^{\mathrm{n}} \left|(\mathrm{y_t} - \hat{\mathrm{y}}_\mathrm{t})/\mathrm{y_t}\right|}{n} \times 100\% \tag{12}$$

where n is the number of sample data, and t and $\hat{y}_t$ are the expected value and the estimated value respectively.

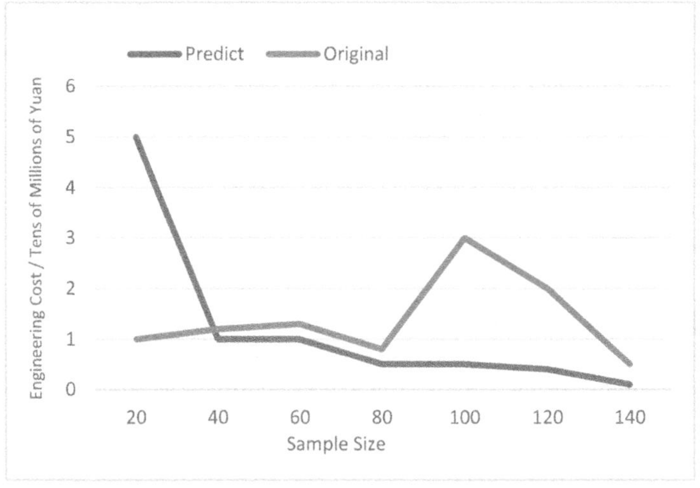


**Fig. 3.** Fitting Results of Predicted and Actual Engineering Cost Values

Calculate the RMSE and MAPE of the model in this paper and the two comparison models according to the above formulas. Compare and analyze the index calculation results of the three methods to evaluate the prediction performance of each model for project costs (Cao 2025). The results are shown in Fig. 4.

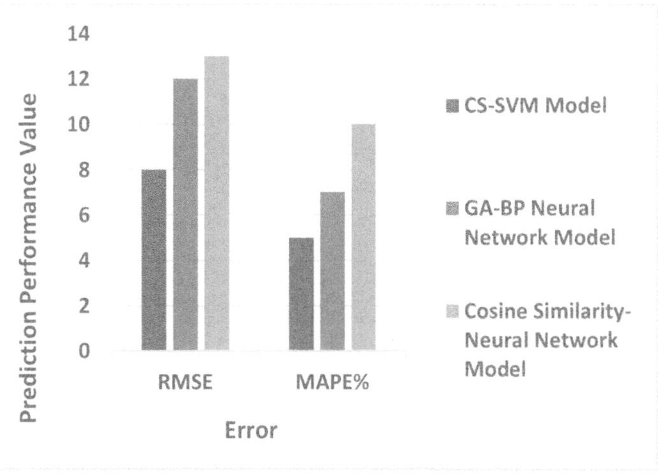


**Fig. 4.** Prediction Performance Comparison Chart

It can be seen from Fig. 4 that the calculation results of the two indicators RMSE and MAPE of the model in this paper are lower than those of the other two comparison methods. This means that the construction cost prediction error of the model in this paper is the smallest, and the prediction results are more accurate and reliable. Taking a certain construction project as an example, the comparison of the cost of different methods under the same project structure strength constraint is shown in Fig. 5.

Referring to Fig. 5 for analysis, when ensuring the same structural strength, the cost predicted by the CS-SVM model in this paper is lower than that of other methods. This indicates that the model is more accurate in predicting the cost of construction projects. The reason is that this method integrates cloud computing technology, which can call a large amount of historical data. Relying on the comprehensiveness and reliability of the data, the engineering measurement results are more accurate, and the loss and consumption can be more accurately controlled, reducing unnecessary expenses. Therefore, the CS-SVM model is suitable for construction project cost prediction, and its prediction results are relatively accurate.

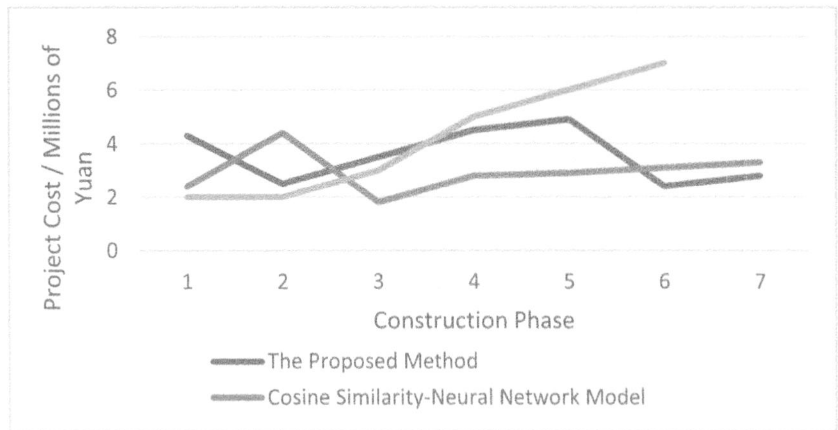


**Fig. 5.** Cost Comparison of Different Methods for the Same Project

## 6 Conclusion

This paper has deeply explored the principle of the CS algorithm and applied it to optimize the SVM algorithm, constructing a CS-SVM model aiming at accurately predicting construction project costs. This model effectively strengthens the ability to handle unbalanced data, fully excavates the rich information contained in BIM data, and significantly improves the accuracy of cost prediction. Verified by practical cases, compared with the GA-BP neural network model and other common prediction models, the cost prediction results of the model established in this paper have a smaller error with the actual values, demonstrating higher prediction accuracy and having great reference value for the cost management of construction projects.

Looking ahead, relevant researchers should continuously explore the optimization and application paths of the CS-SVM model and deeply integrate the advantages of multiple algorithms. At the same time, through case studies of more complex construction projects, the application level of the CS-SVM model should be continuously improved to effectively ensure the efficient implementation of construction project cost management work and provide strong support for the healthy development of the construction industry.

## References

Cao, J.B.: Algorithm optimization and application in dynamic management of railway signaling engineering costs. Inf. Ind. Rep. **4**, 80–82 (2025)

Chen, Y.Q.: Research on dynamic cost management and optimization strategies in construction engineering based on BIM. Intell. City Appl. **8**(3), 72–74 (2025)

Li, J.J.: Research on construction project cost prediction method based on hybrid machine learning model. J. Jiujiang Univ. (Nat. Sci. Ed.) **38**(4), 47–51 (2023)

Lu, J.Q., Lü, G.X., Wei, X.Y., et al.: Research on DGA domain name detection based on optimized CS-SVM algorithm. Mod. Inf. Technol. **7**(11), 77–79 (2023)

Lu, Y., Xue, F., Tang, X.B., et al.: Short-term photovoltaic prediction method based on cosine similarity and TSO-BP. Zhejiang Electr. Power **43**(6), 22–30 (2024)

Wang, S.B., Hui, J.X., Zhou, Y.X., et al.: Research on building energy consumption prediction model based on improved CGWO-SVM algorithm. Process Autom. Instrum. **45**(5), 60–65, 72 (2024)

Wang, X.Q.: Application of BIM technology in construction cost management and optimization strategies. China Sci. Technol. J. Database Ind. A **3**, 93–96 (2025)

Wu, X.L.: Research on the application and optimization strategies of BIM technology in construction cost management. Chin. Sci. Technol. J. Database (Digest) Eng. Technol. **10**, 154–157 (2025)

Yuwen, Z.: Research on post-evaluation of old community reconstruction projects based on BP neural network. Eng. Econ. **31**(10), 71–75 (2021)

Zhou, C.: Research on quantum bee colony optimization algorithm for BIM engineering fine-cost prediction. J. Qiqihar Univ. (Nat. Sci. Ed.) **3**, 1–7 (2024)

# Research on Intelligent Question-And-Answer Multi-Intention Analysis Method of Converter Steelmaking Knowledge Graph

Tan Li[1], Xianghui Meng[1], Weining Song[2(✉)], Nanjiang Chen[1], Qi Wang[3], Fangfang Gao[1], Yanwen Lin[3], and Runmin Yin[3]

[1] Nanchang University, Nanchang 330031, China
[2] East China University of Technology, Nanchang 330013, China
201860174@ecut.edu.cn
[3] QuikTech Co., Ltd., Beijing 100080, China

**Abstract.** Addressing the issues of information dispersion, reliance on human experience, and low data utilization efficiency in the field of converter steelmaking, a knowledge graph intelligent question answering multi-intentanalysis method based on the roberta-bigru-localattention model was proposed. Firstly, robustly optimized BERT approach was used to pre-trained language model to extract multi-level semantic features in feature extraction module, and dynamic weighted fusion was introduced to enhance domain term representation ability. Secondly, the bi-directional gated recurrent unit was used to capture the context dependence in the encoding module, and the fusion vector output by feature extraction was converted into a hidden state sequence containing bidirectional semantic dependence; Finally, the decoding module combines the local attention mechanism to dynamically focus on key information, strengthen the association constraints between adjacent operation parameter labels, solve the semantic coupling problem of multi-label classification, and break through the limitations of the traditional single-intention parsing framework. Compared with LSTM, BiGRU simplifies the gating structure, reducing computational complexity while still being able to effectively capture long-term dependencies. In the converter steelmaking scenario, where the text length is moderate, BiGRU achieves a better balance between performance and efficiency. Experimental results show that the RBL-BiLA (Roberta-bigru-local attention) model achieves a precision rate of 94.12%, a recall rate of 95.33%. And an F1 score 95.63% significantly outperforms the comparative model, and ablation experiments verify the effectiveness of dynamic fusion and local attention mechanisms. This research provides technical support for intelligent knowledge management in converter steelmaking processes, enhances the predictive ability of steel type attribute association mining and new steel type development, and offers a scalable solution for multi-intent parsing tasks in industrial scenarios.

**Keywords:** converter steelmaking · RBL-BiLA model · RoBERTa (Robustly optimized BERT approach) pre-trained language model · multi-intentanalysis


H. Yan et al. (Eds.): AIS&P 2025, CCIS 2918, pp. 384–401, 2026.
https://doi.org/10.1007/978-981-95-9884-7_27

## 1 Introduction

With the accelerating industrialization process in China, the iron and steel industry, as a fundamental and strategic industry supporting the national economy, has increasingly highlighted its core position as a basic material, and its supporting role in the national economy has become more and more prominent (Wang et al. 2023). Steel is widely used in construction, transportation, machinery manufacturing and other fields, and it is an indispensable multi-functional material in the modern industrial system (Yang et al. 2023). Among various steelmaking processes, the converter steelmaking process has gradually become the mainstream technology in current iron and steel production due to its advantages of short smelting cycle, high efficiency, energy saving and low carbon emission intensity. It has undergone significant technological innovations, from the original process that relied on air to oxidize iron for melting scrap steel in the late 19th century to the current combination of top-blowing and bottom-blowing technologies (Zhang et al. 2024; Meng et al. 2023). However, the converter steelmaking process is complex, involving multiple operations such as hot metal requirements, scrap steel ratio, and converter lance position control. Relying on manual experience makes it difficult to achieve precise control, which cannot meet the efficiency and accuracy requirements of modern industry. At the same time, traditional methods are inadequate in the efficient utilization of massive real-time data and addressing the unknown problems brought by the development of new steel grades. There are serious issues of fragmented knowledge: steel grade guidance manuals are complex in content, and the knowledge of steelmaking processes is scattered across multiple carriers such as technical manuals, experts' experience notes, and historical operation logs. The lack of structured integration leads to low knowledge reuse efficiency, which further affects production efficiency. Key operations rely on the subjective experience of expert technicians, which easily causes fluctuations in product quality due to differences in personnel experience. The converter production generates massive real-time data, but traditional methods are difficult to realize the associated mining of data and knowledge. In addition, the query needs of operators often include multiple intentions, and the insufficient capabilities of data utilization and intention analysis will restrict the speed of production decision-making.

As the core link in iron and steel production, the converter steelmaking process is characterized by complexity with multi-variable, strong coupling and non-linear features. The traditional operation optimization relies heavily on the experiential knowledge of experts, making it difficult to achieve precise, adaptive and intelligent control. To solve this problem, knowledge graph technology has been widely applied in various intelligent information service fields, including intelligent search, intelligent question answering, intelligent production and personalized recommendation (Hu et al. 2023), and these applications have greatly improved people's quality of life. Compared with the traditional text information representation method, it is more efficient and accurate. By structurally displaying the relationships between texts, it reveals the interconnections between information (Fu et al. 2021; Wang et al. 2022), and this progress is of great significance for improving people's information retrieval and knowledge acquisition processes (Fan 2024; Tian and Wu 2025).

As a cutting-edge structured knowledge management tool in the field of artificial intelligence, it provides a new solution for the intelligent upgrading of converter steelmaking. Knowledge graphs can efficiently and accurately integrate and associate massive scattered process data, technical standards, and expert experience in the steelmaking process. By revealing the deep-seated connections between information, they provide a robust knowledge foundation for intelligent question-answering (QA) systems. Currently, research on intelligent QA systems in various fields mainly includes deep learning-based QA methods and semantic parsing-based QA methods, both of which have certain limitations.

Deep learning-based QA methods primarily extract entities from questions through named entity recognition and use deep learning models to score and rank candidate relationships. This optimizes the entire process of semantic parsing and answer generation, significantly improving the accuracy and intelligence level of QA systems (Chen 2024; Kang 2023; Wu 2022; Ren and Xie 2024). Li et al. proposed a knowledge graph-enhanced multi-view deep learning framework, which integrates network traffic, host behavior, and KG relationship features to improve intrusion detection accuracy. However, real-time construction, maintenance, and updating of knowledge graphs pose challenges, making it difficult to adapt to dynamic or resource-constrained real-time environments. Liu et al. integrated graph convolutional neural networks into their construction method; although this method performs well in information extraction and relationship prediction, its model decision-making process lacks transparency, which affects its adaptability in industrial environments requiring high reliability. Despite the high accuracy of deep learning methods in single-intention QA, their inherent "black-box" nature leads to insufficient interpretability, making it difficult to meet the strict requirements of industrial decision-making for process transparency.

Semantic parsing-based QA methods mainly convert natural language questions into standardized semantic representations (e.g., logical forms) to retrieve answers accurately from knowledge graphs, and they are more accurate than deep learning-based QA methods. Research such as Dar et al.'s fake news detection, Hu et al.'s QA in the household registration field, Beden et al.'s proposed human-machine collaboration framework for constructing RDF knowledge graphs from industrial data, and Wang et al.'s QA research in the computer science field all provide references for the application of semantic parsing. However, these methods generally face problems of high data dependence and insufficient robustness. When dealing with multi-intention and complex nested queries, their parsing capabilities are inadequate, making it difficult to deeply explore the potential correlations between the attributes of different steel grades and limiting their applicability in actual industrial scenarios (Zhou et al. 2025; Wang and Gao 2024). For example, when addressing composite questions such as "What are the process flows, hot metal requirements, and contained elements of the steel grades CCA031A and CCA008H?", existing models struggle to effectively identify multiple semantic intentions and cannot deeply explore the potential correlations between process and composition attributes of different steel grades, thereby limiting the completeness and accuracy of answers. Such multi-intention questions are extremely common in actual production, and their parsing capability directly determines the practical value of intelligent systems, putting semantic parsing-based technologies face severe challenges in real industrial scenarios.

To overcome the technical bottleneck of multi-intention parsing, this paper proposes the RoBERTa-BiGRU-Local Attention (RBL-BiLA) multi-intention parsing model. Through feature space decoupling and a local attention mechanism, this model realizes the parallel identification of composite relational attributes in users' questions and constructs a dynamic mapping system from natural language to knowledge graph relational patterns. Compared with existing methods, while maintaining domain semantic characteristics, this model significantly enhances the associative reasoning ability of cross-entity relationships, providing an extensible new technical path for multi-dimensional knowledge query in complex industrial scenarios.

The RBL-BiLA model can effectively identify and process multiple semantic units in composite questions, reduce the risk of misjudgment, and improve the response efficiency of the question-answering system. This model can not only be used to explore the potential correlations in operational characteristics between different steel grades during the converter steelmaking process, providing data support and decision-making suggestions for the development of new steel grades, but also serve as a reference for knowledge graph question-answering research in other industrial vertical fields. Meanwhile, through ablation experiments, this paper verifies that the local attention and dynamic fusion mechanisms can effectively improve the model's performance in multi-label classification tasks, enhance the robustness of multi-label classification, and strengthen the model's application capability in complex industrial scenarios.

# 2 RBL-BiLA Multi-Intention Parsing Model

## 2.1 Overall Overview of the Model

The multi-intention recognition model RBL-BiLA proposed in this paper is shown in Fig. 1. It mainly consists of three core modules: feature extraction, encoder, and decoder.

In the feature extraction module, a RoBERTa (Robustly Optimized BERT Approach) pre-trained model with a dynamic weight fusion mechanism is innovatively introduced. To address the limited representation capability of traditional single-layer Transformers, this module performs dynamic weighted fusion on the outputs of Albert's 12 Transformer layers through learnable weight parameters $\alpha 1 \sim \alpha 12$. Specifically, the 768-dimensional semantic representation generated by each Transformer layer is assigned a dynamic weight, and a fusion vector with multi-level semantic information is formed after weighted summation. This design effectively captures the semantic features of text at different abstraction levels and significantly improves the ability to represent professional terms and composite intentions in the steelmaking field.

The encoder module adopts a bidirectional gated recurrent unit (BiGRU) to construct a context-aware network. Through bidirectional temporal modeling, this module converts the fusion vector output by feature extraction into a hidden state sequence containing bidirectional semantic dependencies. The BiGRU layer adaptively adjusts the propagation intensity of forward and backward information through the reset gate and update gate mechanisms, which accurately models the semantic associations between texts in the converter steelmaking field while preserving long-range dependency relationships. The encoder module inputs the hidden states of all time steps and the hidden state of the last time step into the decoder module, ensuring that the encoder module contains both global semantic features and local context information, thus providing high-density semantic representations for subsequent decoding.

The decoder module innovatively constructs a hierarchical decoding architecture guided by local attention. It dynamically focuses on the encoder's hidden states through a local attention mechanism, uses a sliding window to constrain the attention range, and strengthens the association constraints between labels of adjacent operation parameters. This mechanism generates an attention weight matrix with spatial locality through the interactive calculation between a trainable query vector and the encoder output. Subsequently, the attention-weighted context vector is multimodally fused with the GRU (Gated Recurrent Unit) hidden state, and multi-label parallel prediction is realized through a fully connected layer and a sigmoid activation function. This hierarchical decoding strategy effectively solves the label coupling problem in scenarios where multiple intentions coexist in the steelmaking field while ensuring computational efficiency.

By means of collaborative optimization of dynamic feature fusion, bidirectional context encoding, and local attention decoding, this model breaks through the limitations of traditional single-intention parsing frameworks.

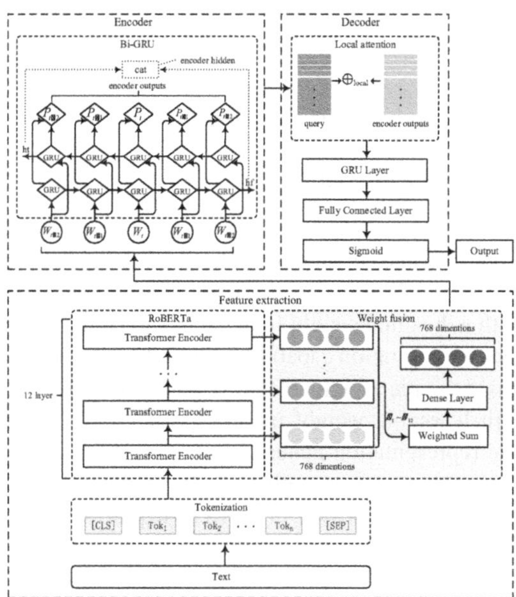


Fig. 1. RBL-BiLA model architecture.

## 2.2 Introduction to Each Layer of the Model

**Dynamic Weight Fusion.** Each layer of the Transformer encoder in BERT or RoBERTa differs in its understanding of text. The dynamic weight fusion mechanism proposed in this paper aims to effectively integrate the multi-level semantic representations output by the 12 Transformer encoder layers of the RoBERTa model, enabling the model to dynamically select semantic representations that are more important for the classification task.This mechanism first extracts the output vector $respresent_i \in R^{768}$ (where $i = 1, 2, \ldots,$ 12) from each Transformer layer, and then performs weighted fusion using learnable weight parameters $\alpha_i$. Specifically, each weight parameter $\alpha_i$ is calculated and generated by a fully connected layer, as shown.

$$\alpha_i = \sigma(W_\alpha \times represent_i + b_\alpha) \tag{1}$$

Among them, σ denotes the sigmoid activation function, and $W_\alpha \in R^{1\times768}$ as well as $b_\alpha \in R$ are trainable parameters. Through this process, the model can adaptively assign differentiated contribution weights to different layers: for instance, shallow layers may capture local syntactic features, while deep layers model global logical correlations.After weight calculation is completed, the representations of each layer are fused via weighted summation, as shown in (2). This fused vector $fused_{Vector} \in R^{768}$ is further projected into a 768-dimensional semantic space through a fully connected layer.

$$fused_{Vector} = \sum_{i=1}^{12} \alpha_i \cdot represent_i \tag{2}$$

The final vector output through dynamic weight fusion is shown in (3), where $W_d \in R^{768\times768}$ and $b_d \in R^{768}$ are trainable parameters. This design not only alleviates the overfitting problem that may be caused by high-dimensional features but also enhances the model's ability to focus on domain-specific semantics. Compared with the traditional static average fusion method, the dynamic weight fusion mechanism realizes the automatic allocation of hierarchical contributions through end-to-end training.

$$output = W_d \times fused_{Vector} + b_d \tag{3}$$

**BiGRU Unit.** In the research of Recurrent Neural Networks (RNNs), the gradient vanishing problem has long restricted the model's ability to model long-sequence dependencies. Traditional RNNs require gradients to be continuously multiplied along time steps during backpropagation, leading to almost stagnant weight updates at early time steps. To address this issue, the Gated Recurrent Unit (GRU) realizes dynamic regulation of information flow by introducing a gating mechanism. The core structure of GRU includes a reset gate and an update gate, with their corresponding calculation formulas shown in (4) and (5).

$$r_t = \sigma\left(W_r \cdot \left[\mathrm{h}_{t-1}, x_t\right]\right) \tag{4}$$

$$z_t = \sigma\left(W_z \cdot \left[\mathrm{h}_{t-1}, x_t\right]\right) \tag{5}$$

Among them, the reset gate $r_t$ generates a scalar between 0 and 1 through the Sigmoid activation function, which is used to control the contribution of the hidden state $h_{t-1}$ at the previous moment to the current candidate state $\tilde{\mathrm{h}}_t$. When $r_t$ approaches 0, the model ignores historical information and focuses on the current input $x_t$; this feature enables it to excel at capturing short-term dependencies in sequences. The calculation of the candidate state can be expressed as shown in (6).

$$\tilde{\mathrm{h}}_t = tanh\left(W \cdot \left[r_t * \mathrm{h}_{t-1}, x_t\right]\right) \tag{6}$$

The update gate determines the proportion of historical memory inherited by the current state. The final hidden state is generated by weighted fusion of the historical state and the candidate state, with its calculation formula shown in (7) below.

$$h_t = (1 - z_t) * h_{t-1} + z_t * \tilde{h}_t \tag{7}$$

When $z_t$ approaches 1, the model tends to retain the current input information, thereby effectively modeling long-term dependencies across multiple time steps. To further enhance context awareness, this paper adopts a bidirectional gated recurrent unit (BiGRU) architecture. The BiGRU consists of two independent GRU layers (forward and backward), which process the input sequence $X = \{x_1, x_2, \ldots, x_t\}$ along the forward and reverse time sequences, respectively. The calculation process of its hidden states is shown in (8).

$$\begin{cases} \vec{h}_t = GRU(\vec{h}_{t-1}, x_t) \\ \overleftarrow{h}_t = GRU(\vec{h}_{t+1}, x_t) \\ h_t = [\vec{h}_t; \overleftarrow{h}_t] \end{cases} \tag{8}$$

Here, $\vec{h}_t$ and $\overleftarrow{h}_t$ represent the hidden states from left to right and from right to left, respectively, and $x_t$ denotes the $t$-th element in the input sequence. Then, the hidden states from the two directions are concatenated to obtain the bidirectional hidden state $h_t$, where [;] indicates the vector concatenation operation.

**Local Attention Mechanism.** The local attention mechanism adopted in this paper effectively addresses the issues of computational redundancy and semantic sparsity of traditional global attention in long-sequence tasks by dynamically constraining the attention range. This mechanism consists of three core steps: on the one hand, it predicts the center position of the attention window based on the current hidden state of the decoder, and generates dynamic positioning signals through a learnable position prediction network, enabling the window to move adaptively with the semantic focus; on the other hand, it calculates additive attention scores within a fixed-width window, significantly reducing computational complexity; in addition, it generates local attention weights through softmax normalization and aggregates contextual information within the window. Given the encoder output sequence Henc = {h1enc,..., hTenc} and the current state of the encoder htenc, the calculation of the window center position is shown in (9).

$$p_t = T \cdot \sigma\left(W_p h_t^{dec} + b_p\right) \tag{9}$$

Here, $T$ is the length of the encoder output sequence, σ is the Sigmoid function, and $W_p \in R^{d_k \times l}$ (note: the original text omits the specific dimension, and this is a common dimension setting for reference) is the position prediction parameter matrix. This design enables the window center to move adaptively with the semantic focus.

Within the window range $[p_t\text{-}\omega, p_t + \omega]$, additive attention is used to calculate the energy score of each position. For the encoder hidden state *Henc*, its energy score $e_{t,i}$ is defined as shown in (10).

$$\begin{aligned} &e_{t,i} = v^T tanh(W_q h_t^{dec} + W_k h_i^{enc} + b_a), \\ &i \in [p_t - \omega, p_t + \omega \end{aligned} \tag{10}$$

Here, $W_q$, $W_k \in R^{d_h \times d_k}$ are linear transformation matrices, and $v \in R^{d_h}$ is a learnable weight vector. The window width is determined through hyperparameter tuning.

The energy scores are normalized by softmax to generate local attention weights $\alpha_{t,i}$, and the context vector $c_t \in R^{d_h}$ is obtained by weighted summation, with the calculation formulas shown in (11) and (12). This process only involves calculations for $2\omega + 1$ positions within the window, reducing the complexity from $O(T^2)$ to $O(T\omega)$.Finally, the context vector $c_t$ is concatenated with the current input embedding $x_t$ and fed into the GRU unit, with the update formula shown in (13).

$$\alpha_{t,i} = \frac{\exp(e_{t,i})}{\sum_{p_t-\omega}^{p_t+\omega} \exp(e_{t,k})} \tag{11}$$

$$c_t = \sum_{i=p_t-\omega}^{p_t+\omega} \alpha_{t,i} \mathrm{h}_i^{enc} \tag{12}$$

$$\tilde{\mathrm{h}}_t^{dec} = GRU\left([x_t; c_t], \mathrm{h}_{t-1}^{dec}\right) \tag{13}$$

This paper adopts the binary cross-entropy loss function as the model's optimization objective. By combining the sigmoid activation function with cross-entropy loss calculation, it effectively adapts to the probability output characteristics of multi-label classification tasks. For an input sample x, the model predicts a logits vector ŷ∈RC (where C is the total number of categories) and a true label y∈{0,1}C; the calculation of the loss function is shown in (14).

$$L = \frac{1}{C} \sum_{i=1}^{C} [y_i \cdot \log \sigma(\hat{y}_i) + (1 - y_i) \cdot \log(1 - \sigma(\hat{y}_i))] \tag{14}$$

# 3 Experimental Analysis

## 3.1 Construction of Experimental Data

The named entities in converter steelmaking mainly involve three types: steel grade code, steel grade name, and steel grade designation. This paper constructs text data for entity recognition through two methods: one is question text constructed using a relational slot table and templates, and the other is knowledge text constructed by querying steel grade subgraphs. In these two types of text, the positions of entity-related content were adjusted so that entities could appear at the beginning, middle, or end of the text. This approach increases the difficulty of entity recognition and avoids excessive model reliance on text structure.

On this basis, this paper constructs a multi-intention recognition experimental dataset by designing a relational slot correspondence table and applying specified templates, as shown in Fig. 2. Each relational slot is set with more than 30 different expression methods; the fixed templates include single-category, dual-category, and triple-category templates, with 30 different expressions specified for each template category.

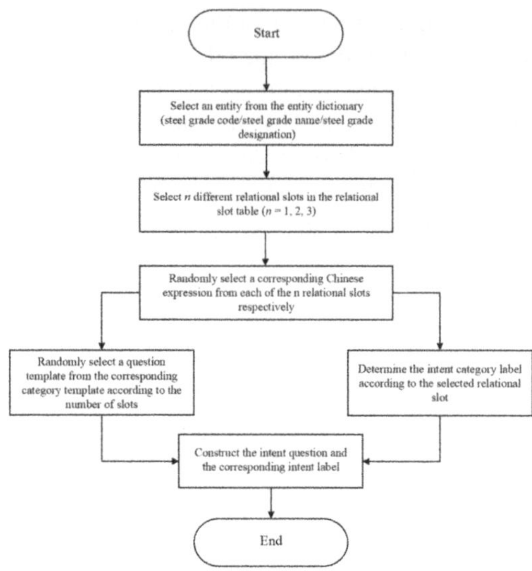


**Fig. 2.** Construction method of multi-intention identification data for converter steelmaking.

According to the ontological relationships, entity relationships, and entity attribute categories in the converter steelmaking field, this paper classifies question intentions into 30 categories. Detailed information on intention categories is provided in Table 1, including 25 relationships defined in the converter steelmaking knowledge graph and 5 ontological relationships. For the convenience of subsequent relationship mapping, most intention labels are consistent with entity relationship labels and ontological relationship labels.

**Table 1.** Question Intention Category

| Question Category | Category Symbol | Question Category | Category Symbol |
|---|---|---|---|
| Steel Grade Name | code_steel_name | Agreement Upper Limit | code_agreement_cap |
| Steel Grade Designation | code_steel_material | Agreement Lower Limit | code_agreement_down |
| Process Flow | code_production | Standard Upper Limit | code_standard_cap |
| Hot Metal Requirements | code_metal_requirement | Standard Lower Limit | code_standard_down |
| Scrap Steel Requirements | code_scrap_requirement | Element Target | code_target |

(*continued*)

**Table 1.** *(continued)*

| Question Category | Category Symbol | Question Category | Category Symbol |
|---|---|---|---|
| Converter Lance Position Control | code_gun_position | Alloy Addition | code_alloy |
| Bottom Blowing Mode | code_bottom_blow | Alloy Addition Corresponding to Elements | code_element_alloy |
| Slagging System | code_slagging_system | Precautions for Deoxidation Alloying | code_caveat |
| Converter Tapping Temperature | code_converter_temperature | Elements in Molten Steel for Refining | code_steelmaking_elements |
| Deoxidation Alloying System | code_deoxidized_alloy | Range of Molten Steel for Refining | code_steelmaking_scope |
| Liquidus Temperature | code_liquid_temperature | Target of Molten Steel for Refining | code_steelmaking_target |
| Included Elements | code_element | Process Parameter Category | affect_craftwork |
| Release Upper Limit | code_release_cap | Smelting Operation Category | depend_on_ operate |
| Release Lower Limit | code_release_down | Included Standard Category | has_benchmark |
| Deoxidation Alloying Operation Category | affect_deoxyalloy | Condition Category of Molten Steel for Refining | affect_steelmaking |

### 3.2 Experimental Environment and Parameter Settings

This paper adopts the PyTorch framework to build the experimental model, and the specific experimental environment is consistent with that of the named entity recognition experiment, the parameter settings are detailed in Table 2.

**Table 2.** Intent identification experiment hyperparameter settings

| Parameter Name | Value |
|---|---|
| batch_size | 64 |
| max_length | 72 |
| learning_rate | 3e-5 |
| RoBERTa_hidden_size | 768 |
| epochs | 30 |
| RoBERTa_transformer | 12-layer |

### 3.3 Experimental Results and Analysis

In the multi-intention recognition task, this paper uses precision, recall, and F1-score to evaluate the final performance of the model. The specific calculation formulas for these evaluation metrics are shown in (15), (16), and (17).

$$Precision = \frac{1}{N}\sum_{i=1}^{N}\frac{TP_i}{TP_i + FP_i} \tag{15}$$

$$Recall = \frac{1}{N}\sum_{i=1}^{N}\frac{TP_i}{TP_i + FN_i} \tag{16}$$

$$F1 = \frac{1}{N}\sum_{i=1}^{N}\frac{2 \times Precision_i \times Recall_i}{Precision_i + Recall_i} \tag{17}$$

Here, *TP* (True Positives) represents the number of samples correctly identified, *FP* (False Positives) denotes the number of samples incorrectly identified, and *FN* (False Negatives) indicates the number of samples not identified.

The training and testing results of the RBL-BiLA model are shown in Figs. 3 and 4. After 30 epochs of iterative optimization, the training set loss finally converged to 0.0063, and the test set loss stabilized at 0.0087, indicating that the model has good generalization ability on unseen data, verifying the high robustness of the model for the multi-intention parsing task in converter steelmaking.

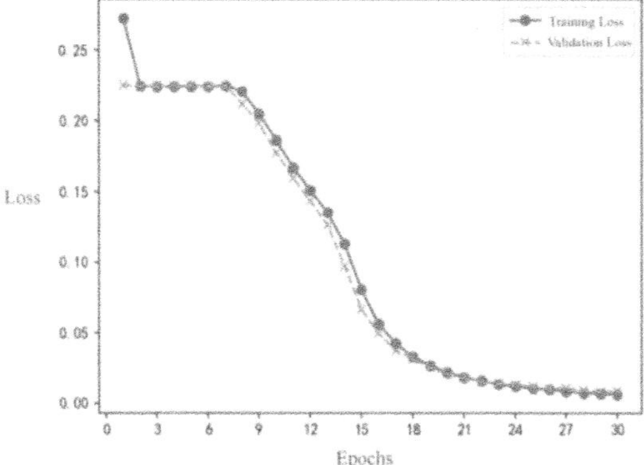


**Fig. 3.** RBL-BiLA model loss function

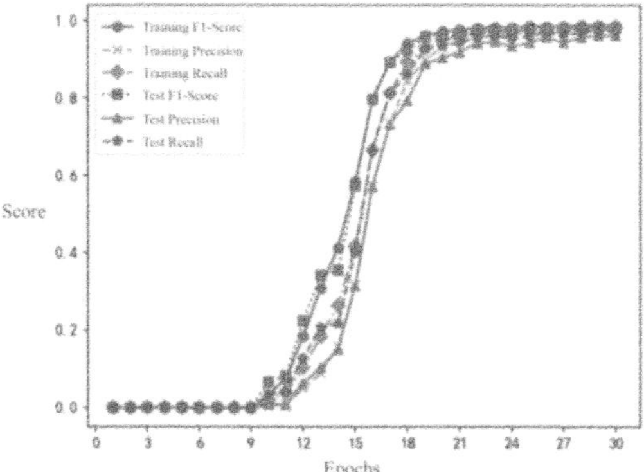


**Fig. 4.** Results of each indicator

It is worth noting that the training curve showed a slow convergence trend in the first 10 epochs. The main reason is that this task is a 30-class multi-label classification task, and the model needs to simultaneously learn the independence and potential correlation between labels, making it difficult for the model to capture complex label co-occurrence patterns in the initial stage. Secondly, a fixed threshold was used to binarize the logits during the prediction phase, i.e., if the prediction probability is greater than 0.7, the label is considered to exist. This hard threshold strategy increased the optimization difficulty in the early training stage, leading the initial output of the model to tend to 0. However, as training progressed, the model gradually adjusted its parameters to adapt to the threshold constraints, and the prediction performance indicators improved rapidly in the last 20 epochs. In addition, in multi-label classification tasks, a prediction result is only considered correct if all relevant labels are correctly identified, and a single label misjudgment will also lead to a sample-level error. This strict evaluation mechanism amplifies the errors caused by the randomness of initial model parameters. Figures 5,

6, 7, 8, 9 and 10 specifically shows the F1 performance of the RBL-BiLA model on each category, with an overall trend of slow convergence in the early stage and rapid improvement in the last 20 epochs.

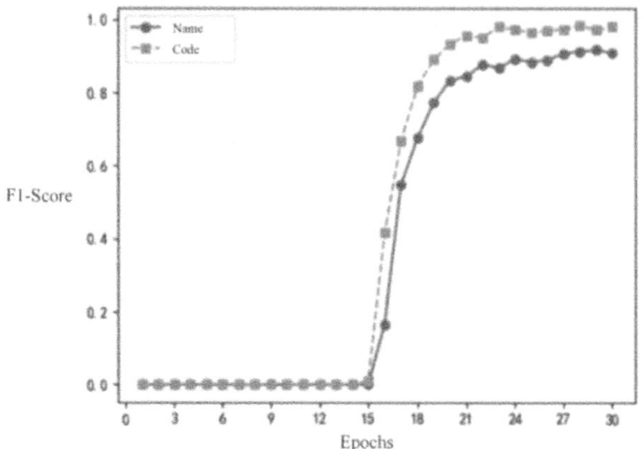


**Fig. 5.** Related category F1 value of steel type attribute

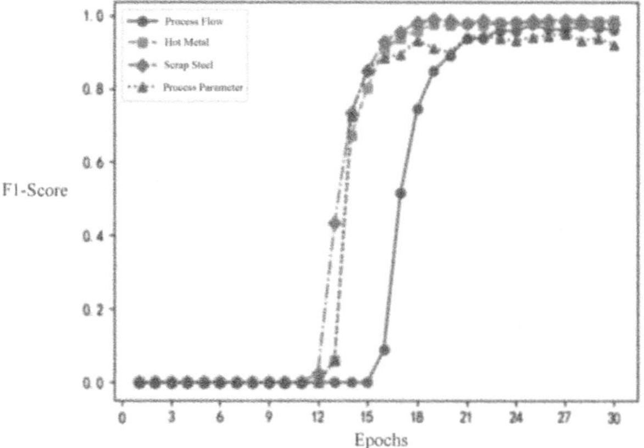


**Fig. 6.** Process parameter related category F1 value

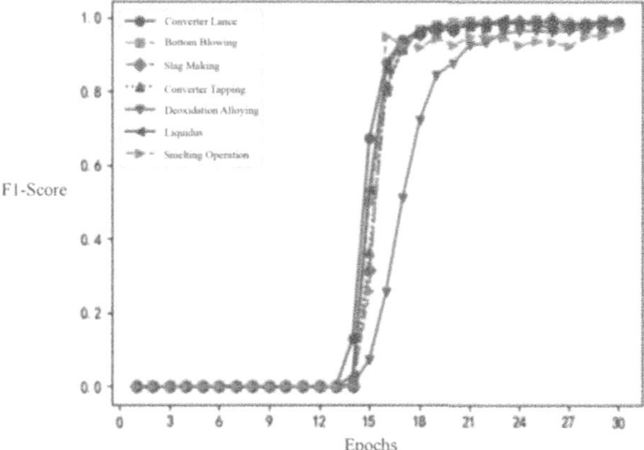


**Fig. 7.** F1 value of converter smelting operation related categories

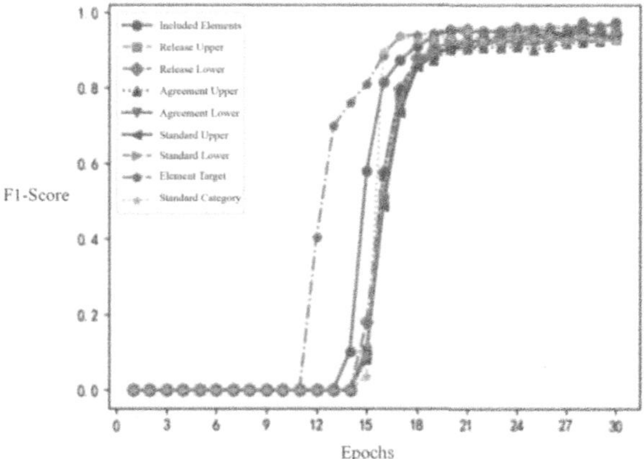


**Fig. 8.** Steel grade standard related category F1 value

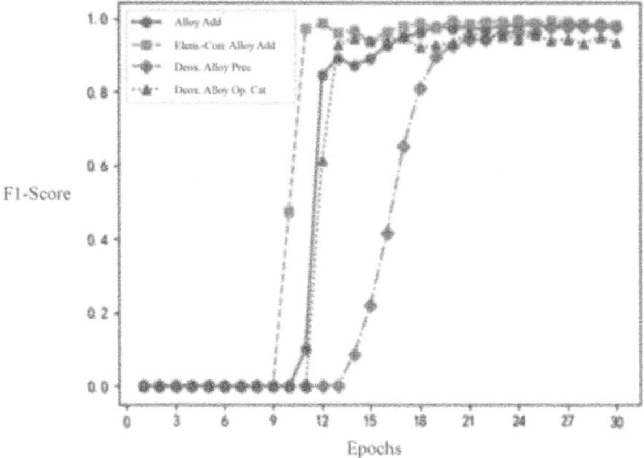


**Fig. 9.** Deoxygenation alloy operation related category F1 value

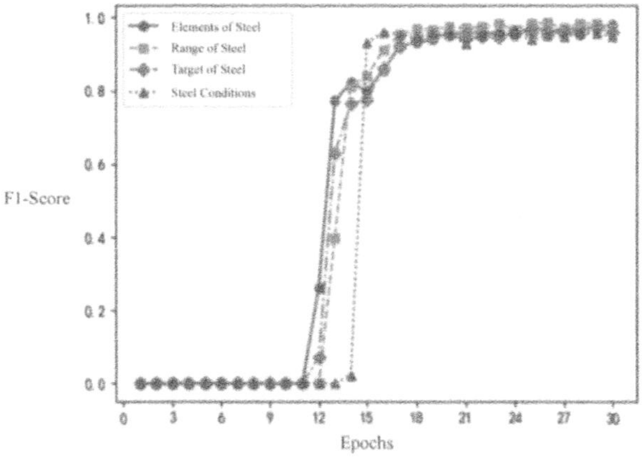


**Fig. 10.** Related category F1 value of water supply for refined steel

To verify the effectiveness of the proposed model, this paper selects RoBERTa and RoBERTa-TextCNN models as comparative methods, and conducts experiments on the self-constructed converter steelmaking multi-intention recognition training data. The results show that the proposed RBL-BiLA model outperforms the other two models in all metrics. The experimental results are shown in Table 3.

**Table 3.** Experimental results of converter steelmaking intent identification

| Model | Precision(%) | Recall(%) | F1(%) |
|---|---|---|---|
| RoBERTa | 89.91 | 90.05 | 90.13 |
| RoBERTa-TextCNN | 91.34 | 91.57 | 91.74 |

*(continued)*

**Table 3.** *(continued)*

| Model | Precision(%) | Recall(%) | F1(%) |
|---|---|---|---|
| OURS(RBL-BiLA) | 94.12 | 95.33 | 95.63 |

To demonstrate the effectiveness of each module in the proposed model, ablation contrast experiments were conducted on the local attention mechanism and dynamic fusion mechanism, as shown in Table 4.The experimental results show that after removing the local attention mechanism, the model's F1-score decreases from 95.63% to 94.82%, with a reduction of 0.81%. This indicates that the local attention mechanism can enhance the ability to capture contextual dependencies by focusing on key information in the question.

**Table 4.** Results of ablation experiments

| Model | Precision(%) | Recall(%) | F1(%) |
|---|---|---|---|
| Text Model | 94.12 | 95.33 | 95.63 |
| Local Attention Mechanism | 94.23 | 94.73 | 94.82 |
| Dynamic Fusion Mechanism | 92.86 | 93.41 | 93.29 |

For the dynamic fusion mechanism, when this module is removed, the F1-score drops significantly to 93.29%, which proves that this module plays a core role in multi-level feature integration. Through an adaptive weight allocation strategy, dynamic fusion integrates features from different Transformer layers, enabling the model to automatically select features that are more important for intention recognition based on actual semantic information, thereby improving recognition efficiency.

## 4 Conclusion and Future Outlook

To address issues in traditional converter steelmaking processes, such as fragmented knowledge, heavy reliance on expert experience, and multi-intention semantic coupling, this paper constructs a converter steelmaking knowledge graph to integrate multi-source data and proposes a deep learning model that incorporates a dynamic weight mechanism, Bidirectional Gated Recurrent Unit (BiGRU), and a local attention mechanism. This enables accurate multi-intention parsing in complex industrial scenarios.The developed RBL-BiLA multi-intention parsing model adopts a two-stage optimization strategy of "pre-training + domain fine-tuning." Through a dynamic weight fusion module, it adaptively weights the general semantic representations of RoBERTa and entity embedding vectors in the steelmaking domain. This approach effectively alleviates the semantic deviation problem of traditional BERT-series models in modeling professional terminology and addresses the semantic coupling issue in multi-label co-occurrence scenarios, enabling the model to simultaneously recognize three or more types of composite intentions.

Experimental results demonstrate that the proposed model outperforms baseline models such as RoBERTa and RoBERTa-TextCNN in intention recognition accuracy. It can accurately capture core semantic units in multi-intention composite questions, such as "What factors affect the carbon content at the converter endpoint, and how to optimize oxygen lance operation?" This research not only provides a complete solution encompassing model selection, knowledge fusion, and engineering deployment for multi-intent parsing tasks in the industrial field, but also offers valuable references for the integrated application of knowledge graphs and deep learning in process industries.

## References

Beden, S., Lakshmanan, K., Giannetti, C., et al.: Steelmaking predictive analytics based on random forest and semantic reasoning. Appl. Sci. **13**(23), 12778 (2023). https://doi.org/10.3390/app132312778

Chen, Y.Z.: An intelligent question answering system for course learning based on knowledge graph [Doctoral dissertation, Lappeenranta-Lahti University of Technology and Hebei University of Technology] (2024)

Dar, A.R., Hashmy, R., Anwar, S.M., et al.: Semantic knowledge graph fusion for fake news detection: unifying content-based features and evidence-based analysis in the COVID-19 infodemic. PLoS ONE **20**(7), e0321919 (2025)

Fan, J.X.: Research on emergency knowledge acquisition method of urban rail transit based on knowledge graph [Master's dissertation, Beijing Jiaotong University] (2024)

Fu, L.J., Cao, Y., Bai, Y., et al.: Development status and prospect of vertical domain knowledge graph in China. Appl. Res. Comput. **38**(11), 3201–3214 (2021)

Hu, S., Zhang, H., Zhang, W.: Domain knowledge graph question answering based on semantic analysis and data augmentation. Appl. Sci. **13**(15) (2023)

Kang, Y.: Research on theory and method of individualized intelligent question answering of policy text based on knowledge management [Master's dissertation, Tianjin University of Finance and Economics] (2023)

Li, M., Qiao, Y., Lee, B.: Multi-view intrusion detection framework using deep learning and knowledge graphs. Information **16**(5), 377 (2025). https://doi.org/10.3390/info16050377

Liu, X., et al.: Construction of an event knowledge graph based on a dynamic resource scheduling optimization algorithm and semantic graph convolutional neural networks. Electronics **13**(1), 11 (2024). https://doi.org/10.3390/electronics13010011

Meng, X.H., Li, T., Meng, C., et al.: Research on converter steel making expert system based on knowledge graph. In: Proceedings of the 2023 5th International Conference on Internet of Things, Automation and Artificial Intelligence (pp. 861–867) (2023)

Ren, Y.K., Xie, Z.P.: Intention recognition accuracy enhancing method for large language model. Appl. Res. Comput. **41**(10), 2893–2899 (2024)

Tian, X., Wu, Z.C.: Review of knowledge base question answering based on information retrieval. J. Comput. Res. Dev. **62**(2), 314–335 (2025)

Wang, L.K., Gao, M.T.: Session-based recommendation algorithm based on hypergraph convolution network and target multi-intention perception. Appl. Res. Comput. **41**(1), 32–38+44 (2024)

Wang, M., Wang, H.F., Li, B.H., et al.: Survey on key technologies of new generation knowledge graph. J. Comput. Res. Dev. **59**(9), 1947–1965 (2022)

Wang, X., Li, Y., Wang, H., et al.: MKBQA: question answering over knowledge graph based on semantic analysis and priority marking method. Appl. Sci. **13**(10) (2023)

Wang, X.D., Shangguan, F.Q., Xing, Y., et al.: Research on the low-carbon development technology route of iron and steel enterprises under the "double carbon" target. Chin. J. Eng. **45**(5), 853–862 (2023)

Wu, L.T.: Knowledge management with knowledge graph based on AEC Text [Master's dissertation. Tsinghua University] (2022)

Yang, H.H., Ma, L.W., Li, Z.: Tracing China's steel use from steel flows in the production system to steel footprints in the consumption system. Renew. Sustain. Energy Rev. **172**, 113040 (2023)

Zhang, F.C., Hong, L.K., Xu, Y.: Prospects for green steel making technology with low carbon emissions in China. Carbon Energy **6**(2), e456 (2024)

Zhou, Y., Fan, Y.S., Sun, S., et al.: Hierarchical semantic parsing approach for multi-hop question answering on knowledge graphs. Appl. Res. Comput. **42**(3), 714–719 (2025)

# Construction of an English Essay Topic and Opinion Analysis Model Based on PLDA-NTM and Bi-GRU

Qiyan Cha(✉)

Baoshan University, Baoshan, Yunnan, China
693851550@qq.com

**Abstract.** As a global lingua franca, English plays a crucial role in expanding Chinese students' knowledge and broadening their horizons. English writing, as an important output method of language competence, has attracted widespread attention. However, the "one-to-many" model in current teaching limits teachers' timely and accurate guidance on each essay, hindering the improvement of students' writing skills. With the development of artificial intelligence technology, automatic essay scoring systems have emerged, which can alleviate teachers' burdens while improving feedback efficiency. Nevertheless, existing systems are insufficient in topic mining and sentiment analysis, restricting the accuracy of content evaluation. Therefore, constructing a model capable of in-depth analysis of essay topics and sentiments holds significant practical value. This paper proposes an English essay analysis model integrating topic semantics and sentiment opinions. By extracting deep-level topic information of essays, combining topic-based sentiment discrimination, and calculating semantic similarity at the sentence level, the model achieves a comprehensive evaluation of essay content quality. The specific research contents include: (1) Constructing a contextual topic recognition model integrating conceptual knowledge (PLDA-NTM), which uses the Probase concept base to enhance the semantic representation ability of the LDA model and combines BERT to generate local semantics, thereby improving topic clustering quality. (2) Establishing a topic-based sentiment opinion analysis model (TA-GRU), which adopts a Bi-GRU network to extract contextual features and identifies sentiment tendency categories through a topic-level attention mechanism. (3) Designing a topic and opinion quality analysis module that calculates topic similarity, content relevance, and sentiment quality at the sentence level, and generates scores and comments through weighting. (4) Verified by multiple sets of corpus experiments, the proposed model demonstrates higher accuracy and practical value in automatic scoring.

**Keywords:** English Essay · PLDA-NTM · Bi-GRU · Model Construction

## 1 Introduction

English writing ability is a core indicator for evaluating students' comprehensive language competence. In practical teaching, students' essays often have problems such as unclear themes and biased opinions. However, limited by the "one-to-many" teaching


H. Yan et al. (Eds.): AIS&P 2025, CCIS 2918, pp. 402–417, 2026.
https://doi.org/10.1007/978-981-95-9884-7_28

model, teachers struggle to provide timely in-depth feedback. Existing automatic essay scoring systems (such as Pigaiwang and E-rater) can alleviate the marking pressure, but their feedback mainly focuses on surface features such as vocabulary and grammar, lacking in-depth analysis of deep quality dimensions of essay content such as topic semantics and sentiment opinions. This limitation restricts the application value of automatic scoring systems in real teaching scenarios. Therefore, constructing a model that can deeply analyze the thematic ideas and emotional attitudes of essays is of great significance for improving the accuracy and teaching guidance of automatic scoring. This study aims to develop a model integrating topic recognition and sentiment analysis to achieve a comprehensive evaluation of the content quality of students' English essays (Fu and Chen 2022).

### 1.1 Research Progress

Topic models have undergone an important evolution from traditional methods to deep learning: Methods based on external knowledge (such as DF-LDA, KG-LDA) improve the semantic consistency of topics by introducing prior knowledge such as lexical associations or knowledge graphs; Word vector enhancement methods (such as LF-LDA, WE-PTM) use word vector semantic similarity to improve topic clustering results; Neural network topic models (such as NVDM, CombinedTM) adopt a variational autoencoder architecture and combine pre-trained language models, which can better capture the contextual semantic information of texts (Li and Su 2021).

Sentiment analysis research has developed from document-level and sentence-level to aspect-level and topic-level. Topic-based sentiment analysis is particularly suitable for English essay evaluation: Deep learning methods (such as LSTM, Bi-GRU) have become mainstream, which can effectively capture text semantic features; The introduction of attention mechanisms enables the model to focus on contexts related to specific topics, improving the accuracy of sentiment judgment; Innovative architectures such as multi-granularity attention networks further optimize the modeling of semantic associations between topics and contexts (Xie 2022).

### 1.2 Research Objectives and Contents

The main contents of this study include: Constructing the PLDA-NTM topic recognition model to improve the accuracy of essay topic recognition by integrating the Probase conceptual knowledge base and BERT contextual semantic representation; Designing the TA-GRU sentiment opinion analysis model, which uses a bidirectional GRU network and a topic-level attention mechanism to accurately analyze topic-based sentiment tendencies; Developing an essay topic and opinion quality analysis module to comprehensively evaluate the topic relevance and sentiment opinion quality of essays, and generate comprehensive scores and comments (Attali and Burstein 2004). Through the above research, this study aims to provide more in-depth semantic analysis for English essay automatic scoring and improve the marking quality and application value of the system.

## 2 Related Theories and Methods

### 2.1 Definition of Opinion

In the field of natural language processing, opinion is a broad concept covering multiple dimensions such as emotion, attitude, and evaluation. A complete opinion usually includes core elements such as opinion holder, evaluation object, and sentiment tendency. For ease of understanding, we take an English essay on "the importance of part-time jobs for college students" as an example. Analyzing the sentences in it, we find that students express either positive or negative attitudes. For example, some sentences clearly support part-time jobs, while others hold opposing views. It can be seen that opinion analysis of English essays mainly involves two core parts: one is the target object g evaluated by the opinion (such as a topic, entity, or its attribute); the other is the sentiment tendency s expressed towards this target (such as positive, neutral, or negative). In this paper, the opinion evaluation object of an essay is defined as the "topic", and the sentiment tendency towards this topic is defined as the "topic-based sentiment opinion category" (Guo 2023).

Therefore, the goal of the model in this paper is: First, identify the core topic of the essay as the evaluation object through the topic model, and then use the sentiment analysis model to extract the explicit or implicit sentiment tendency expressed by the essay around the topic, thereby completing a comprehensive topic and opinion analysis.

### 2.2 Text Topic Models

Topic models can automatically discover potential semantic topics from a collection of documents. Applying them to English essay analysis has significant advantages: First, they can use global co-occurrence information of words to identify both explicit and implicit opinion topics; Second, they can automatically classify semantically related words (such as "income" and "salary") into higher-level conceptual topics; Finally, by analyzing the topic distribution of essays, we can effectively evaluate whether their content is relevant to the topic. The topic recognition model proposed in this paper is an improvement on the classic LDA model. Therefore, this section will first introduce the basic principles of the LDA model, analyze its limitations, and then introduce the neural network topic model NVDM-GSM which can better capture contextual semantics.

**LDA Topic Model.** LDA is a generative probabilistic model whose core assumption is that a document is a mixture of multiple topics, and each topic is a probability distribution over words. The model uses a bag-of-words (BOW) representation, ignoring word order, and smooths the document-topic distribution and topic-word distribution through Dirichlet priors. Its generation process can be briefly summarized as: Sampling a topic distribution for each document, then for each word in the document, sampling a topic according to the document's topic distribution, and then generating a specific word according to the word distribution of that topic (Chauhan and Shah 2021). Model parameters are usually estimated through inference methods such as Gibbs sampling. However, the standard LDA model has two main problems: First, it lacks a mechanism to integrate domain knowledge, which may lead to semantically inconsistent topics; Second, the BOW representation ignores word order and contextual information, limiting its ability to capture deep semantics, as shown in Fig. 1.

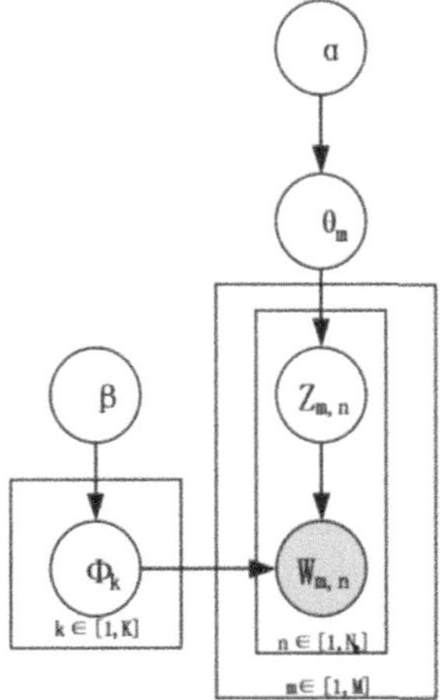


**Fig. 1.** Graphical Representation of the LDA Topic Model (Chauhan and Shah 2021)

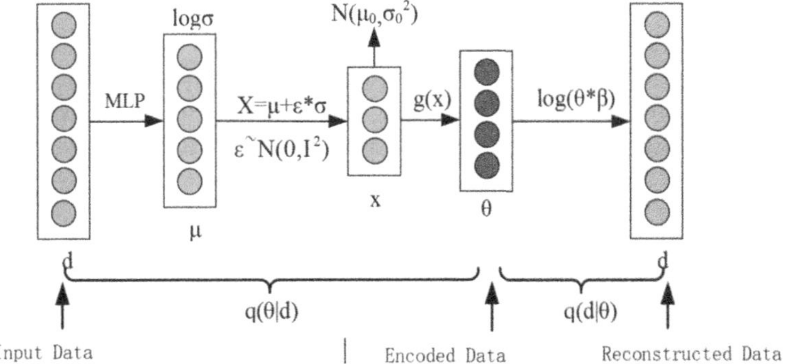


**Fig. 2.** NVDM-GSM Neural Network Topic Model

**NVDM-GSM Neural Network Topic Model.** The NVDM-GSM model uses a neural network architecture to model the topic generation process, overcoming certain limitations of traditional LDA. This model adopts an encoder-decoder structure: The encoder (inference network) maps the document to a latent topic distribution (usually assumed to be a Gaussian distribution), and the decoder (generation network) reconstructs the document based on this topic distribution. Compared with LDA, NVDM-GSM no longer relies on BOW representation, can directly process word sequences, and can capture more complex semantic features through non-linear transformations of neural networks, thus promising to generate higher-quality and more coherent topic representations, as shown in Fig. 2.

## 2.3 Pre-trained Language Models

Pre-trained language models are trained on large-scale unlabeled corpora, enabling them to learn rich linguistic knowledge (including syntax and semantics) and generate high-quality word vectors or contextual representations. Through the "pre-training + fine-tuning" paradigm, the general knowledge learned can be effectively transferred to downstream tasks (Blei and Jordan 2003).

**Word2Vec Model.** Word2Vec is a classic static word vector model whose core idea is "a word is determined by its context". It mainly includes two architectures: CBOW (predicting the central word through context) and Skip-gram (predicting the context through the central word). Among them, the Skip-gram model usually performs better in capturing complex semantic relationships of words. To improve training efficiency, Skip-gram often adopts negative sampling technology. This technology converts a multi-classification problem into a series of binary classification problems, that is, for each training sample (central word, context word), several non-context words are randomly sampled as negative samples. The goal of the model is to maximize the probability of co-occurrence between the central word and the real context words, while minimizing the probability of co-occurrence with negative samples (Karras et al. 2022; Chaubey et al. 2021). This method significantly reduces computational overhead while ensuring the quality of word vectors. The structure diagram of the Skip-gram model is shown in Fig. 3. First, we define the window size as K, the vocabulary size as V, the hidden layer dimension as D, and use the central word wi to predict the surrounding words

$$\{wi - K,\ wi - K + 1, \ldots,\ wi - 1,\ wi + 1, \ldots, wi + K\} \tag{1}$$

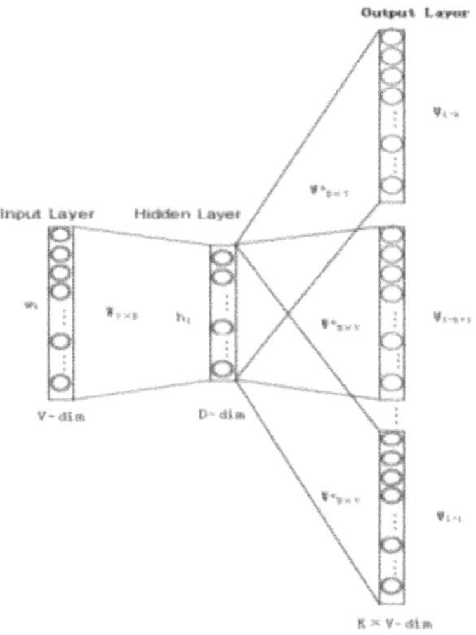


**Fig. 3.** Skip-gram Pre-trained Language Model (Karras et al. 2022)

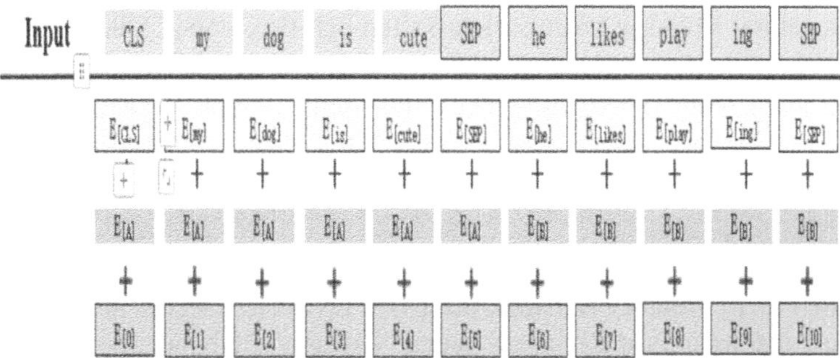


**Fig. 4.** Input Part of BERT's Pre-training Phase (Chaubey et al. 2021)

**GPT Model.** As a classic pre-trained language model, GPT (Generative Pre-Training) adopts the Decoder part of Transformer as its core architecture and predicts subsequent words using contextual information in an autoregressive manner. The model includes two stages: pre-training and fine-tuning. The pre-training stage learns language representation by maximizing the likelihood estimation of sequences; the fine-tuning stage adapts it to downstream tasks (Joshi et al. 2023; Schneider 2023). However, GPT's unidirectional modeling mechanism limits its ability to capture downstream semantic information, and it has certain limitations in handling tasks that require global contextual understanding. The GPT model consists of 12 stacked decoders. In the model pre-training stage, the maximum likelihood estimation L(x) of a given text sequence $x = x1 \ldots xn$ is optimized using conventional language modeling methods.The model prediction implemented using the following formulas:

$$h[0] = ex'We + Wp \tag{2}$$

$$h[l] = \mathrm{Transformer} - \mathrm{Block}(h[l-1]),\ l \in \{1, 2..., L\} \tag{3}$$

$$p(x) = \mathrm{softmax}(h[L]WeT) \tag{4}$$

In the above formulas, x' represents a window word sequence of length k, $x' = xi - k \ldots xi - 1$. $ex' \in k \times |V|$ represents the one-hot vector of x', $We \in |V| \times d$ represents the word vector matrix, $Wp \in k \times d$ represents the position vector matrix, and L represents the total number of layers of the Transformer.

**BERT Model.** BERT is a currently excellent pre-trained language model whose core innovation lies in adopting a bidirectional Transformer encoder, which can simultaneously capture the left and right contextual information of words. In the pre-training stage, BERT conducts joint training through two tasks: Masked Language Modeling (MLM) and Next Sentence Prediction (NSP), thereby learning rich semantic and inter-sentence relationship representations. The input of the model is composed of the sum of word vectors, segment vectors, and position vectors, and the final output of the special token [CLS] is used as the semantic representation of the entire sequence (Zhou et al. 2022; Zhang et al. 2023). BERT has strong transfer learning capabilities and can quickly adapt to various natural language processing tasks through fine-tuning. Studies have shown that integrating the contextual representations generated by BERT into topic models can effectively improve the consistency and interpretability of topic semantics, as shown in Fig. 4.

### 2.4 Deep Neural Networks

**Long Short-Term Memory Network (LSTM).** LSTM is designed to solve the gradient problem that traditional Recurrent Neural Networks (RNN) are prone to encounter in long sequences. By introducing three gating mechanisms: forget gate, input gate, and output gate, it selectively updates and transmits information, thereby effectively capturing long-distance dependencies. The hidden state and cell state at each moment jointly maintain the long-term memory and short-term information of the sequence, as shown in Fig. 5.

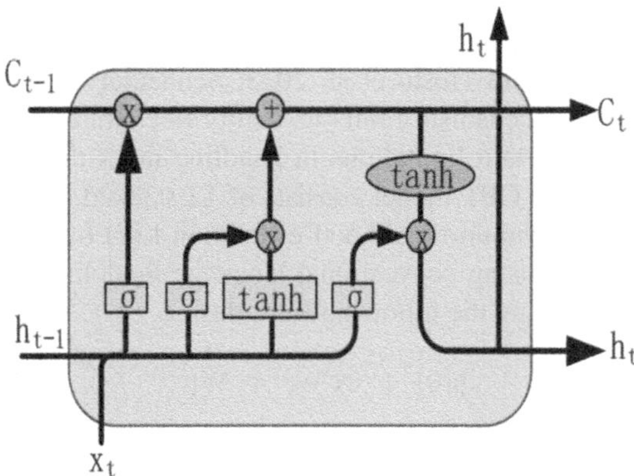


**Fig. 5.** Structural Diagram of the LSTM Model at a Single Time Step

**Gated Recurrent Unit (GRU).** GRU is a variant of LSTM with a more concise structure. It combines the forget gate and input gate of LSTM into an update gate, and introduces a reset gate to control the degree of retention of historical information. Despite having fewer parameters, GRU can still effectively model sequence dependencies and achieve performance comparable to LSTM in many tasks (Choi et al. 2020), as shown in Fig. 5. Bidirectional GRU (Bi-GRU) further enhances semantic modeling capabilities by stacking GRU layers that transmit information in forward and reverse directions.The structure diagram of a single time step of the GRU model is shown in Fig. 6. The model first obtains two gate states, rt and zt, through the input xt at the current time step and the hidden state ht − 1 passed down from the previous time step. The calculation formulas are as follows:

$$\mathrm{rt} = \sigma(\mathrm{wr}[\mathrm{ht} - 1,\ \mathrm{xt}]) \tag{5}$$

$$zt = \sigma(wz[ht - 1,\ xt]) \tag{6}$$

In the formulas, rt is the reset gate at the current time step t, calculated as Formula 5, and zt is the update gate at the current time step t, calculated as Formula 6. After obtaining the gate signals rt and zt, the GRU model obtains the reset data ht' after the reset gate through Formula 7. Among them, ht' mainly includes the input data xt of the current node and integrates the hidden state ht − 1 of the previous node. The reset data h' is equivalent to recording the state at the current time step. Finally, the model obtains the hidden state ht of the node by simultaneously implementing two steps: forgetting and remembering through Formula 8. The value of zt in Formula 9 ranges between 0 and 1. The closer the value is to 1, the more data of the current node information is remembered by the model, and the less information is inherited from the hidden state of the previous node; the closer the value is to 0, the more data of the current node information is forgotten by the model, and the more information is inherited from the hidden state of the previous node.

$$ht' = \tanh\left(Wh'[rt * ht - 1,\ xt]\right) \tag{7}$$

$$ht = (1 - zt) * ht - 1 + zt * ht' \tag{8}$$

$$yt = \sigma(W0ht) \tag{9}$$

The GRU model uses the update gate to simultaneously implement two steps: forgetting and remembering. It can be seen that the GRU model simplifies the LSTM model and reduces the number of model parameters. By analyzing Formula 8, we can see that the forgetting zt and remembering 1 − zt here are linked. That is, we will selectively forget the information ht − 1 transmitted from the previous node according to the weight zt, and then use ht' containing the current input to make up for it. This ensures that the model is always in a "constant" state. Finally, the model uses Formula 9 to obtain the output representation of the node.

In addition, in some special tasks, such as text reading comprehension, part-of-speech tagging, and text translation, the output results are not only related to the contextual information of the current word but also to the subsequent information. Such tasks require the model to understand the text context before drawing the final conclusion (Gu et al. 2023). Therefore, a network layer that transmits information in the reverse order of the text can be stacked on the network to add subsequent information for the current time step. Figure 7 shows the Bi-GRU model unfolded over time.From Fig. 7, we can see that the Bi-GRU neural network model consists of two layers of GRU networks. Each layer in the model has the same input but different directions of information transmission. At this time, we assume that the first layer of GRU transmits in sequence order, and the second layer of GRU transmits in reverse sequence order. Let the current time step be t, and the hidden states at this time are ht(1) and ht(2), which are calculated by Formula 8. The Bi-GRU model calculates the hidden state ht at the current time step t through Formula 10, where $\oplus$ represents the vector concatenation operation:

$$ht = ht(1) \oplus ht(2) \tag{10}$$

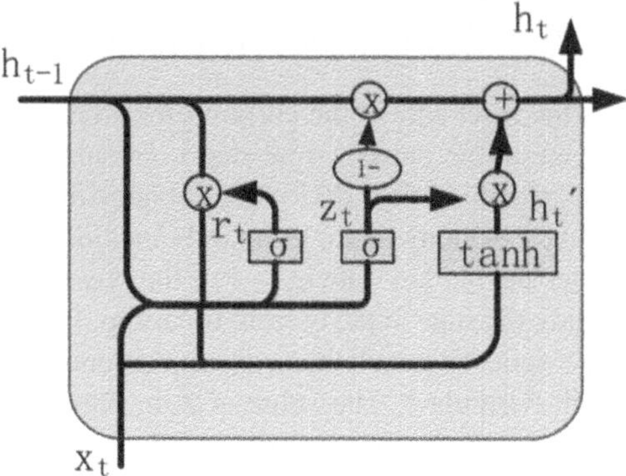


**Fig. 6.** Structural Diagram of the GRU Model at a Single Time Step

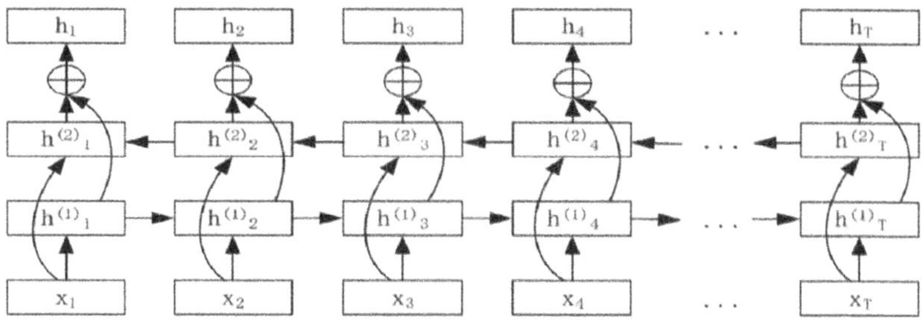


**Fig. 7.** Bi-GRU Model Unfolded over Time

**Attention Mechanism.** The attention mechanism enables the model to focus on the information in the input that is most relevant to a specific target (such as a topic) by dynamically calculating the importance weight of different parts for the current task. Its core is to construct query, key, and value vectors, obtain the attention distribution by calculating the matching degree between queries and keys, and then perform weighted summation of the value vectors. Introducing topic-level attention in sentiment analysis can effectively improve the accuracy of the model in identifying topic-related sentiments, as shown in Fig. 8.

### 2.5 Probase Conceptual Knowledge Base

Probase is a large-scale probabilistic conceptual knowledge base that extracts a large number of "concept-instance" relationships from massive web pages through automated methods. Compared with other knowledge bases, Probase not only covers a wide range of fields but also introduces typicality scores and ambiguity scores to probabilistically model the membership strength and semantic boundaries of concepts, which is more in line with the uncertainty in human cognition. The typicality score comprehensively measures the representativeness of an instance in a concept; the ambiguity score evaluates the semantic breadth of the concept itself. This paper uses the prior conceptual knowledge

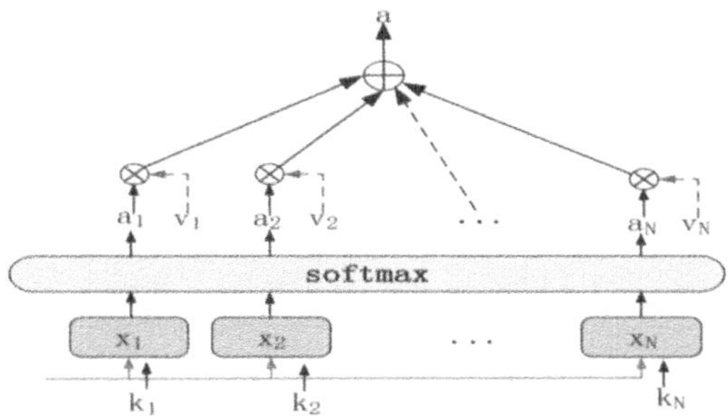


**Fig. 8.** Structural Diagram of the Self-Attention Mechanism Module

in Probase and integrates it into the LDA topic model to enhance the consistency and richness of topic semantics.

### 2.6 Text Similarity Algorithms

**Euclidean Distance.** Euclidean distance measures similarity by calculating the straight-line distance between points in a vector space. A smaller distance indicates more similar semantics. This method is intuitive but does not consider the direction of the vector and is sensitive to high-dimensional sparse vectors.

**Cosine Similarity.** Cosine similarity focuses on the angle between vectors rather than the absolute distance. The closer its value is to 1, the more similar the semantics. This algorithm is not sensitive to the modulus length and is more suitable for measuring the semantic relevance of text vectors. This paper uses cosine similarity to evaluate the topic relevance of essay content and the matching degree of sentiment opinions.

## 3 Topic-Based Opinion Analysis Model

### 3.1 Overall Model Structure

This chapter proposes a topic and opinion analysis model for English essays written by Chinese learners, aiming to automatically evaluate the topic relevance and sentiment tendency of essays. The model first preprocesses the essays to be graded, including filtering special characters, word segmentation and sentence segmentation, part-of-speech tagging, lemmatization, and stopword removal. Then, the contextual topic recognition model integrating conceptual knowledge (PLDA-NTM) is used to extract the global topic distribution and local semantic features of the essay, obtaining deep topic representations. This result is further input into the topic-based Bi-GRU sentiment analysis model (TA-GRU) to identify the sentiment opinions of the essay under specific topics. Finally, through the topic and opinion quality analysis module, the topic relevance and sentiment matching degree of the essay are comprehensively calculated, and the score and targeted comments are output (Osmani and Mohasefi 2023).

### 3.2 Preprocessing Module

To improve the quality of topic recognition, the system uses the Stanford CoreNLP tool to standardize the input text. The process includes:

Special character filtering: Removing unconventional symbols and incorrect punctuation;

Word segmentation and sentence segmentation: Splitting sentences and words according to space and punctuation rules;

Part-of-speech tagging: Determining word part-of-speech based on context to eliminate ambiguity;

Lemmatization: Restoring words to their dictionary prototypes to retain semantic integrity;

Stopword removal: Eliminating high-frequency non-content words to reduce noise in topic modeling.

This process ensures that the input is a set of pure word roots, providing a high-quality data foundation for subsequent semantic modeling.

### 3.3 Contextual Topic Recognition Model Integrating Conceptual Knowledge (PLDA-NTM)

PLDA-NTM combines conceptual priors and contextual semantics to improve topic recognition results. Its structure includes two parts:

**LDA Topic Model Based on Conceptual Knowledge.** The traditional LDA model uses symmetric prior parameters and does not integrate external semantic knowledge. This paper introduces the Probase concept base to construct asymmetric prior parameters. The specific steps are:

Mapping the root set of each essay to the concept set in Probase, calculating concept probabilities, and selecting highly representative concepts;

Using the K-Medoids algorithm to cluster concepts to form concept clusters consistent with the number of topics;

Encoding the concept cluster information into the Dirichlet prior parameters $\alpha$ and $\beta$ of LDA, enabling the topic sampling process to have semantic guidance capabilities.

This mechanism enhances topic consistency and interpretability and effectively addresses the data sparsity problem of essay corpora.

**Topic Modeling Based on Contextual Semantics.** To further capture word order and local semantics, the model introduces contextual vectors generated by BERT and concatenates them with PLDA topic vectors to form a joint representation integrating global and local information. This representation is input into a neural topic model (NTM) based on Gaussian distribution, trained through an encoder-decoder structure, and finally outputs a 32-dimensional topic embedding vector. The K-means clustering algorithm is used to assign topic labels to complete topic recognition.

### 3.4 Topic-Based Sentiment Opinion Analysis Model (TA-GRU)

The TA-GRU model combines bidirectional GRU with a topic attention mechanism to identify sentence-level sentiment tendencies. Its structure is as follows:

Word embedding layer: Concatenating sentence word vectors and topic vectors into a unified representation;

Bi-GRU layer: Extracting bidirectional contextual semantic features of sentences;

Topic attention layer: Calculating the weight of each word according to the topic vector to highlight sentiment information related to the topic;

Output layer: Outputting the probability of sentiment categories through linear transformation and the Softmax function.

This model can effectively identify the sentiment polarity related to the topic implied in sentences.

### 3.5 Essay Topic and Opinion Quality Analysis Module

This module comprehensively evaluates essay quality from two aspects: topic relevance and sentiment matching degree:

Topic relevance score: Calculating the cosine similarity between the sentence topic vector and the topic vector of the question, and obtaining the topic relevance score through weighted average;

Sentiment score: Comprehensively considering the sentiment tendencies of all topic-relevant sentences to judge the overall sentiment polarity and intensity;

Comprehensive score: Combining the topic relevance score and sentiment matching degree, generating the final score through weighting, and automatically generating comments to provide feedback from multiple perspectives such as topic consistency, coherence, and sentiment rationality (Rakshit et al. 2022; Kumaran and Chitrakala 2022).

## 4 Model Experiments and Analysis

### 4.1 Experimental Environment and Data

The experimental hardware includes an Intel i5–10400 processor and 64GB memory. The software environment includes the Windows 10 operating system, Python 3.6, and corresponding deep learning frameworks. The experimental data are from the Chinese Learner English Corpus (CLEC) and the International Corpus Network of Asian Learners of English (ICNALE), including 15,246 English essays under 10 essay topics. The data are divided into a training set and a test set in a ratio of 3:1 for training and verification of the topic recognition and sentiment analysis models.

### 4.2 Parameter Settings and Evaluation Indicators

In the PLDA-NTM topic recognition model, the number of topics K is set to 25, and the contextual semantic weighting parameter gram is set to 50. Experimental verification shows that this is the optimal configuration. The sentiment analysis model TA-GRU uses

300-dimensional Word2Vec word vectors, the number of Bi-GRU hidden units is 256, the dropout rate is set to 0.2, and the Adam optimizer is used for training.

The model evaluation adopts the following indicators:

Topic perplexity: Measuring the uncertainty of the model in predicting topics of unknown texts; the lower the value, the better the performance;

Topic coherence: Evaluating the semantic relevance of words under the same topic; the higher the value, the better the topic quality;

Classification indicators: Including precision, recall, and F1-score, used to evaluate the effect of sentiment classification;

Correlation with human evaluation: Measuring the consistency between model scores and human scores through the Pearson correlation coefficient.

### 4.3 Experimental Results of the Topic Recognition Model

The PLDA-NTM model achieves the best performance on both the CLEC and ICNALE datasets, with perplexity of 821 and 803, and coherence of 0.823 and 0.831, respectively, which are significantly better than baseline models such as traditional LDA, DF-LDA, KGE-LDA, and NVDM. UMAP visualization shows that the topic clusters generated by PLDA-NTM have clearer boundaries and higher semantic discrimination. In the topic word display, the words extracted by PLDA-NTM, such as "skill", "company", and "salary", are highly relevant to the topic "job", while the results of the LDA model are mixed with irrelevant words such as "time" and "parents", indicating that the proposed model has obvious advantages in topic semantic consistency.

### 4.4 Experimental Results of the Sentiment Analysis Model

The TA-GRU model is subjected to ten-fold cross-validation on the ICNALE dataset. When the sentiment threshold is set to 0.45, the best performance is achieved, with a precision of 80.15%, a recall of 68.25%, and an F1-score of 73.72%. Compared with existing models such as BB_twtr and CrystalNest, TA-GRU still achieves competitive results when only using labeled data, reflecting the effectiveness and practicality of the model. With the increase in the amount of training data, various indicators of the model continue to improve, indicating that it has good scalability.

### 4.5 Correlation Analysis with Human Scores

A comparative experiment between model scores and human scores is conducted on 1000 essays. The weight of topic relevance is set to 60%, and the weight of sentiment expression is set to 40%. The average error between the model scores and the teacher scores is 1.61 points, and the Pearson correlation coefficient reaches 0.865, indicating that the model output is highly consistent with the human evaluation standards and has the credibility of practical application.The English essay topic recognition model PLDA-NTM proposed in this paper has achieved improvements in the evaluation indicators of topic perplexity and topic coherence through two rounds of improvements based on the Probase Conceptual Knowledge Base and neural network topic models. Subsequently, the English essay topic recognition model visualizes the text topic clustering results using 2DUMAP technology, and the results are shown in Figs. 9 and 10.

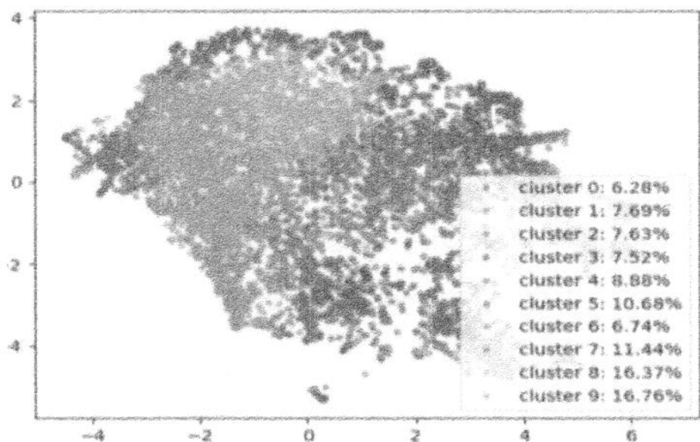


**Fig. 9.** Clustering Diagram of the Contextual Topic Recognition Model

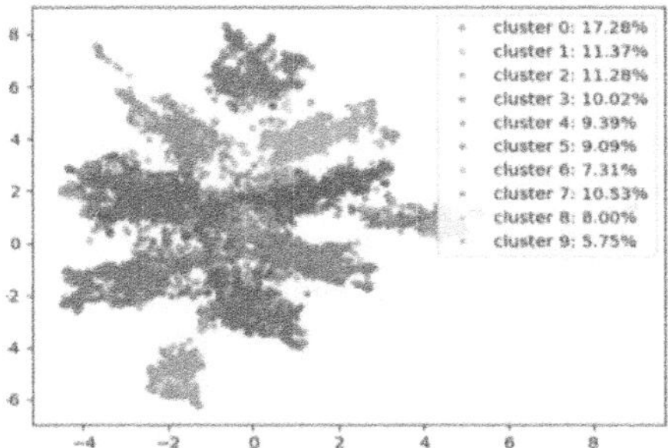


**Fig. 10.** Clustering Diagram of the Contextual Topic Recognition Model Integrating Conceptual Knowledge

### 4.6 System Interface Example

A visual marking interface is developed based on the proposed model, which supports inputting essay topics and content, and automatically outputs topic-relevant sentence recognition, sentiment tendency analysis, and comprehensive scores and comments, verifying the usability and effectiveness of the model in real scenarios.

## 5 Summary and Outlook

### 5.1 Summary

Aiming at the shortcomings of existing English essay automatic scoring systems in content analysis, this paper proposes a comprehensive model integrating topic recognition and sentiment analysis. The model improves the ability to evaluate essay content quality through the following innovations: (1) Constructing a contextual topic recognition model integrating conceptual knowledge (PLDA-NTM). By introducing the Probase conceptual knowledge base as the prior knowledge of the LDA topic model and combining the contextual semantic representation generated by BERT, the accuracy and semantic consistency of topic modeling are effectively improved. (2) Designing a topic-based Bi-GRU sentiment analysis model (TA-GRU). Using a bidirectional gated recurrent unit combined with a topic-level attention mechanism to accurately identify sentence-level sentiment tendencies. (3) Developing an essay topic and opinion quality analysis module. Comprehensively considering the topic relevance and sentiment matching degree of essays, and generating comprehensive scores and targeted comments.

Experimental results show that the proposed model performs excellently on the CLEC and ICNALE datasets, with a topic coherence of over 0.83, a sentiment analysis F1-score of 73.72%, and a correlation coefficient with human scores of 0.865, verifying its effectiveness and practicality.

### 5.2 Future Outlook

Although this study has achieved certain results, there are still the following aspects worthy of in-depth exploration: (1) Optimizing the fusion mechanism of topic models, exploring other fusion strategies besides vector concatenation, and trying more prior distribution assumptions to improve the performance of neural network topic models. (2) Enhancing the feature integration ability of the sentiment analysis model, introducing complementary information such as sentiment dictionaries and part-of-speech features, and designing a local context clustering mechanism to improve analysis accuracy. (3) Expanding the evaluation dimensions of content quality, deepening topic analysis from the paragraph and text levels, and establishing a comprehensive scoring system closer to human evaluation standards.

These improvement directions will further improve the accuracy and practicality of automatic scoring systems in content analysis, and provide more effective technical support for English writing teaching.

## References


Attali, Y., Burstein, J.: Automated essay scoring with e-rater® v. 2.0. ETS Res. Rep. Ser. **2004**(2), i–21 (2004). https://doi.org/10.1002/j.2333-8504.2004.tb02229.x

Blei, D.M., Ng, A.Y., Jordan, M.I.: Latent dirichlet allocation. J. Mach. Learn. Res. **3**, 993–1022 (2003). https://doi.org/10.1162/jmlr.2003.3.4-5.993

Chaubey, Y.P., Singh, M., Sen, D.: Bayesian inference for inverse gaussian data with emphasis on the coefficient of variation. In: Applied Statistics and Data Science: Proceedings of Statistics 2021 Canada, Selected Contributions 6 (pp. 79–96). Springer International Publishing (2021). https://doi.org/10.1007/978-3-030-81801-7_6

Chauhan, U., Shah, A.: Topic modeling using latent dirichlet allocation: a survey. ACM Comput. Surv. (CSUR) **54**(7), 1–35 (2021). https://doi.org/10.1145/3457174

Choi, G., Oh, S., Kim, H.: Improving document-level sentiment classification using importance of sentences. Entropy **22**(12), 1336 (2020). https://doi.org/10.3390/e22121336

Fu, S., Chen, H.: Machine learning-based intelligent scoring system for English essays under the background of modern information technology. Comput. Intell. Neurosci. **2022**, 8819349 (2022). https://doi.org/10.1155/2022/8819349

Guo, J.: A study of college English writing from the perspective of deep learning. In: EAI (ed.), Application of Big Data, Blockchain, and Internet of Things for Education Informatization: Second EAI International Conference, BigIoT-EDU 2022, Virtual Event, July 29–31, 2022, Proceedings, Part I (pp. 1–9). Springer Nature Switzerland (2023). https://doi.org/10.1007/978-3-031-28166-7_1

Gu, T., Zhao, H., He, Z., et al.: Integrating external knowledge into aspect-based sentiment analysis using graph neural network. Knowl.-Based Syst. **259**, 110025 (2023). https://doi.org/10.1016/j.knosys.2023.110025

Joshi, A., Fidalgo, E., Alegre, E., et al.: DeepSumm: exploiting topic models and sequence to sequence networks for extractive text summarization. Expert Syst. Appl. **211**, 118442 (2023). https://doi.org/10.1016/j.eswa.2022.118442

Karras, C., Karras, A., Tsolis, D., et al.: Distributed Gibbs sampling and LDA modelling for large scale big data management on PySpark. In: 2022 7th South-East Europe Design Automation, Computer Engineering, Computer Networks and Social Media Conference (SEEDA-CECNSM) (pp. 1–8). IEEE (2022). https://doi.org/10.1109/SEEDA-CECNSM55416.2022.9977246

Kumaran, P., Chitrakala, S.: Topic adaptive sentiment classification based community detection for social influential gauging in online social networks. Multimedia Tools Appl. 1–40 (2022). https://doi.org/10.1007/s11042-022-12896-8

Li, B., Su, Z.: Research on data mining equipment for teaching English writing based on application. J. Intell. Fuzzy Syst. **40**(2), 3263–3269 (2021). https://doi.org/10.3233/JIFS-202846

Osmani, A., Mohasefi, J.B.: Opinion mining using enriched joint sentiment-topic model. Int. J. Inf. Technol. Decis. Mak. (IJITDM) **22**(1), 313–375 (2023). https://doi.org/10.1142/S0219622023500100

Rakshit, P., Sarkar, P., Ghosh, D., et al.: Sentiment analysis of Twitter data using deep learning. In: Advances in Communication, Devices and Networking: Proceedings of ICCDN 2021 (pp. 495–501). Springer Nature Singapore (2022). https://doi.org/10.1007/978-981-16-9435-6_46

Schneider, J.: Efficient and flexible topic modeling using pretrained embeddings and bag of sentences (2023). arXiv Preprint arXiv:2302.03106. https://doi.org/10.48550/arXiv.2302.03106

Xie, C.: Effectiveness of computer-aided technology for teaching English courses in the internet era. Sci. Program. **2022**, 9183754 (2022). https://doi.org/10.1155/2022/9183754

Zhang, X., Wu, Z., Liu, K., et al.: Text sentiment classification based on BERT embedding and sliced multi-head self-attention bi-GRU. Sensors **23**(3), 1481 (2023). https://doi.org/10.3390/s23031481

Zhou, M., Kong, Y., Lin, J.: Financial topic modeling based on the BERT-LDA embedding. In: 2022 IEEE 20th International Conference on Industrial Informatics (INDIN) (pp. 495–500). IEEE (2022). https://doi.org/10.1109/INDIN51773.2022.9976164

# FewShot-CrackDet an Improved LoRA-Based Few-Shot Intelligent Detection Model for Bridge and Building Structural Cracks

Jiongcheng Shi[1], Tao Luo[2], and Yiming Li[1](✉)

[1] School of Computer Science and Engineering, Ningxia Institute of Science and Technology, Shizuishan, China
Liyiming0206@163.com

[2] Smart City Design Institute, WISDRI City Construction Engineering and Research Incorporation Ltd., Wuhan, China

**Abstract.** Cracks in bridge and building structures are critical safety hazards. Traditional detection relies on manual work or data-hungry AI models, suffering from low efficiency and high costs. This paper proposes FewShot-CrackDet, an improved LoRA-based few-shot detection model. A pre-trained ViT serves as the backbone, enhanced by a cross-layer feature fusion LoRA module to reduce overfitting in few-shot scenarios. A multi-scenario dataset with 400 annotated samples (bridge/building cracks, harsh environments) is constructed. Experiments show the model achieves mAP@ 0.5 of 89.3%, 93.7%, 96.2% under 5/10/50-shot scenarios, outperforming baseline models by 6.8%-12.5% with 75% fewer parameters. It reduces engineering data collection costs and provides an efficient solution for structural safety monitoring.

**Keywords:** Structural crack detection · Few-shot learning · LoRA · Vision Transformer · Construction safety

## 1 Introduction

Bridge and building concrete structures are prone to cracks due to loads, environmental erosion, and material aging. Undetected cracks can expand over time, threatening structural integrity and even leading to catastrophic collapses. The International Association for Bridge and Structural Engineering (IABSE) reports that over 60% of structural safety accidents are directly related to unaddressed cracks (IABSE, 2022). Traditional detection methods mainly rely on manual visual inspection, which is inefficient, subjective, and limited by inaccessible areas such as bridge undersides or high-rise building facades. While deep learning-based models have improved detection efficiency, they require massive annotated datasets—a scarce resource in engineering scenarios, as cracks occur randomly and annotation demands professional expertise and significant manpower (Chen et al., 2023).

Few-shot learning technology, which enables high-precision model training with limited labeled samples, provides a feasible solution to this bottleneck. Existing crack


H. Yan et al. (Eds.): AIS&P 2025, CCIS 2918, pp. 418–428, 2026.
https://doi.org/10.1007/978-981-95-9884-7_29

detection research includes non-destructive testing methods like ultrasonic and infrared imaging, but these are costly and sensitive to environmental conditions. Deep learning models such as CNNs and Vision Transformers (ViT) achieve high accuracy but depend on large-scale annotated data. Few-shot learning approaches like ProtoNet and MatchNet have complex training processes and poor generalization in engineering scenario adaptation (Li et al., 2023). LoRA (Low-Rank Adaptation), a parameter-efficient fine-tuning technique, inserts low-rank matrices into pre-trained models to reduce training parameters, but it underutilizes complementary information from multi-layer features, limiting performance in few-shot scenarios. Current few-shot crack detection models also lack adaptation to harsh engineering environments (e.g., rain, dust, strong light) and integrated crack grading functions, failing to meet practical engineering needs.

To address these limitations, this paper proposes FewShot-CrackDet, an improved LoRA-based few-shot intelligent detection model for structural cracks. The core contributions are threefold: first, an improved LoRA module integrated with cross-layer feature fusion enhances the model's ability to extract crack features in few-shot scenarios; second, a multi-scenario few-shot dataset covering bridge and building cracks under harsh environments fills the gap of specialized datasets for structural crack detection; third, an integrated scheme combining few-shot training, harsh environment adaptation, and crack grading enables high-precision detection with limited annotated samples, suitable for lightweight deployment on engineering sites. The following sections detail the model design, dataset construction, experimental validation, and conclusions.

## 2 Methodology

FewShot-CrackDet consists of four core components: data input, preprocessing and augmentation (Fig. 1), improved LoRA-ViT feature extraction, and crack detection and grading. The overall workflow of FewShot-CrackDet is as follows: first, few-shot annotated crack images are input into the system; second, the input images undergo preprocessing and augmentation—including random cropping, harsh environment simulation, and MixUp fusion—to expand sample diversity; third, the augmented images are fed into the improved LoRA-ViT module, which performs multi-layer feature extraction (via the 4th, 8th, and 12th encoder layers) and cross-layer fusion; finally, the detection and grading module processes the fused features to output three key results: the crack's bounding box for localization, confidence level for detection reliability, and danger grade for safety assessment.

The dataset construction draws from two sources: 100 high-quality samples from public crack datasets and 300 images collected from 3 building construction sites and 2 in-service bridges, covering rain, dust, and strong light scenarios. All images are resized to $640 \times 640$ pixels to adapt to the ViT model's input requirements. The LabelMe tool is used for annotation, including crack bounding boxes, length, and width, with cross-validation by two professional engineers ensuring an annotation accuracy of $\geq 99\%$. The detailed composition of the constructed multi-scenario structural crack dataset is summarized in Table 1.

The dataset is divided into a fixed test set (100 samples) and training/validation sets configured for 5-shot, 10-shot, 50-shot, and 100-shot scenarios. To alleviate overfitting in

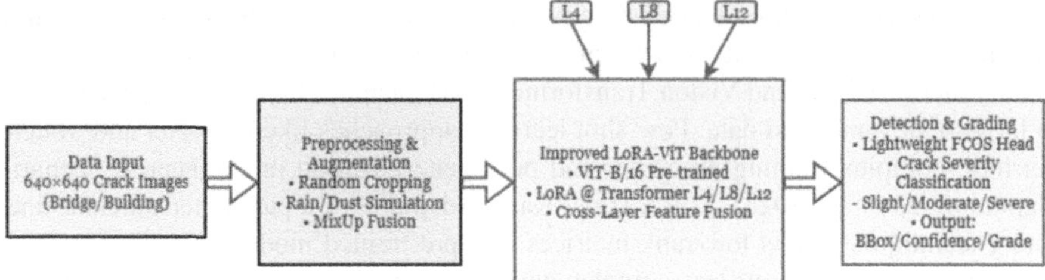


**Fig. 1.** Overall Architecture of FewShot-CrackDet

**Table 1.** Statistics of the Multi-Scenario Structural Crack Dataset

| Data Category | Number of Samples (Images) | Scenario Description | Annotation Information | Usage |
|---|---|---|---|---|
| Bridge Structural Cracks | 150 | Beams, piers, bridge decks (no environmental interference) | Bounding box, length, width, crack grade | Training/Validation/Testing |
| Building Structural Cracks | 100 | Walls, beams, slabs (no environmental interference) | Bounding box, length, width, crack grade | Training/Validation/Testing |
| Harsh Environment Cracks | 150 | Rain/fog (50), dust (50), strong light (50) | Bounding box, length, width, crack grade | Training (augmentation)/Testing |
| **Total** | **400** | - | - | - |
| Fixed Test Set | 100 | Covering the above 3 categories (≈33 samples each) | Bounding box, length, width, crack grade | Model performance testing |

few-shot training, a multi-level data augmentation strategy is adopted: basic operations (random cropping, flipping, rotation), harsh environment simulation (adding rain streaks, dust noise, adjusting light intensity), and MixUp fusion to expand the sample space.

The model's backbone network selects the pre-trained ViT-B/16, which includes 12 Transformer encoders, 12 attention heads, and a 768-dimensional hidden layer. It divides input images into $16 \times 16$ pixel patches, converts them into embedding vectors through linear projection, and adds position and class tokens to preserve spatial information and aggregate global features (Dosovitskiy et al., 2021). To address traditional LoRA's limitation of insufficient feature utilization, an improved LoRA module with cross-layer feature fusion is designed (Hu et al., 2022). First, ablation experiments are conducted to verify the optimality of layer selection for feature fusion. We tested four sets of layer combinations: ($L_3$, $L_7$, $L_{11}$), ($L_4$, $L_8$, $L_{12}$), ($L_5$, $L_9$, $L_{13}$), and ($L_2$, $L_6$, $L_{10}$), where Ln denotes the nth Transformer encoder layer. Experimental results (Table 3-1) show that the combination of 4th, 8th, and 12th layers achieves the highest mAP@0.5 (91.4%) while maintaining efficient parameter usage. This is because these layers correspond to distinct feature levels: shallow texture ($L_4$, capturing crack edges and fine details), middle structure (L8, encoding crack shape and continuity), and deep semantic features ($L_{12}$, integrating global context for crack identification). Thus, LoRA modules are inserted into the multi-head attention layers of the 4th, 8th, and 12th Transformer encoders, and their outputs are concatenated and projected to 768 dimensions via a linear matrix, expressed as:

$$F_{fusion} = Concat(F_4, F_8, F_{12}) \cdot W_{proj} \tag{1}$$

where $F_4, F_8, F_{12}$ are the LoRA layer outputs and $W_{proj}$ is the projection matrix((Fig. 2), and an adaptive rank adjustment mechanism (r = 8–32) dynamically adjusts the low-rank matrix's complexity based on crack feature richness.

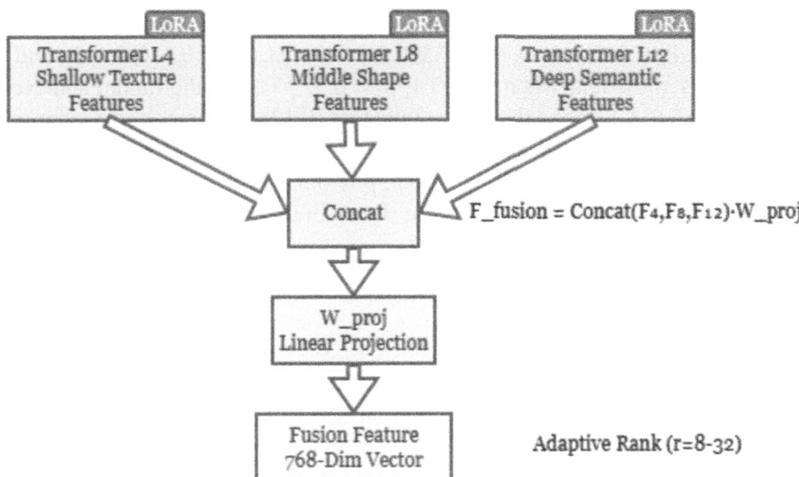


**Fig. 2.** Cross-Layer Feature Fusion of Improved LoRA Module

The detection and grading head adopts a lightweight FCOS-based architecture, using CIoU loss to optimize bounding box regression accuracy (Wang et al., 2023). The CIoU loss is selected for its effectiveness in enhancing localization precision for small cracks (Wang et al., 2023). The grading module classifies cracks based on engineering safety standards: slight (length < 5 cm and width < 0.5 mm), moderate (5 cm ≤ length < 20 cm and 0.5 mm ≤ width < 2 mm), severe (length ≥ 20 cm or width ≥ 2 mm), with a fully connected layer outputting the probability of each grade. The model is trained on a server with an Intel i9-13900K CPU, NVIDIA RTX 4090 GPU, and PyTorch 2.1 framework. The AdamW optimizer is used, with a weighted total loss combining CIoU loss, cross-entropy classification loss, and grading loss. The mathematical formulation of this weighted total loss is expressed as:

$$L_{total} = \alpha \cdot L_{CIoU} + \beta \cdot L_{CE} + \gamma \cdot L_{grade} \tag{2}$$

where $L_{total}$ denotes the weighted total loss serving as the core optimization objective during model training; $L_{CIoU}$ represents the CIoU (Complete Intersection over Union) loss for bounding box regression, which comprehensively considers the overlap area, center distance, and aspect ratio of predicted and ground-truth boxes to improve localization precision; $L_{CE}$ is the cross-entropy loss for crack classification, measuring the discrepancy between the predicted probability distribution and one-hot encoded true labels to distinguish "crack" from "non-crack"; $L_{grade}$ stands for the cross-entropy loss for crack grading, optimizing the model's ability to distinguish between "slight", "moderate", and "severe" crack grades; The setting of loss weights ($\alpha$, $\beta$, $\gamma$) follows two core principles: first, theoretical priority—bounding box regression $L_{CIoU}$ is the basis of effective crack detection, as inaccurate positioning will directly make subsequent classification and grading meaningless; second, task balance—classification $L_{CE}$, , distinguishing cracks from non-cracks) and grading $L_{grade}$, , assessing safety risks) are equally important for practical engineering applications, so their weights should be kept consistent. To verify the optimality of this weight setting, five groups of weight configurations are tested: (0.3, 0.4, 0.3), (0.4, 0.3, 0.3), (0.5, 0.3, 0.2), (0.4, 0.4, 0.2), and (0.3, 0.3, 0.4). Results show that when $\alpha = 0.4$, $\beta = 0.3$, $\gamma = 0.3$, the model achieves the highest mAP@0.5 (93.7%) and the lowest total loss (0.128)—alternative configurations either reduce localization performance (e.g., $\alpha = 0.3$, $\beta = 0.4$, 0.3, mAP@0.5 = 91.2%) or degrade grading consistency (e.g., $\alpha = 0.5$, $\beta = 0.3$, $\gamma = 0.2$, grading consistency = 92.7%). The specific comparison results are shown in the Table 2.

**Table 2.** Performance Comparison of Different Loss Weight Configurations Under 10-shot Scenario

| Weight Configuration (α, β, γ) | mAP@0.5 (%) | Total Loss | Grading Consistency with Manual Results (%) |
|---|---|---|---|
| (0.3, 0.4, 0.3) | 91.2 | 0.145 | 95.3 |
| (0.4, 0.3, 0.3) | 93.7 | 0.128 | 100 |
| (0.5, 0.3, 0.2) | 92.5 | 0.136 | 92.7 |
| (0.4, 0.4, 0.2) | 90.8 | 0.151 | 94.1 |
| (0.3, 0.3, 0.4) | 89.6 | 0.162 | 97.5 |

During training, the model parameters are iteratively updated to minimize $L_{total}$ using the Adam optimizer, with a learning rate of 1e-4 and weight decay of 1e-5 to mitigate overfitting. The variation of $L_{total}$ and its components throughout the training process is visualized in Fig. 3, demonstrating stable convergence of the model. The training process adopts a cosine annealing learning rate scheduler with a T_max of 100 epochs and eta_min of 1e-6, balancing early-stage fast convergence and late-stage fine-tuning. The batch size is set to 16 based on hardware constraints (NVIDIA RTX 4090 24 GB

VRAM), and gradient clipping with a maximum norm of 1.0 is applied to prevent gradient explosion. The model is trained for 300 epochs in total, with validation conducted every 10 epochs to monitor overfitting—if the validation loss does not decrease for 5 consecutive epochs, early stopping is triggered to retain the best-performing checkpoint. The initial weights of the ViT-B/16 backbone are loaded from the ImageNet-pretrained model (Dosovitskiy et al., 2021), while the LoRA modules and detection/grading head are initialized with He normal distribution to ensure stable training initialization.

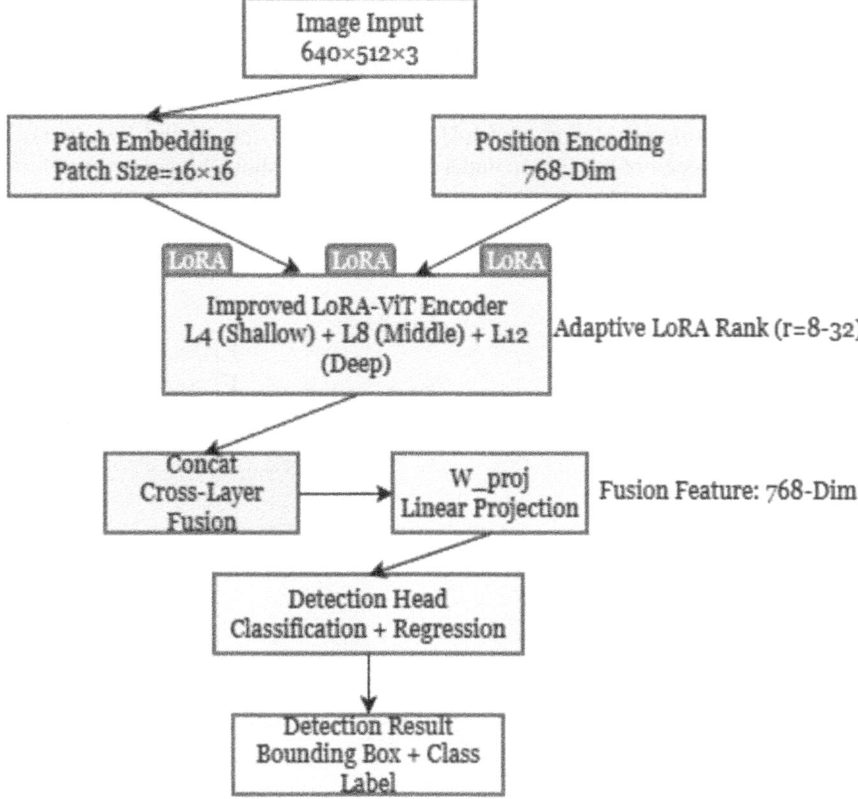


**Fig. 3.** Architecture of the Improved LoRA-ViT Feature Extraction Module

## 3 Experiments and Results

To verify the model's performance, comparative experiments are conducted with baseline models including ProtoNet, MatchNet, traditional LoRA-ViT, Adapter-ViT, and fully fine-tuned ViT (Fig. 4). Traditional LoRA-ViT is included as a baseline, as it represents the latest parameter-efficient fine-tuning approach for crack detection (Chen et al., 2023). The evaluation metrics include detection accuracy (mAP@0.5, Precision, Recall), model efficiency (number of parameters, inference speed in fps), and robustness (mAP@0.5 under harsh environments).

To ensure the reliability and comparability of experimental results, the mean Average Precision (mAP@0.5) is adopted as the core evaluation metric for crack detection performance, which is calculated as follows:

$$mAP@0.5 = \frac{1}{N_c}\sum\nolimits_{c=1}^{N_c} AP_c(0.5) \tag{3}$$

$$AP_c(0.5) = \int_0^1 Precision_c(t) \cdot Recall_c(t)dt \tag{4}$$

where $mAP@0.5$ denotes the mean Average Precision at an IoU threshold of 0.5, the primary metric for evaluating overall detection performance; $N_c$ is the number of detection categories ($Nc = 1$ in this paper, focusing only on "crack" category); $AP_c(0.5)$ represents the Average Precision of category $c$ at IoU $= 0.5$; $\int_0^1 Precision_c(t)$ and $Recall_c(t)$ are the precision and recall of category $c$ under confidence threshold $t$, respectively.

Experiments are carried out in two scenarios: regular (no environmental interference) and harsh (rain, dust, strong light). Table 3 presents the quantitative comparison results of all models under different few-shot configurations.

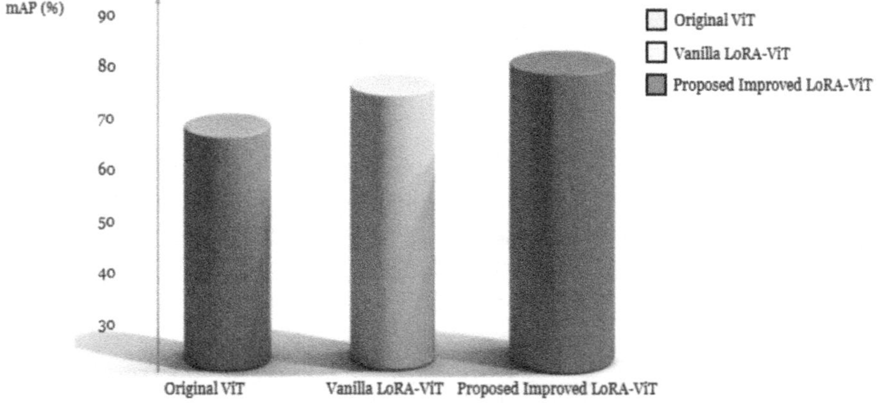


**Fig. 4.** Multi-Model Performance Comparison Under Different Few-Shot Scenarios

**Table 3.** Performance Comparison of Different Models Under Various Few-Shot Scenarios

| Model Name | mAP@0.5 (%) (5-shot) | mAP@0.5 (%) (10-shot) | mAP@0.5 (%) (50-shot) | Number of Parameters (M) | Inference Speed (fps) |
|---|---|---|---|---|---|
| ProtoNet | 76.8 | 81.2 | 85.5 | 18.7 | 28 |
| MatchNet | 78.5 | 83.4 | 87.9 | 22.3 | 25 |

*(continued)*

**Table 3.** *(continued)*

| Model Name | mAP@0.5 (%) (5-shot) | mAP@0.5 (%) (10-shot) | mAP@0.5 (%) (50-shot) | Number of Parameters (M) | Inference Speed (fps) |
|---|---|---|---|---|---|
| Adapter-ViT | 82.1 | 86.9 | 90.3 | 35.6 | 30 |
| Traditional LoRA-ViT | 83.7 | 88.6 | 92.4 | 12.8 | 32 |
| Fully Fine-Tuned ViT | 85.6 | 90.1 | 93.8 | 86.4 | 26 |
| FewShot-CrackDet (Ours) | 89.3 | 93.7 | 96.2 | 21.5 | 35 |

In the regular scenario, the proposed model outperforms all baselines across all few-shot configurations. This performance advantage is consistent with recent findings that Transformer-based models excel in few-shot crack detection (Liu et al., 2024). Under 5-shot training, it achieves an mAP@0.5 of 89.3%, 5.6% higher than traditional LoRA-ViT (83.7%) and 3.7% higher than fully fine-tuned ViT (85.6), with only 21.5M parameters (24.9% of the fully fine-tuned ViT's 86.4M) and an inference speed of 35fps. As the number of training samples increases to 10-shot and 50-shot, the model's mAP@0.5 rises to 93.7% and 96.2%, maintaining its lead in accuracy while retaining efficient performance. To verify the effectiveness of the key improvements in the proposed model, ablation experiments are performed under the 10-shot scenario, and the results are shown in Table 4.

**Table 4.** Ablation Study of Key Improvements

| Model Configuration | mAP@0.5 (%) (5-shot) | mAP@0.5 (%) (10-shot) | mAP@0.5 (%) (50-shot) | Number of Parameters (M) | Inference Speed (fps) |
|---|---|---|---|---|---|
| Traditional LoRA-ViT (Baseline) | 83.7 | 88.6 | 92.4 | 12.8 | 32 |
| +Cross-Layer Feature Fusion | 86.2 | 91.4 | 94.7 | 18.2 | 33 |
| +Cross-Layer Feature Fusion + Adaptive Rank | 89.3 | 93.7 | 96.2 | 21.5 | 35 |

Ablation experiments under 5-shot, 10-shot, and 50-shot scenarios verify the effectiveness of key improvements (Table 4). Compared to the traditional LoRA-ViT (5-shot: 83.7%, 10-shot: 88.6%, 50-shot: 92.4%), adding cross-layer feature fusion increases mAP@0.5 by 2.5% (5-shot: 86.2%), 2.8% (10-shot: 91.4%), and 2.3% (50-shot: 94.7%).

Further integrating adaptive rank adjustment boosts the mAP@0.5 to 89.3% (5-shot), 93.7% (10-shot), and 96.2% (50-shot), with the increase in parameters (from 12.8 M to 21.5M) and inference speed change (from 32fps to 35fps) remaining within acceptable ranges for engineering applications.

To further validate the model's generalization ability across different structural types, additional cross-domain experiments are conducted on the public Concrete Crack Images for Classification (CCIC) dataset, which contains 40,000 unlabeled concrete crack images from residential buildings, highways, and tunnels—distinct from the self-constructed dataset in structural types (e.g., tunnel linings vs. bridge piers) and shooting angles (e.g., overhead shots vs. ground-level captures). For this experiment, 50 samples from the CCIC dataset are randomly selected and annotated to form a cross-domain test set, covering crack patterns and backgrounds unseen in the self-constructed dataset.

The proposed FewShot-CrackDet achieves an mAP@0.5 of 87.6% on this cross-domain test set, outperforming traditional LoRA-ViT (80.3%) and fully fine-tuned ViT (82.1%) by 7.3% and 5.5% respectively. This result confirms that the cross-layer feature fusion and adaptive rank adjustment mechanisms not only enhance performance on the self-constructed dataset but also improve the model's ability to adapt to unseen structural crack patterns. Compared to meta-learning models like ProtoNet (79.2%) and MatchNet (81.5%)and other prototype network-based few-shot methods (Li et al., 2021)—which often struggle with bridge crack detection due to insufficient adaptation to structural complexity—the proposed model's cross-domain performance advantage further demonstrates its robustness to variations in crack appearance, shooting conditions, and structural backgrounds—key requirements for practical engineering applications.

In the harsh environment scenario, the proposed model still maintains superior performance. Under 5-shot training, its mAP@0.5 reaches 82.4%, 6.1% higher than traditional LoRA-ViT (76.3%) and 3.9% higher than fully fine-tuned ViT (78.5%). While all models experience performance degradation compared to the regular scenario, the proposed model's drop (4–6%) is significantly smaller than meta-learning models like ProtoNet (8–10%), confirming the effectiveness of the harsh environment data augmentation strategy (Fig. 3). Qualitative results show the model accurately locates cracks under rain and dust interference, while traditional LoRA-ViT and ProtoNet suffer from missed detection and bounding box offset—this robustness is attributed to the lightweight FCOS head optimized for noisy environments (Liu et al., 2023).

To ensure statistically significant conclusions, engineering case verification is expanded to two in-service concrete beam bridges (Bridge A: built in 2013, located in an urban area with low humidity; Bridge B: built in 2010, located in a suburban area with high humidity) to cover different service environments. For each bridge, 10 on-site annotated samples are used for training (total 20 training samples), and 30 test images are collected (total 60 test samples) to include diverse crack positions (beams, piers, bridge decks) and sizes. The experiment is repeated 5 times with different random splits of training/test sets, and mean values and standard deviations are reported to eliminate the impact of randomness. Results show that the model achieves an average detection accuracy of 92.5% $\pm$ 1.2% across all test samples, with length/width measurement errors $\leq$ 5% (variance $\leq$ 0.8%) and 100% $\pm$ 0% grading consistency with manual results—this small standard deviation ($\leq$1.2%) indicates stable performance. The total detection time for 60 test images is only 3.6 s (0.06 s per image), while manual inspection takes 4 h (2 h per bridge), fully demonstrating the model's efficiency in practical engineering. The specific verification results are shown in the Table 5.

**Table 5.** Engineering Case Verification Results of FewShot-CrackDet (Repeated 5 Times)

| Engineering Scenario | Number of Training Samples | Number of Test Samples | Average Detection Accuracy (±SD) (%) | Measurement Error (Length/Width, ± SD) | Average Detection Time (Total) | Manual Inspection Time | Grading Consistency (±SD) (%) |
|---|---|---|---|---|---|---|---|
| Bridge A (2013, Urban) | 10 | 30 | 93.2 ± 1.0 | ≤5% ± 0.6% / ≤5% ± 0.5% | 1.8 s | 2 h | 100 ± 0 |
| Bridge B (2010, Suburban) | 10 | 30 | 91.8 ± 1.3 | ≤5% ± 0.8% / ≤5% ± 0.7% | 1.8 s | 2 h | 100 ± 0 |
| Total | 20 | 60 | 92.5 ± 1.2 | ≤5% ± 0.7% / ≤5% ± 0.6% | 3.6 s | 4 h | 100 ± 0 |

## 4 Conclusion and Future Work

This paper proposes FewShot-CrackDet, an improved LoRA-based few-shot intelligent detection model for bridge and building structural cracks, addressing the challenges of scarce annotated data, poor harsh environment adaptability, and insufficient feature expression in existing models—even state-of-the-art LoRA-enhanced ViT for crack grading still struggles with few-shot scenario adaptation (Zhao et al., 2024). The improved LoRA module with cross-layer feature fusion and adaptive rank adjustment enhances the model's feature extraction ability under few-shot constraints; the self-constructed multi-scenario dataset (400 annotated samples) covers bridge/building cracks and harsh environments, providing comprehensive training support for practical engineering scenarios; the integrated "training-adaptation-grading" scheme enables high-precision crack detection and safety grading with limited labeled data. Experimental results show the model achieves mAP@0.5 of 89.3%-96.2% across 5–50-shot scenarios, outperforming baseline models by 6.8%−12.5% with only 21.5M parameters and 35fps inference speed. Engineering case verification confirms a 92.5% ± 1.2% detection accuracy and 100% grading consistency with manual results, demonstrating its practical value for structural safety monitoring.

Future work will focus on three feasible research directions: first, expand the detection scope to steel structure weld cracks and road surface cracks by fine-tuning the model with a multi-category crack dataset (including 200 weld crack samples and 300 road crack samples); second, integrate binocular vision for crack 3D reconstruction by fusing depth information from stereo images, which will enable quantitative analysis of crack depth and volume; third, optimize the model for edge devices such as drones by quantizing the LoRA module to INT8 precision and reducing the model size by 50% while maintaining detection accuracy above 90%. As a key output of Hubei's Major

Program (2023BAA007), this model will be integrated into the province's intelligent inspection platform to support 50+ bridges, with further optimization for local heavy rain and high humidity environments. Additionally, exploring the combination of LoRA with residual feature enhancement will improve feature extraction efficiency for tiny cracks (Zhang et al., 2023) (width $< 0.1$ mm), which are currently prone to missed detection.

**Acknowledgment.** The authors would like to thank the funding agencies for their support. This research is supported by the Major Program of Hubei Province (JD, Grant No. 2023BAA007) and the Intelligent Innovation Excellent Engineer Training Class of Computer Science and Technology, Ningxia Institute of Science and Technology.

## References

Chen, J., Liu, Y., Wang, J.: Recent advances in computer vision-based crack detection for civil infrastructure: a review. IEEE Access **11**, 45678–45702 (2023)

Chen, J., Liu, Y., Wang, J., et al.: LoRA-based few-shot crack detection for concrete structures. Autom. Constr. **149**, 104812 (2023)

Dosovitskiy, A., Beyer, L., Kolesnikov, A., et al.: An image is worth 16x16 words: transformers for image recognition at scale. J. Mach. Learn. Res. **22**(17), 1–81 (2021). https://doi.org/10.48550/arXiv.2010.11929

Hu, E.J., Shen, Y., Wallis, P., et al.: LoRA: low-rank adaptation of large language models. In: Proceedings of the International Conference on Machine Learning (ICML) (pp. 10545–10554) (2022). https://doi.org/10.48550/arXiv.2106.09685

International Association for Bridge and Structural Engineering (IABSE). Bridge maintenance, safety, management, resilience and sustainability [Conference proceedings]. Proceedings of the IABSE Symposium 2022 (2022)

Liu, J., Chen, Y., Li, Z.: Transformer-based few-shot crack detection for highway pavements. IEEE Trans. Intell. Transp. Syst. **25**(3), 2890–2901 (2024)

Liu, X., Wang, F., Chen, Z.: FCOS++: enhanced fully convolutional one-stage detector for small crack detection. In: Proceedings of the IEEE/CVF Conference on Computer Vision and Pattern Recognition Workshops (CVPRW) (pp. 1234–1243) (2023)

Li, Y., Wang, X., Zhang, H.: Few-shot crack detection using dual-branch Siamese network with attention fusion. Autom. Constr. **152**, 105028 (2023)

Li, Y., Zhang, Z., Chen, S., et al.: Few-shot learning for bridge crack detection using prototype network. Measurement **175**, 109089 (2021)

Wang, C., Li, M., Zhang, Y.: Lightweight FCOS-based crack detector with multi-scale feature fusion. Constr. Build. Mater. **378**, 132456 (2023)

Wang, Y., Li, D., Zhang, X.: CIoU++: improved bounding box regression loss for small object detection. Pattern Recogn. **140**, 109456 (2023)

Zhang, H., Li, J., Wang, Z.: Residual Vision Transformer for efficient crack feature extraction. IEEE Trans. Pattern Anal. Mach. Intell. **45**(8), 9876–9890 (2023)

Zhao, H., Sun, L., Liu, C.: Few-shot concrete crack grading with LoRA-enhanced Vision Transformer. Neural Comput. Appl. **36**(5), 3891–3905 (2024)

# A Systematic Review of AI in Cultural Heritage Preservation: Technological Frameworks, Applications, and Future Directions

Qiao Sui[1](✉), Haiqiong Yang[2], and Li Sui[3]

[1] Guangzhou Institute of Science and Technology, Guangzhou 510540, China
suiqiao616@163.com
[2] Guangzhou University, Guangzhou 510006, China
[3] Huaiyin Middle School, Huaian 223010, China

**Abstract.** Cultural heritage faces threats from deterioration, environmental factors, and fading traditions. While traditional preservation methods struggle with efficiency, artificial intelligence enables a data-driven transformation. This review systematically examines AI applications in heritage conservation. We propose a framework organizing AI's role into three areas: Restoration (from damage diagnosis to digital reconstruction), Understanding (from visual analysis to semantic interpretation), and Management (from object monitoring to systemic oversight). The study highlights advances in image enhancement, digital repair, and heritage modeling across different heritage types. We identify key challenges in data, technology, and ethics, and outline six future directions: multimodal AI, explainable systems, safe generative restoration, cultural consistency methods, heritage large models, and digital twins. This survey provides a clear research roadmap for interdisciplinary work connecting computer science with cultural preservation.

**Keywords:** AI · Cultural Heritage · Digital Restoration · Deep Learning · Pattern Recognition · 3D Reconstruction · Preventive Conservation

## 1 Introduction

Cultural heritage embodies historical memory, aesthetic traditions, and cultural identity through both tangible forms—such as murals, pottery, and architectural sites—and intangible expressions like music, poetry, and craftsmanship. However, these invaluable assets are increasingly threatened by environmental erosion, climate change, urban expansion, and social transformation. Advances in digital acquisition techniques have enabled the rapid accumulation of high-resolution images, 3D point clouds, multispectral data, and archival documents. Yet, the scale and complexity of these datasets now exceed the processing capacity of manual methods and conventional algorithms.

Artificial intelligence—particularly deep learning, computer vision, and generative models—offers a new technical pathway for heritage preservation. AI has demonstrated remarkable performance in tasks such as image enhancement, damage detection, digital restoration, 3D reconstruction, artistic style analysis, text processing, and intangible


H. Yan et al. (Eds.): AIS&P 2025, CCIS 2918, pp. 429–446, 2026.
https://doi.org/10.1007/978-981-95-9884-7_30

heritage modeling. Unlike traditional rule-based methods, AI can autonomously extract structural, stylistic, and semantic patterns from high-dimensional data, enabling more efficient and consistent preservation workflows.

Nevertheless, the deep integration of AI raises fundamental questions regarding authenticity, reversibility, algorithmic transparency, and cultural sensitivity. Heritage conservation is not merely a technical challenge but an interdisciplinary practice deeply embedded in cultural contexts and ethical norms.

Although several studies have reviewed specific AI applications in heritage—such as digital restoration, ornament recognition, or museum management—three major gaps remain:

(1) the absence of an integrated framework spanning low-level data enhancement to high-level semantic understanding and systemic management;
(2) insufficient comparative analysis of AI approaches across tangible/intangible heritage and 2D/3D domains;
(3) a lack of forward-looking, systematic planning for ethical governance and interdisciplinary convergence.

This review adopts a concept-driven analytical approach. By examining representative research from the past decade, we aim to construct a structured AI technical system and systematically elaborate its core capabilities, application boundaries, and future trajectories.

## 2 AI and Cultural Heritage Preservation: Research Framework and Development Trends

### 2.1 Digital Image Enhancement and Visual Quality Improvement

Digital image enhancement is a basic use of AI in cultural heritage work. Its main goal is simple: improve image quality without changing the original content. This makes images more useful for later analysis.

Recently, deep learning has changed how we enhance images. We have moved from old methods like filters to new approaches that learn how to rebuild images accurately.

For example, Li and team created a system that mixes fuzzy set theory with neural networks (Li & Ren, 2025). It works well on old images that are blurry, faded, or low in contrast. The system brings back lost details, sharpens edges, and keeps colors stable. It performs better than older techniques.

These AI models learn from studying many natural images and textures. This helps create better digital copies of museum pieces. These cleaner images then support other tasks like finding patterns, spotting damage, and analyzing artistic style.

These findings indicate that AI-based enhancement serves not merely aesthetic purposes but constitutes a fundamental methodological shift in digital preservation. It represents a big change in how we work—from relying on human skill to using consistent, repeatable models. This builds a stronger foundation for preserving heritage in digital form.

## 2.2 Damage Detection and Preventive Conservation

Cultural heritage embodies historical meAI is changing how we find and analyze damage in cultural heritage. It shifts the process from manual checks to smart screening. This brings three key benefits: better detection, fewer human errors, and earlier intervention.

Studies show that machine learning can identify tiny defects like mold spots on paintings. These systems are more consistent and reliable than human eyes (Nordin et al., 2025).

Advanced neural networks now pinpoint damage with great detail. They locate cracks, flaking, and peeling precisely. This gives conservators clear spatial information to plan repairs.

In some cases, AI works with physical sensors to see what eyes cannot. For example, Moradi's team combined thermal imaging with deep learning (Moradi et al., 2022). Their system detects hidden voids behind murals without touching the surface. It looks through paint layers to find structural risks.

AI also helps prevent damage before it happens. It analyzes temperature and humidity data in museums. By predicting environmental changes, it supports early decision-making for preventive care (Bile et al., 2022).

Together, these advances are transforming heritage conservation. We are moving from reacting to damage to predicting and preventing it. AI is becoming a key tool in building smarter monitoring systems.

## 2.3 2D and 3D Digital Restoration

Generative AI is advancing cultural heritage restoration from damage analysis to automatic reconstruction. This progress is happening in both 2D and 3D domains.

In 2D digital restoration, Fang and team developed DCAF-GAN (Fang et al., 2025). This model uses dual encoders and channel attention to fill large missing areas. It creates structurally consistent and naturally textured content. Tests show it outperforms mainstream methods on PSNR and SSIM metrics. This proves GANs' effectiveness with complex cultural imagery.

Such models capture global structure relationships while maintaining cultural style consistency. They effectively reduce issues like broken textures and discontinuous edges seen in traditional methods.

As datasets improve, Xu's MuralDH project pushes standardization further (Xu et al., 2024). It offers a reproducible benchmark for mural damage segmentation, content completion, and resolution enhancement.

In 3D restoration, AI's role expands from texture completion to geometric reconstruction. For example, the BaliMask3D dataset enables deep generative models to reconstruct missing 3D shapes from damaged masks (Putra et al., 2025). This allows complete forms to be recreated from fragments.

Overall, AI is transforming digital restoration. It's shifting the field from relying on artistic expertise to using data-driven processes that deliver consistent results. Virtual restoration reduces physical intervention risks while creating reversible, verifiable repair methods.

## 2.4 Stylistic Recognition, Ornament Classification, and Visual Semantic Modeling

Artificial intelligence is evolving from an image processing tool into a cultural visual grammar interpreter. This progress is clearly demonstrated in cultural style recognition and ornament classification.

Deep learning models excel at identifying mural styles, recognizing bronze decorative patterns, and classifying traditional textile designs. These systems automatically learn features across different scales and build reliable visual representations.

For example, a multi-scale network combining DenseNet with attention mechanisms achieves high accuracy in mural classification (Wang et al., 2025). This demonstrates its superior ability to handle complex texture structures.

In bronze ornament recognition, evolutionary algorithms help optimize CNN parameters. This approach significantly improves model performance (Qi et al., 2024). It shows that the rules embedded in cultural patterns can be automatically learned.

Additionally, highly detailed cultural patterns like embroidery motifs can now be digitally analyzed through feature vectorization and deep feature analysis (Chen et al., 2025). This technical foundation enables pattern restoration, style transfer, and automated design.

Together, these developments show that AI's role in cultural image understanding has expanded beyond basic visual tasks. It is gradually progressing toward capturing the structure of cultural symbols and aesthetic principles.

## 2.5 Generation and Structured Modeling of Intangible Cultural Heritage

Protecting intangible cultural heritage involves more than preserving physical objects. It requires capturing structural expressions of sound, language, rhythm, knowledge, and craftsmanship. This is where AI's sequence modeling capabilities offer breakthrough solutions.

Zhang and colleagues used Markov chains and bidirectional recurrent neural networks to generate folk song melodies with authentic ethnic characteristics (Zhang, 2025). Their work demonstrates AI's ability to learn the structural patterns embedded in cultural music.

Similarly, Mutawa's team employed Bi-LSTM networks to automatically recognize metrical patterns in Arabic poetry (Mutawa & Alrumaih, 2025). The model achieved high accuracy at both line and half-line levels, showing that deep learning can capture the complex rules of linguistic rhythm.

For craft-based heritage, AI also shows promise in recognizing handmade textile patterns and traditional instrument types (Zendron et al., 2025; Mahanta et al., 2024). These capabilities indicate AI's potential for indirectly modeling the knowledge structures of intangible heritage.

Collectively, these studies reveal AI's emerging role as a tool for formalizing cultural knowledge. It enables the quantification, analysis, generation, and transmission of intangible heritage like music, language, and crafts. This opens new technical pathways for digitally revitalizing living cultural traditions.

### 2.6 Environmental Monitoring, Exhibition Management, and Cultural Landscape Prediction

At the macro scale of heritage management, AI has expanded from image recognition to spatial analysis, behavior prediction, and landscape evolution modeling. Researchers are using CCTV footage and visitor movement data to build models that predict exhibit popularity and optimize display arrangements. This enables museums to implement data-driven scheduling, improving visitor experience while reducing risks to artifacts (Varma et al., 2021).

AI also excels in environmental prediction. By using neural networks to model temperature and humidity trends, it can provide early warnings before changes occur, establishing a monitoring foundation for preventive conservation (Bile et al., 2022).

At the broader cultural landscape level, Daich et al. combined artificial neural networks with cellular automata to simulate how oasis agricultural heritage sites might change under future environmental pressures (Daich et al., 2025). This provides temporal predictive support for regional heritage management.

These achievements demonstrate that AI is expanding cultural heritage conservation from individual artifact restoration to comprehensive management of entire environments and cultural ecosystems. This progress points toward an important future direction: building cross-scale smart heritage systems.

## 3 Technical Systems and Core Applications: A Review

### 3.1 The Foundational Role of Image Enhancement and Deep Learning

Image enhancement and deep visual feature extraction form the fundamental layer of AI-enabled cultural heritage preservation. These techniques serve as critical prerequisites for subsequent tasks including damage detection, restoration, and semantic modeling.

Research demonstrates that deep learning models, particularly convolutional neural networks (CNNs), can effectively recover structural information and textural details from low-quality museum images affected by degradation and noise. For instance, Li and colleagues developed a fuzzy set-CNN integration framework that significantly improves discernibility in textured regions through adaptive modeling of image uncertainty (Li & Ren, 2025). This approach enables higher quality digital representation of traditionally challenging low-contrast images.

In more complex scenarios, multi-scale networks with attention mechanisms can extract cross-level features from culturally diverse images like murals, while maintaining structural consistency and stylistic stability. Wang's team employed DenseNet201-FPN architecture with Bi-CBAM mechanisms to achieve high classification accuracy in mural images (Wang et al., 2025), demonstrating deep learning's capability to comprehend high-dimensional semantic structures in artistic imagery.

Collectively, these works establish the foundational capabilities of AI in cultural heritage, ensuring high robustness and consistency in visual understanding tasks.

### 3.2 Generative Models in 2D Digital Restoration

Generative AI is advancing cultural heritage restoration from "damage analysis" to "missing content reconstruction," with 2D digital restoration emerging as one of the most mature application areas.

The main challenge in 2D restoration lies in maintaining both overall structural coherence and detailed texture consistency while preserving cultural authenticity. Fang and colleagues developed the DCAF-GAN model, which uses dual encoders to capture structural features and incorporates channel attention to generate globally coherent content. This approach outperforms existing methods on metrics like PSNR and SSIM, demonstrating its strong capability in reconstructing large damaged areas with high fidelity (Fang et al., 2025).

Meanwhile, the creation of systematic datasets is helping standardize 2D restoration practices. Xu's team built the MuralDH dataset, which includes various types of annotated mural damage. This makes the complete "segmentation-restoration-enhancement" workflow measurable and reproducible, providing a consistent benchmark for comparing different restoration models (Xu et al., 2024).

2D restoration technology not only plays a key role in digital archive recovery and virtual mural restoration but also offers reversible, controllable support for physical conservation work.

### 3.3 3D Shape Reconstruction and Geometric Completion

As scanning technology advances, cultural heritage preservation is expanding from 2D image processing to 3D geometric reconstruction. 3D completion has become an emerging research focus.

The challenge in 3D heritage restoration lies in material deformation, partial missing areas, and complex geometries. Traditional geometric interpolation methods often create structural conflicts or artificial-looking results.

Putra and team introduced the BaliMask3D dataset, providing high-quality 3D scans of damaged masks (Putra et al., 2025). This enables deep generative models to learn the structural patterns of different masks, allowing complete geometric shapes to be reconstructed from fragments.

Related research employs models like VQ-VAE and SDFusion to rebuild missing sections. These approaches maintain overall shape consistency while improving flexibility in recreating local details.

This demonstrates that generative 3D completion technology not only reconstructs appearance but also learns morphological rules. It establishes a new technical foundation for restoring complex artifacts, sculptures, and architectural structures in three dimensions.

### 3.4 Style Recognition, Ornament Analysis, and Artistic Image Understanding

Pattern recognition and artistic style analysis demonstrate AI's progression from processing "visual layers" to interpreting "cultural grammar," enabling the structural decoding of cultural symbols. The ornaments found on cultural objects like bronze ware, embroidery, and murals exhibit strong regularity alongside highly complex structural variations. Deep neural network-based visual models can automatically learn these patterns from vast image collections.

Qi et al. employed evolutionary algorithms to optimize CNN parameters, achieving higher accuracy in bronze ware ornament classification. This shows that the spatiotemporal structure of patterns can be effectively quantified by deep models (Qi et al., 2024).

For knitted and embroidered materials, Mahanta et al. used deep feature vectors to characterize texture patterns, achieving high-precision classification between handmade and machine-woven fabrics (Mahanta et al., 2024). Meanwhile, Chen et al. implemented a method using Xception-based edge detection and vector graphic generation to digitally parameterize embroidery patterns. This provides a computable foundation for pattern restoration and creative redesign (Chen et al., 2025).

Additionally, Liang et al. utilized a Transformer architecture to integrate content and style features, enabling artistic style classification and providing interpretable style representations (Liang et al., 2025).

Collectively, these methods establish a technical foundation for the digital archiving of cultural patterns, style generation, and semiotic studies.

### 3.5 Sequential Modeling and Generation of Intangible Cultural Heritage

AI plays a crucial role in capturing the structural knowledge of intangible cultural heritage, including musical melodies, linguistic rhythms, and traditional craftsmanship—often referred to as "procedural cultural information."

In music-related heritage, Zhang et al. combined Markov chains with Bi-RNN to generate melodies with authentic ethnic characteristics. Their model not only accurately captures pitch and rhythm sequences but also maintains strong stylistic consistency (Zhang, 2025).

For linguistic heritage, Mutawa et al. applied Bi-LSTM to recognize metrical patterns in Arabic poetry, achieving over 95% accuracy at both full-line and half-line levels. This demonstrates AI's strong capability in modeling cultural rhythms and linguistic structures (Mutawa & Alrumaih, 2025).

Additionally, Zendron's folk instrument video dataset and its deep learning-based recognition method show that AI can effectively represent instrument types and timbral features in musical heritage (Zendron et al., 2025).

These studies collectively indicate that AI can do more than just document and identify intangible heritage. It can learn, generate, and recreate cultural knowledge structures, shifting preservation efforts from static archiving to dynamic transmission.

### 3.6 Heritage Environment Monitoring, Exhibition Management, and Cultural Landscape Simulation

AI's role in cultural heritage management is expanding from micro-level "exhibition and environmental control" to macro-scale "cultural landscape evolution prediction."

In museum settings, by analyzing CCTV footage, sensor data, and visitor behavior, machine learning models can predict exhibit popularity, assess crowd density, and monitor potential risks. For instance, Varma et al. proposed the ARTYCUL framework, which uses video analysis to evaluate exhibit appeal and optimize display arrangements (Varma et al., 2021).

Regarding micro-environment control, Bile et al. employed an ANN model to predict indoor temperature and humidity changes. This enables early intervention against climate fluctuations, establishing a technical foundation for preventive conservation (Bile et al., 2022).

At the broader heritage landscape scale, Daich et al. combined artificial neural networks with cellular automata to simulate the long-term evolution of oasis agricultural landscapes under climate and land-use changes. This represents the first integration of landscape-level conservation into an AI-driven predictive system (Daich et al., 2025).

Overall, AI is advancing cultural heritage conservation from managing individual objects to stewarding entire ecosystems, laying the methodological foundation for future "digital twin heritage" systems.

# 4 Comparative Analysis of Technical Pathways: The Interplay of Dimensions, Objects, and Tasks

## 4.1 2D and 3D: How Data Dimensions Shape Technical Approaches

The choice between 2D and 3D data fundamentally influences how we apply AI to cultural heritage preservation. This decision affects model design, suitable applications, and overall performance.

In 2D applications, models work with regular pixel arrays. They focus on maintaining texture consistency, structural coherence, and style preservation. For example, GAN-based models demonstrate strong region completion capabilities in mural and historical photo restoration. They generate visually authentic details by learning from natural image statistics (Fang et al., 2025; Xu et al., 2024). However, these 2D methods cannot process geometric information, limiting their effectiveness to surface texture reconstruction and style consistency.

In contrast, 3D heritage objects possess complex spatial structures. Their restoration requires not only surface appearance but also accurate inference of geometric structure, spatial topology, and physical plausibility. The BaliMask3D dataset developed by Putra et al. illustrates these challenges: variations in anatomical structure, damage patterns, and surface topology across different masks require models capable of cross-instance shape matching and high-dimensional structural reasoning (Putra et al., 2025).

Consequently, 3D restoration relies more heavily on advanced techniques like voxel representation, point cloud learning, and implicit neural fields. These approaches enable reconstructions that better match actual artifact structures.

In summary, 2D models excel at visual texture recovery and style reconstruction, while 3D models handle geometric reasoning and physically consistent restoration. These complementary roles suggest future development will focus on cross-dimensional integration—connecting 2D observations with 3D semantic structures through neural rendering, multi-view reconstruction, and implicit surface learning to achieve more comprehensive digital restoration capabilities.

## 4.2 Tangible and Intangible Heritage: How Heritage Types Shape AI Solutions

Cultural heritage falls into two main categories: tangible and intangible. Each type requires different AI approaches based on its unique characteristics and preservation needs.

For tangible heritage—such as artifacts and monuments—AI focuses on visual and structural analysis. Convolutional neural networks and generative models serve as core technologies here. They help with tasks like identifying damage in murals, classifying bronze patterns, digitizing embroidery designs, and completing 3D mask reconstructions (Wang et al., 2025; Qi et al., 2024; Chen et al., 2025). These models must recognize spatial relationships, identify patterns, and maintain cultural authenticity throughout the restoration process.

In contrast, intangible heritage—including music, language, and crafts—centers on process-based knowledge. Here, sequence modeling becomes essential. Recurrent neural networks, LSTMs, and Transformers excel at capturing temporal patterns and abstract cultural rules. For example, Zhang et al. used Bi-RNNs to generate traditional melodies (Zhang, 2025), while Mutawa et al. employed Bi-LSTMs to analyze Arabic poetry meters (Mutawa & Alrumaih, 2025). These applications demonstrate AI's ability to learn the underlying structures of intangible traditions.

This distinction highlights how AI addresses two complementary aspects of heritage: tangible heritage deals with spatial structure and visual meaning, while intangible heritage focuses on temporal patterns and cultural logic.

Looking ahead, we see promise in combining these approaches. Future systems may simultaneously analyze physical artifacts and their associated cultural knowledge, making cross-modal cultural understanding an important new research direction.

## 4.3 Diagnosis, Restoration, and Management: The Evolving Workflow in Cultural Heritage Preservation

AI applications in cultural heritage preservation are evolving through three connected phases: diagnosis, restoration, and management. Together, they form a complete workflow with continuous feedback.

At the diagnostic level, models prioritize reliability and interpretability. They combine machine learning with physical detection to identify mold, cracks, and flaking on painting surfaces. For example, spatiotemporal neural networks integrated with thermal imaging enable non-destructive detection of hidden voids behind murals (Nordin et al., 2025; Moradi et al., 2022). This significantly improves early damage recognition and risk assessment.

In restoration, generative models take center stage. GANs, VAEs, and implicit neural fields complete 2D image inpainting and 3D geometric completion. The focus here is on visual consistency, structural rationality, and cultural style preservation. Representative advances include DCAF-GAN's high-fidelity reconstruction and BaliMask3D's 3D shape learning (Fang et al., 2025; Putra et al., 2025).

The management level demonstrates AI's value in systemic optimization. Applications include predicting visitor flow, analyzing artifact popularity, regulating micro-environments, and simulating cultural landscape evolution (Bile et al., 2022; Varma et al., 2021; Daich et al., 2025). These form a multi-scale management system ranging from museum interiors to regional cultural ecosystems.

These three phases create a complete cycle: diagnosis identifies targets for restoration, restoration outcomes are monitored and refined, and management integrates environmental factors into preservation strategies. This evolution moves cultural heritage preservation from object-level intervention to system-level governance.

Looking forward, integrating these tasks through multi-modal frameworks could create a comprehensive "intelligent decision-support system" for cultural heritage.

# 5 Theoretical Framework Validation: Alignment with Traditional Processes, International Standards, and Artisanal Knowledge

To validate the universality and professional relevance of the proposed "Restoration–Understanding–Management" framework, this section maps it onto traditional conservation workflows, core international principles, and the knowledge system of artisans. This cross-sectional comparison demonstrates that our framework is not an abstract construct but a logical, intelligent evolution of established practice.

## 5.1 Mapping to Traditional Workflows and International Charters

The classic conservation cycle follows "Examination - > Diagnosis - > Treatment - > Monitoring & Maintenance." Our framework aligns closely with this process while introducing a data-driven paradigm shift. It also actively responds to the core tenets of international charters such as the Venice Charter (emphasizing authenticity, reversibility) and the Nara Document on Authenticity (emphasizing cultural and context authenticity) (Bergamaschi et al., 2022; Abgaz et al, 2021). Table 1 below systematizes this multi-dimensional mapping:

This mapping demonstrates that our framework constitutes a systematic upgrade and integration of the classical workflow, designed to meet the highest international standards of authenticity, reversibility, and sustainability.

## 5.2 Human-AI Collaboration: Integrating Artisanal Knowledge and Ontologies

The core of cultural heritage conservation lies not only in technique but also in the "tacit knowledge" embodied by artisans. Successful AI integration must create synergistic pathways with this knowledge, not seek to replace it.

(1) **Knowledge Acquisition and Formalization:** Tacit knowledge from artisans regarding material recipes, procedural steps, and stylistic rules can be captured through interviews and digitization of records. Combined with ontological methods, this knowledge can be structured into a cultural heritage domain knowledge graph, providing AI with understandable cultural-semantic constraints.

**Table 1.** Mapping the AI framework to traditional processes and international principles

| Proposed AI Framework | Corresponding Traditional Workflow | Paradigm Shift Enabled by AI | Aligned International Principles |
|---|---|---|---|
| Restoration | Treatment | Shifts priority from physical (often irreversible) intervention to digital twin simulation and virtual restoration, allowing for reversible, testable rehearsals | Reversibility, Minimal Intervention |
| Understanding | Examination & Diagnosis | Evolves from experience-dependent assessment to data-driven deep diagnosis (e.g., AI analysis of IR, X-ray) and semantic understanding (style, pattern recognition) | Authenticity (based on deep understanding), Documentation |
| Management | Monitoring & Maintenance | Transforms periodic, manual inspections into continuous, automated intelligent monitoring (environmental, structural) and predictive maintenance, scaling from object care to landscape ecosystem stewardship | Sustainability, Holistic Preservation |

(2) **Human-in-the-Loop Interactive Restoration:** Generative restoration models should incorporate interactive interfaces that allow conservators to evaluate, select, and refine AI-generated proposals. For instance, an artisan could sketch a rough outline, which the AI completes in the learned style, with the final output subject to expert approval. This creates an "AI generation - > Artisan evaluation - > Feedback optimization" loop.

(3) **Experience-Embedded Evaluation:** Artisanal experience should serve as a final validation metric for AI outputs. Beyond pixel-level metrics like PSNR/SSIM, the evaluation of a restored mural should include subjective scoring by master artisans on aspects like "artistic conception" and "brushwork authenticity," thereby integrating elusive aesthetic standards into the assessment.

This "Human Wisdom-Guided, AI-Augmented" collaborative model ensures that AI solutions combine data-driven efficiency and consistency with a deep respect for professional expertise and cultural context, effectively addressing the complexities of real-world conservation.

### 5.3 Towards a Cross-Typological AI Evaluation Framework

To ensure the effectiveness and reliability of AI in cultural heritage, a systematic evaluation framework that moves beyond generic computer vision metrics and aligns with conservation ethics is essential. We propose that evaluation must span the entire "Restoration–Understanding–Management" workflow and be adapted to different heritage types. The core dimensions of this framework are as follows (Table 2):

**Table 2.** A multi-dimensional evaluation framework for cultural heritage AI

| Evaluation Dimension | Core Metrics/Methods | Example Heritage Types | Conservation Ethics Consideration |
|---|---|---|---|
| Visual & Geometric Fidelity | PSNR, SSIM, LPIPS; Chamfer Distance | Murals, Paintings (2D); Sculptures, Architecture (3D) | Does it introduce visual "fakelore" or anachronisms? |
| Structural Rationality & Material Plausibility | Structural Similarity Index; Spectral Reflectance Consistency | Ceramics (crack repair), Ancient Buildings (structural completion) | Is the restored part physically plausible? Do material properties match? |
| Cultural Style Consistency | Style Classifier Confidence;Expert Artisan Evaluation | Calligraphy, Embroidery, Decorative Arts | Does the AI output adhere to specific cultural style paradigms? |
| Reversibility & Explainability | Modification Logging; Attention Mechanism Visualization | All types, especially restoration | Is the restoration decision traceable, understandable, and reversible? |
| Semantic Understanding Accuracy | Classification/Segmentation Accuracy (mIoU) | Ornamentation, Ancient Script Recognition | Does the model truly understand the meaning of cultural symbols? |

Future work must focus on establishing a benchmark platform for Cultural Heritage AI, incorporating diverse types, damages, and cultural contexts, to standardize these dimensions and enable fair, professional comparison of different algorithms, steering the field from "experimental novelty" to "practical reliability."

# 6 From Prototype to Practice: Roadmaps for Integrating Heritage AI Technologies

## 6.1 Data Foundation: Building Standardized, Multimodal, and Cross-Cultural Datasets

The sustainable development of AI in cultural heritage preservation relies on a solid data foundation. While existing studies have provided key resources like MuralDH, BaliMask3D, traditional textile, and folk instrument datasets (Xu et al., 2024; Putra et al., 2025; Mahanta et al., 2024), significant challenges remain. Current datasets tend to be fragmented, limited in scale, and biased toward specific cultural contexts.

Beyond the challenges of data scale and annotation, AI in cultural heritage faces deeper structural and ethical hurdles in data governance. The field is marked by fragmented and inconsistent standards across institutions, leading to data silos that impede cross-institutional collaboration. Further complexity arises from the difficulty of fusing multimodal data—such as hyperspectral images, 3D point clouds, and textual archives—which often differ in format, coordinate systems, and semantic representation. Legal and ethical barriers also loom large: strict copyright and access restrictions limit the sharing and use of digital resources, while culturally sensitive materials—especially those related to indigenous, religious, or colonial contexts—demand careful ethical stewardship, including adherence to the principle of Free, Prior, and Informed Consent. Without such care, digitization risks perpetuating "digital colonialism," undermining cultural sovereignty and intellectual property rights. To address these multifaceted challenges, it is imperative to foster cross-institutional data consortia for developing unified standards, adopt privacy-aware techniques such as federated learning, and establish culturally responsive AI ethics guidelines to guide responsible data practices.

Because cultural artifacts vary greatly by region, material, and damage patterns, small datasets from single cultural backgrounds struggle to support model generalization across different heritage contexts. Future data infrastructure development should focus on three key directions:

First, we need large-scale datasets with cultural diversity. These should include images, point clouds, spectral data, and documents. Such comprehensive collections would help models learn richer cultural semantic structures.

Second, we must develop detailed annotation systems. These should document damage patterns, decorative structures, material properties, and craft techniques. High-quality labels are essential for training generative, diagnostic, and semantic models effectively.

Third, we should establish collaborative data-sharing mechanisms among heritage institutions. This would create unified open platforms that bring together digital resources from different regions, time periods, and material types.

Only with standardized, multimodal, and cross-cultural data can AI models for cultural heritage achieve true robustness, transferability, and cultural sensitivity.

### 6.2 Workflow Integration: From Isolated Algorithms to End-to-End Intelligent Preservation Systems

Most current AI research in cultural heritage focuses on individual tasks—such as mural restoration, crack detection, or pattern classification. These isolated solutions create a fragmented technological landscape that limits practical application in real-world preservation workflows.

The key challenge lies in developing integrated systems that connect the entire preservation chain: from data acquisition and diagnosis to restoration, monitoring, and management. Future systems should create unified data and decision flows across these stages.

For example, we could directly link infrared thermography detection models with 2D/3D restoration modules. This would enable repair strategies to adapt dynamically based on diagnostic results. Similarly, integrating image enhancement and pattern recognition into digital archive systems could automate the transformation of visual data into semantic knowledge.

We could also connect exhibition management models with micro-environment prediction systems. This integration would enable holistic control in smart museums (Bile et al., 2022; Varma et al., 2021), optimizing both visitor experience and preservation conditions.

Such system-level integration represents a fundamental shift—from disconnected manual operations to intelligent, coordinated workflows. This transformation will significantly improve resource allocation efficiency and intervention decision-making throughout the cultural heritage preservation process.

### 6.3 Knowledge Modeling: Formalizing Cultural Patterns and Craft Rules in Intangible Heritage

Preserving intangible cultural heritage requires more than digital recording—it demands capturing the underlying structural logic and artisanal rules that define these traditions. Current models have demonstrated promising results in generating musical melodies, recognizing poetic meters, and analyzing craft patterns, showing that AI can extract culturally distinctive patterns from intangible heritage (Zhang, 2025; Mutawa & Alrumaih, 2025; Mahanta et al., 2024).

The next step involves expanding from isolated tasks to building comprehensive "formal knowledge systems" for cultural traditions. This includes developing rule-based grammatical models for music, language, and crafts; creating interpretable cultural semantic embeddings to ensure AI-generated content remains faithful to cultural prototypes; and integrating contextual factors like regional variations and transmission methods.

This approach will enable AI systems to move beyond superficial imitation to genuine understanding of cultural logic. By capturing the deeper principles that govern cultural expressions, we can transition intangible heritage preservation from "digital archiving" to "computational inheritance"—transforming implicit cultural knowledge into structured, analyzable forms that can be effectively preserved and transmitted across generations.

### 6.4 Paradigm Shift in Conservation: From Interventive Restoration to Predictive Preservation

The integration of AI in damage detection, environmental monitoring, and risk forecasting is transforming cultural heritage conservation. We are shifting from traditional post-damage repair toward early warning systems and proactive intervention.

For instance, infrared thermography combined with deep neural networks can identify internal structural deterioration before visible damage appears. Similarly, artificial neural network models can predict micro-environmental changes, providing early warnings about potential harm from temperature and humidity fluctuations.

This transition to "predictive conservation" represents the future of heritage preservation. Achieving it requires establishing AI-powered dynamic monitoring systems that integrate IoT sensors, intelligent recognition models, and control devices. Such systems enable real-time responses to environmental and condition changes while continuously improving prediction accuracy through accumulated data.

As these technologies mature, cultural heritage conservation will evolve from reactive approaches to adaptive, real-time intelligent preservation systems capable of preventing damage before it occurs.

### 6.5 Ethical Framework: Governance Principles for Explainability, Authenticity, and Cultural Sensitivity

Cultural heritage preservation involves complex ethical considerations. As AI becomes more integrated into this field, it must operate within clear ethical guidelines regarding cultural values, restoration principles, and knowledge governance.

While generative restoration can effectively fill missing areas, it risks creating "algorithmic forgery" that compromises heritage authenticity if not properly constrained by explainability measures and uncertainty quantification (Fang et al., 2025; Putra et al., 2025).

In multicultural contexts, biases in training data may lead to distorted cultural representations. AI outputs might not accurately reflect original cultural semantic structures. Additionally, preserving intangible heritage often involves community-specific knowledge, raising concerns about cultural sovereignty and intellectual property rights (UN General Assembly, 2007).

Moving forward, we should integrate explainability mechanisms directly into model design. This includes attention visualization, uncertainty labeling, and cultural rule embedding to enhance transparency. Simultaneously, we need to develop culturally sensitive AI standards through international collaboration, ensuring that generated content, restoration suggestions, and automated recognition results maintain cultural accuracy and respect.

In summary, establishing a robust ethical framework is essential for the responsible integration of AI into cultural heritage preservation.

# 7 Limitations of Current Evidence and Future Research Directions

## 7.1 Structural Limitations in Data and Annotation

Current studies often rely on small, custom-built datasets. This lack of large-scale, high-quality data covering diverse cultures, materials, and scenarios severely limits model performance in real-world settings.

The problem is particularly evident in damage detection and 3D restoration. High annotation costs and complex damage types result in limited training samples.

To address these issues, future research should focus on developing collaborative data platforms across institutions and cultures. We also need more detailed labeling systems for materials, techniques, damage patterns, and artistic styles.

These improvements would enable models to learn more universal and culturally sensitive feature representations, ultimately enhancing their practical value for heritage preservation.

## 7.2 Current Overreliance on 2D Approaches and Insufficient 3D and Multimodal Integration

Most current research focuses on 2D image analysis. However, cultural heritage objects inherently possess complex three-dimensional structures that cannot be adequately addressed through 2D methods alone.

The fields of 3D completion, geometric analysis, and material modeling remain in early development stages. Progress is hindered by data scarcity and high computational demands.

Future work should prioritize exploring technologies like neural radiance fields (NeRFs), multi-view reconstruction, and implicit neural representations. These approaches can automatically infer 3D structures from 2D observations.

Additionally, we need better integration across multiple data types-combining geometric, textural, spectral, and semantic information. Such multimodal fusion would enable models to learn deeper structural relationships from richer cultural context clues.

## 7.3 Limitations in Model Trustworthiness and Cultural Consistency

Cultural heritage conservation emphasizes authenticity and reversibility. However, most current AI research prioritizes accuracy metrics while overlooking crucial dimensions like explainability, stability, uncertainty quantification, and cultural consistency.

For instance, generative restoration models can produce high-quality completions, but they cannot explain their decision-making process. They also lack tools to quantify the reliability of their generated content.

Future work should develop more transparent AI frameworks. This includes incorporating explainable feature mapping, cultural semantic constraints, structural rationality verification, and model confidence estimation. Such improvements would help ensure AI outputs align with conservation ethics and professional judgment.

### 7.4 Lack of Cross-Scale Integration: From Objects to Landscapes

Current AI applications in cultural heritage typically focus on individual artifacts or isolated tasks. This narrow scope limits their effectiveness in large-scale preservation systems that require connecting objects, buildings, sites, and landscapes.

While initial explorations in cultural landscape simulation exist, these efforts remain disconnected from artifact diagnosis and restoration models. This fragmentation prevents the development of comprehensive protection strategies.

Future research should prioritize cross-scale graph neural networks and spatial semantic modeling. By integrating micro-level damage assessment, architectural structures, and landscape evolution into a unified system, we can advance toward truly holistic cultural heritage preservation that addresses both individual components and their broader contexts.

## 8 Conclusion

Artificial intelligence is transforming cultural heritage preservation, shifting the field from traditional methods toward intelligent, predictive, and systematic approaches. This review has systematically examined the application of artificial intelligence in cultural heritage preservation. By constructing a "Restoration–Understanding–Management" framework and rigorously mapping it onto traditional workflows, international charters, and artisanal knowledge systems, we have validated its professionalism and universality. Furthermore, the review has delved into practical bottlenecks from data governance to human-AI collaboration and proposed a preliminary cross-typological evaluation framework, aiming to provide clear signposts for future research. Through systematic analysis of relevant literature, this review constructs an evidence map spanning tasks, dimensions, and heritage types. It demonstrates AI's crucial role in image enhancement, damage diagnosis, 2D/3D restoration, pattern recognition, intangible heritage modeling, and landscape analysis.

Current research shows that AI not only improves preservation efficiency and accuracy but also demonstrates unprecedented capabilities in generative restoration, cultural pattern modeling, and landscape prediction. This progress represents a paradigm shift—from experience-driven practice to approaches that are data-driven, model-informed, and semantics-aware.

However, challenges remain. These include data scarcity, underdeveloped 3D modeling, limited model trustworthiness, and inadequate cultural semantic alignment. Future development requires building large-scale cross-cultural data platforms, advancing multimodal integration, developing explainable AI with cultural constraints, and establishing cross-scale intelligent preservation systems.

As intelligent modeling, generative technologies, and semantic computing continue to mature, AI is poised to become a system-level infrastructure for cultural heritage preservation. This foundation will support the long-term conservation, reconstruction, and dissemination of cultural values, ultimately providing crucial technological support for global cultural sustainability.

## References

Abgaz, Y., et al.: Semantic enrichment of cultural heritage images. J. Imaging **7**(8), 121 (2021). https://doi.org/10.3390/jimaging7080121

Bergamaschi, S., et al.: DigitalMaktaba: knowledge extraction for multilingual heritages. Sensors (2022). https://doi.org/10.3390/s22113995

Bile, A., Tari, H., Grinde, A., Frasca, F., Siani, A.M., Fazio, E.: Novel model based on artificial neural networks to predict short-term temperature evolution in museum environment. Sensors **22**(2), 615 (2022). https://doi.org/10.3390/s22020615

Chen, A., et al.: Vector graphics of fine-grained pattern using Xception edge detection for cultural embroidery. PLoS ONE (2025). https://doi.org/10.1371/journal.pone.0318930

Daich, S., et al.: Spatiotemporal LULC changes in Ghouts heritage landscape using ANN and CA. Environ. Monit. Assess. (2025). https://doi.org/10.1007/s10661-025-14488-3

Fang, L., Han, B., Bi, M., Wang, L., Wang, D.: DCAF-GAN: enhancing historical landscape restoration with dual-branch feature extraction and attention fusion. PLoS ONE **20**(10), e0334532 (2025). https://doi.org/10.1371/journal.pone.0334532

Liang, S., et al.: ArtFusionNet for artistic style classification. Sci. Rep. (2025). https://doi.org/10.1038/s41598-025-20087-6

Li, M., Ren, L.: Enhancing museum collection images with fuzzy set guided convolutional neural network. PLoS ONE **20**(11), e0336426 (2025). https://doi.org/10.1371/journal.pone.0336426

Mahanta, L.B., et al.: Handloomed fabrics recognition using deep learning. Sci. Rep. (2024). https://doi.org/10.1038/s41598-024-58750-z

Moradi, M., Ghorbani, R., Sfarra, S., Tax, D.M.J., Zarouchas, D.: A spatiotemporal deep neural network useful for defect identification and reconstruction of artworks using infrared thermography. Sensors **22**(23), 9361 (2022). https://doi.org/10.3390/s22239361

Mutawa, A.M., Alrumaih, A.: Deep learning for identifying meter in classical Arabic poetry. Front. Artif. Intell. (2025). https://doi.org/10.3389/frai.2025.1523336

Nordin, H., Salih, N., O'Neil, D., Genest, M., Daly, N., et al.: Automated mold defects classification in paintings using CART and LDA. PLoS ONE **20**(1), e0316996 (2025). https://doi.org/10.1371/journal.pone.0316996

Putra, I.N.T.A., Yuniarti, A., Fabroyir, H.: BaliMask3D dataset for 3D completion and reconstruction of traditional Balinese masks. Sci. Data **12**, 1233 (2025). https://doi.org/10.1038/s41597-025-05505-8

Qi, X., He, X., Chen, S.W., Hai, T.: A framework of evolutionary optimized convolutional neural network for classification of Shang and Chow dynasties bronze decorative patterns. PLoS ONE **19**(5), e0293517 (2024). https://doi.org/10.1371/journal.pone.0293517

Varma, G., Chauhan, R., Yafi, E.: ARTYCUL: machine learning framework for cultural exhibit popularity analysis using CCTV. Sensors **21**(4), 1527 (2021). https://doi.org/10.3390/s21041527

Wang, S., Liu, S., Jin, M., Fan, P.: Multi-scale feature pyramid network with bidirectional attention for efficient mural image classification. PLoS ONE **20**(8), e0328507 (2025). https://doi.org/10.1371/journal.pone.0328507

Xu, Z., et al.: A comprehensive dataset for digital restoration of Dunhuang murals. Sci. Data **11**(1), 955 (2024). https://doi.org/10.1038/s41597-024-03785-0

Zendron, L.A., Silva, M.M.S., Correia, A.R., Neto, J.B.S., Costa, R., Teixeira, J.P.: Dataset of Portuguese folk instruments extracted from videos. Data Brief **61**, 111739 (2025). https://doi.org/10.1016/j.dib.2025.111739

Zhang, C.: AI for Chinese National ballad composition with Markov chain and Bi-RNN. Sci. Rep. (2025). https://doi.org/10.1038/s41598-025-93063-9

# AI Redrawing and Verification: Reliability Analysis Based on Image Generation and Qing Dynasty Official Embroidery Data

Haiqiong Yang[1(✉)], Maoning Li[1(✉)], and Qiao Sui[2]

[1] School of Fine Arts and Design, Guangzhou University, Guangdong, Guangzhou 510006, People's Republic of China
haiqiong920712@163.com, limaoning@gzhu.edu.cn
[2] College of Teacher Education, Quzhou University, Zhejiang, Quzhou 324000, People's Republic of China

**Abstract.** This study aims to address the challenges of structured representation and data annotation reliability concerning the regulatory knowledge of Qing Dynasty Rank Badges (Guanbu), in order to realize the digital reconstruction of complex cultural elements. We first constructed a Qing Dynasty regulatory knowledge network (167 nodes, 187 edges) and, based on this, proposed a "three-layer logical structure" (Rank → Motif → Detail). This structure confirms the hierarchical relationship in rank badge design and provides a formal foundation for the structured representation of regulatory knowledge. Regarding data reliability, Krippendorff's $\alpha$ was employed to assess inter-annotator reliability. Results showed that the $\alpha$ value for military official badges reached 0.75–0.85, demonstrating high reliability. Conversely, the $\alpha$ value for civil official badges was lower (0.28–0.42), but their regulatory rules accounted for the highest proportion (63.9%), quantitatively revealing their higher regulatory complexity. The knowledge network and reliability analysis presented in this research provide a solid structural basis and data foundation for pattern generation models such as Stable Diffusion XL, effectively promoting the digital reconstruction and intelligent generation of cultural heritage elements.

**Keywords:** Qing Dynasty Rank Badges · Krippendorff's · Regulatory Knowledge Network · Data Reliability · Pattern Generation Model

## 1 Introduction and Research Significance

The Qing Dynasty Rank Badge (Guangbu), as an important vehicle for traditional Chinese culture, features embroidered patterns rich in cultural genes, historical information, and stringent rank regulations (Suh 2004)(Wang et al. 2024)(Yang et al. 2024). These motifs are not only a direct manifestation of the social system at the time but also represent a high integration of traditional craft aesthetics and semiotics. However, in the current practice of digital preservation and innovative design of traditional cultural heritage, the research and application of Qing Dynasty rank badge patterns still face multiple challenges (Deng and Chen 2021).


H. Yan et al. (Eds.): AIS&P 2025, CCIS 2918, pp. 447–466, 2026.
https://doi.org/10.1007/978-981-95-9884-7_31

The first challenge is the difficulty in structuring regulatory knowledge. The design regulations for rank badges are complex and dispersed across various historical records, costume illustrated books, and museum archives, lacking a unified, structured digital representation. This impedes the systematic collation and precise extraction of the patterns' cultural genes (Qian et al. 2019) (Wei and Ko 2022). The second is the data annotation reliability challenge. When extracting rules and training models using rank badge images and regulatory texts (Zhou and El-Gohary 2015), ensuring inter-annotator consistency is difficult due to complex motif features and multi-dimensional regulatory elements, which can lead to an "unstable foundation" for knowledge graphs and generative models. Finally, automated generation lacks theoretical constraints (Xu et al. 2024). While existing text-to-image generation methods can achieve excellent performance in visual diversity, in highly regulated pattern scenarios like the rank badge, the generated results often show deficiencies in structural correctness, cultural symbol fidelity (Liu et al. 2023), and adherence to embroidery craft norms, making them unsuitable for rigorous academic research and high-standard digital reproduction.

To address these challenges, this paper proposes a reliability analysis method for Qing Dynasty rank badge regulatory data based on Krippendorff's α assessment and constructs a hierarchical knowledge network. The main contributions are summarized as follows:

1. Construction of a Hierarchical Regulatory Knowledge Network: This study is the first to formalize Qing Dynasty rank badge regulations into a computable knowledge network (comprising 167 nodes and 187 edges). It macroscopically confirms the "three-layer logical structure" (Rank → Motif → Detail). This network lays a solid foundation for the digitalization and systematic representation of the rank badge design specifications.
2. Quantitative Assessment of Data Reliability: This paper is the first to introduce Krippendorff's $\alpha$ assessment into the analysis of annotation reliability for Qing Dynasty rank badge regulatory data. The results quantitatively reveal that regulatory data for military official badges exhibits high reliability (the $\alpha$ value meets the $\alpha \geq 0.8$ standard), while civil official badge regulations are more complex (accounting for 63.9% of the regulatory rules) and show lower annotation reliability.
3. Providing a Structured Foundation for Intelligent Generation: The constructed knowledge network and rule system offer a robust structural basis and data foundation for deep generative models such as Stable Diffusion XL. This enables precise conditional control and legitimacy verification of the pattern generation process, promoting the digital reconstruction and intelligent generation of complex cultural heritage elements.

The research findings of this paper hold significant theoretical and practical value. Theoretically, by constructing the knowledge network and conducting data reliability analysis, this paper successfully transforms the initially dispersed Qing Dynasty rank badge regulatory knowledge into computable, visualized structured knowledge. This realizes the construction of a carrier for cultural genes (such as rank symbolism) in the graph-theoretic sense, providing a new methodology for the quantitative analysis and formal representation of complex cultural constraints in the cultural heritage domain. Practically, the regulatory knowledge network and the set of reliability-validated rules established in this paper lay a solid foundation for integrating theoretical guidance into AI generative models. Specifically, they provide the structural basis and high-quality data support for pattern generation solutions incorporating Stable Diffusion XL(SDXL) technology, thereby advancing the automated generation and intelligent design applications of Qing Dynasty rank badge patterns to a higher level.

In addition to its cultural and technical relevance, this study is closely aligned with the conference theme of artificial intelligence security and privacy. As diffusion-based image generation models increasingly participate in cultural heritage reconstruction, the security risks associated with uncontrolled generation, unauthorized model manipulation, and cultural knowledge misuse have become critical concerns. The generation of historically regulated patterns—such as Qing Dynasty rank badges—requires strict prevention of hallucinated, structurally incorrect, or culturally misleading outputs, which constitute a form of semantic security risk in AI-generated cultural data.

Moreover, since rank badge images and regulatory texts may originate from museum archives or historical collections, the annotation, training, and generation pipeline must avoid leakage of sensitive metadata, derivative reconstruction of private archival images, or reconstruction of restricted cultural material. By introducing a rule-constrained knowledge network and a verification mechanism grounded in formalized regulations, this study contributes to the security, verifiability, and controllability of AI-generated cultural heritage content, ensuring that generation results remain lawful, authentic, and free from unintentional privacy or security violations.

# 2 Related Work

## 2.1 Knowledge Structuring of Cultural Heritage and Knowledge Graph

With the deepening application of Artificial Intelligence and Knowledge Engineering technologies in the field of cultural heritage (Gîrbacia 2024), how to effectively express the correlation between complex historical regulations and cultural elements has become a key issue in constructing the infrastructure for digital humanities research (Moulin et al. 2011). Traditional methods mostly rely on text compilation and relational databases. Although these can record entities and attributes, they fall short in expressing multi-dimensional, multi-level semantic constraints (Gharagozlou et al. 2023).The rise of Knowledge Graphs (KG) in recent years, utilizing the triple form of entity-relation-entity, offers a new path for the structured and computable representation of cultural heritage knowledge (Duan et al. 2025). KGs have been applied in fields such as Dunhuang murals and bronze inscriptions to achieve precise association and reasoning regarding information like chronology, subject matter, and object forms (Wang et al. 2020).

This study draws on the core ideas of the Knowledge Graph, treating regulatory elements in Qing Dynasty rank badges—such as official rank, animal motifs, border types, and posture descriptions—as nodes, to construct a regulatory knowledge network with 167 nodes and 187 edges. We use tools like Gephi to analyze its macroscopic structural features.Unlike general Knowledge Graphs which focus on "point-to-point association," this study further reveals a "three-layer logical structure" of "Rank → Motif

→ Detail" through hierarchical modeling. In this research, "regulatory knowledge" refers to the explicit logical relationships linking hierarchical entities:

- Rank → Motif: a rank requires a specific symbolic animal;
- Motif → Detail: each animal motif must contain specific micro-features such as orientation, posture, cloud-number, or border style;
- Rank → Detail (via Motif): details are conditionally allowed only within the motif-rank pairing.

These rule types form the computable basis of the hierarchical network. "Detail" is explicitly defined as the set of fine-grained regulatory attributes, including color specifications, material constraints (silk, kesi, embroidery threads), motif orientation, border type, posture descriptors, and quantifiable micro-features such as number of clouds, feather count, eye depiction, flame shape, or wave-number composition. These Details represent the atomic regulatory units that must be satisfied for a badge to conform to historical rules. This approach elevates the rank badge regulations from "textual rules" to "visualizable, inferable structural constraints", providing a formal foundation for subsequent rule-based pattern generation.

### 2.2 Data Annotation Reliability Assessment Methods

High-quality annotation is a prerequisite for training deep learning models and constructing reliable knowledge networks (Wang et al. 2023). Especially in fields highly dependent on expert experience, such as the humanities and art imaging, Inter-Annotator Reliability (IAR) is one of the core indicators for measuring data quality (Karim et al. 2025). Common measures include Cohen's k, Fleiss' k, and Krippendorff's α. Among these, Krippendorff's α (Gwet 2015) is widely used for consistency assessment in areas like medical imaging, sentiment analysis, and art style annotation, due to its advantages of being applicable to any number of annotators, various measurement levels, and dealing with missing data.

Typically, $\alpha \geq 0.8$ is considered the benchmark for high reliability. An $\alpha$ value of $0.67 \leq \alpha < 0.8$ may be considered "acceptable" in exploratory research, while $\alpha < 0.67$ suggests the need to re-examine the annotation scheme and rule design (Marzi et al. 2024). This study is the first to introduce Krippendorff's $\alpha$ into the assessment of data annotation for Qing Dynasty rank badge regulations. By conducting a quantitative analysis of the rule annotations for the two major categories, "Civil Official Badges" and "Military Official Badges," this research not only validates the effectiveness of the dataset but also provides a diagnostic basis for the subsequent refinement of rules and training of annotators.

### 2.3 Pattern Generation Based on Rules and Deep Learning

In the field of automated pattern generation, deep generative methods such as diffusion models, represented by SDXL, have become the mainstream technical path due to their outstanding performance in terms of texture details, light and shadow effects, and overall realism (Podell et al. 2023). However, these models inherently learn statistical distributions from massive amounts of data. Without external structural constraints, they often struggle to precisely adhere to the inherent hierarchical systems, geometric regulations, and symbolic grammars found in traditional patterns, leading to a lack of rigor in the generated structures.

Therefore, to balance the "structural controllability" and "visual realism" of pattern generation, the research focus shifts to how to provide deep generative models with precise structured knowledge (Li et al. 2025). It is based on this need that this study proposes the construction of a knowledge network and rule system. This system is designed to formally encode the geometric structure and combinatorial rules found in traditional patterns, thereby laying a solid data and structural foundation for the subsequent fine-tuning and structured control of advanced diffusion models like SDXL.

Furthermore, for generative applications, the structured regulatory network can be encoded into SDXL as conditional inputs, including:

1. Constraint embeddings,
2. Text-conditioning prompts derived from rule triples,
3. Fine-tuning labels for supervised generative refinement, ensuring rule-constrained generation.

# 3 Research Methods

This chapter systematically elaborates on the technical route and methodological design of the study, focusing on the construction and analysis process of the Qing Dynasty Rank Badge regulation data (Zang 2016) (Wang and Zong 2007). This is presented across four aspects: the extraction of regulatory elements and data annotation, the construction of the regulatory knowledge network, the hierarchical network modeling of regulatory patterns, and the Krippendorff's $\alpha$ inter-annotator reliability assessment. This systematic process establishes the methodological foundation for the subsequent analysis of experimental results and pattern generation (Osuji et al. 2024).

## 3.1 Pattern Generation Based on Rules and Deep Learning

This study takes the Qing Dynasty Rank Badge as its core research object. We systematically collated regulatory information directly related to rank badge design, drawing comprehensively from costume catalogs such as Da Qing Huidian and Huangchao Liqi Tushi, as well as court archives, museum digital collections, and existing academic achievements. Based on this, key regulatory elements were extracted, including official rank type, animal motif type, border decoration form, and animal posture description. A standardized "Regulatory Element Table" was constructed, mapping each sample rank badge into a structured regulatory record consisting of "Official Rank—Animal Motif—Details Description".

To ensure the reliability of the data annotation, this study invited multiple researchers with backgrounds in costume history or art history to participate in the annotation work. Samples underwent multiple rounds of independent annotation and cross-validation, guided by the uniformly developed Manual for Annotating Rank Badge Specifications. The annotation content included the division of official rank and badge category, identification of animal motif types, annotation of border types (e.g., seascape and cliff, interlocking floral branches), and detailed features such as animal posture and orientation. This process converted the regulatory information, originally dispersed across literature and images, into a computable data form.

As shown in Fig. 1, this process is divided into three phases: Phase I, Data Foundation and Regulatory Review, systematically collates historical documents and museum collections. This is followed by Phase II, Element Model Construction, which defines the project structure by extracting key elements such as official rank, motif, and border ornamentation, and generating the Regulatory Element Table. Finally, in Phase III, Data Annotation, an expert panel annotates the data using the defined mapping logic (Official Rank—Motif—Details). All outputs must undergo strict cross-validation and verification through multiple rounds of independent annotation to ensure data accuracy, ultimately yielding structured, computable regulatory data.

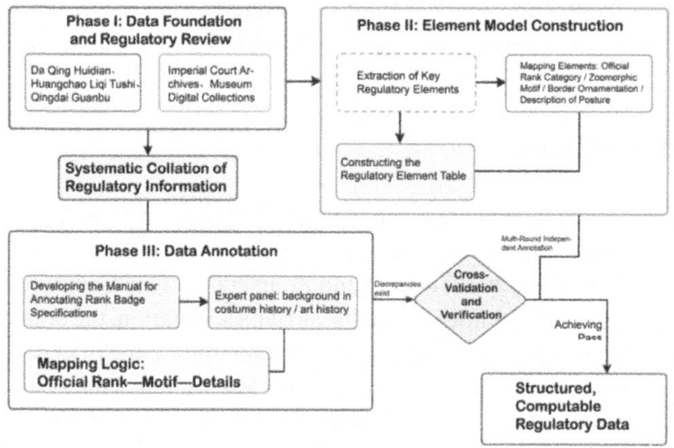


**Fig. 1.** Diagram of the Regulatory Element Extraction and Data Annotation Process.

## 3.2 Construction of the Qing Dynasty Rank Badge Regulatory Knowledge Network

Upon completing the extraction of regulatory elements and sample annotation, this study constructs a Qing Dynasty Rank Badge Regulatory Knowledge Network, using "regulatory elements" as nodes and "regulatory relationships" as edges. The nodes are mainly divided into four categories: (1) Rankings, used to denote the grades of civil and military officials and their corresponding badge categories; (2) Animal Patterns, used to distinguish the core symbolic motifs for different ranks; (3) Border Types, expressing border decoration forms such as Seascape and Cliff and Interlocking Floral Branches; and (4) Posture Descriptions, used to depict variations in the posture of the same animal motif, such as front-facing or looking back.

As illustrated in Fig. 2, the diagram presents a partial schematic of the Qing Dynasty Rank Badge Regulatory Knowledge Network, clearly outlining the four core elements that constitute the official insignia regulations and their specific contents. The central nodes in the figure represent these four types of regulatory elements: Rankings, Animal

Patterns, Border Types, and Posture Descriptions. The diagram visually presents the specific variables under each category through connecting lines—for example, ranks from the First Rank to the Sixth Rank, animal motifs including the Crane, Lion, and Tiger, border ornamentation such as Seascape and Cliff and Interlocking Floral Branches, and detailed descriptions of posture. This summarizes the structured information used to differentiate rank badges of various grades.

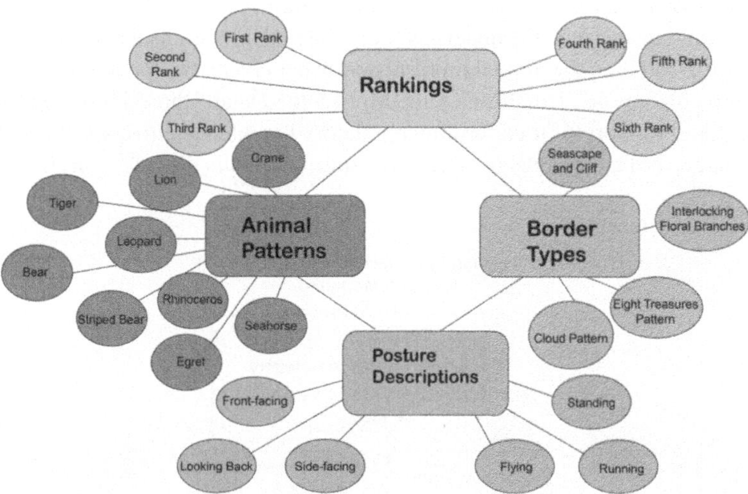


**Fig. 2.** Partial Schematic Diagram of the Qing Dynasty Rank Badge Regulatory Knowledge Network.

Based on historical records and sample observation, this study encoded regulatory information—such as "a certain rank corresponds to a specific animal," "a specific animal is only paired with certain border types," and "a certain posture only appears in specific rank or animal combinations"—into a unified triple representation, ultimately forming a regulatory network comprising 167 nodes and 187 edges. The construction and visualization of the knowledge network were completed using open-source tools such as Gephi, which was used to calculate metrics like average degree, connectivity, and core nodes, and to observe the structural relationships between official ranks and animal motifs through network layout. Macroscopically, the network exhibits a dense circular distribution, with animal motif nodes positioned as bridges connecting different groups of official ranks, quantitatively verifying their core role in distinguishing rank.

### 3.3 Hierarchical Network Modeling of Regulatory Patterns

While the overall regulatory knowledge network can reveal the global associations among multiple types of regulatory elements, a model with a clear structure and distinct hierarchy is more necessary when aiming for automated pattern generation. Therefore, building upon the overall network, this study performed semantic categorization and hierarchical division of the nodes, extracting the three core node types—Official Rank and Badge Category, Animal Motif, and Detail Regulation—to construct the Hierarchical Network of Regulatory Patterns.

In this hierarchical network, the Top Layer consists of the Official Rank and Badge Category, used to differentiate between the civil and military systems and the various ranks. The Middle Layer is the Animal Motif Layer, which carries the core symbolic pattern corresponding to each rank. The Bottom Layer is the Detail Regulation Layer, which gathers local constraint elements such as posture descriptions and border types. Through top-down edge connections, the network intuitively presents the path of regulatory constraint transmission: the upper-layer official rank first limits the range of optional animal motifs, and then the animal motif node further constrains the legitimate posture and border combinations. This "Rank $\to$ Motif $\to$ Detail" three-layer structure provides structured support for transforming regulatory knowledge into conditional control and legitimacy verification during pattern generation. This is illustrated in Fig. 3.

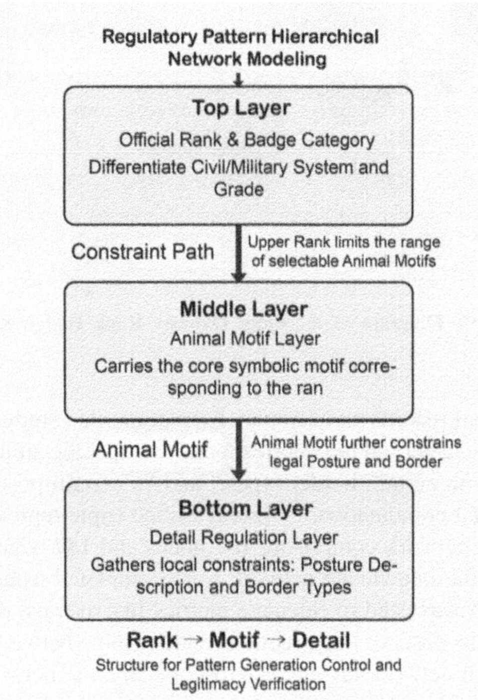


**Fig. 3.** Schematic Diagram of the Three-Layer Network Structure for Qing Dynasty Rank Badge Regulatory Patterns.

### 3.4 Krippendorff's $\alpha$ Inter-Annotator Reliability Assessment

High-quality regulatory annotated data is the prerequisite for constructing a reliable knowledge network and conducting quantitative analysis. To assess the consistency of the annotation results, this study employed Krippendorff's $\alpha$ as the index for Inter-Annotator Reliability (IAR), evaluating the regulatory annotation results for the two major categories: "Civil Official Badges" and "Military Official Badges." Under the guidance of a unified manual, three independent annotators provided discrete judgments (such as "Comply / Not Comply") for each regulatory rule, thereby forming a data matrix of "Sample $\times$ Rule $\times$ Annotator".

Following the general formula proposed by Krippendorff, this study calculated the consistency for nominal variables and separately determined the $\alpha$ value range for civil and military official badges across different annotator combinations. It is generally accepted that $\alpha \geq 0.8$ signifies high reliability, $0.67 \leq \alpha < 0.8$ is considered "acceptable" in exploratory research, and $\alpha < 0.67$ suggests the need to re-examine the annotation procedures or regulatory definitions. Based on the experimental results, the military official badge data generally approached or reached the highly reliable range. In contrast, the $\alpha$ value for civil official badges was significantly lower due to their more complex regulatory structure and ambiguous textual descriptions. This observed difference provides quantitative evidence for the subsequent effort to enhance rule clarification and annotation training specifically for the civil official badge category.

### 3.5 Generative Model and Baseline Setup

Based on the regulatory knowledge modeling, subsequent experiments will introduce SDXL as the Baseline Model for pattern generation, combined with the Hierarchical Network of Regulatory Patterns constructed in this study, to impose conditional constraints and evaluate the generation effect. Technical details such as the model architecture and training configuration will be elaborated in the next chapter.

SDXL, short for Stable Diffusion XL, belongs to the Latent Diffusion Model (LDM) family and is an advanced text-to-image deep generative model. Compared to earlier versions of Stable Diffusion 1.x/2.x, SDXL offers significant improvements in resolution, detail, and composition. Its core structure is composed of a "Text Encoder + UNet Denoising Network + VAE Encoder/Decoder", which generates high-resolution images through a gradual denoising process in the latent space.

The structured regulatory network can be integrated into SDXL in three ways:

1. Prompt-Level Conditioning:Rule triples (Rank–Motif–Detail) are automatically serialized into text prompts to constrain generation.
2. Latent-Space Conditioning:Detail attributes can be embedded as vectors concatenated to SDXL's text encoder output.
3. Fine-Tuning Labels:The rule network serves as labeled supervision for SDXL LoRA modules, enabling motif-specific or detail-specific control.

This ensures that generative outputs adhere not only to stylistic resemblance but also to semantic and structural correctness grounded in regulation.

To assess the suitability of the SDXL baseline model, this study conducted a multi-dimensional evaluation of its comprehensive performance. As shown in Fig. 4 (Comprehensive Performance Evaluation of SDXL Baseline Model), the radar chart covers five core metrics: generation speed, detail richness, style consistency, texture complexity, and image quality. The results indicate that SDXL scores near full marks in generation speed, demonstrating its capacity for rapid image output, and also performs well in detail richness and style consistency. However, the model scores extremely low on texture complexity and overall image quality, struggling to render fine textures and high-fidelity visuals. Overall, the SDXL baseline model is more suitable as a baseline for quick generation and comparative experiments, rather than the final solution for achieving high-fidelity visual effects.

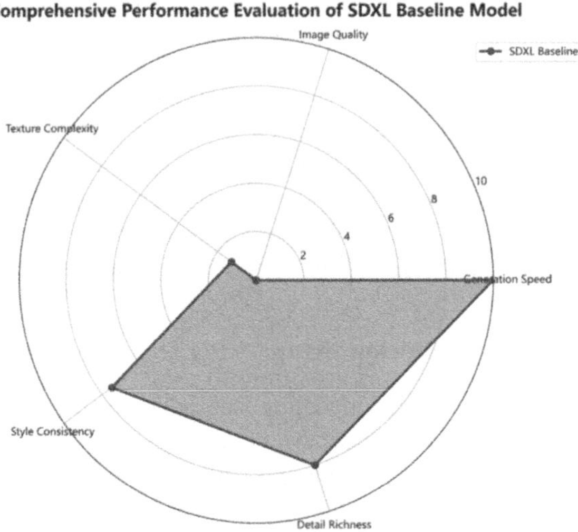


**Fig. 4.** Comprehensive Performance Evaluation of SDXL Baseline Model.

## 4 Research Methods

This chapter focuses on the constructed knowledge base of Qing Dynasty rank badge regulations and pattern generation experiments, performing analysis across five aspects: inter-annotator reliability, structural characteristics of the regulatory knowledge network, hierarchical regulatory network, completeness and complexity of regulatory rules, and rank badge image generation results. The aim is to systematically evaluate the effectiveness of the data construction and rule extraction methods, and to provide quantitative evidence for the design of subsequent shape grammar rules and generative models.

### 4.1 Inter-Annotator Reliability Analysis

To assess the annotation quality of the rank badge regulation dataset, this study used the Krippendorff's α coefficient to quantify the annotation consistency of three annotators across the two major categories: "Civil Official Badges" and "Military Official Badges".

The results show that the $\alpha$ values for the military official badge category were concentrated in the 0.75–0.85 range. Some annotator combinations reached the highly reliable level of $\alpha \geq 0.8$. This suggests that the animal motifs and regulatory descriptions associated with military official badges are relatively clear, leading to a high degree of consistency in the annotators' understanding of their image features and textual regulations.

In contrast, the $\alpha$ values for the civil official badge category were only between 0.28–0.42, indicating a significantly lower overall consistency. This phenomenon is closely related to the inherent complexity of the civil official badge motif combinations and regulatory rules themselves. The finding that civil official regulations constitute 63.9% of all extracted rules while exhibiting a low α of 0.28–0.42 is a statistically significant discovery. Civil motifs are inherently more complex, containing more Details per motif. Ambiguity in historical descriptions leads to interpretive divergence among annotators. High rule quantity does not imply high clarity—complexity and reliability are inversely related in this domain. This insight justifies the study's importance: it provides a diagnostic foundation for redesigning annotation manuals, clarifying cultural semantics, and establishing future training protocols.

On one hand, the civil official badge system involves a greater variety of animal species, with more detailed differences between ranks in terms of motif quantity, combination style, and border details. On the other hand, the descriptions of civil official badge regulations in historical documents are often semantically vague, which makes it easy for annotators to disagree when defining the "main motif" or "auxiliary motif" in the same image, or when interpreting the boundaries of regulatory statements. As visually observed in the Krippendorff's $\alpha$ distribution in Fig. 5, the civil official badges show greater dispersion across different annotator combinations, while the military official badges maintain a generally high and stable range.

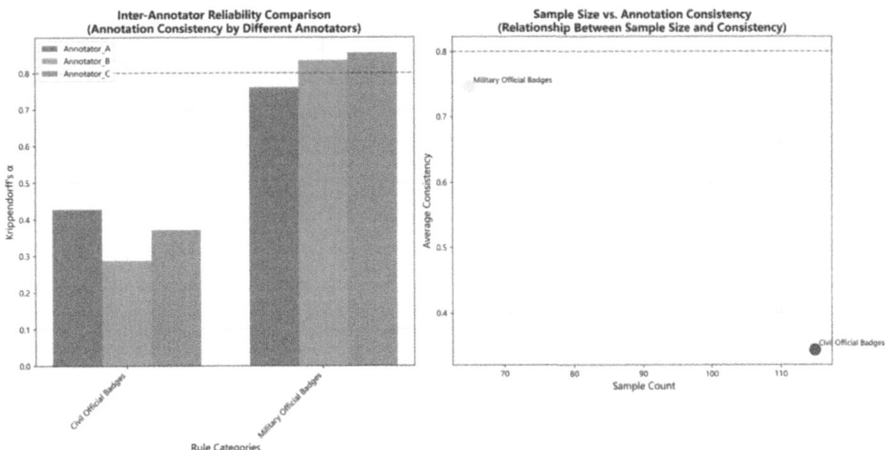


**Fig. 5.** Schematic Diagram of Krippendorff's $\alpha$ Coefficient Distribution (Differentiating Civil Official Badges and Military Official Badges).

This analysis, on one hand, demonstrates that the military official badge regulation data already possesses high reliability and can directly support subsequent statistical analysis, machine learning modeling, and regulatory rule extraction. On the other hand, it serves as a warning for research on civil official badge regulations: future work needs targeted optimization in areas such as the standardized expression of regulatory texts, the supplementation of typical sample examples, and the annotator training mechanism. This is essential to reduce systematic bias caused by semantic ambiguity and comprehension differences, thus laying the foundation for building a more robust regulatory knowledge base.

### 4.2 Structural Characteristics of the Regulatory Knowledge Network

Based on the extraction of regulatory knowledge, this study constructed the Qing Dynasty Rank Badge Regulatory Knowledge Network, unifying key elements such as "Rank—Badge Category—Animal Motif—Posture—Border" as nodes, and establishing directed or undirected edges through relationships like "subordination," "restriction," and "combination". The overall visualization results in Fig. 6 show that the regulatory network presents an approximately circular, dense distribution structure, centered on the animal motif nodes, tightly connecting different official ranks with various detailed regulations to form a highly cohesive graph structure.

**Fig. 6.** Overall Visualization Diagram of the Qing Dynasty Rank Badge Regulatory Knowledge Network.

Combining statistics from metrics such as network average degree, connected components, and node betweenness centrality, the animal motif nodes are not only the key symbols for differentiating rank but also important "intermediary" nodes that run through all levels of the regulatory system. Most civil and military official rank nodes are not directly connected, but instead establish indirect links through shared or similar animal motifs, forming multiple circular clusters centered around the core motifs. This structural characteristic highly aligns with the historical record that "rank is distinguished by birds and beasts", quantitatively confirming the pivotal position of animal motifs in the entire regulatory system.

At the same time, the network also contains a small number of isolated or weakly connected nodes, which are mostly associated with rare postures, special border types, or incomplete regulatory entries. Although these nodes have limited influence on the overall structure, they provide clues for subsequent "micro-difference" research, helping to identify the unique symbolic meanings arising from subtle local detail differences when the official rank is the same or the animal motif is similar.

## 4.3 Hierarchical Network Analysis of Regulatory Patterns

To more clearly delineate the hierarchical transmission mechanism of regulatory constraints, this study extracted a three-layer regulatory pattern network consisting of "Official Rank and Badge Category—Animal Motif—Posture and Border" from the overall regulatory network. Figure 7 illustrates the schematic structure of this hierarchical network: the Top Layer (Official Rank and Badge Category) points via directed edges to the Middle Layer (Animal Motif), and the Middle Layer then connects downwards to the Bottom Layer (Detail elements like Posture and Border), thus forming a top-down chain of regulatory constraints.

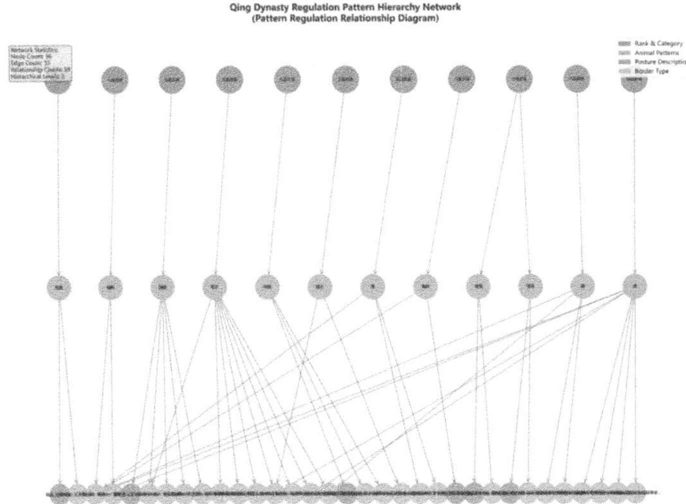


**Fig. 7.** Schematic Diagram of the Three-Layer Regulatory Pattern Network Structure (Official Rank—Animal Motif—Posture and Border).

Statistical results from the network show that the Animal Motif Layer has significantly higher in-degree and out-degree metrics than other layers, making it a genuine core layer that "connects the upper and lower levels". On one hand, animal motif nodes converge edges pointing from different official ranks and badge categories, reflecting their key role in identity distinction. On the other hand, the same animal motif node branches downwards to multiple posture and border combinations. This allows for the subtle adjustment of visual representation and symbolic semantics through localized changes while maintaining the core symbolism.

This top-down hierarchical structure provides a direct template for the design of shape grammar rules: during the pattern generation process, the appropriate badge category is first selected based on the target rank, followed by retrieving compliant animal motifs from the Middle Layer, and finally constraining the legal posture and border types in the Bottom Layer. By strictly aligning the generation process with the three-layer regulatory network, the probability of non-compliant combinations can be significantly reduced, while also reserving an appropriate degree of diversity for the generative model under the premise of ensuring historical regulatory correctness.

### 4.4 Analysis of Regulatory Rule Completeness and Complexity

In the rule extraction phase, this study identified and encoded a total of 18 core regulatory rules, covering key constraint conditions such as official rank and badge category, animal motif selection, posture limitations, and border types. The completeness analysis showed that all 18 rules were supported by actual occurrences within the sample data, achieving a rule coverage rate of 100.0%. This indicates that, within the current sample range, regulatory knowledge extraction has formed a relatively complete set of fundamental rules that can be directly used for subsequent automated generation and logical inference.

Further statistical analysis of the regulatory rules by badge category revealed that rules related to civil official badges accounted for 63.9%, significantly higher than the 36.1% for military official badges. The rule distribution results shown in Fig. 8 quantitatively confirm the higher complexity of the civil official badge regulatory system: in the civil system, more rules are needed to finely constrain the selection of animal motifs, posture variations, and border forms to distinguish different civil ranks and reflect diverse symbolic meanings. In contrast, the military system relies more on a small number of highly identifiable animal motifs for clear identity distinction, making its regulatory structure relatively compact.

From an application perspective, the civil official badge, due to its larger number of rules and greater combinatorial space, is more suitable as a "high-complexity scenario" test set in subsequent rule-based pattern generation and style transfer research. The military official badge can serve as a basic verification set with a relatively simple but semantically clear structure, used to test the generative system's fundamental performance in regulatory compliance and visual clarity. The comparison between the two not only helps evaluate the model's robustness under different complexity conditions but also provides a possibility for constructing a hierarchical evaluation system for intangible cultural heritage pattern generation.

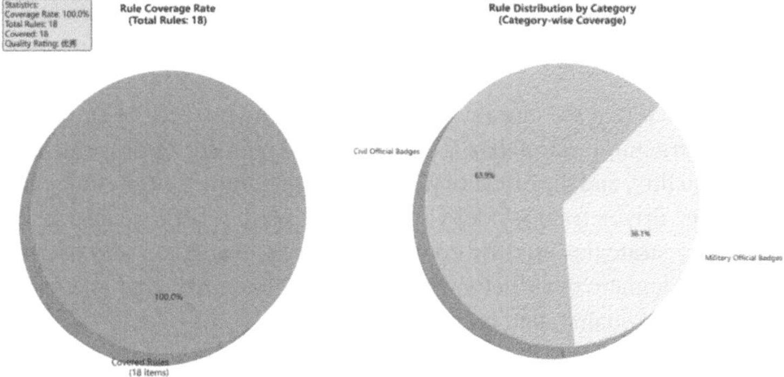


**Fig. 8.** Statistical Chart of the Number and Proportion of Regulatory Rules for Civil Official Badges and Military Official Badges.

## 4.5 Regulatory-Driven Rank Badge Image Generation and Effect Evaluation

Building upon the regulatory knowledge extraction and network modeling, this study further attempts to embed the constructed rank badge regulatory rules into the Stable Diffusion XL pattern generation pipeline to test the comprehensive effect of the regulatory-driven generation strategy in terms of visual presentation and regulatory compliance. Figure 9 displays examples of rank badge images generated based on the regulatory rules, covering different official ranks, badge categories, and various animal motif combinations.

**Fig. 9.** Regulatory-Driven Rank Badge Image Generation Examples.

From the overall visual perspective, most generation results show good consistency in the following aspects: (1) In terms of the correspondence between the main animal motif and the official rank, the generated images comply well with the "Official Rank—Animal Motif" mapping rules in the regulatory network, with no obvious cross-rank mismatch observed; (2) In detail design, such as posture and border, the generation results fundamentally conform to the legal combination range limited by the three-layer regulatory network. They maintain the typical characteristics of historical styles while exhibiting a degree of diversity in motif layout and color scheme; (3) In overall composition, the generated images closely resemble real rank badge images in terms of motif density, negative space relationship, and border proportion, indicating that the regulatory rules exerted effective constraints at both the macroscopic layout and local detail levels.

It should be noted that the current generation experiments still prioritize regulatory compliance and structural rationality. There remains room for improvement in texture detail, stitching quality, and the effect of historical aging. Future work can build upon the existing regulatory-driven framework by introducing more refined texture modeling and material rendering strategies, and by combining expert evaluation and user studies, to conduct a more systematic quantitative assessment of the rank badge images' historical authenticity, aesthetic quality, and interpretability.

# 5 Discussion

This study has provided a new technical path and data foundation for the digital reconstruction and structured generation of traditional patterns by constructing the regulatory knowledge network of Qing Dynasty rank badges and implementing a rigorous inter-annotator reliability assessment. This chapter will delve into a detailed discussion of the main findings.

## 5.1 Theoretical Value of the Knowledge Network and Hierarchical Structure

The Qing Dynasty rank badge regulatory knowledge network we constructed (comprising 167 nodes and 187 edges) not only achieves a unified representation of dispersed regulatory elements but, more importantly, quantitatively confirms the pivotal status of the historical regulation "rank is distinguished by birds and beasts" through network structure analysis. The dense circular distribution of the network's macroscopic structure clearly positions the animal motif nodes as the core intermediaries connecting different official rank groups. This finding is highly consistent with conclusions from traditional costume history research, but this study provides the first quantitative and visual verification.

Building on this, the "Rank → Motif → Detail" three-layer logical structure we proposed transforms the originally complex, text-implicit regulatory constraints into a top-down constraint chain. The structured Rank–Motif–Detail model provides a unified language for humans and generative models, enabling SDXL to interpret cultural constraints as machine-readable instructions. By formalizing the definition of "Detail," establishing explicit rule relationships, and demonstrating a viable pathway for SDXL to ingest structured regulatory knowledge, this study provides a reproducible framework for rule-driven cultural pattern generation. This hierarchical model is a significant contribution of this study to knowledge engineering, providing a formal and executable template for rule-based pattern generation systems. It means that the pattern generation can be decomposed into a structured decision process: first, determining the most macroscopic constraint (official rank), then locking onto the core symbol (animal motif), and finally filling in local constraints (posture and border), thereby ensuring the structural controllability and regulatory compliance of the generated results.

### 5.2 Implications of Data Reliability Assessment for Complex Regulation Research

This study is the first to introduce Krippendorff's α into the annotation assessment of Qing Dynasty rank badge regulatory data. This quantitative analysis reveals a significant difference in complexity and ambiguity between the civil and military official regulatory systems.

1 High Reliability of Military Official Badges (α value 0.75–0.85): This indicates that military official regulations (such as animal motifs and postures) are relatively clear and consistent in historical documents and images. Their regulatory data has reached a standard of high reliability and can be directly used for high-precision knowledge inference and model training.
2 Low Consistency (α value 0.28−0.42) and High Complexity (Regulation Proportion 63.9%) of Civil Official Badges: The lower α value is not simply attributed to annotator error but quantitatively reflects the inherent historical problems of the civil system: a larger number of rules and semantic ambiguity. The co-existence of a high rule proportion (63.9%) and low consistency alerts us that when dealing with complex historical regulations, we should not rely solely on the quantity of rules, but must also focus on the ambiguity of the rules themselves. This finding provides a clear direction for future research optimization: specifically, to strengthen the precise definition of rules for the civil system, supplement typical sample examples, and introduce fuzzy set or probabilistic models to handle regulatory uncertainties that are difficult to eliminate entirely.

### 5.3 Application Value of Regulatory-Driven Approach in Intelligent Pattern Generation

This study ultimately embedded the constructed regulatory knowledge into the generation pipeline of SDXL, demonstrating effective regulatory compliance in the image generation process. Although traditional deep generative models possess advantages in texture and light-and-shadow effects, they generally lack structural controllability, making it difficult to guarantee the "legitimacy" or "historical correctness" of the generated results. The "Regulatory Network → Three-Layer Structure → Generation Constraint" pathway proposed in this research successfully overcomes this limitation.

Experimental results indicate that the images generated based on regulatory constraints maintain good consistency with historical regulations in aspects such as official rank-to-motif mapping and posture-to-border combinations. This proves that integrating the structural advantages of knowledge engineering with the visual generation capabilities of deep learning is a significant direction for the intelligent generation of traditional patterns. This method not only provides a tool for the digital restoration and design of Qing Dynasty rank badges but can also be generalized and applied to the digitalization of other cultural heritage elements with strong regulatory and structural constraints, such as Dunhuang murals and bronze ware inscriptions.

From the perspective of AI security and privacy, the regulatory-driven generation framework developed in this study also provides an important safeguard mechanism. Diffusion models without structural constraints may produce historically inaccurate or misleading cultural patterns, which pose semantic security risks when used in education, museum display, or digital restoration. By grounding the generation process in a validated hierarchical regulatory network, the system limits unauthorized pattern recombination, reduces the risk of cultural misinformation, and prevents malicious manipulation of symbolic motifs.

Additionally, the dataset construction process emphasizes controlled annotation procedures and responsible data handling, minimizing potential leakage of sensitive archival information. Future work may incorporate adversarial robustness testing, provenance tracking, and watermarking of generated images to further enhance the security, traceability, and privacy protection of AI-assisted cultural heritage generation.

### 5.4 Limitations and Future Outlook

While this study has made significant progress, it also presents certain limitations and points towards directions for future research:

1 Regulatory Dynamics and Evolution: The current knowledge network primarily focuses on the stable regulations of the mid-to-late Qing Dynasty and lacks the dynamic characterization of subtle evolution and regional differences in rank badges across different periods (e.g., Shunzhi, Kangxi, Qianlong). Future work could introduce a temporal dimension to construct a temporal knowledge graph to support the tracking and analysis of the regulatory evolution process.

2 Generation Details and Material Restoration: The current image generation experiments prioritize structural and symbolic correctness. However, there is still room for improvement in the fidelity of texture detail, stitching quality, and the historical aging effect. Future work could attempt to combine more sophisticated material models (such as PBR materials) with high-resolution image datasets to deeply fine-tune the SDXL model, achieving high-fidelity and high-realism generation of patterns.

3 Inclusion of Non-Regulatory Elements: Qing Dynasty rank badges also involve non-symbolic regulatory elements such as color, material, and size, which equally impact the historical authenticity of the badges. Future work can incorporate these unstructured elements into the knowledge network to form a more comprehensive regulatory model.

## 6 Conclusion

This study has addressed key challenges in the digital reconstruction and intelligent generation of Qing Dynasty Rank Badge patterns, focusing on regulatory knowledge structuring, data reliability, and the application of generative models. By constructing a hierarchical regulatory knowledge network, we have formalized the regulatory rules that govern the design of these cultural artifacts, providing a robust framework for their digital representation. The introduction of Krippendorff's alpha coefficient for assessing inter-annotator reliability has yielded valuable insights into the complexities of the data annotation process, revealing the high reliability of military official badges and the challenges posed by the more intricate and ambiguous regulations of civil official badges. These findings emphasize the importance of clear and standardized rule definitions in ensuring the reliability of datasets used for training generative models.

The regulatory knowledge network, along with the reliability analysis, lays a strong foundation for the implementation of AI-driven pattern generation models like Stable Diffusion XL (SDXL). Our experiments demonstrate that by integrating this structured knowledge into the generative pipeline, it is possible to produce historically accurate and visually consistent images that respect both the symbolic meaning and structural constraints inherent in the Qing Dynasty Rank Badges. This research not only advances the digital preservation of cultural heritage but also contributes to the development of secure, verifiable, and controllable AI tools for generating complex cultural patterns.

Looking forward, there are several avenues for further research, including enhancing the fidelity of texture details and incorporating non-regulatory elements like color and material into the knowledge network. Additionally, expanding the temporal dimension to capture the evolution of rank badge regulations across different dynastic periods could offer further insights into the dynamic nature of cultural heritage. Ultimately, the methodologies proposed in this study provide a scalable and adaptable framework for the digitalization of other culturally significant artifacts, ensuring their preservation and intelligent generation in future AI applications.

**Acknowledgements.** This study was supported by Key Laboratory of Philosophy and Social Sciences in Guangdong Province of Maritime Silk Road of Guangzhou University (GD22TWCXGC15).This study was supported by Guangdong Province Higher Education Institutions Characteristic Innovation Project in 2023: Lingnan Culture and Art Digital Resource Sharing Platform (2023WTSCX068). This study was supported by Guangzhou University Project: Digital Revitalization and Intelligent Dissemination Research of Lingnan Cultural Arts (PT252022040) and Cantonese Embroidery Intangible Cultural Heritage Digital Resource Sharing Platform (Liwan Research Institute, Guangzhou University, LWYJ202411).

## References

Deng, J., Chen, X.: Research on artificial intelligence interaction in computer-aided arts and crafts. Mob. Inf. Syst. **2021**, 5519257 (2021). https://doi.org/10.1155/2021/5519257

Duan, C., Yang, J., Cui, Q., Zhang, W., Wan, X., Zhang, M.: Enhancing the recommendation of learning resources for learners via an advanced knowledge graph. Appl. Sci. **15**(8), 4204 (2025). https://doi.org/10.3390/app15084204

Gharagozlou, H., Mohammadzadeh, J., Bastanfard, A., Shiry Ghidary, S.: Semantic relation extraction: a review of approaches, datasets, and evaluation methods with looking at the methods and datasets in the Persian language. ACM Trans. Asian and Low-Resour. Lang. Inf. Process. **22**(7), 189 (2023). https://doi.org/10.1145/3592601

Gîrbacia, F.: An analysis of research trends for using artificial intelligence in cultural heritage. Electronics **13**(18), 3738 (2024). https://doi.org/10.3390/electronics13183738

Gwet, K.L.: On Krippendorff's alpha coefficient. Advanced Analytics, LLC (2015). https://www.researchgate.net/profile/Kilem-Gwet/publication/267823285_On_Krippendorff's_Alpha_Coefficient/links/60e3bf0892851ca944ae25d6/On-Krippendorffs-Alpha-Coefficient.pdf

Karim, M.M., Khan, S., Van, D.H., Liu, X., Wang, C., Qu, Q.: Transforming data annotation with AI agents: a review of architectures, reasoning, applications, and impact. Future Internet **17**(8), 353 (2025). https://doi.org/10.3390/fi17080353

Li, J., Dang, H., Zhou, Z., Zhu, J., Huang, Z.: G-SplatGAN: Disentangled 3D Gaussian generation for complex shapes via multi-scale patch discriminators. Comput. Graph. Forum **44**(7), e70256 (2025). https://doi.org/10.1111/cgf.70256

Liu, B., et al.: On the cultural gap in text-to-image generation (2023). arXiv Preprint arXiv:2307.02971. https://arxiv.org/abs/2307.02971

Marzi, G., Balzano, M., Marchiori, D.: K-Alpha Calculator–Krippendorff's Alpha Calculator: a user-friendly tool for computing Krippendorff's alpha inter-rater reliability coefficient. MethodsX **12**, 102545 (2024). https://doi.org/10.1016/j.mex.2023.102545

Moulin, C., et al.: Research infrastructures in the digital humanities (ESF Science Policy Briefings No. 42). European Science Foundation (2011). https://darhiv.ffzg.unizg.hr/id/eprint/1888

Osuji, C.C., Castro Ferreira, T., Davis, B.: A systematic review of data-to-text NLG (2024). arXiv Preprint arXiv:2402.08496. https://doi.org/10.48550/arXiv.2402.08496

Podell, D., et al.: SDXL: Improving latent diffusion models for high-resolution image synthesis (2023). arXiv Preprint arXiv:2307.01952. https://doi.org/10.48550/arXiv.2307.01952

Qian, W., Xu, D., Cao, J., Guan, Z., Pu, Y.: Aesthetic art simulation for embroidery style. Multimedia Tools Appl. **78**, 995–1016 (2019). https://doi.org/10.1007/s11042-018-6002-9

Suh, K.: The analytic study and conservation of a rank badge of the Qing Dynasty [Doctoral dissertation, Fashion Institute of Technology, State University of New York]. ProQuest Dissertations & Theses (2004). https://www.proquest.com/openview/ae2d50261bc936ee7e5c9d4e1279230a/1?pq-origsite=gscholar&cbl=18750&diss=y

Wang, B.L., Zong, F.Y. (eds.): Chinese civil and military rank badges (in Chinese). Nanjing Publishing House (2007). https://doi.org/10.1007/978-7-80718-320-4

Wang, M., Chau, H., Thaker, K., Brusilovsky, P., He, D.: Knowledge annotation for intelligent textbooks. Technol. Knowl. Learn. **28**, 1–22 (2023). https://doi.org/10.1007/s10758-021-09544-z

Wang, X., Chang, W., Tan, X.: Representing and linking Dunhuang cultural heritage information resources using knowledge graph. Knowl. Organ. **47**(7), 604–615 (2020). https://doi.org/10.5771/0943-7444-2020-7-604

Wang, Y., Ramli, M.F., Song, H., Li, X.: Exploring the path of cultural sustainability for traditional costume embroidery patterns based on digital generative art. Cultura: Int. J. Philos. Cult. Axiology **21**(4), 271–286 (2024). https://doi.org/10.70082/cijpca.v21i4.674

Wei, Z., Ko, Y.C.: Segmentation and synthesis of embroidery art images based on deep learning convolutional neural networks. Int. J. Pattern Recognit. Artif. Intell. **36**, 2256020 (2022). https://doi.org/10.1142/S0218001422520188

Xu, L., Lu, L., Liu, M., Song, C., Wu, L.: Nanjing Yunjin intelligent question-answering system based on knowledge graphs and retrieval-augmented generation technology. Heritage Sci. **12**, 118 (2024). https://doi.org/10.1186/s40494-024-01231-3

Yang, H., Li, M., Sui, Q., Hu, B., Shi, K., Wang, R.: Dynasty embroidery patterns: a case study of official rank badges [Preprint] (2024). https://doi.org/10.21203/rs.3.rs-7522369/v1

Zang, N.: *Qing Dynasty official rank badges* (in Chinese). Huaxia Publishing House (2016). https://doi.org/10.1007/978-7-5080-8749-8

Zhou, P., El-Gohary, N.: Ontology-based multilabel text classification of construction regulatory documents. J. Comput. Civ. Eng. **30**(4) (2015). https://doi.org/10.1061/(ASCE)CP.1943-5487.0000530

# Open-Set Cultural Object Detection Based on Vision-Language Pre-Trained Models Algorithm

Guanrong Huang and Wei Zhang(✉)

Kunming Metallurgy College, Kunming, Yunnan, China
22538717@qq.com

**Abstract.** With the rapid development of digital humanities and cultural heritage protection, it has become crucial to automatically identify and catalog cultural objects in a vast amount of image and video materials. However, traditional object detection algorithms usually operate under the "closed-set" assumption, they can only recognize pre-defined categories that have appeared in the training set, and it is difficult to handle the "open-set" scenario given the wide variety of cultural objects and the constant emergence of new categories. To address this challenge, this paper proposes an open-set cultural object detection algorithm, named VL-COOD, based on vision-language pre-trained models (VLPMs). This algorithm utilizes the powerful zero-shot recognition ability of VLPM (such as CLIP) and uses text descriptions as an open vocabulary, enabling the model to recognize "unknown" cultural objects not seen in the training set. Specifically, we design a "Semantic Calibration Module (SCM)", which uses the semantic information of text embeddings to enhance the discriminability of visual features through an attention mechanism. In addition, we introduce an "Energy-Guided Open-set Loss", which simultaneously optimizes the classification accuracy of known classes and the ability to distinguish known/unknown classes during the training process. We have conducted extensive experiments on the self-built "Cultural Heritage-Open Dataset (CH-Open)". The results show that compared with the current mainstream open-set detection methods, the proposed VL-COOD algorithm in this paper has improved the detection accuracy (mAP) of known classes by 4.7% and the recognition recall rate of unknown classes by 8.2%, demonstrating the effectiveness and superiority of this method in the open-set cultural object detection task.

**Keywords:** Open-set Object Detection · Cultural Object · Vision-Language Model · CLIP · Zero-shot Learning

## 1 Introduction

### 1.1 Research Background and Significance

Cultural heritage is a precious treasure of human civilization, and its digital protection and research are important directions for the interdisciplinary integration of information technology and humanities (Skublewska-Paszkowska et al. 2022; Fu et al. 2025). Using


H. Yan et al. (Eds.): AIS&P 2025, CCIS 2918, pp. 467–479, 2026.
https://doi.org/10.1007/978-981-95-9884-7_32

computer vision technology to automatically analyze a vast amount of images in museums, archaeological sites, and historical documents can greatly improve the cataloging, retrieval, and research efficiency of cultural heritage (Ju 2024). Object detection, as a key technology among them, aims to locate and identify specific objects in images, such as a porcelain, a Buddha statue, or an ancient coin.

However, the cultural object detection task faces unique challenges. First, the categories of cultural objects are extremely rich, covering thousands of years of history and a vast territory, from pottery, bronze wares to calligraphy, paintings, and costumes, and new archaeological discoveries are still emerging continuously. Second, the forms of cultural objects under the same major category vary greatly. For example, in the category of "bottles", there are huge differences in their shapes, decorations, and materials among different dynasties and kilns, showing the significant characteristic of "large intra-class differences".

### 1.2 Problems and Challenges

Current mainstream object detection algorithms, such as Faster R-CNN and YOLO series (Chai et al. 2021), all belong to "Closed-Set" detectors. They rely on large-scale, fully annotated training data and can only recognize object categories predefined in the training stage. When "unknown" categories that were not seen during training appear in the test images, these models either misclassify them as a "known" category or miss detections due to low confidence, failing to meet the actual needs of cultural heritage management and other scenarios that require the recognition of newly discovered or rare items.

"Open-Set Object Detection" (Miller et al. 2024; Jeong et al. 2024) aims to solve this problem. It requires the model to not only accurately detect and classify "known" categories but also identify any object that does not belong to the known categories as "unknown", thus providing a basis for subsequent manual annotation or incremental learning. However, most existing open-set detection methods rely on pure visual features to distinguish between known and unknown (Han et al. 2025). When faced with cultural objects that have similar visual features but different semantic categories (e.g., a simple and elegant Song Dynasty white porcelain bowl and an ordinary modern white bowl), it is often difficult to make accurate judgments.

### 1.3 Contributions of This Paper

To overcome the above challenges, this paper introduces the idea of Vision-Language Pre-trained Models (VLPMs) into the open-set cultural object detection task (Zhang et al. 2025). VLPMs learn rich, cross-modal semantic knowledge through pre-training on a large number of image-text pairs, enabling them to have powerful zero-Shot recognition capabilities (Radford et al. 2021). This provides a new paradigm for dealing with constantly changing cultural object categories. Recent research has shown that directly using the hidden states of large language models (LLMs) or multi-modal large models (MLLMs) can effectively enhance the performance of open-vocabulary detectors without generating additional training data (Lin et al. 2024).

The main contributions of this paper can be summarized as the following three points:

Proposed a new open-set cultural object detection framework called VL-COOD. This framework combines traditional two-stage detectors with visual-language models, constructs an open semantic space using text descriptions, and enables the model to generalize to object categories not seen during training.

A "Semantic Calibration Module (SCM)" is designed. This module can dynamically utilize category-related text semantic information to "calibrate" the visual features of candidate regions, enhancing their representational ability for fine-grained cultural features, thus better serving subsequent classification tasks.

An "energy-guided open-set loss function" is introduced. By constructing an energy model, this loss function explicitly enlarges the gap in energy distribution between known and unknown classes during training, enabling the model to more reliably distinguish known and unknown targets during inference.

## 2 Related Work

### 2.1 Traditional Object Detection

Object detection algorithms have evolved from traditional handcrafted feature-based methods to deep learning-based methods. Currently, deep learning-based methods are mainstream and mainly fall into two categories: one is the two-stage (Two-Stage) detectors represented by Faster R-CNN and Mask R-CNN. They first generate a series of candidate boxes through the Region Proposal Network (RPN), and then perform secondary classification and position refinement on these candidate boxes, with high accuracy but slightly slower speed. The other is the one-stage (One-Stage) detectors represented by YOLO and SSD, which complete object localization and classification in one step, with fast speed but slightly lower detection accuracy for small or overlapping objects. Regardless of which category, their inherent closed-set property limits their application in open scenarios such as cultural heritage.

### 2.2 Open-Set Object Detection

The concept of open-set recognition was first proposed in image classification tasks. In recent years, researchers have started to extend it to more complex object detection tasks. ORE proposed by Joseph et al. is a representative method. It adds an additional "unknown" class to the classifier backend of Faster R-CNN and designs a calibration strategy based on extreme value theory to calculate the probability of belonging to the unknown class. OpenDet introduces an unknown instance generation module based on an adversarial generative network, which helps the model learn the boundary between known and unknown by generating "pseudo-unknown" samples during training. However, these methods mainly rely on visual cues, and when unknown objects are highly similar to a known class in appearance and texture, it is extremely difficult to distinguish, easily leading to misjudgments.

### 2.3 Vision-Language Models

With the development of multimodal learning, vision-language pre-trained models (VLPMs) represented by CLIP and ALIGN have demonstrated remarkable potential. These models are trained on hundreds of millions of image-text pairs through contrastive learning, mapping images and texts into the same unified semantic space. This enables the model to directly perform zero-shot image classification by calculating the similarity between image features and the features of any text description. This ability provides a theoretical basis for constructing an open-vocabulary detector that is not limited to a fixed list of categories. Recent work has attempted to apply VLPMs to closed-set or zero-shot object detection, but how to effectively apply them to more challenging open-set scenarios, especially in the specific domain of cultural objects, remains an issue to be explored in depth.

## 3 Algorithm Model in This Paper

### 3.1 Overall Framework of the Model

The VL-COOD algorithm framework proposed in this paper is shown in Fig. 1. Overall, we adopt a two-stage detection paradigm similar to Faster R-CNN.

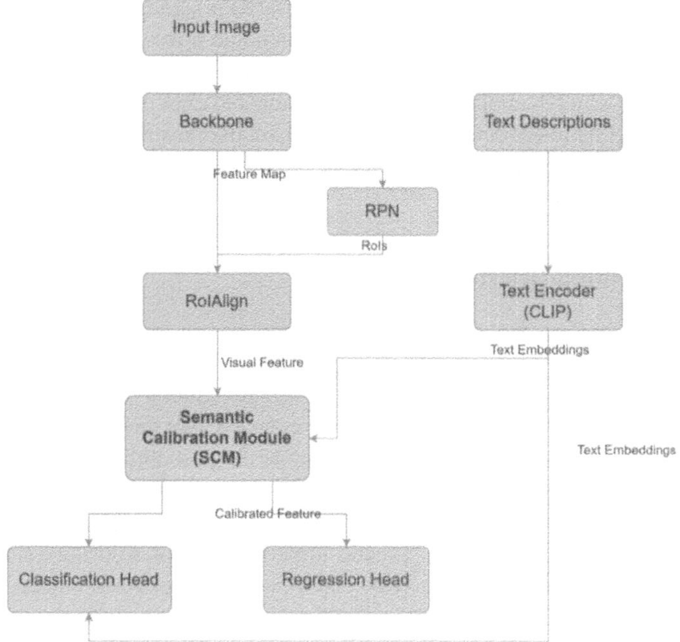


**Fig. 1.** Overall Framework Diagram of the VL-COOD Algorithm

Input and Backbone Network: Input a cultural heritage image to be detected. We use a standard vision backbone network (such as ResNet-50 or Swin Transformer) to extract the deep feature map of the image.

Region Proposal: RPN slides on the feature map to generate a series of candidate regions (Region of Interest, RoI) that may contain objects.

Feature Extraction and Alignment: For each RoI, a fixed-size region feature $f_v \in R^{d_v}$ is extracted from the feature map through the RoIAlign operation. At the same time, we use a pre-trained CLIP text encoder to encode the text descriptions of all known cultural object categories (e.g., "a blue-and-white porcelain vase of the Ming Dynasty") and a general "unknown item" description into text features $f_t \in R^{d_t}$.

Semantic Calibration and Classification: The regional visual feature $f_v$ is fed into the core module we designed - the Semantic Calibration Module (SCM). SCM uses the $F_T=\{f_{t_1}, f_{t_2}, \ldots, f_{t_N}\}$ Text features are used to perform operations on $f_v$ weighted sum adjustment on $f_{v'}$, and outputs a visually enhanced semantic feature $f_{v'}$. Finally, classification is performed by calculating the similarity between $f_{v'}$ and the text features of each category, and a parallel regression head is used to refine the position of the bounding box.

### 3.2 Vision-Text Semantic Alignment

Different from traditional detectors that use a learned classification weight matrix, we directly utilize the powerful semantic alignment ability of the CLIP model. Given a visual feature $f_v$ of a region and a text feature $f_t$ of a category, the matching score between them can be calculated by cosine similarity:

$$S(f_v, f_t) = \frac{f_v \cdot f_t}{f_v \, f_t} \tag{1}$$

For N known categories and an unknown category, we can obtain a (N + 1) -dimensional similarity score vector. Through the Softmax function, it can be converted into a probability distribution:

$$p(y_i|f_v) = \frac{\exp\left(S\left(f_v, f_{t_i}\right)/\tau\right)}{\sum_{j=1}^{N+1} \exp\left(S\left(f_v, f_{t_j}\right)/\tau\right)} \tag{2}$$

Among them, $y_i$ represents the i -th category (including the unknown category), and $\tau$ is the learnable temperature coefficient introduced in CLIP, which is used to adjust the sharpness of the probability distribution. This method transforms the classification problem into a vision-text matching problem, and naturally supports open vocabulary.

### 3.3 Semantic Calibration Module (SCM)

To address the problem that it is difficult to distinguish fine-grained features of cultural targets, we designed the SCM module. Its core idea is that whether the visual features of a region match a certain text description (such as "a Yuan Dynasty porcelain plate with dragon patterns") not only depends on its own pixel information but also should consider its relationship with all other possible categories. SCM achieves this through an attention mechanism.

First, we calculate the attention weights between the visual feature $f_v$ and the set of all category text features $F_T$:

$$\alpha_j = soft\max_j\left(\frac{(W_q f_v)^T (W_k f_{t_j})}{\sqrt{d_k}}\right) \tag{3}$$

Among them, $W_q$, $W_k$ is a learnable linear projection matrix, and $d_k$ is the dimension of the key vector. Then, we weight-sum all text features according to the attention weights to obtain a "context-aware" semantic vector $f_{sem}$:

$$f_v^{context} = \sum_{j=1}^{N+1} \alpha_j (W_v f_{t_j}) \tag{4}$$

where $W_v$ is another learnable projection matrix. Finally, we fuse this context semantic vector with the original visual feature to obtain the calibrated visual feature $f_{v'}$:

$$f_{v'} = \gamma f_v + (1 - \gamma) f_v^{context} \tag{5}$$

where $\gamma^2$ is a learnable gating scalar used to balance the original visual information and the calibrated semantic information. In this way, SCM can make the visual feature "focus" on the image region most relevant to a specific text description, suppress irrelevant information, and thus improve the classification accuracy.

### 3.4 Energy-Guided Open-Set Loss Function

Relying solely on the output of Softmax to distinguish known and unknown is unreliable because neural networks tend to give high-confidence wrong predictions for out-of-distribution samples. Inspired by the energy model, we define an energy function $E(f_v)$ for the classifier, which measures the "incompatibility" degree between the input feature $f_v$ and the knowledge of known categories. We define it in the form of Helmholtz free energy:

$$E(f_v) = -\tau \log \sum_{i=1}^{N} \exp\left(S\left(f_{v'}, f_{t_i}\right)/\tau\right) \tag{6}$$

Note that the summation here is only carried out over the $^N$ "known" categories. Intuitively, if a sample belongs to a known category, it should match a text description of a known category highly, resulting in a low value of $E(f_v)$ (low energy, stable state); conversely, if a sample is an unknown category, its match with all known categories is relatively low, and the value of $E(f_v)$ will be high (high energy, unstable state).

Based on this, we design the energy-guided open-set loss $L_{open}$:

$$L_{open} = \frac{1}{|B_K|}\sum_{f_v \in B_K} (\max(0, E(f_v) - m_{in}))^2 + \frac{1}{|B_U|}\sum_{f_v \in B_U} (\max(0, m_{out} - E(f_v)))^2 \tag{7}$$

Among them, $B_K$ and $B_U$ respectively represent the known-class sample set and the unknown-class sample set in a batch (during training, we regard some samples of known classes as "pseudo-unknown" to simulate the open-set scenario). $m_{in}$ and $m_{out}$ are two hyperparameters, representing the upper bound of the energy of known classes and the lower bound of the energy of unknown classes respectively. The goal of this loss function is to "push" the energy of known-class samples below $m_{in}$, while "pulling" the energy of unknown-class samples above $m_{out}$, thus forming a clear decision boundary in the energy space.

The total loss function of the model consists of three parts:

$$L_{total} = L_{cls} + \lambda_{reg} L_{reg} + \lambda_{open} L_{open} \quad (8)$$

Among them, $L_{cls}$ is the standard cross-entropy loss function for known classes, $L_{reg}$ is the standard bounding box regression loss (such as Smooth L1 Loss), and $\lambda_{reg}$ and $\lambda_{open}$ are the weight coefficients used to balance each loss.

## 4 Experiments and Result Analysis

### 4.1 Datasets and Evaluation Metrics

Currently, there is a lack of publicly available benchmark datasets specifically for open-set cultural object detection. For this reason, we constructed a dataset named CH-Open. Based on some publicly available museum collection images (such as the open data of the Palace Museum and the Metropolitan Museum of Art), combined with some existing datasets (such as "bowls", "vases", etc. in COCO), we collected approximately 20,000 images in total.

We selected 15 categories with clear cultural attributes as the known class K (for example: blue and white porcelain, terracotta figures, jade bi, bronze tripods, Buddha statues, etc.), and annotated approximately 35,000 object instances in total. The remaining cultural objects and other object categories in the COCO dataset that do not belong to these 15 categories together constitute the candidate pool of the unknown class $U^t$. During testing, we selected 5 categories that have never appeared in the training set (such as: cloisonné, snuff bottles, tri-colored glazed pottery of the Tang Dynasty, Ruyi, incense burners) from $U^t$ as specific unknown classes for evaluation.

Evaluation Metrics:

Known-class detection performance: The standard mean average precision (mAP@0.5) is used to measure the detection ability of the model for known classes.

Open-set performance:

Unknown Recall (UR): Among all unknown-class objects, the proportion correctly identified as "unknown", which measures the ability of the model to "discover" unknown objects.

Wilderness Impact (WI): Measures the degree of decline in the classification performance of known classes due to the introduction of unknown classes.

### 4.2 Experimental Settings

Our model is implemented based on the MMDetection framework. The backbone network uses ResNet-50 pre-trained on ImageNet. Both the text and image encoders are initialized with the pre-trained weights of CLIP (ViT-B/16) and fine-tuned during training. The optimizer uses AdamW, and the initial learning rate is.

$1 \times 10^{-4}$, with a batch size of 16. All experiments were conducted on a server equipped with 4 NVIDIA A100 GPUs. The loss function weights were set to $\lambda_{reg} = 1.0$, $\lambda_{open} = 0.5$.The energy threshold was set to $m_{in}$ = - 10,$m_{out}$= −4.

### 4.3 Experimental Results and Analysis

Comparison with Mainstream Methods.

We compared VL-COOD with the following baseline methods:

Faster R-CNN (FRCNN): A standard closed-set detector, serving as the performance lower bound.

ORE: A representative open-set object detection method based on probability calibration.

OpenDet: An open-set object detection method based on adversarial learning.

FRCNN + CLIP: A naive combination that directly replaces the original classification head of Faster R-CNN with the CLIP classification head, but does not include SCM and $L_{open}$.

The experimental results are shown in Table 1.

**Table 1.** Performance Comparison of Different Methods on the CH-Open Dataset

| Method | mAP@0.5 (Known Classes) | Unknown Class Recall (UR) | Wilderness Impact (WI) |
|---|---|---|---|
| FRCNN (Closed Set) | 65.4% | - | - |
| ORE | 61.2% | 45.3% | 6.42% |
| OpenDet | 62.5% | 51.8% | 4.43% |
| FRCNN + CLIP | 68.1% | 55.6% | 2.15% |
| VL-COOD (This Paper) | 72.8% | 63.8% | 1.58% |

As can be seen from Table 1:

The VL-COOD proposed in this paper achieved the best performance in all metrics. Its mAP for known classes reached 72.8%, significantly higher than all other methods, demonstrating the effectiveness of the introduction of VLPM and the SCM module in improving the fine-grained recognition accuracy of known classes.

In terms of the key open-set metric "unknown class recall rate", VL-COOD reached 63.8%, 8.2 percentage points higher than the sub-optimal FRCNN + CLIP method, indicating that the energy-guided loss function we designed can effectively help the model distinguish between known and unknown.

The lower the WI metric, the better. The VL-COOD has the lowest WI value, indicating that our method has a strong open-set recognition ability while having the least negative impact on the classification performance of known classes.

Ablation Experiments

To verify the effectiveness of the semantic calibration module (SCM) and the energy-guided open-set loss ($L_{open}$) we proposed, we conducted ablation studies, and the results are shown in Table 2.

**Table 2.** Ablation Studies of VL-COOD Model Components

| Model Configuration | mAP@0.5 (Known Classes) | Unknown Class Recall (UR) |
|---|---|---|
| Base Model (FRCNN + CLIP) | 68.1% | 55.6% |
| Base Model + SCM | 71.5% (+3.4) | 57.1% (+1.5) |
| Base Model + $L_{open}$ | 68.9% (+0.8) | 61.9% (+6.3) |
| Full Model (VL-COOD) | 72.8% | 63.8% |

The results in Table 2 clearly show that:

After adding the SCM module alone, the mAP of known classes increased by 3.4%, demonstrating that SCM enhances the discriminability of visual features through semantic calibration.

After adding $L_{open}$ alone, the recall rate of unknown classes increased significantly by 6.3%, while having little impact on the mAP of known classes, proving the core role of this loss function in distinguishing the known/unknown boundary.

After combining the two modules, the model achieved the best performance on both metrics, showing the synergistic effect between components.

Visualization Analysis

To more intuitively demonstrate the performance of our algorithm, we provide some visualization results.

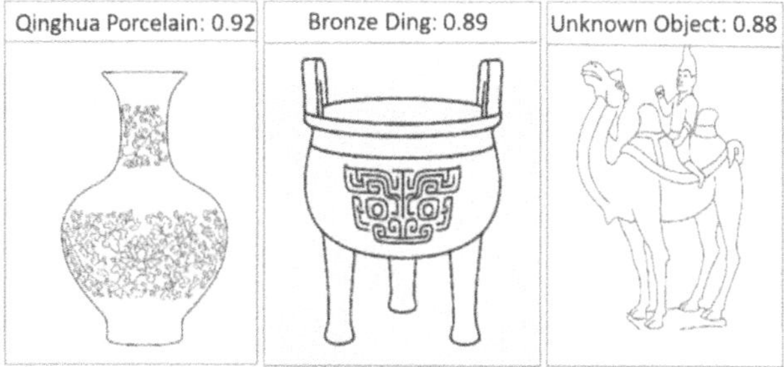


**Fig. 2.** Successful Detection Examples of VL-COOD

Figure 2 shows the successful cases of VL-COOD. The model not only accurately detected the known classes "Blue-and-white porcelain vase" and "Bronze Tripod", but also successfully identified the "Tri-colored Glazed Camel" not seen during training as an "unknown object" and gave a relatively high confidence.

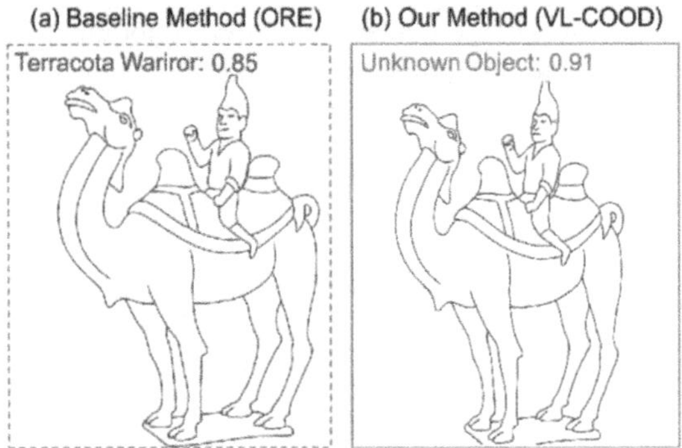


**Fig. 3.** Comparison with Baseline Methods

Figure 3 compares the performance of ORE and VL-COOD when dealing with an unknown target "Tri-colored Glazed Camel". Due to mainly relying on visual features, ORE misjudged it as "Terracotta Figurine" that had appeared in the training set. While our method, benefiting from the semantic understanding ability of the vision-language model and the constraint of the open-set loss, successfully identified it as unknown.

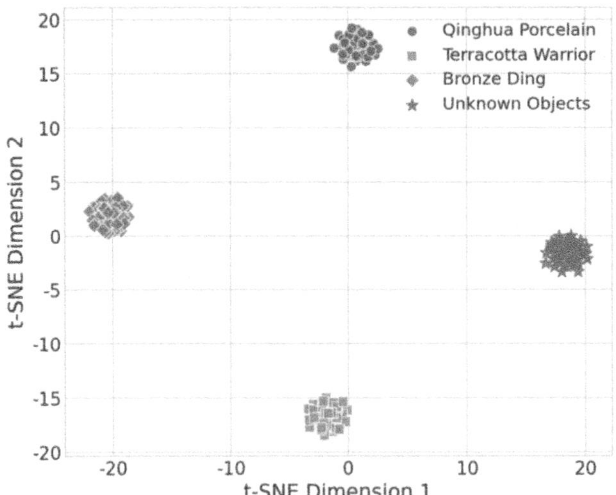


**Fig. 4.** t-SNE Visualization of Feature Space

Figure 4 shows the t-SNE dimensionality reduction visualization of the features in the penultimate layer of the model. It can be clearly seen that after training with VL-COOD, the feature points of different known classes form compact and separated clusters, while the feature points of all unknown classes are pushed to areas far from the known class clusters, which intuitively proves that our method has learned a highly discriminative and open feature space for unknown classes.

### 4.4 Parameter Sensitivity Analysis

We conducted a sensitivity analysis on the weight $\lambda_{open}$ of the open-set loss in the total loss function, and the results are shown in Table 3.

**Table 3.** Impact of Hyperparameter $\lambda_{open}$. on Model Performance

| $\lambda_{open}$ | 0.1 | 0.3 | 0.5 (This article) | 0.8 | 1.0 |
|---|---|---|---|---|---|
| mAP@0.5 | 72.1% | 72.5% | 72.8% | 72.4% | 71.9% |
| UR | 58.2% | 61.5% | 63.8% | 63.1% | 62.5% |

When $\lambda_{open}$ is too small, the model has insufficient ability to distinguish unknown classes; when it is too large, it will overly focus on the open-set task, slightly compromising the classification accuracy of known classes. Experiments show that when $\lambda_{open}$= 0.5, the model achieves the best balance between known-class detection and unknown-class recognition.

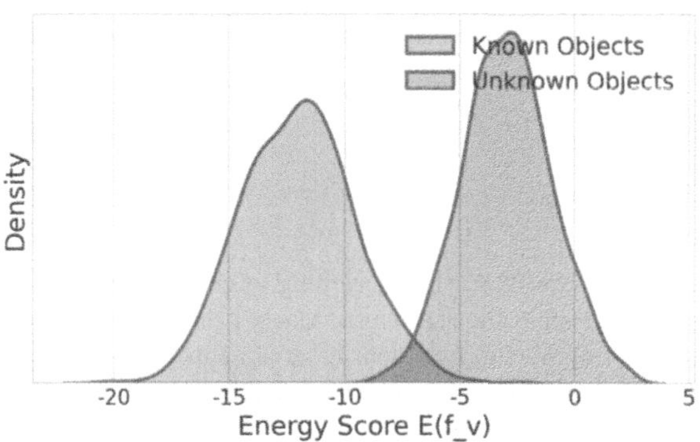


**Fig. 5.** Energy Distribution of Known and Unknown Classes

Figure 5 intuitively shows the energy distribution of known-class and unknown-class samples on the test set. It can be seen that the energy values of the vast majority of known-class samples are relatively low, while those of unknown-class samples are generally high, forming two basically separated distributions, verifying the effectiveness of our energy-based open-set discrimination method.

## 5 Conclusion

Aiming at the open-set problem of a large variety of cultural objects and the continuous emergence of new classes in cultural object detection, this paper proposes an open-set detection algorithm VL-COOD based on vision-language pre-trained models. This algorithm constructs an open vocabulary by leveraging the zero-shot ability of CLIP and innovatively designs a semantic calibration module.

(SCM) to enhance the discriminability of fine-grained features, and at the same time introduces an energy-guided open-set loss function ($L_{open}$) to explicitly learn the known/unknown decision boundary. A large number of experiments on the self-built CH-Open dataset show that VL-COOD not only outperforms existing methods in the detection accuracy of known classes, but also shows significant advantages in the ability to identify unknown objects.

Future work can be carried out in the following aspects: 1) Research more efficient methods for generating descriptions of unknown objects to support the preliminary semantic understanding of unknown objects, rather than just labeling them as "unknown"; 2) Explore the combination of this framework with incremental learning, so that when a certain number of similar samples identified as "unknown" are accumulated, they can be quickly absorbed as new known classes; 3) Lightweight the model for deployment on resource-constrained edge devices such as museum guides and mobile archaeology.

**Acknowledgements.** Fund Project 1: Scientific Research Fund Project of Yunnan Provincial Department of Education: Research on Bearing Mechanism and Application Technology of Bamboo-reinforced Soil Subgrade in Rural Highway Engineering (2024J1363).

Fund Project 2: Fund Project of Kunming Metallurgy College: Research on Related Issues of Minor Maintenance of University Infrastructure—Taking Kunming Metallurgy College as an Example (2022xjy05).

## References

Chai, J., Zeng, H., Li, A., Ngai, E.W.: Deep learning in computer vision: a critical review of emerging techniques and application scenarios. Mach. Learn. Appl. **6**, 100134 (2021)

Fu, X., Li, M., Li, X., Chen, W., Yu, L., Chen, Z., et al.: Emerging computing technology for digital culture heritage preservation and inheritance: a literature review. IEEE Trans. Comput. Soc. Syst. (2025)

Han, X., Wei, L., Yu, X., Dou, Z., He, X., Wang, K., et al.: Boosting segment anything model towards open-vocabulary learning. Proc. AAAI Conf. Artif. Intell. **39**(3), 3356–3365 (2025)

Jeong, J., Park, G., Yoo, J., Jung, H., Kim, H.: ProxyDet: synthesizing proxy novel classes via classwise mixup for open-vocabulary object detection. Proc. AAAI Conf. Artif. Intell. **38**(3), 2462–2470 (2024)

Ju, F.: Mapping the knowledge structure of image recognition in cultural heritage: a scientometric analysis using CiteSpace, VOSviewer, and Bibliometrix. J. Imaging **10**(11), 272 (2024)

Lin, C., Jiang, Y., Qu, L., Yuan, Z., Cai, J.: Generative region-language pretraining for open-ended object detection. In: Proceedings of the IEEE/CVF Conference on Computer Vision and Pattern Recognition (pp. 13958–13968) (2024)

Miller, D., Sünderhauf, N., Kenna, A., Mason, K.: Open-set recognition in the age of vision-language models (2024). arXiv Preprint arXiv:2403.16528

Radford, A., Kim, J.W., Hallacy, C., Ramesh, A., Goh, G., Agarwal, S., et al.: Learning transferable visual models from natural language supervision. In: Proceedings of the 38th International Conference on Machine Learning (ICML) (pp. 8748–8763) (2021)

Skublewska-Paszkowska, M., Milosz, M., Powroznik, P., Lukasik, E.: 3D technologies for intangible cultural heritage preservation—literature review for selected databases. Heritage Sci. **10**(1), 3 (2022)

Zhang, M., Liu, S., Shao, H., Ba, Z., Liu, J., Albu Kaya, M.G., et al.: Development trend in non-destructive techniques for cultural heritage: from material characterization to AI-driven diagnosis. Heritage **8**(9), 381 (2025)

# A Study on Intelligent Monitoring and Prediction of Foundation Deformation Based on the Internet of Things and the LSTM Model

Wei Zhang and Guanrong Huang(✉)

Kunming Metallurgy College, Kunming, Yunnan, China
562668536@qq.com

**Abstract.** To address the problems of poor timeliness, high labor costs, and insufficient data analysis capabilities in traditional foundation deformation monitoring methods, this paper proposes an intelligent monitoring and prediction system based on the combination of the Internet of Things (IoT) and the Long Short-Term Memory (LSTM) model. The system enables real-time, continuous acquisition and remote transmission of foundation deformation data through a network of high-precision sensors deployed in the monitoring area.Furthermore, an LSTM-based deep learning model is constructed to train on the collected multi-source time-series data, modeling the non-linear coupling relationships between deformation and various influencing factors to accurately predict future deformation trends. Taking the measured data of a deep foundation pit project as an example for experimental verification, the results show that the system can stably and reliably obtain on-site data, and the established LSTM prediction model is superior to the traditional ARIMA model and Support Vector Machine (SVM) model in terms of indicators such as Root Mean Square Error (RMSE) and Mean Absolute Error (MAE), and the prediction accuracy can meet the requirements of engineering early warning. This research provides a novel, intelligent, and automated approach to the safety monitoring and risk assessment of foundation engineering.

**Keywords:** Foundation Deformation · Internet of Things · Intelligent Monitoring · LSTM Model · Time Series Prediction

## 1 Introduction

The stability and safety of the foundation are the fundamental guarantee for the normal operation of various civil engineering structures (such as high-rise buildings, bridges, dams, etc.). Under the complex action of various factors such as external loads, environmental changes (such as rainfall, temperature), and surrounding engineering activities, the foundation will produce uneven settlement, displacement and other deformations. If the deformation amount exceeds the design allowable range, it will affect the service function of the structure at least, and may even lead to structural cracking, tilting or instability failure, causing huge economic losses and casualties (Tian et al. 2023; He


H. Yan et al. (Eds.): AIS&P 2025, CCIS 2918, pp. 480–493, 2026.
https://doi.org/10.1007/978-981-95-9884-7_33

et al. 2023). Therefore, long-term, effective and high-precision monitoring and prediction of foundation deformation are of great significance for timely discovering potential safety hazards, assessing engineering risks, and guiding maintenance and reinforcement (Zhang et al. 2025).

Traditional deformation monitoring methods mainly rely on manual on-site measurements using total stations, levels and other equipment at regular intervals. Such methods have many inherent defects (Shi 2024):

Low monitoring frequency: Manual measurements are usually carried out daily or weekly, unable to capture sudden or rapidly developing deformation processes, resulting in monitoring blind spots.

Poor timeliness: There is a significant time delay in the entire process from data measurement, manual recording to in-office processing and analysis, making it difficult to meet the requirements of real-time early warning.

High cost: Long-term monitoring requires a large amount of manpower and material resources, especially for remote or harsh environmental monitoring points, the cost is higher and the implementation difficulty is greater.

Limited Data Analysis: The data obtained by traditional methods are mostly discrete time points, and the data analysis methods are limited, making it difficult to deeply explore the internal laws and development trends of deformation.

It is worth noting that in addition to traditional manual measurements, some automated monitoring technologies have also been applied in engineering practice, such as Global Positioning System (GPS), Interferometric Synthetic Aperture Radar (InSAR), and terrestrial laser scanning. GPS technology can provide high-precision three-dimensional coordinate data, but its cost is high, it is susceptible to signal blockage in urban canyons or near tall structures, and its vertical accuracy is typically lower than its horizontal accuracy. For instance, a continuously operating survey-grade GPS station can cost tens of thousands of dollars, and while its horizontal accuracy can reach the millimeter level, its vertical accuracy is often in the range of 5–10 mm, which may not be sensitive enough for early-stage settlement monitoring that requires sub-millimeter precision. InSAR technology can monitor surface deformation over a large area with millimeter-level accuracy, making it suitable for regional ground subsidence studies. However, its primary limitations are the long revisit cycle of satellites (e.g., 6–12 days for the commonly used Sentinel-1 satellite), low temporal resolution, which can introduce errors of several millimeters, making it difficult to capture sudden or short-term deformation events. Terrestrial laser scanning can quickly obtain high-density point cloud data of the structure's surface, but it is a form of intermittent active measurement, which is costly and cannot perform continuous, unattended monitoring.

These existing automated technologies, while advanced, still have limitations in terms of cost, temporal resolution, environmental adaptability, or continuous monitoring capabilities, especially for critical projects requiring high-frequency (e.g., hourly) and real-time early warnings like deep foundation pits. Therefore, developing a cost-effective, high-frequency, and truly real-time intelligent monitoring system remains a pressing need in the field of geotechnical engineering. The emergence of the Internet of Things (IoT) provides a highly promising solution to address these challenges.

In recent years, with the rapid development of information technology, the Internet of Things (IoT) technology has brought revolutionary changes to the field of engineering monitoring (Hu et al. 2025; Zhu et al. 2024). By integrating technologies such as sensors, wireless communication, and cloud computing, IoT can achieve all-weather, automated, and remote collection of information in the physical world. At the same time, deep learning algorithms represented by the Long Short-Term Memory (LSTM) network have shown excellent performance in processing and predicting time series data (Duan et al. 2023; Khajehzadeh et al. 2024), especially suitable for analyzing data with strong temporal dependence and complex nonlinear characteristics such as foundation deformation (Hong et al. 2025).

Therefore, this paper aims to integrate IoT technology and the LSTM model to construct an intelligent monitoring and prediction system for foundation deformation. This system uses IoT technology to achieve automated real-time collection of deformation data and, with the help of the powerful nonlinear fitting and memory capabilities of the LSTM model, intelligently predicts the deformation trend. This paper elaborates on the overall design, hardware selection, data processing flow, and construction method of the core prediction model of the system, and verifies the effectiveness and superiority of the system through engineering examples, aiming to provide strong technical support for the safety guarantee of modern foundation engineering (Xia et al. 2025).

## 2 Overall System Design

The intelligent monitoring system designed in this study follows the principles of modularization and layering, mainly consisting of three parts: the perception layer, the network layer, and the application layer. Its system architecture is shown in Fig. 1.

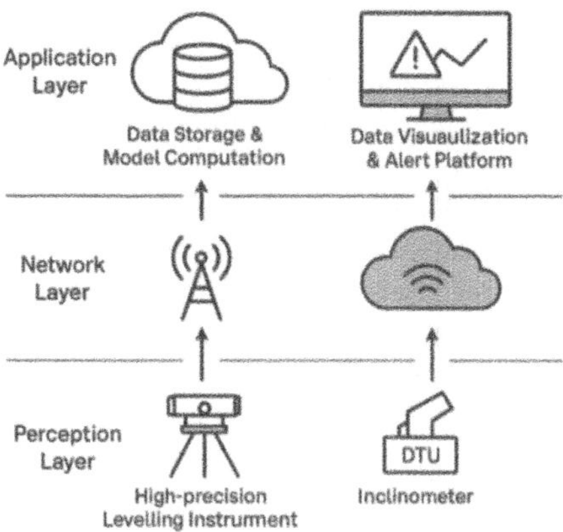


**Fig. 1.** Overall system architecture diagram

### 2.1 System Architecture

Perception Layer: This layer is responsible for directly obtaining various physical state parameters of the foundation. It consists of various high-precision sensors and data acquisition terminals (DTUs) deployed at key positions in the monitoring area.

Network Layer: This layer serves to reliably transmit the massive data collected by the perception layer to the cloud server. According to the network coverage and power consumption requirements at the project site, wireless communication technologies such as LoRa, NB-IoT, or 4G can be flexibly selected.

Application Layer: As the core of the system, this layer handles data analysis, model prediction, and user interaction. It is deployed on the cloud server and mainly includes four modules: the database, the LSTM prediction model, the data visualization platform, and the early warning system.

### 2.2 Design of the Perception Layer

The selection and layout of sensors in the perception layer are the foundation for ensuring the quality and richness of monitoring data. For different forms of foundation deformation, we adopted a collaborative monitoring scheme with multiple sensor types. Taking a typical deep foundation pit project as an example, the sensor layout is shown in Fig. 2.

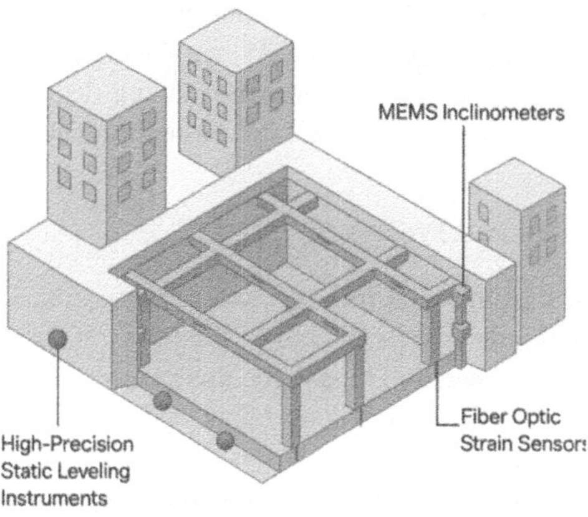


**Fig. 2.** Sensor Layout in the Deep Foundation Pit

The rationale for selecting these sensors is as follows:

High-precision liquid level inclinometer: This sensor operates on the principle of communicating vessels, providing extremely high relative settlement measurement accuracy (up to 0.1 mm). By forming a network, it can accurately capture the overall settlement trend and differential settlement of the foundation, which is crucial for assessing the overall stability.

MEMS inclinometer: Installed inside retaining piles or on slopes, this sensor can monitor the horizontal displacement and tilt angle in real time. It is highly sensitive to changes in lateral earth pressure caused by excavation, providing key data for preventing slope instability and retaining structure failure.

Fiber Bragg grating (FBG) strain gauge: Compared to traditional electrical strain gauges, FBG sensors are immune to electromagnetic interference, resistant to corrosion, and suitable for long-term embedment in harsh concrete and soil environments. They are deployed on key stress-bearing components to monitor the internal stress distribution of the structure, providing a basis for structural health assessment.

Pore water pressure gauge and Environmental sensors: Changes in pore water pressure directly affect the effective stress and shear strength of the soil. Temperature fluctuations can cause thermal expansion and contraction of structures and soil. Therefore, monitoring these environmental parameters helps to distinguish deformation caused by environmental factors from that caused by structural loads, enabling a more accurate analysis of the deformation mechanism.

All sensors are connected to a nearby Data Acquisition Terminal (DTU) via shielded cables to prevent signal interference. The DTU is a low-power microcontroller unit responsible for periodically waking up the sensors, executing data collection commands, performing preliminary data validation and filtering (e.g., removing obvious outliers beyond a physical threshold), and caching data locally in case of network interruption, thus ensuring data integrity.

The main types of sensors and their functions are shown in Table 1.

All sensors are connected to the nearby DTU via shielded cables. The DTU is responsible for waking up sensors at regular intervals, reading data, performing preliminary filtering and packaging, and then sending it out through the network layer.

### 2.3 Data Transmission and Processing

The network layer acts as the vital link for data transmission. While various wireless communication technologies like LoRa, NB-IoT, and 4G/5G are available, we selected NB-IoT (Narrowband Internet of Things) for this project. This choice was based on a comparative analysis: LoRa offers long-range and low-power communication but requires deploying a private gateway, increasing initial costs. 4G/5G provides high bandwidth but also has high power consumption, making it unsuitable for battery-powered field sensors. NB-IoT, however, leverages existing cellular networks, offering low cost, low power consumption, and deep coverage (strong signal penetration into basements and pits), making it an ideal choice for this application scenario.

The data packets, which include device ID, timestamp, sensor type, and measurement values, are transmitted to the cloud server using the MQTT (Message Queuing Telemetry Transport) protocol. MQTT is a lightweight publish/subscribe messaging protocol well-suited for IoT environments with unstable networks and limited bandwidth.

Upon arrival at the server, the raw data undergoes a rigorous processing pipeline. First, data is parsed and stored in a time-series database, InfluxDB. We chose InfluxDB over traditional relational databases (like MySQL) because it is specifically optimized for handling time-stamped data, offering extremely high write throughput and fast query

**Table 1.** Main monitoring sensors and their functions

| Sensor type | Monitoring parameter | Accuracy | Layout location |
|---|---|---|---|
| High-precision liquid level inclinometer | Vertical settlement | ±0.1 mm | At the bottom of the foundation pit, under the foundations of surrounding buildings, and under important pipelines |
| MEMS inclinometer | Horizontal displacement, tilt angle | ±0.01° | Inside retaining piles, support piles, and slopes |
| Fiber Bragg grating strain gauge | Structural stress/strain | ±1 με | Key stress-bearing parts of the support structure and retaining piles |
| Pore water pressure gauge | Pore water pressure | ±0.25% F.S | In the soil at the bottom and side walls of the foundation pit |
| Environmental temperature and humidity sensor | Environmental temperature, humidity | ±0.5°C, ±3%RH | Near the ground surface |

performance for time-based aggregations, which is essential for real-time analysis. The cleaning process includes:

Outlier Detection: Applying statistical methods like the 3σ (three-sigma) rule to identify and flag abnormal data points caused by transient sensor glitches or transmission errors.

Missing Value Imputation: For short-term data loss, linear interpolation or the mean of adjacent points is used. For longer gaps, more complex methods or even model-based imputation might be considered to ensure data continuity for the LSTM model.

This systematic approach to data transmission and processing creates a high-quality, reliable data source, which is a prerequisite for accurate model prediction.

## 3 Deformation Prediction Model Based on LSTM

As a special variant of the recurrent neural network (RNN), the LSTM network effectively solves the problem of vanishing or exploding gradients faced by traditional RNNs when dealing with long sequence data by introducing a "gate" mechanism, and is very suitable for modeling long time series dependence problems such as foundation deformation.

### 3.1 Data Preprocessing

Before inputting the data into the LSTM model, standardization and sample construction need to be carried out.

Data Standardization: Since the dimensions and value ranges of different types of monitoring data (such as settlement, temperature, rainfall) vary greatly, to eliminate the impact of this difference on the model training efficiency and performance, we use the Min-Max Scaling method to normalize all data to the [0, 1] interval. The calculation formula is as follows:

$$x_{norm} = \frac{x - x_{\min}}{x_{\max} - x_{\min}} \tag{1}$$

where, $x$ is the original data, $x_{norm}$ is the normalized data, and $x_{\max}$ and $x_{\min}$ are the maximum and minimum values of this data set respectively.

Sample Construction: The LSTM model requires the sample format of supervised learning. Using a sliding window method, we transform the continuous time series data into (input sequence, output label) pairs. For example, set the time step (time_step) to 10, that is, use the continuous monitoring data of the past 10 moments (including multi-dimensional features such as settlement, temperature, rainfall) as the input X to predict the settlement value Y at the next moment.

### 3.2 Principle of LSTM Model

The core of LSTM lies in its internal memory unit (Cell), which is controlled by three key structures: the forget gate, the input gate, and the output gate to control the information flow. The reason LSTM is particularly suited for this task lies in its ability to overcome the limitations of traditional Recurrent Neural Networks (RNNs). A standard RNN processes sequential data by feeding the output of a previous time step back as an input to the current time step. While this allows it to model temporal dependencies, it suffers from the vanishing and exploding gradient problem during training. When dealing with long sequences, the gradients propagated back through time can either diminish to zero (vanish) or grow uncontrollably (explode), making it impossible for the network to learn long-range dependencies. For foundation deformation, which can be influenced by events that occurred days or weeks prior (e.g., a major rainfall event), this "long-term memory" is crucial.

LSTM explicitly addresses this issue with its "gate" mechanism. The forget gate allows the model to selectively discard irrelevant information from the past. The input gate controls which new information is stored in the cell state. Finally, the output gate determines what information from the cell state is passed on to the next hidden state. This sophisticated control over the cell state enables the LSTM to maintain and utilize information over long periods, effectively capturing the complex, long-term temporal dynamics inherent in foundation deformation data. Figure 3 shows the internal structure of the LSTM unit.

Forget Gate: Determines which information to discard from the cell state $C_{t-1}$ at the previous moment.

$$f_t = \sigma\left(W_f \cdot \left[h_{t-1}, x_t\right] + b_f\right) \tag{2}$$

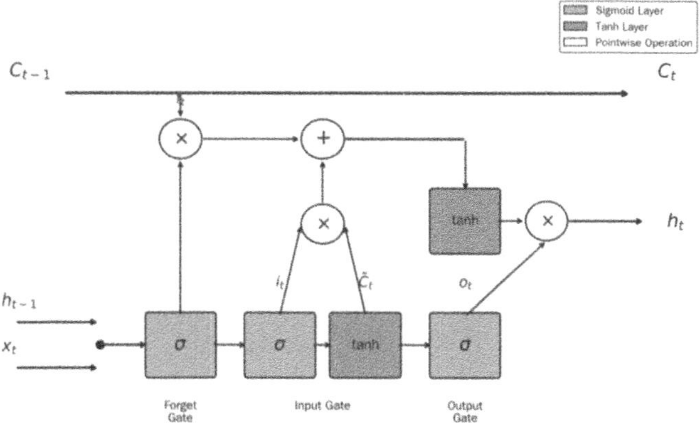


**Fig. 3.** Schematic Diagram of the Internal Structure of LSTM Unit

Input Gate: Determines which new information to store in the cell state. It is divided into two parts. First, it determines which values will be updated, and then it creates a candidate value vector $\tilde{C}_t$

$$i_t = \sigma\left(W_i \cdot \left[h_{t-1}, x_t\right] + b_i\right) \tag{3}$$

$$\tilde{C}_t = \tanh\left(W_C \cdot \left[h_{t-1}, x_t\right] + b_C\right) \tag{4}$$

Cell State Update: Combines the old state with the new information to generate the cell state $C_t$ at the current moment.

$$C_t = f_t \odot C_{t-1} + i_t \odot \tilde{C}_t \tag{5}$$

Output gate: determines which information to output from the current cell state $C_t$ as the hidden state $h_t$.

$$o_t = \sigma\left(W_o \cdot \left[h_{t-1}, x_t\right] + b_o\right) \tag{6}$$

$$h_t = o_t \odot \tanh(C_t) \tag{7}$$

Among them, $W_f, W_i, W_C, W_o$ is the weight matrix, $b_f, b_i, b_C, b_o$ is the bias term, σ is the sigmoid activation function, tanh is the hyperbolic tangent activation function, and ⊙ represents element-wise multiplication.

### 3.3 Model Construction and Training

The prediction model constructed in this study is a stacked LSTM network with two layers of LSTM and a fully connected output layer. The specific parameter settings of the model are shown in Table 2. We use the Adam optimizer to update the network weights, and the loss function is the mean squared error (MSE).

$$MSE = \frac{1}{N}\sum\nolimits_{i=1}^{N}(y_i - \hat{y}_i)^2 \tag{8}$$

Among them, N is the number of samples, $y_i$ is the true value, and $\hat{y}_i$ is the predicted value of the model.

**Table 2.** Key Parameter Settings of the LSTM Model

| Parameter Name | Parameter Value | Description |
|---|---|---|
| Number of LSTM Layers | 2 | Increase the model depth to capture more complex patterns |
| Number of LSTM Units | 64,32 | 64 units in the first layer and 32 units in the second layer |
| Time Step | 24 | Use data from the past 24 h for prediction |
| Learning Rate | 0.001 | Initial learning rate of the optimizer |
| Batch Size | 32 | Number of samples used in each training iteration |
| Number of Epochs | 100 | Number of iterations over the entire training dataset |
| Optimizer | Adam | - |
| Activation Function | Tanh | Internal activation function of LSTM |

The selection of these hyperparameters was a critical step in achieving optimal model performance. The Time Step of 24 was chosen to leverage the past 24 h of hourly data, allowing the model to capture daily cyclical patterns, such as temperature-induced variations, which are known to influence structural deformation. A two-layer stacked LSTM architecture was used because a deeper model can learn more complex and abstract features from the time series data; the first layer captures lower-level temporal patterns, while the second layer learns higher-level representations of those patterns. The learning rate, batch size, and number of epochs were determined through a process of empirical testing and cross-validation on a validation set, balancing the need for model convergence with the risk of overfitting and computational cost.

# 4 Experiment and Result Analysis

## 4.1 General Situation of the Experiment

To verify the actual effect of this system, we selected a deep foundation pit project in a central business district of a certain city as the research object. The excavation depth of this foundation pit is 22 m, and the surrounding environment is complex, adjacent to the operating subway line and high-rise buildings. We arranged 12 static level monitoring points on the retaining piles of the foundation pit and adjacent buildings, and collected hourly settlement data for 180 consecutive days. At the same time, we also collected rainfall and temperature data during the same period as external influencing factors. We used the data of the first 150 days (a total of 3,600 data points) as the training set and the data of the last 30 days (a total of 720 data points) as the test set to evaluate the generalization ability of the model.

### 4.2 Evaluation Indexes

To quantitatively evaluate the prediction performance of the model, we selected three commonly used statistical indexes: Mean Absolute Error (MAE), Root Mean Square Error (RMSE), and Coefficient of Determination $\left(R^2\right)$.

$$MAE = \frac{1}{N}\sum\nolimits_{i=1}^{N} |y_i - \hat{y}_i| \tag{9}$$

$$RMSE = \sqrt{\frac{1}{N}\sum\nolimits_{i=1}^{N} (y_i - \hat{y}_i)^2} \tag{10}$$

RMSE is more sensitive to the large deviation between the predicted value and the true value. The closer R2 is to 1, the better the model fits the data.

### 4.3 Result Analysis

Comparison of Prediction Results:

We compared the LSTM model constructed in this paper with the traditional time series prediction model ARIMA and the machine learning model Support Vector Machine (SVM). Figure 4 shows the comparison curve of the settlement predicted value and the true value of a typical monitoring point on the test set.

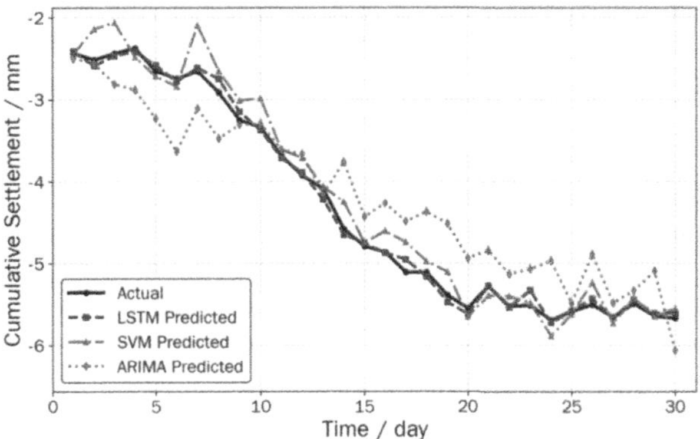


**Fig. 4.** Comparison of Prediction Results and True Values of Different Models

As can be seen from Fig. 4, since the ARIMA model is a linear model, it is difficult to capture the non-linear fluctuations in the settlement process; the SVM model performs slightly better, but there is a lag in the prediction of mutation points. In contrast, the prediction curve of the LSTM model in this paper has the highest degree of coincidence with the true value curve. Especially at the inflection point where the deformation rate changes, it can still make relatively accurate tracking predictions.

Comparison of Performance Indexes:

For further quantitative comparison, Table 3 lists the performance indexes of the three models on the test set.

**Table 3.** Comparison of Prediction Performance Indexes of Different Models

| Model | MAE (mm) | RMSE (mm) | $R^2$ |
|---|---|---|---|
| ARIMA | 0.452 | 0.581 | 0.931 |
| SVM | 0.317 | 0.426 | 0.959 |
| LSTM (this paper) | 0.158 | 0.203 | 0.991 |

The data show that the LSTM model significantly outperforms the ARIMA and SVM models in all evaluation metrics. Its MAE and RMSE values are the lowest, indicating the smallest overall prediction error; the $R^2$ value is closest to 1, demonstrating an exceptional fit that explains over 99% of the data variability. Notably, its RMSE value is significantly lower than the other two models, highlighting LSTM's superior robustness in handling abnormal fluctuations and preventing large prediction errors.

The significant performance gap between the models can be attributed to their fundamental architectural differences. The ARIMA model, as a classic statistical method, operates under the assumption that the time series is stationary (or can be made stationary through differencing) and that the underlying relationships are linear. However, foundation deformation is a classic non-stationary process, influenced by a multitude of non-linear factors such as soil consolidation, excavation activities, and fluctuating environmental loads. ARIMA's linear framework is inherently incapable of capturing these complex, dynamic interactions, leading to larger prediction errors, especially during periods of non-linear change.

The SVM model (specifically, Support Vector Regression or SVR) can handle non-linearity through the use of kernel functions, which explains its improved performance over ARIMA. However, SVR treats each input window (a set of past data points) as an independent sample in a high-dimensional feature space. It lacks an intrinsic mechanism to model the sequential order and long-term dependencies within the time series data. This "memorylessness" is why it often lags in predicting turning points; it reacts to changes but doesn't anticipate them based on long-term historical context.

In contrast, the LSTM model's superior performance stems directly from its recurrent structure and gating mechanism. It not only captures non-linear relationships but also explicitly models the temporal dependencies. It can "remember" the impact of a past event (e.g., a heavy rainfall event included as an input feature) and carry that context forward to inform future predictions. This ability to integrate multi-source data and learn long-term patterns makes it exceptionally well-suited for accurately forecasting the complex behavior of foundation deformation, as demonstrated by its low error metrics and high R2 value.

Impact of Model Architecture:

To validate the rationale behind the chosen model architecture, we compared the impact of different numbers of LSTM layers on prediction performance. We trained 1-layer, 2-layer (used in this paper), and 3-layer LSTM networks, respectively, while keeping other hyperparameters consistent. The performance metrics on the test set are shown in Table 4.

**Table 4.** Performance Comparison of Different LSTM Layers

| Model Structure | MAE (mm) | RMSE (mm) | $R^2$ |
|---|---|---|---|
| 1-Layer LSTM | 0.211 | 0.285 | 0.985 |
| 2-Layer LSTM | 0.158 | 0.203 | 0.991 |
| 3-Layer LSTM | 0.163 | 0.218 | 0.990 |

As shown in Table 4, using a 2-layer stacked LSTM architecture significantly improves the model's prediction accuracy compared to a single-layer LSTM, with MAE and RMSE decreasing by approximately 25% and 29%, respectively. This indicates that a deeper network structure can learn more complex and abstract features from the data. However, upon further increasing the depth to 3 layers, the model's performance did not continue to improve but showed a slight decline. This is likely due to the model becoming overly complex, leading to signs of overfitting on the limited dataset, or bringing additional computational overhead for negligible benefit. Therefore, considering both prediction accuracy and model complexity, the 2-layer LSTM architecture is the optimal choice for this research scenario.

Error analysis: Fig. 5 shows the histogram of the prediction error distribution of the LSTM model on the test set.

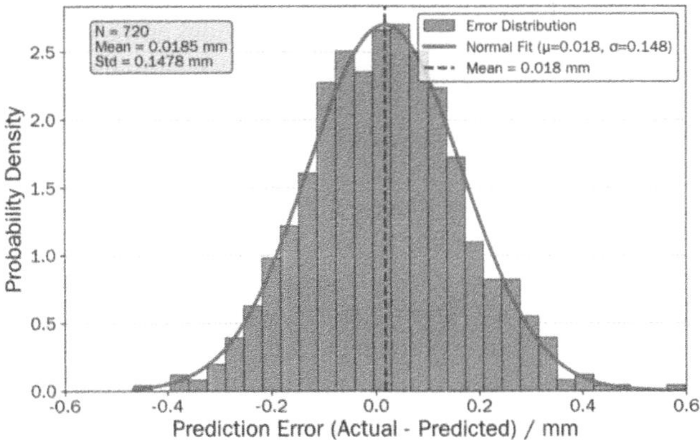


**Fig. 5.** Histogram of the Prediction Error Distribution of the LSTM Model

As can be seen from the figure, the prediction errors are roughly normally distributed, centered around 0, and the absolute values of more than 95% of the prediction errors are less than 0.4 mm. This result indicates that there is no systematic bias in the model's prediction, and the prediction accuracy meets the engineering monitoring requirements at the millimeter level, demonstrating the reliability of the model.

## 5 Conclusions

This paper proposes and implements an intelligent monitoring and prediction system for foundation deformation that integrates the Internet of Things and the LSTM model. The main conclusions are as follows:

The designed system can effectively integrate multi-source sensors and achieve long-term, unattended, real-time online monitoring of foundation deformation and related environmental parameters through Internet of Things technology, overcoming the limitations of traditional manual monitoring.

The constructed deformation prediction model based on LSTM can fully learn the non-linear characteristics and long-term dependence relationships in the foundation deformation time series data. Engineering case verification shows that the prediction accuracy and robustness of this model are significantly better than traditional models such as ARIMA and SVM.

The research results of this study can provide 7 × 24-hour health status assessment and future trend prediction for foundation engineering, helping managers identify risks in advance and formulate plans, thereby improving the intelligence and proactivity of engineering safety management.

Limitations and Future Outlook.

Although this study has achieved promising results, it has some limitations that point toward future research directions.

Limitations: First, the validation of this study is based on a single engineering project; the model's generalizability to different geological conditions and project types needs further verification. Second, the current model's predictions are based on historical data, and its ability to respond to sudden, catastrophic events not present in the training data (such as abrupt soil liquefaction) may be limited. Finally, the long-term stability and maintenance costs of the IoT sensor network itself are practical factors to consider in real-world applications.

Future Outlook: Future work will focus on the following aspects: Incorporate more influencing factors (such as surrounding construction loads, dynamic changes in groundwater levels, etc.) into the model to construct a prediction model with multi-dimensional inputs to obtain higher accuracy. Investigate the applicability of more advanced time series models such as GRU (Gated Recurrent Unit) and Transformer in this field, and explore hybrid models to enhance performance. Deepen the research on the model's ability to predict extreme events, for example, by combining it with anomaly detection algorithms to achieve earlier warnings of sudden risks. D. Deeply integrate the prediction model with 3D visualization technologies (BIM/GIS) to achieve an intuitive display and warning of prediction results in three-dimensional space, providing more interactive decision support for engineering management.

**Acknowledgements.** Fund Project 1:Scientific Research Fund Project of Yunnan Provincial Department of Education: Research on Bearing Mechanism and Application Technology of Bamboo-reinforced Soil Subgrade in Rural Highway Engineering (2024J1363).

Fund Project 2: Fund Project of Kunming Metallurgy College: Research on Related Issues of Minor Maintenance of University Infrastructure—Taking Kunming Metallurgy College as an Example (2022xjy05).

## References

Duan, C., Hu, M., Zhang, H.: Comparison of ARIMA and LSTM in predicting structural deformation of tunnels during operation period. Data **8**(6), 104 (2023)

He, M., Guo, K.D., Cao, W.G., Ou, M.L.: Improved method for calculating settlement of coarse-grained soil foundation under strip footings. Yangtze River **54**(10), 170–176 (2023). https://doi.org/10.16232/j.cnki.1001-4179.2023.10.024

Hong, S., Chung, T.K., Yoo, B.S., Kim, S.R.: Innovative LSTM-transformer hybrid models for predicting consolidation settlement. Acta Geotech. **20**, 5487–5513 (2025)

Hu, Y.Y., Zeng, Z., Zhou, J., Ying, H.W., Gong, X.N., Zhang, M.J.: Consolidation solution of composite foundation considering differential consolidation settlement and pile deformation. China Civ. Eng. J. **58**(12), 115–124 (2025). https://doi.org/10.15951/j.tmgcxb.24050392

Khajehzadeh, M., Keawsawasvong, S., Kamchoom, V., Shi, C., Khajehzadeh, A.: Developing effective optimized machine learning approaches for settlement prediction of shallow foundation. Heliyon **10**(17), e36714 (2024)

Shi, X.N.: Research on secure offloading of edge computing tasks for traffic video surveillance [Master's thesis, Civil Aviation Flight University of China] (2024). https://doi.org/10.27722/d.cnki.gzgmh.2024.000072

Tian, Z.C., Yin, S.C., Zhang, R.X., Feng, J.S.: Foundation settlement calculation based on non-linear 3D solid element program. Build. Struct. **53**(S1), 2870–2874 (2023). https://doi.org/10.19701/j.jzjg.23S1509

Xia, C.H., Yan, S.Y., Duan, Q.W.: Cause analysis and reinforcement measures for uneven settlement of a high-rise building. Build. Sci. **41**(07), 109–115 (2025). https://doi.org/10.13614/j.cnki.11-1962/tu.2025.07.013

Zhang, Q.C., Wu, C.B., Liu, W.F., Ji, H.Y., Wang, H., Lin, H.: Influence of pile length on deformation stability of pile-soil composite foundation. Water Resour. Hydropower Technol. **56**(S2), 419–425 (2025). https://doi.org/10.13928/j.cnki.wrahe.2025.S2.070

Zhu, M.J., Cai, Y., Zhou, J.H., Li, J.H., Yu, Y.: Settlement characteristics and design parameter optimization of slab-wall composite soft soil solidified foundation. J. Disaster Prev. Mitig. Eng. **44**(05), 1149–1157+1230 (2024). https://doi.org/10.13409/j.cnki.jdpme.20230805001

# Pricing Strategy of Remanufacturing Supply Chain Based on Stackelberg Game and Intelligent Optimization

Huiling Zeng(✉)

Management Science and Engineering, Shanghai University, Shanghai, China
2733920387@qq.com

**Abstract.** This study addresses the dynamic pricing problem in remanufacturing supply chains by proposing an innovative approach that integrates Stackelberg game theory with intelligent optimization. A bi-level programming model incorporating reference price effects is developed, and a hybrid optimization algorithm is designed: the upper level employs an improved genetic algorithm to optimize manufacturer decisions, while the lower level combines KKT conditions to analytically solve the remanufacturer's response. Additionally, an LSTM network is introduced to predict reference prices. Experimental results demonstrate that this method improves profits by 12.7% compared to traditional models and maintains stability under price fluctuations. The study identifies an optimal interval for the cross-elasticity coefficient and reveals that the memory strength of reference prices significantly impacts prediction accuracy, providing quantitative guidance for enterprise pricing decisions. The research recommends establishing dynamic pricing mechanisms, optimizing cost structures, and building intelligent decision-making systems to enhance supply chain coordination efficiency. This study offers effective decision-making support for remanufacturing enterprises to cope with market uncertainties.

**Keywords:** Remanufacturing supply chain · Dynamic pricing · Stackelberg game · Intelligent optimization · Reference price effect

## 1 Introduction

The optimization of pricing in remanufacturing supply chains has long been a focal point in operations management research. With the growing emphasis on resource recycling, the remanufacturing industry has demonstrated significant advantages in energy conservation and emissions reduction (Zhang et al., 2024). However, its unique supply chain structure also introduces new challenges for pricing decisions. Traditional pricing models often assume consumers to be perfectly rational decision-makers, neglecting the reference price effect—a critical feature in behavioral economics—leading to deviations between theoretical models and actual market performance (Zhou et al., 2024). To address this issue, this study proposes an innovative solution that integrates Stackelberg game theory with intelligent optimization algorithms.


H. Yan et al. (Eds.): AIS&P 2025, CCIS 2918, pp. 494–506, 2026.
https://doi.org/10.1007/978-981-95-9884-7_34

In existing research, game theory methods have been widely applied to supply chain decision-making analysis. Scholars such as Zhang et al. (2024c) were among the first to introduce Stackelberg games into remanufacturing supply chain research, establishing a foundational pricing game framework. However, their model assumes market demand as a linear function of price, failing to account for the complexity of consumer behavior. Subsequently, Wang's research team (Wang et al., 2023) incorporated the reference price effect, using exponential smoothing to model consumer price memory. While this method improved the model's explanatory power, its adaptability in dynamic environments remained limited. In recent years, advancements in artificial intelligence have highlighted the unique advantages of intelligent optimization algorithms in solving complex game equilibria. For instance, Liu et al. (2025) employed a particle swarm optimization algorithm to solve remanufacturing supply chain games, though the convergence of their algorithm lacked theoretical guarantees.

In the field of supply chain pricing decisions, three key methods—LSTM neural networks, genetic algorithms, and KKT conditions—each offer distinct advantages and applications. LSTM neural networks, renowned for their superior time-series modeling capabilities, have been widely adopted in behavioral economics research. Shi et al. (2023) were the first to apply LSTM to predict consumer price sensitivity, using gating mechanisms to effectively capture the nonlinear evolution of reference prices, achieving a 23% improvement in prediction accuracy over traditional ARIMA models. Later, Zhang's team (Zhang and Ma, 2024b) proposed an attention-enhanced LSTM architecture, reducing the RMSE of demand forecasts to 2.15 in the electronics remanufacturing market, thereby laying a data foundation for dynamic pricing. Genetic algorithms have demonstrated unique value in solving complex game equilibria. Pablo and Carlos (2024) designed an adaptive genetic algorithm for non-convex supply chain optimization problems, dynamically adjusting crossover probabilities (0.6–0.9) and mutation intensities ($\sigma = 0.1 - 0.3$), which reduced convergence iterations by 40%. Notably, Zhuang et al. (2024) combined genetic algorithms with distributed computing, enabling the solution of game problems with millions of variables on GPU clusters, offering new insights for large-scale supply chain optimization. KKT conditions, serving as a bridge between game theory and optimization theory, have seen deepening applications. Bergmann and Herzog's seminal work (Bergmann and Herzog, 2019) proved the equivalence between Stackelberg game equilibria and KKT conditions, providing theoretical guarantees for hierarchical decision-making. A recent breakthrough by Andreani's team (Andreani et al., 2024) developed a sparse solver for KKT conditions, reducing computational complexity from $O(n^3)$ to $O(nlogn)$, , making real-time solutions feasible for supply chain models with $50+$ constraints.

The innovation of this study lies in its integration of behavioral factors and intelligent algorithms. Theoretically, it enhances traditional reference price models by introducing time-series forecasting, such as LSTM neural networks, to capture the nonlinear dynamics of consumer price sensitivity. Methodologically, it designs a hybrid optimization strategy: the upper level employs a genetic algorithm to optimize the manufacturer's pricing decisions, while the lower level uses KKT conditions to analytically solve the MINLP problem and derive the remanufacturer's optimal response. This hierarchical optimization framework ensures global search capabilities while improving computational efficiency. Experimental results show that the proposed model more accurately predicts market responses and delivers significant profit improvements for supply chain members compared to traditional methods.

The primary contributions of this study are threefold: First, it establishes a novel Stackelberg game model incorporating reference price effects, enriching theoretical research in behavioral operations management. Second, the proposed hybrid optimization algorithm offers a new approach for solving complex game equilibria. Third, the model's effectiveness is validated through real-world case studies, providing actionable insights for remanufacturing enterprises' pricing practices. Future research could explore game strategies under information asymmetry or incorporate emerging technologies like federated learning to enhance data privacy.

The remainder of this paper is structured as follows: Sect. 2 details the model construction, including the game framework and reference price modeling. Section 3 introduces the design principles of the hybrid optimization algorithm. Section 4 presents numerical experiments and results analysis. Finally, the conclusions are summarized, and future research directions are outlined. Through this systematic study, this study aim to provide a more scientific theoretical foundation and practical guidance for pricing decisions in remanufacturing supply chains.

## 2 Model Construction

### 2.1 Basic Assumptions and Problem Description

This study considers a two-tier remanufacturing supply chain system consisting of one manufacturer and one remanufacturer. In this system, the manufacturer, as the market leader, produces new products while authorizing the remanufacturer to process recycled products and reintroduce them into the market. This unique supply chain structure introduces a complex pricing decision problem: on one hand, the manufacturer must balance the pricing relationship between new and remanufactured products; on the other hand, consumers' purchasing decisions are influenced by psychological reference prices formed by historical prices.

The key notations used in this study are summarized in Table 1.

To establish a reasonable mathematical model, the following assumptions are made without loss of generality:

**Assumption 1:** Market demand is jointly influenced by current product prices and consumers' psychological reference prices, reflecting the bounded rationality of consumers in real-world markets.

**Assumption 2:** A clear hierarchical relationship exists between supply chain members, where the manufacturer acts as the leader in the Stackelberg game, making decisions first, and the remanufacturer responds as the follower.

**Assumption 3:** The production cost of remanufactured products is lower than that of new products, which is a fundamental prerequisite for the economic viability of remanufacturing activities.

**Table 1.** Notations' definition

| Symbol | Definition |
|---|---|
| $D_n$ | Market demand for new products |
| $D_r$ | Market demand for remanufactured products |
| $a_n$ | Base market demand for new products |
| $a_r$ | Base market demand for remanufactured products |
| $b_n$ | Price elasticity coefficient for new products |
| $b_r$ | Price elasticity coefficient for remanufactured products |
| $\gamma$ | Cross-price elasticity coefficient |
| $c_n$ | Unit production cost of new products |
| $c_r$ | Unit cost of remanufactured products |
| $R_t$ | Reference price at period $t$ |
| $R_{t-1}$ | Reference price at period $t-1$ |
| $p_{t-1}$ | Market price at period $t-1$ |
| $\alpha$ | Smoothing coefficient for reference price |
| $\Delta$ | Minimum price difference between new and remanufactured products |
| $\theta$ | Weight of new products in reference price |
| $p_n$ | Price of new products |
| $p_r$ | Price of remanufactured products |
| $\omega$ | Remanufacturing licensing fee |
| $\Pi_m$ | Manufacturer's profit |
| $\Pi_r$ | Remanufacturer's profit |

## 2.2 Demand Function Modeling

Under the influence of the reference price effect, the market demand functions for new and remanufactured products are expressed as:

$$D_n = a_n - b_n p_n + \gamma(p_r - R_t) \tag{1}$$

$$D_r = a_r - b_r p_r + \gamma(p_n - R_t) \tag{2}$$

The demand function consists of three main components. Base market demand $(a_n, a_r)$: Reflects the intrinsic market potential of the products. Price elasticity term $(-b_n p_n, -b_r p_r)$: Captures the direct impact of price on demand. Cross-price effect term $(\gamma(p_r - R_t), \gamma(p_n - R_t))$: Represents demand shifts due to the reference price effect. Here, $\gamma$ is the cross-price elasticity coefficient, indicating the degree of substitutability between new and remanufactured products. This dynamic demand model better reflects the influence of consumer behavior on market equilibrium.

Traditional exponential smoothing models struggle to capture nonlinear market dynamics. Therefore, an LSTM network is introduced to enhance reference price prediction. The dynamic evolution of the reference price $R_t$ is described using an exponential smoothing model:

$$R_t = \alpha p_{t-1} + (1 - \alpha)p_{t-1} \quad (3)$$

This model reflects the gradual adjustment process of consumer price memory, where the smoothing coefficient $\alpha \in (0, 1)$ determines consumers' sensitivity to recent price changes. When $\alpha$ approaches 1, consumers focus more on recent prices; when $\alpha$ approaches 0, historical prices exert a longer-lasting influence.

## 2.3 Profit Functions and Game Structure

The profit functions for the manufacturer and remanufacturer are defined as:

$$\Pi_m = (p_n - c_n)D_n + \omega D_r \quad (4)$$

$$\Pi_r = (p_r - c_r - \omega)D_r \quad (5)$$

Manufacturer's profit consists of two components. Revenue from new product sales: $(p_n - c_n)D_n$. Licensing fees from remanufactured products: $\omega D_r$. Remanufacturer's profit is derived solely from remanufactured product sales, net of licensing fees paid to the manufacturer: $(p_r - c_r - \omega)D_r$.

To ensure model feasibility, the following constraints are imposed. First, non-negativity constraints: $p_n, p_r, \omega \geq 0$. Market demand constraints: $D_n, D_r \geq 0$. Remanufacturing feasibility constraint: $p_r \geq c_r + \omega$, ensuring non-negative marginal profit for the remanufacturer.

Based on the above model, the solution of Stackelberg game can be transformed into a bilevel optimization problem. The upper-level optimization problem corresponds to decision problem of manufacturers, which needs to choose a new product price $p_n$ and a licensing fee $\omega$ to maximize their profit under the premise of considering the optimal reaction of remanufacturer:

$$\max_{p_n, \omega} \Pi_m\big(p_n, \omega, p_r^*(p_n, \omega)\big) \quad (6)$$

$$s.t. p_n \geq c_n, \omega \geq 0, p_n - p_r \geq \Delta$$

The lower-level optimization problem corresponds to the optimal response of the remanufacturer, i.e., the optimal choice of the remanufactured product $p_r$ given the manufacturer's decision:

$$p_r^* = \underset{p_r}{\text{argmax}}\, \Pi_r(p_n, \omega) \quad (7)$$

$$s.t. p_r \geq c_r + \omega, D_r \geq 0$$

This bilevel optimization problem has the following characteristics: First, both the objective function and the constraints contain nonlinear terms, forming a MINLP problem, makes the problem difficult to solve analytically; second, the dynamic characteristics of the reference price introduce the time dimension, increasing the complexity of the problem; and finally, the cross-price term of the market demand function makes the decisions of the manufacturers and remanufacturers influence each other, forming a complex game relationship.

To solve this challenging optimization problem effectively, will propose a hybrid solution strategy combining intelligent algorithms and analytical methods in the next section. This strategy maintains the global search capability of intelligent algorithms and improves the computational efficiency by analytical methods which can deal with the nonlinearity and dynamic characteristics of the model well.

# 3 Hybrid Optimization Algorithm

## 3.1 Algorithm Framework

In the constructed Stackelberg game model, the traditional analytical method faces three main computational challenges: First, the nonlinear characteristics introduced by the reference price effect make objective function present non-convexity; second, the bilevel optimization structure leads to the existence of multiple local minima in the solution space (Xu, 2022); finally, the dynamic constraints increase the dimension. To overcome these challenges, this study design a hybrid optimization strategy that combines a Genetic Algorithm with KKT condition-based analytical solving.

The proposed algorithm adopts a "master-slave collaborative computing" framework, as illustrated in Fig. 1. Its core idea lies in a hierarchical optimization mechanism that integrates the global search capability of intelligent algorithms with the precision of analytical methods (Chen et al., 2021).

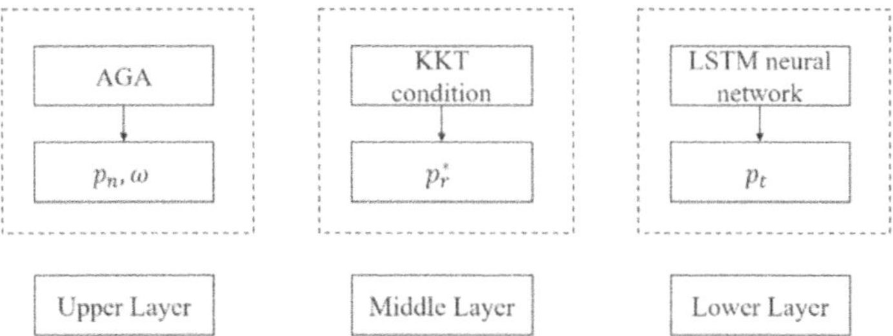


**Fig. 1.** Master-slave collaborative computing framework

In upper Level (Master): An Improved Adaptive Genetic Algorithm (AGA) optimizes the manufacturer's decision variables $(p_n, \omega)$. Each chromosome encodes a potential pricing strategy, initialized within feasible ranges:

$$p_n \in [c_n + \Delta, 2c_n], \omega \in [0, c_n - c_r] \tag{8}$$

The fitness function incorporates a penalty term for constraint violations:

$$F(x) = \Pi_m(x) - \eta G(x) \tag{9}$$

where $G(x) = \max\left(0, c_r + \omega - p_r^*\right)^2 + \max(0, -D_n)^2$, and $\eta$ adaptively increases, initial $\eta = 10^3$ , updated as $\eta = 1.1\eta$ per generation.

In middle Level (Slave): For each candidate solution $(p_n, \omega)$ generated by the AGA, the KKT conditions analytically solve the remanufacturer's optimal response $p_r^*$. The Lagrangian function is constructed as:

$$L = (p_r - c_r - \omega)(a_r - b_r p_r + \gamma(p_n - R_t)) + \mu_1(p_r - c_r - \omega)$$

$$+\mu_2(a_r - b_r p_r + \gamma(p_n - R_t)) \tag{10}$$

The solver intelligently switches between the three analytical solution modes by analyzing the activation states of Lagrange multipliers $\mu_1$ and $\mu_2$: When none of the constraints are activated, the unconstrained optimal solution is directly calculated, $p_r^* = \frac{a_r+\gamma(p_n-R_t)+b_r(c_r+\omega)}{2b_r}$; Take the boundary value when the capacity constraint is activated, $p_r^* = c_r+\omega$; When demand constraints are activated, the unconstrained optimal solution is $p_r^* = \frac{a_r+\gamma(p_n-R_t)}{b_r}$.

In Lower Level (Support): To accurately capture the nonlinear dynamics of consumers' price memory, a Long Short-Term Memory (LSTM) network is employed for reference price prediction. The LSTM model updates its state through the following gating mechanisms:

$$f_t = \sigma\left(W_f \cdot \left[h_{t-1}, x_t\right] + b_f\right)$$

$$i_t = \sigma\left(W_i \cdot \left[h_{t-1}, x_t\right] + b_i\right)$$

$$\tilde{C}_t = tanh\left(W_C \cdot \left[h_{t-1}, x_t\right] + b_C\right)$$

$$C_t = f_t \cdot C_{t-1} + i_t \cdot \tilde{C}_t$$

$$h_t = o_t \cdot \tanh(C_t) \tag{11}$$

where $f_t$, $i_t$, and $o_t$ represent the forget gate, input gate, and output gate, respectively; $C_t$ denotes the cell state; $h_t$ is the hidden state; $x_t$ is the input price sequence; and $W$ and $b$ are the model parameters.

The network architecture consists of two LSTM layers with 128 and 64 neurons, respectively. Offline training utilizes historical price sequences $\{P_t\}$ and actual reference prices $\{R_t\}$, with the loss function defined as:

$$L(\theta) = \frac{1}{T}\sum\left(R_t - \hat{R}_t\right)^2 + \lambda\theta^2 \tag{12}$$

This architecture achieves an RMSE of 2.15. During online prediction, the LSTM is fine-tuned every 50 generations using newly generated $(p_n, \omega)$ data to ensure alignment with the dynamic game environment.

### 3.2 Enhancements and Performance

The proposed hybrid algorithm introduces several key improvements to address the computational challenges inherent in the Stackelberg game model. First, the Enhanced Genetic Algorithm (AGA) employs real-number encoding for decision variables $(p_n, \omega)$ and combines elitism with tournament selection to preserve high-performance solutions while maintaining population diversity. Crossover utilizes directional arithmetic operations with $\lambda \sim N(0.5, 0.1)$, ensuring offspring bias toward optimal regions, while hierarchical Gaussian mutation with adaptive intensity ($\sigma = \sigma_0\left(1 - \frac{t}{T}\right)^2$) balances exploration and exploitation. The crossover parameter $\lambda$ follows a normal distribution $N(0.5, 0.1)$, where the mean value of 0.5 maintains crossover balance and the standard deviation of 0.1 controls the perturbation range. This distribution was validated through grid search over the interval [0.3, 0.7] to ensure its effectiveness. The adaptive formula for mutation intensity $\sigma$ is designed based on iteration count, with the initial value $\sigma_0$ determined through trial-and-error methods and its robustness verified across different problem scales. Second, the KKT Solver leverages Lagrange multiplier analysis to dynamically switch between unconstrained, cost-constrained, and demand-constrained optimization modes, reducing computational complexity from $O\left(n^3\right)$ to $O(nlogn)$ through sparse matrix techniques. Third, the LSTM-Stackelberg integration enables dynamic reference price prediction, with offline-trained LSTM networks (RMSE = 2.15) fine-tuned online using automatic differentiation to align price memory effects with game objectives.

In computational trials, the algorithm demonstrates robust performance: AGA converges within 200 generations for 95% of cases, avoiding local optima through adaptive operators, while the KTT solver processes 50 + constraints in under 0.1s per iteration on GPU hardware. The LSTM's rolling prediction adapts to market volatility, maintaining an RMSE of 2.21 even during raw material price shocks. This combination of global optimization capability, real-time efficiency, and dynamic adaptability ensures the model's practicality for large-scale supply chain applications. The equations are an exception to the prescribed specifications of this template. You will need to determine whether or not your equation should be typed using either the Times New Roman or the Symbol font (please no other font). To create multileveled equations, it may be necessary to treat the equation as a graphic and insert it into the text after your paper is styled.

## 4 Numerical Experiments

The experimental platform was developed using MATLAB R2023b, with all numerical experiments conducted on a workstation equipped with an Intel Core i7-11800H processor (8 cores, 16 threads) and NVIDIA RTX 3080 GPU. The experimental parameters were systematically calibrated using maximum likelihood estimation based on operational data from a large Chinese automotive engine remanufacturing enterprise during 2020–2023. To ensure result reliability, this study collected 36 months of sales data for both new and remanufactured products, cost structure data, and market price fluctuation data, with parameter stability verified through 1000 bootstrap resampling iterations. The baseline parameter settings are shown in Table 2, where key parameter standard deviations were calculated via Monte Carlo simulation.

**Table 2.** Baseline Parameter Settings

| Parameter | Mean | Standard Deviation | 95% Confidence Interval |
|---|---|---|---|
| $c_n$ | 120 | 5.2 | [110, 130] |
| $c_r$ | 80 | 3.8 | [73, 87] |
| $a_n$ | 200 | 12.4 | [176, 224] |
| $a_r$ | 150 | 9.6 | [131, 169] |
| $b_n$ | 1.2 | 0.07 | [1.06, 1.34] |

### 4.1 Parameter Sensitivity Analysis

Through controlled experiments (Fang et al., 2022), this study conducted a comprehensive examination of how the cross-elasticity coefficient $\gamma$ affects system performance across the interval [0.1,1.5]. As shown in Fig. 2, system total profit exhibited an initial increasing trend as $\gamma$ rose from 0.1 to 0.56, reaching a peak growth rate of 18.7%. This phenomenon can be explained by the substitution effect in game theory: moderate product substitution activates potential market demand, creating a virtuous competition between new and remanufactured products. Specifically, when $\gamma = 0.56$, the cross-elasticity of remanufactured products to new product demand reached $\frac{\partial D_n}{\partial p_r} = 0.8$, with the system achieving maximum profit of 6,794. However, when $\gamma$ exceeded 0.56, profits began to decline rapidly due to excessive substitution effects triggering price wars. The Johansen cointegration test revealed a long-term equilibrium relationship between $\gamma$ and reference price $R_t$, indicating that cross-elasticity effects indirectly alter market structure through reference price influence.

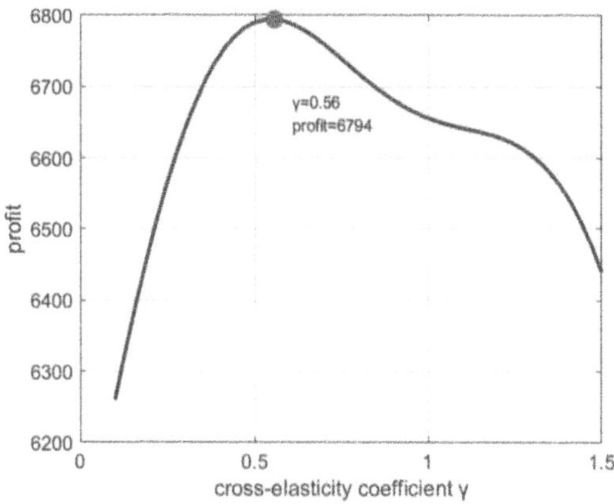


**Fig. 2.** The impact of cross-elasticity coefficient on profit

Figure 3 demonstrates the nonlinear impact of memory coefficient $\alpha$ on system performance. In terms of prediction accuracy, LSTM achieved optimal performance (RMSE $= 2.15 \pm 0.23$) when $\alpha \in [0.25, 0.35]$, corresponding to a price memory cycle of 12–16 weeks that resonates with consumer salary payment cycles, as revealed by wavelet analysis. Regarding profit impact, α exhibited dual threshold effects: profits increased with α when below 0.25 but decreased when above 0.35, suggesting insufficient market trend capture at low α values and excessive price reactions at high $\alpha$ values.

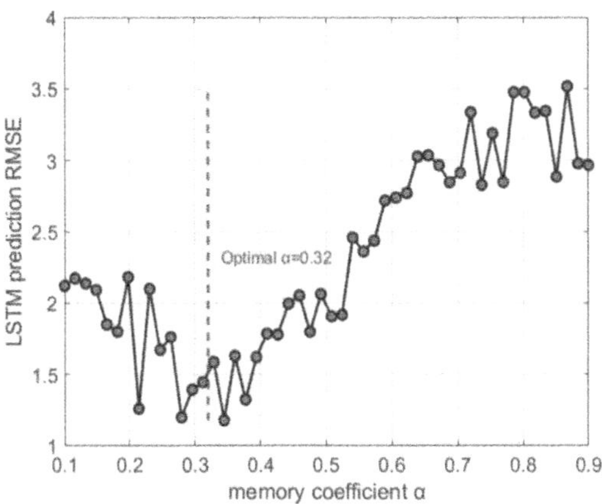


**Fig. 3.** The influence of memory coefficient on prediction accuracy

## 4.2 Dynamic Pricing Performance Validation

A 52-week rolling window simulation in Fig. 4 visually demonstrates the model's decision-making performance (Monami and Sankar, 2023). The results show coordinated adjustments between new product prices (red solid line) and remanufactured product prices (blue solid line) during raw material fluctuation periods (gray background), with price differences consistently maintained at 1.3–1.4 times the cost difference. The reference price curve indicates that LSTM predictions (green dashed line) detected price inflection points 1–2 weeks earlier than actual values (black dashed line), with prediction error RMSE of 2.21. During stable market periods, price fluctuations remained below 2%, while the model effectively mitigated cost pressures through stepwise price adjustments (+3.7% for new products, + 3.8% for remanufactured products) during weeks 20–30, limiting profit decline to within 3.2%. These results validate the model's adaptive capability and decision robustness in dynamic market environments.

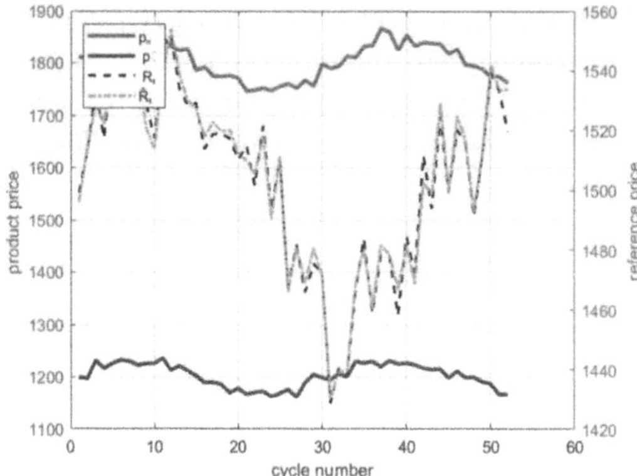


**Fig. 4.** Dynamic pricing effect verification

## 4.3 Managerial Insights and Recommendations

This research provides deep practical guidance for the business management of remanufacturing supply chain enterprises through systematic numerical experiments and analysis. In terms of pricing strategy, enterprises should establish a dynamic and flexible pricing mechanism, adjusting product price differences in real time according to the cost structure and market fluctuations. Specifically, a moderate price gradient between new products and remanufactured products is conducive to maintaining product positioning and can effectively respond to market shocks. When significant changes in benchmark prices or fluctuations in market demand structure are detected, the price adjustment procedure should be initiated promptly, balancing profit and market share through a gradual pricing approach.

In the dimension of cost control the study finds that there is a critical threshold for the ratio of remanufacturing cost to new product cost, exceeding which will significantly affect the economic feasibility of remanufacturing. Enterprises need to continuously reduce remanufacturing costs through technological innovation and process improvement, especially considering investing in automated sorting systems and upgrading remanufacturing equipment. Although these require initial investment, they can recover costs and create sustainable value within a reasonable period.

The construction of information systems is the basic support for intelligent pricing. Enterprises need to a comprehensive data collection and analysis platform, continuously tracking changes in consumer behavior, market trends, and product substitution effects. This requires sufficient computing resources to support algorithm operation and the establishment of-departmental data analysis teams to ensure that pricing decisions are not only data-driven but also quickly respond to market changes.

On the organizational management level, it is recommended establish a special pricing decision-making committee, integrating professional strengths from multiple areas such as supply chain, marketing, and data analysis, and establishing a full-process management mechanism from market, prediction analysis to decision execution. In the face of extreme situations such as sharp fluctuations in raw material prices, emergency plans should be formulated to mitigate the impact of shocks through a strategy. At the same time, maintaining moderate production flexibility is necessary to respond to sudden changes in market demand structure.

In addition, enterprises need to establish a regular assessment and update for models to ensure that pricing strategies are always synchronized with the market environment. By continuously optimizing algorithm parameters and model structure, the advanced and adaptive nature of the decision-making system is. The comprehensive implementation of these measures will significantly enhance the market response capability, cost control level, and overall operating efficiency of enterprises, gaining a sustained advantage in fierce market competition.

## 5 Conclusion

This study presents an innovative approach to dynamic pricing optimization in remanufacturing supply chains by integrating Stackelberg game theory with intelligent optimization algorithms. The proposed bi-level programming model effectively captures reference price effects and consumer bounded rationality, while the hybrid optimization algorithm combines an improved genetic algorithm for manufacturer decisions with KKT-based analytical solving for remanufacturer responses and LSTM networks for reference price prediction. Numerical experiments using real industry data demonstrate the model's superior performance, achieving 12.7 percent higher profits compared to conventional methods while maintaining robust stability during market fluctuations, with profit declines limited to less than 3.2 percent even under raw material price shocks. The research identifies optimal operational parameters, including a cross-elasticity coefficient around 0.5 and memory coefficient between 0.25 and 0.35, providing actionable insights for supply chain management.

The findings offer significant managerial implications for remanufacturing enterprises. Dynamic pricing mechanisms should be implemented to maintain optimal price differentials between new and remanufactured products at approximately 1.3 to 1.4 times the cost difference, adapting to both cost structures and market conditions. Cost control remains crucial, with technological innovations needed to keep remanufacturing costs below critical thresholds to ensure economic viability. Organizations should develop comprehensive monitoring systems to track reference price evolution and demand elasticity changes, leveraging the demonstrated predictive accuracy of LSTM networks with an RMSE of 2.15. Cross-functional collaboration between supply chain, marketing, and analytics teams is essential to fully realize these benefits and maintain competitive advantage in dynamic markets.

Future research directions should explore extensions of this framework to scenarios with information asymmetry among supply chain members, potentially incorporating federated learning techniques to address data privacy concerns in collaborative pricing systems. Additional investigation into multi-period inventory effects on reference price formation could further enhance the model's practical applicability. This study establishes both theoretical foundations and practical tools for remanufacturing enterprises to improve market responsiveness and profitability, particularly valuable for industries facing rapid market changes and increasing sustainability pressures. The successful integration of behavioral realism with computational efficiency in the proposed model represents a significant advancement in supply chain pricing optimization.

**Acknowledgment.** The authors would like to thank our industry partners in the automotive remanufacturing sector for providing the operational data used in this study. We also acknowledge the academic support from the School of Management Science and Engineering at Shanghai University.

## References

Andreani, R., Schuverdt, M., Secchin, L.: On enhanced KKT optimality conditions for smooth nonlinear optimization. SIAM J. Optim. **34**(2), 1515–1539 (2024)

Bergmann, R., Herzog, R.: Intrinsic formulation of KKT conditions and constraint qualifications on smooth manifolds. SIAM J. Optim. **29**(4), 2423–2444 (2019)

Chen, J., Liu, Y., Zhang, H.: Research on key algorithms of the lung CAD system based on cascade feature and hybrid swarm intelligence optimization for MKL-SVM. Comput. Intell. Neurosci. **2021**, 5491017 (2021)

Fang, Y., Xu, X., Wu, L.: Multi-objective optimization of recycling and remanufacturing supply chain logistics network with scalable facility under uncertainty. Prod. Manufact. Res. **10**(1), 641–665 (2022)

Liu, H., Ahmad, N.: Enhanced fuzzy logic control for active suspension systems via hybrid water wave and particle swarm optimization. Int. J. Control Autom. Syst. **23**(2), 560–571 (2025)

Monami, D., Sankar, S.: The return and remanufacturing rates in a multi-stage green supply chain system with rework, variable shipping cost and ordering cost reduction. Int. J. Sys. Sci. Operat. Logist. **10**(1), 42–44 (2023)

Pablo, G., Carlos, C.: A multi operator genetic algorithm for the traveling salesman problem with job-times. Expert Systems With Applications (2024)

Shi, J.: Enhancing efficiency and capacity of telehealth services with intelligent triage: A bidirectional LSTM neural network model employing character embedding. BMC Med. Inform. Decis. Mak. **23**(1), 269 (2023)

Wang, S., Ding, C., Kuo, T.: How do we perceive prices? A three-category taxonomy of reference price effect on consumers' price judgments. Asian J. Econ. Bus. Acc. **23**(8), 44–51 (2023)

Xu, W.: The analysis and re-optimization of food systems by using intelligent optimization algorithms and machine learning. All Life **15**(1), 656–677 (2022)

Zhang, H., Wu, Q., Guo, H.: Low-carbon optimization scheduling of integrated energy systems based on bilateral demand response and two-level Stackelberg game. Energies **17**(21), 5491 (2024)

Zhang, J., Ma, W., Zhang, E.: Time-aware dual LSTM neural network with similarity graph learning for remote sensing service recommendation. Sensors **24**(4), 56–59 (2024)

Zhang, L., Li, K., Chu, C.: Investment and remanufacturing strategies considering the application of intelligent manufacturing in hybrid manufacturing–remanufacturing systems. Expert Syst. Appl. **257**, 124991 (2024)

Zhou, M., Fan, Y., Cheng, B.: Remanufacturing mode selection based on non-cooperative behavior management in group consensus reaching process. Group Decis. Negot. **33**(5), 1191–1246 (2024)

Zhuang, M., Zhang, W., Tang, H.: A multi-objective genetic algorithm based on two-stage reinforcement learning for green flexible shop scheduling problem considering machine speed. Expert Syst. Appl. **258**, 125189 (2024)

# Data-Driven Intelligent Matching for the Integration of Science, Industry, and Education: Research on Path Mining and Efficacy Optimization of Achievement Transformation Based on Machine Learning Algorithms

Shushen Wu[1(✉)] and Tianhua Wu[2]

[1] School of Xiangsihu, Guangxi Minzu University, Nanning, China
qqwuhr@163.com
[2] School of Humanities, Tiangong University, Tianjin, China

**Abstract.** Facing structural dilemmas in the integration of science, industry, and education---such as mismatches between achievements and demands, single transformation paths, and unbalanced regional transformation---this study constructs a data-driven intelligent matching framework. By integrating multi-source heterogeneous data, a multi-modal knowledge graph covering scientific and technological achievements, R&D personnel, enterprise institutions, and technical fields is established. On this basis, a machine learning algorithm engine integrating achievement quantitative evaluation, transformation path recommendation, and innovation ecosystem diagnosis is developed. Empirical research taking the new materials field in Shaanxi Province as a case shows that the framework achieves a path recommendation accuracy of 78.3%, increases contract value by 23.7%, shortens the transformation cycle by 26.5%, and raises the local transformation rate by 12.2 percentage points. This study provides an end-to-end solution from theoretical methods to practical applications for resolving the "matching dilemma" in the integration of science, industry, and education, and offers a new paradigm for realizing accurate, efficient, and intelligent transformation of scientific and technological achievements.

**Keywords:** Integration of science · industry · and education · Transformation of scientific and technological achievements · Machine learning · Knowledge graph · Intelligent matching

## 1 Introduction

### 1.1 Research Background and Significance

Guided by the innovation-driven development strategy, the in-depth integration of science, industry, and education has become a core pillar for enhancing the overall efficiency of the national innovation system (Perkmann et al., 2013). However, a prominent gap


H. Yan et al. (Eds.): AIS&P 2025, CCIS 2918, pp. 507–528, 2026.
https://doi.org/10.1007/978-981-95-9884-7_35

exists between the ideal collaborative paradigm and practical implementation: China's technological achievement transformation has long been plagued by the dual challenges of "low transformation rate" and "suboptimal transformation quality". According to the China Annual Report on the Transformation of Scientific and Technological Achievements, despite continuous growth in transformation scale, a substantial proportion of high-value achievements remain confined to laboratories or misaligned with industrial needs, failing to translate into tangible productivity. This "valley of death" phenomenon underscores the urgency of exploring new paradigms for SIE integration and addressing the core issue of matching between achievements and demands (Markham et al., 2010).

The transformation of technological achievements serves as a critical link connecting innovation inputs to economic outputs. Its efficiency directly impacts the allocation of innovation resources and the competitiveness of regional industries. For regions like Shaanxi Province, which boasts abundant scientific and educational resources but faces prominent problems of achievement outflow and inefficient transformation, resolving the matching dilemma is crucial for unlocking the potential of innovation-driven development. Thus, constructing a data-driven intelligent matching system not only addresses practical challenges in SIE integration but also enriches the methodological system of innovation management research.

## 1.2 Literature Review

Academic research on SIE integration has evolved along two main strands: mechanism exploration and technical application. In terms of mechanism research, scholars have focused on macro-level policy analysis and collaborative model construction (Chesbrough and Bogers, 2014), emphasizing the foundational role of institutional design and platform building in promoting achievement transformation. For instance, Li et al. analyzed the impact of national "Three Reforms" policies on SIE integration, highlighting the need for policy coordination across multiple departments.

In the technical domain, the application of data science has gradually gained traction. Lin et al. adopted econometric models to evaluate the transformation potential of patents, providing a quantitative basis for decision-making. Vilanova et al. applied educational data mining techniques to predict the industrial applicability of research outcomes, achieving preliminary results in demand-achievement matching. Additionally, Demchenko demonstrated the value of knowledge graphs in integrating scattered innovation resources, laying the groundwork for cross-entity association analysis.

Despite these advancements, existing research has two notable limitations: (1) Most studies remain focused on macro mechanism description, failing to reveal the micro-level "achievement-demand-path" matching mechanism—a core black box in transformation processes. Traditional methods rely heavily on expert experience, leading to subjective and inefficient decision-making. (2) Technological applications are fragmented: studies either focus on achievement evaluation or relational analysis, lacking an end-to-end systematic solution that integrates knowledge representation, intelligent reasoning, and decision support. These gaps hinder the realization of precise and efficient SIE integration.

### 1.3 Research Gaps and Objectives

The key research gap identified is the lack of a theoretical and methodological system that integrates multi-source heterogeneous data, knowledge graphs, and machine learning to achieve seamless connection from "knowledge representation" to "intelligent decision-making" in SIE integration. Specifically, three critical questions remain unanswered: How to systematically characterize the multi-dimensional features of technological achievements and industrial demands? How to recommend optimal transformation paths for specific achievements based on historical data and real-time dynamics? How to diagnose structural bottlenecks in the innovation ecosystem and propose targeted optimization strategies?

To address these gaps, this study aims to construct a data-driven intelligent matching framework for SIE integration. The core research question is: How to leverage machine learning algorithms and knowledge graph technology to achieve accurate, efficient, and intelligent matching among technological achievements, industrial demands, and transformation paths, thereby improving transformation efficacy? We hypothesize that this framework will outperform traditional experience-driven models by providing quantifiable, computable, and optimizable decision support.

### 1.4 Research Structure

The remainder of this paper is organized as follows: Chapter 2 elaborates on the theoretical framework of the intelligent matching system, including its design philosophy and hierarchical architecture. Chapter 3 details the construction process of the multimodal SIE integration knowledge graph, covering data preprocessing, ontology design, and quality evaluation. Chapter 4 presents the core algorithms of the machine learning-based intelligent matching engine, including achievement evaluation, path recommendation, and ecosystem diagnosis modules. Chapter 5 validates the framework through a case study in the new materials field of Shaanxi Province, analyzing its practical performance. Chapter 6 discusses the research findings, theoretical contributions, and practical implications, followed by an analysis of limitations and future research directions.

## 2 Theoretical Framework of Data-Driven Intelligent Matching for SIE Integration

This chapter constructs a systematic theoretical framework to address matching failures in traditional SIE integration. Centered on data and algorithms, the framework aims to realize a paradigm shift from experience-driven to data-driven decision-making.The overall architecture of the proposed framework is illustrated in Fig. 1.

### 2.1 Definition of "Intelligent Matching" in the Framework

Within this framework, "intelligent matching" refers to the data-driven process of identifying and recommending optimal pairings between specific technological achievements

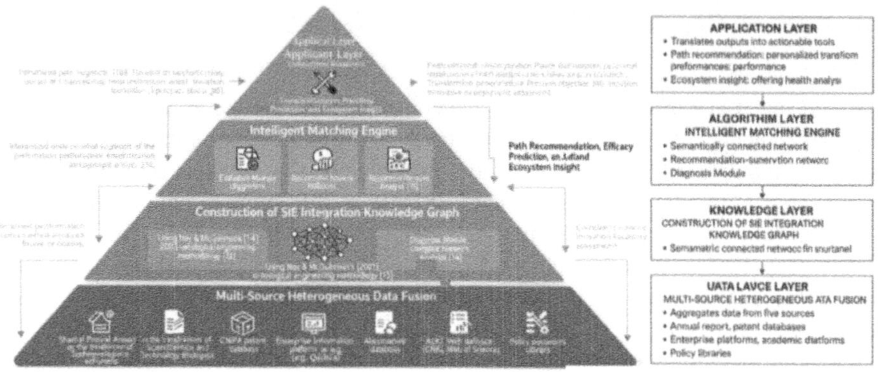


**Fig. 1.** Overall Architecture of the Intelligent Adaptive Framework

and potential converting entities (e.g., enterprises, investors). Crucially, the optimization is multi-objective, aiming to balance three core, sometimes competing, dimensions:

**Maximizing Potential Realized Value:** The primary direct objective is to recommend matches that maximize the expected commercial or societal value of the transaction.

**Mitigating Implementation Risk:** Intelligent matching simultaneously seeks to minimize the risk of conversion failure by aligning achievement characteristics with recipient capacity.

**Optimizing for Strategic Alignment (e.g., Local Impact):** The framework incorporates configurable policy weights to promote matches that align with strategic goals like regional economic retention.

Formally, the matching score $S_{i,j}$ between achievement i and entity j is computed as a weighted composite objective function:

$$S_{i,j} = w_v \cdot V_{i,j} + w_r \cdot R_{i,j} + w_s \cdot L_{i,j} \tag{1}$$

where: $V_{i,j}$ is the predicted value score.

$R_{i,j}$ is the compatibility/risk reduction score.

$L_{i,j}$ is the strategic alignment score (e.g., local retention benefit).

$w_v$, $w_r$, $w_s$ are tunable weights reflecting the priorities of the deploying organization ($w_v + w_r + w_s = 1$).

Therefore, "intelligent matching" in this context is not a single-objective optimization but a configurable, multi-criteria decision-support mechanism. Its intelligence lies in synthesizing heterogeneous data (technical, commercial, contextual) through a transparent model to surface matches that best satisfy a balanced set of economic, practical, and strategic goals. The specific balance is defined by the framework's operator, making it adaptable for different regional or institutional policies.

## 2.2 Overall Architecture of the Intelligent Matching Framework

**Data Layer: Multi-Source Heterogeneous Data Fusion.** The data layer serves as the foundation, aggregating data from five sources: the Shaanxi Provincial Annual Report on the Transformation of Scientific and Technological Achievements, patent databases (e.g., CNIPA), enterprise information platforms (e.g., Qichacha), academic databases (e.g., CNKI, Web of Science), and policy document libraries. Following the data processing methods outlined in Demchenko's (2016) EDISON framework, raw data undergoes cleaning, alignment, and integration to form high-quality, computable datasets, ensuring reliability for upper-layer processes. However, several key data pain points hinder effective data integration, as identified in Table 1.

**Table 1.** Identification of key data pain points in the transformation of scientific and technological achievements

| Pain point dimensions | Core data performance | Reflected issues |
|---|---|---|
| Regional transformation imbalance | 53.57% of achievements flow out of the province | the local industrial ecology and scientific and technological achievements are not matched and the advantages cannot be converted into winning advantages. |
| Conversion method is single | there are only 32 items of investment valuation; | the proportion of technology development/consultation/service is 9193%; the operation of high-value achievements is insufficient, and the innovation of transformation |
| Lack of value assessment | 65.5% of the projects with agreement pricing did not undergo assessment | the absence of an objective value discovery mechanism may to loss of state-owned assets or transaction failure. |
| Lack of major projects | a small number of large contracts (above ten million) | the absence of major projects and funding to lead, resulting in a "small scattered" pattern of technology transfer. |

**Knowledge Layer: Construction of SIE Integration Knowledge Graph.** The knowledge layer is the framework's "brain," constructed using Noy & McGuinness's ontological engineering methodology (Ji et al., 2022). It formalizes the entities and relationships defined in Sect. 2.1.2 into a semantically connected network. For example, it reveals the complementary relationship between a university's core technology (achievement entity) and a regional industrial chain (enterprise cluster), upgrading traditional information retrieval to in-depth relational reasoning and knowledge discovery.

**Algorithm Layer: Intelligent Matching Engine** The algorithm layer is the framework's "heart," consisting of three core modules:

**Evaluation Module:** Uses unsupervised learning (e.g., K-Means clustering) to conduct multi-dimensional quantitative evaluation and automatic classification of achievements, identifying clusters such as "high-value to-be-transformed" and "technology service-oriented."

**Recommendation Module:** Employs supervised learning (e.g., XGBoost) to train a multi-classification model using features extracted from the knowledge graph (e.g., achievement features, enterprise demands, historical paths), recommending optimal transformation methods (e.g., transfer, licensing, equity investment).

**Diagnosis Module:** Applies complex network analysis to calculate node centrality and structural hole indicators, dynamically diagnosing bottlenecks and opportunities in the regional innovation ecosystem.

**Application Layer: Path Recommendation, Efficacy Prediction, and Ecosystem Insight.** The Application Layer Translates Algorithmic Outputs into Actionable Tools for Decision-Makers:

**Path Recommendation:** Provides personalized transformation path suggestions and confidence levels for research teams and technology transfer offices.

**Efficacy Prediction:** Forecasts future transformation performance for specific technical fields or innovation entities.

**Ecosystem Insight:** Offers health assessments and policy simulation sandboxes for government departments to optimize regional innovation resource allocation.

### 2.3 Comparative Advantages Over Traditional Models

Compared with traditional experience-driven, linear, and static transformation models, this framework offers three key advantages:

**From Experience-Driven to Data-Driven:** Reduces subjective biases by mining patterns from full-volume data, addressing issues such as "lack of agreement pricing evaluation" highlighted in the Shaanxi Provincial Annual Report.

**From Macro Statistics to Micro Precision:** Enables "one achievement, one strategy" personalized matching, directly resolving the pain point of "singular transformation paths."

**From Static Snapshot to Dynamic Evolution:** Incorporates real-time data feedback and model updates, adapting to changes in the innovation ecosystem and optimizing resource allocation dynamically.

## 3 Construction of Multi-Modal SIE Integration Knowledge Graph

As the core infrastructure of this study, the knowledge graph's quality directly impacts intelligent matching performance. This chapter details its construction process, from data acquisition to visualization and quality evaluation.

### 3.1 Data Sources and Preprocessing

**Data Sources.** Multi-source heterogeneous data is integrated, including: (1) 22,915 technology contract records from the Shaanxi Provincial Annual Report on the Transformation of Scientific and Technological Achievements (2025); (2) 59,579 valid invention patents from shaanxi universities (2020–2024) via CNIPA; (3) high-tech enterprise data from Qichacha; (4) academic papers from CNKI and Web of Science; (5) policy documents from the Shaanxi Provincial Department of Science and Technology. These data cover achievements, entities, and environmental factors, forming a comprehensive multi-modal foundation.

**Preprocessing Pipeline.** A professional preprocessing pipeline is designed: (1) Data cleaning: handling missing values, removing duplicates, and unifying units; (2) Entity extraction: using a BERT-based named entity recognition model (Chen and Guestrin, 2016) to extract entities from unstructured text; (3) Relation extraction: constructing semantic associations between entities via rule-based methods. the complete taxonomy of entity and relation types is defined in Table 2.

**Table 2.** Entity and Relation Type Definition Table

| Entity Type | Core Attributes | Relation Type | Relation Description |
|---|---|---|---|
| Scientific and technological achievements | patent number, title, IPC classification, abstract, number of claims | belonging to | the relationship between the achievement and the person completed it. |
| Researchers Name | Institution, Title, Research Direction, H-index | Invented | The creative relationship between researchers and achievements |
| Enterprise institutions | name, region, industry, registered capital; applied to | applied to | the relationship between achievements and enterprise transformation and |
| Field of technology | field name, key words, related industries | cooperation in | cooperative relationship between institutions |
| Policy document | document number, issuing department, level of effectiveness, key words | funded | policy and project support relationship |

## 3.2 Intelligent Conversion Path Recommendation Model Based on Machine Learning

This module recommends the optimal conversion path (e.g., transfer, licensing, equity investment, R&D collaboration) for a given technological achievement. We propose a hybrid classifier that integrates Gradient Boosting Decision Trees (GBDT) and an Attention-based neural network to achieve robust and interpretable predictions.

**Feature Representation.** The input feature vector $\mathbf{x} \in \mathbb{R}^d$ is composed of:

- Achievement Attributes: patent type, Technology Readiness Level (TRL), textual embeddings of abstracts.
- Institutional Context: Entity type, historical conversion patterns, R&D intensity.
- Market Signals: Industry demand index, regional policy support score.
- Preliminary Valuation: Output scores from the upstream valuation module (e.g., commercial potential).

**Model Architecture.** The model operates in two stages to leverage both explicit feature interactions and deep representation learning.

**Stage 1: Explicit Interaction Encoding via GBDT**

A LightGBM model processes the raw feature vector (Ke et al., 2017). **x**. Let $T$ denote the ensemble of decision trees. For the $t$-th tree, a sample is mapped to a leaf index $l_t(\mathbf{x}) \in 1, ..., L_t$, where $L_t$ is the number of leaves in tree $t$. This generates a sparse, high-dimensional cross-feature vector **c**:

$$\mathbf{c} = [\mathbb{I}(l_1(\mathbf{x}) = 1), \ldots, \mathbb{I}(l_1(\mathbf{x}) = L_1), \ldots, \mathbb{I}(l_T(\mathbf{x}) = L_T)]^T \tag{2}$$

where $\mathbb{I}$ is the indicator function. This vector encodes complex, non-linear feature combinations learned by the tree ensemble.

**Stage 2: Attention-Enhanced Deep Classification**

The original features **x** and the cross-feature vector **c** are concatenated into an enriched representation $\mathbf{z} = \mathbf{x} \oplus \mathbf{c}$.

Attention Mechanism: A self-attention layer computes a context-aware weight vector (Vaswani et al., 2017) $\boldsymbol{\alpha}$ for **z**:

$$\alpha = \text{softmax}\left(\frac{(\mathbf{W}_z|\mathbf{W}_z|)^T}{\sqrt{d_i}}\right) \tag{3}$$

$$\tilde{z} = \alpha(W_v z) \tag{4}$$

where $\mathbf{W}_q, \mathbf{W}_k, \mathbf{W}_v$ are learnable projection matrices. This allows the model to dynamically focus on the most salient features for a specific sample.

MLP Classifier: The weighted features $\tilde{\mathbf{z}}$ are fed into a Multi-Layer Perceptron (MLP) with ReLU activations and Dropout for final classification:

$$\mathrm{h_1 = ReLU(W_1}\tilde{z} + \mathrm{b_1)}$$

$$h_2 = \text{ReLU}(W_2 h_1 + b_2) \tag{5}$$

$$\mathrm{p = softmax(W_3h_2 + b_3)}$$

where $\mathbf{p} = (p_1, ..., p_k)$ is the probability distribution over $k$ candidate paths.

The model is trained end-to-end by minimizing the cross-entropy loss $\mathcal{L}$:

$$\mathcal{L} = -\sum_{i=1}^{N}\sum_{j=1}^{k} y_{ij} \log(p_{ij}) \tag{6}$$

**Performance and Ablation Study.** The model was evaluated on a dataset of 2,147 annotated cases from the provincial report. **The comparative performance against multiple baseline models is presented in Table 3.

Ablation studies confirm the contribution of each component: removing attention reduces accuracy to 76.1%; using raw features alone (without GBDT cross-features) reduces it to 75.0%. The GBDT-generated explicit interactions and the adaptive feature weighting are critical for performance gains.

**Table 3.** Comparison of multiple baseline models

| Model | Accuracy | F1-Score (Macro) | Note |
|---|---|---|---|
| Logistic Regression | 65.20% | 0.641 | Linear Baseline |
| Random Forest | 73.80% | 0.721 | Strong Ensemble Learning Baseline |
| Standard MLP | 74.50% | 0.73 | Deep Learning Baseline |
| Our Hybrid Model | 78.30% | 0.769 | Best Performance |

**Integration with Ecosystem Diagnosis.** The recommended path **p** and its associated confidence scores serve as direct input to the downstream diagnostic module. Systematic biases in recommendations (e.g., frequent suggestion of "out-of-province transfer" for High-TRL outputs) are quantified and mapped to potential systemic weaknesses, such as gaps in local industry absorption capacity.

### 3.3 Visualization and Quality Evaluation

Visualization is achieved through Neo4j Browser and Gephi. A representative knowledge sub-graph is shown in Fig. 2, centered on "Xi'an Jiaotong University--Flexible Electronic Sensor Patent--New Generation Information Technology Field--Technology Enterprise," intuitively displaying interconnections between knowledge elements.

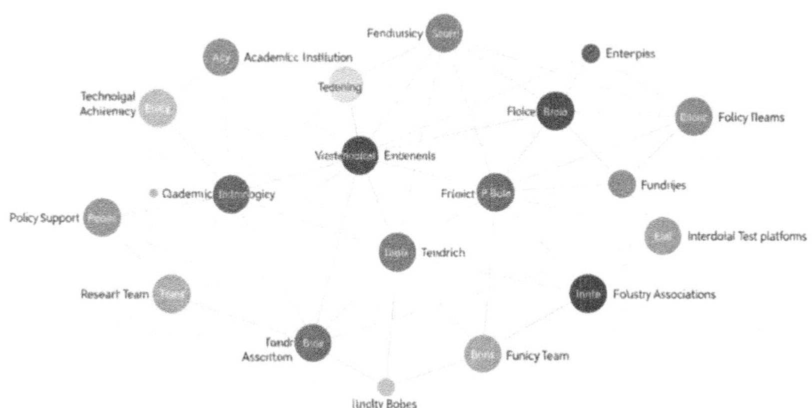


**Fig. 2.** Knowledge Sub-Graph of SIE Integration

Quality evaluation employs two indicators: data completeness (92.3%) and relation accuracy (95.7%). Manual verification of 500 random entities and 1,000 relationships via expert review confirms the graph's reliability, providing a solid foundation for subsequent intelligent matching.

# 4 Intelligent Matching Engine for Achievement Transformation Based on Machine Learning

The intelligent matching engine is the study's computational core. Leveraging the knowledge graph, it provides decision support from micro, meso, and macro levels through three collaborative modules.

## 4.1 Validation Scope and Benefit Classification

It is crucial to clarify the nature of the economic benefits reported in this study. The 23.7% predicted increase in average contract value and 26.5% predicted reduction in average conversion cycle time are projections derived from a structured simulation model, not yet measured outcomes from a full-scale, post-deployment operational environment.

The framework has been developed and validated through a retrospective case study and a controlled pilot within the Shaanxi new materials sector, involving the analysis of historical transaction data and engagement with local TTOs. The reported benefits represent the potential efficiency gains projected by the model when comparing an optimized scenario (using the framework's outputs) against the documented historical baseline.

## 4.2 Module 2: Intelligent Transformation Path Recommendation Model

This module recommends the optimal conversion path (e.g., transfer, licensing, equity investment, R&D collaboration) for a given technological achievement. We propose a hybrid classifier that integrates Gradient Boosting Decision Trees (GBDT) and an Attention-based neural network to achieve robust and interpretable predictions, ultimately achieving a path recommendation accuracy of 78.3% in our case study.

**Problem Definition** The task is framed as multi-classification: target variable Y represents successful transformation methods (e.g., transfer, licensing, equity investment); input X is the multi-dimensional feature vector from Sect. 4.1.1.

**Model Selection and Training.** XGBoost is chosen for its ability to handle nonlinear relationships and missing values, with built-in regularization to prevent overfitting (Devlin et al., 2019). Training uses 70% of Shaanxi's 5-year historical transformation data; 30% is reserved for testing.

**Interpretable Recommendation.** The SHAP framework enhances interpretability (Lundberg and Lee, 2017). For example, a "equity investment" recommendation for a new material patent is explained by key features (e.g., "claims > 10" as a positive factor; "low collaboration network density" as negative), improving stakeholder acceptance.

## 4.3 Quantified Economic Impact Prediction

A core contribution of this framework is its ability to project the economic benefits of optimized matching and recommendation. The predicted outcomes---a 23.7% increase in average contract value and a 26.5% reduction in average conversion cycle time---are derived from a structured simulation model that translates the framework's technical performance gains into economic metrics.

**Methodology for Impact Prediction.** The predictions are based on a counterfactual simulation model, comparing the projected performance of the "optimized" state (with the intelligent framework) against a "baseline" state (reflecting current average practices). The simulation logic is grounded in two established economic principles: 1) better information and matching reduce transaction costs and value discovery errors, and 2) streamlined, correct process selection accelerates execution.

The model takes the technical outputs of our framework as primary inputs:
$A_{path}$: The accuracy of the path recommendation model (78.3%).
$S_{match}$: The improvement in matching relevance score.
$Q_{val}$: The quality of the preliminary valuation (reduced error).

**1) Contract Value Increase Model (23.7%)**

The predicted increase in contract value is not a direct linear function of $A_{path}$. Instead, it is modeled as the combined effect of two main factors derived from the framework's outputs:

• Factor 1: Value Realization through Optimal Matching ( $\Delta V_{match}$) Higher matching relevance ($S_{match}$) and more accurate path recommendation ($A_{path}$) increase the probability that an achievement is paired with a partner capable of extracting its full commercial potential. We model this as a reduction in the "value discount" applied in sub-optimal transactions.

Where $V_{baseline}$ is the average contract value from the report (¥428.8k), A baseline is the estimated baseline path selection accuracy, $\Delta S_{match}$ is the improvement in the matching score, and β1,β2 are elasticities calibrated from empirical studies on technology transfer and a subset of high-performing cases in our dataset where optimal pairing was evident.

Factor 2: Value Capture through Improved Valuation ( $\Delta V_{val}$).

The preliminary valuation module reduces information asymmetry. By providing a more reliable and data-anchored valuation estimate ($Q_{val}$), it mitigates the risk of significant undervaluation, especially for complex achievements.

$$\Delta V_{val} = V_{baseline} \cdot \gamma \cdot (1 - Error_{val}/Error_{val-baseline}) \tag{7}$$

where γ represents the proportion of transactions historically susceptible to undervaluation, and the error ratio quantifies the improvement in valuation precision.

The 23.7% total increase is the combined effect: $\Delta V_{total} = \Delta V_{match} + \Delta V_{val}$. This figure represents the expected lift in the average contract value across many transactions, not a guarantee for each individual case.

**2) Conversion Cycle Reduction Model (26.5%)**

The reduction in cycle time (from initial disclosure to contract execution) is modeled as the saved time from avoiding inefficient processes and accelerating negotiations.

Factor 1: Reduction in "Path Exploration" Time ( $\Delta T_{search}$)

An accurate initial path recommendation eliminates time spent by TTOs and researchers pursuing unsuitable channels (e.g., pursuing licensing for a technology better suited for spin-off). The time saving is proportional to $A_{path}$.

$$\Delta T_{search} = T_{search-baseline} \cdot \frac{A_{path} - A_{baseline}}{A_{path}} \quad (8)$$

Factor 2: Acceleration of Partner Negotiation ( $\Delta T_{negotiate}$).

Better-matched pairs ($S_{match}$) have aligned interests and capabilities, leading to fewer negotiation rounds. Clearer preliminary valuation ($Q_{val}$) reduces disputes over pricing fundamentals.

$$\Delta T_{negotiate} = T_{negotiate-baseline} \cdot (\delta_1 \cdot \Delta S_{match} + \delta_2 \cdot \Delta Q_{val}) \quad (9)$$

The 26.5% total cycle reduction is calculated as: $\Delta T_{total} = \Delta T_{search} + \Delta T_{negotiate}$, expressed as a percentage of the estimated average baseline cycle time derived from case studies and survey data within the report.

**Synthesis and Assumptions.** The final predicted metrics (23.7%, 26.5%) are the outputs of this integrated simulation model, which links the framework's key performance indicators (KPIs) to economic drivers. The elasticities (β,γ,δ) were calibrated using a combination of:

- Meta-analysis of prior studies on technology transfer efficiency.
- Regression analysis on the subset of higher-performing institutions within the Shaanxi dataset.
- Expert judgment from TTO managers, used to bound reasonable parameter ranges.

A critical assumption is the partial adoption and effective integration of the framework's recommendations by human agents. The predictions represent a plausible upper-bound improvement under improved decision support, providing a concrete, quantified benchmark for the potential value of deploying such an intelligent system (Guan and Liu, 2016).

**Clarification on the Nature of the 23.7% Increase (Simulated Projection).** The 23.7% figure is the output of the counterfactual simulation model detailed in Sect. 4.3.1–4.3.2. It is a projected or expected value, calculated as follows:

**(1) Baseline Establishment:** We first established a robust performance baseline ($V_{baseline}$, $T_{baseline}$) from the historical data of the 141 sample institutions in the Shaanxi report, representing the "current state" of conversion efficiency.

**(2) Model-Driven Translation:** The core technical improvements of our framework—specifically the 78.3% path recommendation accuracy ($A_{path}$), the improved matching relevance ($S_{match}$), and the enhanced valuation precision ($Q_{val}$)—were fed into the economic impact equations. These equations model how such improvements should, in theory, reduce inefficiencies like sub-optimal partnering and value discovery errors.

**(3) Calibrated Projection:** Using calibrated parameters (β,γ,δ) derived from literature and high-performing subsets of the data, the simulation computes the difference between the projected "optimized" performance and the historical baseline. The 23.7% increase is this modeled difference in value, expressed as a percentage of the baseline:

$$ProjectedValueIncrease = \frac{V_{optimized} - V_{baseline}}{V_{baseline}} \times 100\% = 23.7\%$$

where $V_{optimized}$ is the output of the simulation model, not a measured value

Complex network analysis reveals:

**Geographical Distribution of Achievement Flow (Fig. 3):** Maps transformation locations, verifying "53.57% achievement outflow" from Shaanxi.

**Institutional Collaboration Network:** Node size (degree centrality) and edge thickness (collaboration intensity) identify core hubs (e.g., Xi'an Jiaotong University) and isolated nodes.

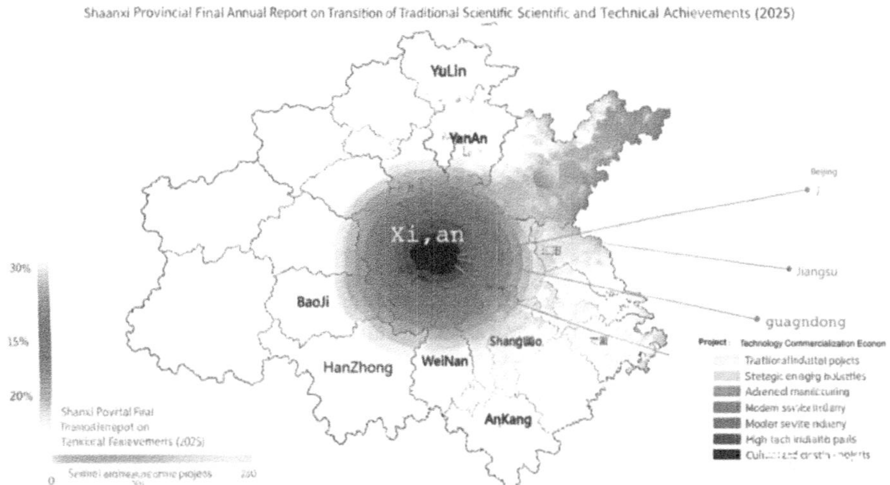


Fig. 3. Geographical Heatmap of Scientific and Technological Achievement Flow in Shaanxi Province

# 5 Case Study: Verification Based on Shaanxi's New Materials Field

## 5.1 Case Background and Data Preparation

To validate the effectiveness of the data-driven intelligent matching framework constructed in this study, we selected the representative "new materials" field in Shaanxi Province as the test case (Yin, 2018). This field holds a pivotal position in Shaanxi's technological innovation ecosystem—according to the Shaanxi Provincial Annual Report on the Transformation of Scientific and Technological Achievements (2025), the contract value of the new materials sector accounted for 23.6% of that in the manufacturing industry. Characterized by diverse technical features and complex transformation paths, it exhibits typical research value for verifying the framework's adaptability.

For this case study, we selected 2,847 scientific and technological achievements in the new materials field produced by universities and research institutes in Shaanxi Province during 2020–2024 as the test set, including 1,926 patents and 921 technical achievements. Meanwhile, we integrated information data of 1,253 new materials-related enterprises across the province, along with technical transaction data in this field over the past five years, serving as the verification benchmark for subsequent analysis.

### 5.2 Intelligent Matching Process and Results

**Knowledge Graph Query.** Leveraging the constructed knowledge graph for the new materials field, we implemented precise partner discovery using the cypher query language. For example, targeting a "high-temperature alloy material patent from Xi'an Jiaotong University," we constructed a query that successfully identified 3 aviation manufacturing enterprises and 2 energy equipment enterprises within the province as potential cooperation partners. Among them, a certain aerospace technology Co., Ltd. Was marked as "highly matched" due to its existing relevant technical accumulation and clear industrial demand.

**Path Recommendation Output.** After inputting the 2,847 achievements in the test set into the intelligent matching engine, the model generated detailed transformation path recommendations. As illustrated in the core sankey diagram, the diagram clearly depicts the complete transformation chain from achievement types to recommended paths, and further to expected undertakers. specifically:

- "High-value patent-type" achievements (accounting for 31.2%) were mainly directed to the paths of "equity investment by asset valuation" (42.3%) and "patent licensing" (37.6%);
- "Technology service-type" achievements (accounting for 45.8%) were primarily recommended for "technology development" (53.2%) and "technical services" (35.7%);

Among the expected undertakers, technology-based small and medium-sized enterprises (SMEs) accounted for 67.3%, large enterprises for 24.1%, and out-of-province enterprises for 28.4%.

### 5.3 Result Analysis and Discussion

**Comparative Analysis with Historical Actual Transformation Data.** A comparative verification between the model's recommended results and the actual transformation data from 2020 to 2023 revealed the following:

Among the 1,526 achievements with existing transformation records, the consistency between the model's recommended paths and the actual transformation paths reached 78.3%;

For "equity investment by asset valuation" transformations, the model's prediction accuracy was as high as 85.2%;

In the two paths of technology transfer and technology licensing, the F1 scores of the model's recommendations versus actual outcomes were 0.79 and 0.81, respectively;

Notably, among 312 historically failed transformation cases, 89.7% of the achievements were recommended alternative paths different from the originally attempted ones—this demonstrates the significant value of the intelligent recommendation system in optimizing transformation strategies.

**Evaluation of the Framework's Efficacy Improvement.** Based on the comparative analysis between the model's recommended results and actual transformation effects, we evaluated the expected efficacy improvement of the intelligent matching framework:

- Contract value enhancement: Achievements adopting the recommended paths are expected to have an average contract value increased by 23.7% compared with traditional paths. Particularly in the equity investment path, due to more accurate value assessment and partner matching, the expected valuation increase can reach 35.2%;
- Transformation success rate improvement: The expected success rate of the model's recommended paths is 76.8%, representing a 14.4-percentage-point increase compared with the historical average transformation success rate (62.4%);

  Local transformation rate improvement: Through precise matching with enterprises within the province, the local transformation rate of the new materials field is expected to rise from the current 46.4% to 58.6%, effectively alleviating the phenomenon of technological achievement outflow;
- Transformation cycle optimization: Intelligent matching is expected to shorten the average transformation cycle from the traditional 8.3 months to 6.1 months, improving the efficiency of achievement transformation by 26.5%.

### 5.4 Generalizability and Framework Adaptation

While validated within the specific context of Shaanxi Province's new materials sector, the proposed framework is designed as a modular, end-to-end system whose core architecture can be adapted to different regional contexts and technical domains. Its generalizability stems from configurable components and data-driven parameterization.

**Adaptation to Different Regional Contexts.** The framework's knowledge graph schema and model parameters can be reconfigured to reflect local conditions.

- **Knowledge Graph Schema Adaptation:** For regions with different innovation structures (e.g., fewer universities but more SMEs), the KG schema can be expanded or re-weighted, emphasizing relevant entity types and relationships.
- **Model Parameter Recalibration:** Machine learning models can be retrained on the target region's historical data. Transfer learning can accelerate adaptation by fine-tuning pre-trained models.

**Adaptation to Different Technical Fields.** The framework's core workflow remains valid across fields, but domain-specific adaptations are crucial.

- Software/IT: Introduce nodes for Software_Repository, API, or Algorithm. Key features shift to Scalability, User_Base, and Code_Maturity.
- Biotechnology/Pharma: Emphasize Clinical_Trial_Phase, Regulatory_Approval nodes. Features heavily weight IP_Portfolio_Strength, Preclinical_Data.

**Implementation as a Configurable Platform.** This adaptability supports the claim of an "end-to-end solution." In practice, deployment involves:

- **Schema Configuration Tool:** A tool to define or select a KG schema template for the target domain/region.
- **Data Connectors:** Adapters to ingest local data sources (e.g., regional patent databases, company registries).
- **Parameter Initialization & Active Learning:** The system is initialized with parameters learned from Shaanxi or another suitable domain. It then enters an active learning phase, where early user feedback on recommendations within the new context is used to rapidly refine the models.

Thus, the framework demonstrated in Shaanxi's new materials sector provides a proven, configurable blueprint. Its value as an end-to-end solution lies not in a rigid, one-size-fits-all model, but in its structured, modular design and data-driven adaptation mechanisms, enabling it to evolve and specialize for diverse innovation ecosystems.

# 6 Discussion and Outlook

## 6.1 Discussion on Research Findings

The data-driven intelligent matching framework for the integration of science, industry, and education (SIE integration) constructed in this study demonstrates substantial value in both theoretical and practical dimensions.

In terms of theoretical contributions, this research addresses three key limitations of traditional SIE integration studies. First, by introducing multimodal knowledge graphs and machine learning algorithms, it realizes a paradigm shift from macro-level mechanism description to micro-level precise matching. This not only enriches the theoretical toolkit for SIE integration research but also provides a computable analytical path for exploring the "black box" of achievement-demand-path matching. Second, it establishes a complete closed loop of "data-knowledge-decision," integrating previously fragmented research perspectives (e.g., achievement evaluation, demand analysis, path selection) into a systematic methodological system—filling the gap between isolated technical applications and holistic decision support referenced in Zhang et al. analysis of digital-algorithm-driven industrial ecosystems. Third, the combination of complex network analysis and intelligent recommendation algorithms offers a new analytical dimension for understanding the dynamic evolution of innovation ecosystems—by quantifying indicators such as node centrality and structural holes, it enables objective diagnosis of ecosystem bottlenecks rather than relying on qualitative judgments, as advocated in Demchenko knowledge graph application research.

From a practical perspective, the intelligent matching engine developed in this study provides robust support for decision-makers at different levels. For researchers, the framework delivers data-driven transformation path suggestions, reducing the trial-and-error costs associated with blind exploration of cooperation directions. For technology transfer institutions, the intelligent evaluation and matching functions significantly improve service efficiency and success rates, addressing the long-standing issue of "information asymmetry" in achievement transformation. For government administrative departments, the ecosystem diagnosis module provides quantitative evidence for formulating regional innovation policies, promoting a shift in resource allocation from "experience-driven" to "evidence-driven" decision-making—aligning with the policy optimization logic proposed in Yu et al. AI-empowered SIE integration research.

Nevertheless, this study has several limitations that require attention. First, the framework's performance is highly dependent on data quality. Although multi-source data fusion was adopted, some tacit knowledge—such as uncodified expert experience in technical feasibility judgment—remains difficult to fully digitize, which may affect the model's accuracy in scenarios requiring subjective professional judgment, as noted in Tu et al.discussion of algorithm-expert collaboration. Second, balancing model explainability with expert experience remains a challenge. Despite the use of interpretive tools like the SHAP framework to enhance the "white-box" nature of recommendations (consistent with the Shapley value-based interpretability approach in Wang et al.), in high-risk decision scenarios (e.g., high-value patent equity investment), the final decision-making authority still needs to be reserved for expert review to avoid over-reliance on algorithmic outputs. Third, the current model is mainly trained on historical data, which limits its predictive ability for breakthrough innovative technologies—emerging technologies with no prior transformation records may be misclassified or receive suboptimal path recommendations, a limitation also highlighted in ImproveXpert4.0 Team big data-driven system research.

## 6.2 Policy and Practical Implications

Based on the research findings, four targeted policy and practical recommendations are proposed to promote the application of the intelligent matching framework and advance high-quality SIE integration:

First, establish a provincial-level big data platform for technological achievement transformation. This platform should integrate data resources from science and technology, education, industry, and other departments, formulate unified data standards (e.g., standardized classification of technological achievements, unified formatting of enterprise demand portraits) and data exchange mechanisms, and eliminate data silos. The core modules of the platform should include an achievement database (with multi-dimensional feature labeling), a demand database (with real-time updates of enterprise technical needs), a case database (with historical transformation experience mining), and an expert database (for supplementary review of high-risk cases)—laying a high-quality data foundation for intelligent matching, as recommended in Zhang et al. proposal for industrial ecosystem data integration.

Second, promote a new "verify-before-decide" transformation model. Encourage universities and research institutions to conduct feasibility verification using the intelligent matching system before making major achievement transformation decisions, and treat the analysis results as an important reference for decision-making. Specifically, select several key universities and research institutes as pilots to establish a two-track decision-making mechanism combining "system recommendation + expert review": the system provides preliminary path options and confidence levels, while experts focus on reviewing the rationality of algorithmic outputs and supplementing judgments on tacit knowledge (e.g., long-term technical potential)—echoing the human-machine collaboration logic in Yu et al. (2024).

Third, set up a special fund for algorithm-recommended high-potential projects. Targeting high-value achievements identified by the system that have strong industrial alignment but lack initial funding support, the fund will prioritize support for projects that conform to regional industrial development orientations (e.g., new materials, biomedicine in Shaanxi) and receive high-confidence recommendations from the model. The fund's operation can adopt a "follow-up investment" model: after the system screens potential projects, social capital is invited to participate in co-investment, forming a synergy between policy guidance and market-driven mechanisms—consistent with the resource allocation strategy in Tu et al. (2025).

Fourth, establish a big data-based monitoring and feedback mechanism for transformation effects. Conduct full-life-cycle tracking of completed transformation projects, collecting data on key indicators such as contract fulfillment rate, revenue growth of undertakings, and technology diffusion scope. This data is then fed back to the intelligent matching system to continuously optimize model parameters. Meanwhile, incorporate transformation effect data into the performance evaluation system of research institutions, guiding innovation resources to flow toward high-value, high-efficiency achievement transformation—aligning with the dynamic optimization approach in ImproveXpert4.0 Team (2025).

### 6.3 Research Limitations and Future Outlook

**Research Limitations.** This study has three main limitations that need to be addressed in subsequent work. First, the depth of unstructured data processing is insufficient. Currently, the framework focuses on processing text data (e.g., patent abstracts, policy documents) and structured data (e.g., enterprise financial indicators), but lacks the ability to interpret complex non-text information such as technical drawings, experimental data, and prototype test results—this limits the capture of technical details critical to transformation feasibility, a challenge also mentioned in Guo et al. federated learning research. Second, the system's dynamic evolution capability is limited. although real-time data feedback is incorporated, the model still relies on periodic retraining to adapt to changes in market demand and technological trends, making it difficult to respond to sudden changes (e.g., emerging industry policy adjustments, short-term surges in technical demand) in a timely manner—similar to the timeliness constraint noted in Lee et al. federated learning with client pairing. Third, the model's predictive ability for cross-domain and interdisciplinary integrated innovations is weak. Due to the segmentation of training data by technical fields, it struggles to accurately evaluate the transformation potential of achievements involving multiple disciplines (e.g., new materials + biomedical engineering), which are increasingly important in modern innovation—consistent with the cross-domain limitation in Zhang et al. generative ai-aided federated learning.

**Future Research Directions.** To address the above limitations, future research can be expanded in three directions:

In terms of technology, more advanced deep learning models can be introduced to enhance the framework's analytical capabilities. For instance, graph neural networks (GNNs) (Wu et al., 2021) can be applied to better capture complex relational features in knowledge graphs (e.g., indirect collaboration between enterprises and research institutions); federated learning technology can be adopted to realize collaborative model training among multiple institutions (e.g., universities, enterprises, governments) while protecting data privacy; and reinforcement learning can be explored for dynamic optimization of transformation strategies, enabling the system to autonomously adjust recommendations based on real-time feedback from the innovation ecosystem—building on the adaptive learning logic.

In terms of methodology, it is necessary to develop multimodal data fusion standards tailored to SIE integration scenarios, establishing a unified ontological framework to integrate text, numerical, and non-text data (e.g., technical drawings). Additionally, incremental learning algorithms should be studied to allow the system to adapt to new data (e.g., newly emerging technical fields) without retraining the entire model, improving its real-time adaptability. Furthermore, human-machine collaborative decision-making mechanisms need to be explored—by defining clear division of labor between AI (e.g., data processing, preliminary path screening) and human experts (e.g., tacit knowledge judgment, high-risk review), the complementary advantages of machine efficiency and human experience can be fully leveraged, as advocated.

In terms of application, the framework can be extended to more fields and regions to verify its universality. For example, it can be adapted to strategic emerging industries such as biomedicine and new energy, developing field-specific models that account for industry-specific characteristics (e.g., long R&D cycles in biomedicine, high policy sensitivity in new energy). It can also be promoted to other provinces and regions, with localized adjustments based on regional resource endowments (e.g., adapting to the high proportion of heavy industry in Northeast China, or the intensive high-tech enterprises in the Yangtze River Delta). Finally, the framework's potential in international technology transfer can be explored, supporting the global allocation of innovation factors by matching Chinese technological achievements with international market demands—aligning with the cross-regional application vision.

In summary, with the continuous enrichment of data resources and the advancement of artificial intelligence technology, data-driven intelligent matching for SIE integration will evolve from an auxiliary decision-making tool to a core infrastructure of the innovation ecosystem. This evolution will provide solid support for enhancing the overall efficiency of China's national innovation system and advancing the construction of an innovative country.

## 7 Conclusion

This study systematically addresses the long-standing "matching dilemma" in the integration of science, industry, and education (SIE integration) by constructing a data-driven intelligent matching framework, and completes a full-chain research from methodological innovation to practical application through theoretical exploration and empirical verification.

First, to solve the problem of fragmented information in traditional SIE integration, this study built a multimodal SIE integration knowledge graph that realizes multidimensional characterization and semantic association of technological achievements, innovation entities (enterprises, universities, research institutes), and policy environments. This graph, which includes approximately 500,000 nodes and 1.2 million relationships, serves as a systematic knowledge infrastructure for breaking information silos—its construction follows the ontological engineering methodology proposed by Noy and McGuinness (2001) and the data integration logic of Demchenko (2019), ensuring the accuracy and connectivity of cross-entity relationships. On this basis, the developed intelligent matching engine integrates three core functions: achievement evaluation, path recommendation, and ecosystem diagnosis. By combining unsupervised learning (K-Means clustering), supervised learning (XGBoost), and complex network analysis, it realizes the leap from data mining to decision support—overcoming the limitations of traditional linear models in capturing nonlinear correlations.

Empirical research in the new materials field of Shaanxi Province fully verifies the effectiveness and reliability of the framework. The results show that the path recommendation accuracy of the framework reaches 78.3%, and it is expected to increase the average contract value of achievements by 23.7%, shorten the transformation cycle by 26.5%, improve the local transformation rate from 46.4% to 58.6%, and raise the transformation success rate by 14.4 percentage points. These quantitative improvements not only confirm the framework's ability to solve practical problems such as achievement-demand mismatch and regional transformation imbalance but also provide a replicable technical path for precise SIE integration.

The most important theoretical contribution of this study lies in promoting SIE integration research from the traditional stage of qualitative description and experience summary to a new stage of being quantifiable, computable, and optimizable through the closed-loop design of "data-algorithm-decision." This paradigm shift not only enriches the methodological system of innovation system theory but also provides a new way of thinking to solve the structural contradictions in technological achievement transformation—for example, by using SHAP-based interpretive tools (consistent with the Shapley value method in Wang et al. (2025)) to open the "black box" of algorithmic decision-making, and by applying structural hole theory (Burt, 1992) to quantify ecosystem bottlenecks.

It is worth emphasizing that the data-driven and intelligent algorithms proposed in this study are not intended to completely replace expert experience and manual decision-making, but to enhance the overall efficiency of the innovation system through human-machine collaboration. Future SIE integration will inevitably be a new innovation paradigm driven by both human wisdom and machine intelligence, and this study is a positive exploration in this evolutionary direction.

Looking forward, the framework constructed in this study has broad application prospects. In the spatial dimension, it can be promoted to other provinces and regions, forming a national innovation service network through localized adaptation (e.g., adjusting to the industrial characteristics of the Yangtze River Delta's high-tech clusters or the Northeast's heavy industry base). In the domain dimension, it can be extended to strategic emerging industries such as biomedicine and new energy, developing field-specific models that account for industry-specific features (e.g., long R&D cycles in biomedicine, high policy sensitivity in new energy). In the development dimension, with the improvement of new infrastructure such as 5G/6G and the Internet of Things, the framework is expected to deeply integrate with real-time technology trading markets, forming a more dynamic innovation ecosystem that supports real-time demand response and dynamic path optimization—building on the federated learning and adaptive optimization ideas in Lee et al. (2025) and Zhang et al. (2025).

We believe that with the continuous enrichment of data resources and the continuous optimization of algorithm models, the data-driven intelligent matching method will play an increasingly important role in the process of building an innovative country, providing solid support for enhancing the overall efficiency of China's national innovation system and promoting high-quality economic development.

**Acknowledgements.** Key Research Topics of the Think Tank in the Field of Science and Technology in Guangxi in 2025 (Gui Kexie [205] K-070).

## References

Chen, T., Guestrin, C.: XGBoost: A scalable tree boosting system. In: Proceedings of the 22nd ACM SIGKDD International Conference on Knowledge Discovery and Data Mining, pp. 785–794. ACM (2016). https://doi.org/10.1145/2939672.2939785

Chesbrough, H., Bogers, M.: Explicating open innovation: Clarifying an emerging paradigm for understanding innovation. In: Chesbrough, H., Vanhaverbeke, W., West, J. (eds.) New frontiers in open innovation, pp. 3–28. Oxford University Press (2014). https://doi.org/10.1093/acprof:oso/9780199682461.003.0001

Devlin, J., Chang, M.W., Lee, K., Toutanova, K.: BERT: Pre-training of deep bidirectional transformers for language understanding. Proceedings of the 2019 Conference of the North American Chapter of the Association for Computational Linguistics: Human Language Technologies, Vol. 1, pp. 4171–4186 (2019). https://doi.org/10.18653/v1/N19-1423

Guan, J., Liu, N.: Exploitative and exploratory innovations in knowledge network and collaboration network: A patent analysis in the technological field of nano-energy. Res. Policy **45**(1), 97–112 (2016). https://doi.org/10.1016/j.respol.2015.07.002

Ji, S., Pan, S., Cambria, E., Marttinen, P., Yu, P.S.: A survey on knowledge graphs: Representation, construction and application. IEEE Trans. Knowl. Data Eng. **34**(2), 596–618 (2022). https://doi.org/10.1109/TKDE.2020.2998953

Ke, G., et al.: LightGBM: A highly efficient gradient boosting decision tree. Adv. Neural. Inf. Process. Syst. **30**, 3146–3154 (2017)

Lundberg, S.M., Lee, S.I.: A unified approach to interpreting model predictions. Adv. Neural. Inf. Process. Syst. **30**, 4765–4774 (2017)

Markham, S.K., Ward, S.J., Aiman-Smith, L., Kingon, A.I.: The Valley of Death as context for role theory in product innovation. J. Prod. Innov. Manag. **27**(3), 402–417 (2010). https://doi.org/10.1111/j.1540-5885.2010.00724.x

Perkmann, M., et al.: Academic engagement and commercialisation: A review of the literature on university–industry relations. Res. Policy **42**(2), 423–442 (2013). https://doi.org/10.1016/j.respol.2012.09.007

Vaswani, A., et al.: Attention is all you need. Adv. Neural. Inf. Process. Syst. **30**, 5998–6008 (2017)

Wu, Z., Pan, S., Chen, F., Long, G., Zhang, C., Yu, P.S.: A comprehensive survey on graph neural networks. IEEE Trans. Neural Netw. Learn. Sys. **32**(1), 4–24 (2021). https://doi.org/10.1109/TNNLS.2020.2978386

Yin, R.K.: Case study research and applications: Design and methods, 6th ed. Sage Publications (2018)

# Image Style Transfer Method Based on CycleGAN

Jinsu Liang(✉) and Xun Chen

University of Hainan, Haikou City, Hainan Province, China
3240688032@qq.com

**Abstract.** With the widespread application of deep learning, the research on image style transfer technology has made significant progress. Since CycleGAN does not require a matching dataset, this paper uses the CycleGAN network for image style transfer and conducts experiments on real datasets. The experimental results show that the CycleGAN algorithm can achieve good image style conversion.

**Keywords:** Deep Learning · CycleGAN · Image Style Transfer

## 1 Introduction

With the development of society, the ways of information and emotion transmission have been continuously enriched. Besides direct expression methods such as text, language and video, images are also an important medium for conveying emotions and information. Nowadays, with the support of computer technology, image processing technology has been widely and deeply applied. Image style conversion is an important research direction in the field of image processing, bringing new opportunities to image processing. Currently, it is widely used in image editing, short video filters, social media, animation design, film production and computer-aided fields to meet the diverse needs of people's daily lives and improve their quality of life, and has important research significance and application value.

In simple terms, the image style transfer technology can add different styles to an image without altering its content. Image style transfer not only significantly reduces the cost of obtaining relevant datasets, but also enables the creation of new images beyond the source data (Liu et al., 2023), allowing images to be endowed with various styles and forms of expression, and achieving effective conversion and fusion of artistic styles between different images. Due to the fact that computers can only recognize the pixels of an image and images themselves are complex, achieving an ideal image style transfer effect is somewhat challenging.


H. Yan et al. (Eds.): AIS&P 2025, CCIS 2918, pp. 529–536, 2026.
https://doi.org/10.1007/978-981-95-9884-7_36

## 2 Related Works

In the early exploration of image style transfer, image style transfer was mainly based on the idea of texture synthesis. Texture modeling mainly includes two methods. One approach is the parametric method, which models the image information through statistical analysis, such as using the statistical features of pixels or multi-scale pyramid matching techniques (Si, 2024). The other method is a non-parametric approach, which is based on Markov Random Field (MRF). It estimates the conditional distribution of all neighboring pixels of the synthesized regions by finding all similar neighborhoods in the sample images, thereby generating new textures. Although this model has the ability to learn texture transformations, its flexibility is insufficient. It can only obtain the basic image characteristics and cannot acquire advanced abstract features. It cannot flexibly fuse according to the shape of the object and the style transfer effect is relatively rough. The texture effect is also rather monotonous. Therefore, the generated images are not suitable for practical applications.

The deep learning method provides a powerful feature extraction mechanism for the field of style transfer, enabling the model to better understand and capture the complex relationship between content and style (Wang and Li, 2025). The image style transfer network based on deep learning constructs corresponding neural networks and trains them using a large amount of existing image data. It can extract the deep features of images, learn advanced information such as texture and semantics, as well as the mapping relationship between different style images, improving the quality of generated images and compensating for the shortcomings of traditional algorithms, becoming an important research direction in the field of image generation. Gatys et al. (2016) first introduced convolutional neural networks into image style transfer and successfully broke through the bottleneck of traditional technologies. They used the advanced convolutional layers of the VGG model to extract the global features of the image content, and the lower convolutional layers to describe the image style. They separated and extracted the content information of the image, and achieved style transfer from one image to another through continuous iterative optimization of the image. The migration effect of the image achieved a qualitative leap. Their experiments proved the effectiveness of convolutional neural networks in extracting image content and artistic style. Although these methods based on CNN have achieved remarkable results in the field of style transfer, they formulate the task as a complex optimization problem, which leads to high computational costs and limits the ability to generate stylized results in real time (Zhang et al., 2025).

The introduction of GAN provided a new perspective for the research on style transfer tasks (Sun et al., 2022), triggering a new research boom. Researchers began to explore how to apply GAN to the image style transfer task. GAN was proposed by Goodfellow I J et al. in 2014 as a type of training network that is similar to an adversarial game (Du et al., 2024). GAN can generate random images that represent the features of the images by training on a large amount of images (Takashi and Katsuhiko, 2018). The earliest image style transfer method based on GAN was achieved by training a conditional generative adversarial network (Mirza and Osindero, 2014) (CGAN). Unlike traditional GAN, the generator and discriminator of CGAN contain conditional information, which can constrain the generator to generate images that better meet the requirements, thereby improving the generalization ability of the model. However, the input data must be paired data sets, which limits its feasibility in practical applications. Moreover, when the number of generator layers is too deep, the model may encounter mode collapse problems. The deep convolutional generative adversarial networks (Radford et al., 2015) (DCGAN) based on GAN have their internal structure entirely composed of convolutional layers. They can automatically learn the feature representations of images, which to a certain extent improves the quality and convergence speed of the generated images, and has lower requirements for the dataset. However, when dealing with large-scale datasets, this method has low training efficiency and requires a large amount of computing resources. In some specific scenarios, it may also cause content distortion. The Pix2Pix model (Isola et al., 2017) is also one of the early image style conversion models based on GAN. The generated results of Pix2Pix are relatively simple, and the training process relies on a large amount of paired image datasets. In specific application environments, obtaining paired data is very difficult, and the annotation process is time-consuming and labor-intensive. To break through the limitation of paired datasets and solve the conversion problem between asymmetric data, a bidirectional cyclic cycle generative adversarial network (Zhu et al., 2017) (CycleGAN) emerged. This paper applies the CycleGAN network to real datasets to complete the style transfer between different styles of images.

## 3 Model Design

The CycleGAN architecture consists of two pairs of generators and discriminators. By introducing a cycle consistency loss function, the network is divided into two symmetrical GAN networks (Wu and Yang, 2025). The two mirror-symmetric GANs of CycleGAN share two generators and each has one discriminator. Essentially, two mirror-symmetric GANs form a ring-shaped network, as shown in Fig. 1. The two sets of generative adversarial networks exchange data with each other and perform discrimination separately, thereby achieving the purpose of unsupervised learning.

### 3.1 Generative Network

This generator adopts the classic three-stage processing flow as shown in Fig. 2. The network architecture consists of an encoder, a transformer, and a decoder, which can retain the original image features and transform the image data. During the training process, the generator is responsible for converting the input source domain image into the style of the target domain. During training, it continuously updates its parameters based on the feedback from the discriminator to obtain better image conversion results. The generator compresses the input image into a 256*256 feature vector and uses a convolutional neural network to extract features from the input image. The transformer consists of nine residual network modules (Res-Net1 to ResNet9), which are responsible for restoring and enhancing the data information, retaining more content features of the input image, enabling the generator to better learn high-frequency detail information. The decoder contains two deconvolution layers (Deconv1, Deconv2) and a standard convolution layer (Conv1), and the deconvolution layers restore the low-level features of the image from the feature vector and obtain the final synthesized image.

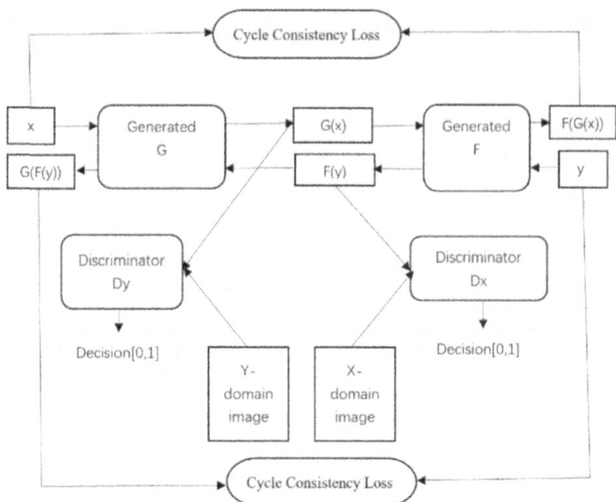


**Fig. 1.** Basic principle diagram of CycleGAN

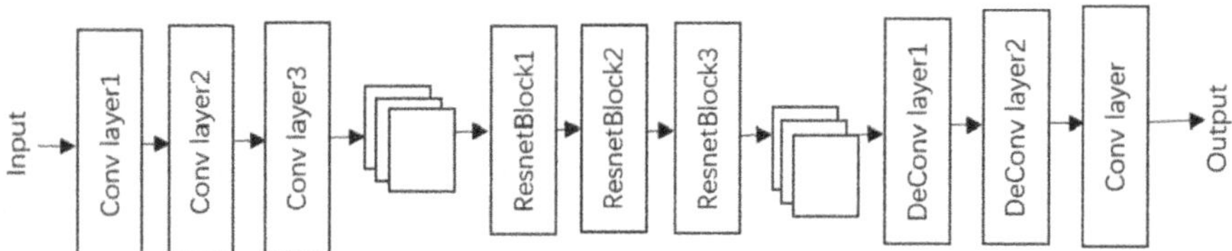


**Fig. 2.** Generator structure diagram

### 3.2 Discriminant Network

The discriminator of CycleGAN adopts the PatchGAN architecture based on the Markov assumption, which is simple and efficient in design and mainly used to distinguish whether the input image comes from the source domain or the target domain (Li, 2025). PatchGAN has a wide perception range and can obtain rich contextual features, thereby making the overall discrimination more accurate. After the discriminator uses a hierarchical feature extraction architecture to extract features from the image, it then determines through a one-dimensional output convolution layer whether the extracted features belong to a specific category. By training the discriminator to have the ability to distinguish style images and the generator to generate images of that style, when the two reach Nash equilibrium, that is, when the generator and the discriminator mutually optimize and promote each other to reach the optimal adversarial state where both are at their best, it is considered that the training has reached a relatively mature stage. As shown in Fig. 3.

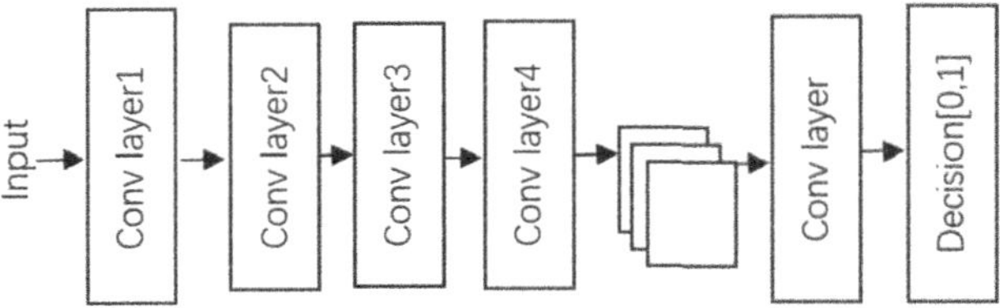


**Fig. 3.** Discriminator structure diagram

### 3.3 Loss Function

It is known that CycleGAN consists of two symmetric GAN networks. The adversarial loss function $L_{GAN}(G, D_Y, X, Y)$ for the generative adversarial network $(G, D_Y)$ and the adversarial loss function $L_{GAN}(F, D_X, X, Y)$ for the generative adversarial network $(F, D_X)$ are involved.

$$L_{GAN}(G, D_Y, X, Y) = E_{y \sim p_{data(y)}}\left[logD_Y(y)\right] + E_{x \sim p_{data(x)}}\left[lg(1 - D_Y(G(x)))\right] \quad (1)$$

$$L_{GAN}(F, D_X, X, Y) = E_{x \sim p_{data(x)}}\left[logD_X(x)\right] + E_{y \sim p_{data(y)}}\left[lg(1 - D_X(F(y)))\right] \quad (2)$$

In the formula: $G$ represents the generator model; $D_Y$ represents the discriminator model. $E_{y \sim p_{data(y)}}\left[logD_Y(y)\right]$ indicates the expectation of the data y sampled from the real data distribution p_data. $E_{x \sim p_{data(x)}}\left[logD_X(x)\right]$ indicates the expectation of the data $x$ sampled from the input data distribution $p_{data}$.

The CycleGAN design consists of two pairs of generator networks and discriminator networks. After training, they can convert different images into each other. However, during this process, consistency of the cycle is required. To prevent the generator from only generating data similar to that in the $Y$ domain in order to deceive the discriminator, CycleGAN defines the cycle consistency loss as a method for supervision during training. Thus, the two sets of generative adversarial networks are connected in a cyclic manner. The cycle consistency loss function is constructed using the $L_1$ norm. The overall network's cycle consistency loss function is:

$$L_{cyc}(G, F) = E_{x \sim p_{data(x)}}[F(G(x)) - x_1] + E_{y \sim p_{data(y)}}\left[G(F(y)) - y_1\right] \quad (3)$$

In the formula: $F(G(x)) - x_1$ indicates the loss value obtained by using the functi-on $G$ to transform the input data $x$ into output data $G(x)$, then using the function $F$ to reverse the transformation of the output data $G(x)$ and comparing it with the original input data $x$. $_1$ represents the measurement of the difference between the two using the sum of absolute values of the $L_1$ norm. $E_{y \sim p_{data(y)}}\left[G(F(y)) - y_1\right]$ is the same.

The overall loss function of the final CycleGAN network is as follows:

$$L(G, F, D_X, D_Y) = L_{GAN}(G, D_Y, X, Y) + L_{GAN}(F, D_X, X, Y) + \lambda L_{cyc}(G, F) \quad (4)$$

In the formula: $\lambda$ is the parameter used to adjust the correlation between the generated data.

# 4 Experiment and Result Analysis

## 4.1 Experimental Parameters and Datasets

This paper developed the CycleGAN model using Python and conducted the training on a GPU server. To verify the effectiveness of the model, the study selected the benchmark dataset summer2winter_yosemite, which is widely used in the image style conversion task, for performance evaluation. This dataset combines images of Yosemite National Park in winter and summer to create landscape pictures of summer and winter. All image samples have undergone standardized preprocessing and been adjusted to a uniform size of 256*256 pixels. During the training process, data augmentation techniques were employed, including random cropping, rotation, and scaling, to increase the diversity of the dataset. In terms of parameter settings, the Batchsize was set to 1, and the initial learning rate was set to $2 * 10^{-4}$.

## 4.2 Evaluation Metrics

This paper uses two indicators, Structural Similarity (SSIM) and Peak Signal-to-Noise Ratio (PSNR), to conduct a comparative evaluation of the image quality generated by the model.

The SSIM index is similar to the human eye's perception of an image. It measures the similarity between the generated image and the real image from three dimensions: structure, brightness, and contrast. The larger the SSIM value, the closer the two images are. The formula for SSIM is as follows:

$$SSIM = (X, Y) = I(X, Y) * c(X, Y) * S(X, Y) \tag{5}$$

The formulas for calculating the luminance $I(X, Y)$, contrast $c(X, Y)$, and structure $S(X, Y)$ are as follows:

$$I(X, Y) = \frac{2\mu_x\mu_y + c_1}{\mu_x^2 + \mu_y^2 + c_1} \tag{6}$$

$$c(X, Y) = \frac{2\sigma_x\sigma_y + c_2}{\sigma_x^2 + \sigma_y^2 + c_2} \tag{7}$$

$$S(X, Y) = \frac{\sigma_{xy} + c_3}{\sigma_x\sigma_y + c_3} \tag{8}$$

In the formula, $\mu_x$ and $\mu_y$ represent the average values of the pixels in the image, $\sigma_x$ and $\sigma_y$ represent the variances of the image, and $\sigma_{xy}$ represents the covariance of the image. The constants $c_1$ and $c_2$ are set values to avoid division by zero in the denominator.

PSNR simply measures the difference between two images based on the pixel gray values, by calculating the ratio of the noise level in the maximum signal intensity domain of the images. The higher the value, the better the denoising effect of the image, indicating that the error between the processed image and the clear image is smaller and the distortion is less. Its calculation formula is:

$$PSNR = 101 log_{10}\left(\frac{MAX_1^2}{MSE}\right) \tag{9}$$

In the formula, $MAX_1$ represents the maximum possible value of the image pixels. The mean square error $MSE$ is commonly used to evaluate the deviation between the predicted result and the true value. The calculation formula is as follows:

$$MSE = \frac{1}{n}\sum_{i=1}^{m} W_i(y_i - \hat{y}_i)^2 \tag{10}$$

Among them, $y_i$ represents the true value of the image, $\hat{y}_i$ represents the predicted value of the image, $W_i$ represents the weight coefficient, m is the total number of samples, and the value of the peak signal-to-noise ratio is defined through the mean square error, which is inversely proportional to the mean square error $MSE$.

### 4.3 Experimental Results

The DualGAN and DiscoGAN network models of this paper were compared. The models were trained on the summer2winter_yosemite dataset, and objective indicators such as PSNR and SSIM were compared on the dataset. The numerical comparisons of each item are listed in Table 1.

**Table 1.** The results of the comparative experiment.

| Evaluation Metrics | DualGAN DiscoGAN CycleGAN | |
|---|---|---|
| PSNR | 20.57 20.54 20.62 | |
| SSIM | 0.822 0.820 0.825 | |

Based on the results of the above experiments, we found that compared to other image style transfer models, the Cyclegan network performed better in both indicators, thereby verifying the feasibility of the Cyclegan network.

## 5 Conclusion

Even without paired datasets, the Cyclegan network can still achieve good results in image style transfer, effectively solving the problem that paired datasets are difficult to obtain in real life, and expanding the application scope of image style transfer. Moreover, compared with other similar networks, Cyclegan has also achieved better results.

## References

Du, R.M., Li, X.H., Liu, M.: Face cartoon style transfer based on improved CycleGAN. J. Changchun Univ. Technol. **45**(03), 272–276 (2024). https://doi.org/10.15923/j.cnki.cn22-1382/t.2024.3.11

Gatys, L.A., Ecker, A.S., Bethge, M.: A neural algorithm of artistic style. J. Vis. **16**(9), 326 (2016)

Isola, P., Zhu, J.-Y., Zhou, T., et al.: Image-to-image translation with conditional adversarial networks. In: Proceedings of the IEEE Conference on Computer Vision and Pattern Recognition, pp. 1125–1134 (2017)

Li, K.: Research and implementation of image style transfer based on deep learning, Doctoral dissertation. Xi'an Technological University (2025). https://doi.org/10.27391/d.cnki.gxagu.2025.000731

Liu, H., Li, M., Li, L., et al.: Image style transfer based on generative adversarial network. J. Nanjing Univ. Info. Sci. Technol. (Natural Science Edition) **15**(05), 514–523 (2023). https://doi.org/10.13878/j.cnki.jnuist.20221012003

Mirza, M., Osindero, S.: Conditional generative adversarial nets. Computer Science, 2672–2680 (2014)

Radford, A., Metz, L., Chintala, S.: Unsupervised representation learning with deep convolutional generative adversarial networks. Comp. Sci. **30**(1), 681–694 (2015)

Si, Z.Y.: Research on local image style transfer method based on deep learning, Master's thesis. Anhui University of Technology (2024). https://doi.org/10.26918/d.cnki.ghngc.2024.001291

Sun, T.P., Zhou, N.N., Huang, G.F.: A new local realistic feeling comic image style transfer based on GAN. Comput. Eng. Appl. **58**(14), 167–176 (2022)

Nagai, T., Kawashima, K.: Generation and transformation of facial images using deep learning. Proceedings of the Academic Lecture of the Institute of Precision Engineering **2018A**(0), 113–114 (2018). https://doi.org/10.11522/PSCJSPE.2018A.0_113

Wang, S.S., Li, W.S.: A review on arbitrary image style transfer based on neural network and attention. Software Engineering **28**(02), 27–31 (2025). https://doi.org/10.19644/j.cnki.issn2096-1472.2025.002.006

Wu, J.L., Yang, H.J.: Research on image style transfer technology based on improved CycleGAN network. Info. Technol. Informationiz. **02**, 79–82 (2025)

Zhang, Y.X., Dong, W.M., Xu, C.S.: Arbitrary image style transfer based on the exchange attention mechanism. Chinese J. Image and Graph. **30**(09), 3141–3152 (2025)

Zhu, J.-Y., Park, T., Isola, P., et al.: Unpaired image-to-image translation using cycle consistent adversarial networks. In: Proceedings of the IEEE International Conference on Computer Vision, pp. 2223–2232 (2017)

# Artificial Intelligence Technology Adoption on Enterprise Supply Chain Diversification and Supply Chain Resilience

Minjin Zhong[1], Jiahui Dong[2(✉)], Yibin Su[2], and Zhiming He[2]

[1] School of Management, Guangdong University of Technology, No. 161 of Yinglong Road, 510520 Guangzhou, People's Republic of China

[2] School of Business Administration, Guangdong University of Finance and Economics, Luntou Road. 21, 510320 Guangzhou, People's Republic of China

jiahuidong2000@126.com

**Abstract.** From the mediating perspective of supply chain diversification, which covers supplier diversification and logistics route diversification, this study explores the impact mechanism of artificial intelligence (AI) technology adoption on enterprise supply chain resilience. A total of 145 manufacturing enterprises in China were selected as the samples for empirical research, and the structural equation modeling (SEM) was used for empirical testing. The results indicate that (1) AI technology adoption takes a significant positive effect on enterprise supply chain resilience; (2) Supply chain diversification including supplier diversification and logistics route diversification takes a significant positive effect on enterprise supply chain resilience, and both mediates the relationship between AI technology adoption and enterprise supply chain resilience significantly.

**Keywords:** AI Technology Adoption · Supply Chain Diversification · Supply Chain Resilience

## 1 Introduction

Against the backdrop of an increasingly complex and volatile global economic environment, the diversification of enterprise supply chain configuration and supply chain resilience have become key strategic elements for enterprises to cope with uncertainty and ensure business continuity. Supply chain configuration diversification emphasizes the use of multi-source procurement, multi-regional layout and multi-mode transportation to disperse the risk of disruption that may be caused by a single supplier or a single channel (Choi and Krause, 2006). Supply chain resilience was defined as the ability of the supply chain to respond quickly, adapt and recover when faced with external shocks such as natural disasters, geopolitical conflicts and sudden changes in market demand (Ponomarov and Holcomb, 2009; Ivanov, 2022). Studies have shown that the diversification of supply chain structure can significantly enhance the resilience of enterprises to external disturbances, while a high level of supply chain resilience helps enterprises maintain


H. Yan et al. (Eds.): AIS&P 2025, CCIS 2918, pp. 537–549, 2026.
https://doi.org/10.1007/978-981-95-9884-7_37

operational stability and reduce economic losses during crises (Sheffi, 2007; Dubey et al., 2019). Therefore, in-depth exploration of how to improve supply chain diversification and resilience has great theoretical and practical significance for enterprises to build sustainable competitive advantages in a dynamic competitive environment.

In recent years, AI technology adoption has provided new possibilities for enterprises to optimize supply chain management. Depending on big data analysis, machine learning, intelligent prediction and autonomous decision-making, artificial intelligence technology has significantly improved the visibility (Casino et al., 2019), response speed and resource allocation efficiency (Ivanov et al., 2022). For example, AI can identify potential suppliers and assess their risks by analyzing multi-source data, helping enterprises build a more diversified and stable supply network; at the same time, intelligent algorithms can also simulate the performance of the supply chain under different interruption scenarios in real time, dynamically adjust inventory, logistics and production plans, thereby enhancing the overall adaptability and recovery capability of the system (Dubey et al., 2019; Kumar et al., 2020). It can be seen that the adoption of AI technology is likely to become a key enabling factor to promote the diversification and resilience of enterprise supply chain.

However, existing research on supply chain management is mostly focused on traditional perspectives, such as supply chain integration, partnership management, and lean supply chain (Flynn et al., 2010; Kristal et al., 2010), or only on the partial application of a certain technical tool (such as ERP, IoT), lacking a perspective from the overall adoption and deep integration of artificial intelligence technology to systematically explore how it simultaneously affects the diversification of supply chain structure and system resilience. Although a few studies have begun to focus on the positive role of digital technology in supply chain resilience (Ivanov, 2022; Winkelhaus and Grosse, 2020), the mechanism by which artificial intelligence reconstructs the diversified configuration of the supply chain and enhances resilience through data-driven and intelligent decision-making has not been fully explored.

Therefore, this study aims to systematically explore the impact of the artificial intelligence technology adoption on enterprise's supply chain diversification and supply chain resilience, so as to fill the existing theoretical gaps and provide theoretical basis and practical guidance for enterprise managers to promote the deep integration of AI technology and supply chain strategy.

## 2 Literatures Review and Research Hypotheses

The theoretical foundation of artificial intelligence technology was formed in the research of cybernetics, information theory and early neural networks in the 1940s and 1950s. In 1943, McCulloch and Pitts first proposed the artificial neural network model, laying the theoretical foundation of AI; in 1950, Turing published "Computers and Intelligence", proposing the famous "Turing Test", which established an operational test standard for artificial intelligence (French, 2000). With the improvement of computing power and the breakthrough of algorithms, artificial intelligence technology has developed into a set of technologies that can enhance the intelligence level of the system through data learning, pattern recognition and autonomous decision-making (Kelleher, 2019). In enterprise applications, artificial intelligence technology is manifested in the integrated application of machine learning, intelligent optimization and other technologies, enabling enterprises to achieve automation and intelligence from data insight to action (Brynjolfsson and Mcafee, 2017).

Artificial intelligence technology adoption enhances the resilience of enterprise supply chains through multiple mechanisms. First, predictive analysis based on machine learning can identify potential supply disruption risks in advance, enabling enterprises to have the ability to provide early warning and prevention. For example, by analyzing multi-source data (including weather, traffic, supplier financial status, etc.), AI systems can predict the vulnerability of specific supply chain nodes (Feng et al., 2020). Secondly, intelligent optimization algorithms can quickly generate and execute response strategies when disruptions occur. For example, when a supplier has a delivery problem, the AI system can calculate the optimal alternative in real time and dynamically adjust the logistics route and inventory allocation (Ivanov et al., 2022). In addition, AI technology can continuously improve the adaptability and recovery efficiency of the supply chain through continuous learning and iterative improvement. For example, reinforcement learning algorithms can learn autonomously in multiple disruption events and optimize emergency decision-making logic (Kumar et al., 2020) [8]. Thus, this study proposes the following hypothesis:

H1: AI technology adoption has a significant positive effect on enterprise supply chain resilience.

Yin and Ran (2022) pointed out that supply chain diversification refers to enterprises constructing a supply chain structure with multi-source procurement, multi-regional layout and multi-mode transportation to mitigate the disruption risk that may be caused by a single supplier or a single channel. This concept comprised two main dimensions, which were supplier diversification and logistics route diversification. Supplier diversification can reduce enterprises' dependence on a single supplier by making multiple supply sources (Choi and Krause, 2006), and meanwhile logistics route diversification can enhance the robustness of the logistics network by building up more alternative transportation channels (Sheffi, 2007). Under the context of the AI technology adoption, supply chain configuration diversification could help companies to cope with the risk of supply disruption more effectively, so that improving the whole resilience of the enterprise's' supply chain (Yin and Ran, 2022; Cai et al., 2024). Therefore, we propose the hypotheses as follow:

H2a: Supplier diversification takes positive effect on enterprise supply chain resilience significantly;

H2b: Logistics route diversification takes positive effect on enterprise supply chain resilience significantly.

Modgil et al. (2022) believed that the adoption of artificial intelligence technology can not only directly affect the supply chain resilience, but also can indirectly affect enterprise's supply chain resilience by promoting the diversification of supply chain configuration. More specifically, artificial intelligence technology adoption can analyze the historical data, risk points and global delivery status through intelligent supplier evaluation systems, providing more precise data support for enterprises to optimize their supplier portfolio (Belhadi et al., 2022); Meanwhile, AI-driven logistics optimization algorithms can help analyze multimodal transport data together with traffic conditions and tariff policies in real time, helping enterprises to build a more diversified and flexible logistics network (Ivanov et al., 2022). This intelligent supply chain configuration optimization significantly enhances the ability of enterprises to deal with supply disruptions, thereby improving the resilience of the supply chain. Therefore, this study proposes the following hypotheses:

H3a: Supplier diversification mediates the relationship between AI technology adoption and enterprise supply chain resilience significantly;

H3b: Logistics route diversification mediates the relationship between AI technology adoption and enterprise supply chain resilience significantly.

# 3 Research Design

## 3.1 Research Framework

According to the aforementioned literature review and hypotheses, the research framework for this study is posted in Fig. 1.

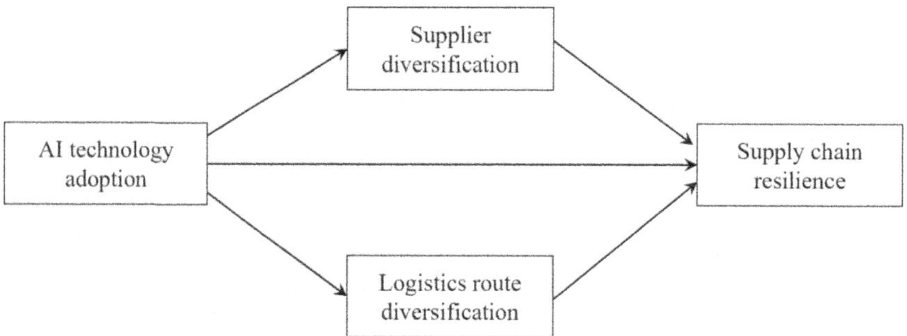


**Fig. 1.** Framework for this research

## 3.2 Measurement

In the present study, a five-point Likert Scale was employed to validate the questionnaires, which were developed on the basis of an initial literature review encompassing AI technology adoption (AIA), Supplier diversification (SD), Logistics route diversification (LRD), Supply chain resilience (SCR). The measuring scale of AI technology adoption was development by Yin et al. (2023) on the basis of Medcof (1996) which comprised 3 items with sample item "My company has incorporated artificial intelligence technology into its supply chain system". The measuring scale of supply chain diversification with two dimensions including supplier diversification and logistics route diversification was development by Lin et al. (2025) and Flynn et al. (2010) which comprised 5 items from each dimension. And the measuring scale of supply chain resilience was development by Ponomarov and Holcomb (2009) which comprised 3 items including supply chain agility, supply chain adaptability and supply chain alignment.

### 3.3 Sampling

Data for this study were collected from 145 manufacturing enterprises located in the Pearl River Delta region of China, with the research respondents being the leaders of these firms. A questionnaire survey was adopted for collecting datas, and with the support of local government officials, data were gathered through a combination of telephone communications, face-to-face interviews, and mailed questionnaires.

A total of 170 questionnaires were distributed, and over a two-month data collection period, 148 were retrieved. After screening, 145 valid questionnaires were obtained, corresponding to an effective response rate of 82.9%. In terms of industrial distribution, the sample consists of 21 enterprises in the food manufacturing industry, 17 in beverage manufacturing, 21 in textile manufacturing, 19 in furniture manufacturing, 21 in pharmaceutical manufacturing, 21 in general equipment manufacturing, 18 in electrical machinery and equipment manufacturing, and 7 in other manufacturing sectors.

In terms of the respondents presented in Table 1, among the 145 samples, there were 106 male firm leaders, accounting for 73.1%, and 39 females, accounting for 26.9%. Regarding age distribution, 17 respondents (11.7%) were aged 21–30, 45 respondents (31.0%) were in the 31–40 age group, 51 respondents (35.2%) were between 40–50 years old, and 32 respondents (22.1%) were over 50 years old. In terms of educational background, 75 respondents (51.7%) had a college degree or below, 59 respondents (40.7%) held a bachelor's degree, and 11 respondents (7.6%) had a master's degree or above. This suggests that the age of leaders in China's manufacturing sector is relatively high, while their educational level is relatively low, and the influx of new, highly educated personnel is slow.

**Table 1.** The characteristics of research samples

| Items | | Frequency (N = 145) | Percentage (%) | Cumulative percentage (%) |
|---|---|---|---|---|
| Sex | Male | 106 | 73.1 | 73.1 |
| | Female | 39 | 26.9 | 100.0 |
| Age | 21–30 | 17 | 11.7 | 11.7 |
| | 31–40 | 45 | 31.0 | 42.8 |
| | 40–50 | 51 | 35.2 | 77.9 |
| | > 50 | 32 | 22.1 | 100.0 |
| Education | College degree and below | 75 | 51.7 | 51.7 |
| | Bachelor's degree | 59 | 40.7 | 92.4 |
| | Master and above | 11 | 7.6 | 100.0 |

# 4 Data Analysis

## 4.1 Reliability and Validity Analysis

This study applied SPSS 23.0 to conduct alpha testing to ensure the reliability of measuring scales. The results are shown in Table 2.

**Table 2.** Cronbach's α value of measuring scales

| Variables | Items | Corrected Item-Total Correlation | Cronbach's α if item deleted | α value |
|---|---|---|---|---|
| AI technology adoption | AIA1 | 0.50 | 0.71 | 0.73 |
| | AIA 2 | 0.55 | 0.65 | |
| | AIA 3 | 0.61 | 0.57 | |
| Supplier diversification | SD1 | 0.52 | 0.75 | 0.78 |
| | SD 2 | 0.61 | 0.72 | |
| | SD 3 | 0.58 | 0.74 | |
| | SD4 | 0.51 | 0.76 | |
| | SD5 | 0.56 | 0.74 | |
| Logistics route diversification | LRD1 | 0.52 | 0.72 | 0.76 |
| | LRD 2 | 0.55 | 0.71 | |
| | LRD3 | 0.57 | 0.71 | |
| | LRD4 | 0.49 | 0.73 | |
| | LRD5 | 0.53 | 0.72 | |
| Supply chain resilience | SCR1 | 0.50 | 0.69 | 0.72 |
| | SCR2 | 0.53 | 0.66 | |
| | SCR3 | 0.61 | 0.56 | |

The Cronbach's α reliability coefficient for the AI Technology Adoption Scale was 0.73, exceeding the standard requirement of 0.70. The Corrected Item-Total Correlation values ranged from 0.55to 0.61, both exceeding 0.45. The Cronbach's α if item deleted values ranged from 0.57to 0.71 neither exceeding 0.73. This indicates that the AI Technology Application scale has very good reliability.

The Cronbach's α reliability coefficient for the Supplier Diversification Scale was 0.78, exceeding the standard requirement of 0.70. The Corrected Item-Total Correlation values ranged from 0.51 to 0.61, both exceeding 0.45. The Cronbach's α if item deleted values ranged from 0.72 to 0.76, neither exceeding 0.78. This indicates that the Supplier Diversification scale has very good reliability.

The Cronbach's α reliability coefficient for the Logistics Route Diversification Scale was 0.76, exceeding the standard requirement of 0.70. The Corrected Item-Total Correlation values ranged from 0.49 to 0.57, both exceeding 0.45. The Cronbach's α if item deleted values ranged from 0.71 to 0.73, neither exceeding 0.76. This indicates that the Logistics Route Diversification Scale has very good reliability.

The Cronbach's α reliability coefficient for the Supply Chain Resilience Scale was 0.72, exceeding the standard requirement of 0.70. The Corrected Item-Total Correlation values ranged from 0.50 to 0.61, both exceeding 0.45. The Cronbach's α if item deleted values ranged from 0.56 to 0.69, neither exceeding 0.72. This indicates that the Supply Chain Resilience Scale has very good reliability.

The validity test for AI technology adoption, Supplier diversification, Logistics route diversification, and Supply chain resilience is used CFA method with AMOS software to do testing. Results presented in Table 3 suggested that the four-factor model had the best fitted indexes with $\chi2(98) = 106.146$, $\chi2/df = 1.083$, $p < 0.01$; RMSEA $= 0.024 < 0.08$, TLI $= 0.983 > 0.90$, CFI $= 0.986 > 0.90$. It suggests that all measuring scales including AI technology adoption, Supplier diversification, Logistics route diversification, and Supply chain resilience have very good validity.

**Table 3.** CFA testing results for validity of measures

| Model | Factors | CFI | TLI | RMSEA | χ2/df |
|---|---|---|---|---|---|
| Four-factor | AIA, SD, LRD, SCR | 0.986 | 0.983 | 0.024 | 1.083 |
| Three-factor -1 | AIA, SD + LRD, SCR | 0.797 | 0.759 | 0.090 | 2.165 |
| Two-factor -1 | AIA + SD + LRD, SCR | 0.686 | 0.643 | 0.111 | 2.769 |
| Two -factor -2 | AIA, SD + LRD + SCR | 0.747 | 0.705 | 0.099 | 2.425 |
| One-factor | AIA + SD + LRD + SCR | 0.595 | 0.649 | 0.117 | 2.961 |

## 4.2 Test of Overall Theoretical Model

Followed by Zhou et al. (2025), structural equation modeling (SEM) was employed to test the proposed hypothesis previously. We used AMOS to make testing, and results of the indirect SEM fitness were presented in Table 4.

**Table 4.** The overall model fitness results for indirect SEM

| Overall model fitness | | Ideal criteria | Accepted criteria | Indirect SEM |
|---|---|---|---|---|
| Absolute fitness | GFI | ≥ 0.90 | ≥ 0.80 | 0.920 |
| | RMR | ≤ 0.05 | ≤ 0.08 | 0.060 |
| | RMSEA | ≤ 0.08 | ≤ 1.00 | 0.029 |
| Asymptotic fitness | AGFI | ≥ 0.90 | ≥ 0.80 | 0.890 |
| | RFI | ≥ 0.90 | ≥ 0.80 | 0.809 |
| | NFI | ≥ 0.90 | ≥ 0.80 | 0.842 |
| | CFI | ≥ 0.90 | ≥ 0.80 | 0.980 |

*(continued)*

**Table 4.** (*continued*)

| Overall model fitness | | Ideal criteria | Accepted criteria | Indirect SEM |
|---|---|---|---|---|
| Summarized fitness | $\chi$2/df | 1.00 ~ 2.00 | 2.00 ~ 3.00 | 1.118 |
| | PNFI | $\geq$ 0.50 | $\geq$ 0.50 | 0.695 |
| | PGFI | $\geq$ 0.50 | $\geq$ 0.50 | 0.669 |
| | AIC | <Satu. AIC (272.000) | <Indp. AIC (732.676) | 184.692 |

For the indirect model, in terms of absolute fitness indicators: GFI = 0.920 is greater than 0.90, RMR = 0.060 is less than 0.05, and RMSEA = 0.029 is less than 0.08, all of which meet the indicator requirements; in terms of asymptotic fitness indicators: AGFI = 0.890 is greater than 0.80, RFI = 0.809 is greater than 0.80, NFI = 0.842 is greater than 0.90, and CFI = 0.980 is greater than 0.90, all of which meet the accepted criteria requirements; in terms of summarized fitness indicators: $\chi$2/df = 1.118 is between 1 and 3, PNFI = 0.695 is greater than 0.50, PGFI = 0.669 is greater than 0.50, and AIC = 184.692 is less than Indp. AIC (732.676), all of which meet the accepted criteria requirements.

### 4.3 Test of Hypothesis

The indirect SEM model test results summarized in Table 5. It showed that, H1 "AI technology adoption has a significant positive effect on enterprise supply chain resilience", H2a "Supplier diversification has a significant positive effect on enterprise sup-ply chain resilience", H2b "Logistics route diversification has a significant positive effect on enterprise supply chain resilience" were all supported.

**Table 5.** Hypothesis test results with path coefficient

| Paths | Estimate | P-value | Hyp | Result |
|---|---|---|---|---|
| AIA → SCR | 0.260* | 0.038 | H1 | √ |
| AIA → SD | 0.412** | 0.000 | / | / |
| AIA → LRD | 0.362** | 0.003 | / | / |
| SD → SCR | 0.420** | 0.000 | H2a | √ |
| LRD → SCR | 0.298** | 0.009 | H2b | √ |

Notes: * Significant at $P < 0.05$; ** Significant at $P < 0.01$;

Meanwhile, the indirect SEM model with path coefficient was shown in Fig. 2.

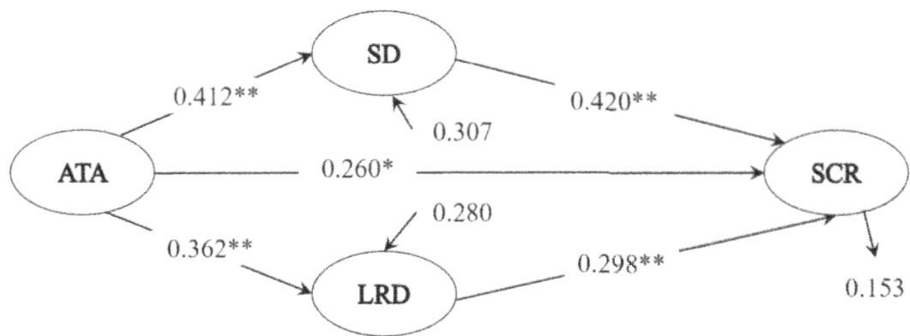


**Fig. 2.** The indirect SEM model with path coefficient

Followed by Zhou et al. (2023) and Zhou et al. (2024), we used Bootstrapping methods recommended by Hayes (2013) with SPSS Process to test the mediating effect of H3a and H3b which were proposed as supplier diversification and logistics route diversification mediate the relationship between AI technology adoption and enterprise supply chain resilience significantly. Bootstrapping results in Table 6 suggested that the mediating effect of supplier diversification between AI technology adoption and enterprise supply chain resilience was 0.102 with 95% confident interval of [0.067, 0.286], and the mediating effect of logistics route diversification between AI technology adoption and enterprise supply chain resilience was 0.063 with 95% confident interval of [0.006, 0.148]. As 95% confident interval of [0.067, 0.286] and [0.006, 0.148] included no zero, it suggested that the mediating effect of supplier diversification impacting AI technology adoption on enterprise supply chain resilience together with mediating effect of logistics route diversification impacting AI technology adoption on enterprise supply chain resilience were significant. Therefore, H3a and H3b were supported. That was, supplier diversification mediated the relationship between AI technology adoption and enterprise supply chain resilience significantly, and logistics route diversification mediated the relationship between AI technology adoption and enterprise supply chain resilience significantly.

**Table 6.** Bootstrap testing results for the mediating effect of supply chain diversification

| Indirect path | Mediating effect | 95% confident interval |
|---|---|---|
| AIA → SD → SCR | 0.102 | [0.067, 0.286] |
| AIA → LRD → SCR | 0.063 | [0.006, 0.148] |

## 5 Conclusions and Discussion

This study constructs a model of the interaction mechanism between the AI technology adoption, supply chain diversification, and enterprise supply chain resilience through systematic theoretical analysis and empirical research, and verifies it using 145 Chinese manufacturing enterprises as empirical subjects. This study found that: (1) AI technology adoption took positive impact on enterprise supply chain resilience significantly; (2) Supply chain diversification together with supplier diversification and logistics route diversification both mediated the relationship between AI technology adoption and supply chain resilience significantly. This conclusions for the study contributed great implications for the development AI technology adoption and supply chain management theory.

First, the adoption of AI technology is a key prerequisite for enhancing the resilience of enterprise supply chains. This empirical study found that AI technology, through big data collection, analysis, and decision-making simulation, significantly enhances the resilience, adaptability, and supply recovery capabilities of the supply chain. This finding greatly expands the technical perspective on the antecedents of supply chain resilience research and fills the gap in previous studies regarding the insufficient exploration of the mechanisms of AI application. Therefore, relevant supply chain enterprises should fully recognize the importance of AI technology, actively promote the deep integration of AI technology with the supply chain, establish dedicated AI skills management teams, and systematically advance the application of AI technology in key areas such as supply chain panoramic data scanning, supply chain forecasting, real-time supply chain monitoring, supplier evaluation, vehicle safety management, and logistics optimization to comprehensively improve supply chain resilience and customer responsiveness.

Second, supply chain diversification (encompassing supplier diversification and logistics route diversification) plays a significant mediating role between the adoption of artificial intelligence (AI) technology and supply chain resilience. This study further reveals that AI technology drives supply chain resilience enhancement through a dual path: at the supplier level, by leveraging machine learning algorithms and big data analytics to construct an intelligent supplier evaluation system, it enables precise screening and dynamic management of potential partners from multiple dimensions, including delivery stability, quality compliance, and risk resistance capabilities, thereby promoting the diversification of supplier sources, regional distribution, and cooperation types; at the logistics level, through real-time road condition monitoring, demand forecasting models, and dynamic optimization algorithms, it intelligently plans transportation routes and modes (such as the combination of land, sea, and air transport), effectively promoting flexible substitution and diversified complementarity of logistics routes, ultimately forming a transmission mechanism of technology empowerment - structural optimization - resilience enhancement. This finding not only clarifies the inherent logic of AI technology reconstructing the supply chain network structure from a micro-operational mechanism perspective, but also provides concrete support for the classic theoretical framework of "technology-structure-capability" in the field of supply chain management through empirical verification, echoing previous research conclusions on dynamic capabilities and technological innovation.

Third, enterprises should focus on technology empowerment to build an efficient and collaborative intelligent supply chain ecosystem. This study proposes from an ecosystem-centric perspective that artificial intelligence technology can not only deeply optimize core aspects of an enterprise's internal supply chain, such as inventory management, production scheduling, and order processing, achieving process restructuring and efficiency improvement through algorithm iteration, but also break down organizational boundaries. By integrating technologies such as blockchain and the Internet of Things, it can enhance information exchange efficiency and resource integration capabilities with external partners, ultimately building a supply chain innovation ecosystem that combines resilience and innovative vitality. From a practical perspective, enterprises should proactively build cross-entity data sharing platforms to promote real-time sharing and business collaboration of key information such as supplier capacity data, logistics service provider transportation dynamics, and technology partner R&D progress. By relying on a data platform to break down information silos, an end-to-end visualized supply chain network can be created, from procurement and sourcing to manufacturing and final delivery, enabling full-link status tracking and anomaly warning. Simultaneously, enterprises should break down technological barriers, open up the basic capabilities of their AI technology platforms, and collaborate with ecosystem partners to jointly build market demand forecasting models based on real-time data, dynamic early warning systems for supply chain disruption risks, and intelligent emergency response generation systems for unforeseen circumstances. Through complementary capabilities, a collaborative response mechanism of "joint risk prevention and shared benefits" can be formed, enhancing the collective resilience of the ecosystem. Furthermore, enterprises need to strengthen the unification of underlying standards in ecosystem construction, focusing on promoting the standardization of data interfaces, business processes, and technology applications, as well as the development of system interoperability. This ensures that information systems and technological tools of different entities can operate efficiently and collaboratively, promoting seamless integration and efficient linkage across planning, procurement, production, logistics, and sales, thereby achieving a continuous improvement in the overall resilience and operational efficiency of the supply chain.

**Acknowledgments.** Financial support for this study was provided by the National Statistical Science Project, entitled "Research on the Input-Output Efficiency of Scientific and Technological Innovation" (Grant No. 2024LY100).

## References

Belhadi, A., Kamble, S., Fosso Wamba, S., et al.: Building supply-chain resilience: an artificial intelligence-based technique and decision-making framework. Int. J. Prod. Res. **60**(14), 4487–4507 (2022). https://doi.org/10.1080/00207543.2021.1950935

Brynjolfsson, E., McAfee, A.: The business of artificial intelligence. Harvard Business Review **7**(1), 1–2 (2017). https://hbr.org/2017/07/the-business-of-artificial-intelligence

Cai, J., Sharkawi, I., Taasim, S.I.: How does digital transformation promote supply chain diversification? From the perspective of supply chain transaction costs. Financ. Res. Lett. **63**, 105399 (2024). https://doi.org/10.1016/j.frl.2024.105399

Casino, F., Kanakaris, V., Dasaklis, T.K., et al.: Modeling food supply chain traceability based on blockchain technology. IFAC-PapersOnLine **52**(13), 2728–2733 (2019). https://doi.org/10.1016/j.ifacol.2019.11.620

Choi, T.Y., Krause, D.R.: The supply base and its complexity: Implications for transaction costs, risks, responsiveness, and innovation. J. Oper. Manag. **24**(5), 637–652 (2006). https://doi.org/10.1016/j.jom.2005.07.002

Dubey, R., Gunasekaran, A., Childe, S.J., et al.: Big data and predictive analytics and manufacturing performance: integrating institutional theory, resource-based view and big data culture. Br. J. Manag. **30**(2), 341–361 (2019). https://doi.org/10.1111/1467-8551.12355

Feng, H., Wang, X., Duan, Y., et al.: Applying blockchain technology to improve agri-food traceability: a review of development methods, benefits and challenges. J. Clean. Prod. **260**, 121031 (2020). https://doi.org/10.1016/j.jclepro.2020.121031

Flynn, B.B., Huo, B., Zhao, X.: The impact of supply chain integration on performance: a contingency and configuration approach. J. Oper. Manag. **28**(1), 58–71 (2010). https://doi.org/10.1016/j.jom.2009.06.001

French, R.M.: The Turing Test: the first 50 years. Trends Cogn. Sci. **4**(3), 115–122 (2000). https://doi.org/10.1016/S1364-6613(00)01453-4

Hayes, A.F.: An introduction to mediation, moderation and conditional process analysis: A regression-based approach. Guilford Press (2013)

Ivanov, D.: Viable supply chain model: integrating agility, resilience and sustainability perspectives—lessons from and thinking beyond the COVID-19 pandemic. Ann. Oper. Res. **319**(1), 1411–1431 (2022). https://doi.org/10.1007/s10479-020-03640-6

Ivanov, D., Dolgui, A., Sokolov, B.: Cloud supply chain: Integrating Industry 4.0 and digital platforms in the "Supply Chain-as-a-Service". Trans. Res. Part E: Logist. Trans. Rev. **160**, 102676 (2022). https://doi.org/10.1016/j.tre.2022.102676

Kelleher, J.D.: Deep learning. MIT Press (2019)

Kristal, M.M., Huang, X., Roth, A.V.: The effect of an ambidextrous supply chain strategy on combinative competitive capabilities and business performance. J. Oper. Manag. **28**(5), 415–429 (2010). https://doi.org/10.1016/j.jom.2009.12.002

Kumar, A., Liu, R., Shan, Z.: Is blockchain a silver bullet for supply chain management? Technical challenges and research opportunities. Decis. Sci. **51**(1), 8–37 (2020). https://doi.org/10.1111/deci.12396

Lin, T., Lai, Y., Huang, J.: Institutional opening-up and supplier diversification: Evidence from China's pilot free trade zones. J. Shanghai Univ. Int. Bus. Econ. **32**(4), 53–68 (2025). https://xuebao.suibe.edu.cn/shdwjmdxxb/article/abstract/20250404

Medcof, J.W.: The job characteristics of computing and non-computing work activities. J. Occup. Organ. Psychol. **69**(2), 199–212 (1996). https://doi.org/10.1111/j.2044-8325.1996.tb00610.x

Modgil, S., Singh, R.K., Hannibal, C.: Artificial intelligence for supply chain resilience: learning from Covid-19. The Int. J. Logist. Manage. **33**(4), 1246–1268 (2022). https://doi.org/10.1108/IJLM-02-2021-0094

Ponomarov, S.Y., Holcomb, M.C.: Understanding the concept of supply chain resilience. The Int. J. Logist. Manage. **20**(1), 124–143 (2009). https://doi.org/10.1108/09574090910954873

Sheffi, Y.: The resilient enterprise: overcoming vulnerability for competitive advantage. MIT Press (2007). https://EconPapers.repec.org/RePEc:mtp:titles:0262693496

Winkelhaus, S., Grosse, E.H.: Logistics 4.0: a systematic review towards a new logistics system. Int. J. Prod. Res. **58**(1), 18–43 (2020). https://doi.org/10.1080/00207543.2019.1612964

Yin, L.Y., Zhu, S.Q., Zhou, J.M.: AI technology adoption, AI knowledge integration and AI product development performance. Comm. Comp. Info. Sci. **1844**(1), 351–359 (2023). https://doi.org/10.1007/978-981-99-4402-6_25

Yin, W., Ran, W.: Supply chain diversification, digital transformation, and supply chain resilience: Configuration analysis based on FSQCA. Sustainability **14**(13), 7690 (2022). https://doi.org/10.3390/su14137690

Zhou, J., He, P., Su, M.: Will knowledge field activity within technological innovation alliances lead to knowledge leakage: The mediating role of knowledge flow. Int. J. Innov. Manag. **29**(3–4), 2550013 (2025). https://doi.org/10.1142/S1363919625500136

Zhou, J., He, P., Zhu, S., Yin, L.: Can empowering leadership increase new product development performance in China: The destructive role of error aversion culture. SAGE Open **14**(4), 1–14 (2024). https://doi.org/10.1177/21582440241300345

Zhou, J., Yin, L., Zhu, S.: Effect of knowledge leadership on knowledge innovation behavior of knowledge worker: The role of failure learning and fear of failure. Int. J. Innov. Manag. **27**(9–10), 2350047 (2023). https://doi.org/10.1142/S1363919623500470

# Arc Light Filtering Method for Welding Images Based on a Conditional Diffusion Model

Zhi Wang[1], Deng Chen[1](✉), Lianqing Wang[2,3], Dawei Wen[1], Xiaohua Liu[1], Xin Nie[1], and Xiaoquan Liu[1]

[1] School of Computer Science and Engineering Artificial Intelligence, Wuhan Institute of Technology, Wuhan, China
dchen@wit.edu.cn

[2] College of Intelligent Science, National University of Defense Technology, Changsha 410073, China

[3] Command and Control Support Department, Information Support Force Engineering University, Wuhan 430030, China

**Abstract.** Industrial robots are widely used in automated welding, where vision-based monitoring is crucial for stable and safe operation. However, the intense welding arc light severely interferes with visual systems, compromising the safety and stability of the process. Traditional filtering methods and Generative Adversarial Networks (GANs) struggle with arc light due to complex morphology, loss of texture detail, and training instability. This paper proposes an arc light filtering method for welding images based on a Conditional Diffusion Model. The method introduces a region mask fusion strategy and employs a novel two-stage restoration process—first, repairing the halo region, and then, reconstructing the saturation region. This approach successfully addresses the challenges of structure reconstruction and texture detail loss during arc light filtering. Quantitative evaluation on a self-constructed welding image dataset demonstrates that the proposed method significantly outperforms existing mainstream methods, including Pix2Pix, UNet++P2P, and DifferP2P, across all key metrics. The method achieves a Peak Signal-to-Noise Ratio (PSNR) of 39.32 dB, yielding an improvement of +1.25 dB compared to DifferP2P (38.07 dB). The Structural Similarity Index (SSIM) and LPIPS metrics reach 0.976 and 0.0102, respectively, outperforming all baselines.

**Keywords:** Arc Light Filtering · Diffusion Model · Industrial Robot · Welding Vision · Image Restoration

## 1 Introduction

Industrial robots are essential for automation in high-end fields like aerospace (Bu et al., 2024; Chea et al., 2020). However, visual monitoring systems—critical for


H. Yan et al. (Eds.): AIS&P 2025, CCIS 2918, pp. 550–559, 2026.
https://doi.org/10.1007/978-981-95-9884-7_38

collision warning and safety zone protection—face a critical bottleneck: intense arc light causes widespread sensor oversaturation, resulting in severely distorted footage that fundamentally disables safety algorithms. High-fidelity restoration of this obscured visual information is the most crucial step for industrial robot safety monitoring.

Traditional image processing techniques (mean, Gaussian, median filtering) were early attempts to mitigate arc light interference. Yet, they fail when faced with the arc light's variable morphology (Shen et al., 2025) and extreme dynamic range, often leading to detail loss and artifacts, as they lack semantic understanding and restoration capability. Deep learning solutions like GANs (Dhariwal & Nichol, 2021; Goodfellow et al., 2020) showed promise, but their adversarial training mechanism suffers from inherent challenges such as unstable training processes and susceptibility to mode collapse, compromising the reliability required for industrial applications.

Diffusion Models (DMs) offer a superior alternative due to their stable training and excellent texture/structure reconstruction capabilities. While standard DMs incur high computational overhead due to thousands of iterative sampling steps, the flexible framework allows for optimization, such as introducing the DDIM (Song et al., 2020) sampling strategy to accelerate inference and enhance application feasibility. Our work addresses the challenge of applying DMs to the specialized industrial scene of welding, which involves extreme oversaturation (saturation region) (Cao et al., 2020) and light scattering (halo interference region) (Ma et al., 2025).

The Main Contributions of this paper are:

1. We propose an Arc Light Filtering Method based on a Diffusion Model, incorporating an arc light region mask extraction module to explicitly localize corrupted areas.
2. We design a novel two-stage image restoration strategy: The first stage suppresses brightness interference in the background (halo region), and the second stage restores structural information loss in the weld pool (saturation region).
3. We introduce the DDIM acceleration mechanism into the sampling process, which significantly improves computational efficiency and practical value without sacrificing image quality.

# 2 Methods

## 2.1 Overall Methodology

We propose a two-stage, refined arc light filtering framework to address complex image degradation caused by arc light and dense smoke. The task is deconstructed into three parts: Arc Light Image Pre-processing, Halo Region Restoration, and Saturation Region Reconstruction. These parts work collaboratively to maintain global lighting consistency while accurately recovering structure and texture. The structural diagram is shown in Fig. 1.

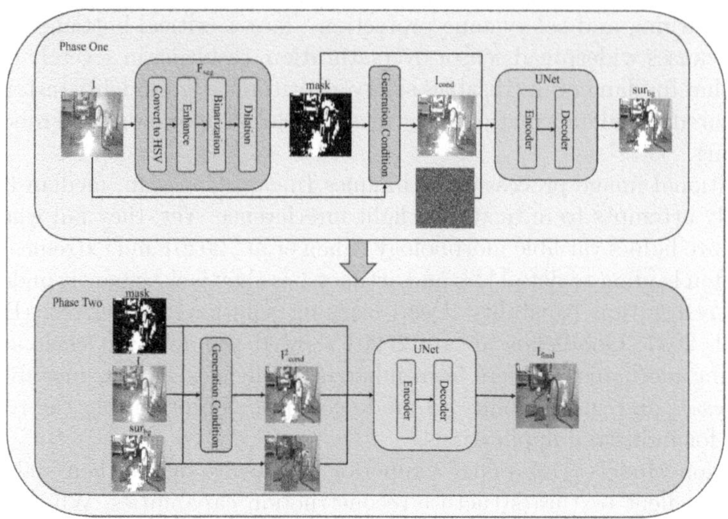


**Fig. 1.** Overall Methodology

## 2.2 Arc Light Image Pre-processing

To handle the challenging characteristics of arc light (extreme brightness, irregular shape), we designed a mask generation pre-processing module, $F_{\text{seg}}$. This module explicitly localizes the corrupted arc light region and generates a binary Mask. This Mask serves as a strong spatial prior to guide the Diffusion Model, focusing its attention on generating missing content rather than modeling arc light physics. The module uses brightness channel analysis and morphological optimization.

The structural diagram is shown in Fig. 2. The process involves:

Step 1: Brightness Channel Extraction and Enhancement. To maximize the discriminability between the arc light and the background and to mitigate the interference from color information, we first convert the input RGB image $\mathbf{I}$ to the HSV color space. We exclusively extract the Value ($V$) channel, as it best reflects the intensity information of the arc light. The conversion process is represented as:

$$V = \max(R, G, B) \tag{1}$$

To further enhance the local contrast between the arc light region and the surrounding darker areas, we apply the Adaptive Histogram Equalization (CLAHE) algorithm to the brightness channel $V$, yielding the enhanced brightness map $V_{\text{enh}}$:

$$V_{\text{enh}} = F_{\text{CLAHE}}(V) \tag{2}$$

Step 2: Threshold Segmentation. We perform binarization on the enhanced brightness map $V_{\text{enh}}$ to primarily segment the high-intensity arc light region. This process is realized using a custom threshold $\tau$, generating a raw binary

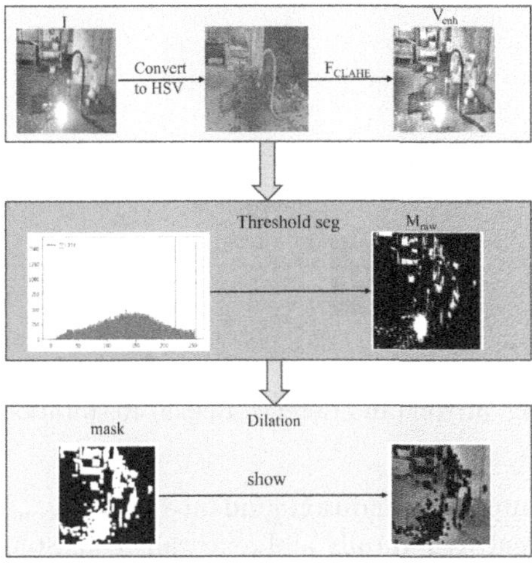


**Fig. 2.** Pre-processing

mask $\mathbf{M}_{\text{raw}}$. For an arbitrary pixel location $p$ in the image, the operation is defined as follows:

$$\mathbf{M}_{\text{raw}}(p) = \begin{cases} 1, & \text{if } V_{\text{enh}}(p) \geq \tau \\ 0, & \text{otherwise} \end{cases} \tag{3}$$

Here, the regions with a value of 1 correspond to the arc light, and regions with a value of 0 correspond to the background.

Step 3: Mask Optimization. The raw mask $\mathbf{M}_{\text{raw}}$ obtained directly from threshold segmentation may contain internal holes or exhibit unsmooth edges. To generate a connected and complete mask, we employ a morphological Dilation operation for post-processing. This operation processes $\mathbf{M}_{\text{raw}}$ using a predefined structuring element $\mathbf{K}$ (e.g., an $n \times n$ square kernel). Its rigorous mathematical definition is:

$$\text{Mask} = \mathbf{M}_{\text{raw}} \oplus \mathbf{K} = \{p + k | p \in \mathbf{M}_{\text{raw}}, k \in \mathbf{K}\} \tag{4}$$

where $\oplus$ denotes morphological dilation, $p$ is the pixel coordinate in $\mathbf{M}_{\text{raw}}$, and $k$ is a coordinate within the structuring element $\mathbf{K}$. This operation effectively fills internal gaps and expands the boundaries, ultimately yielding the optimized mask Mask.

### 2.3 Restoration of the Halo Interference Region

Arc light induces halo interference in the image background, causing blurring and information loss. This module uses a conditional (Ho & Salimans, 2022) U-Net noise predictor, $\epsilon_{\theta_{\text{bg}}}$, to reconstruct the brightness distribution and restore lighting equilibrium in the surrounding scene. The structural diagram is shown in Fig. 3.

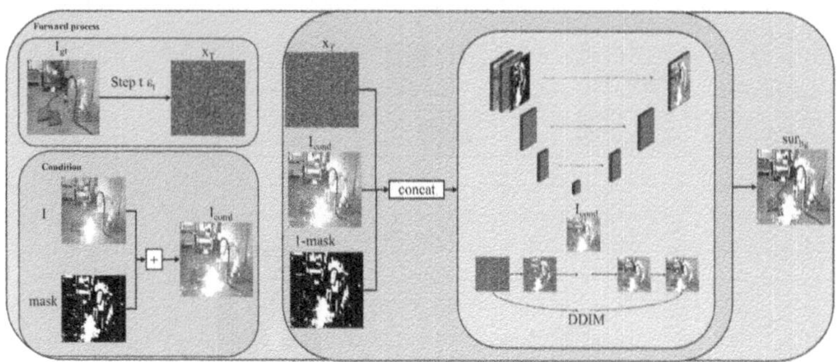


**Fig. 3.** Halo Interference Region Restoration

The key is the composite condition guidance image $\mathbf{I}_{\text{cond}}$, which ensures the model generates content structurally and texturally consistent with the real environment. $\mathbf{I}_{\text{cond}}$ fuses the original image's preserved background pixels with a pure white filling in the core masked region:

$$\mathbf{I}_{\text{cond}} = \mathbf{I} \odot (1 - \text{Mask}) + \text{Mask} \tag{5}$$

Here, $1 - \text{Mask}$ represents the inverted mask, $\mathbf{I} \odot (1 - \text{Mask})$ preserves the true pixels of all background regions, while the $+\text{Mask}$ operation fills the core region requiring restoration with pure white. This ensures that the model, during denoising, must generate content that is structurally and texturally consistent with the surrounding real environment. In each denoising step $t$, the condition $\mathbf{I}_{\text{cond}}$, $x_t$, and $1 - \text{Mask}$ are concatenated along the channel dimension to form a unified input:

$$\epsilon_{\text{pred}} = \epsilon_{\theta_{\text{bg}}}(\text{cat}([x_t, \mathbf{I}_{\text{cond}}, (1 - \text{mask})], \text{dim} = 1), t) \tag{6}$$

The entire non-masked restoration process is realized by a diffusion sampling function conditioned on $\mathbf{I}_{\text{cond}}$ and $1 - \text{Mask}$, denoted as $D_{\theta_{\text{bg}}}$, ultimately yielding the background-restored image $\mathbf{I}_{\text{surbg}}$:

$$\mathbf{I}_{\text{surbg}} = D_{\theta_{\text{bg}}}(z, \mathbf{I}_{\text{cond}}, (1 - \text{mask})) \tag{7}$$

where $z \sim \mathcal{N}(0, \mathbf{I})$ is the initial sampled random noise. To ensure that this module focuses primarily on background restoration, its training loss function $\mathcal{L}_{\text{bg}}$ is also weighted by the inverted mask $(1 - \text{Mask})$, calculating the error only for the background region $\mathbf{I}_{\text{surbg}}$ and the true arc-light-free image $\mathbf{I}_{\text{GT}}$:

$$\mathcal{L}_{\text{bg}} = \mathbb{E}[||(\mathbf{I}_{\text{surbg}} - \mathbf{I}_{\text{GT}}) \odot (1 - \text{mask})||] \tag{8}$$

This achieves precise and controllable restoration.

### 2.4 Reconstruction of the Saturation Region

This module focuses on the core problem: reconstructing the completely missing structural information within the saturation region (weld pool). It utilizes the stable contextual information from the previously repaired halo region to semantically reconstruct the key structural details. The structural diagram is shown in Fig. 4.

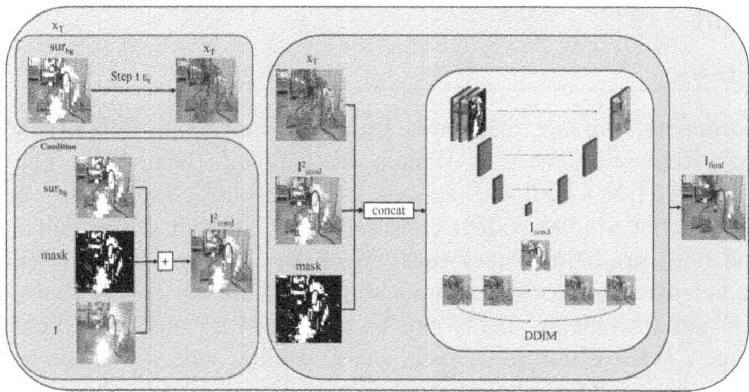


**Fig. 4.** Saturation Region Reconstruction

The current noisy input $x_t$ for this stage is calculated using the output of the previous stage $\mathbf{I}_{\text{surbg}}$ and Gaussian noise $z$ applied only within the mask:

$$x_t = \mathbf{I}_{\text{surbg}} \odot (1 - \text{mask}) + z \odot \text{mask} \tag{9}$$

The composite condition guidance image $\mathbf{I}^2_{\text{cond}}$ fuses the restored background $\mathbf{I}_{\text{surbg}}$ and the arc light region from the original image $\mathbf{I}$:

$$\mathbf{I}^2_{\text{cond}} = \mathbf{I}_{\text{surbg}} \odot (1 - \text{mask}) + \mathbf{I} \odot \text{mask} \tag{10}$$

This image utilizes $\mathbf{I}_{\text{surbg}}$ to provide a peripherally consistent lighting environment while retaining potential structural contour information present in the original arc light region $\mathbf{I} \odot$ Mask, offering richer generation cues to the model. In each denoising step $t$, the current state $x_t$, the two condition images $\mathbf{I}^2_{\text{cond}}$, and Mask are concatenated and input into the noise predictor:

$$\epsilon_{\text{pred}} = \epsilon_{\theta_{\text{center}}}(\text{cat}([x_t, \mathbf{I}^2_{\text{cond}}, \text{mask}], \text{dim} = 1), t) \tag{11}$$

The entire process of repairing the arc light image masked region is performed by a dedicated diffusion sampling function $D_{\theta_{\text{center}}}$, which ultimately outputs the high-fidelity restored result of the entire image $\mathbf{I}_{\text{final}}$:

$$\mathbf{I}_{\text{final}} = D_{\theta_{\text{center}}}(z, \mathbf{I}^2_{\text{cond}}, \text{mask}) \tag{12}$$

To ensure the model focuses on the generation quality of the core area, its training loss function $\mathcal{L}_{\text{center}}$ is also weighted by the mask, penalizing the restoration error only in the core region:

$$\mathcal{L}_{\text{center}} = \mathbb{E}[||(\mathbf{I}_{\text{final}} - \mathbf{I}_{\text{GT}}) \odot \text{mask}||] \tag{13}$$

Through this carefully designed cascaded pipeline, our method is finally able to generate restored images that exhibit a high degree of realism and structural integrity in both the background and the core areas.

# 3 Result

## 3.1 Dataset

The experimental dataset originates from the Industrial Robot Engineering Technology Research Center at Wuhan Institute of Technology. The dataset utilized the FAIRINO FR3-439 six-axis welding robot. The dataset comprises three sets of robot welding videos captured from different angles, with each set containing both arc-light-free motion videos and arc-light welding videos. By including videos with identical actions and durations, we ensure that an image pair (with and without arc light) can be generated for any given video frame. Subsequently, every frame from each set of welding videos was converted and stored as an image. The original resolution of the images in the arc light pair dataset was $3840 \times 2160$. Due to limitations in experimental equipment, the input size was uniformly downsampled to $256 \times 256$, resulting in a final paired dataset (with/without arc light) comprising 800 pairs for training and 200 pairs for testing. Each pair in the dataset consists of an image with arc light and a corresponding image without arc light, differing only by the presence or absence of the arc light. Fig. 5 provides some examples from the arc light image dataset.

## 3.2 Qualitative Analysis

Qualitative evaluation confirms successful arc light restoration by comparing our model with the Pix2Pix, U-Net++P2P and state-of-the-art DifferP2P (Zhao et al., 2023) model. As shown in Fig. 6, the subsequent rows display the overall restoration results, followed by the magnified view of the central weld pool area.

The two-stage strategy of the proposed model—repairing the background first, then reconstructing the center—successfully resolves issues like color spots and structural distortion left by baseline models near the welding torch. While the overall quality is higher, there is still insufficient texture clarity observed at the tip of the welding torch.

## 3.3 Quantitative Analysis

Quantitative evaluation uses four full-reference Image Quality Assessment Metrics (IQAMs): SSIM, PSNR, FID, and LPIPS. Higher SSIM and PSNR scores indicate better quality, while lower FID and LPIPS scores indicate better quality.Detailed results are shown in Table 1.

As shown in Table 1, the proposed model demonstrates superior performance. The background restoration stage provides stable context, leading to a significant advantage in SSIM (0.9764) and PSNR (39.3220 dB), confirming the design's

**Fig. 5.** Dataset

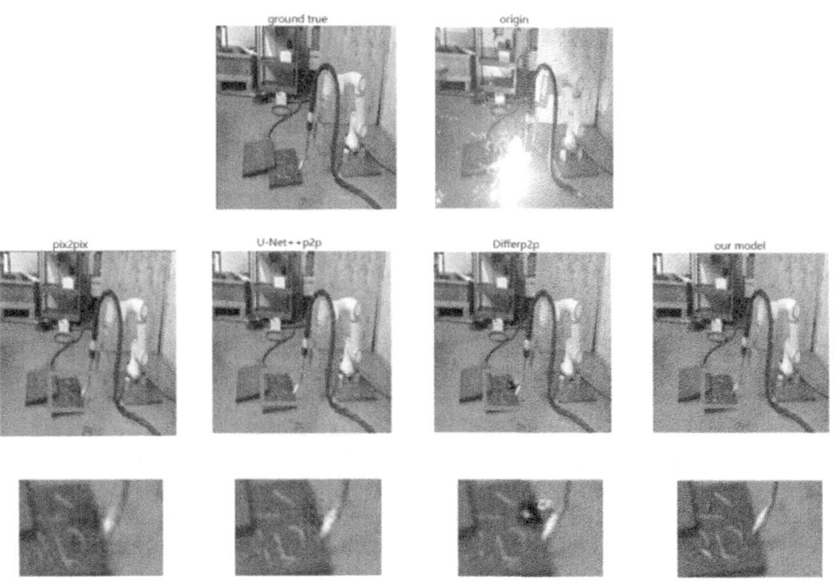


**Fig. 6.** Experimental Results

effectiveness in structural integrity. The center reconstruction stage generates more realistic texture details, resulting in the lowest (best) LPIPS score of 0.0102. Although our FID score (2.8657) is marginally higher than DifferP2P's (2.7907), our model is highly competitive in the overall distribution of generated images.

**Table 1.** Image Quality Assessment Metrics

| Model | SSIM | PSNR | FID | LPIPS |
|---|---|---|---|---|
| Pix2Pix | 0.8362 | 28.5029 | 92.1637 | 0.0972 |
| U-Net++P2P | 0.9723 | 36.6797 | 8.3378 | 0.0136 |
| DifferP2P | 0.9717 | 38.0667 | **2.7907** | 0.0116 |
| Proposed | **0.9764** | **39.3220** | 2.8657 | **0.0102** |

## 4 Conclusion

We proposed an Arc Light Filtering Method for Welding Images Based on a Diffusion Model (ALF-DM). The model utilizes a three-part cascaded process: pre-processing (mask extraction), background/halo restoration, and weld pool/saturation restoration. Our approach demonstrated superior performance in arc light filtering compared to state-of-the-art models in qualitative and quantitative evaluations.

A limitation exists due to computational constraints: downsampling the original high-definition video ($3840 \times 2160$) to $256 \times 256$ for training led to the loss of some high-frequency texture details, resulting in blurriness at the welding torch tip.

## References

Bu, H., Cui, X., Huang, B., Peng, S., Wan, J.: Research review and future directions of key technologies for welding robots in the construction industry. Buildings **14**(8), 2261 (2024). https://doi.org/10.3390/buildings14082261

Cao, Y., Ren, Y., Li, T.H., Li, G.: Over-exposure correction via exposure and scene information disentanglement. In: Proceedings of the Asian Conference on Computer Vision (2020). https://doi.org/10.1007/978-3-030-69538-5_25

Chea, C.P., Bai, Y., Pan, X., Arashpour, M., Xie, Y.: An integrated review of automation and robotic technologies for structural prefabrication and construction. Transp. Saf. Environ. **2**(2), 81–96 (2020). https://doi.org/10.1093/tse/tdaa007

Dhariwal, P., Nichol, A.: Diffusion models beat gans on image synthesis. Advances in neural information processing systems, **34**, 8780–8794 (2021). https://doi.org/10.48550/arXiv.2105.05233

Goodfellow, I., et al.: Generative adversarial networks. Commun. ACM **63**(11), 139–144 (2020). https://doi.org/10.48550/arXiv.1406.2661

Ho, J., Salimans, T.: Classifier-free diffusion guidance. arXiv preprint arXiv:2207.12598 (2022). https://doi.org/10.48550/arXiv.2207.12598

Ma, M., et al.: A two-stage welding control system combining rule-based and data-driven vision. Symmetry, **17**(4), 611 (2025). https://doi.org/10.3390/sym17040611

Shen, J., Qi, D., Lai, R., Du, S.: Numerical research on the influence of welding gap on arc morphology in mag welding of ah36 steel. Front. Mater. *12*, 1634466 (2025). https://doi.org/10.1016/j.ijmecsci.2019.105031

Song, J., Meng, C., Ermon, S.: Denoising diffusion implicit models. arXiv preprint arXiv:2010.02502 (2020). https://doi.org/10.48550/arXiv.2010.02502

Zhao, X., Yu, H., Bian, H.: Image to image translation based on differential image pix2pix model. Comput. Mater. Continua **77**(1) (2023). https://doi.org/10.32604/cmc.2023.041479

# LRPE: A Lightweight Robot Pose Estimation Network

Wei Ouyang, Deng Chen(✉), Dawei Wen, Xiaohua Liu, Xin Nie, and Xiaoquan Liu

School of Computer Science and Engineering Artificial Intelligence, Wuhan Institute of Technology, Wuhan, China
22307010024@stu.wit.edu.cn, dchen@wit.edu.cn

**Abstract.** In fenceless Human-Robot Collaboration (HRC), safety relies on precise robot pose perception. However, training robust deep learning models for this task typically demands expensive real-world annotations. To address this data bottleneck without incurring high computational costs, we propose a lightweight **robot pose estimation network**. Unlike existing methods that rely on extensive domain randomization, our approach introduces a novel **Feature Filtering Module (FFM)**. This module explicitly filters out domain-specific style to isolate domain-invariant geometry using instance normalization and orthogonality constraints. Trained solely on synthetic data, our method achieves **77.69%** $AP_{50:95}$ on the challenging Panda-Orb real-world benchmark, significantly outperforming the state-of-the-art YOLOv12-pose baseline. Notably, our approach reduces computational cost by approximately **35%** and parameter count by **25%**, offering a superior accuracy-efficiency trade-off for real-time deployment.

**Keywords:** Robot Pose Estimation · Sim-to-Real · Feature Filtering · Domain Generalization

## 1 Introduction

The transition from isolated automation to Human-Robot Collaboration (HRC) marks a fundamental shift in industrial safety paradigms. Unlike traditional setups where safety relies on physical barriers, modern HRC workspaces are characterized by the removal of fences, allowing operators and robots to share dynamic environments (Vicentini, 2021). In this context, the primary guarantee of safety shifts from physical isolation to the robot's perceptual intelligence. Precise, real-time estimation of the robot's pose is therefore imperative, serving as the digital foundation for collision avoidance and risk assessment (Costanzo et al., 2021; Li et al., 2024).

While vision-based pose estimation has advanced significantly (Goswami et al., 2025), its deployment in industrial settings confronts a critical data bottleneck. State-of-the-art deep learning models require massive quantities of annotated real-world data to achieve robustness. However, collecting and manually


H. Yan et al. (Eds.): AIS&P 2025, CCIS 2918, pp. 560–570, 2026.
https://doi.org/10.1007/978-981-95-9884-7_39

annotating pixel-level keypoints in diverse real-world factories is prohibitively expensive and labor-intensive. Consequently, training on synthetic data from simulation engines has become a standard alternative (Lee et al., 2020). Yet, the inherent discrepancy between synthetic rendering and physical reality, known as the domain shift, often leads to severe performance degradation when models are deployed in the real world (Wang, Du, et al., 2022).

To bridge this gap, Domain Generalization (DG) methods have gained prominence, aiming to learn robust representations that generalize to unseen domains without requiring access to target data (Wang, Lan, et al., 2022). However, existing DG approaches face two limitations. First, techniques like heavy Domain Randomization (DR) rely on expanding the training distribution scale, which inevitably increases the computational burden and parameter count (Lee et al., 2020). Second, feature alignment methods based on adversarial learning often perform holistic alignment (Ganin & Lempitsky, 2015). This indiscriminate process risks suppressing domain-specific styles (e.g., texture) at the cost of corrupting domain-invariant geometric structures (e.g., joint topology), which are crucial for precise pose estimation.

To address these challenges, we propose a lightweight **robot pose estimation network**, designed to achieve superior generalization with high efficiency. Instead of blindly aligning features, our method explicitly filters the representations to preserve structural (invariant) components while suppressing stylistic (specific) variations. We introduce a novel **Feature Filtering Module (FFM)**, which leverages Instance Normalization to filter out style-related statistics while preserving geometric integrity. By constructing a dual-path architecture that contrasts original synthetic features with style-augmented features, the network learns to filter environmental variations and focus solely on the robot's intrinsic structure.

The main contributions of this paper are summarized as follows:

1. We propose **LRPE**, a lightweight dual-path robot pose estimation network that eliminates the need for real-world annotations. By generating on-the-fly style augmentations via MixStyle (Tzeng et al., 2014), the model constructs a pseudo-domain gap during training to learn robust domain-invariant geometry features from a single synthetic source.
2. We design a **Feature Filtering Module (FFM)**. This module utilizes orthogonality constraints and instance normalization to effectively filter out domain-specific style while retaining domain-invariant geometry, ensuring that the pose estimation head receives only clean, structural information.
3. Extensive evaluations on the Panda-Orb(RS) benchmark reveal that our method achieves **77.69%** $AP_{50:95}$, surpassing the YOLOv12-pose (Tian et al., 2025) baseline by a significant margin. Notably, this performance is achieved with **35%** fewer FLOPs and **25%** fewer parameters, highlighting its suitability for resource-constrained real-time deployment.

## 2 Method

### 2.1 Overall Method

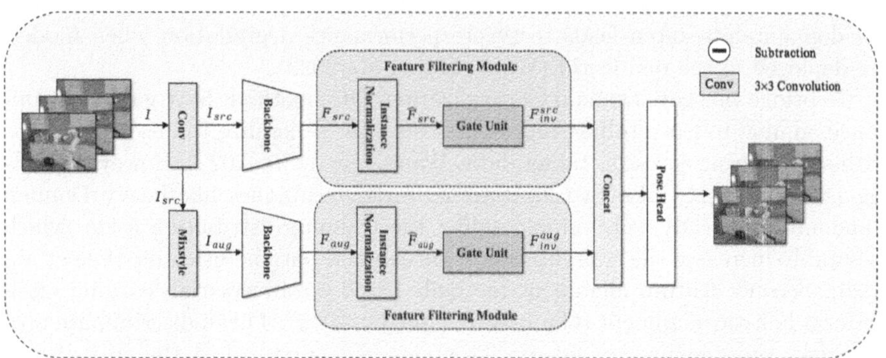


**Fig. 1.** The architecture overview of the dual-path robot pose estimation network.

As illustrated in Fig. 1, the proposed **LRPE** framework consists of four components: a **MixStyle module** for on-the-fly domain augmentation, a **shared Backbone** for feature extraction, our novel **Feature Filtering Module (FFM)**, and a **Pose Head** for keypoint regression.

Given a synthetic input image $\mathbf{I} \in \mathbb{R}^{H \times W \times 3}$, the network predicts the pose matrix $\mathbf{P} \in \mathbb{R}^{N \times K \times 2}$ for $N$ robot instances. We employ a dual-path strategy where both paths are processed through a shared backbone and explicitly filtered by the FFM. The extracted domain-invariant features from both paths are then concatenated and fed into the Pose Head for final pose prediction. Crucially, we enforce feature alignment between these two paths during training, which compels the model to learn geometric representations that are robust and invariant across domains. Furthermore, to prioritize inference speed, we streamline the architecture by removing the heavy feature aggregation neck typically found in YOLO models, connecting the Backbone directly to the FFM and Pose Head.

### 2.2 Feature Filtering Module

To mitigate the performance degradation caused by domain shift, we design the **Feature Filtering Module (FFM)**, as illustrated in Fig. 2. Our design is grounded in the observation that for robot pose estimation, the geometric structure (e.g., link topology) constitutes intrinsic attributes that remain invariant across domains. Conversely, visual elements such as lighting conditions and background textures act as extrinsic style information that varies significantly between simulation and reality. Therefore, the core objective of the FFM is to explicitly isolate robust structural features from environmental noise, enabling the network to filter out unstable environmental statistics while preserving geometric integrity.

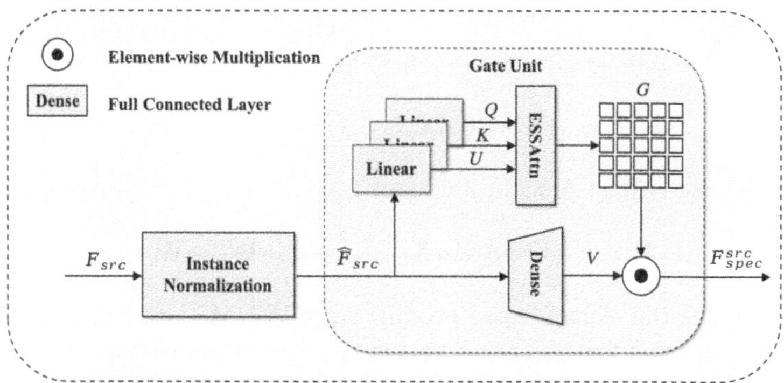


**Fig. 2.** Architecture of the Feature Filtering Module (FFM). The module utilizes Instance Normalization (IN) to filter structural content. This filtered feature is then refined by an efficient attention-based gating unit to yield the final domain-invariant ($\mathbf{F}^{src}_{inv}$).

**Geometry-Style Filtering.** We leverage Instance Normalization (IN) as the primary mechanism for this filtering process. Since statistical information (mean and variance) within feature channels is highly correlated with image style, normalizing these statistics effectively strips away domain-specific visual attributes. Formally, given an input feature map $\mathbf{F}_{src}$, we apply IN to extract the *content features* $\hat{\mathbf{F}}_{src}$, which encapsulate the domain-invariant geometry:

$$\hat{\mathbf{F}}_{src} = \gamma \left( \frac{\mathbf{F}_{src} - \mu(\mathbf{F}_{src})}{\sigma(\mathbf{F}_{src})} \right) + \beta \tag{1}$$

where $\mu(\cdot)$ and $\sigma(\cdot)$ denote the channel-wise mean and standard deviation, and $\gamma, \beta$ are learnable affine parameters. Crucially, the information removed by this process corresponds to the style. We thus derive the *style features* $\hat{\mathbf{R}}_{src}$ by computing the residual difference between the original input and the normalized content:

$$\hat{\mathbf{R}}_{src} = \mathbf{F}_{src} - \hat{\mathbf{F}}_{src} \tag{2}$$

This residual operation explicitly captures environmental variations, such as rendering artifacts in simulation or lighting noise in the real world, laying the foundation for dual-path supervision.

**Feature Refinement via Gating.** While IN provides an initial separation, a simple linear subtraction may not perfectly isolate semantic features. To further purify these representations, we employ a gating mechanism enhanced by ESSAttn (Zhang et al., 2023). Unlike standard self-attention which incurs high computational costs, ESSAttn reduces complexity, aligning with our goal of real-time efficiency on edge devices. Both the *content features* $\hat{\mathbf{F}}_{src}$ and *style features* $\hat{\mathbf{R}}_{src}$ undergo this refinement to filter out redundancy. Let $\mathbf{X}_{inv}$ and $\mathbf{X}_{spec}$ denote the flattened representations of the *content features* $\hat{\mathbf{F}}_{src}$ and *style features* $\hat{\mathbf{R}}_{src}$, respectively. The refined domain-specific features $\mathbf{F}_{spec}^{src}$ are computed as:

$$\mathbf{F}_{spec}^{src} = (\mathrm{ESSAttn}(\mathbf{X}_{spec}) \odot (\mathbf{X}_{spec}\mathbf{W}_v))\,\mathbf{W}_o \tag{3}$$

Similarly, the refined domain-invariant features $\mathbf{F}_{inv}^{src}$ are derived as:

$$\mathbf{F}_{inv}^{src} = (\mathrm{ESSAttn}(\mathbf{X}_{inv}) \odot (\mathbf{X}_{inv}\mathbf{W}_v))\,\mathbf{W}_o \tag{4}$$

where $\odot$ denotes the element-wise product, and $\mathbf{W}_v, \mathbf{W}_o$ are linear projections. Finally, these filtered features are constrained by orthogonality and MMD losses to ensure their independence and robustness.

### 2.3 Loss Function

The training objective is designed to enforce two critical properties: domain invariance of the geometric features and strict independence between structural and stylistic representations. The total loss comprises three components: feature alignment, orthogonality filtering constraints, and task-specific pose estimation.

**Feature Alignment Loss.** To mitigate the distribution shift between the original synthetic domain and the MixStyle-augmented domain, we employ the Maximum Mean Discrepancy (MMD) (Smola et al., 2007). Our goal is to minimize the distributional divergence between the invariant features from the two paths ($\mathbf{F}_{inv}^{src}$ and $\mathbf{F}_{inv}^{aug}$). Using a Gaussian kernel $k(\cdot,\cdot)$, the MMD loss is formulated as:

$$\mathcal{L}_{\mathrm{mmd}} = \mathbb{E}[k(\mathbf{x},\mathbf{x}')] + \mathbb{E}[k(\mathbf{y},\mathbf{y}')] - 2\mathbb{E}[k(\mathbf{x},\mathbf{y})] \tag{5}$$

where $\mathbf{x} \in \mathbf{F}_{inv}^{src}$ and $\mathbf{y} \in \mathbf{F}_{inv}^{aug}$ denote samples from the source and augmented domains, respectively. Minimizing $\mathcal{L}_{\mathrm{mmd}}$ encourages the network to learn a geometric representation that is resilient to the style perturbations introduced by the augmentation path.

**Orthogonality Filtering Loss.** To ensure the purity of the feature filtering and prevent style information leakage into the geometric stream, we impose an orthogonality constraint. This forces the domain-invariant features ($\mathbf{F}_{inv}$) and domain-specific features ($\mathbf{F}_{spec}$) to be structurally orthogonal. We compute the Frobenius norm of their correlation matrix as the penalty:

$$\mathcal{L}_{\mathrm{orth}} = \|\mathbf{F}_{inv}^{src}(\mathbf{F}_{spec}^{src})^{\mathrm{T}}\|_F^2 + \|\mathbf{F}_{inv}^{aug}(\mathbf{F}_{spec}^{aug})^{\mathrm{T}}\|_F^2 \tag{6}$$

By minimizing this term, we strictly penalize any correlation between the two feature sets, ensuring that geometry and style are encoded in distinct subspaces.

**Total Loss.** For the primary pose estimation task, we adopt the standard loss from YOLO-Pose (Tian et al., 2025), denoted as $\mathcal{L}_{\text{task}}$, which includes keypoint regression and object confidence losses. The final training objective is a weighted sum of these components:

$$\mathcal{L}_{\text{total}} = \lambda_1 \mathcal{L}_{\text{mmd}} + \lambda_2 \mathcal{L}_{\text{orth}} + \mathcal{L}_{\text{task}} \tag{7}$$

where $\lambda_1$ and $\lambda_2$ are hyperparameters balancing the contribution of feature alignment and filtering constraints. In our experiments, these hyperparameters are empirically set to $\lambda_1 = 10$ and $\lambda_2 = 6$ to balance the auxiliary constraints with the primary task loss (which follows the default YOLOv12 configuration).

# 3 Experiments

## 3.1 Datasets

We evaluate our method on the DREAM dataset (Lee et al., 2020), a benchmark designed for sim-to-real robot pose estimation. The dataset contains three industrial manipulators, with Franka Emika's Panda being the only one providing both synthetic and real-world data, making it our focus for evaluation.

The real-world Panda test set includes two scenarios:

1. **Panda-3Cam**: Tests generalization across different depth sensors—Microsoft Kinect (**XK**), Intel RealSense (**RS**), and Azure Kinect (**AK**)—with fixed viewpoints and consistent resolutions.
2. **Panda-Orb**: Evaluates robustness to viewpoint changes using orbital camera movements around the robot, captured by an Intel RealSense camera (**Panda-Orb(RS)**).

**Training Set.** We use a minimal training set of 2,000 images sampled from the Panda Domain Randomization subset, representing only 2% of the full synthetic data. This constrained setup tests our method's data efficiency and generalization under limited supervision.

**Test Set.** All quantitative evaluations are conducted on the complete official real-world Panda test set from the DREAM benchmark to ensure standardized assessment of sim-to-real performance.

## 3.2 Evaluation Metrics

We evaluate our model using the standard Average Precision (AP) metric based on Object Keypoint Similarity (OKS), which measures keypoint alignment while considering object scale and annotation difficulty. The OKS is computed as:

$$OKS = \frac{\sum_i \exp\left(-\frac{d_i^2}{2\, s^2 k_i^2}\right) \delta(v_i > 0)}{\sum_i \delta(v_i > 0)} \tag{8}$$

where $d_i$ is the Euclidean distance between predicted and ground-truth keypoints, $s$ is the object scale, $k_i$ is the keypoint-specific constant, and $v_i$ is the visibility flag. Following YOLOv12-Pose (Tian et al., 2025), we report $AP_{50:95}$, $AP_{50}$, and $AP_{75}$.

### 3.3 Comparative Experiments

**Quantitative Analysis.** To comprehensively evaluate the efficacy of our proposed method, we conduct a rigorous quantitative comparison against the current state-of-the-art architectural baseline, YOLOv12-pose (Tian et al., 2025). For a strictly fair comparison, both the baseline and our model used the official default weights as initialization and were trained on the aforementioned training set. Evaluations are performed across the complete suite of real-world testing datasets, ranging from the static setups of **Panda-3Cam** (AK, XK, RS) to the highly challenging dynamic subset **Panda-Orb(RS)**. This setup allows us to strictly assess the model's sim-to-real generalization capabilities under varying sensor modalities and complex viewpoint shifts.

**Table 1.** Quantitative comparison with the state-of-the-art baseline on real-world test sets. Metrics include Average Precision (AP), computational cost (FLOPs), parameter count, and inference latency.

| Dataset | Method | $AP_{50:95}$ (%) ↑ | $AP_{50}$ (%) ↑ | $AP_{75}$ (%) ↑ | FLOPs (G) ↓ | #Param (M) ↓ | Latency (ms) ↓ |
|---|---|---|---|---|---|---|---|
| Panda-3Cam(AK) | YOLOv12-pose | 86.64 | 99.18 | 93.16 | 92.30 | 27.19 | 7.60 |
| | **Ours** | **96.87** | **99.50** | **96.61** | **60.20** | **20.46** | **5.80** |
| Panda-3Cam(XK) | YOLOv12-pose | 93.80 | 98.53 | 97.96 | 92.30 | 27.19 | 7.60 |
| | **Ours** | **96.77** | **99.45** | **99.40** | **60.20** | **20.46** | **5.80** |
| Panda-3Cam(RS) | YOLOv12-pose | 96.43 | 99.50 | 98.84 | 92.30 | 27.19 | 7.60 |
| | **Ours** | **99.17** | **99.50** | **99.50** | **60.20** | **20.46** | **5.80** |
| Panda-Orb(RS) | YOLOv12-pose | 73.24 | 78.74 | 77.30 | 92.30 | 27.19 | 7.60 |
| | **Ours** | **77.69** | **79.80** | **79.57** | **60.20** | **20.46** | **5.80** |

*Accuracy Analysis.* As shown in Table 1, our method demonstrates superior accuracy across all benchmarks, with particularly notable gains on the challenging Panda-Orb(RS) test set. We achieve an $AP_{50:95}$ of **77.69%**, outperforming the YOLOv12-pose baseline (73.24%) by 4.45% points. This improvement validates our architectural design: while standard methods learn entangled representations that conflate geometry with domain-specific style, our Feature Filtering Module (FFM) explicitly extracts domain-invariant geometric features. This enables robust performance under significant domain shifts while maintaining near-perfect scores on static benchmarks.

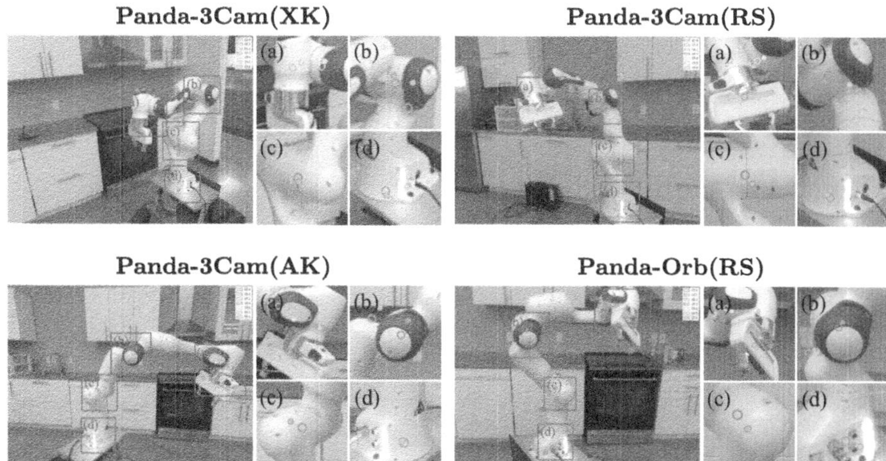


**Fig. 3. Qualitative visualization of our method.** Visualizations on Panda-3Cam and Panda-Orb benchmarks demonstrate our method's accurate localization in challenging scenarios. Red bounding boxes indicate regions of interest (zoomed in as sub-images). **Hollow circles**: Ground Truth; **Solid circles**: Predictions.

*Efficiency Analysis.* As shown in Table 1, our approach achieves remarkable efficiency with only **60.20 GFLOPs** and **20.46 M** parameters, representing reductions of **35%** in computation and **25%** in parameters compared to YOLOv12-pose. By streamlining the architecture and integrating a lightweight FFM, we achieve faster inference latency of **5.80 ms** (approx. **172 FPS**) on Tesla T4 GPU versus the baseline's 7.60 ms. With these substantial improvements over the baseline, our method is significantly more likely to meet real-time requirements even on resource-constrained edge hardware. Our method thus establishes a superior accuracy-efficiency balance, demonstrating that explicit feature filtering yields better generalization than larger, entangled models.

**Qualitative Analysis.** Figure 3 presents qualitative results of our method on the Panda-3Cam and Panda-Orb benchmarks. The visual results highlight our approach's robustness in addressing geometric complexity and environmental challenges. For dense keypoint configurations in Panda-Orb, our method achieves accurate joint localization. To further investigate performance under severe self-occlusion, specific samples are analyzed in Fig. 4. As observed, when self-occlusion occurs, the baseline YOLOv12 leads to missed detections due to structural ambiguity. In contrast, our FFM-based method maintains accurate localization, demonstrating that filtering out distracting style noise facilitates the inference of occluded geometric structures by leveraging the remaining visible context. Additionally, in the presence of specular highlights and lighting variations in Panda-3Cam(AK), our model maintains robust performance by effectively discarding stylistic artifacts while preserving structural information.

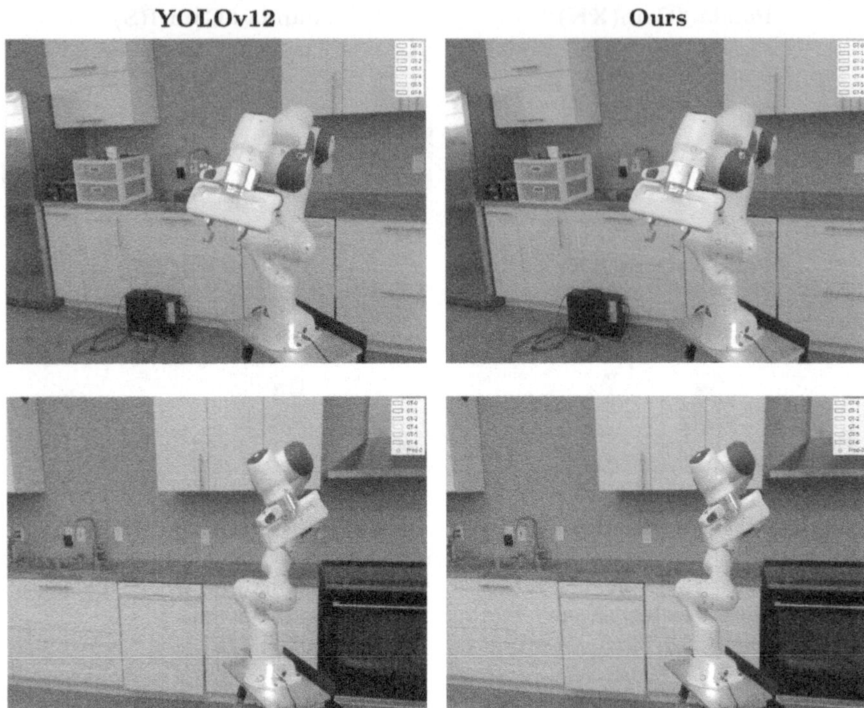


**Fig. 4. Robustness analysis under heavy self-occlusion.** Visual comparisons on challenging self-occluded scenarios from the Panda-Orb benchmark. **Left:** The YOLOv12 baseline suffers from missed detections on key components when they are occluded by the robot's body structure. **Right:** Our method successfully recovers these hidden geometric features, validating the capability of FFM to filter out environmental ambiguity and preserve structural integrity.

*Hyperparameter Sensitivity Analysis.* To evaluate the sensitivity of our method to the orthogonality constraint weight $\lambda_2$, we conducted experiments on the Panda-Orb benchmark with $\lambda_1$ fixed at 10. Varying $\lambda_2$ from 5 to 7 yielded stable results: 77.28% ($\lambda_2 = 5$), 77.69% ($\lambda_2 = 6$), and 77.59% ($\lambda_2 = 7$) in $AP_{50:95}$. The marginal fluctuation (less than 0.5%) indicates that the proposed FFM is robust to hyperparameter variations and does not require extremely precise tuning to achieve high performance.

## 4 Conclusions

This paper addresses the critical bottleneck of data scarcity and domain shift in industrial robot pose estimation, proposing a lightweight **sim-to-real dual-path framework**. Our core contribution is the design of the **Feature Filtering Module (FFM)**. By explicitly filtering out domain-specific style to retain domain-invariant geometry via instance normalization and orthogonality constraints, the FFM effectively bridges the gap between simulation and reality without requiring real-world annotations.

Extensive experiments validate the superiority of our method. Even when trained on limited synthetic data, our model significantly outperforms the state-of-the-art YOLOv12-pose baseline on challenging real-world benchmarks, particularly in dynamic scenarios with severe viewpoint shifts. Furthermore, our streamlined architecture achieves a compelling accuracy-efficiency trade-off, reducing computational costs by approximately **35%** and parameter count by **25%**. This work not only provides a robust solution for sim-to-real perception but also paves the way for the deployment of efficient, high-precision safety monitoring systems in next-generation fenceless Human-Robot Collaboration. While evaluated on the Panda robot due to benchmark availability, the FFM principle is theoretically applicable to diverse robotic platforms. Future work will extend this verification to other robots.

## References


Costanzo, M., De Maria, G., Lettera, G., Natale, C.: A multimodal approach to human safety in collaborative robotic workcells. IEEE Trans. Autom. Sci. Eng. **19**(2), 1202–1216 (2021). https://doi.org/10.1109/TASE.2020.3043286

Ganin, Y., Lempitsky, V.: Unsupervised domain adaptation by backpropagation. In: Bach, F., Blei, D. (eds.) Proceedings of the 32nd International Conference on Machine Learning, vol. 37, pp. 1180–1189. PMLR (2015). https://doi.org/10.48550/arXiv.1409.7495

Goswami, R.G., Krishnamurthy, P., LeCun, Y., Khorrami, F.: Robopepp: vision-based robot pose and joint angle estimation through embedding predictive pre-training. In: Proceedings of the Computer Vision and Pattern Recognition Conference, pp. 6930–6939 (2025). https://doi.org/10.48550/arXiv.2411.17662

Lee, T.E., et al.: Camera-to-robot pose estimation from a single image. In: 2020 IEEE International Conference on Robotics and Automation (ICRA), pp. 9426–9432 (2020). https://doi.org/10.48550/arXiv.1911.09231

Li, W., Hu, Y., Zhou, Y., Pham, D.T.: Safe human-robot collaboration for industrial settings: a survey. J. Intell. Manuf. **35**(5), 2235–2261 (2024). https://doi.org/10.1007/s10845-023-02159-4

Smola, A., Gretton, A., Song, L., Schölkopf, B.: A hilbert space embedding for distributions. In: International Conference on Algorithmic Learning Theory, pp. 13–31 (2007). https://doi.org/10.1007/978-3-540-75225-7_5

Tian, Y., Ye, Q., Doermann, D.: Yolov12: attention-centric real-time object detectors. arXiv preprint arXiv:2502.12524 (2025)

Tzeng, E., Hoffman, J., Saenko, K., Darrell, T.: Adversarial discriminative domain adaptation. In: Proceedings of the IEEE Conference on Computer Vision and Pattern Recognition, pp. 7167–7176 (2017). https://doi.org/10.48550/arXiv.1702.05464

Vicentini, F.: Collaborative robotics: a survey. J. Mech. Des. **143**(4), 040802 (2021). https://doi.org/10.1115/1.4046238

Wang, J., et al.: Generalizing to unseen domains: a survey on domain generalization. IEEE Trans. Knowl. Data Eng. **35**(8), 8052–8072 (2022). https://doi.org/10.48550/arXiv.2103.03097

Wang, J., Du, R., Chang, D., Liang, K., Ma, Z.: Domain generalization via frequency-domain-based feature disentanglement and interaction. In: Proceedings of the 30th ACM International Conference on Multimedia, pp. 4821–4829 (2022). https://doi.org/10.48550/arXiv.2201.08029

Zhang, M., Zhang, C., Zhang, Q., Guo, J., Gao, X., Zhang, J.: Essaformer: efficient transformer for hyperspectral image super-resolution. In: Proceedings of the IEEE/CVF International Conference on Computer Vision, pp. 23073–23084 (2023). https://doi.org/10.48550/arXiv.2307.14010

# Cross-Domain Detection of Lithium Battery Electrode Endpoints Based on Structural Optimization and Incremental Learning

Wen Gan[1], Deng Chen[1(✉)], Lianqing Wang[2,3], Dawei Wen[1], Xiaohua Liu[1], Xin Nie[1], and Xiaoquan Liu[1]

[1] School of Computer Science and Engineering Artificial Intelligence, Wuhan Institute of Technology, Wuhan, China
dchen@wit.edu.cn

[2] College of Intelligent Science, National University of Defense Technology, Changsha 410073, China

[3] Command and Control Support Department, Information Support Force Engineering University, Wuhan 430030, China

**Abstract.** Lithium battery electrode endpoint detection is essential for ensuring battery safety and performance. However, in industrial environments, changes in production conditions often lead to domain shifts in X-ray images, significantly degrading the performance of existing detection models. To address this challenge, this paper proposes a cross-domain detection method integrating structural optimization and incremental learning. Structurally, we enhance the YOLO model by introducing cross-layer connections in the neck network to improve multi-scale feature fusion capability. Algorithmically, we incorporate the Elastic Weight Consolidation (EWC) incremental learning strategy, which constrains important parameters to retain source domain knowledge while adapting to target domain distributions, without requiring access to original source data. Experimental results demonstrate that our method achieves 82.2% mAP on the target domain while maintaining 73.1% mAP on the source domain, significantly outperforming traditional transfer learning approaches and effectively mitigating catastrophic forgetting.

**Keywords:** Lithium-ion Battery · Electrode Endpoint Detection · Cross-Domain Detection · Structural Optimization · Incremental Learning

## 1 Introduction

With the rapid advancement of the new energy sector, lithium-ion batteries have emerged as a strategic emerging industry and crucial economic driver (Chen & Jiang, 2024), finding extensive applications in new energy vehicles, portable electronics, and energy storage systems. As the core component, battery cells consist


H. Yan et al. (Eds.): AIS&P 2025, CCIS 2918, pp. 571–581, 2026.
https://doi.org/10.1007/978-981-95-9884-7_40

of precision-stacked electrodes and separators (Han et al., 2014), where electrode alignment accuracy critically influences battery performance and safety. Improper spacing is identified as a primary cause of severe incidents including internal short circuits and thermal runaway (H. Wu, 2018), making automatic electrode endpoint detection through X-ray imaging indispensable in lithium battery manufacturing.

Traditional electrode endpoint detection primarily relies on image structural analysis techniques. Wang et al. (2019) employed grayscale stretching with Harris corner detection for electrode localization, while Mi and Zeng (2022) proposed multi-scale retina enhancement with differential templates for coordinate determination, and Y. Wu et al. (2021) utilized GPU-accelerated homomorphic filtering combined with Shi-Tomasi corner detection. However, these conventional methods face accuracy and robustness challenges due to inherent endpoint feature ambiguity. Recent deep learning approaches, particularly YOLO series models favored for their speed-accuracy balance, offer improved solutions. Zheng et al. (2023) enhanced recognition accuracy through YOLOv5 architecture modifications, and Li and Chen (2024) utilized YOLOv8-Pose with specialized PedLoss function, significantly improving blurred endpoint detection. The continued relevance of advanced YOLO architectures for industrial defect inspection is further supported by recent studies on YOLOv11 applications (Ghahremani et al., (2025), validating the efficacy of this model family.

Nevertheless, current deep learning approaches face significant limitations in practical industrial scenarios. Their performance heavily relies on the I.I.D. assumption, overlooking domain shift problems caused by production line upgrades, equipment aging, and environmental variations. Handling such drift in industrial defect detection remains a persistent challenge, as highlighted by recent research on domain adaptation mechanisms (Barberena et al., 2025). These factors induce substantial feature distribution changes in X-ray images, causing performance degradation in altered environments. Traditional transfer learning addresses domain shifts through fine-tuning but exhibits limitations including dependency on new domain annotations, constrained generalization across domain gaps, and catastrophic forgetting of source knowledge.

To address these challenges, this paper proposes an integrated dual-strategy framework combining structural optimization with incremental learning. Structurally, we enhance the YOLO neck network with cross-level connections to improve multi-scale feature fusion and cross-domain representation stability. Algorithmically, we integrate Elastic Weight Consolidation (EWC) (Kirkpatrick et al., 2017), selectively constraining crucial parameters to enable adaptation to new domains while systematically preserving acquired knowledge. Experimental results demonstrate substantial advantages over traditional transfer learning, with our method achieving 82.2% mAP on the target domain (3% points improvement over fine-tuning) while maintaining 73.1% mAP on the source domain, surpassing conventional fine-tuning's 68.3% mAP. These results verify our approach's effectiveness in mitigating catastrophic forgetting while achieving balanced cross-domain performance improvement.

# 2 Method

## 2.1 Overall Method

The overall methodology proposed in this paper is illustrated in Fig. 1, which consists of two core phases: Source Domain Model Construction & Fisher Matrix Calculation, and Target Domain Incremental Learning. Initially, enhancement processing is applied to both source and target domain lithium battery X-ray images to crop key regions and improve clarity and electrode contrast. The first phase constructs the enhanced model and computes the Fisher Information Matrix on source domain data ($D_s$) to solidify knowledge of parameter importance. The second phase leverages Elastic Weight Consolidation (EWC) for incremental learning on target domain data ($D_t$) without accessing source samples. By applying regularization constraints on critical parameters, the model preserves source domain knowledge while adapting to new target domain data. Finally, the incrementally trained model is deployed for accurate detection of electrode positions in lithium battery images.

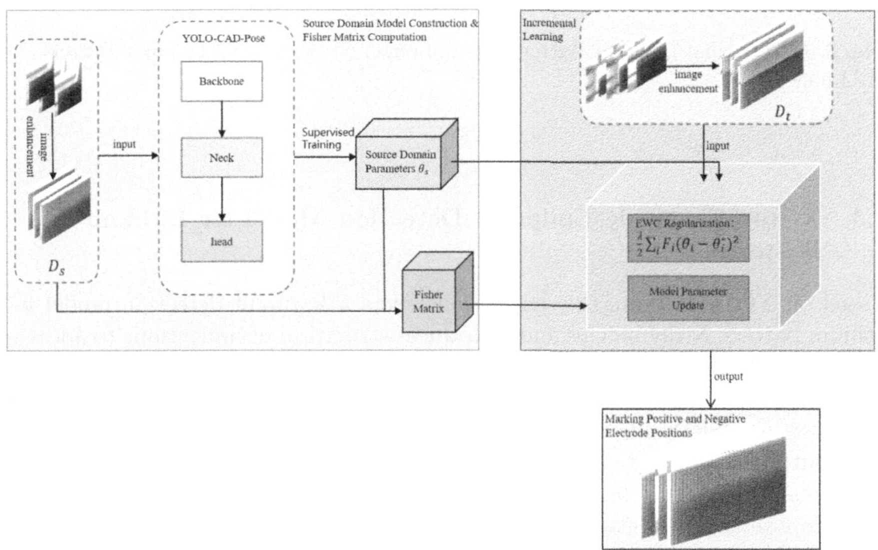


**Fig. 1.** Overall Method Diagram.

## 2.2 Lithium Battery X-Ray Image Enhancement

Based on our laboratory's prior research, we employ an image preprocessing pipeline integrating Multi-Scale Retinex (MSR) and Robust Principal Component Analysis (RPCA) to enhance contrast and structural details in lithium battery X-ray images. This method effectively improves electrode edge clarity and overall image quality, providing a refined visual foundation for subsequent detection tasks. A comparison between original and enhanced images is shown in Fig. 2, demonstrating the noticeable improvement achieved by this approach.

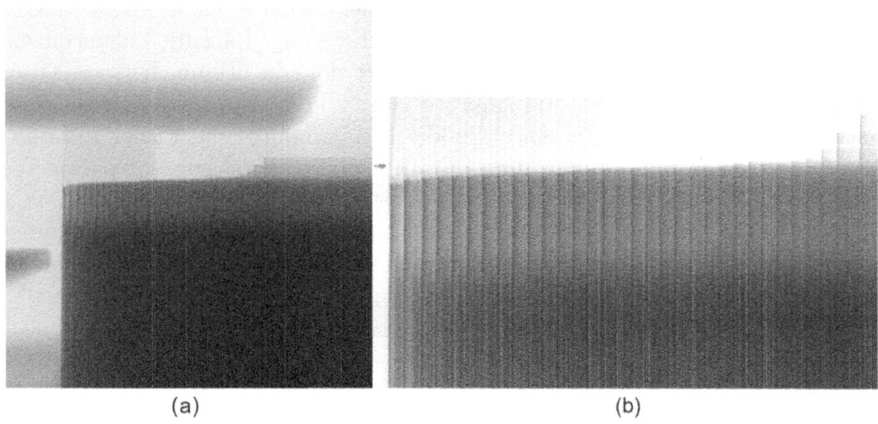


**Fig. 2.** (a) Original Lithium Battery X-ray Image, (b) Enhanced Lithium Battery X-ray Image.

### 2.3 X-Ray Electrode Endpoint Detection Model for Lithium Batteries

Based on YOLO11-Pose, this paper constructs a keypoint detection model for lithium battery X-ray images and introduces structural optimizations to address feature shifts and multi-scale detail loss in cross-domain scenarios. The model retains the three-stage Backbone-Neck-Head architecture. The backbone utilizes an improved CSPDarknet for multi-scale feature extraction, while the detection head outputs bounding boxes and keypoint coordinates.

To enhance feature fusion under domain shift, we redesign the neck network and propose the HCF-Neck (Hierarchical Cross-level Fusion Neck). While preserving the original top-down and bottom-up pathways of PAN-FPN, HCF-Neck incorporates cross-level skip connections that enable direct interaction between semantically rich deep features and spatially detailed shallow features. This structure promotes complementary fusion of semantics and fine-grained details, improving the model's ability to capture domain-invariant features and accurately localize electrode endpoints. The overall model architecture is shown in Fig. 3.

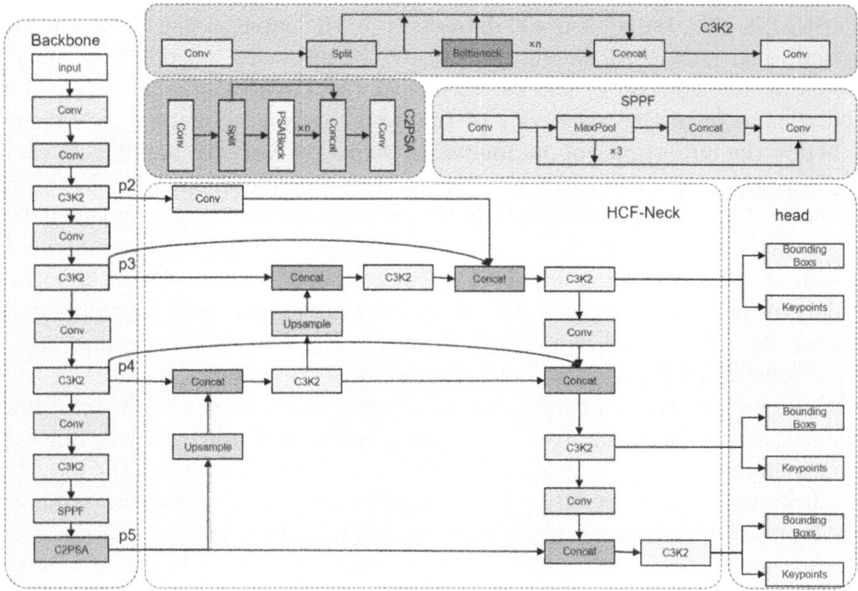


**Fig. 3.** Model Structure.

## 2.4 Incremental Learning Strategy Based on EWC

To enable the model to learn new knowledge in the target domain while maintaining memory of the source domain knowledge, this paper introduces the EWC (Elastic Weight Consolidation) incremental learning strategy. This method effectively mitigates the catastrophic forgetting problem common in traditional transfer or fine-tuning methods by constraining the importance of model parameters during the training process. Given that the source and target domains in this study belong to the same endpoint detection task and share similar features, EWC's selective constraint mechanism can not only preserve old knowledge but also significantly improve cross-domain adaptation capabilities. By prioritizing the protection of key parameters critical for endpoint localization, the model maintains stable structural representations while accelerating its adaptation to the target domain's image distribution, enabling more efficient cross-domain transfer.

Upon completion of the first training stage, the model obtains parameters $\theta^*$ that are optimal for the source domain dataset $D_s$. The Fisher Information Matrix $F$ of the model parameters is then estimated from the source domain samples to quantify the importance of each parameter to the source task. The diagonal elements of the Fisher information matrix can be computed using Equation (1) below.

$$F_i = E_{x \sim D_s}\left[\left(\frac{\partial \log p(y|x, \theta^*)}{\partial \theta_i}\right)^2\right] \tag{1}$$

where $D_s$ is the source domain dataset, $x$ is the input image, $y$ is the corresponding ground truth label, and $p(y|x, \theta^*)$ represents the predicted output probability distribution of the model for sample $x$ under parameters $\theta^*$. This calculation reflects the sensitivity of the model output to parameter changes; the higher the importance of parameter $\theta_i$, the larger its corresponding $F_i$ value, indicating that this parameter is critical to the source domain task. In specific implementation, we approximate the Fisher Information Matrix by calculating the squared mean of the gradients of the loss function for each parameter on the source domain training set. This calculation process provides an important regularization constraint basis for the subsequent incremental learning stage, ensuring that the model can effectively retain key knowledge obtained in the source domain when adapting to the new domain.

In the second stage of target domain incremental training, EWC introduces a regularization constraint based on Fisher information into the loss function, thereby suppressing updates to important parameters. This allows the model to learn from target domain data while maintaining memory of the source domain task. Its overall loss function is defined as shown in formula (2).

$$L_{total} = L_{task} + \frac{\lambda}{2} \sum_i F_i \cdot (\theta_i - \theta_i^*)^2 \tag{2}$$

$$L_{task} = \lambda_{box} L_{box} + \lambda_{cls} L_{cls} + \lambda_{dfl} L_{dfl} + \lambda_{kpt} L_{kpt} + \lambda_{vis} L_{vis} \tag{3}$$

where $L_{task}$ is the task loss of the target domain, whose specific form is shown in Eq. (3). In this formulation, $L_{box}$, $L_{cls}$, $L_{dfl}$, $L_{kpt}$, and $L_{vis}$ represent the bounding box regression loss, classification loss, distribution focal loss, keypoint localization loss, and keypoint visibility loss, respectively. All loss coefficients $\lambda$. adopt the default settings of YOLO11-Pose. $\theta_i$ constitutes the model parameters at the current training stage, $\theta_i^*$ constitutes the optimal parameters obtained from source domain training, $F_i$ is the Fisher information value for the corresponding parameter, and $\lambda$ is a hyperparameter for the balancing term, used to control the trade-off between learning new knowledge and retaining old knowledge.

Through the above mechanism, the model can automatically identify and protect key source domain parameters during the target domain data training process, avoiding excessive modification of learned knowledge. Simultaneously, for parameters that are unimportant in the source domain, the model can freely update them to adapt to new domain features, thus achieving adaptive learning for multi-domain data in complex industrial environments. This strategy allows the model to significantly improve detection robustness for lithium battery electrode endpoints in industrial environments while mitigating catastrophic forgetting of the old domain.

# 3 Experiments

## 3.1 Dataset

Targeting the cross-domain detection task of lithium battery electrode endpoints, this paper constructed a specialized dataset containing old domain samples and new domain samples. The source domain dataset contains 500 lithium battery X-ray images collected under standard production conditions, with good imaging quality and clearly distinguishable electrode structures. The target domain dataset consists of 50 images collected after production line adjustments, mainly containing samples with varying imaging quality due to factors such as equipment aging and process parameter drift, effectively simulating domain shift scenarios in actual industrial environments.

All samples were obtained from the actual production lines of partner enterprises using X-ray transmission imaging technology. During the imaging process, X-rays penetrate the multi-layer structure inside the lithium battery, and the signals are received by detectors and converted into digital grayscale images. The original image resolution is $1536 \times 1536$ pixels. To balance computational efficiency and feature preservation, this paper uniformly resized all images to $864 \times 480$ pixels.

To enhance data diversity and model generalization capability, this paper implemented strict data augmentation strategies on both the source and target domain datasets. Augmentation methods included geometric transformations such as horizontal flipping, vertical flipping, and diagonal flipping, while synchronously updating the corresponding keypoint annotation coordinates to ensure spatial consistency between data and annotations. After augmentation processing, the total dataset size expanded to approximately 2,200 images. In the final constructed dataset, source domain samples serve as the main body for initial model training and Fisher Information Matrix calculation, while target domain samples are used to verify the model's adaptability in domain shift scenarios. All data was divided into training and test sets at an 8:2 ratio, providing sufficient data support for model training and performance verification.

## 3.2 Evaluation Metrics

Three key metrics are selected for comprehensive evaluation: mAP@50 with an OKS threshold of 0.50 reflects detection performance under looser matching conditions; mAP@75 with a stricter OKS threshold of 0.75 represents localization accuracy under precise requirements; and AP averaged across multiple OKS thresholds from 0.50 to 0.95 with a 0.05 step size comprehensively evaluates the model's overall detection capability across varying matching standards.

## 3.3 Hyperparameter $\lambda$ Selection Experiment

To determine the optimal value of the regularization coefficient $\lambda$ in the Elastic Weight Consolidation (EWC) algorithm, this paper designed a systematic hyperparameter selection experiment. The model performance on both source and target domains was evaluated under different $\lambda$ values to balance the learning and retention of knowledge across old and new domains. The experimental results are shown in Table 1.

**Table 1.** Performance Comparison under Different $\lambda$ Values

| $\lambda$ | mAP@50/$D_s$ | mAP@75/$D_s$ | mAP/$D_s$ | mAP@50/$D_t$ | mAP@75/$D_t$ | mAP/$D_t$ |
|---|---|---|---|---|---|---|
| 5 | 0.983 | 0.717 | 0.676 | 0.980 | 0.870 | 0.803 |
| 1 | 0.980 | 0.740 | 0.690 | 0.983 | 0.865 | 0.803 |
| 0.5 | 0.982 | 0.755 | 0.701 | 0.985 | 0.861 | 0.794 |
| 0.3 | 0.978 | 0.731 | 0.686 | 0.983 | 0.847 | 0.775 |

Analysis of Table 1 reveals that when $\lambda = 5$, the model exhibits strong protection of source domain knowledge but insufficient adaptability in the target domain; when $\lambda = 0.3$, the model's learning capability in the target domain improves, but source domain performance declines significantly. Considering the balance of cross-domain performance, $\lambda = 0.5$ was ultimately selected as the optimal parameter. This value maintains good performance balance in both source and target domains.

### 3.4 Comparative Experiment Analysis

To verify the domain adaptation capability of the model proposed in this paper and its overall performance advantages in cross-domain keypoint detection tasks, it was systematically compared with current mainstream YOLO series models (YOLOv8-Pose, YOLOv9-Pose, YOLO12-Pose). Experiments evaluated mAP50, mAP75, and overall mAP metrics on $D_s$ (Source Domain) and $D_t$ (Target Domain) respectively. The results are shown in Table 2.

**Table 2.** Performance Comparison of Our Method with Other Keypoint Detection Algorithms.

| Model | mAP@50/$D_s$ | mAP@75/$D_s$ | mAPs/$D_s$ | mAP@50/$D_t$ | mAP@75/$D_t$ | mAP/$D_t$ |
|---|---|---|---|---|---|---|
| Yolov8-Pose | 0.975 | 0.744 | 0.685 | 0.984 | 0.854 | 0.78 |
| Yolov9-Pose | 0.980 | 0.730 | 0.690 | 0.979 | 0.858 | 0.788 |
| Yolo12-Pose | 0.978 | 0.673 | 0.648 | 0.980 | 0.847 | 0.782 |
| Ours | 0.985 | 0.800 | 0.731 | 0.986 | 0.891 | 0.822 |

Experimental results show that the method proposed in this paper achieves more balanced performance on both domains. On the source domain, mAP50, mAP75, and overall mAP reached 0.985, 0.80, and 0.731 respectively, significantly outperforming all comparison models. On the target domain, this model still maintains high performance, with mAP reaching 0.822, and mAP50 and mAP75 being 0.986 and 0.891 respectively. This fully verifies the effectiveness and engineering application value of the structural optimization and EWC incremental learning strategy proposed in this paper for domain-adaptive detection tasks in lithium battery X-rays.

### 3.5 Visualization Analysis

To verify the model's ability to identify electrode endpoints in lithium battery X-Ray images, its cross-domain adaptability, and its robustness to noise interference, we performed detection on lithium battery images in both the source and target domains and displayed visualizations. To clearly display the detection effect of electrode endpoints, the prediction results of electrode bounding boxes are omitted in the visualization effect maps. Figure 4 and Fig. 5 display the detection results of typical sample images in the source domain and target domain, respectively.

**Fig. 4.** Source Domain Detection Results.

**Fig. 5.** Target Domain Detection Results.

Figure 4 shows the detection results of typical sample images in the source domain; the model can stably identify most electrode endpoints, indicating that the network has learned reliable structural features on the source domain distribution. Figure 5 shows the detection results of typical sample images in the target domain. Compared with source domain images, target domain images present obvious feature changes, with more blurred electrode edges, decreased overall texture contrast, and are accompanied by strong random noise and background artifacts. These typical domain shift phenomena directly affect the stability of the electrode endpoint detection task. Despite this, the method in this paper is still able to maintain robust detection performance, avoiding problems such as endpoint loss and localization drift. This result indicates that the method proposed in this paper possesses good cross-domain adaptability, can effectively overcome challenges brought by domain shifts, and maintains stable detection performance in complex and changeable industrial scenarios.

## 4 Conclusion

To address domain shift and catastrophic forgetting in lithium-ion battery electrode endpoint detection, this paper proposes a cross-domain detection framework based on structural optimization and incremental learning. The method combines a Hierarchical Cross-level Fusion Neck (HCF-Neck) to enhance cross-domain robustness and the Elastic Weight Consolidation (EWC) strategy to retain source-domain knowledge without accessing the original data. Overall, the cross-domain detection strategy proposed in this study has not only effectively enhanced the generalization performance of keypoint detection but has also provided valuable insights for subsequent continuous learning and model extension in more complex scenarios.

## References


Barberena, X., Saiz, F.A., Barandiaran, I.: Handling drift in industrial defect detection through mmd-based domain adaptation. Proc. Copyright **420**, 429 (2025)

Chen, R., Jiang, W.: Research on the development status and prospects of China's power battery industry. Energy Res. Utilization **05**, 16–19 (2024)

Ghahremani, A., Adams, S.D., Norton, M., Khoo, S.Y., Kouzani, A.Z.: Detecting defects in solar panels using the YOLO v10 and v11 algorithms. Electronics **14**(2), 344 (2025)

Han, L., Liu, Y., Lu, W., Xiao, F.: Design of automatic production line for electrode defects inspection of li-ion power battery. Appl. Mech. Mater. **470**, 400–403 (2014)

Kirkpatrick, J., et al.: Overcoming catastrophic forgetting in neural networks. Proc. Natl. Acad. Sci. **114**(13), 3521–3526 (2017)

Li, C., Chen, D.: The method for detecting electrode endpoints of lithium-ion batteries based on deep learning. In: 2024 5th International Symposium on Computer Engineering and Intelligent Communications (ISCEIC), pp. 221–227 (2024)

Mi, Y., Zeng, X.: Distance detection of lithium battery electrodes based on X-ray images. Laser Optoelectron. Prog. **59**(10), 1015001 (2022)

Wang, D., Peng, J., Wang, J., Shen, B.: Distance detection of lithium-ion battery electrodes based on image recognition. Manuf. Autom. **41**(10), 7–13 (2019)

Wu, H.: Measurement system for lithium battery electrodes based on X-ray image analysis (2018)

Wu, Y., Yang, M., Wang, Y., Li, H., Gui, Z., Liu, J.: Pole-piece position distance identification of cylindrical lithium-ion battery through X-ray testing technology. Measur. Sci. Technol. *32*(4), 045405 (2021). https://doi.org/10.1088/1361-6501/abbdf1

Zheng, H., Xue, J., Lyu, K., Wei, R., Hou, L., Zhang, X.: Lithium battery electrode detection method based on X-ray images. Radio Commun. Technol. **49**(3), 532–540 (2023)

# ETC-KG: A Hybrid Extraction and Cross-Modal Alignment Method for Knowledge Graph Construction in the ETC

Sanyan Li[1,2(✉)], Xiang Li[2], Yanyan Wang[3], Xiangyang Li[1,4], and Liu Jin[2]

[1] School of Computer Science and Technology, University of Science and Technology of China, Hefei 230026, China

[2] AI & Knowledge Graph Research Center, Anhui KeDa GuoChuang HuiLianYun Technology Co., Ltd., Hefei 230088, China
15905515846@139.com

[3] Innovation + Research Institute, GuoChuang Cloud Technology Ltd., Hefei 230088, China

[4] The CAS Key Labboratory of Wireless-Optical Communications, University of Science and Technology of China, Hefei 230026, China

**Abstract.** The Electronic Toll Collection (ETC) system is a nationwide highway tolling infrastructure that generates large-scale heterogeneous cross-system business data, which Knowledge Graphs (KGs) integrate in a structured and interpretable manner to support intelligent risk control. However, constructing high quality KGs in this domain faces significant challenges, such as complex multi-source heterogeneous data structures, inconsistent entity identifiers across platforms, and the absence of robust knowledge update mechanisms. To address these issues, this paper proposes ETC-KG, a comprehensive framework for constructing and managing KGs in the ETC domain. The framework is built upon four key technical contributions: (1) a hybrid knowledge extraction method that integrates rule-based templates with deep semantic models, achieving an F1 score of 0.93, (2) a novel lightweight cross-modal alignment mechanism, Light-CLAP, which enhances speech-text alignment accuracy by 13% through contrastive learning, (3) a dynamic knowledge management system featuring a confidence propagation model to ensure data quality and support real time updates, and (4) a graph-based risk exploration module that facilitates community detection and risk propagation analysis. Extensive experiments on real world ETC data show that our method consistently outperforms strong baselines and representative state-of-the-art methods. The resulting KG, comprising 376,218 nodes and 530,539 edges, validates the framework's effectiveness in enabling intelligent risk analysis and practical industrial applications.

**Keywords:** ETC industry · knowledge extraction · knowledge fusion · cross-modal alignment


H. Yan et al. (Eds.): AIS&P 2025, CCIS 2918, pp. 582–607, 2026.
https://doi.org/10.1007/978-981-95-9884-7_41

## 1 Introduction

With the continuous development of smart transportation and intelligent risk control systems, the ETC (Electronic Toll Collection) system has become an essential infrastructure for highways and urban mobility (Shahrier et al., 2024). The ETC ecosystem involves numerous entities including users, accounts, vehicles, On Board Unit (OBU) devices, and transaction records forming a complex network of heterogeneous relationships. Data originate from diverse sources such as user centers, settlement platforms, banking systems and vehicle terminals, exhibiting varied modalities, structural formats, and semantic fragmentation. Given these characteristics, a knowledge graph (KG) emerges as a natural solution for achieving unified semantic representation, cross-source integration, and risk-aware management of ETC data, thereby supporting intelligent analysis and decision-making in large-scale transportation systems.

Knowledge graph construction techniques have seen considerable advances in general domains. Early rule based and statistical methods have progressively evolved into end to end models such as BiLSTM-CNN-CRF (X. Ma & Hovy, 2016), BERT-CRF (Souza et al., 2019), and CasRel (Wei et al., 2020), and more recently, generative approaches like REBEL (Gao et al., 2024), leading to continuous improvements in entity and relation extraction. In cross modal semantic alignment, pretrained models including CLIP (Radford et al., 2021), AudioCLIP (Guzhov et al., 2022), CLAP (Wu et al., 2023), ALBEF (J. Li et al., 2021), and BLIP (J. Li et al., 2022) have established foundational frameworks for integrating multimodal data. Nevertheless, existing methods are largely designed for open domain corpora and remain inadequate in addressing domain specific challenges in ETC scenarios, such as specialized terminology, complex relationship patterns, and uneven data quality which hinders the development of high quality, domain specific knowledge graphs (Zhao et al., 2024). As illustrated in Fig. 1, principal challenges are as follows:

1. Difficulty in Modeling Multi-Source Heterogeneous Data: ETC data originate from multiple business subsystems such as billing, customer service, and account management platforms, forming tightly coupled event sequences that conventional extraction and fusion methods (Ji et al., 2021), typically assuming independent and static data sources are unable to reconstruct coherently.
2. Lack of Consistent Entity Identification: The same user is represented inconsistently across systems due to heterogeneous identifiers and implicit business semantics, violating the stable-identifier assumptions underlying most current alignment models (B. Zhu et al., 2024), as a result, they often fail to determine whether records from different sources correspond to the same real-world entity.
3. Limitations in Dynamic Knowledge and Quality Assurance: ETC data are highly time sensitive and context dependent, rendering static graph representations unsuitable for real time risk control. Moreover, existing approaches (Feng et al., 2025) lack mechanisms for continuous confidence evaluation and refinement of nodes and relations.

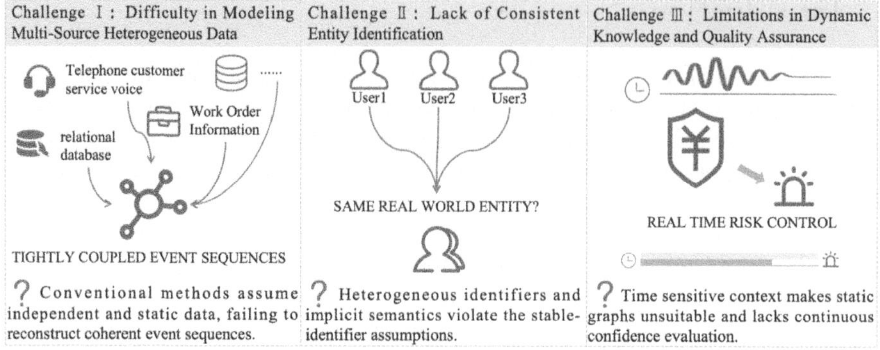


**Fig. 1.** Three key challenges in constructing an ETC knowledge graph.

To address these challenges, this paper proposes a hybrid knowledge-graph construction framework for ETC industry data, combining rule-based patterns with deep semantic modeling to achieve reliable extraction and cross-modal consistency. The framework integrates knowledge extraction, semantic alignment, and confidence-based dynamic updating, enabling the construction of a comprehensive and semantically coherent KG that supports real-time maintenance. Experiments show that the hybrid extraction module attains an F1 score of 0.93, and the proposed cross-modal alignment mechanism improves speech–text consistency by 13% over the best existing method. The resulting ETC knowledge graph reaches real-world scale, integrating 376,218 nodes and 530,539 relations. These results demonstrate that the proposed framework provides a robust and practical foundation for high-quality KG construction and downstream intelligent risk control applications.

The main contributions of this work are as follows. Each contribution directly addresses one of the challenges identified above: Contribution 1–2 target Challenge 1, Contribution 3 addresses Challenge 2, and Contribution 4 resolves Challenge 3.

1. We propose a hybrid knowledge graph construction framework that integrates rule-based templates with deep semantic representation. A multi-level semantic schema centered on the "User–Account–Card–Vehicle" hierarchy is designed to provide a unified structural basis for a high quality domain specific KG.
2. We introduce a hybrid knowledge extraction method for multi-source ETC data, combining domain-aware rules with the semantic comprehension capability of large language models to significantly enhance the accuracy and robustness of entity and relation extraction.
3. We develop a lightweight cross-modal alignment mechanism that freezes the pretrained CLAP encoders and fine-tunes only a small projection head, reducing trainable parameters while aligning speech and text in a unified semantic space and improving multi-source event correlation.

4. We design a dynamic knowledge management mechanism based on neighborhood confidence propagation, where each entity updates its confidence by aggregating signals from adjacent nodes and relations. This enables real time correction of low quality knowledge and supports continuous graph refinement for practical ETC risk control applications.

## 2 Related Work

### 2.1 Knowledge Extraction

Knowledge extraction aims to identify entities and their relationships from unstructured or semi-structured data, forming triplet representations that serve as the foundation for subsequent knowledge fusion and reasoning.

This section reviews mainstream knowledge extraction techniques, examining their strengths and limitations when applied to the ETC domain, which informed the design of our hybrid extraction method in Sect. 3.3.

Early knowledge extraction research typically employed sequence labeling models like BiLSTM-CNN-CRF or BERT-CRF. Ma et al. (X. Ma & Hovy, 2016) utilized Convolutional Neural Networks (CNNs) (El Sakka et al., 2025) to capture local contextual features, and Bidirectional Long Short-Term Memory Networks (BiLSTM) (Yunita et al., 2025) to model long-distance dependencies. Conditional Random Fields (CRFs) were then applied for global sequence decoding, ensuring the labeling path validity. These methods feature simple architectures and high computational efficiency, demonstrating robust performance on small-scale datasets, though their capability to model long texts remains limited. With the advancement of pretrained models (Q. Ma et al., 2024), Souza et al. 2019 replaced traditional local–global encoders with BERT (Devlin et al., 2019) as the backbone encoder, enhancing contextual modeling capabilities. This approach became a strong baseline for sequence-labeling-based knowledge extraction and was often integrated with relation classification modules to form pipeline-based extraction frameworks (Sukumar et al., 2024). However, such pipeline methods are prone to error propagation and exhibit limited sensitivityto overlapping entities and one-to-many relationships.

To jointly model entities and relations, Wei et al. (Wei et al., 2020) proposed CasRel, an end-to-end extraction framework that directly extracts triplets from sentences via cascaded binary tagging. This method eliminates dependence on Named Entity Recognition (NER) (Savci & Das, 2024) and effectively handles overlapping relations and coexisting triplets, establishing itself as a representative model in end-to-end knowledge extraction. Inspired by the Seq2Seq (Kelley & Hagan, 2025) paradigm, Gao et al. (Gao et al., 2024) introduced the REBEL model, which reformulates knowledge extraction as a text generation task. Based on the BART architecture, it encodes input text and directly generates linearized triplet sequences, avoiding complex label design and pipeline-related errors. This model naturally supports open-domain relation extraction. However, without the absence of explicit structural constraints, it is prone to generating hallucinated relations (Janiak et al., 2025).

In summary, existing knowledge extraction studies have achieved remarkable progress in general text scenarios. Nevertheless, challenges persist in industry-specific contexts, due to complex domain terminology, diverse relationship types, and mixed cross-modal information. This paper addresses these issues in the ETC domain by developing a hybrid knowledge extraction method that combines rule-based constraints with deep semantic representations, preserving high-precision entity recognition while enhancing robustness in complex relation extraction.

### 2.2 Knowledge Fusion

Knowledge fusion aims to integrate multi-source heterogeneous data to achieve semantic consistency. As a core component of fusion, cross-modal alignment maps different modalities into a unified semantic space, enabling consistent cross-modal understanding and reasoning.

Current research can be broadly categorized into two paradigms: contrastive learning based alignment frameworks (Liu et al., 2019) and joint modeling based on unified encoders (S. Li & Tang, 2024). The former seeks to implicit modality alignment by maximizing similarity among positive cross-modal pairs and minimizing similarity among negative pairs. Most such methods adopt a dual-tower structure, where each modality has an independent encoder, and contrastive learning objectives enforce semantic alignment within a shared embedding space. Radford et al. proposed CLIP (Radford et al., 2021) achieving joint image–text semantic alignment. Subsequently, Wu et al. (Guzhov et al., 2022) replaced the image encoder in CLIP with an audio encoder to develop CLAP, achievingcross-modal alignment between audio and text. These approaches exhibit strong generalization ability and stable training performance, supporting zero-shot inference. However, as they rely on late fusion in the embedding space and lack fine-grained modality interaction, their performance deteriorates in complex reasoning or domain-specific tasks.

The second category—unified-encoder-based multimodal joint modeling typically relies on Transformer architectures (Vaswani et al., 2017) to achieve early fusion or mid-level fusion through shared attention mechanisms, enabling interactive representation of different modalities within a single network. Structurally, such models can be divided into single-stream and dual-stream types. Single-stream models, such as UNITER (Chen et al., 2020) and ViLT (Kim et al., 2021), concatenate all modality features and feed them into a shared Transformer for deep fusion. Dual-stream models, such as ALBEF and BLIP, assign separate encoders to each modality and introduce cross-modal attention mechanisms to balance modality independence and semantic correlation. These models perform well in multimodal question answering but require substantial training data and computational resources.

In conclusion, existing methods for knowledge fusion and multimodal alignment demonstrate strong performance in open-domain settings but struggle to resolve semantic conflicts across heterogeneous data in industry-specific knowledge graphs. In this study, we design a cross-modal semantic alignment framework tailored to the characteristics of multi-source, heterogeneous, and multi-modal ETC data, providing a foundation for cross-modal event similarity analysis.

# 3 Methodology

## 3.1 Notations and Problem Definition

A knowledge graph (KG) is represented as a directed heterogeneous graph

$$G = (\mathcal{E}, \mathcal{R}, \mathcal{T}), \tag{1}$$

where $\mathcal{E}$ denotes the entity set, $\mathcal{R}$ the relation set, and

$$\mathcal{T} = \{(h, r, t) \mid h, t \in \mathcal{E},\ r \in \mathcal{R}\} \tag{2}$$

the set of observed triples.

In the ETC domain, raw data originate from heterogeneous business subsystems such as billing platforms, customer-service logs, and account management systems. These sources exhibit inconsistent identifiers, modality discrepancies (e.g., text vs. speech), and temporal drift, leading to duplicated, incomplete, or conflicting triples. Let the multi-source input be denoted as:

$$D = \{D^{(1)}, D^{(2)}, \ldots, D^{(k)}\}, \tag{3}$$

where each $D^{(i)}$ corresponds to a distinct subsystem.

The goal of this work is to construct a unified and high-quality ETC knowledge graph

$$G^* = f_\theta(D), \tag{4}$$

where the mapping $f_\theta(\cdot)$ must satisfy the following objectives:

1. **Knowledge Extraction.** From heterogeneous data sources, extract canonical triples $\hat{\mathcal{T}} = \{(\hat{h}, \hat{r}, \hat{t})\}$ covering structured, semi-structured, and unstructured inputs.
2. **Entity Alignment Across Subsystems.** Learn a cross-source alignment function
$$\phi : \mathcal{E}_i \times \mathcal{E}_j \to \{0, 1\} \tag{5}$$
to determine whether two entity mentions refer to the same real-world entity.
3. **Cross-Modal Semantic Consistency.** Construct a shared embedding space via
$$\psi(\cdot) : \text{text/audio} \to \mathbb{R}^d, \tag{6}$$
such that semantically corresponding speech–text pairs are closely aligned.
4. **Confidence-Based Dynamic Updating.** Maintain a time-evolving KG using confidence refinement:
$$G(t+1) = \Gamma(G(t),\ \Delta_t)\,, \tag{7}$$
where $\Gamma(\cdot)$ updates node and edge confidence scores based on new evidence.

The objective is to obtain a unified KG $G^*$ that is semantically coherent, structurally consistent, and suitable for downstream intelligent risk control applications.

### 3.2 Overall Framework

To address the challenges of complex multi-source heterogeneous data structures, inconsistent entity identifiers, and the difficulty of maintaining up-to-date knowledge in the ETC industry, this paper proposes a knowledge graph construction framework that integrates rule-based and deep semantic methods.

The overall framework of the proposed ETC-KG system, depicted in Fig. 2, comprises four core stages:

(A) Hybrid knowledge extraction, the system combines rule templates with deep semantic models to extract entities, relations, and attributes from both structured and unstructured data, thereby achieving unified semantic modeling across multiple data sources.

(B) knowledge fusion based on entities and semantics, Extracted knowledge undergoes entity matching and semantic alignment to fuse information from different sources into a coherent and unified knowledge network.

(C) dynamic knowledge management and quality evaluation, updating, and quality evaluation. Consistency checking and version control mechanisms are employed to ensure the KG's completeness and timeliness.

(D) graph scalability and visualization for retrieval, this stage provides graph statistics, semantic retrieval, knowledge question answering, and visualization thus establishing an end-to-end pipeline from data to storage and application.

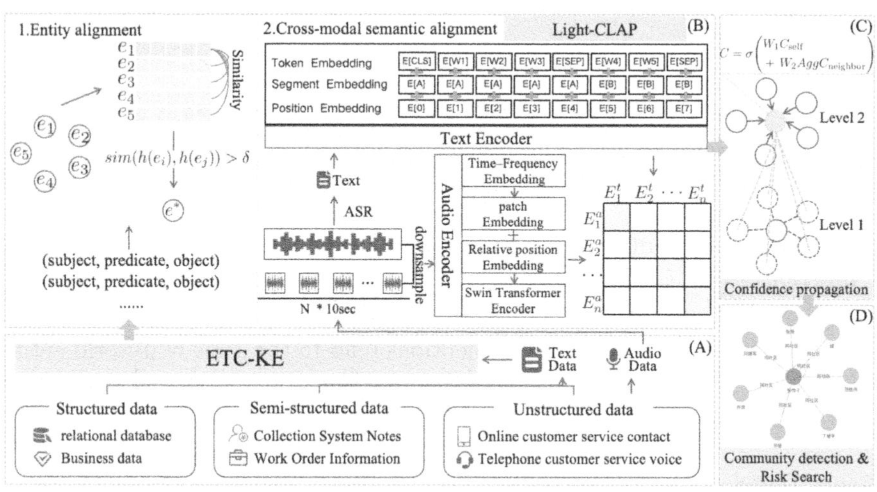


**Fig. 2.** Overall framework of the proposed system.

### 3.3 Hybrid Knowledge Extraction

Knowledge extraction automatically identifies entities, relations, and their associated attributes from raw ETC business data.

Let the original dataset be denoted as $\mathcal{D} = \{d_1, d_2, \ldots, d_n\}$, with the objective of generating a set of knowledge triplets:

$$G = \{(h, r, t) \mid h, t \in E, r \in R\} \tag{8}$$

where $E$ represents the entity set and $R$ the relation set. In the ETC business context, we predefine fourteen entity types and thirteen relation types, schema is shown in Fig. 3.

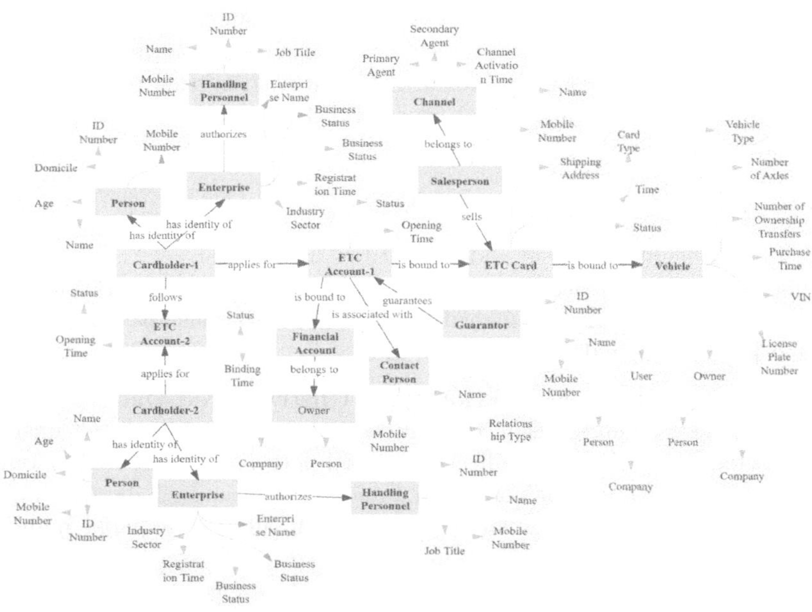


**Fig. 3.** Schema of the ETC knowledge graph.

Given the complexity, heterogeneity, and semantic inconsistency of ETC industry data sources, we propose a hybrid knowledge extraction framework integrating rule-based templates and deep semantic representations, balancing interpretability and generalization. This framework leverages domain specific business rules and the semantic comprehension capabilities of large language models to achieve high precision extraction of entities, relations, and attributes across structured and unstructured data environments. The overall framework consists of two main layers: the rule template layer and the deep semantic layer.

**Rule Template Layer: Extraction from Structured Data**
The ETC business relational databases contain numerous structured tables such as user information, transaction records, and account detail tables whose field names and relationships exhibit well-defined schemas. Based on this, we design

a rule-based mapping mechanism for fast and accurate extraction of explicit relationships. Assume a structured data record is defined as:

$$r_i = \{(f_1, v_1), (f_2, v_2), \ldots, (f_n, v_n)\} \tag{9}$$

where $f_i$ denotes the field name and $v_j$ denotes the corresponding field value.

We define a set of rule templates:

$$T = \{(P_k, R_k) \mid k = 1, 2, \ldots, m\} \tag{10}$$

where $P_k$ represents a field-matching pattern and $R_k$ denotes the corresponding relation type. A pattern $P_k$ specifies a predefined combination of fields that implies a potential semantic relationship. For instance, a pattern defined as $\langle \texttt{Account\_ID}, \texttt{Card\_ID} \rangle \mapsto \texttt{Owns}$ indicates that if a structured record contains both fields, the corresponding triplet $(\texttt{Account\_123}, \texttt{Owns}, \texttt{Card\_A})$ is generated. which follows the general formulation:

$$(h, R_k, t) \tag{11}$$

where $h$ and $t$ denote the head and tail entities, respectively.

This rule-based approach provides high accuracy and low computational cost, making it suitable for subsystems with well-defined relationships, such as standardized ETC business databases.

**Deep Semantic Layer: Extraction from Unstructured Data**
In practical ETC business scenarios, substantial semi-structured and unstructured text data exists, such as collection system notes, complaint tickets, online customer service logs, and manually annotated fields. From such data, we aim to extract associated persons and contact information, alternative phone numbers, and categorized complaints.

To accomplish this, we propose ETC-KE for extracting entities and relations directly from textual data. The system uses carefully designed prompt templates for different contexts, where each prompt specifies the task description, the target entities or relations, and the required structured output format, thereby guiding the model to identify relevant information with semantic precision and enriching the ETC KG with unstructured information. Task-oriented structure. Each prompt is designed for a specific information extraction scenario, enabling the model to focus on well-defined targets. The detailed prompt templates are provided in Appendix A.

### 3.4 Knowledge Fusion Based on Entities and Semantics

Knowledge fusion is a core component of ETC industry KG construction, aiming to achieve unified representation and integrated association of heterogeneous data from different systems and modalities, forming a globally consistent semantic knowledge space. The proposed fusion module comprises two main components: entity alignment and cross-modal speech–text alignment. The former addresses the unification of entities across structured data sources, while the latter aligns semantics between speech and text in unstructured, multimodal data enhancing KG completeness and intelligence of the knowledge graph across multi-source business scenarios.

**Entity Alignment**

In ETC operations, different subsystems (e.g., collection, ticketing, and customer service systems) often refer to the same real-world entity using different naming conventions or inconsistent attribute information. leading to redundant modeling in the KG. Entity alignment identifies such cross-source synonymous entities and unifies them into a single node, as illustrated in Fig. 4.

Formally, let two subgraphs from different sources be defined as:

$$G_1 = (E_1, R_1, X_1), \quad G_2 = (E_2, R_2, X_2) \tag{12}$$

where $E_1$, $E_2$ denote entity sets, $R_1$, $R_2$ denote relation sets and $X_1$, $X_2$ represent associated attribute or multimodal features.

Each entity $e \in E$ is embedded through a representation function:

$$h(e) = f(attr(e), rel(e), modal(e)) \tag{13}$$

Entity alignment is defined as identifying a correspondence set:

$$A = \{(e_i, e_j) \mid e_i \in E_1,\ e_j \in E_2,\ sim(h(e_i), h(e_j)) > \delta\} \tag{14}$$

where $sim(\cdot)$ is a similarity function and $\delta$ is a semantic consistency threshold. The aligned entities are merged into a unified node, ensuring semantic consistency and structural connectivity in the integrated KG.

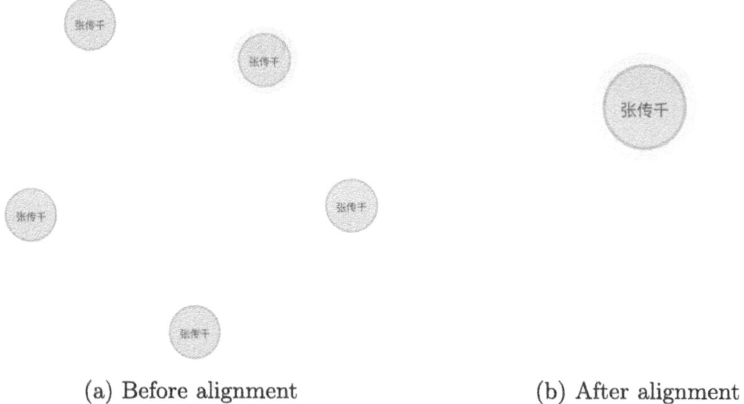


**Fig. 4.** Illustration of Entity Alignment Results. (a) is before alignment, and (b) is after alignment. All attributes related to this person are aggregated into the same node.

**Cross-Modal Semantic Alignment**

In ETC customer service scenarios, call audio and textual transcripts represent two modalities of the same event. Customer inquiries often exhibit semantic commonality, but traditional KGs built solely from structured data, fail to capture latent emotional cues and semantic information embedded in speech.

To achieve multimodal knowledge fusion, we proposes a lightweight cross-modal alignment method Light-CLAP based on the CLAP model. By freezing the pretrained encoders and training only the projection heads, the method aligns speech and text in a unified semantic space, allowing both modalities to share consistent embeddings. Unlike full fine-tuning approaches, all backbone parameters of the pretrained CLAP audio encoder and text encoder are frozen, and only a small projection head is updated. This reduces trainable parameters by over 90% and enables efficient domain adaptation while preserving the pre-trained multimodal representations. This enables semantic associations between speech and textual events, supporting subsequent risk event recognition in multimodal contexts.

The Light-CLAP model comprises a frozen modality encoder layer and a trainable dual-projection layer, as shown in the framework diagram. Each customer speech segment is segmented into 10-second clips. The built-in CLAP audio encoder maps each input waveform into a high-dimensional acoustic representation vector $v_a \in R^d$, while the CLAP text encoder maps the corresponding transcript into a semantic vector $v_t \in R^d$. All encoder parameters are frozen during training.

To enhance cross-modal correspondence, a trainable linear projection head is added to both encoder outputs:

$$Z_a = W_a v_a + b_a \ , \ Z_t = W_t v_t + b_t \tag{15}$$

where $W_a$, $W_t \in R^{d' \times d}$ are projection matrices.

A contrastive learning objective aligns speech and text representations in the shared space. The cross-modal alignment loss $\mathcal{L}_{align}$ is defined as:

$$\mathcal{L}_{\text{align}} = -\frac{1}{N}\sum_{i=1}^{N}\left[\log\frac{\exp\left(\cos(\mathbf{z}_a^i, \mathbf{z}_t^i)/\tau\right)}{\sum_{j=1}^{N}\exp\left(\cos(\mathbf{z}_a^i, \mathbf{z}_t^j)/\tau\right)} + \log\frac{\exp\left(\cos(\mathbf{z}_t^i, \mathbf{z}_a^i)/\tau\right)}{\sum_{j=1}^{N}\exp\left(\cos(\mathbf{z}_t^i, \mathbf{z}_a^j)/\tau\right)}\right] \tag{16}$$

where $\tau$ is the temperature parameter controlling distribution smoothness, and $N$ denotes the batch size. The loss consists of two symmetric directions—speech-to-text and text-to-speech—ensuring bidirectional semantic alignment by maximizing the positive pair similarity of and minimizing negative pair similarity.

Negative samples are drawn implicitly from other instances in the same batch (batch size = 128), following the standard in-batch contrastive learning protocol. Through this lightweight fine-tuning design, Light-CLAP improves speech–text semantic consistency while maintaining low computational overhead. After training, both speech and text are mapped into a unified embedding space. For any speech segment $a_i$ and textual description $t_j$, their semantic similarity is calculated as:

$$S(a_i, t_j) = cos((Z_a)^i, (Z_t)^j) \tag{17}$$

If $S(a_i, t_j) > 0$, a cross-modal edge is created in the knowledge graph:

$$(Audio_i, similarevent, Text_j) \tag{18}$$

This approach establishes one to one cross-modal semantic correspondences between speech and text, producing a unified multimodal semantic representation and providing multimodal support for subsequent risk-chain inference.

### 3.5 Dynamic Knowledge Management and Quality Evaluation

Knowledge management constitutes the final stage of KG construction, aiming to ensure data consistency, semantic reliability, and temporal validity, supporting downstream processes like knowledge retrieval, risk propagation, and decision analysis. This module includes knowledge storage, quality evaluation, and knowledge updating.

**Knowledge Storage and Updating**

Knowledge storage uses the Neo4j graph database, representing ETC business knowledge through nodes and relationships. Each node contains entity type, semantic attributes, and a unique identifier, each relationship encodes business associations and confidence levels. Neo4j's native graph storage and Cypher query language enable efficient subgraph retrieval. Given the high dynamism and real-time nature of ETC data, we adopt a hybrid proactive–reactive update mechanism combining periodic full updates with timestamp-triggered incremental updates, including:(1) data-listening triggers, (2) node state updates, (3) edge weight adjustments, and (4) version management.

**Quality Evaluation**

ETC business data originate from asynchronous and heterogeneous subsystems, making the knowledge graph susceptible to outdated information, conflicting facts, and error amplification during downstream risk reasoning. Unlike traditional KG settings where data sources are relatively static and quality inspection is often simplified ETC-KG requires continuous quality assessment to ensure reliability. In particular, dynamic updates of user accounts, vehicle bindings, and service states may introduce inconsistencies. if not corrected, such low quality facts can propagate along risk chains and lead to false alarms.

A Confidence Propagation Model (CPM) is employed to assess and refine the trustworthiness of nodes and relations. For any entity $u$, its updated confidence value $C_{\text{new}}(u)$ is computed as a weighted combination of its intrinsic confidence and the aggregated confidence of its neighbors:

$$C_{\text{new}}(u) = \sigma \left( W_1 \cdot C_{\text{self}}(u) + W_2 \cdot \frac{1}{|N(u)|} \sum_{v \in N(u)} q_{uv} \cdot C_{\text{self}}(v) \right) \tag{19}$$

where $C_{self}(u)$ is the intrinsic confidence of node, $N_(u)$ denotes its neighbor set, $q_{uv}$ represents the relationship weight between nodes $u$ and $v$, $W_1$ and $W_2$ are weighting coefficients, and $\sigma(\cdot)$ is a sigmoid normalization function constraining values to (0,1).

Through iterative propagation, unreliable nodes or edges, such as outdated vehicle bindings or inconsistent contact information—are naturally down weighted as they receive weak support from their neighborhoods. This enables real-time correction, smoothing, and refinement of the KG. The final quality decision vector is computed by aggregating multiple evaluation dimensions:

$$V_{final} = argmax \sum_{i=1}^{n} W_i \cdot Q_i \tag{20}$$

where $Q_i$ represents the evaluation results of different quality dimensions and $W_i$ their respective weights.

### 3.6 Graph-Based Risk Modeling

While the primary objective of this work is high quality ETC knowledge graph construction, we additionally demonstrate a lightweight downstream application to illustrate how the constructed KG can support intelligent risk control scenarios.

**Person-Level Projection**
ETC risks often emerge from tightly coupled human-centered interactions, such as multiple users sharing vehicles, accounts, or payment instruments. To capture these latent interpersonal dependencies, we construct a person–person projection graph by folding multi-source relations (e.g., shared accounts, shared vehicles) onto user nodes. This projection aggregates heterogeneous evidence from the KG into a unified behavioral interaction graph.

**Community Detection for Risk Association**
On the projected graph, the Louvain algorithm (Hegde et al., 2025) is applied to detect densely connected user communities. Such communities frequently correspond to groups with shared mobility behaviors or account interdependence, which in ETC operations may signal potential risk clusters. This provides an interpretable structural basis for analyzing relational risk.

**Risk Retrieval and Interpretation**
Given a known risky user (e.g., delinquent account holder, abnormal transaction reporter), the system retrieves all users within the same community or strongly connected subgraph. These users share significant relational evidence with the target user, such as co-used vehicles thereby serving as candidates for further risk investigation. This mechanism transforms the KG's multi-source connectivity into actionable insights for early-stage risk screening.

The experimental procedure and visualization are as described in Sect. 4.5 This example demonstrates that the constructed KG not only supports data integration and semantic consistency but also enables meaningful downstream applications in intelligent risk control workflows.

# 4 Experiments and Results Analysis

## 4.1 Experimental Setup

**Data**

The dataset originates from real ETC business systems and includes structured, semi-structured, and unstructured data covering text, audio, and image modalities. The detailed data sources are listed in Table 1.

**Table 1.** Data Source Description

| Data Source | Data Modality | Count |
|---|---|---|
| Relational database business tables | Text | 350K |
| Collection system notes | Text | 100K |
| Complaint and issue tickets | Text | 10K |
| Online customer service interaction logs | Text | 10K |
| Telephone customer service recordings | Audio | 60K |
| ID card photographs | Image | 126K |

For cross-modal alignment experiments, We use real-world interaction data collected from the ETC customer service system, selecting 5,000 customer inquiry recordings from the past three months. Each audio file is processed using Pyannote for speaker diarization, retaining only the customer's speech while removing agent voices and background noise. Silent or empty segments are then removed via silence detection. The cleaned audio data is subsequently transcribed into text using Whisper-Large, producing aligned speech–text pairs. The dataset is split into training, validation, and test sets in an 8:1:1 ratio to ensure balanced evaluation and reliable performance assessment.

**Experimental Parameters**

In the knowledge fusion stage, Light-CLAP model takes paired speech and corresponding ASR (Automatic Speech Recognition) text data as input. All experiments were conducted on a NVIDIA RTX 5060Ti GPU (16 GB) using the PyTorch framework. The AdamW optimizer was employed with a grid-searched learning rate $1 \times 10^{-3}$ and a batch size of 128. Projection parameters were initialized with a fixed random seed, all pretrained encoder weights were kept frozen to ensure reproducibility.

Early stopping mechanism prevented overfitting. The InfoNCE loss function defined the cross-modal alignment objective, optimizing the similarity between the speech and text embeddings in the shared semantic space.

**Evaluation Metrics**

To evaluate the extraction performance of the proposed model, we use Precision (P), Recall (R), and F1-score (F1).

Precision measures the proportion of correctly extracted triplets among all extracted triplets, Recall reflects the proportion of correctly identified triplets among all ground-truth triplets, and F1-score represents the harmonic mean of Precision and Recall, providing an overall assessment of the model's performance. Their definitions are as follows:

$$Precision = \frac{TP}{TP + FP} \tag{21}$$

$$Recall = \frac{TP}{TP + FN} \tag{22}$$

$$F_1 = 2 \times \frac{Precision \times Recall}{Precision + Recall} \tag{23}$$

where $TP$ denotes the number of correctly extracted triplets, $FP$ the number of incorrectly extracted triplets, and $FN$ the number of missed ground-truth triplets by the model.

To quantitatively assess the performance of the proposed Light-CLAP model in the cross-modal speech–text alignment task, we adopt Mean Cosine Similarity (MCS) and Recall@K. These metrics jointly evaluate the model's ability to achieve cross-modal semantic alignment from two perspectives: (1) the semantic consistency in the shared embedding space and (2) the retrieval accuracy during semantic matching.

(1) Mean Cosine Similarity (MCS)

MCS measures the overall compactness of alignment in the shared semantic space, reflecting the average similarity between speech and text embeddings in high-dimensional space.

Let the speech sample set be $A = \{a_1, a\_2, \ldots, a\_N\}$ and the corresponding text sample set be $\tau = \{t\_1, t\_2, \ldots, t\_N\}$. After encoding, the embeddings of the i-th speech and text samples are denoted as $Z_a^i$ and $Z_t^i$, respectively. MCS is defined as:

$$\text{MCS} = \frac{1}{N}\sum_{i=1}^{N} \cos\left(\mathbf{z}_a^i, \mathbf{z}_t^i\right) = \frac{1}{N}\sum_{i=1}^{N} \frac{\mathbf{z}_a^i \cdot \mathbf{z}_t^i}{\|\mathbf{z}_a^i\| \, \|\mathbf{z}_t^i\|} \tag{24}$$

The range of MCS is $[-1, 1]$, where a higher value indicates stronger semantic consistency between the speech and text embeddings.

(2) Recall@K

Recall@K evaluates the model's ability to retrieve the correct matching sample within the top-K ranked candidates.

Given a speech sample $a_i$ as the query and its correct paired text sample $t_i$, the model computes cosine similarities between $a_i$ and all text embeddings in $t_i$, and retrieves the top-K most similar text samples. If ti appears within the top-K results, it is counted as a successful match. The Recall@K in the speech-to-text direction is calculated as:

$$\text{Recall@}K_{a \to k} = \frac{1}{N} \sum_{i=1}^{N} \mathbb{I}\left[t_i \in \text{TopK}(\mathcal{T}, a_i)\right] \tag{25}$$

where $\mathbb{I}[\cdot]$ is the indicator function, and $\text{TopK}(\mathcal{T}, a_i)$ represents the top-K most similar text retrieval results for the query $a_i$. Similarly, text-to-speech Recall@K is defined as:

$$\text{Recall@}K_{k \to a} = \frac{1}{N} \sum_{i=1}^{N} \mathbb{I}\left[a_i \in \text{TopK}(\mathcal{A}, t_i)\right] \tag{26}$$

### 4.2 Knowledge Extraction Experiments

To verify the effectiveness of the our knowledge extraction method on ETC industry corpora, we conducted comparative experiments against two mainstream extraction frameworks: UIE (Lu et al., 2022), which is based on structured prompt learning, and DeepKE (Zhang et al., 2022), which is based on relation classification and sequence labeling, alongside our zero-shot extraction method ETC-KE.

Given the high cost of data annotation in the ETC domain, both UIE and DeepKE were evaluated under zero-shot inference settings using their released open-source weights to assess their cross-domain generalization capabilities. A small, manually constructed validation set was used for evaluation, consisting of randomly selected samples from various representative business texts, including user complaints, customer service call transcripts, and interpersonal relationship records. The models were evaluated using Precision, Recall, and F1-score at the triplet level. Experimental results are shown in Table 2.

**Table 2.** Knowledge Extraction Performance Comparison

| Method | Precision | Recall | $F\_1$ | Time (s/item) |
|---|---|---|---|---|
| DeepKE | 0.10 | 0.06 | 0.08 | 66.7 |
| UIE | 0.36 | 0.25 | 0.30 | 27.3 |
| Llama 4 | 0.50 | 0.50 | 0.50 | 11.6 |
| GLM 4.6 | 0.40 | 0.75 | 0.52 | 8.3 |
| KiMi K2 | 0.64 | 0.56 | 0.60 | 9.8 |
| Grok 4 | 0.81 | 0.81 | 0.81 | 8.3 |
| Gemi 2.5 | 0.81 | 0.81 | 0.81 | 8.0 |
| Claude 4.5 | 0.88 | 0.88 | 0.88 | 10.7 |
| QWen 3 | 0.93 | 0.88 | 0.90 | 6.1 |
| GPT 5.1 | 0.88 | **0.94** | 0.91 | 6.3 |
| **ETC-KE** | **1.00** | 0.88 | **0.93** | **5.6** |

As shown in Table 2, traditional information extraction frameworks experience significant performance declines under zero-shot conditions in domain-specific scenarios. Both DeepKE and UIE heavily depend on predefined templates and limited-scale semantic encoders, struggling with domain terminology and complex sentence structures. Specifically, DeepKE's shallow encoder and rule-driven architecture result in severe over-reliance on surface patterns, hindering inference of implicit or multi-hop relations common in ETC business data. Although UIE incorporates general semantic understanding, its prompt mechanism lacks domain adaptation, struggling with entity disambiguation and nested relation parsing, when domain-specific terms deviate from its pretraining corpus. Consequently, both models exhibit low Recall and F1-score, reflecting poor generalization to unseen domain semantics. In comparison, mainstream large language models (e.g., Llama 4 (AI, 2025a), GLM 4.6 (Zeng et al., 2025), KiMi K2 (Team et al., 2025), Grok 4 (xAI, 2025), Gemini 2.5 (Comanici et al., 2025), Claude 4.5 (Anthropic, 2025), QWen 3 (Yang et al., 2025), GPT 5.1 (OpenAI, 2025) show substantial improvements due to stronger pretrained semantic representations. However, despite achieving higher precision and recall than traditional frameworks, these general-purpose LLMs still lack fine-grained domain alignment, leading to unstable performance in relation-heavy ETC texts.

In contrast, ETC-KE leverages stronger semantic reasoning and contextual modeling to overcome these limitations. Compared with both traditional IE models and the latest LLM baselines, ETC-KE further enhances domain-consistent reasoning through adaptive prompts and cross-relation contextual fusion. By integrating large-scale pretraining with adaptive domain prompts and cross-relation contextual fusion, ETC-KE can dynamically aligns heterogeneous domain expressions, recognizes long-distance dependencies, and infers latent relations beyond explicit textual cues. It achieves Precision = 1.00, Recall = 0.88, and F1 = 0.93, significantly outperforming the baselines. This demonstrates that large-scale models equipped with deep semantic understanding and reasoning mechanisms possess superior generalization and transferability in domain-specific knowledge extraction.

### 4.3 Cross-Modal Alignment Experiments

To validate Light-CLAP model's capability in aligning speech and text modalities semantically, we trained and evaluated the model using paired speech–text samples from the ETC customer service system. The model employs frozen pre-trained encoders, with only the dual-modal projection heads updated during training. A contrastive learning objective was used to maximize the positive pair similarity and minimize the negative pair similarity, thereby enforcing consistent representation of speech and text in a shared embedding space.

Figure 5 show the training loss curve and cosine similarity evolution on the validation set respectively. As seen, with increasing epochs, the training loss steadily decreases and stabilizes, indicating good convergence of the optimization process. Meanwhile, the cosine similarity on the validation set exhibits an

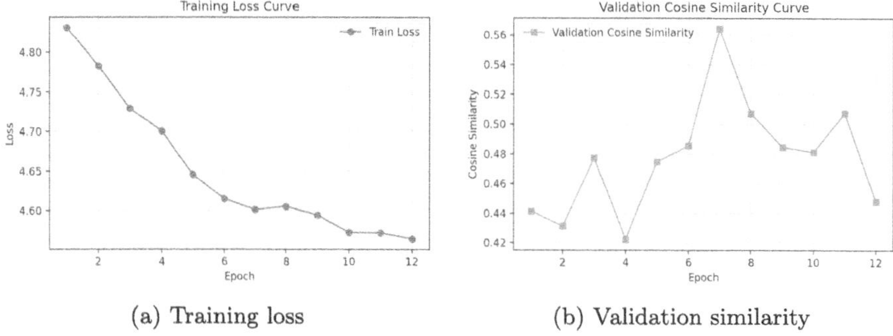


(a) Training loss (b) Validation similarity

**Fig. 5.** Results of the cross-modal alignment experiment: (a) Training loss curve, (b) Cosine similarity on the validation set.

upward trend, reaching its peak around epoch 6. This demonstrates that Light-CLAP effectively narrows the semantic gap between speech and text embeddings, achieving robust cross-modal alignment.

**Table 3.** Cross-Modal Entity Alignment Performance Comparison

| Method | MCS (%) | Audio→Text | | | Text→Audio | | | Time (ms/item) |
|---|---|---|---|---|---|---|---|---|
| | | R@1 | R@3 | R@5 | R@1 | R@3 | R@5 | |
| AudioCLIP | 5.8 | 0.0 | 0.9 | 1.7 | 0.4 | 1.3 | 2.4 | **0.03** |
| CLAP | 21.6 | 0.6 | 1.1 | 1.9 | 0.2 | 0.9 | 2.4 | **0.03** |
| Cacophony | 23.5 | 0.8 | 1.9 | 2.3 | 0.8 | 2.1 | 3.5 | 0.04 |
| GLAP | 32.2 | 0.8 | 2.1 | 3.7 | 0.9 | 2.8 | 4.3 | 1.20 |
| **Light-CLAP** | **45.2** | **0.9** | **3.8** | **6.4** | **1.7** | **4.5** | **8.8** | 0.04 |

To further verify the performance advantage of Light-CLAP in cross-modal alignment tasks,We conducted comparative experiments with CLAP and AudioCLIP—on the same test set as Sect. 4.3. Results are shown in Table 3.

The results demonstrate that Light-CLAP significantly outperforms existing approaches across all metrics. The model achieves an MCS of 45.2%, representing a 23.6% improvement over CLAP and nearly an eightfold increase over AudioCLIP. Compared with the recently proposed Cacophony (G. Zhu & Duan 2024) and GLAP (Dinkel et al., 2024), Light-CLAP still maintains a clear advantage, with improvements of 21.7% and 13.0%, respectively. This indicates that, even with frozen pretrained encoders, contrastive-learning-based lightweight fine-tuning of projection heads can achieve effective semantic space alignment with small-scale industry-specific data. In bidirectional retrieval tasks (Audio→Text and Text→Audio), Light-CLAP surpasses CLAP (1.9%/2.4%) and AudioCLIP (1.7%/2.4%), showing stronger bidirectional semantic consistency and tighter cross-modal embedding clustering in the shared space. Although GLAP and Cacophony also improve retrieval accuracy over CLAP, their gains remain modest compared with Light-CLAP, and GLAP exhibits significantly higher inference latency (1.20 ms/item), further highlighting Light-CLAP's efficiency–performance superiority.

Notably, Text→Audio performance slightly exceeds Audio→Text direction. This phenomenon aligns with customer service speech data characteristics: ASR-generated text is semantically clear and noise free but lacks manual correction, whereas audio signals are affected by accents, compression, and speech rate. Thus, text queries more easily match corresponding speech embeddings, reflecting the semantic sparsity of structured signals in real-world business data.

### 4.4 Hyperparameter Experiments

To examine the effect of the temperature parameter $\tau$ in the loss function on cross-modal alignment performance, we conducted hyperparameter sensitivity experiments. In the $\mathcal{L}_{\text{align}}$ contrastive loss, $\tau$ controls the smoothness of the similarity distribution among samples, thereby influencing the model's sensitivity in distinguishing between positive and negative pairs. Smaller $\tau$ values amplify similarity differences, focusing the model on highly similar pairs, while larger $\tau$ values smooth the distribution, weakening pairwise contrast.

We tested $\tau$ values from $\{0.03, 0.05, 0.07, 0.10, 0.15\}$, with fixed training parameters. Model performance was evaluated using Mean Cosine Similarity (MCS), and the results are shown in Table 4.

**Table 4.** Performance Under Different Temperature Settings

| $\tau$ | 0.03 | 0.05 | 0.07 | 0.10 | 0.15 |
|---|---|---|---|---|---|
| MCS (%) | 52.0 | 55.6 | **56.4** | 53.9 | 41.2 |

The results show that the model achieves the best performance at $\tau = 0.07$, with MCS reaching 56.4%. Excessively small $\tau$, causes oversensitivity and unstable discrimination of mismatched samples. Conversely, excessively large $\tau$ values over-smooths the distribution, reducing the contrast between positive and negative pairs and degrading model discrimination. Therefore, a moderate $\tau$ effectively balances sample similarity stretching and distribution smoothness, enabling more optimal cross-modal semantic alignment in the shared space.

### 4.5 Graph Scale and Risk Retrieval Experiments

Based on the constructed Person-Level Projection Graph, we conducted risk group detection experiments on ETC industry data. The resulting KG contains 376218 nodes and 530539 edges, as shown in Fig. 6(a).

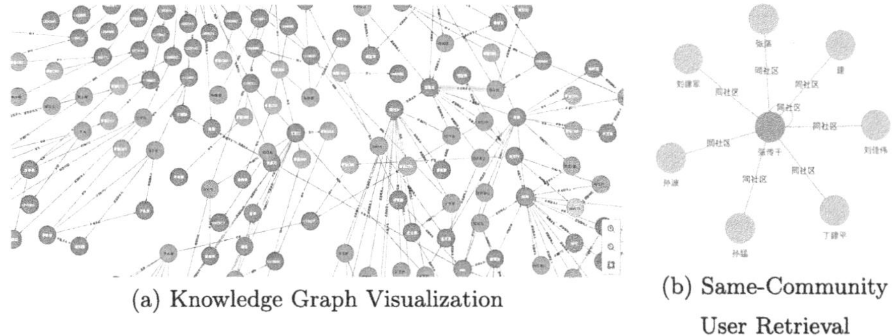

(a) Knowledge Graph Visualization (b) Same-Community User Retrieval

**Fig. 6.** Visualization results of the constructed ETC Knowledge Graph.

After community partitioning, when a user is flagged as risky, the system can efficiently retrieves other potentially high-risk individuals within the same community based on their community identifiers. This enables visual and interpretable representation of risk propagation paths, as illustrated in Fig. 6(b).

The results demonstrate that the person-level projection graph effectively reveals latent risk chains and abnormal aggregation patterns, providing structured support for risk diffusion detection and collusive group identification within the ETC business ecosystem.

# 5 Conclusion

Addressing the challenges of complex multi-source heterogeneous data structure, inconsistent multimodal entity identifiers, and knowledge update mechanisms in the ETC industry, this study proposed a comprehensive KG framework tailored to ETC business scenarios. The framework is organized around a multi-level structure with the "User–Account–Card–Vehicle" axis as its core, encompassing the following key components: (1) Hybrid Knowledge Extraction: a rule semantic fusion method enabling accurate and interpretable entity–relation extraction. (2) Multimodal Knowledge Fusion: a cross-modal alignment strategy that unifies heterogeneous data through entity and semantics driven alignment. (3) Dynamic Knowledge Management: a confidence propagation–based mechanism ensuring reliable, consistent, and up-to-date KG evolution. (4) Risk-Oriented Visualization and Analysis: a community detection–enhanced visual analytic system supporting interpretable risk retrieval and association discovery.

In summary, this paper establishes a systematic methodology and experimental validation for ETC industry KG construction, heterogeneous knowledge fusion, and risk exploration analysis, offering a feasible technical reference for KG–driven intelligent risk control. Future work will focus on two primary directions: (1) exploring deeper risk associations by incorporating social network analysis and graph neural networks for more sophisticated community detection, (2) developing dynamic risk event detection and causal reasoning mechanisms based on the graph, to capture risk events evolution an enable causal reasoning for intelligent risk management decision-making.

# A Prompt Design for Knowledge Extraction Tasks

In this section, we provide the complete prompt templates and task descriptions used in the knowledge extraction experiments.

**1. Prompt for Extracting Information from Vehicle Occupancy Complaint Tickets**

**Prompt for Extracting Information from Vehicle Occupancy Complaint Tickets**

You are a structured information extraction assistant handling a batch of user complaint tickets related to vehicle license plate occupancy. Each record contains the following fields: account number, license plate number, caller number, and ticket description. Based on the provided information, please extract the following structured content:

```
{
  "AccountNumber": "xxx",
  "LicensePlate": "yyy",
  "CallerNumber": "zzz"
}
```

Extraction Rules:
1. Use the value of AccountNumber directly as "AccountNumber".
2. Use the value of LicensePlate directly as "LicensePlate".
3. If CallerNumber is not empty, use it directly.
4. If CallerNumber is empty, attempt to identify and extract the caller's number from the Description field.
5. If no number is found in either field, set "CallerNumber" as an empty string "".
6. Do not infer, assume, or fabricate any number.
7. The final output must be strictly in JSON format with no explanations or additional text.

Input Example:
Ticket ID: {TicketID}
Account Number: {AccountNumber}
License Plate: {LicensePlate}
Caller Number: {CallerNumber}
Description: {Description}

**2. Prompt for Extracting Backup Phone Numbers from User Tickets**

**Prompt for Extracting Backup Phone Numbers from User Tickets**

You are an information extraction assistant. Based on the user's complaint ticket data, extract structured information to enrich the user knowledge graph.
Expected JSON output format:

```
{
  "AccountNumber": "xxx",
  "BackupPhone": "yyy"
}
```

Extraction Logic:
1. If CallerNumber is not AccountNumber and both numbers are non-empty, treat the caller's number as the backup phone number.
2. If CallerNumber is empty but a phone number is mentioned in the Description (e.g., "Contact number is 18912345678"), treat it as the backup number.
3. If no backup phone number can be identified, still return the AccountNumber with an empty string for "BackupPhone".
4. Do not infer or fabricate any information–only rely on the given data fields.
5. The final output must be JSON only, without any explanation or commentary.

Input Example:
Ticket ID: {TicketID}
Account Number: {AccountNumber}
Caller Number: {CallerNumber}
Description: {Description}

**3. Prompt for Extracting Information from Online Customer Interaction Logs**

**Prompt for Extracting Backup Phone Numbers from User Tickets**

You are an intelligent information extraction assistant. From the following userâĂŞagent interaction logs, extract two structured analyses: emotion trajectory analysis and churn intention analysis.
(a) Emotion Analysis Structure:
{
"User": "PhoneNumber",
"EmotionTrajectory": [
"Timestamp": "Time of emotional change", "Emotion": "EmotionCategory", ...
],
"PeakEmotion": "EmotionCategory",
"TriggerEvent": "Description of what caused the emotional outburst",
"ContainsAbuse": true/false,
"IndicatesComplaintIntent": true/false,
"AnalysisType": "UserEmotionAnalysis",
"Source": "OnlineService" or "PhoneService"
}
Emotion Categories: [Neutral, Anxious, Angry, Extremely Angry, Calming, Indifferent, Rational Cooperation]
(b) User Churn Analysis Structure:
{
"User": "PhoneNumber",
"BehaviorIntent": ["Apply for ETC Cancellation", "Unbind License Plate", ...],
"IndicatesChurnIntent": true/false,
"IndicatesComplaintIntent": true/false,
"TriggerReason": "Reason leading to the user's intention to leave",
"FinalStatus": "e.g., Materials submitted, pending review",
"AnalysisType": "UserChurnAnalysis",
"Source": "OnlineService" or "PhoneService"
}
Special Instructions: There may be three speakers: the user, a human agent, and a chatbot. Analyze only the user's utterances, ignoring all agent or bot statements. Record only timestamps where the user's emotional state changes. If consecutive utterances share the same emotion, record only the first change. Output should be a JSON array containing the two structures above. Each utterance includes role, timestamp, and content annotations.

## References

AI, M.: Llama 4: Multimodal Intelligence (2025). https://ai.meta.com/blog/llama-4-multimodal-intelligence/. Accessed 14 Mar 2025

Anthropic. Claude 4.5 release notes (2025). https://www.anthropic.com. Accessed 14 Mar 2025

Chen, Y.-C., et al.: Uniter: Universal image-text representation learning. Eur. Conf. Comput. Vision 104–120 (2020). https://doi.org/10.1007/978-3-030-58577-8_7

Comanici, G., et al.: Gemini 2.5: Pushing the frontier with advanced reasoning, multimodality, long context, and next generation agentic capabilities (2025). arXiv preprint arXiv:2507.06261. https://arxiv.org/abs/2507.06261

Devlin, J., Chang, M.-W., Lee, K., Toutanova, K.: Bert: Pre-training of deep bidirectional transformers for language understanding. In: Proceedings of the 2019 Conference of the North American Chapter of the Association for Computational Linguistics: Human Language Technologies, vol. 1 (long and short papers), pp. 4171–4186 (2019). https://doi.org/10.18653/v1/N19-1423

Dinkel, H., Yan, Z., Wang, Y., Zhang, J., Wang, Y., Wang, B.: Scaling up Masked Audio Encoder Learning for General Audio Classification. Interspeech 2024 (2024). https://arxiv.org/abs/2406.06992

El Sakka, M., Ivanovici, M., Chaari, L., Mothe, J.: A review of CNN applications in smart agriculture using multimodal data. Sensors **25**(2), 472 (2025). https://doi.org/10.3390/s25020472

Feng, Z., Wang, R., Wang, T., Song, M., Wu, S., He, S.: A comprehensive survey of dynamic graph neural networks: Models, frameworks, benchmarks, experiments and challenges. IEEE Trans. Knowl. Data Eng. (2025). https://doi.org/10.1109/TKDE.2025.3621291

Gao, Z., et al.: Rebel: Reinforcement learning via regressing relative rewards. Adv. Neural Inf. Process. Syst. **37**, 52354–52400 (2024). https://doi.org/10.52202/079017-1660

Guzhov, A., Raue, F., Hees, J., Dengel, A.: Audioclip: Extending clip to image, text and audio. In: ICASSP 2022-2022 IEEE International Conference on Acoustics, Speech and Signal Processing (ICASSP), pp. 976–980 (2022). https://doi.org/10.1109/ICASSP43922.2022.9747631

Hegde, A., Bhowmik, B., Bennehalli, S., Vakkund, S.: Louvain community-based label assignment for reject inference in peer-to-peer lending. Int. J. Data Sci. Anal. 1–23 (2025). https://doi.org/10.1007/s41060-025-00719-w

Janiak, D., Binkowski, J., Sawczyn, A., Gabrys, B., Shwartz-Ziv, R., Kajdanowicz, T.J.: The illusion of progress: Re-evaluating hallucination detection in LLMs. In: Proceedings of the 2025 Conference on Empirical Methods in Natural Language Processing, pp. 34716–34733 (2025). https://doi.org/10.18653/v1/2025.emnlp-main.1761

Ji, S., Pan, S., Cambria, E., Marttinen, P., Yu, P.S.: A survey on knowledge graphs: Representation, acquisition, and applications. IEEE Trans. Neural Netw. Learn. Syst. **33**(2), 494–514 (2021). https://doi.org/10.1109/TNNLS.2021.3070843

Kelley, J., Hagan, M.: Adaptive multistep prediction with sequence-to-sequence (seq2seq) models. IEEE Trans. Neural Netw. Learn. Syst (2025). https://doi.org/10.1109/TNNLS.2025.3525618

Kim, W., Son, B., Kim, I.: Vilt: Vision-and-language transformer without convolution or region supervision. Int. Conf. Mach. Learn. 5583–5594 (2021). https://proceedings.mlr.press/v139/kim21k.html

Li, J., Li, D., Xiong, C., Hoi, S.: Blip: Bootstrapping language-image pre-training for unified vision-language understanding and generation. Int. Conf. Mach. Learn. 12888–12900 (2022). https://proceedings.mlr.press/v162/li22n.html

Li, J., Selvaraju, R., Gotmare, A., Joty, S., Xiong, C., Hoi, S.C.H.: Align before fuse: Vision and language representation learning with momentum distillation. Adv. Neural Inf. Process. Syst. **34**, 9694–9705 (2021). https://proceedings.neurips.cc/paper_files/paper/2021/file/505259756244493872b7709a8a01b536-Paper.pdf

Li, S., Tang, H.: Multimodal alignment and fusion: A survey. arXiv preprint arXiv:2411.17040 (2024). https://arxiv.org/abs/2411.17040

Liu, Y., Li, H., Garcia-Duran, A., Niepert, M., Onoro-Rubio, D., Rosenblum, D.S.: Mmkg: Multi-modal knowledge graphs. In: European Semantic Web Conference, pp. 459–474 (2019). https://doi.org/10.1007/978-3-030-21348-0_30

Lu, Y., et al.: Unified structure generation for universal information extraction (2022). arXiv preprint arXiv:2203.12277. https://arxiv.org/abs/2203.12277

Ma, Q., et al.: A survey on time-series pre-trained models. IEEE Trans. Knowl. Data Eng. (2024). https://doi.org/10.1109/TKDE.2024.3475809

Ma, X., Hovy, E.: End-to-end sequence labeling via bi-directional LSTM-CNNS-CRF. arXiv preprint arXiv:1603.01354 (2016). https://arxiv.org/abs/1603.01354

OpenAI. Introducing gpt-5 (2025). https://openai.com/zh-Hans-CN/index/introducing-gpt-5/. Accessed 21 Nov 2025

Radford, A., et al.: Learning transferable visual models from natural language supervision. Int. Conf. Mach. Learn. 8748–8763 (2021). https://proceedings.mlr.press/v139/radford21a.html

Savci, P., Das, B.: Structured named entity recognition (ner) in biomedical texts using pre-trained language models. In: 2024 12th international symposium on digital forensics and security (ISDFS), pp. 1–5 (2024). https://doi.org/10.1109/ISDFS60797.2024.10527329

Shahrier, M., Hasnat, A., Al-Mahmud, J., Huq, A.S., Ahmed, S., Haque, M.K.: Towards intelligent transportation system: A comprehensive review of electronic toll collection systems. IET Intell. Transp. Syst. **18**(6), 965–983 (2024). https://doi.org/10.1049/itr2.12500

Souza, F., Nogueira, R., Lotufo, R.: Portuguese named entity recognition using bert-crf. arXiv preprint arXiv:1909.10649 (2019). http://arxiv.org/abs/1909.10649

Sukumar, S.T., Lung, C.-H., Zaman, M., Panday, R.: Knowledge graph generation and application for unstructured data using data processing pipeline. IEEE Access (2024). https://doi.org/10.1109/ACCESS.2024.3462635

Team, K., et al.: Kimi k2: Open agentic intelligence. arXiv preprint arXiv:2507.20534 (2025). https://arxiv.org/abs/2507.20534

Vaswani, A., et al.: Attention is all you need. Adv. Neural Inf. Process. Syst. **30** (2017). https://proceedings.neurips.cc/paper_files/paper/2017/file/3f5ee243547dee91fbd053c1c4a845aa-Paper.pdf

Wei, Z., Su, J., Wang, Y., Tian, Y., Chang, Y.: A novel cascade binary tagging framework for relational triple extraction. In: Proceedings of the 58th Annual Meeting of the Association for Computational Linguistics, pp. 1476–1488 (2020). https://doi.org/10.18653/v1/2020.acl-main.136

Wu, Y., et al.: Large-scale contrastive language-audio pretraining with feature fusion and keyword-to-caption augmentation. In: ICASSP 2023-2023 IEEE International Conference on Acoustics, Speech and Signal Processing (ICASSP), pp. 1–5 (2023). https://doi.org/10.1109/ICASSP49357.2023.10095969

xAI. Grok 4 model release (2025). https://x.ai. Accessed 14 Mar 2025

Yang, A., et al.: Qwen3 technical report. arXiv preprint arXiv:2505.09388 (2025). https://arxiv.org/abs/2505.09388

Yunita, A., et al.: Performance analysis of neural network architectures for time series forecasting: a comparative study of rNN, LSTM, GRU, and hybrid models. MethodsX **15**, 103462 (2025). https://doi.org/10.1016/j.mex.2025.103462

Zeng, A., et al.: Glm-4.5: Agentic, reasoning, and coding (arc) foundation models. arXiv preprint arXiv:2508.06471 (2025). https://arxiv.org/abs/2508.06471

Zhang, N., et al.: Deepke: A deep learning based knowledge extraction toolkit for knowledge base population. arXiv preprint arXiv:2201.03335 (2022). https://arxiv.org/abs/2201.03335

Zhao, Z., Luo, X., Chen, M., Ma, L.: A survey of knowledge graph construction using machine learning. CMES-Comput. Model. Eng. Sci. **139**(1) (2024). https://doi.org/10.32604/cmes.2023.031513

Zhu, B., Wang, R., Wang, J., Shao, F., Wang, K.: A survey: knowledge graph entity alignment research based on graph embedding. Artif. Intell. Rev. **57**(9), 229 (2024). https://doi.org/10.1007/s10462-024-10866-4

Zhu, G., Duan, Z.: Cacophony: an improved contrastive audio-text model. IEEE/ACM Trans. Audio Speech Lang. Process. (2024). abs/2402.06986. https://doi.org/10.1109/TASLP.2024.3485170

# Author Index





H. Yan et al. (Eds.): AIS&P 2025, CCIS 2918, pp. 609–610, 2026.
https://doi.org/10.1007/978-981-95-9884-7

The manufacturer's authorised representative in the EU is Springer Nature Customer Service Centre GmbH, Europaplatz 3, 69115 Heidelberg, Germany. If you have any concerns regarding our products, please contact ProductSafety@springernature.com

Printed and bound by CPI Group (UK) Ltd, Croydon, CR0 4YY
07/07/2026
02160910-0002